AF528587

Petroleum Geology of India

THE AUTHOR

Dr. G.N. Rao was born in Vizianagaram on 17th April 1953 and graduated in Geology from the Maharajah's College. He obtained his Masters in Applied Geology from Andhra University, Visakhapatnam in 1975 with first rank. He joined in Oil and Natural Gas Commission in 1976 as Geologist. He served in the organization for more than 34 years in different capacities which made him obtain experience in oil Exploration and Development right from basic geological data collection in an exploratory well to generation of a drillable prospect. His experience started in Assam and in Bengal Mahanadi, Krishna Godavari and Cauvery Basins. He obtained M.Tech., in Petroleum Exploration from IIT (ISM) in 1984.

He was the person for complete understanding the Basin analysis of Krishna Godavari Basin which also made him to devise the Litho-startigraphic nomenclature of the Basin.

He worked in foreign basin wing in the KDMIPE and also obtained experience in technical Docket preparation for offered exploration Blocks by the Government. He obtained Doctorate on the Sub-surface Geology of the Krishna-Godavari Basin from the Andhra University in 1994.

He was awarded Energy award by The Geological Society of India in 2001 and published paper award by ONGC. He was trained in the Asian Institute of Management in Manila and in the operation of mud-logging unit of M/s Geoservices in France.

He worked also as Country Manager of ONGC-Videsh Limited in Libya to take care of exploratory activity. He worked as General Manager (Exploration and Production) in Star Oil Company in Sudan. Later he was consultant for the National Oil Company of Angola to make plan and monitor the Brown fields of the nation.

Presently on professional front he is teaching for post graduate students in Petroleum Exploration as Honorary Professor and guest Lecturer in Andhra University, Aditya College and JNTU, Kakinada. He is also Board of Studies Member for Petroleum Engineering in Aditya College (Autonomous)

His Technical publications are about 25 in different national and International Journals, which includes one in AAPG about KG Basin.

Petroleum Geology of India

Exploration, Prospect Generation, Drilling, Evaluation Assessment and Indian Sedimentary Basins

Dr. G.N. Rao

Former General Manager (Geology)
ONGC and Honorary Professor,
Andhra University,

2025

Daya Publishing House®

A Division of

Astral International Pvt. Ltd.

New Delhi – 110 002

© 2025 AUTHORS
ISBN 978-93-90435-84-5 (HB)

Publisher's Note:

Every possible effort has been made to ensure that the information contained in this book is accurate at the time of going to press, and the publisher and author cannot accept responsibility for any errors or omissions, however caused. No responsibility for loss or damage occasioned to any person acting, or refraining from action, as a result of the material in this publication can be accepted by the editor, the publisher or the author. The Publisher is not associated with any product or vendor mentioned in the book. The contents of this work are intended to further general scientific research, understanding and discussion only. Readers should consult with a specialist where appropriate.

Every effort has been made to trace the owners of copyright material used in this book, if any. The author and the publisher will be grateful for any omission brought to their notice for acknowledgement in the future editions of the book.

All Rights reserved under International Copyright Conventions. No part of this publication may be reproduced, stored in a retrieval system, or transmitted in any form or by any means, electronic, mechanical, photocopying, recording or otherwise without the prior written consent of the publisher and the copyright owner.

Published by : **Daya Publishing House®**
A Division of
Astral International Pvt. Ltd.
– ISO 9001:2015 Certified Company –
4736/23, Ansari Road, Darya Ganj
New Delhi-110 002
Ph. 011-43549197, 23278134
E-mail: info@astralint.com
Website: www.astralint.com

Digitally Printed at : **Replika Press Pvt. Ltd.**

Dedicated to my Grand-daughters

Ria, Sonia and Esha

Foreword

I have great pleasure in writing the foreword to this book which has been brought out by the author who served as geologist in the Industry more than four decades, with marking "Kelly" even in odd hours, while drilling for oil, to learn more on the hidden treasure of the Earth. From first look I understood his vision, hard-work and dedication to the subject. He also worked in different countries to have global experience to fulfil his thirst for knowledge. But his confidence and zeal with first love to this oil and gas job made him to climb slowly to the citadel of knowledge.

His interest in academics along with duty, made him to acquire the first multi-disciplinary degree in petroleum exploration in 1984. His pioneering work in the then virgin basin Krishna –Godavari Basin helped to analyse the basin for prospectivity along with real cause for abnormal pressures in the basin. This in turn helped the management to avoid super pressure zones for economic exploitation of gas in the basin. His work in the K G Basin also helped him to obtain Doctoral degree, which added another milestone to see the same as AAPG publication with cover page of the K G Basin.

He also obtained needed knowledge and technology in Upper Assam, Bengal, Mahanadi and Cauvery Basins. Some new pools in Assam and few fields in KG basin and Mahanadi Basin with producible hydrocarbons are few results of his result oriented hard work.

Within the company, he got opportunity to associate with international operators and which lead to start operations in Libya alone by him which show his spirit to work for the company.

Though he became senior citizen, his thrust for sharing the knowledge made him to teach the upcoming generation on integrated petroleum exploration in Post-graduate course. By seeing the gap in teaching with data, the author worked hard to integrate all data on petroleum geology of India with latest available knowledge in this book shape, which help the students enormously. Though he teaches regulary in Andhra University, he creates interest in Petroleum Engineering students of our Colleage on invited lectures, which make them understand the subject with more clarity.

The author also gave very briefly the geopolitics of the oil game, technical and statistical illustrations where ever possible.

The chosen canvas of the author is so large that even though only the principal elements to the extent possible for focussing the learner's mind to learn. However, the more than century old knowledge with dynamic episodes of technology, touched more pages.

The result is a pleasant blend of the technical data in the desired narrative form for the knowledge seekers and wish him to be success in his endeavour.

N. Sesha Reddy

Chairman
Aditya Colleges of Engineering and Learning
Surampalli, Andhra Pradesh

Preface

Prior to my attempt to place before the learners about my experience as geologist in oil industry, introduce saying that I worked over forty years in the industry. It all strating with by collecting geological samples while drilling, to locate oil and gas in the farthest jungles of Assam during late seventies of last century. As time passes, encountered with wide variety of people working in various capacities and at diverse levels of knowledge in developing our nation's oil industry. Such individuals helped me to shape the ideas on which this treatise is based. The core notions came from reflections over the years and experience of the people with whom my work in the industry was carried out.

There is natural tendency to view the world's oil and gas potential from our own vision. But the view point needs to be unbiased and the presented data with a non-commercial perspective, may help the people who want to learn with clear vision. Individual's background, knowledge, overview and experience shapes such presentation. This is true for all those involved in the oil game. Though many intellectuals present the view point in a particular part of this vast subject, it is an attempt to present all available data in a simple way for the benefit of new-comers in the game to understand the latest status in India.

My travel through many areas of exploration and production of oil and gas with in the country and also some countries in the world, during four decades, has provided chance to understand many divergent views. This experience has provided a panorama to distil a multi-faceted account of the oil game.

Like oil was found in seeps in the North-Eastern corner of India, my initial opportunity was in that area for an understanding to find new oil in this game of high stakes.

My next landing was in the Bengal Basin during early eightees, similar to that of "Stanvac", where all ingredients needed to find oil or gas are visualised and deepest technically challenging wells were drilled. But all those efforts were remained as only knowledge bank till now. This experience involved wide exposure to theories and ideas that have been developed to understand the complex shadow side of the up-stream business. This gave me an opportunity to express before the technical intellectuals about my thoughts for not making the basin as a commercial one.

The next step was in the Krishna- Godavari Basin. By then, the basin was proved gas bearing within the technically challenging depth in land areas and less volume of oil in deep waters of offshore. So there were problems in justifying the investment in the basin, when imported oil was cheaper. The team work of some of colleagues with fair knowledge of understanding the unseen geology has made us to reduce the risk in drilling complex areas. There by,the quick results of finding new oil and gas in onland and more oil in the shallow waters in offshore not only made the basin commercially viable but also added more techno-management perspective to understand the basin for obtaining returns on investment. Some higher technocrats gave me invaluable opportunity to devise the startigraphic nomenclature vis–a-vis the basin analysis. Since then, the idea of sharing the knowledge to others was initiated by way of publications. This knowledge driven-path has now shaped as this comprehensive version to make new- learners more comfortable.

Then I was asked to move to research and development activity of the company, where in, an over view the activity of the foreign basins first and then to Assam basin. But this time not only exploration game but also how to take care of development drilling with unbiased perspective to see the seals are really sealing oil, where it was viewed, such that the faults are not faulty ones. This has added the third dimension to the knowledge reservoir.

In the hydrocarbon game, the economics will play a major role for the investor. I was fortunate to be associated in marking the blocks offered in the first round of the country and prepared the related documentation to help the prospective bidders for investment in the Indian exploration scenario.

The oil industry of India has travelled a long way from the jungles of Digboi to the deep blue waters of the "Bombay High" to the shallow waters of of the "Ravva". It has survived the economic, social and political turbulence in the country as well as in the world. Being a small traveller in this uncertain journey to win, I was asked to take care of complete exploration activity of Mahanadi offshore. In the new regime of Exploration and Production activity in India some down-stream players have become partners of the upsteam game.This situation added new skills to the individual, in budgeting finance and to convince the partners for approval of the investment.

With growing international activity by the company, the author, was given an opportunity to find oil in the awarded acreage in the southern desert of Libya. This opportunity gave immense exposure to build desired infra-structure for oil and gas in a new environment. The discussions with many international operating companies has also helped in shaping thoughts for the better survival in the competitive world of the oil game.

My last leg in the first nation oil company of India (ONGC) was to look after the exploration activity in the southern Cauvery Basin. This gave an opportunity for understanding the complex sub-surface tectonic activity concealed in the basin to locate "oil-finds" in the matured basin.

These thirty-five years of varied exposure gave me an opportunity to take responsibility in a private operator's exploration work in Sudan. The knowledge base has helped to find new oil in that land also. Then another opportunity came, to represent the Angola's national oil company in the blocks operated by many international Oil and Gas operators.

This four decades of the knowledge obtained has helped me to accept an opportunity to teach students who want to pursue their interest in this complex oil game by obtaining fair knowledge.

Like many technical-writers, I have gone through much of the related work of many individuals of different deciplines, who published in different forums. All such knowledge sharing of many were referred and mentioned in this attempt. I heartfully thank all such individuals.

Right from the first "rig-man" who helped me to mark "Kelly" with grease to know the pros and cons of drilling in search of oil and now to use the "static model" in computors for the same objective, at various stages, many persons have helped, I express my gratitude to all of them.

The person who gave me the opportunity to be a player in this thrilling and challenging oil exploration, Mr. S N Talukdar in 1976 can never be forgotten and all my big "namaskar" to him. I express my deep gratitude to Mr. P K Chandra, for his continuous help from the day of my joining in the company.

I express my thanks to Dr. Giri Prasad, HOD Petroleum Technology of Aditya Engineering College who right from start, not only encouraged in re-shaping the thoughts which can help now the students on the subject. Similarly I am thankful to Dr. Karuna Karudu for ideas to formulate the module of the content for the benefit of the students on the subject, also going through the some part of the text and for sugesstions. The help extended by Mr. YER Sastry earlier GM (Geophysics) for critically verifying the chapte-5 and corrections made. Similarly the help extended by Mr. S L N Rao electrologging expert for chapter-7 on the subject can never be forgotten.

Again, like history, this book is an attempt with love for the subject and an intention to help the students who learn this complicated interlinked subject of oil and gas exploration and production. In this attempt, many persons' published work on related topics were referred and thank all those countless hearts in achieving the goal, to help the knowledge thirst youngesters. I thank, publishers for all help particularly the unknown reviewer.

There are many others who also helped me at various stages of my growth, though I fail to mention, for all of them I express my heart-full thanks.

Dr. G.N. Rao

Contents

Chapter 1

Introduction

The Petroleum Exploration and Production technology explains the stages in the life of an oil or gas field, from exploration, through appraisal, development, planning and production. The objective is to provide a comprehensive methodology of the upstream industry, useful for those who learn to be professionals in the industry and who wish to be better informed about the basic methods, concepts and technology used in oil and gas exploration and production. It is also intended to provide fair knowledge to the readers who are not directly working in the oil and gas field activities, but are providing related support services.

Specifically, this attempt intends to help the reader to understand the major technical and business considerations which are required at every stage in the life of an oil/gas field, and also need to demonstrate the link between the many disciplines involved in the process. Exploration activities are aimed at finding more volumes of hydrocarbons, thus replacing the volumes being produced.

Several conditions need to be satisfied for the existence of a hydrocarbon accumulation. The first of these is an area in which a suitable sequence of rocks have accumulated oil and gas over geologic time called the sedimentary basin. One of the geo-scientific breakthrough has been the acceptance of the concept of plate tectonics. The plate tectonic model postulates that the positions of the oceans and continents are gradually changing through geologic times, which likely generate a sedimentary basin.

If two elements hydrogen and carbon are ideally combined in nature to form either liquid or gas, that combination which generates the energy is called hydrocarbons. However, since it is easy to take and carry liquids, first it was concentrated to search for oil. In the earth all things are available within the rocks, this "rock oil" is called Petroleum.

The first oil well was in Pennsylvania in 1859 (Drake well up to 20 m). In the next, an elephant helped in finding oil in thick Assam jungles by the sticky greasy substance seen on its leg while crossing an oil seep.

This rock oil is generally of straw, green, brown or black in colour with varying densities from 0.79 to 0.95. American Petroleum Institute (API) has devised a gravity formula to classify the quality of crude oil. It is reverse of the normal gravity, that is, high API gravity crude means low density and high quality. So the API gravity of oil will directly related to the quality of rock oil. API gravity of various crudes ranges from 5 to 61°.

Since some oils show optical activity which is the unique property of biological things, the biogenic origin is popular for the generation of rock oil. Because it is black in color, sometimes called as "Black Gold". It occurs within the pore space available in rocks, such rocks are termed as sedimentary rocks. The generation of this "Rock oil" takes some millions of years and occurs as oil pools in porus rocks in the area of occurrence. Further, the quantum of occurrence is also depends on quantum of source rock.

Petroleum geology is the study of origin, occurrence, movement, accumulation, and exploration of hydrocarbon fuels. It refers to the specific set of geological disciplines that are applied to the search for hydrocarbons. Petroleum geology is principally concerned with the evaluation of these seven key elements in a sedimentary basin. Sedimentary basins are regions where considerable thicknesses of sediments have accumulated which are widespread both onshore and offshore.

The sedimentary basins have formed on earth during the past 600 million years are by lithospheric extension which is a flexure or bending in the earth's crust. The isostasy is a vertical deflections in the region where the load is applied in one side of the area it sinks and the other side rises.

The potential of hydrocarbon generation and accumulation in the sedimentary basins depends upon a number of factors such as thickness of sediments, rates of sedimentation, nature of sediments and their organic content, tectonic and thermal history of basins. The prospectivity of a basin is determined by analysing these factors. Further, the results of earlier exploration, if any, also provide necessary input in these analyses. A prognosis of the hydrocarbon resource potential is thus made through probabilistic models based on statistical concepts tempered with sound geological reasoning. Classification of sedimentary basin in the context of plate tectonic evolution is used as an analogy-model to identify the basins of interest. The parameters that go into the identification of a sedimentary basin are important for initiation of exploration activities which hold good for selection of a tectonic block or sector within a basin. The exploration cycle starts with a reconnaissance survey followed by semi detailed data acquisition and analysis. The Generation, Migration and Entrapment (GME) model is to be build up which will be followed by drilling.

Petroleum Reserves

The reserves are those quantities of petroleum which are anticipated to be commercially recovered from known accumulations in the reservoirs below. These reserves cannot be measured directly but only estimated on the basis of geological, engineering knowledge and principles. Like all estimates, reserve estimates are subject to uncertainty.

The uncertainty depends chiefly on the amount of reliable geo-scientific data available at the time of the estimate and the interpretation of these data. The relative degree of uncertainty may be conveyed by placing reserves into one of two principal classifications, either proved or unproved. Unproved reserves are less certain to be recovered than proved reserves and may be further sub-classified as probable and possible reserves to indicate the increasing uncertainty in their economic recoverability. The disclosure of the quantities classified as unproved reserves is left to the discretion of the countries or companies involved.

Furthermore, many are confidential or subject to certain limitations or rules of the land. Companies use different numbers for internal and external purposes, and different companies involved in a particular field may report different estimates. It is accordingly very difficult to obtain possibly real estimates, and even more difficult to determine their status over a period. It is also possible that, sometimes some fields may lack reserve data, meaning that, announced discovery was not substantiated with reliable data. Fields may change their names where they straddle concession or national boundaries.

The method of estimation is called deterministic if a single best estimate of reserves is made based on known geological, engineering, and economic data. The method of estimation is called probabilistic when the known data are used to generate a range of estimates and their associated probabilities. Identifying reserves as proved, probable, and possible has been the most frequent classification method and gives an indication of the probability of recovery (Figure 1.1).

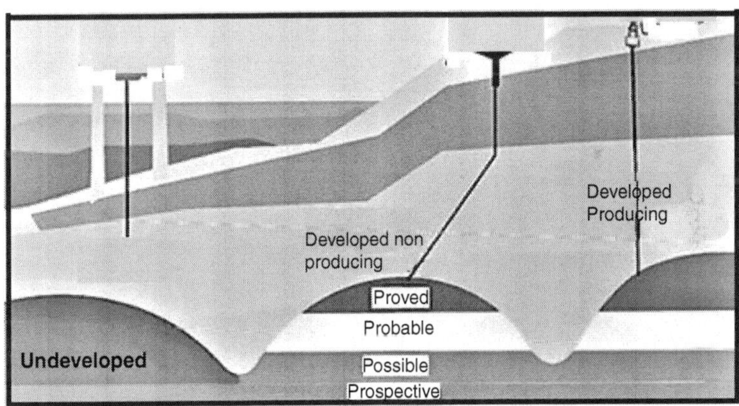

Figure 1.1. Categories of Oil Reserves.

Because of potential differences in uncertainty, caution should be taken when aggregating reserves of different classifications. Reserves estimates will generally be revised, as additional geological or engineering data made available or as global economic conditions change. Improved recovery methods include all methods for supplementing natural energy or altering natural forces in the reservoir to increase ultimate recovery. Examples of such methods are pressure maintenance, cycling, water-flooding, thermal methods, chemical flooding, and the use of miscible and immiscible displacement fluids.

Recovery Factors

The growth of reserves is also can be taken due to the improvement in the percentage of the oil-in-place to be recovered. The uncertainty relating to the size of the reserves is large at the outset but reduces progressively during production stage, until the abandonment of the field. The oil-in-place is estimated at an early stage from seismic and other data, and the uncertainty does not improve over the life of the field unless additional data are added to find new quantum of reserves are known either in area or new reservoir in depth. Recovery factor is used to estimate reserves only at the beginning of the project, later estimates come from well performance and stimulation techniques. The recovery factors are commonly indicated as percentage of in-place reserves which is as a factor by percentage indicates for its recoverable estimates. This quantum of hydrocarbons (oil and/or gas) is likely to be produced from a pool/or accumulation, considering the envisaged rock properties and reservoir dynamics. However, the average recovery factors for different types of reservoirs vary widely because of geological character of the reservoir coupled with type of oil which defines its mobility.

Exploration Stage

Although a basin analysis is usually part of the first study, a company conducts prior to moving into an area for exploration, it is also conducted or updated during the exploration phase as and when new information is obtained. Exploration geology comprises all the activities and studies necessary for finding new hydrocarbon occurrence. Usually seismic (or 3D seismic) studies are shot if needed and old exploration data (seismic lines, well logs, reports) are used to the studies. Sometimes, gravity and magnetic studies are conducted, and oil seeps/spills are mapped to find potential areas for hydrocarbon occurrences. As soon as a significant hydrocarbon occurrence is found by an exploration, the appraisal stage starts.

Appraisal Stage

The appraisal stage is used to delineate the extent of the discovery. Hydrocarbon reservoir properties, connectivity, hydrocarbon type and gas-oil and oil-water contacts are determined to calculate potential recoverable volumes. This is usually done by drilling more appraisal wells around the initial exploration well. Production tests may also give insight in reservoir pressures and connectivity. Geochemical and petro-physical analysis gives information on the type of the oil, like viscosity, composition, API gravity and the nature of the reservoir like, porosity and permeability.

Production Stage

After a hydrocarbon occurrence has been discovered and appraisal has indicated it as a commercial find, the production stage is initiated. This stage focuses on extracting the hydrocarbons in a controlled way without damaging the formation, within commercial favorable volumes. Production wells are drilled and completed in strategic positions. 3D seismic is usually available by this stage to target wells precisely for optimal recovery. Sometimes enhanced recovery methods like steam injection, pumps are used to extract more hydrocarbons or to redevelop aged or brown fields.

Analysis of Reservoir

The possible occurrence of a reservoir rock is determined through a combination of regional studies, stratigraphy and sedimentology and seismic interpretation. Once a possible hydrocarbon reservoir is identified, the key physical characteristics of a reservoir rock are focused to know its bulk rock volume, net-to-gross ratio, porosity and permeability.

Bulk rock volume or gross volume of the rock above hydrocarbon-water contact, is determined by mapping and correlating sedimentary packages. The net-to-gross ratio, typically estimated from analogues and wire-line logs, is used to calculate the proportion of the sedimentary packages that contains permeable reservoir rocks. The bulk rock volume multiplied by the net-to-gross ratio gives the net rock volume of the reservoir. The net rock volume multiplied by porosity gives the total hydrocarbon pore volume, *i.e.*, the volume within the sedimentary package that fluids can occupy. The summation of these volumes for a given exploratory prospect will allow to determine whether the production can be viable.

The important parameters of the reservoir like porosity and permeability are determined through the study of rock specimens collected from well or contiguous parts of the reservoir rock that outcrop at the surface, if so available. If not, the technique of formation evaluation using wire-line tools recorded down the well.

After the discovery and quantification of reserves, the optimum exploitation rate depends on a number of factors which are analysed/reviewed/evaluated in techno-economic analysis with time progressive based on estimated life. Based on the knowledge of reservoir fluid and rock characteristics, production testing and reservoir studies, technological schemes for exploitation of oil/gas fields are prepared for primary recovery and secondary and enhanced oil recovery phases using full scale reservoir simulator.

By a rationale behind the development of an oil field means, the one with which the reservoir pool/field is developed by minimum possible number of wells ensuring the envisaged optimum rates of oil/gas production and ultimate high yield with the least possible capital investment. The basic objectives of such development plan include:

- ☆ Effective well spacing and pattern for production/injection and their sequence of drilling.
- ☆ Fixing the output of the well at different stages of the producing life.
- ☆ Determine productive life of individual wells and there by the field.

Geologists are often asked: How much oil and gas is left in the world? There is no simple answer to this question. Geology, economics, politics and technology play a role in assessing our resources and determining the economic viability of those resources.

It is necessary, to periodically re-assess petroleum resources, not only because new data become available and better geologic models are developed, but also because many non-geologic factors like logistics and dynamic price scenario, the production will be economical and acceptable over the foreseeable future.

The vision of an oil and gas company is dominantly influenced firstly by prevailing market forces and demand, secondly by changing policies of the nation. Since many countries in the world are energy starved particularly with hydrocarbons, demand is increasing. World oil demand averaged 97.20 mb/d in 2017, up by 1.7 per cent over a year, with increase taking place in Asia and Pacific region, particularly China and India.

For developing countries, the huge investment is required for exploration and exploitation of hydrocarbons and also the outflow of foreign exchange is major concern. In order to reduce such burden, the government is changing its energy policy by inviting private participation in the petroleum industry and diversification in energy sources. Due to prevailing low oil prices, the E and P companies world over are planning for non-domestic prolific areas. In this competitive business world, any company can survive by adjusting to the changing environment.

The consumption level of petroleum products in the next decade indicates marginal increase in demand pattern consistent with requirements of country's economic development efforts. Our country India has to depend on oil and gas as energy source even beyond 2030.

World Outlook

The availability of world's recoverable oil consists of cumulative production, remaining reserves, reserve growth and undiscovered resources is estimated at about 3 trillion barrels of oil. Of this, about 24 percent has been produced and an additional 29 percent has been discovered and called as reserves. The natural gas potential is estimated at 15.4 quadrillion cubic feet (2.5 trillion barrels of oil equivalent).

More or less the present Organization of Petroleum Exporting Countries (OPEC) output was flat, and the group's share of global production is reducing (!). Declines in Libya and Angola were offset by gains in Iraq, Saudi Arabia and Iran. World natural gas consumption increased by just 0.4 per cent annual, well below the 10-year average of 2.4 per cent.

Total world proven crude oil reserves stood at 1,483 billion barrels at the end of 2017, sufficient to meet 52 years of global necessity. The largest addition to reserves are from Saudi Arabia. OPEC countries continue to hold the majority of the world's reserves.

If the reasons behind the current decline in prices seen as an interesting mix of demand and supply-side factors. On the supply side, North American production continues to boom as a result of the shale-oil. Global consumption increased by 840,000 b/d, with emerging economies accounting for all of the growth. Since July 2014, oil prices have fallen. There is an upward revision for future oil use in the petro-chemical sector for India and China but the advent of new technologies has led to downward pressures on oil demand.

The world will have to depend primarily on fossil fuels in the foreseeable future, and hence hydrocarbons play a vital role. OPEC has hiked the export prices of oil in stages up to 2008. Because of these high import prices many countries undertook energy efficiency drives and enhanced exploratory ventures even hither to uneconomical ventures earlier. This price induced conservation has reduced the consumption of oil which in turn resulted in less capacity utilization of production in case of major oil exporting countries. Realizing the problem, the OPEC has devised a system called net-back price which drastically reduced the world oil prices.

These low oil prices have dampened the industry's enthusiasm for exploration activity. However, by adopting new technology, the hydrocarbon exploration and production is becoming economical in nineties. But in the case of developing countries, the rapid economic growth, urbanization and increase in the transport needs will contribute to increase the oil demand. Considering the growing population levels in case of India, even a small increase in per capita consumption though lowest, it will lead to a sharp quantum jump. The share of hydro-electric and nuclear power in the overall energy scene has reduced in the years due to prevailing low oil prices.

The natural gas is an environment friendly fuel and it is expected to play a major role in the emerging energy scenario and also seen as an attractive substitute even in transport sector. Presently gas consumption is mainly restricted to the oil producing country due to requirement of high capital cost for transportation.

A surge in oil production from shale reservoirs replaced the need for LNG imports of USA. In Europe, gas demand stagnated as a result of the economic slowdown. Despite this, dependence on increasingly costly North Sea fields and long distance pipe imports kept prices relatively high. Demand growth has slowed in key countries particularly, China for whom an economic slowdown and lagging industrial production have affected the need for gas. Longer term, the outlook is more uncertain, because the demand will be governed in large part by Asia's increasing needs. Global natural gas demand is likely to grow by 1.9 per cent p.a. to approximately 4,700 BCM by 2030.

Demand

At present, the share of petroleum products in the world's energy demand is nearly 38 per cent. The world's oil demand was 4,220 MMT in 2014. The world's energy demand has been raising at the rate of 2.5 per cent on average per year, which is expected not to increase much in future up to 5,050 MMT by year 2040 due to shift towards non-fossil energy resources.

Oil demand in industrialized countries is projected to increase at only about 1 percent per year over the next two decades. In all regions of the world, the largest increases in oil consumption are expected to result from oil's use as transportation fuel.

Supply

The estimated oil reserves in the world are nearly 239 BMT in 2017 and with present rate of consumption of 4.397 BMT/year, these reserves may last for 52 years more. The Reserves/Production ratio of world zones are: - MiddleEast-95, LatinAmerica-43, Africa 25, Europe and Asia-Pacific 17 and North America-10 years.

The gas reserves are estimated to be 193.5 TCM and annual production was 3.165 TCM, which may last for about 65 years. Hence, the reserves are sufficient to meet the energy requirements for quite some time, but what matters now is, the areas of supply are less than the centers of consumption which make energy demand nations to depend more on imports.

With present price scenario, if oil companies reduce exploration, to save money now, future supplies may be effected. Oil companies have to think about the fact that, any oil development takes between 5-10 years from exploration to full production.

Price-Scenario

Crude oil prices have been in downtrend since 2013. The initial fall was rapid and unexpected. This was because production grew faster than projections and demand reduced faster than expected, resulting in an excess of oil for sale in the world. Recent events, such as the fall in Chinese growth projections and the end of sanctions against Iran, have given economists a reason to downgrade their projections for future crude oil price scenario.

OPEC is a permanent, inter-governmental organization, established in Baghdad, Iraq, in 1960. It comprises 12 Members: Algeria, Angola, Ecuador, Iran, Iraq, Kuwait, Libya, Nigeria, Qatar, Saudi Arabia, the United Arab Emirates and Venezuela. The Organization has its headquarters in Vienna, Austria. OPEC has always recognized to bring stability of the oil market and thereby to the global economy.

A main factor for the sharp price drop, is the continuing unwillingness of OPEC, to intervene to stabilize markets as over-supplied. Iran, Venezuela and Algeria have been pressing the cartel to cut production to firm up prices, but Saudi Arabia, the United Arab Emirates and other gulf allies are refusing to do so. At the same time, Iraq is actually pumping more. Saudi officials have thought if they cut production and prices go up, they will lose market share and merely benefit their competitors.

In the year 1960, the Convention on the Organization for Economic Co-operation and Development (OECD) was also formed and now 16 countries have become members of the Organization. Member countries include: Australia, Austria, Belgium, Canada, Chile, Czech Republic, Denmark, Estonia, Finland, France, Germany, Greece, Hungary, Iceland, Ireland, Israel, Italy, Japan, Korea, Luxembourg, Mexico, the Netherlands, New Zealand, Norway, Poland, Portugal, Slovak Republic, Slovenia, Spain, and Sweden.

The per capita oil use will remain lower in developing countries than in the OECD. On the supply side, the last few years have seen significant growth from nonOPEC countries. In addition, in terms of the global energy mix, renewables from hydropower, wind and solar are expected to continue to grow at a fast pace. Oil production is not declining fast enough in the United States and other countries. Demand for fuels is recovering in some countries, and that could help crude prices recover.

There are couple of other global and geo-political factors which will force the crude oil price to remain lower, which include, the evolving Syria situation and the consequent US-Russia strain and the longevity of extreme steps Saudi Arabia is willing take to consolidate its position.

The oil industry, with booms and downs, is in a new downturn. Earnings are down for companies that have made record profits earlier, leading to rig count and sharp decline in investments in exploration and production. The fall in oil prices since mid-2014, largely driven by the strength of supply, as nonOPEC production grew and OPEC maintained its production levels in order to protect its market share. So it was of great change and task of the industry is to meet present challenges while continuing to invest to meet future demand, safely and sustainably.

World Oil Price Turbulence

The crude oil market was wracked with turbulence during 1998, as prices fell and it was the lowest since 1973. The declining oil prices were influenced by an unexpected slowdown in the growth of energy demand worldwide. The growth in nonOPEC oil supplies played a significant role as competitor for OPEC's market share. North America dominated nonOPEC supply in the early 1970s, the North Sea and Mexico evolved as major producers into the 1980s, and much of the new production in the 1990s has come from the developing countries of Latin America and the nonOPEC Middle East as well as China.

By mid-1998, declining prices set in motion renewed efforts to manage oil supplies of OPEC. Agreements were sought to restrict production within OPEC, and initiatives were launched to gain cooperation from nonOPEC producers.

OPEC producers are expected to be the major beneficiaries of increased production requirements, but nonOPEC supply is expected to remain highly competitive with major increments to supply coming from deep-water resources, especially in West Africa. Several West African producers like, Angola, Cameroon, Chad, Congo, Gabon, Ivory Coast could eventually reap the benefits of substantial offshore exploration activity when the oil price environment became more favourable.

For the development of oil industry, it is the projected price scenarios will help to new investors in the field. Due to such optimistic projections in 2008, the oil price has reached nearly $150 per barrel, so the period has seen lot of new players and acquisitions of marginal fields and redevelopment of aged/brown fields.

Despite all available geological expertise, much oil and gas is still found by drilling. Gas is assuming more prominent role in the petroleum industry. Much needs to be learned about the viability of unconventional gas deposits such as coal-bed methane, shale gas, gas hydrates.

Many industrialized nations had reduced their dependence on oil where practical alternatives were available, such as electric utility and industrial boilers and home heating. The United States led the way in this transformation. European countries are expected to continue moving away from oil for non-transportation uses.

The result is, since the year 2013, huge fall of oil prices were seen which fallen to $42 per barrel by mid-2016. Though the annual production of the world has stable its demand has not increased as projected earlier due to quick development of alternative energy resources coupled with efficient consumption of energy.

America extracted more amounts of oil from the less expensive shale plays even as Russia and Saudi Arabia withheld supplies to revive the commodity. The Alberta oil sands contain 1.6 trillion barrels of oil, of which 311 billion barrels are recoverable with current technology. Further, a significant amount of U.S. natural gas production now comes from sources once considered unconventional. As far as overall production is concerned, the United States has been the world's top producer since 2012.

Oil edged lower due to pressure by expectations that Saudi Arabia and Russia could pump more crude to compensate for a potential supply shortfall the Brent crude futures were down at $75.24 a barrel by mid-2018. One analysis points to a possible energy future of mass electrification, digitization, and new technologies, in which the rise in electric vehicles and continued improvements in fuel efficiency lead to ultimately some displacement of future views on oil prices.

Complex Oil Game in the World

The complex global oil game has always been there and there were many historical studies of the hydrocarbon business, oil crisis and oil related wars. During last century seen corporate oil collusions and collapse of Imperial Russia into many small nations in the southern part and around Caspian Sea having hydrocarbon reserves.

The discovery of new oil reserves in the Middle-East, Venezuela and Brazil transformed the scale and geography of the oil game. After the last world war, America faced dependence on foreign oil especially from the Middle-East. Since Suez is canal not for oil movement, contributed to the redrawing the world oil map as concern over the security of supply moved up to the west.

Great oil price shocks during 1970s and 1980s realigned the game of "oil-chess". Conflicts of Iran with hostile America were intensified. The dramatic plunge in the price of crude in 1986 forced the industry into restructuring. The crisis involving political troubles in Iraq, Iran and Kuwait have brought new problems for strategy and corporate portfolio management.

Europe and America have diverged in interest and alignment. Some aspects of this trans-Atlantic divergence had an impact on the oil game, possibly because many oil companies are of Europe and America's dependence more on oil. Some companies discounted sanctions imposed on Iraq, Myanmar, Libya and Sudan. The cartel trend relations between Saudi Arabia and America is well known to world oil game.

Fundamental rationale behind overthrow of earlier regime of Iraq is also known in the game. The development of oil industry in Syria is being retarded since the unrest started cannot attract major investments in the oil exploration there.

The new nations around Caspian Sea like Kazakhstan, the exploration of the promising area retarded due to lack of pipe line to the open world market. Though some new nations Uzbekistan, Kirgizstan and Turkmenistan want to develop their oil and gas resource could not get positive results because major part is gas which need costly pipe line to the demand area? On the way Afghanistan with military control by America coupled with rugged and difficult terrain became the retarding element for such massive investment. Similarly though India need gas from Iran and/or Qatar either by sea route or land route pipe-line, the peace troubled Pakistan became point of concern for construction of such pipeline.

The OPEC is projected as important, but the role of Nigeria, Angola, Brazil and Venezuela is to be strategically known in the organization. There have been some ethnic violence keeping in mind the oil location within the country like Sudan. This conflict in the south for over log period ultimately lead to Independent nation. When sanctions applied in Sudan by the America, the Asian oil companies are benefitted to be in the game. Similarly in Nigeria two different ethnic groups from north and south creating unrest in the oil rich Nigeria.

China asserted a new prominence as key player to take calculative lead where western companies are in low profile due sanctions or prevailing inter-state turmoil. Recently China has shifted focus to South China Sea where some oil resource nations are known like Vietnam and Brunei.

Further, the low oil prices since 2013, the oil rich nations in South America, Brazil and Venezuela became financially weak lead to much unrest both politically and economically.

Historical Journey of Oil and Gas Exploration in India

The petroleum industry in India started since 1866, that is only 7 years after the discovery made by the Drake Well in USA. Oil was found by seepages near Digboy during construction work. The Makum discovery had a small production with a rapid decline. Digboi had a more sustained production, which was about 100,000 tpa of oil by 1920 (Viswanath., SN 1997).

At around 1901, another shallow oil field was discovered near Badarpur in the Cachar area of Assam, which was developed by the Burmah Oil Company in 1915 and was producing around 60,000 tpa of oil by 1920. Digboi field was also taken over by BOC as operator and the production was increased to 300,000 tpa by the late 1930s. Badarpur meanwhile started to decline and had to be abandoned in 1933. At the time of independence, India had a small production of 0.27 mtpa from only Digboi oil field. The discovery of Nahorkatiya was in 1953 and drilled a deep well in Moran which resulted in to discovered oil field in 1956 (Das Gupta, A.B, 1976).

Since Independence the Government of India felt the need to build in some petroleum geology competence within the Geological Survey of India and it was followed by creating Oil and Natural Gas Division within it in 1955. This was converted to a Directorate in May, 1956, and then to a Commission in August, 1956. ONGC was converted in to a statutory body in 1959 which became a commercial Corporation in 1994. The GOI was also able to organize massive support for ONGC from the then USSR since inception.

Oil India Private Limited (OIL) was incorporated by Government of India in 1961, in which AOC and BOC became equal partners by transforming OIL into a JV company. OIL discovered Kusijan oilfield in 1969 and Jorajan oilfield in 1972. Later, gas was discovered by OIL in Tengakhat field of Assam in 1973. OIL also ventured from Assam to Odissa both in onshore and offshore. During 1979-89 it went to Andaman offshore and Rajasthan onshore. The Kharsang oilfield was discovered by OIL in 1976. Government of India took over OIL and it became full-fledged government enterprise in 1982.

ONGC systematically started its geophysical surveys in all areas which were considered prospective on the basis of global analogy. Further, thrust was given for survey in area of Himalayan foothills and adjoining Ganga plains, alluvial tracts of Gujarat, upper Assam and Bengal. The exploratory drilling carried out in Himalayan foothill during 1957, remained unsuccessful. ONGC discovered oil at Cambay the giant Ankleshwar field in 1960, Kalol in 1961, Lakwa in 1964, Geleki in 1968 and gas discovery-Manhar tibba in Rajasthan in 1969. Thus, Gujarat region turned out to be a major petroleum bearing area with production increasing rapidly to reach peak about 6.4 mtpa by 1991.

ONGC also found a lot more of oil and gas in the known upper Assam petroleum province. ONGC moved into Tripura Cachar area and found number of gas fields. However, its next major success was the discovery of the offshore super giant Bombay High oil field in 1974. Many other discoveries have followed in the East Coast basins and Rajasthan. As a result, the country has been able to reach a production level of 34 mtpa by 1990.

Offshore exploration was initiated by ONGC in the form of experimental seismic survey in 1962 in Gulf of Cambay and later in western offshore. The exploration was continued in entire western off-shore including Kerala-Konkan basin and eastern offshore area. This led to large discovery of Bassein and Neelam in western offshore and PY-3 and Ravva in Eastern offshore. ONGC's geo-scientific survey spread out to UP, Bihar, Tamil Nadu, Rajasthan, J&K, Kutch and Andhra Pradesh. By mid 1980s ONGC successfully discovered prospects in Cauvery and KG basin.

The government efforts during early Nineties gave thrust for opening up Oil and Gas sector in India to private players also. Hindustan Oil Exploration Company (HOEC) was among such private players. In view of the liberalized policy adopted by GoI, a need for an independent upstream regulatory body called the Directorate General of Hydrocarbons (DGH) was formed in 1993. Opening up of deepwater exploration has led to discovery of the recent large gas fields in Krishna-Godavari deep water.

Hydrocarbon Potential of India

India is endowed with a vast sedimentary cover in both onland and offshore. The estimated areal extent of the sedimentary area is about 3.14 million square kilometres. Now the oil and gas contribute 39.2 per cent to primary energy consumption however, natural gas constituted about 7.8 per cent of the energy mix.

Oil production is more or less stagnant and India is expected to meet over 78 percent of demand for crude oil through imports. Initiatives to increase domestic production could not be much successful.

India has balance of reserves of 763 MMT of oil (333 MMt in onshore and 430 MMt in offshore areas, which would last for about 20 years at a production rate of 37 MMT per year. Those of natural gas are placed at 1,354.76 BCM (353.34 BCM in onshore and 1001.42 in offshore areas).

It is well known that Exploration and Production is a highly capital-intensive business and it requires huge investment on a continuous basis – not only to accrue new resources but also to sustain production by improving recovery rates of mature fields and for converting reserves into producing properties.

India uses variety of energy sources and now coal is the primary energy source and meets nearly 56.4 per cent of the needs, the next one are hydrocarbons, with oil 32 per cent and gas 7.5 per cent. Since gas is environmental friendly fuel, the concentrated efforts in last decade has made its presence in nation's energy scene.

The net realized price per barrel of oil produced in India is having a thin margin for the Indian producing companies when compared to low selling prices of crude in international markets. In India, the cost of operations has increased by almost 7-8 percent every year because of production is mainly from the aged fields.

Though gas reserves are more comfortable than that of oil, the higher capital cost for pipelines to transport gas may affect its production. Keeping in mind the matured fields and reducing success in oil find, may further reduce the level of present self-sufficiency in future.

Import of Crude Oil during 2014-15 was 189.432 MMT. Indian oil and gas companies are encouraged to invest overseas and to build strong relations with strategically important countries. India aims to produce 60 MT by 2025 from other countries.

Government's Offer of Blocks for Finding more Oil with Foreign Help

Between 1980 and prior to implementation of NELP in 1996 some exploration blocks were awarded to private companies, where ONGC and OIL had the rights for participation in the blocks after hydrocarbon discoveries. During first round, Chevron, USA and 3 wells were drilled without success. During 1992 and 1993 in the two development rounds, offered to some companies, the marginal fields.

In 1993, GoI offered blocks for geophysical and other surveys to update the information on hydrocarbon potential of India's unexplored sedimentary basins. The second speculative survey round was launched in 1994 and the third round in 1995. The third round was called as Joint Venture Speculative Survey Round (JVSSR) with a provision of risk participation/cost sharing by DGH. As a result of exploratory activities, several unexplored and poorly explored areas, in particular offshore and deep-water areas have been appraised through geophysical surveys and exploratory drilling.

The nine rounds of NELP resulted in the discovery of some gas fields in deep-water Krishna Godavari basin offshore. Though, in Cauvery offshore, small oil and small to marginal gas fields in Mahanadi offshore were found, most of them may not be commercially viable in comparison to the present international price scenario.

The new method called HELP (Hydrocarbon Exploration Licensing Policy), replaces New Exploration Licensing Policy. Companies will now be able to produce any energy source under a single license. Explorers will also be free to bid for an area of their choice instead of waiting for the government to carve out blocks and auction them. Contracts for all these acreages will follow revenue-sharing model instead of profit share.

Energy Security

The prognosticated conventional hydrocarbon resources in 15 sedimentary basins and deep-water areas of the country are of the order of 28.1 billion tones. In-place hydrocarbon volume of about 11,350 million tons of oil and oil equivalent gas could be established through exploration by ONGC, OIL and Private/JV companies. So, more of resources are under "yet to find category".

Although, yet-to-find hydrocarbon exists, India has to look for properties abroad for energy security. ONGC Videsh Limited (OVL), the overseas arm of ONGC, at present produces over 165,000 bopd from its producing fields spread across some countries.

OVL has spread its wings in Russia, CIS countries, West Asia, Southeast Asia, Latin America and Africa. ONGC Videsh has also developed strong partnership alliance with both national upstream companies and international operating companies. BRPL is actively engaged in exploration and development in Mozambique and, along with Videocon, in deep-water Brazil where discoveries have been made.

A National Data Repository (NDR) is created which comprise of reliable exploration and production data for India. Online data management would be drawn up which would be an essential part of an Open Acreage Licensing Policy (OALP).

Since India's energy demand is expected to grow, GOI has adopted a strategy to enhance availability of natural gas in the country through: Coal Bed Methane Exploration and Production activities, developing underground coal gasification and targeting Unconventional gas sources like Shale Gas, Gas Hydrates.

Because the marginal oil and gas fields of ONGC and Oil India Ltd., could not be monetized for years due to various reasons such as isolated locations, size, more development costs, technological constraints, Government of India is with an objective to bring marginal fields to augment the domestic production of oil and gas. It had announced Discovered Small Field (DSF) Policy with the objective to bring discovered small fields to the production at the earliest.

Chapter 2

Sedimentary Environments-Basin Classification

The hydrocarbons are generated below the Earth's surface and trapped in the pores of sedimentary rocks. Crude oil is thought to form from the protected organic matter, that settle to the bottom of marine basins in which sediments are being deposited in bulk and are rapidly buried within sequences of sedimentary rocks. After their formation, oil and natural gas migrate from fine grained source rocks to reservoir rocks which are permeable sedimentary rocks. A high proportion of open pore spaces facilitates the migrating hydrocarbons to move till entrapped. So in all means the formation of sedimentary rocks is important to search for hydrocarbons (Pettijohn, 1984).

Sedimentation is the process for particles in suspension to settle out of the fluid in which they are entrained and get compacted to become sedimentary rock. In geology, sedimentation is the terminal end of sediment transport. Sedimentation may pertain to objects of various sizes, ranging from large rocks in flowing water to suspensions of dust and pollen particles to cellular suspensions to solutions of single molecules such as proteins.

Depositional Processes

The hard rocks of mountains will become pieces of different sizes (boulders to fine clay) due to weathering and transportation. Similarly rivers are also originate from hills which provide transport media for such weathered material. The process of transportation is facilitated by high surface gradient provided by mountainous terrain and once rivers enter the plains, the energy of transport media is reduced. Finally rivers bifurcate to unload the sediments at the shore area (deltas) before meeting the sea. Based on the gradient, the fluvial processes can be categorized as foot hill areas with high gradient, Braided Streams with moderate gradient and finally less gradient Meandering Streams. During transportation the grain size of sediments are also reduced because of breaking in to pieces due to friction. Bigger particles of the sediments are carried by rivers as bed load and very fine particles as suspended load.

Alluvial Fans

A depositional cone formed at the base of a hilly terrain is called alluvial fan. This environment is characterized high energy processes like flash floods, mud flows and debris flows which deposit the coarser and most ill–sorted sediments. In this environment the deposition occurs directly adjacent to the areas of erosion and deposits are of maximum possible size range. The size of the individual fans range in radius from hundreds of meters to tens of kilometres, which is controlled by the area of deposition. Coalescing fans can occur in liner belts of mountain ranges. The thickness of deposits is maximum at apex of the fan which gradually decreases away from the apex. Tectonic activity and climate have influence on the source, transportation and deposition of these fan deposits. Uplift of mountains may result in intensive erosion of rocks and high gradient for drainage system.

Sometimes, sands on the down-stream parts of alluvial fans, along the braided streams, point-bars, distributary channels are picked up by the wind and re-deposited as dune sands. Similarly sands deposited along beaches of the coastal and inter-deltaic regions are re-deposited by winds as sand dunes on Barrier Island or on the main land.

Many types of ***sand dunes*** are known: -

Fore-dune Ridges are adjacent and parallel to the shore-line. Parabolic Dunes are with open end towards beach (wind ward), middle part moves forward in relation to sides. Barchans have steep slope on concave side facing away the beach.

Transverse Dunes are parallel or oblique to the shore, and elongated perpendicular to wind direction. Gentle slope on windward side and steep slope on leeward side. Several Barchans join to linear ridge which look-like Transverse dunes. Longitudinal Dunes are elongated parallel to wind direction and perpendicular to the coast–line.

Dune deposits commonly contain both shells and plants characterized by high angle cross-bedding and usually well sorted. On the contrary the beach deposits are generally horizontal with low angle cross-bedding. Desert sand bodies differ from coastal aeoline sands in the distribution sometimes controlled by two wind directions showing cross-bedding in two directions.

The clastic materials are transported in mountains through stream flows and at lesser gradient it become debris/mud flows and facilitates to form alluvial fans. The mud flows occur due to presence of unconsolidated clay which became slippery when wet, steep gradients, short periods of abundant water and sparse vegetation.

Generally, the alluvial fan environment become braided stream towards down dip with lesser physiographic gradient. In some areas where mountains are located very near to oceans or at large inland lakes, alluvial fans are formed both sub-areally and in sub-merged conditions. Such fans are called 'Gilbert' deltas.

Braided Stream

When rivers enter plains with reduced gradient and moderate energy, the area is called braided stream environment. In this environment the rivers are less sinuous and more in number which merge further downstream to form meandering stream. Due to coarse nature of sediments, bulk of the sediment load of most braided stream is transported as bed load. The combination of steep gradients and high discharge rates of rivers result in deposition of large amounts of coarse material. However, the grain size of deposits are finer than that of alluvial fan deposits.

If mountain areas are located nearer to the sea, the braided stream environment can extend directly to coastline without having third phase called meandering stream (Figure 2.1).

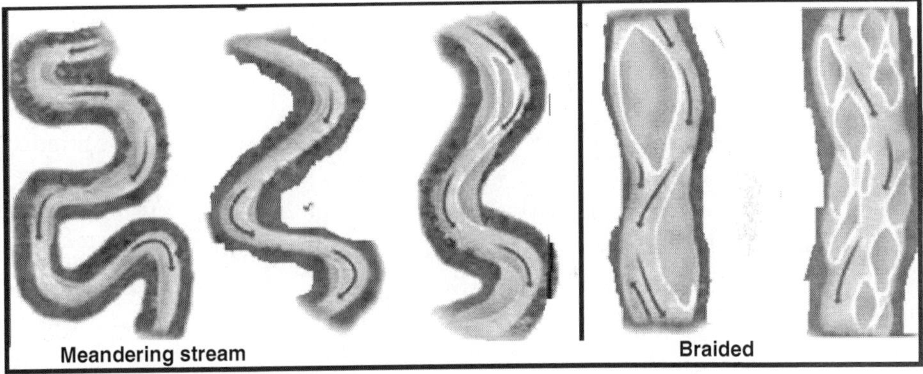

Figure 2.1. Meandering and Braided Stream.

The depositional landforms include longitudinal bars (aligned along the river channel) in upstream and transverse (along the width of river) bars are more common in downstream direction.

Generally the coarser fraction of sediments are carried at the center of the river and due to local hydraulic conditions if part of this coarser fraction is deposited it forms a central ridge with in the channel. Progressively it grows and form a longitudinal bar and the river channel becomes wider due to bank erosion. During high water stage, these longitudinal bars may be dissected by small streams flowing over the bars. These transfer channels can also be filled with time resulting into transverse bars.

Meandering Stream

It is the end part of a flowing river and occur in coastal plain areas between the braided streams and deltas in down-dip. If mountainous terrain is adjacent to shore line meandering stream do not exist. It is characterized by single channel stream which is deeper than the multi-channel braided stream, but more sinuous. Clays and fine silts are transported in suspension and coarser sediments such as gravel and pebbles are transported as bed load. Maximum sediment transport occurs during rising flood stage and deposition during falling flood stage. The important processes include channel migration and channel abandonment.

When current velocity is maximum and directed towards concave side of the channel, bank caving occurs in bends and deposition occur on the convex side which is termed as point bar. In this process the channel migrates further and became more sinuous.

If the meander-stream become more curved and coupled with more velocity at upstream, part of the curved stream can be abandoned. If a single meander loop is abandoned it is called neck or "chute cut off".

Due to frequent channel migration, the coalescing point bar sands can form blanket like sand body of very large regional extent. Another important type of clastic sedimentary process is aeolian deposits in which the process of sedimentation is transportation and deposition by the wind. Two basic conditions for the deposits are large supply of dry sand and sufficient wind velocity. These conditions are commonly present little away along coast-line and in deserts.

Deltas

As a geological term the word delta includes a triangular shaped land area at the mouth of rivers when enter the sea. The delta is a lobate body consisting of sediment that has been transported to the end of a river channel and deposited mostly sub-aqueous (below water) but partly sub-aerially (exposed to land area).

A delta is formed when a river develops a confluence point with a standing water body, like bays and lakes where dumps the sediment load. The sediments of the river through the confluence point are dispersed by the hydro-dynamic factors of waves and currents of the standing water body. The growth and evolution of the delta depends largely on the quantity of the deposits transported and distributed by the river through the confluence point.

Small rivers, which bring less volume of sediments, cannot produce the deltas and termed as estuaries with a character of single channel with funnel shaped nature. When a river is able to discharge huge volume of sediment with good distributary channels, the possibility of the evolution of the delta and its growth would be more. As the deltaic areas are zones of excess sediment supply than the neighbouring coastal zones, the coastline near the delta is protruding into the sea, in case syn-sedimentary subsidence is not in tune with sediment load brought by the river.

The deltaic sediments are partly re-worked by marine processes like waves, currents and tides. A delta landscape usually shows wide range of various fluvial coastal and near shore landforms. The combined effect of factors like fluvial dynamics, waves, currents, tides, climate and biological parameters would promote the variability within the delta in different shapes produced in space and time (Figure 2.2).

Coleman (1985) distinguishes three basic types of distributary network:

- ☆ The first one consists of many seaward bifurcating channels and the delta shows a large number of river mouths. They are mostly associated with low wave action, low offshore slope, small tidal range, and a fine grained sediment load.

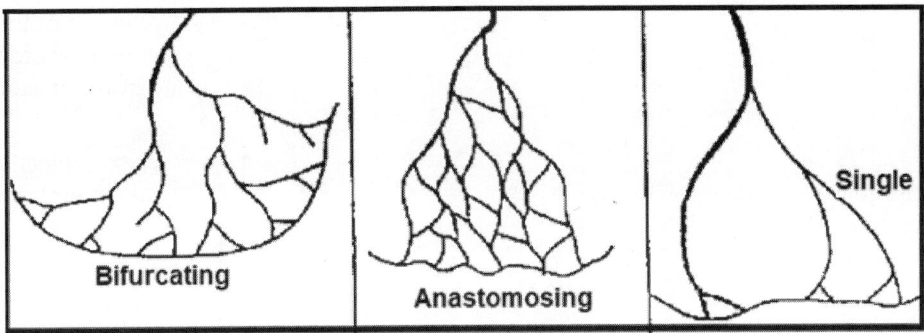

Figure 2.2. Delta Types.

- ☆ The second shows anastomosing branching depicting many branching and re-joining channels; though there are only a few active river mouths. This is due to erratic fluvial discharges, intermediate wave energy, high tide range and relatively steep offshore slope.
- ☆ The third one shows single or few channels, mostly originating from a common point at the head of a delta. This is associated with high wave action, high tidal range and steep offshore slope (Figure 2.3).

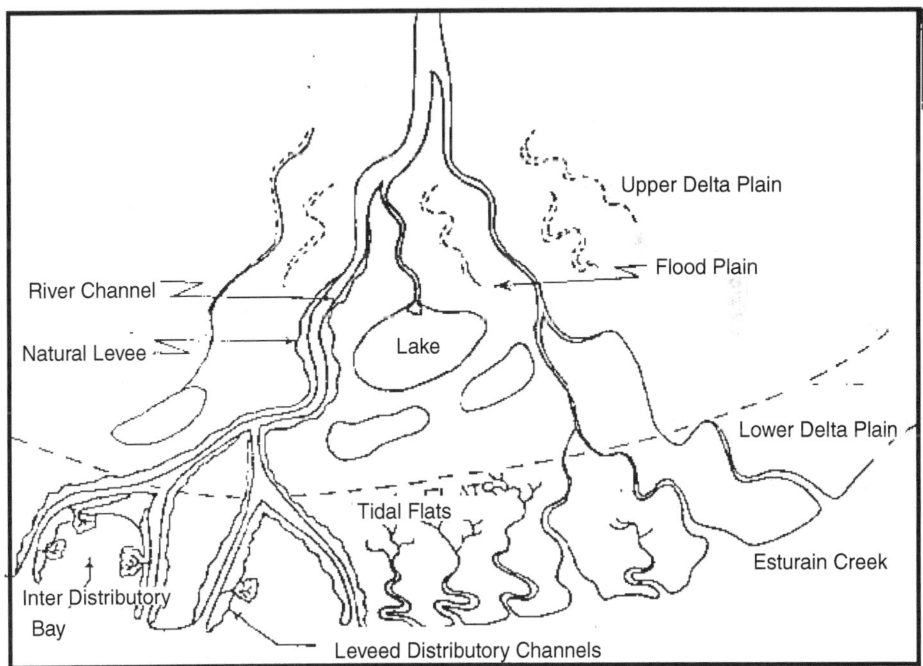

Figure 2.3. Delta Geometry.

The ultimate shape of delta and distribution pattern of sediment at river mouth is also affected by a number of factors, like, water discharge, sediment influx/load, water depth, wave action and tidal currents (Figure 2.4).

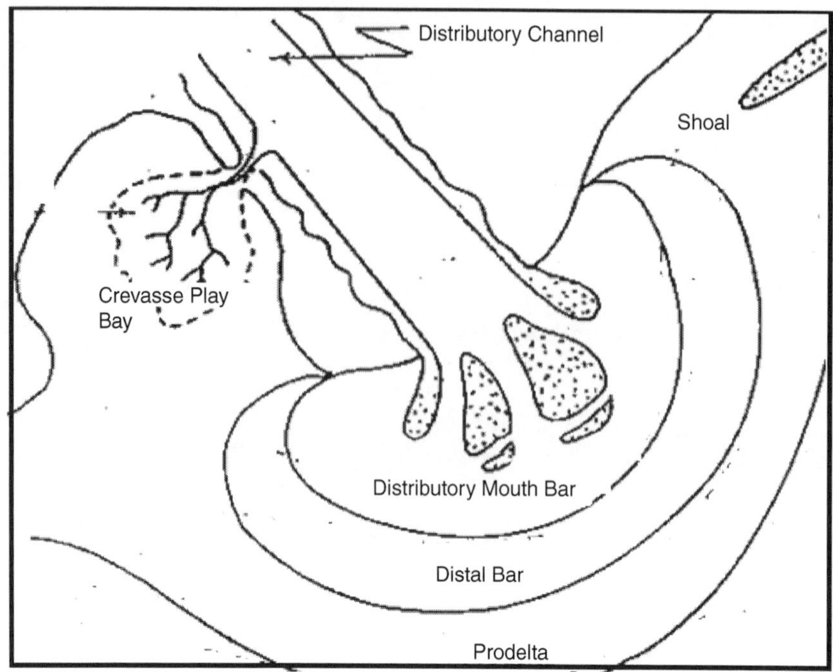

Figure 2.4. Delta – Depositional Patterns.

Since deltas are important for hydrocarbon exploration one has to understand the various depositional processes in deltas. The area towards land-ward part of the delta is called the fluvial depositional area where sediments are carried by rivers flowing with fresh water. Many deltas have built new land along the open sea coast and sometimes big lakes might have filled with deltaic sediments.

In many basins a wide variety of clastic depositional systems may exist. They may range from on-land alluvial fan, lakes, aeolian (transported by wind) systems and dip related fluival, deltaic, shelf and slope systems. Alluvial fan deposits located adjacent to mountainous regions also look like triangular shaped bodies. Both fans and deltas are lobate accumulations of sediment built at the end of river channels. The main difference is that fans are deposited away from the influence of free body of water like lakes or sea (Figure 2.5).

The term Fan-Delta refers to the triangular shaped deposit occur away from the open marine conditions with less transportation. Hence deltas are formed by the combination of fluvial and marine processes (Passega, R., 1964).

The distribution, orientation and internal geometry of deltaic deposits are controlled by climate, morphology, vegetation, water discharge, sediment load, river mouth process, waves, tides, winds, current shelf/slope and the tectonics and geometry of the receiving basin. To accommodate this complex set of variables the deltas are classified as:

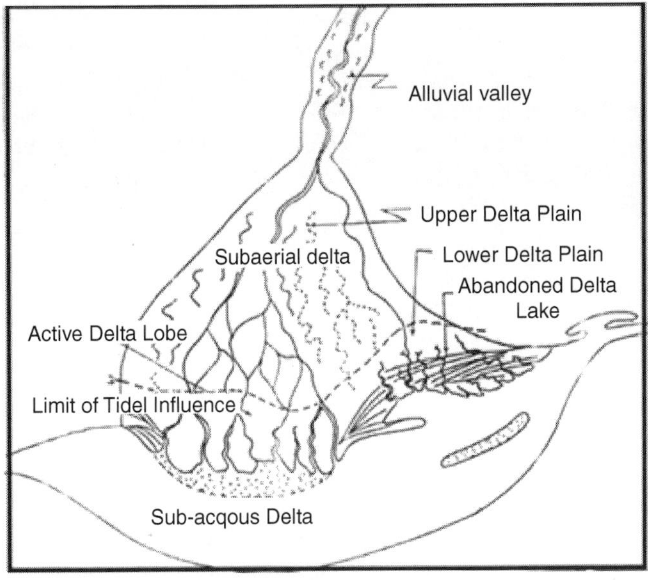

Figure 2.5. Delta Influence.

(i) River dominated deltas, referred as high constructive;

(ii) Wave and tide dominated deltas, referred as high destructive (Figure 2.6).

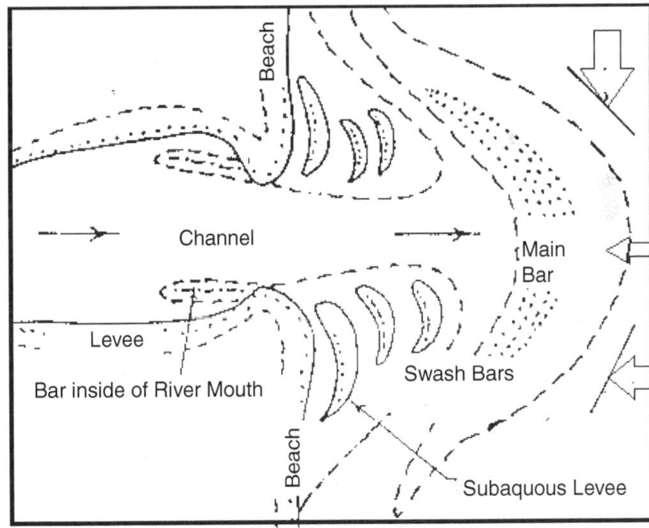

Figure 2.6. Wave Dominated Delta.

River Dominated River Mouth

The river dominated areas are characterized by sediment discharge, and high flow velocities, and with low wave energy. The distributary channels scour into underlying distributary mouth bar sand and runs as a narrow, deep straight channel into the sea. Upon abandonment the channel is primarily filled by muddy sediments. The distributary mouth bar shows a coarse grained bar crest, seawards of it is an extensive gentle sloping finer grained bar front. Seaward of the bar front is distal bar of mainly silty sediments, which ultimately grades into pro-delta clays. A characteristics feature is gradual grain size increase from pro-delta, through distal bar, into distributary mouth bar.

The river mouth sand bodies are essentially thin, laterally wide spread sheet like showing a gradational lower base and also a few bar finger sand type, running across the coast line.

Wave Dominated River Mouth

Many river mouths are exposed to prominent wave action and the wave processes produce complex changes in the depositional processes. The seaward flowing effluent causes shoaling of the waves and breaking of waves in effluent region causes intense vertical mixing, breaking down of density stratification. As a result of seaward decrease in effluent velocity, accompanied by shoreward reworking by waves, much sand is deposited at a short distance from river mouth. Lateral redistribution of sediment of distributary mouth bar by wave induced transport produces an arcuate shape. Rapid deposition of sand also takes place along the flanks of the effluent producing broad shallow subaqueous levees, which is reworked in the form of swash bars. All these sandy deposits join together to be attached to the coast and cause construction of the river mouth.

Tide Dominated River Mouth

Many rivers make their deltas in sea with macro-tidal setting and strong tidal currents. The tides cause intense mixing between effluent and ambient waters destroying the density stratification. It causes bi-directional sediment transport and the range of position of land-sea interface and of fluvial marine interaction is extended both vertically and horizontally (Figure 2.7).

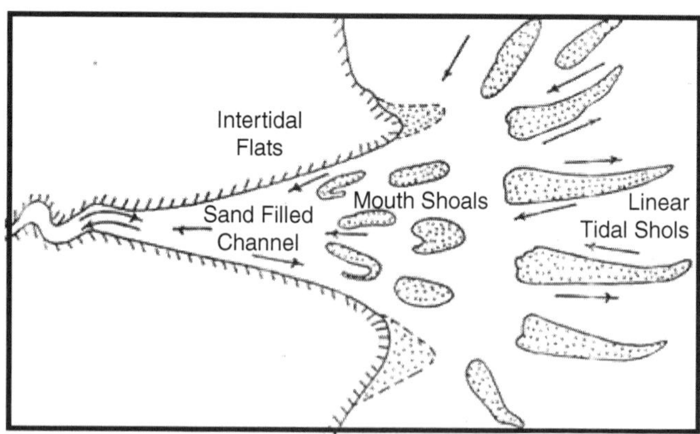

Figure 2.7. Tide Dominated Delta.

The results of strong tidal activity at the river mouth is flaring of the river mouth bar. The distributary mouth bar and distal bar is replaced by subaqueous linear ridges which are partly exposed at low tide. These ridges alternate with channels, and both are made up of sand. Landwards, these linear tidal ridges - channel merge into inter tidal flats and tidal channels and creeks. There is upstream bed load transport in the river mouth and extensive sand accumulation in the channel. The distributary channels are mostly sand filled (McClay, K. R., *et al.* 1998).

The inner continent shelves in front of deltas are characterized by extremely gentle slopes (Coleman, 1976), which are always depositional having a considerable apron of sediments. The outer continental shelves are steepened, and often receive slumped masses of pro-delta regions.

Syn-depositional Tectonics of the Basin

Most of the deltas are located in tectonically active areas, undergoing subsidence, so that, thick sedimentary column is produced. However, many deltas are formed where there is little subsidence, hence only thin section of sediment is produced. Further, the rate of subsidence also changes through time and in a given basin position of depocentres which are likely zones of maximum rate of subsidence. In areas of pronounced subsidence synchronous to sedimentation, thick sequences of same facies are produced. If the subsidence can keep pace with sedimentation, there would be little progradation. If the subsidence is high in delta plain area thick fluvial, marshy sequences are formed and in the delta front region thicker delta front deposits are formed.

The *bottom-set* deposits are made up of fine-grained clayey sediments, deposited offshore of the delta from suspension, showing sub-horizontal disposition. The *fore-set* deposits are made up of steeply inclined layers of sandy sediments making bulk of the deltaic deposit. The *top-set* deposits are again sub-horizontally layered sediments, rather thin, but made up of coarser-grained sandy sediments. They cap the deltaic sequences, and are partly sub-aerial deposit.

Pro-delta is essentially the area where deposition of fine-grained suspended load takes place from the effluent coming out of the river mouth. The sediments are typically silty clay. Due to rapid rate of sedimentation marine fauna is scarce. The laminations are related to colour and grain size differences. Bio-turbation is usually absent, but may be present as distinct thin horizons. In rare cases thin fine sand, silt intercalations may be present.

In deltas, where high amount of un-consolidated mud is deposited and the subsidence can result in to the expulsion of water, the sediments show mass movement and deformation within the sediments called shale-diaperism.

The sub-areal (exposed) area of a delta is termed as deltaic plain and that portion which forms under water is called delta front, delta platform and pro-delta. Depth of water in the basin affects the delta configuration. If other factors are equal, the same amount of sediment builds a delta which forward rapidly in shallow water but less rapid in deep water. Similarly sea level changes effect the delta sediments. In case of rise in sea level, called marine transgression, most of the land area of delta inundates due to which sediment influx decreases and deposit occur near the shore not far into continental shelf. In the case of transgression, since shore line shifts landward the waves rework the top of the delta lobe and will be buried under new non-deltaic sediments. With fall in sea level, called regressive phase, the increased river gradient with more sediments which were initially deposited in the delta may re-transported to down slope and build and sub-aqueous fan in deeper water (Daniel A. Busch AAPG Momor, 21).

Tectonic movements can affect both the supply of sediment and the area of delta growth in the basin. As the land surface gradually subside, many rivers are able to deposit sediments in their channel floors and its banks called "natural levees". Sometimes, due to sediment load of delta itself, the basinal area subsides depending on the nature of the underlying crust of the basin floor. The critical element for smooth delta build up is the dip angle of the basin floor. In case of more angle, the sediments brought by the rivers will be dumped/dispersed into deep marine areas without proper delta build up.

A depositional sedimentary environment in a geomorphic unit named as landforms. The Barrier Island complex comprises three depositional environments, the Barrier Island, the Lagoon behind the barrier and tidal channel between two barrier islands. The small deltaic deposits built within these tidal channels are called Tidal Delta or Ebb Delta. It is aligned more or less perpendicular to the coastline. Beach Ridges are attached to the land and aligned parallel to the shore-line. The outward continental extension of the delta is restricted by beach ridges. The reworked sediments due to wave action can get re-deposited like ridge along the coast called Spit. The tidal flats are feature less plain dissected by a network of tidal channels. The very fine grained sediments in this environs are called mud flats.

Basic Types of Deltas

The types of deltas formed at the mouths of streams are determined by the relative densities of the effluent water and the water of the embayment. If the turbid water of the stream has a density exceeding that of the water of the embayment, in such case the sediment-bearing fluid slides under the water of the embayment and tends to erode a submarine channel along the shelf of the basin and to carve a submarine canyon in the bethel part of a continental slope.

A submarine delta will be deposited just basin-ward from the point of abrupt decrease in gradient. Such deltas are formed far offshore in deep water as the result of density or turbidity currents. The classification based on shape is more diagnostic.

In the constructive phase of delta formation it is characterized by deposits of active building process. When distributaries are over extended, the river is diverted to a new course and initial delta lobe is abandoned. The destruction phase is characterized by reworking process by waves, tides and currents within the basin. Based on the geometry of deltas they are termed as Bird Foot, Lobate, Arcuate, Cuspate and Estuarine.

Arcuate Delta

It can recognized the similarity in shape of the subaerial part of the Greek letter (Δ) referred to accretionary area of a delta. The shape (Figure 2.8) reveals a three sided outline with the shore-ward margin convex (or arcuate) in a seaward direction. **Godavari Delta is an arcuate** delta. All typical deltaic landforms are marked in the area conspicuous one being Kakinada Spit. Based on sedimentological and landsat studies, progradation of the delta towards south-east are marked in the quaternary sediments. Based on our exploratory data it can be said that, good delta build up is seen in the basin since Late Eocene with major marine transgression during Middle Miocene followed by a regression.

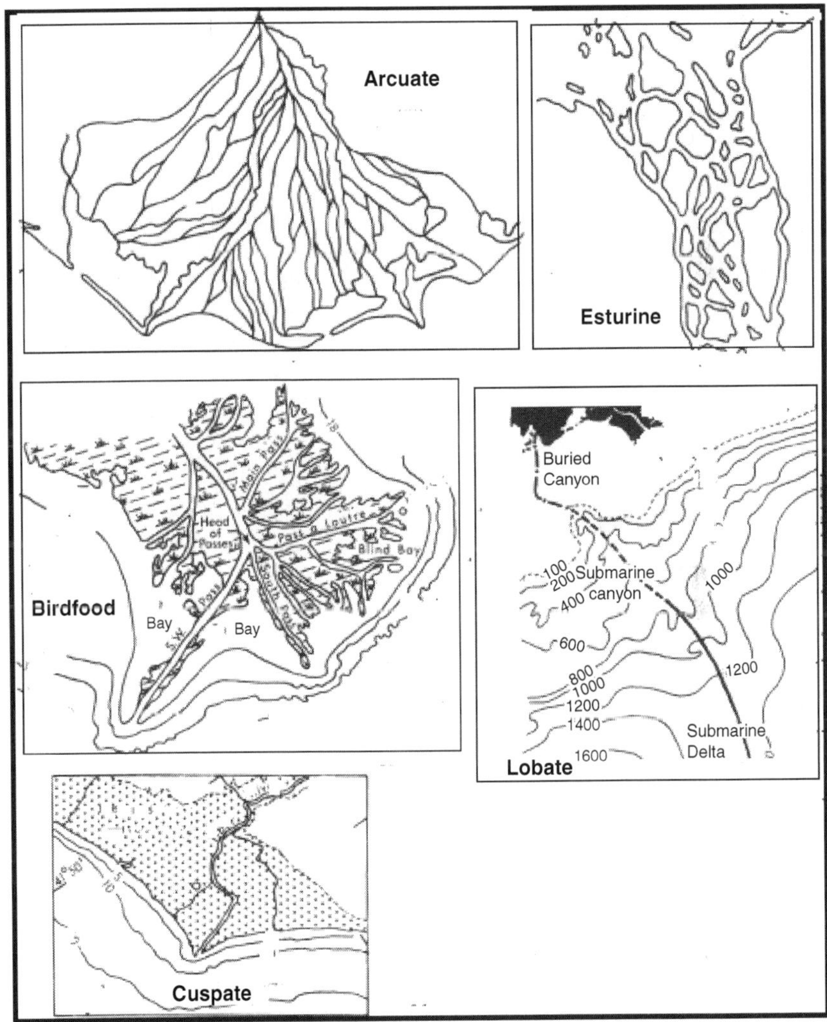

Figure 2.8. Types of Deltas by Deposition.

Estuarine Delta

The estuarine delta is funnel-shaped and has a length generally several times greater than its maximum width. The distributaries are braided and are separated by sand-bars or islands, which may be silty at places. Estuarine sediments may accumulate in certain river mouths along shorelines of submergence. Estuarine deltas, however, do not accumulate in all estuaries. They form only when the opposing forces of stream and tidal energies are nearly equal and the streams are transporting finer clastics like sand and silty sediments with clay.

Birdfoot Delta

This type of delta has the shape of a bird's foot. The distributaries are relatively few in number and the width of the individual channels is great compared to their depth. Such deltas are built into deeper waters, either near the edge of a continental shelf or in a lacustrine environment where there is a comparatively abrupt download of sediment at the mouth of a stream draining into a lake.

Lobate Delta

In *lobate* delta, the sediments are distribution in a basin-ward direction. Initially, at lower sea level, the river eroded a canyon across the sub-aerially exposed shelf. It consists mainly of red shales into which were incised deep channels that subsequently were filled with sands. The main channel did not split, but during time many side channels form at various points along its course. ***Krishna Delta*** is a lobate delta, which conspicuously protrudes into the sea, whereas apex of the delta is far away from the coast. The action of river is more predominant over the waves and currents. The modern distributaries branch out from the main river.

Cuspate Delta

A cuspate delta is one in which triangle-shaped deposits accumulate on either side of the main channel as it accretes seaward from the original shoreline. This type of delta presents a striking contrast to the arcuate delta in that it converges basinward almost to a point. Further, it is split down the middle by a single channel without any significant distributaries.

The canyons mostly possess prominent V-shaped valleys and invariable located in front of the river mouths. Canyons of different magnitudes have also contributed to the total sediment piles in the deep water parts of the basin. The sediments near the canyon mouth are silty sand, while with in the canyon sediments are silty clay and clayey silt, with moderate sorting.

Origin of canyons may be varied, but most of them probably originate during sea-level low stand as fluvial channels on the mid to outer shelf. The most spectacular canyon is the one in front of the Ganges delta, the "Swatch of No Ground". There is Chilaka canyon in front of Mahanadi delta, Godavari and Goutami canyons in front of Godavari delta, Vasista trough, Machili and Nagarjuna canyons in Krishna delta region. Coleroon and Cauvery canyons are present near Cuddalore canyon (Mohapatra G P, *et al.*, 2002).

Bengal Fan is the largest deep sea fan occupied a large part of the Bay of Bengal with a thick pile of sediments mostly of young Neogene and Quaternary age. However, the distal part of the fan extend very far in the Bay.

Deltas and Hydrocarbon Exploration

Deltas are prime centres for hydrocarbon exploration because these areas provide all ingredients required for the task. The luxurious vegetation in Mangrove Swamps (coastal vegetation) and rich organic matter coupled with continuous subsidence of the strata provide good source material for hydrocarbon generation. During growth of a delta, the bottom organisms of in the basin floor are affected by the influx of sediments. However, the planktons (floating organisms) get adjusted according to the ways in which the water from rivers mixes in to open sea basin.

The interplay of sand and shale bodies in cycles of deposition provide both reservoir and cap facies. The continuous subsidence provide overburden load by deposition of younger sediments over the source which help in the expulsion of hydrocarbons from source to reservoir which is called the primary migration. The syn-sedimentary subsidence provide entrapment conditions during movement of hydrocarbons within the reservoir (secondary migration) by growth faults and related roll over structures. In growth faults the throw grows with sedimentation in down-thrown side of the fault plane due to syn-sedimentary nature and is in the process of compaction. Finally the thick clastic wedges form against the land will provide stratigraphic pinch outs for accumulation of hydrocarbons even within thin sand bodies (Bates, C. C., 1953).

Marine Deposits

The deposition in the marine environment occur under a wide range of geological and hydrological conditions ranging from coastal shallow marine area to very deep water area. Two factors which control the sedimentation in marine environment are topographic profile of the sea bed and waves including bottom currents which redistribute the sediments in different areas. The sea bed is divided into four areas based on water depth and profile of the sea floor. Very shallow continental shelf area, steeply dipping continental slope, gently dipping continental rise and abyssal plain at very deep water (Figure 2.9).

The area where shelf turns in to slope it is called hinge line or shelf-edge. The physiography of shelf area is depend on basement/tectonic features like highs and lows. The continental slopes are characteristic of submarine canyons (cut valleys along the slope) which will carry the sediments into deep marine part.

Transgression and Regression

Sea-level changes relative to the land surface, control the succession of sediments in a basin. At times, sea level rise by as much as a couple of hundred meters, creating shallow seas that submerge the interiors of continents. At other times, sea level fall by hundred meters, exposing the continental shelves. Global sea-level changes may be due to a number of factors, including climate change, which controls the amount of ice stored in polar ice caps and causes changes in the volume of ocean basins. Sea level at a location may also be due to the local uplift or sinking of the land surface. When relative sea level rises, the shoreline migrates landward, the process is called transgression. When relative sea level falls, the coast migrates seaward that process is regression.

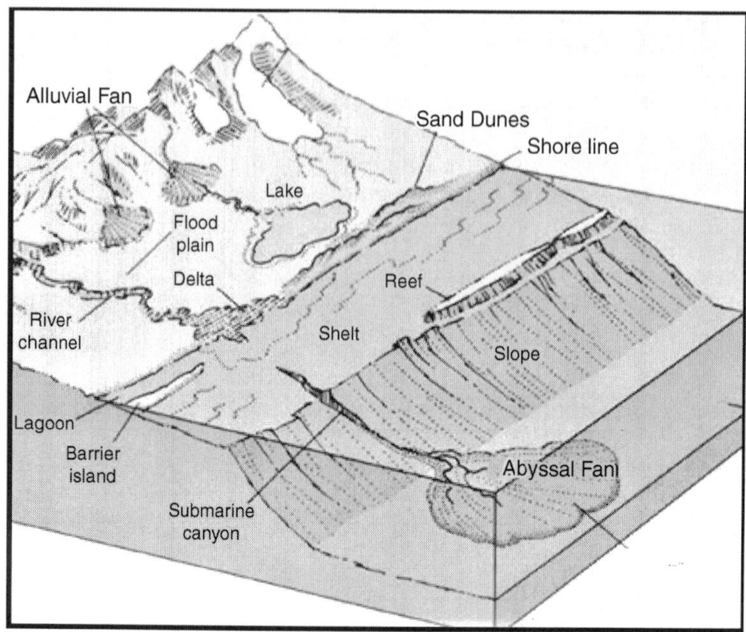

Figure 2.9. Depositional Environments.

If the quantum of sediment supply is more, the delta (land-part) grows beyond the coastline it is termed as progradation of delta, due to which the marine conditions move farther towards sea side. It is called marine regression because of which the area previously covered by marine sediments may become deltaic.

It is also known that the coarser clastics are carried by water through streams, channels, canyons, *etc.* and very fine grained silts and clays are carried by suspension in water. The limit beyond which only clays/silts prevail is termed as strand line meaning more or less the sand limit line. Generally the strand line demarcate the pro-delta regime where mainly clays are deposited.

The width of the continental shelf also controls the areal extent of submarine deltaic and shallow water deposits. Every continental land mass has its own characteristic width of the shelf. If width of the shelf is more it facilitates the smooth delta progradation, otherwise voluminous sediments brought by fluvial systems are dispersed into deep water through submarine canyons.

In an attempt to place sedimentary facies in a tract extending from the continental, shallow marine to slope and deep basin environment, they are described as *undaform* where shallow marine and non-marine facies are deposited, *clinoform* where thick progressively deeper water facies are deposited and *fondaform* where in deep water facies are deposited.

The marine sediments are deposited under three principal types of environmental conditions.

- On the outer (seaward) shelf, the slope and the continental rise as a result of slumping, sliding and tectonic activity.
- In submarine canyons, fan valleys and fans by both bottom and density currents
- In abyssal plains by density (turbidity) and bottom currents.

The sedimentation processes in shelf area may be constructive or destructive. In the constructive process, the deposition occurs either in suspension or in traction. Deposition through suspension occurs off active deltas and in pro-delta regime. Storm ebb currents may transport some suspended sediment seaward. Traction deposition is tidal flow. Where tides are dominant, some of the near shore deposits are transported across the shelf by tidal currents and deposited into submarine canyons and density flows. In the area of abandoned deltaic regime destructive process include reworking of older sediments where sediments are bypassed to the slope.

The near-shore or transition environment deposits are called paralic which are of fluvio-marine origin. These are linear submarine sand bars deposited under sub-tidal environment occurring at some distance from and extend generally parallel to shore line built chiefly by tidal, oceanic, storm or wave generated currents. The Sabkha is a surface of deflation formed by the removal of dry loose particles below the ground water level where evaporite minerals precipitate. The local depressions near the coast may be temporarily inundated by water to form Playa Lake. The water lasts until evaporation dries up the lakes and leaves behind the suspension and solution lodes.

Sometimes a delta is abandoned because of upstream diversion of fluvial channel. In such area if marine transgression occurred, the older deltaic sediments are rework due to marine activity. The sediments are winnowed and re-deposited along the advancing shore-line and final sediments are deposited farther offshore. Over a period of time, as the transgression proceeds, shallow marine sands are deposited over the older deltaic sediments in turn overlain by marine silts and clays.

Sediments deposited over the shelf can be grouped as biogenic with organic origin composition of which varies with climate. Calcareous biogenic are common in low latitudes in the warm oceans and siliceous are at high latitudes in cold water regions. Authigenic that is, *in situ* origin, when an area is starved of fresh sediments, the upwelling phosphatic rich waters will directly precipitate as phosphates within the fossil debris. Terrigenous means clastic sediments which include both suspended and reworked by waves and deposited in shallow waters.

Formation of carbonates is mainly by bio-chemical processes in shallow waters where conditions for biological and physico-chemical fixation of carbonate are optimum. As a result of carbonate accumulation build up to sea level and above (atoll) resulting in a characteristic sequence of deposits in which each unit is deposited in progressively shallow water.

Deposition within the slope systems principally takes place by same variety of gravity or density flow transport and deposition. Submarine canyons and fans are common features associated with continental shelves slopes and rises. The canyons are v-shaped and have steep walls and gradient. Though the origin of deep sea canyon is uncertain yet, it is believed that, it is related to paleo-rivers and formed during low sea-level stage. The coarser and poorly sorted sediments occur in canyons. Sands are common in distributary channels and fan valley and on the lower parts of the open fan. The sand bodies are usually thin and very lenticular and are inter-bedded with fine grained sediments. Relatively intense burrowing by benthic organism occur on the continental slopes. Upper and middle slope are the sites of more intense bio-turbation. The sediments transported through canyons are deposited at base of the slopes or on the continental rise are called submarine fans. The lower parts of fans merge with the abyssal plain

The surfaces of the fans are characterized by lower gradient, distributary channels with natural levee and intra-channel area. The manner in which these clastics are transported is believed to be turbidity currents. The bottom currents play vital role in sediment transportation.

In abyssal plains, the sediments in suspension are deposited which are rich fossiferous oozes containing microscopic calcareous/siliceous organisms that inhibit the upper layers of deep ocean basins. These are called pelagic sediments and become dominant in deep basinal area.

In restricted deep marine basins the evaporites (salt dome) are deposited due to precipitation of minerals by evaporation of saline brines.

The sedimentary unit is the one type of depositional element; units of this rank are classified as channel-complex or floodplain, and represent large-scale features. Each depositional element can also be subdivided into minor litho variation-scale units and are called facies associations interpreted as sub-environments.

Where sedimentary rocks can be observed at outcrop or in cores, a facies can be defined on the basis of colour, bedding, composition, texture, fossils and sedimentary structures. If the biological content of the rock is its dominant aspect, then it should be called a bio-facies, an ichno-facies being distinguished by its suite of trace fossils. If fossils are absent or not considered the emphasis is to be placed on the physical and chemical characteristics of the rock, then it should be called a litho-facies.

Sedimentary Basins-Classification

Basin is a bowl shaped area with rock strata dipping inwards. The word is used in structural geology to signify a syncline that is circular or elliptical in plan (Billings, 1954 and Sloss 1949) defined basins ascertain areas of the cratons which are subject to sinking to an extent greater than that of the surrounding areas. The sedimentary basins are segments of the earth's crust which have been down warped for a considerable geological time (Landes 1951). A basin represents a unit of geological structure which received sedimentary succession unique to it during a given span of geological time (North 1971).

Simply the sedimentary basin is a low area in the Earth's crust, of tectonic origin, in which sediments accumulate. Sedimentary basins range in size from as small as hundreds of meters to global scale called ocean basins.

The tectonic frame work and structural style of sedimentary basins cannot be viewed in isolation confined to the small area of a basin, but should be analysed as a mega event of the earth's crustal evolution during which, vast regions were subject to deformation resulting in the formation of a basin.

Types of Basins

Sedimentary basin is that part of the earth's crust where significant amount of sediments are deposited and preserved, irrespective of size, shape and structural attitude. In the initial concepts these are known as troughs accumulating thick pile of sediments which may result into mountain building through iso-static readjustments. Thus the earlier ideas include geosynclinal theory and orogeny (mountain building activity). These geosynclins are classified based on their geographical location and geometry (Aubouin, J., 1965).

A **geosyncline** is large trough like down warping of the crust in which rocks are accumulated. In other words a geosyncline is a mega basin with the existence of thick successions of sedimentary rocks in orogenic belts. Such accumulations are deformed during later phase of the geological cycle.

I. Intra-cratonic: Auto-geosyncline-elliptical with highlands. Zeugo-geosyncline -adjoining faulted high lands, Exo-geosyncline – clastic wedge from mountain belt. Exo-geosynclines mean sedimentary basins situated at cratonic edges essentially filled by deltaic complex. Auto-geosynclines situated within the cratonic regions called as interior basins. Zeugo-geosynclines also situated within the craton with detritus derived from intra-cratonic mountains

II. Bordering craton: Mio-geosyncline -non volcanic sediment fill. Eugo-syncline – volcanic clastic sediments. It is called Mio-geosyncline if the prograding wedge of mostly shallow-water sediment at a continental margin (Kay, M., 1947). These are more peripheral to the continents and folded only after their neighbouring eugeosynclinal zones have been folded.

III. Post-orogenic: Epi-geosyncline – adjoining complex uplifts. Taphro-geosyncline – Rift. Paralic-geosyncline – coastal basin. Epi-eugeosynclines refer to foredeep and inter-mountain basins. Taphrogeosynclines refer to faulted graben and half graben type basins. Paralic-geosynclines which refer to successor type marginal basins superimposed on earlier geosynclinal basins.

The sedimentary basins are also referred as *Mesogeosyncline* in between continental zone and Mono, Poly, and *Para-geosynclines* along the continental margins with emergent border lands. The *Ortho or true geosynclinal* basins are characterised by gentle folds. Among the true geosynclines are eugeosynclines characterised by paleo-magmatism both at the initial stage of basin formation and later stage by plutonic intrusions (Schuchert 1923).

The important effects of tectonics on sedimentation include: - formation of a basin where sediments can be deposited, nature of sediment, rate of sediment supply, rate of deposition and depositional environment.

Sedimentary basins are the regions of long-term subsidence creating space for infilling by sediments. The subsidence can result from a variety of causes that include: the thinning of underlying crust, sedimentary, volcanic, and tectonic loading, and changes in the thickness or density of adjacent lithosphere. Lithosphere is considered as a continuous spherical layer, which is broken in to number of rigid fragments called "Plates". Movement of such parts of the crust is termed as "Plate tectonics". So the primary classifications to distinguish basins formed in various plate tectonic regime like: divergent, convergent, transform, intra-plate and the proximity of the basin to the active plate margins, and whether oceanic, continental or transitional crust underlies the basin (Figure 2.10).

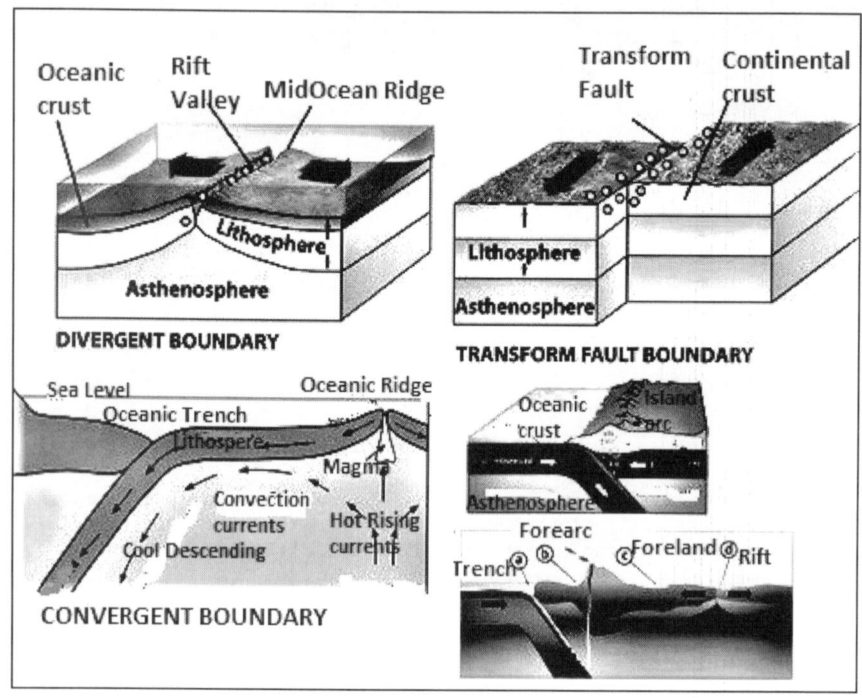

Figure 2.10. Tectonics-Basin Evolution.

Continental rifts: The most common basins associated with continental rifts are half grabens developed on the hanging walls of normal faults. The summarized conceptual models for the tectono-sedimentary evolution of continental rift basins, including their three-dimensional development. Strike-slip deformation*:* Deformation of the lithosphere such that faults are vertical and occur as a result of near horizontal maximum and minimum principal stresses. The resulting zones of subsidence are known as strike-slip or pull-apart basins. Basins formed through strike-slip action occur where a vertical fault plane curves. When the curve in the fault plane moves apart, a region of transition results, creating a basin

In fact, plate-tectonics also affect the climate as the distribution of oceans and continents, change in latitudinal position due to movements and creation of mountain chains in the case of continental collisions. In some cases, sedimentation itself affects the tectonics, to lesser extent by increasing the lithospheric loading in the basin.

Sedimentary loading dominates along continental-oceanic crustal boundaries which are supplied by major rivers and deltas (*e.g.*, continental embankments and remnant ocean basins). Tectonic loading dominates in settings where crustal shortening dominates (*e.g.*, trenches and foreland basins).

Along strike-slip faults can produce small pull-apart basins more on them later (Figure 2.11). Relief of this kind is on a small scale that may not be isostatically compensated (Weeks, L.G., 1952).

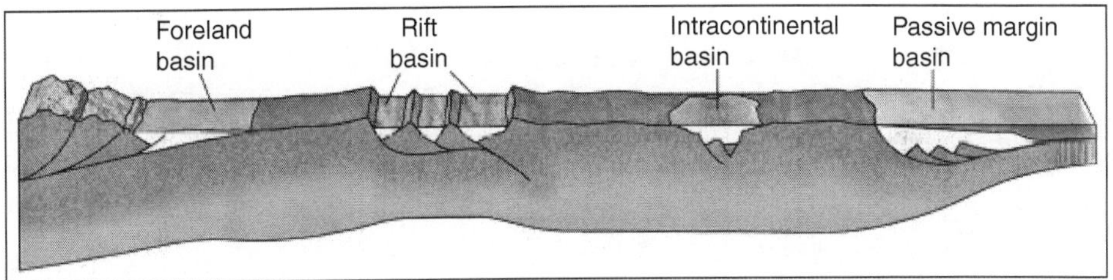

Figure 2.11. Geological Setting of Basins.

However, this classification did not account for the genesis of (basins) or forces responsible for their modification with geological time due to crustal movements.

Plate Tectonics and Basin Evolution

The crust of the Earth constitutes the outer shell of the earth. This is known as the Mohorovicic discontinuity or Moho. The outer-shell of the earth is called the lithosphere which is a brittle solid. The lithosphere can be divided into two layers the lighter continental crust and the denser oceanic crust. The parts of the outer crust (plates) are bounded by ocean ridges, subduction zones and transform faults. The transform faults are those lineaments along which plates slide by each other.

The configuration of plates has no direct relationship to the distribution of oceans and continents. About seven mega plates are marked of which only one is wholly oceanic (Pacific) and others are partly of continental crust and partly of oceanic crust (Africa, America, Antarctica, India, Eurasia and China). Six intermediate plates are also known which include: Somalia, Arabia, Philippines, Nazca, Cocos and Carribean.

Three types of relative motion between plates are possible; separation (divergence/Passive) closing together (convergence/active) and friction. Transform faults are sites of friction where there is no creation or destruction of plates.

Passive continental margin occur along the edge of the open ocean basins and such margins are characterized by minimum tectonic and igneous activity. These margins are also known as divergent margins. Triple junctions or tri-junctions are places from where three continental plates divergence in different directions. If only one plate movement is active in a tri-junction, the rift created between other two plates is termed as failed arm. Divergent margins are characteristic of prominent continent rises gentle continental slope broad continental shelves and broad coastal plains. Along passive continental margins, approximately over the transition from continental to oceanic crust formed by rifting and opening of a full-scale ocean basin which opens by spreading, the zone of heating and extensional thinning of continental crust on either side of the ocean basin subsides slowly by cooling.

Sediments, either siliciclastics derived from land or carbonates generated in place, cover this subsiding transition from continental crust to oceanic crust with a wedge of sediment to build. The subsidence is accentuated by loading of the deposited sediments. Passive-margin basins form along the edges of continents that are not plate boundaries.

Convergent Continental margin include deep oceanic trenches, steep continental slopes, narrow continental shelves and coastal plains. Due to convergence, the compression will occur in marine sediments resulting into uplift of sediments as islands and the zone is called Island Arc. The material of the subducted plate will again come-up as magma which erupt through mountain chains associated with island arcs, so these are also called as volcanic arcs.

So the evolution of sedimentary basin is inter-linked with global plate movements which develop in divergent, intra-plate, convergent, transform, hybrid, and miscellaneous settings. Within each setting are several variants, dependent on type(s) of underlying crust, structural position and sediment supply.

Subsidence of sedimentary basins results from (1) thinning of crust (2) thickening of mantle lithosphere (3) sedimentary and volcanic loading (4) tectonic loading (5) sub-crustal loading (6) asthenospheric flow, and (7) crustal densification (Figure 2.12).

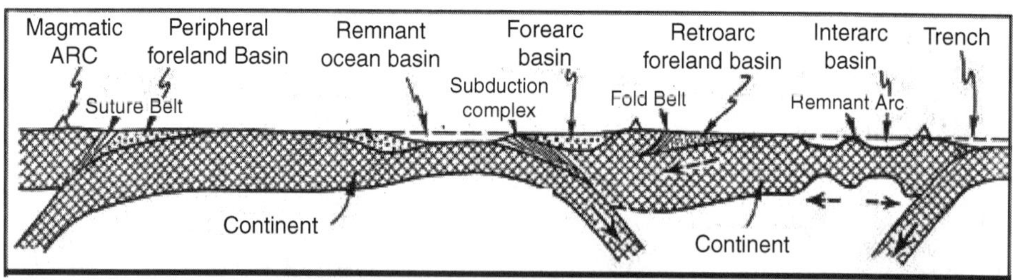

Figure 2.12. Classification of Basins-based on Tectonics.

Continental rifts may evolve into active ocean basins bordered by intra-plate continental margins with three types of configurations: shelf-slope-rise, transform, and embankment. Continental rifts that do not evolve into oceans become fossil rifts, which later become sites for development of intra-cratonic basins and aulacogens. If all plate boundaries within and around an ocean basin become inactive, a dormant ocean basin develops. Sites for sedimentary basins in convergent settings include trenches, trench slopes, forearcs, intra-arcs, backarcs, and retroarcs.

Intra-arc rifting in highly extensional arcs commonly evolves into back-arc spreading to form new oceanic crust. Highly compressional arcs develop retro-arc fold thrust belts and related retro-foreland basins, and may develop hinterland basins, broken retro-forelands may develop in former craton areas.

The continental collision generally initiates at one point and the resulting suture propagates away from this point. Remnant ocean basins form on both sides of the initial impact point, and rapidly fill with sediment derived from the suture zone. With this collision, pro-foreland basins form on continental crust of the subducting plate and collisional retro-foreland basins form on the overriding plate.

In transform settings and along complex strike-slip fault systems related to convergent settings, changing stress regimes related to irregularities in fault trends, rock types, and plate motions result in short-lived sedimentary basins.

Ocean Basins and Continental Margins

During the transition from continental rifting to seafloor spreading, transitional crust forms, as stretched continental crust. Continental rifting evolves into seafloor spreading only in the absence of significant sediments of that oceanic crust. A significant width of transitional crust typically forms on the margins of ocean basins prior to initiation of true seafloor spreading. As these transitional types of crust form and the two continental margins move apart, a new ocean basin develop. Intra-plate continental margins that originate along transform boundaries rather than rift boundaries have narrower sediment prisms and transitional crust.

Intra-cratonic Basins

Most intra-cratonic basins overlie fossil rift renewed periods of subsidence in cratonic basins can generally be correlated with changes in lithospheric stress related to orogenic activity. Subsidence occurs when lithospheric rigidity lessens, allowing uncompensated mass in the upper crust (remnants of fossil rifts) to subside over a broad area.

In contrast to intra-cratonic basins, platforms accumulate sediment of uniform thickness over continental scales. The platform stratigraphic sequences are transitional into continental margins, intra-cratonic basins, foreland basins, and other tectonic settings along continental margins

Active Ocean Basins

As oceanic crust subsides with time and distance from spreading ridges, systematic pelagic and hemipelagic deposits accumulate. The result is a dynamic and predictive stratigraphy relating the age, depth, and paleo-altitude of oceanic crust to oceanic depositional facies. Volcano-clastic and turbidite deposits continental margins makes difficult to predicte stratigraphic sequences on oceanic plates.

Oceanic Islands, Seamounts, Aseismic Ridges, and Plateaus

Islands seamounts, and ridges thermally subside as oceanic plates migrate away from spreading ridges. Thermal anomalies independent of spreading ridges (*e.g.*, hot spots) create new islands, ridges, and plateaus. The post-volcanic history of seamounts can be divided into three sequential stages: subaerial, shallow water, and deep water or bathyal. As an island is eroded and subsides, fringing reefs and atolls may form, depending on latitude, climate, and relative sea level. Dormant ocean basins are floored by oceanic crust, which is neither spreading nor subducting. Following cessation of plate activity within and around the basin, sediment loading is the dominant subsidence mechanism.

The basin forming tectonic style depends on (i) composition of the crust on which basin rests (ii) type of plate movement and (iii) position of the basin on the plate:

- ☆ Intra-continental: Mid plate rift, normal faulted basins, superimposed grabens, depressions. *Interior Sag, Interior Fracture*
- ☆ Divergent margin: Proto-margin rift, margin rift aborted segment, *Marginal Sag*
- ☆ Convergent margin: Fore arc, Fore-deep, Back arc, Foreland, Back Arc rift, *Trench, Oceanic Sag*
- ☆ Transform: Wrench Fore deep, *Oceanic wrench*

Interior Sag is located entirely within the continental crust and simply sags with minor or no faulting. More or less circular/oval in shape and generally do not accumulate great thickness of sediments.

Interior Fractures are caused by tensional faults within the continental block. Vertical faulting and subsidence are the dominant features.

Marginal Sags are located on the outer edges of continental crust. The basin areas lie parallel with the plate boundary and sediments may down-lap on to oceanic crust. When interior fracture is activated, the marginal sag is initiated as the spreading begins. Basement subsides as one block and entire edge of the continent sinks.

Wrench or shear basins are strike slip shear basins which are generated due to lateral movements. Though wrenching is known as extensional movement, they generally exhibit some divergence or convergence. Small angles of convergence show up as wrenching or fold belts and small angles of divergence appear as normal faulting or sagging.

Oceanic Sags: cooling of oceanic crust may result into a little sink. Thus sag areas that accumulate a significant thickness of sediments are called oceanic sag which is also known as inter-arc basin. These are small ocean basins floored by oceanic or transitional crust and are believed to have been formed by a large divergent wrench couplet.

Arc-trench systems may be categorized into three fundamental types: extensional, neutral and compressional. Arc-trench systems with significant strike slip may be considered a fourth type. Strike-slip faults may occur in all types of arc-trench system, but they are especially common in strongly coupled systems experiencing oblique convergence.

Intra-arc basins are defined as thick volcanic-volcano-clastic and other sedimentary accumulations along the arc platform, which is formed of overlapping or superposed volcanoes. A more general term, "arc massif," refers to crust generated by arc magmatic processes, and arc crust may underlie a much broader region than the arc platform.

There are at least three major types of depocenters for volcanic and sedimentary accumulations within arcs. Depocenters may occur in low regions between volcanoes and along their flanks, although these have high preservation potential only below sea-level. They also referred to "fault-bounded basins," which are rapidly subsiding basins where tectonic structures, rather than constructional volcanic features, account for relief along the basin margin.

Trench: Trench and Trench associated basins are located at convergent margins and are of two types. When oceanic plate over-ride the trench depth is less with little sediment fill on the other hand, when a continental plate over-ride, the trenches accumulate thick deep water deposits.

Dynamic nature of sedimentation and tectonics in active trenches. With increasing transverse supply of sediment to the trench, the axial channel of the trench wedge is forced seaward and the trench wedge widens. Turbidites are within trench-slope basins. Average width, sediment thickness, and age of basins increase up slope due to progressive uplift of deformed material and widening of fault spacing. In ancient subduction complexes, trench-slope basins are filled with relatively un-deformed, locally derived turbidites.

Fore-arc Basins

Factors controlling fore-arc geometry include initial setting, sediment thickness on subducting plate, and rate of sediment supply to trench. The net result of widening of the arc-trench gap is the general tendency for fore-arc basins to enlarge through time. Residual and constructed basins tend to evolve into composite basins and is accompanied by filling and shallowing of forearc basins.

Back-arc Basins

There are two types of backarc basins, one is oceanic basins behind intra-oceanic magmatic arcs, and the other is continental basins behind continental-margin arcs that lack foreland fold thrust belts. Many back-arc basins are extensional in origin, forming by rifting and seafloor spreading.

Remnant Ocean Basins

Where continental collisions occur, sutures are formed and along such sutures Remnant Ocean Basin is developed. Upon collision the peri-cratonic basins may be uplifted and become fold belts. Intense deformation occurs in suture belts during the subduction. Suture belts can involve rifted continental margins or various combinations of arcs and continental margins. Colliding continents tend to be irregular, and great variability of timing, structural deformation, sediment dispersal patterns and preservation occurs along strike submarine fans in remnant ocean basins represent the largest accumulations of sediment on Earth.

The Bengal Fan is the largest sediment body in the Bay of Bengal and the Indus Fan located in Arabian Sea is second largest; both are derived from the greatest uplifted area on Earth, the Tibetan Plateau and Himalayas.

These are within suture zones formed by continent-continent collision. Continental margins and subduction zones are with further subduction and suturing, create isolated basins still floored by residual oceanic crust, which receive abundant sediment from adjacent strongly uplifted crust.

Hinterland Basins

There are two classes of hinterland basins, those formed in non-collisional retroarc orogens (*e.g.*, Andes) and collisional orogens (*e.g.*, Himalaya-Tibet). Hinterland basins record nonmarine sedimentation, usually at high elevations, that formed on thickened continental crust.

Aulacogens

There are linear sedimentary troughs at high angles to orogens and extending from the margins toward the interiors of cratons. During continental rifting, three rifts commonly form at approximately 120 degrees, probably because this is a least-work configuration. Regardless of whether initiating processes are "active" or "passive", in the majority of cases, two rift arms proceed through the stages of continental separation, whereas seafloor spreading fails to develop in the third arm, resulting in a fossil rift. The Pranahita Godavari Graben with cross trend against peri-cratonic Krisha-Godavri basin is a failed arm of an Aulacogen.

Rifts that evolve into ocean basins generally are overlain by nascent-ocean and shelf-slope-rise continental margins, whereas fossil rifts adjoining continental margins evolve into re-entrants that capture main drainages of continental interiors; major deltas that form at these re-entrants commonly construct continental embankments (*e.g.*,Godavari Delta).

Rift Basins are within continental lithosphere on cratons. Lithospheric extension on a craton, presumably by regional sub-lithospheric heating, causes major rifts. Some such rifts continue to open and eventually become ocean basins floored by oceanic rather than continental crust. In other cases, the rifts fail to open fully into ocean basins, so they remain floored by thin continental crust rather than new oceanic crust. Rift basins form in continental rifts, regions where the lithosphere is stretching horizontally, and therefore thins vertically. As the rift grows, slip on faults drops blocks of crust down, producing low areas bordered by narrow mountain ridges. These troughs fill with sediment.

Oceanic Rift Basins

This category of basins is transitional between intra-continental rift basins and passive-margin basins. Basins have opened wide enough to begin to be floored with oceanic crust but are still so narrow that the environment is either still non-marine or, if marine, has restricted circulation.

Foreland Basins

There are two kinds of foreland basins, *retro-arc foreland basins*, which are formed on stable continental crust by loading by thrust sheets moving toward the continental interior as a result of compression and crustal shortening in an ocean-continent subduction zone, and *peripheral foreland basins*, formed after continent-continent collisions by loading of the continental crust of the subducted plate by development of thrust sheets in the continental crust of the subducted plate directed back away from the subduction zone. Both kinds tend to be asymmetrical, with their deepest parts nearest the emplaced thrust sheets. They tend to migrate away from the arc or suture zone with time. They are filled by sediments derived from the mountainous terrain associated with the compression and thrusting. Sediment fill is coarse, fluvial, siliciclastics, mainly as alluvial fans, thinning and fining away from the arc or suture, often passing into shallow-marine sandstone-shale successions if sea level is high enough to flood the basin. Thickness may up to many thousands of meters. The classic molasse facies, thick non-marine conglomerates, is deposited in foreland basins.

Pull-Apart Basins

Locally along major strike-slip or transform faults, either in continental crust or in oceanic crust. If a strike-slip fault is stepped or curved rather than straight, movement along it tends to produce tension, where the sense of the curvature and movement are such that the walls of the fault are pulled apart from one another. In the tensional segments, gaps or basins are produced which are filled with sediment from adjacent high crust. Sediment fill is thick coarser to fine clastics, often as alluvial fans passing into lake deposits or into deposits of restricted marine environments. In some cases thick marine turbidites fill the distal parts of the basin. Deposition is concurrent with elongation of the basin.

Isostasy

If faulting ceases and the sediment sheets are eroded away, the foreland basin will be rapidly uplifted as the plate regains isostatic equilibrium. Note that erosion is an important form of unloading which itself drives further uplift. Basins formed by extension of the plates are quite different since isostatic equilibrium is maintained throughout their development, from the rifting stage right through to the end of thermal subsidence.

Orogenic Uplift

The uplift and erosion of sedimentary basins is called basin inversion. The term is most commonly applied to extensional sedimentary basins which have been shortened, showing normal faults which have been reactivated as thrust faults. Epeirogenic uplift besides shortening caused by horizontal plate movements (known as orogeny), there is another important way in which sedimentary basins become uplifted and eroded known as epeirogeny.

Basin Analysis

It is the study which involves application of sedimentological, stratigraphic and tectonic principles. It aims at developing a full understanding of the rocks that fill the sedimentary basin for the purpose of understanding their geological history and evaluating their economic importance if any (Boggs 1995).

The basin modelling/analyses based upon structure, isopach (sediment thickness), lithofacies, paleogeologic and the geophysical maps. So the basin analysis requires a complex reconstruction of basin dynamics on the basis of physical principles and mathematical modelling (Allen, P.A. and Allen, J.R., 1990). Not only the shape of basins and the rate of sedimentation should be reconstructed, but also heat flow, diagenesis, fluid flow,, *etc.*

Indian Scenario

The total sedimentary cover in India up to 400 m isobath, is 3.14 million square kilometres. Many sedimentary basins have so far been identified based on geologic and tectonic considerations (Figure 2.13). Out of the total sedimentary area, only about 20 per cent is moderately to well explored, and around 36 per cent remains less explored to unexplored (Ajay Dwivedi, 2016).

Biswas, *et al.* (1982 and 1987) published classification of Indian sedimentary Basins using the Dickinson (1976) scheme of Basin classification based on the structural style, geometry and size, nature of stratigraphic fill and thermal history and grouped them in three main settings: (i) intracratonic, (ii) rifted, (iii) orogenic settings (Table. 2.1). About 38 sedimentary basins in plate tectonic set up were recognized in onshore and offshore areas of the Indian plate (Raju., A.T.R. 1974) including Western and Eastern deep (>1000m bathymetry) ocean basins (Purnachandra Rao, V. and Kessarkar, M. P. 2001). Most of the basins are peri-cratonic. In the south, the basins are coastal basins extending from onshore to offshore (Bastia, R., et. al. 2006). In the northern part the basins are bordered by orogenic belts of Himalaya and Assam Arakan. These the basins are peripheral foreland basins subducting below the orogenic belt. Peri-cratonic basins in the north are separated by ridges which are first order basement arches (Prabhakar, K. N. and Zutshi, P. L. 1993). In the southern coastal region the basins are rifted basins separated by horsts (Fuloria, R. C., *et al.*, 1992). The intra-cratonic basins are rifted basins like Gondwana basins or structurally controlled depressions such as Vindhyan. Cuddapah. Kaladgi and Bhima basins.

Petroleum occurs mainly in marginal and peri-cratonic rift basins in the southern coastal belt. Occurrence of petroleum in the northern peripheral foreland basin is yet to be discovered excepting in the Upper Assam inter orogenic foreland basin. The basin is situated on the cratonic wedge between two mountain belts, Himalaya and Assam-Arakan Mountains. Tectonic setting and collision history at this north-eastern corner of the Indian plate created favourable habitat for generation and accumulation of hydrocarbons. Interior and intra-cratonic rift basins are also devoid of hydrocarbon excepting a few gas shows in Gondwana rift basins and a single show in the Vindhyan basin. These intra-cratonic and northern foreland basins are categorized as "Frontier Basins" for exploration.

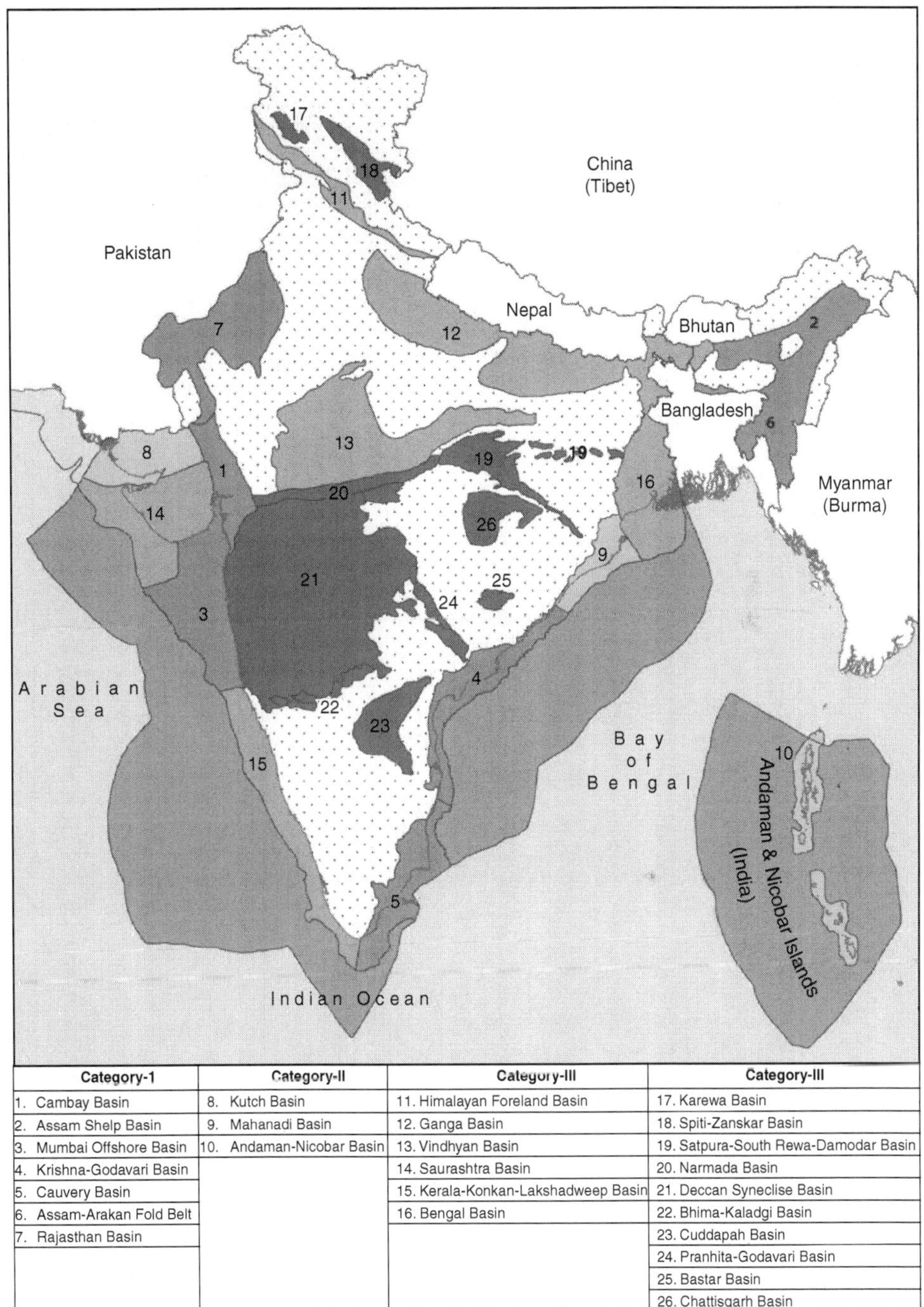

Category-1	Category-II	Category-III	Category-III
1. Cambay Basin	8. Kutch Basin	11. Himalayan Foreland Basin	17. Karewa Basin
2. Assam Shelp Basin	9. Mahanadi Basin	12. Ganga Basin	18. Spiti-Zanskar Basin
3. Mumbai Offshore Basin	10. Andaman-Nicobar Basin	13. Vindhyan Basin	19. Satpura-South Rewa-Damodar Basin
4. Krishna-Godavari Basin		14. Saurashtra Basin	20. Narmada Basin
5. Cauvery Basin		15. Kerala-Konkan-Lakshadweep Basin	21. Deccan Syneclise Basin
6. Assam-Arakan Fold Belt		16. Bengal Basin	22. Bhima-Kaladgi Basin
7. Rajasthan Basin			23. Cuddapah Basin
			24. Pranhita-Godavari Basin
			25. Bastar Basin
			26. Chattisgarh Basin

Figure 2.13. Sedimentary Basins of India (After Ajay Dwivedi, 2016).

The ***Mumbai offshore basin*** is situated on the western continental terraces of the Indian landmass. The eastern portion is bounded by the exposure of the Deccan Trap, while the western portion is marked by a basement high arch, corresponding to Kori Prathap Ridge (Harbison and Bassinger 1973; Naini and Talwani 1983; Bhattacharya and Chaubey 2001). Drilling in the Bombay High field, India has revealed large volume of oil and gas, within the basement rocks composed of basaltic and granitic gneiss. The Bombay High, the largest oil field in India, is situated around 161 km west-northwest of the Mumbai coast. Most of the oil and gas is contained in Miocene limestone reservoirs and a smaller proportion in the basal clastic sandstone reservoirs, which overlie the fractured metamorphic rocks. The main limestone reservoir is encountered at a depth of approximately 1300 m but the basement hydrocarbons occur at 1900 m, on the crest of the structure (Akbar 1993). It has been found that the Bombay High basin shows a deep crustal root of about 25 km. while in the eastern and western portion, the basement becomes uplifted. The basement topography/crustal root in the Bombay High basin is relatively smooth (Bhattacharyya., R and T. J. Majumdar., 2013).

The ***Cambay basin*** derives its importance by virtue of its being one of the two known oil bearing provinces of India. The basin is located in the alluvial plains of Gujarat and extends into the Gulf of Cambay. Tectonically, it is situated at the rim of the Deccan crystalline shield on the north-western edge of the Peninsula (Srinivasan, S. and Khar, B.M., 1995).

The basin is of intra-cratonic type formed by discontinuous normal faults. While the northern part of the basin is a graben, the southern part is mainly a half graben. It is known to be a Cenozoic basin with basic lava flows (known as the Deccan Traps) forming the basement of the overlying Paleogene and Neogene sediments. The presence of sporadic Mesozoic outcrops on the flanks of the basin points to the likelihood that the basin may have existed even in the Mesozoic Era. Further north, the basin is connected with the Kutch Mesozoic basin and with Jaisalmer-Barmer basin, which is the shelf region of the large Indus geosyncline.

The mild tectonism of the basin has probably played a minor part in controlling the sedimentary facies. Eocene sediments, thicker than the other time stratigraphic units, are overlain by Oligocene sediments, which have been laid down in a regressive phase. A prominent disconformity separates the Paleogene sequence from the Neogene.

It is thought that a deep seated fault, formed probably during the end of the Gondwana period, runs parallel to the western coast of India and into the Cambay basin. This fault zone is cut across into two parts by an important transverse basement fault running along the Narmada River and joining up with the southern coast of Saurashtra peninsula. The area to the south of the Narmada River forms the north-eastern corner of the southern part of the fault zone. The other part of the fault zone comprises the rest of Cambay basin (Mathur.,L P.*et al.*, 1968).

Godavari Delta

In geology the dictum says "present is key to the past". So many geologists have studied the present Godavari delta. The extensive deltaic plain with an area of 6,322 sq.km formed by the Godavari River and in the adjoining areas. The coast line along the delta extends 150 km, from Uppada in the north to Narasapur in the south and the delta is arcuate in shape and wave dominated. The Neogene deltaic sequence of Godavari offshore, in particular, is over 3,000 km, thick sedimentary pile, extensively deformed by growth faults and rollover features (Rao, G. N.2001). One of the factors contributing to the development of thick deltaic sequence in the Krishna-Godavari basin is the presence of a long, linear Gondwana Rift valley basin in the hinterland (Choudhari, R. S. 1992).

Godavari River receives its sediments from a variety of geological formations. Deccan traps of late Cretaceous-Eocene period covers 55 per cent of the drainage basin. Descending from the Deccan plateau the river first traverses through a small terrain composed of Archean granites and unclassified crystallines called Dharwars. The river also crosses the formations of Cuddapahs, Vindhyans and Gondwanas. In the lower reaches, the river traverses through the Eastern Ghats and at Rajahmundry the river cuts across the Deccan basalts again and Rajahmundry sandstones of Miocene (Bastia, R. 2004).

The sedimentary environments in the Delta include river channels and estuary, the river splits at Dowleswaram into Gautami (90km), Vainteyam (35km) and Vasista (96km) which flows away from each other. The maximum width of the Gautami is 2km., whereas width of the Vasista is 1km. During the high floods, sediment load is transported to the sea through the distributaries. The Godavari has well developed flood plains on either side of the river channels. Natural levees of Godavari delta include, 500m, wide near Doleswaram and project up to 2m, above the flood plain. The sediments show alternate layers of sand and silts. Mangroves are extensively developed in the Godavari delta in the Gautami estuarine system. The sediments are mud, silt and clay. Tidal channels and narrow tidal creeks are fully developed in the delta plain of the Godavari River. These channels are flooded with tidal water, throughout the year except in the months of high floods.

In and around the Kakinada Bay extensive tidal flats are developed along the gently dipping seacoasts with significant tidal rhythms, where enough sediment is available under low wave energy conditions. Tidal flats are also occur in the Gautami estuary, Vasista estuary. There are three lagoons in Godavari delta. It is a vast, shallow stretch of water which is separated from the open sea in the south by a narrow beach and in the east by a composite system of swamps, tidal creeks and beach. Lagoonal sediments contain poorly sorted silts and fine sands (Bastia, R., *et al.*, 2006b).

Kakinada bay appears to be compound and consists of estuaries adjoining lagoon (Subba Rao, 1958). Sediments brought into the bay are mostly silty clays which should be reworked for providing the finest clay particles to the tidal flats. The finest and purest muds are deposited in the protected and deeper parts of bay where the currents are weakest. The rapid deposition of sediments at the confluences of the distributaries of Godavari in general and that of the Gautami in particular resulting in the formation of spits (Lafond,, *e.g.*, and Prasada Rao, R., 1956). The spit has lengthened continuously northward and is composed of fine sands (Bastia, R. 1999).

Assam Arakan Basin

It was the first basin where oil was found by seepages and exploration and development of oil fields since past 150 years. Oil occurring in Paleocene to Miocene sections and in subsurface anticlines to thrust belts.

Cauvery Basin

The exploratory efforts by ONGC were started in 1958, which resulted in few oil and gas fields. The basin is of east-west aligned horst and graben morphology and all three grabens are found to be commercially producing oil and gas.

Rajastan Basin

This tectonically northward extended basin found to be producing heavy oil very shallow depths in the oldest reservoirs.

Andaman Nicobar Basin

The geological mapping in the area was started in 1965. Some offshore structures were drilled in which only one was found to be producible with hydrocarbon gas. But further delination wells proved to be dry. So due to commercial viability coupled with strategic location the exploration activity is reduced.

Kucth Basin

The basin with western part of Gujarat with extension to southern deep waters of Arabian sea the exploration efforts resulted in producible hydrocarbons but the development and bringing production is still awited.

Mahanadi Basin

This basin extends from onland Odisha through flow area of River Mahanadi extends into deep waters of Bay of Bengal. The deep drilling both in some onland and offshore areas flowed mainy gas and still awaiting to establish their commercial viability.

The sub-surface data of ***Ganga basin*** defined it as a major platform depression and classified into seven tectonic zones, *viz.*, Monghyr-Saharsa ridge, East Uttar Pradesh shelf, Gandak depression, Faizabad ridge, West Uttar Pradesh shelf, Sarada depression and Delhi-Hardwar ridge. This classification is based on the continuation of major tectonic trends from the Peninsular shield into the Ganga basin, the variations in the total thickness of the sedimentary cover, and the basement configuration as deduced from different surveys (Bhattacharjee, D., *et al.*, 2006).

Table 2.1. Indian Basins: Type and Status

Category	Type	Basins	Status
I. Interior Basins	Intra Cratonic Sags	1. Vindhyan Basin 2. Chattisgarh Bastar Basin 3. Cuddapah Basin 4. Kaladgi Basin	Gas Indications

Category	Type	Basins	Status
II. Rifted Setting	IIA Marginal Aulacogen	1.Cambay Basin	Oil and Gas producing Basin
	IIB Pericratonic Rift	1.Kutch Basin 2.Saurashtra Shelf 3.Surat Basin 4.Bombay Ratnagiri Basin 5. Konkan Basin 6. Kerala Basin 7. Western Bengal Basin 8. Mahanadi Basin 9. Krishna Godavari Basin 10. Palar Basin 11. Cauvery Basin 12. Mannar Basin	Oil and Gas – Emerging Basin Oil and Gas Indications Gas discoveries – Emerging Basin Oil and Gas producing Basin Oil and Gas producing Basin Oil and Gas producing Basin
	IIC Intra Cratonic RM	1. Pranhita Godavari Basin 2. Mahanadi Graben 3. Purnea Graben	
	IID Intra-Cratonic Transtensional Basins	1.Satpura Basin 2. Son Basin 3. South Rewa Basin 4. Damodar Basin 5. Narmada Basin	
	IIE Miogeoclinal Prism	1. Deep Water Basins of West coast 2. Deep water Basins of East coast	 Oil and Gas Producing Basins
III. Orogenic Settings	IIIA Superposed Basins	1. Karewa Basin	
	IIIB Peripheral Foreland Basin	1. Rajasthan Basin 2. Punjab Basin 3. Ganga Basin 4. Lower Assam Basin	Oil and Gas Producing Basin Oil and Gas Producing Basin
	IIIC Inter orogenic foreland Basin	1. Upper Assam Basin	Oil and gas Producing Basin
	IIID Forearc Basins	1. Kohima forearc Basin 2. Tripura Mizoram Basin 3. Andaman Basin	 Gas producing Basin Gas Indications
	IIIE Remnant Ocean Basin	1. East Bengal Basin	
After S K Biswas			

The sedimentary cover over most part of the Ganga basin is essentially composed of two main stratigraphic cum structural sequences representing the two main sedimentary stages in the geological evolution of the basin. The oldest, probably corresponding to the Vindhyans, is represented by stable to unstable shelf sediments composed of quartz-arenite-limestone-shale alternations. The younger sequence, unconformably overlying the Vindhyans, corresponds to the Neogene terrigenous clastics (Siwaliks). The presence of a profound unconformity between these two groups of sediments representing a considerable time gap ranging from (?) late Palaeozoic to Paleogene is an important factor in deciphering the tectonic evolution of the Himalaya. However, towards the northernmost depressed parts of the Ganga Basin, the age of the additional thickness of sediments intervening between the two above mentioned groups remains uncertain (Sastri., V.V., *et al.*, 1971).

Vindhyan Basin

This basin is the southern limit of northerly dipping Ganga Basin. The exploratory efforts in recent times indicated good flow of gas from older reservoirs but not yet firmed up.

Bengal Basin

It is westerly extension or basin margin area of mega geosynclinal basin located in the Indian Territory whose major part is situated in Bangladesh area. In Bangladesh part, many commercially producing major gas fields are known. The exploratory drilling about 45 wells since 1958 so far resulted in commercially viable hydrocarbons. However, some wells showed generation of oil and gas without much surface flowing hydrocarbons.

Chapter 3

Petroleum Generation and Expulsion

It is well believed that, oil is formed from organic material available in sediments which include both plant and animal life. The terrestrial organic material is being carried by rivers and deposited into ocean basins where bulk marine organic matter also available with in sediments.

Petroleum forms by the breaking down of molecules of fats, oils and waxes that contributed to the formation of kerogen. As marine life died, it settled at the sea bottom and became buried in layers of clay, silt and sand. The gradual decay by the effect of heat and pressure resulted in the formation of many chemical compounds.

For any commercial accumulation of hydrocarbons, the optimal availability of the following elements is very essential:

- ☆ Source rocks for generation.
- ☆ Maturity of organic matter.
- ☆ Expulsion of hydrocarbons from the source to reservoir (Primary migration).
- ☆ Migration of oil and gas within the reservoir (Secondary migration).
- ☆ Entrapment by both vertical and lateral seals (Trapping mechanism).

It is a well-known fact that, if oxygen is present which may destroy the organic matter by burning. In ocean basins up to the level where light penetrates (called photic zone) and above the wave base oxygen is available. So to protect organic matter from destruction, the first requisite is the sediments should reach reducing environment very fast. Unless faster rate of sedimentation prevails, sediments may not reach reducing environment quickly. In turn, if sedimentation rate is more, there could be dilution of organic matter in the sediments because for a limited supply of organic matter if quantum of sediments are more its percentage of concentration is likely to reduce. On the other hand if rate of sedimentation is slow, most of the organic matter may be destroyed. So, the first requisite is to have optimal percentage of organic matter called total organic carbon percentage (TOC per cent) in the sediments. If good concentration of TOC implies that bottom oxygen is minimum with intermediate rate of sedimentation. TOC with more than 1.5 per cent can be considered as optimal percentage for generation of hydrocarbons in commercial quantities.

Though carbon and hydrogen available in organic residue, its transformation into petroleum needs specific pressure and temperature conditions, otherwise the available carbon can be fixed as coal.

Petroleum geochemistry has been primarily an exploration-oriented subject and provides understanding of origin, migration, accumulation and alteration of petroleum in the earth's crust. With the advancement of various analytical techniques many oil companies have started applying petroleum geochemistry to understand various problems associated with exploration and reservoir management (Kuldeep Chandra and S.N. Bhattacharya 1966).

Source Rocks

Delineation and identification of potential source rocks depend on the studies of the local stratigraphy, paleo-geography and sedimentology to determine the likelihood of organic-rich sediments having been deposited. If the likelihood of existence of a source rock is thought to be high with optimal percentage of organic content, the next subject to address is the state of thermal maturity of the source material, and the timing of maturation. Maturation of source rocks depends on temperature and pressure.

Petroleum source beds are fine grained, clay-rich siliclastic rocks (mudstones, shales) or dark colored carbonate rocks (limestones, marlstones), which might have generated hydrocarbons. A petroleum source is characterized by three essential conditions: it must have a sufficient content of finely dispersed organic matter of biological origin; this organic matter must be of a specific composition, *i.e.*, hydrogen-rich; and the source rock must be buried at certain depths and subjected to proper subsurface temperatures in order to initiate the process of petroleum generation by the thermal degradation of kerogen.

If the bottom waters don't have enough oxygen, there are no living organisms on the bottom. The organic stuff is layered in the sediment and gets buried. When enough overburden of sediments piles on top of the organic layers, the temperature starts to increase due to geothermal gradient. The combination of pressure of the pile of sediment and the increased temperature within the earth cause the organic material to change into petroleum trapped in the sediment. Oil and gas result mostly from the rapid burial of dead micro-organisms in environments where oxygen is so scarce that they do not decompose. This lack of oxygen enables them to maintain their hydrogen-carbon bonds, a necessary ingredient for the production of oil and gas.

Three elements control the deposition of organic-rich sediments: productivity, dilution, and preservation. Biological productivity determines the amount of organic matter that is contributed to sediments. Dilution refers to the amount of inorganic minerals that mixes with the organic matter. Once deposited, the organic matter must be preserved in a form that may later generate petroleum.

Living organisms are mostly composed of proteins, nucleic acids, and lipids. Structural materials, such as lignin and cellulose, and various resins. Most of the organic carbon is bound in a condensed, insoluble macro-molecular material, termed kerogen. It is either directly related to biopolymers in living organisms or formed during diagenesis.

Sometimes, specific bacteria may activate the hydrogenation from organic compounds. The catalytic action will accelerate these reactions. Certain elements like Vanadium, molybdenum, nickel and silicates may act as catalysts in the process. The radioactive bombardment within the sedimentary sequence may further facilitate the transformation of organic matter into petroleum. Above all, optimal pressure and temperature are the primary requisites for the conversion of organic matter in to petroleum. Sometimes though the temperatures are less it can be compensated by exposure of sediments to the process for more geological period. Similarly if geothermal heat supply is more than the normal even younger sediments may generate hydrocarbons by early maturation of organic matter. The minimum time required for the transformation is about one million years.

Kerogen quality is mainly a function of hydrogen content – kerogens with high H/C ratios (> 1.2) are oil-prone, while those with lower H/C ratios (0.5 to 0.8) tend to generate mostly gas (Figure 3.1). Biogenic input of organic matter and the manner of its preservation determine the hydrogen content of kerogen. Some algae and bacteria are extremely rich in membrane lipids and aliphatic biopolymers, such as algaenan. Sediments that receive this input have the potential to be hydrogen-rich (kerogen H/C >1.5). In case of land-plants lignins have low H/C ratios and sediments that received only this input are hydrogen-poor (kerogen H/C ratios < 1.0). The hydrogen, sulfur, and oxygen content of kerogen can be modified greatly during deposition and diagenesis. Anoxic conditions conserve, or even enhance, the initial H/C ratio of the primary organic matter via selective preservation and/or contributions from secondary biota. Euxinic conditions promote the incorporation of sulphur, resulting in high S/C ratios.

From the data the oxygen Index (mgCO_2/g TOC) can also be calculated. The cross-plot of Hydrogen Index and Oxygen Index can indicate type of organic matter along with generation potential (Peters, K.E. 1986).

There are many methods to classify kerogen types by their morphology, maceral and palynological assemblages, and/or bulk chemical composition. The most widely used is a modification of a method developed for coals, whereby the H/C and O/C ratios are plotted. Kerogens are classified as Type I, II, III, or IV depending where they fall on the plot. Type I and II are oil-prone, Type III is gas-prone, and Type IV is inert carbon.

Nature of the kerogens can be most reliably characterized on the basis of the hydrogen indices. Hydrogen index is indicative of quality of source rocks. More the hydrogen Index it can generate hydrocarbon.

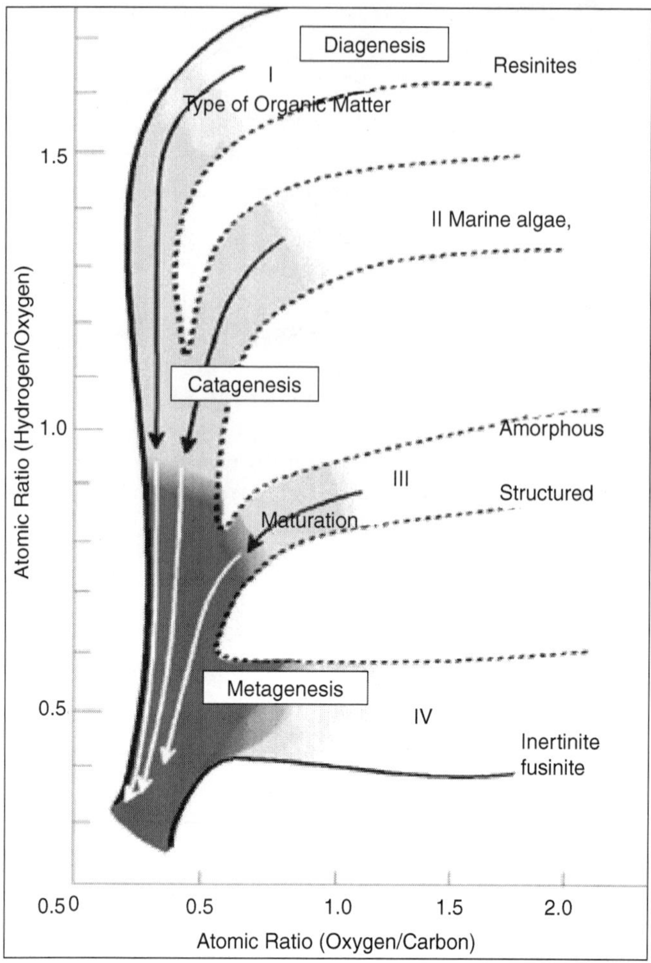

Figure 3.1. Maturation of Organic Matter.

Since geological processes are dynamic in nature, it is difficult to have evidences of transformation of organic matter into petroleum. However, certain analytical techniques to identify the source material may help in understanding impressions left in the sediments.

Based on empirical evidence, minimum concentration levels of 1.5 per cent and 0.5 per cent of Total Organic Carbon (TOC) in source rocks of siliclastic and carbonate lithologies respectively have been established (Hunt, 1996). The organic carbon concentration is an approximate measure of the organic matter content of a rock. Most source rocks which have effectively generated and expelled commercial quantities of petroleum have TOC concentrations in the order of 2-10 per cent. While assessing hydrocarbon potential of a basin even iso-TOC map can be prepared to know prospective areas. Most petroleum source rocks display dark brown to black colors. This is due to the presence of finely disseminated organic matter.

The formation of sediments rich in organic matter is restricted to certain conditions of the depositional environment. These sediments are deposited in aqueous environments which receive a certain contribution of organic matter, *i.e.*, residues of dead organisms, along with the sediment particles. In subaerial environments organic matter is readily destroyed by chemical and microbial oxidation shortly after deposition. Good quality petroleum source rocks can be deposited in marine or lake environments as organic-matter-rich muds providing that bottom waters are oxygen-deficient, *i.e.*, that reducing conditions prevail.

Generation and Expulsion of Oil and Gas

The Hydrogen Index-HI (Hg mg/g TOC) is not influenced by any other geological factor and reflects generating capacity of kerogen in each stage of location of the source bed.

Type I-kerogens are rare in terms of worldwide occurrence and mostly restricted to oil shales. The generation of oil and gas in source rocks is a natural consequence of the increase of subsurface temperature during geologic time. The process of kerogen transformation with increasing temperatures is called maturation, which is subdivided into the catagenesis and metagenesis stages (Tissot and Welte, 1984).

With respect to the stage to which petroleum generation has advanced, the organic matter is called immature prior to the onset of hydrocarbon generation, mature if hydrocarbon generation is in progress, or over-mature when these reactions have been terminated. Heat is the main driving force in maturation and petroleum generation reactions. In this context, the maximum temperatures reached, *i.e.*, the heating rate as well as the exposure time to maximum temperatures, are important. The temperature interval where oil generation is in progress is referred to as the 'oil window'. It extends over the temperature interval of about 80-150°C. For petroleum exploration, it is essential to determine the precise stage at which hydrocarbon generation reactions have progressed in a particular source rock.

This is done by measuring the so-called maturation parameters which the organic matter of a source rock has experienced. These are indirect measures of paleo-temperatures and their exposure duration. The most commonly used maturation parameters are **vitrinite reflectance** (VR_0) and biomarker maturity ratios. Vitrinite particles are widespread trace components in many source rocks which can be identified under a reflectance microscope (Morrow, D.W. and D.R. Issler, 1993).

The **Vitrinite reflectance** is a measurement by microscope of optical properties of vitrinite, a form of organic matter, contained in rock samples. The measurement is expressed in terms of the percentage of incident light reflected from a vitrinite sample. Larger measured values indicate higher levels of thermal maturity. Vitrinite is a material composed of tissues produced in the decay of plants, *etc.* Reflectance of a material is the ratio of reflected light intensity to the incident light intensity. The medium of reflection is also oil and parameter is known as VR_0 (Vitrinite Reflectance in Oil). If VR_0 value is more than 0.6 the sediments are rated as source material for generation of hydrocarbons and if it is more than 1.0 sediments are located beyond the oil generative window (Hubson., GD and E N Tiratsoo 1985).

Sometimes the Vitrinite reflectance (VR_o) of sapropelic organic matter can underestimate or overestimate maturity due to suppression or misidentification of semifusinite as vitrinite, and much care is needed to understand the maturity of source facies (Hasan, N, K.B. Kohli and A. Banerjee.2003).

T_{max} is the temperature which corresponds the top of the peak (S_2) formed by the hydrocarbon compounds expelled during the cracking of the un-extractable organic matter. If T_{max} is more than 440 °C, it indicates the maturation of organic matter for petroleum generation (Banerjee, A, *et al.*, 1988).

The crossplot of Hydrogen Index and T max may help to know maturity in relation to type of organic matter (Figure 3.2).

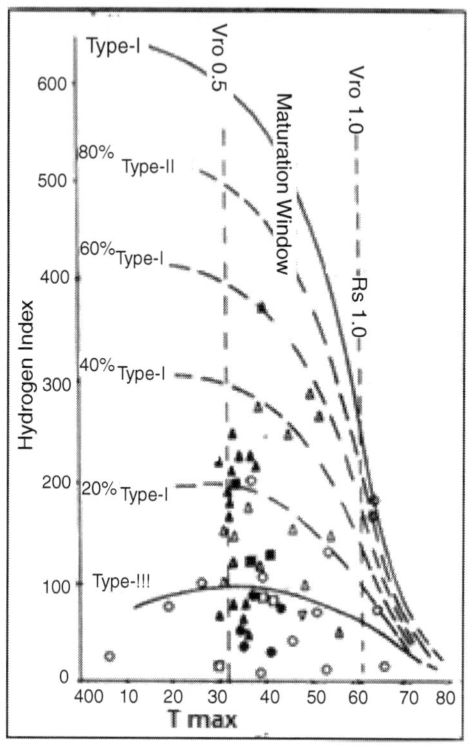

Figure 3.2. Crossplot of HI-Tmax to show Maturity Window.

Maturity modelling is a widely used tool in petroleum exploration to assess the maturity of the source rock where maturity data are not available or are of poor quality, estimate the timing of hydrocarbon generation and migration and check the subsidence and temperature histories, which are utilised in quantitative basin modelling. The most commonly used method for maturity modelling is the **Time-Temperature Index** (TTI) method of Waples (1980), based upon the concept of Lopatin (1971). The Waples method is based upon the assumption that organic reaction rates increase approximately by a factor of two with each 10°C rise of temperature.

The Waples (1980) method is widely used by geoscientists (Tissot, *et al.*, 1987; Sweeney and Burnham, 1990; Waples, *et al.*, 1992), (Kalkreuth and McMohan, 1988; Kantsler, *et al.*, 1983; Samanta and Shukla, 1988; Thomas and Thomas, 1988; Schmoker and Herter, 1990). Morrow and Issler (1993) concluded that, the TTI method generally overestimates maturity as measured by reflectance of the vitrinite maceral.

Implementation of the Lopatin's method begins with a reconstruction of the depositional and tectonic history of the geologic section of interest. This was best accomplished by plotting depth of burial versus geological age (Figure 3.3) and to specify its temperature history (Figure 3.4).

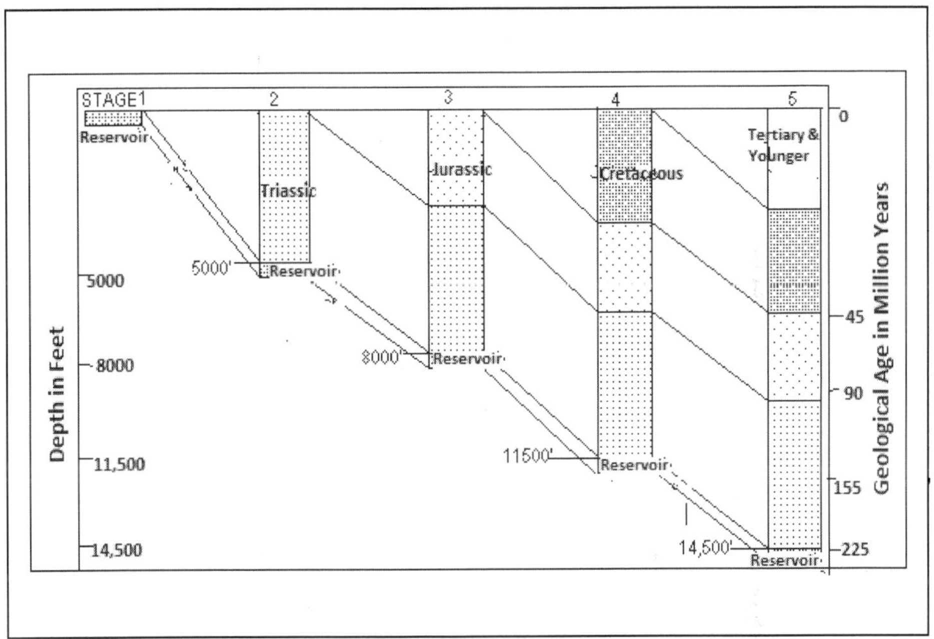

Figure 3.3. Subsidence Curve of a Geological Formation.

The basic equation for the computation is given as:

$$TTI = \sum_{n_{\min}}^{n_{\max}} (\Delta T_n)(r^n)$$

where ΔTi = length of time in millions of years spent by the sediment in the temperature interval; n-max and n-min are the n – values of the highest and lowest temperature interval encountered; r = 2.

The second aspect of the geologic model is the temperature-history. Subsurface temperature was specified for every depth throughout the relevant geologic past. This was done by computing the present geothermal gradient measured from continuous temperature log.

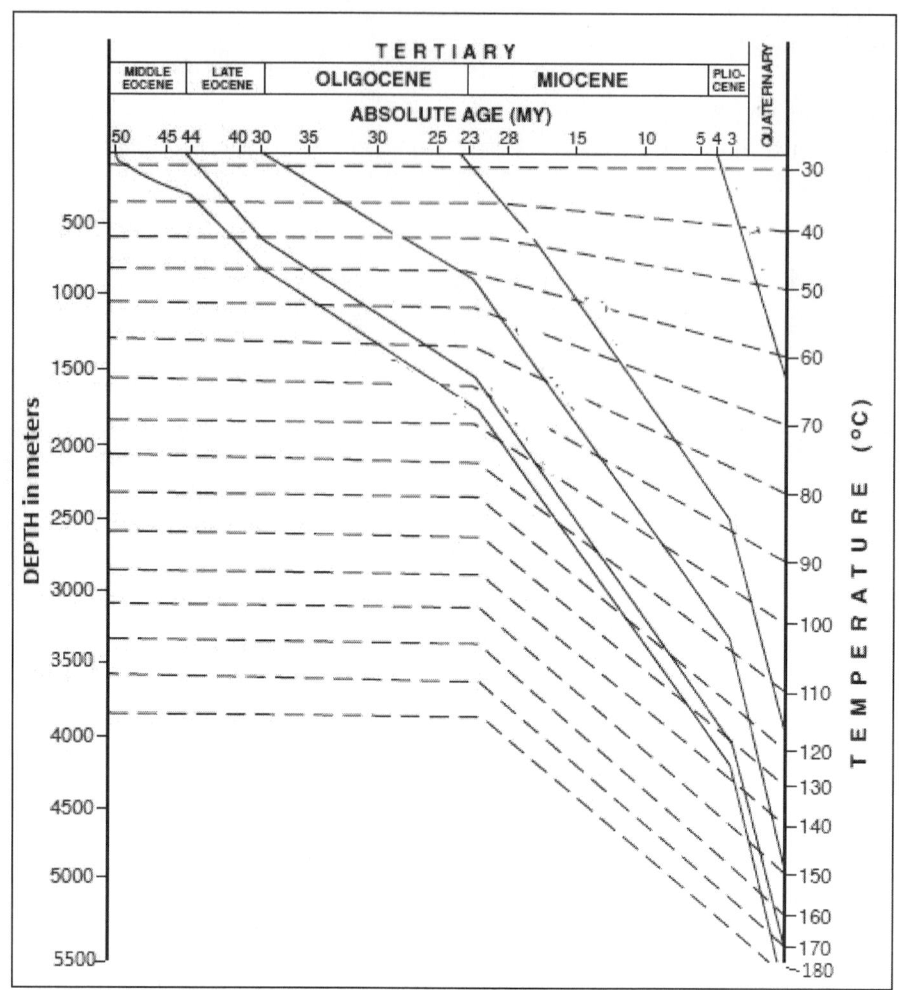

Figure 3.4. Subsidence History with Iso-Temperature Lines.

In case no laboratory data available the source potential and geological time of maturation can be estimated using the stratigraphic column of a drilled well and its bottom temperature. The procedure is called Time Temperature Index (TTI). First the burial history plot of the sedimentary column is to be prepared by positioning the formation by adding overburden formation on geological time scale.

By deducing the surface temperature from the bottom temperature and dividing the geological column in to 10°C increment intervals. In each increment interval calculate $\Delta t^* r^n$ where Δt is the duration of geological period spent by the formation with in the temperature increment interval and r is the temperature factor which is considered as 2. For temperature increment between 100-110°C the value of n is considered as 1. If it is less increment zone reduce value of n by 1 and if it is more increase the value of n by 1. Sum up all such incremental values to get TTI value. If TTI value is more than 15 the strata is supposed to be matured to generate hydrocarbons. The geological age of the top of maturation level can also be known.

During catagenesis, kerogen thermally cracks and produces bitumen. With heating (~90-140°C), C–C bonds break within the evolved polar compounds and from residual kerogen, yielding a hydrocarbon-rich fluid that is then expelled from the source rock's mineral matrix. With additional thermal stress, kerogen yields primarily condensate and then wet gas (C1-C6).

In the initial phase of generation of hydrocarbons is dispersed in minute particles, and even might have been soluble or miscible in water. As the over-burden increase these discrete particles may form patches and accumulate as separate phase and continue to develop buoyancy. The oil is immiscible in water and has lower density than the water. Catagenesis is complete when the kerogen has expended its capacity to generate C2+ hydrocarbons (~150-175°C). Metagenesis may then take place under still more severe thermal alteration, where only methane is produced (Tissot and Welte, 1984).

Pyrolysis is a rapid assay method in which 100 grams of powdered rock is heated at 25 degrees per minute. The helium carries vaporized material generated by this. The first (S1) is attributed to hydrocarbons within the source rock and the second one to hydrocarbons generated by the thermal breakdown, the third (S3) represent CO_2. The S1+S2 is the measure of the genetic potential of the source rock and S1/(S1+S2) is the transformation ratio (Lewan, M.D. 1993).

The chemical analysis and cross plot of Pristane (nC17) and Phytane (nC18) can help in knowing the type of organic matter which help to know indirectly the type of hydrocarbons likely generated in that part of the basin (Figure 3.5) (Rashid, M. A.1985).

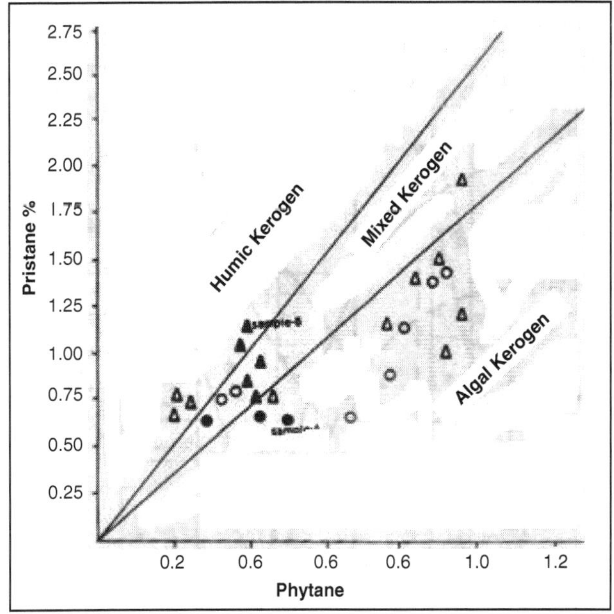

Figure 3.5. Pristane-Phytane Ratio.

Sometimes, spores and pollens of plant derived matter will alter its coloration due to prevailing temperatures in the basin. This can be studied and the parameter of the study is known as Thermal Alteration Index (TAI). If TAI value is more than 2.5 it is known as good source material for hydrocarbon generation.

Once the generation potential of hydrocarbon from a given volume of source rock is known the possible quantum of generated oil or gas in the matured source beds in the basin can be mathematically known. With logical and reasonable assumptions of expelled hydrocarbons in the percentage of source rock volume can also be quantified by estimates.

Geochemical Prospecting

The primary objective of a geochemical exploration survey is to establish the presence and movement of hydrocarbons in the area and to determine the probable hydrocarbon charge to specific exploration leads and prospects. The conventional geochemical techniques have included direct soil analysis, active soil gas measurement. The limitations include, the poor adsorptivity of soils, sampling difficulty due to poor soil permeability, low analytical sensitivity. The surface geochemical prospecting clearly identified anomalously high surface occurrences of hydrocarbon compounds and such anomalies can be linked to a migration of hydrocarbons into scattered accumulations. Hydrocarbons generated and trapped beneath the surface seep or leak to the surface in varying but detectable quantities (Figure 3.6). These phenomena occur because processes and mechanisms such as diffusion, effusion, and buoyancy allow hydrocarbons to escape from reservoirs and migrate to the surface where they may be retained in the sediments and soils or diffuse into atmosphere or water columns

Surface geochemical prospecting technique can be used to map hydrocarbon micro-seeps from subsurface active petroleum and natural gas reservoirs at depth and may reduce the exploration risk of the area. The ratios of the light hydrocarbon gases are sensitive to thermal maturity and they indicate whether an area is more oil and gas prone. Fault and fracture, however, can affect the migration and strongly influence seepage magnitude. There is no relationship between the magnitude of a seep and the economy of the deposit (Jones 2003). Magnitudes are controlled by subsurface pressure and permeability of the migration pathway whereas economics are controlled by the size of the accumulation. The microseeps reflect migration pathways from deeper and sometimes laterally distant locations. Shallow reservoirs give best seep-accumulation relationship. Seeps are difficult when traps are buried beneath thick undisturbed sediments.

Seismic exploration detects structural and stratigraphic traps but geochemical prospecting can help in determining if there is charge of petroleum in the area. Proper integration of geochemical results with geologic and seismic data leads to a better evaluation and risk assessment.

Adsorbed gas anomalies of light hydrocarbons (C1 to C5) are further substantiated by fluorescence studies of higher aromatic hydrocarbons, which are essential constituents of crude oils. These higher hydrocarbons too migrate to surface and are not affected by any microbial activity. Aromatic hydrocarbons are soluble in formation water and thus can be carried vertically in water solution through fractures too small to permit movement of in place oil. The aromatics tend to accumulate on the surface over time at a point of maximum vertical migration rather than in halo position. Integration of all studies may give better indication of prospective areas (Tiwari, R R and Dolly Pal Rana 2003).

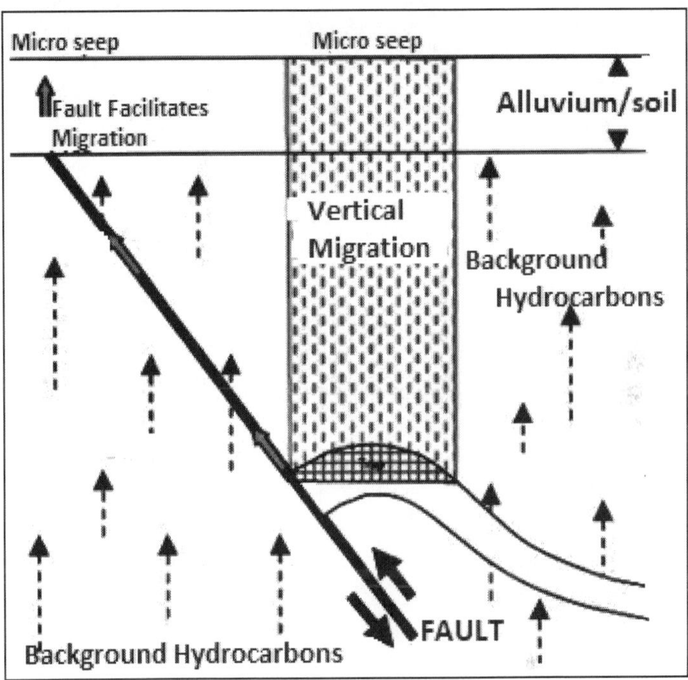

Figure 3.6. Surface Geochemical Survey (After Mithilesh Kumar, *et al.*, 2010).

Geochemical Prospecting-Indian Scenario

AAFB

The survey was carried out in Mizoram where the field samples were acquired using the passive soil gas collector. After interpretation of the geochemical survey the prospects are identified. Two leads are located on the Aizawl anticline and one is on the Keifang anticline, and other is on the Seling anticline (Mithilesh Kumar, *et al.*, 2010).

Cauvery Basin

The ***adsorbed gas survey*** analysis carried out in *Kottur – Mettupalayam* area on many soil samples was carried out and the study established that the samples are thermogenic in nature and have been generated from a similar source input. The data also revealed that the area is oil prone. The findings, supported by superimposing gas information from deep wells of the same area over adsorbed gas data, confirmed the nature of subsurface hydrocarbons. The composite contour map of methane, ethane and propane exhibit that many wells producing from Cretaceous are falling in halo anomalous area, suggesting more potential is available for future exploration (Vyas, D., *et al.*, 2005).

Western Offshore

The surface geochemical surveys were carried out in ***Ratnagiri*** area and about twenty three distinct hydrocarbon anomalies have been delineated and ranked from the three blocks based on the relative magnitude of the anomalies.

The geochemical studies were carried out in the continental shelf, slope, rise and deep sea off ***Dwaraka coast of Arabian Sea***. Bottom samples were characterized by low TOC values as compared to that of Top sediments, which may be due to early diagenetic processes in the sedimentary regime. Good quality organic matter with better enrichment is found in the sediments of bathymetry range 200m to 1500m. This may be attributed to the occurrence of oxygen minimum layer (OMZ) at this depth. The slope sediments are characterized by anoxic environment of deposition (high HI and low OI values) associated with OMZ in the slope region. Most of sediments are characterized as having thermogenic gases indicated by the adsorbed gas studies. The prospective areas have been delineated based on the anomalies in the contours of C1-C4. The yield of hydrocarbon is generally higher in both top portion and bottom portion of the sediments collected from the slope region (Arora H.K and N.J. Thomas., 2003).

Surface geochemical investigation over some selected area in ***Kerala–Konkan offshore*** was carried out as another parameter to evaluate this high-risk high reward area. Many seabed core samples were taken from the area and subjected to adsorbed gas and fluorescence studies. Reinterpretation of data by integration of both techniques has indicated that the area has gas/condensate prospects.

Bengal Basin

Surface prospecting through near surface soil gas survey confirmed by hydrocarbon ratios as well as the stable carbon isotopic study carried out on adsorbed soil gases. In a limited adsorbed soil gas survey in Galsi area did not indicate concentration of light hydrocarbon gases (Kuldeep Chandra, *et al.*, 1975). A systematic reconnaissance adsorbed gas survey in near surface soil samples were carried out in Diamond Harbour, Barasat, Basirhat and Bodra area (Kuldeep Chandra, *et al.*, 1982).This study indicated presence of light hydrocarbon gases from C1 to C4 in a good number of samples. Subsequently the full scale adsorbed survey was taken up in Krishnanagar -Ranaghat area (Ramesh chand and A C Sarkar, *et al.*, 1989), Baharampur-Jalangi-Karimpur (A C Sarkar, *et al.*, 1993), Ranaghat-Bishnupur-Jaguli (Hajra, *et al.*, 1997) and Contai (Hajra, *et al.*, 2000) areas.

All these areas have shown some positive anomaly of light hydrocarbon gases. In Contai area a halo anomaly was observed surrounding the Paleocene wedge-out feature when geochemical near surface soil gas survey anomaly map was superimposed on the subsurface geological feature of the area and later a drilled well for Paleocene wedge prospect, water and gas flow was observed during production testing. The total all the light hydrocarbon constituent concentration for the combined area was made it shows a regional anomaly in and around Eocene hinge zone in the basin. The ratio indicated them of thermogenic origin (Malay Rudra. et. al 2005).

Source Rock Studies- Indian Scenario

Krishna-Godavari Basin

In the basin thick argillaceous beds are known in Permo-Carboniferous, Lower and Upper Cretaceous, Paleocene and Lower Eocene sections in East Godavari sub-basin and also Lower Miocene sequences in offshore part of the basin. Most of these sequences are found to be good source potential. High geothermal gradients are known in the basin, which might have facilitated the early maturation in some of the sequences. The Paleocene and Upper Cretaceous sequences in East Godavari sub-basin have generated hydrocarbons during Lower Eocene. The type of organic matter is mostly type-III, except in Ravva area where type-II was identified (Samanta, U. and R.K. Shukla, 1988).

In *West Godavari Sub Basin* the Gajulapadu Shale hosts (100 – 400m) organic rich oil/gas prone source rocks having good hydrocarbon potential in Gajulapadu-1, area with about 460m thick source facies, Geochemical studies indicate medium density, paraffinic oils of Lingala – Kaikaluru to have been derived from mixed source input comprising of dominantly terrestrial organic matter, whereas Suryaraopeta oil shows dominantly terrestrial organic matter imprint. In the eastern part of the Bantumilli Graben thick source sediments within Raghavapuram Shale were observed (Venkanna, P., *et al.*, 2003).

Source potentials in these sediments observed to be good to excellent based on S2 values which exhibit a range of 2 to 13mg HC/gm., in general. Organic matter in these shales is of type II. But source sediments comprising of type-III organic matter in the southern part, *i.e.*, Nandigama - Pendurru area. The formation exhibits TOC 1 per cent encountered in the area (Samanta,U and RK Shukla 1985).

The Vadaparru sediments of the Oligocene and Eocene age in the ***Yanam area*** have excellent organic matter richness and fair to good hydrocarbon generation potential. Organic matter is predominantly of terrestrial nature (Type III) with contribution from flowering plants and also from marine organisms (Type II). These sediments are deposited in marginal marine conditions as in inner to middle shelf. Maturity results reveal that the Eocene sediments are at low maturity (VRo 0.48 per cent) and Oligocene sediments are immature (VRo 0.44 per cent) and cannot generate liquid hydrocarbon. Rock-Eval pyrolysis and TOC studies indicate that poor source rock characteristics (av. S2= <1mg HC/g of rock; hydrogen indices <50 mg HC/g TOC; av. TOC <1) but the lower part of the Oligocene sequences is marked by excellent organic matter enrichment and fair to good hydrocarbon generation potential (av. S2= 5.2 mg HC/g rock; av. TOC= 3.45 per cent) (Janardhanan, Mala, *et al.*, 2005).

Sediments of Eocene age exhibit TOC, ranging between 1.26-11.2 per cent. The entire Eocene sequence has very good hydrocarbon generation potential (av. S2= 6.35mg HC/g rock). Hydrogen indices (HI) values vary between 112-224 mg HC/g of TOC, indicating that these sediments have good generation potential for gaseous hydrocarbons with subordinate oils. However, the T_{max} data indicate that both Oligocene and Eocene sediments are immature in the area (Goswami, B.G., *et al.*, 2005).

Cambay Basin

The source rock studies in **Kutch offshore** area is fair to poor in Oligocene to Lower Miocene in carbonate sediments and hence these sediments are devoid of any potential source rocks. Argillaceous sediments with higher TOC (1-25.8) and S2 (3.77- 13.68) values with adequate maturity in the Mesozoic to Lower Eocene sections are the potential source rocks. Good potential source rocks are present in Paleocene to Lower Eocene section.

The effect of over estimation of maturity in Cambay basin was studied further by comparison between newly generated VRo on coal/coaly shale and old VRo on mixed lithologies. It revealed that, the maturity was overestimated in some wells where old VRo predicts the onset, peak and end of oil generation at 2400m, 2900m and 3500m respectively while new VRo predicts the same at much deeper depths. VRo measurements should, therefore, be limited to coal/coaly shale eliminating caved samples. Oxidized coals, solid bitumen and pseudo virtinite macerals, inherent in the samples or incorporated from mud additives should also be recognized and discarded. Cavings generally have low reflectance since the samples are coming from shallower depths. Mud additive are more of a problem, since some organic materials cannot be positively distinguished from primary vitrinite. The reflectance of most oxidized vitrinite ranges between 1.0 per cent and 1.4 per cent. Vitrinite that has been oxidized in the subsurface is observed in the red shales and inter bedded red and green shales (Pahari, S., *et al.*, 2003).

Miocene sequences from wells of ***Mehsana block of North Cambay Basin*** where studies indicate S1 value ranges from 0.00 to 4.84 mg. HC/g rock, TOC from 0.12 to 36.71 per cent, PI from 0.00 to 0.61 and EOM from 0.6 to 32.0 mg. HC/g rock, respectively. The geochemical analyses of selected samples from Jhagadia, Kand and Babaguru formations suggest the presence of two types of hydrocarbons, one migratory and the other syngenetic in nature (Kohli K.B., *et al.*, 1994). The migration of accumulated hydrocarbons of Kalol Formation to Babaguru Formation at some locales. Rock-Eval data suggest that the interval 670m to 760m in almost all the wells is comparatively rich and might have accumulations of gas (Kohli K.B., *et al.*, 2003).

Fatty acid distribution in source rock from ***Gandhar field*** was studied to understand the nature of source organic, paleoenvironment and its possible contribution to the Hazad sands. The fatty acid distribution of younger Cambay Shale shows the presence of unsaturated and hydroxy fatty acids are present which indicate algal and microbial input in source organic. C18: 1/C16 ratio is less than 1, indicates anoxic environment of deposition, the older Cambay Shale/

Olpad indicates terrestrial organic matter, input, is indicative of oxic environment of deposition.

The sedimentary sequences from wells situated in northern part and eastern flank of ***Patan depression of Cambay Basin*** are lean in organic matter concentration, whereas organic matter increases in wells of southern part of Patan depression. In southern part, Kalol-Kadi formation is adequately rich in organic matter within Cambay Shale with average TOC of 1.41 per cent, but the grey member is richer in TOC as compared to the red clay member. Sequences in the northern part have been found to be immature with vitrinite reflectance values ranging from 0.3-0.4. Only lower part of Kalol/Kadi formation at some places is at the threshold of hydrocarbon generation whereas Cambay Shale is within the oil window. The quality of organic matter showed the mature Cambay Shale of South Patan-1, 200 m thick section between 2000-2500 m of grey member has fair gas potential whereas source rock facies are absent in the red member (Neelam Niranjan and K.L. Mehrotra-1991).

Source rocks in Cambay Shale and Olpad Formations (lower Eocene and older sediments) are significantly mature and upper Olpad source rocks are in peak hydrocarbon generation and expulsion phase in Wamaj low (Banerjie, V., *et al.*, 1995). Oligocene sediments in the western flank of Tankari depression have entered into catagenic phase. The entire Dadhar Formation has entered the oil window in the western flank area. In the basinal part, where most of the wells were terminated in Cambay Shale Formation, oil window was noticed in Ankleshwer Formation. Intense hydrocarbon generation and expulsion was observed in Nada-Bharkhodra area and deeper basinal part of the basin.

In the Wamij low, the beginning of oil window ranges between 1985-2015 m depth in the area within Cambay Shale. Deeper Olpad sediments beyond about 4075m depth have passed the onset of dry gas window. The Tankari depression has revealed that the entire Dadhar Formation has entered oil window on western flank, Nada- Bharkhodra- Malpur area. The beginning of oil window in basinal part of the Gandhar area was noticed between 2400-2950 m in Ankleshwar Formation (Harvinder Singh, *et al.*, 2005).

Upper Assam Basin

In Rudrasagar, source rocks have upto ~35 per cent Type-II organic matter with potential for early generation. Elsewhere the potential source rocks contain Type III kerogen. Borholla-Champang and Khoraghat oils are from early catagenetic stage whereas Rudrasagar oils are from late diagenetic to very early catagenetic stage. Borholla-Champang gases were generated from primary cracking of hydrocarbons at VRo < 0.9 per cent. Rudrasagar gases were generated during primary cracking (0.9-1.1 per cent VRo) and few at very early stage of secondary cracking (1.3 per cent VRo).

The sediments of *Sylhet Formation* generally has poor organic matter richness with HI below 60 and source potential index (SPI) < 0.5 ton HC/m^2 and indicates that the organic matter is terrestrial (Type III), deposited in oxic environment. Sediments are in oil window in the southern part of the basin around Dimapur, Tenephe and Chumukedima areas.

The *Kopili* sediments of Dhansiri valley area indicate poor reducing depositional environment for organic matter and is dominantly Type III. In Borholla-Champang and Khoraghat areas, the SPI ranges from 0.22 to 1.57. The formation is in early phase of maturation with maximum of 0.80 per cent VRo at 4481m in Tenephe-1.

The sediments of *Barail Formation* have high Pr/Ph (avg>3) and Pr/nC17 (avg>4) and indicate Type III organic matter which was deposited mainly in oxic environment. Barail coal shale sequences have good organic matter richness. In, Borholla-Champang area average TOC per cent ranges from 1.10 to 9.57 with HI < 100 in shales In Rudrasagar, average TOC for coal and shale sequences range from 18.67 to 55.81 per cent

Maturity show that oil window (0.5 per cent VRo) starts at 2500m in Borholla and Khoraghat areas and marginally earlier in Champang area. In Borholla area, Sylhet Formation entered oil window at around 2-3Ma and in Khoraghat area, at around 1-2 Ma. In Rudrasagar area Barail Formation is immature (VRo per cent < 0.5 per cent). It indicates that, oils were expelled at low maturity (Tmax 402°C~407°C) corresponding to VRo 0.42 to 0.45 per cent. (Anuja Mishra, et. al. 2003)

The gas chromatographic analysis of ***Furkating area of Assam Arakan fold belt*** indicate, the presence of Methane, Ethane, Propane, Butanes and Pentane in the proportion of 90.22 per cent, 9.77 per cent, 3.21 per cent, 1.77 per cent and 0.6 per cent respectively. The presence of C2+ hydrocarbons indicates thermal origin with some contribution from biogenic gas. When the concentration contour maps were superimposed on structure contour map at Bokabil Top, it is observed that anomalous areas are corresponding to faults (Kaul, S.K., *et al.*, 2005).

Cauvery Basin

The source rock maturity studies of gas-prone sediments in the well in Kuthalam area, of Cambay Basin has a major contribution from humic material while in the oil-prone unit the algal contribution is around 60 per cent. The oil window ranges around 2800m in Vijayapuram-Andimanam (VR0 range from 0.40-053 per cent) in Kuthalam-Bhuvanagiri (VRo-0.57-0.67 per cent). The Thermal Alteration Index data is in agreement with the other maturity data. TAI 2.5 is attained at 2,800m and 2230m (Pahari, S., *et al.*, 2005).

Pranhita-Godavari Graben

Two coal bands were observed in the drilled sequence in the graben (Aswraopeta) at 1,346 m and 2,634m. The results indicate that the TTI method has overestimated the VRo values, while the calculated VRo values were found to be very close at 1,346m but it overestimates value at 2,634m.

Kutch

The Palyno-stratigraphic and source rock studies of the well Lakhpat-1 located in western part of the mainland were carried out with samples of cores and cuttings. Detailed palynofacies study has indicated that, organic matter is humic/sapropelic and adequately mature in zones I and II upto 2380m with TAI values ranging from

2.5 to 2.75; Representation of phytoplankton in fairly high frequency and suggests shallow marine conditions of deposition.

Bengal Basin

In deeper Sylhet and Jalangi formations major source rock development has taken place in the north-east of the basin, *i.e.*, around Radha and south-east part around Diamond Harbour (Figure 3.7). Significant maturation (R_0=0.65 per cent) of source observed around 4000 m. Jaguli-1 has revealed mature source rock sequence in the Oligocene at 3915-4457 m having wet gas with subordinate oil prone type IIIA kerogens. The lipids from the kerogens had mixed terrestrial-algal source inputs (Chawla, B.R., *et al.*, 1985).

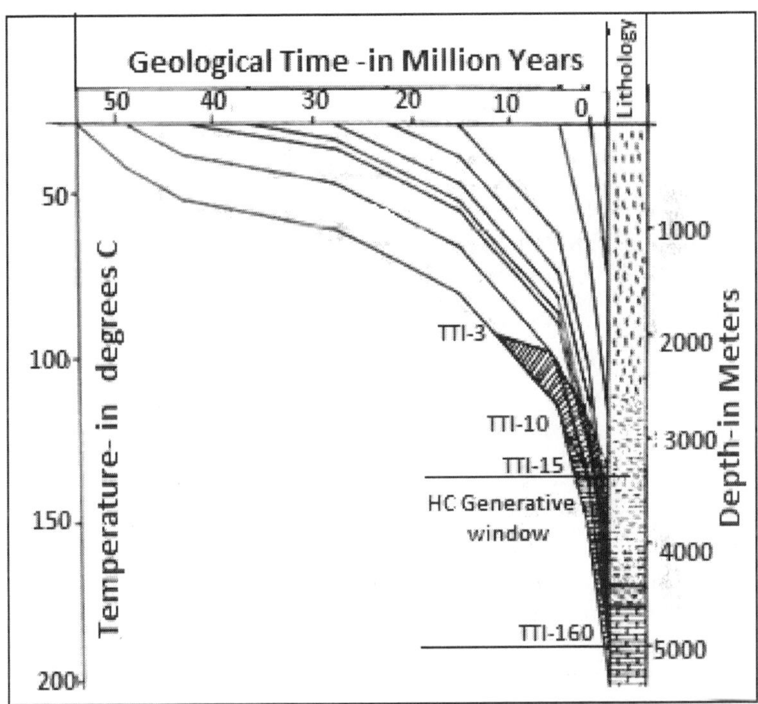

Figure 3.7. Maturation of Source Facies-West Bengal Basin.

The Pliocene and the Miocene have type IIIC kerogens except has a good source rock unit which is inadequately mature. In the Paleocene sequence, Dhananjaypur-1 has type IIIB+IIIA kerogens in the depth interval of 3020-3353 m. (Jalangi Formation) but Bolpur-Ghatal formations have poor type IIID dry gas prone kerogens (Mukherjee, Mrs J. and A.S. Chopra 1990).

Expulsion

After the generation of hydrocarbons, the next important element is the expulsion from the source to the reservoir beds. Generally source beds are argillaceous and poorly permeable beds. This initial migration of petroleum from shales to permeable reservoir beds is called the Primary migration. During this period, the direction of fluid migration is mainly vertical and across the bedding planes. Such vertical migration requires strong vertical fluid potential difference.

The effective hydrocarbon expulsion in their own phase may start at the critical depth, when the viscosity of water within the source sequence exceeds the viscosity of hydrocarbons. The viscosity of water is likely to increase with compaction of sediments. The depth at which the hydrocarbons are expelled in preference over the water is called *critical compaction depth*. During compaction of sediments with geological time, the associated water is also expelled out to overlying horizons. These expelled fluids will facilitate the hydrocarbons to move from source facies to reservoir rock.

In petroliferous basins, the quantum of the source bed may need to be sufficient than that of the thickness of the reservoir beds. These argillaceous facies contain more water at the time of deposition than that of the sandstone. Hence the variations caused in fluid potential in shale compaction is more than that of the sandstone. However, by the time hydrocarbons generated in these shales, much of the compaction might have been completed, which reduces the fluid potential during later phases of compaction. Hence the time of generation of hydrocarbons and the time of expulsion of fluid content from source beds are two important factors which are need to be studied in detail to assess the hydrocarbon potential of an area.

In case, there was no effective expulsion of hydrocarbons, most of them might have been retained within the source beds. Hence the source beds with high concentration of hydrocarbons cannot be considered as potential source beds for accumulations. If depth range of oil and gas generation overlap with migration depth range, critical shale compaction depth, those areas are best suited for effective petroleum migration and entrapment.

Migration and Entrapment

Migration of hydrocarbons dissolved in the gas phase can facilitate the movement of hydrocarbons through the source rock, as the gas phase migrates into shallower regions where temperature is lower, the liquid hydrocarbons come out of solution. However, the gas/oil ratio of most oil fields is too low for the gas to be the only means of transporting the oil out of the source rock. Also at the onset of generation most the kerogen produces little gas, most gas is generated late during the maturation history.

Primary migration of petroleum follows pressure gradients from the mature source rock intervals towards their contacts with reservoir strata. Several processes control the generation of pressures. One of the main driving forces for primary migration is sediment compaction due to overburden load. Compaction is achieved by the reduction of pore spaces due to the expulsion of pore waters. So the good quality source rocks experience further compaction with the expulsion of petroleum. Most source rocks are black shales which have very low permeability. Most hydrocarbons probably are expelled from the source rock as liquids. The expulsion of the oil out of the source rock is a dynamic process driven by the oil generation itself (Figure 3.8).

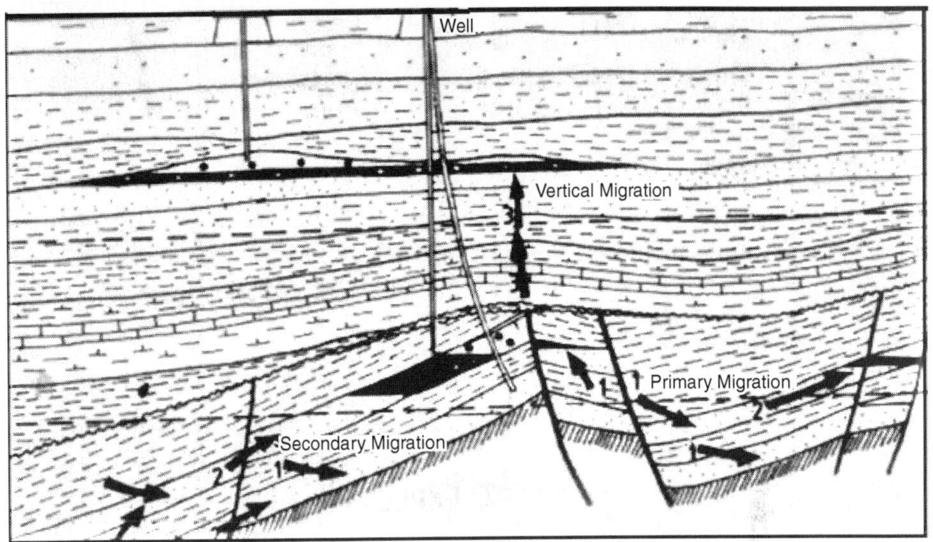

Figure 3.8. Primary and Secondary Migration.

Secondary Migration

The movement within the reservoir beds is called secondary migration. During secondary migration the beds, through which the movement occurs until they are trapped are called carrier beds which are permeable. Significantly higher permeability and sufficiently large pore sizes allow for the formation of oil droplets and small continuous oil stringers that is a network of interconnected oil-filled pores. Migration of petroleum in sandstone reservoir beds is driven by buoyancy and hence directed upwards into inclined carrier rock strata. This movement is referred to as up-dip migration. The rate of this movement depends on the volume of the oil, its buoyancy, as well as on the pore-size distribution and permeability of the reservoir rock. Controlled by capillary forces, oil droplets have a tendency to move into the largest-size pores available (England, *et al.*, 1987). As soon as saturation of a sufficiently dimensioned interconnected network of large-size pores is achieved, breakthrough occurs and the up-dip movement of the oil begins. Within sandstone reservoir strata, the distribution of pore sizes is very heterogeneous.

To produce hydrocarbons in commercial quantities there must be large quantities of oil and gas in permeable beds to form hydrocarbon pools. For this concentration the hydrocarbons move within the reservoir beds unless they are trapped on all sides. Since gas and oil are lighter than water they occupy highest place within the reservoir beds and likely to escape/move vertically in a lithological sequence. The basic requirement to have commercial pools is to stop the vertical migration of hydrocarbons.

Accumulation of Petroleum

The updip migration of petroleum along inclined carrier beds continues as long as it does not encounter structural configurations where the reservoir strata form traps. Traps are areas in the subsurface where petroleum accumulates. The most common traps are culminations of folds, called anticlines. The contact between the oil-saturated and the water saturated pore spaces is always sharp and, in most cases, horizontal, this is referred to as the oil-water-contact (Selley, 1998).

Areally spread impervious beds like shales and basalts will act as seal rocks and are also known as regional cap facies if their areal spread is sufficiently more. The lateral seal will be provided if the reservoir rock become impervious abruptly or came in contact with such rocks. Faults, folds and litho-facies variations provide lateral seals. The passage of movement of hydrocarbons within the carrier beds and or reservoir beds is known as path way of migration.

The guiding principle for establishing the pathway of migration is the fluid pressure gradient. This gradient in turn depends on difference in subsidence and compaction which in turn controlled by overburden of varying thickness. Hence the knowledge on compaction fluid migration will help petroleum explorationists for analysing the prospectivity of an area.

If the quantity of free water in the reservoir/carrier beds is less, the movement of petroleum is presumably less and hence accumulation will be close to source beds. Similarly, if there is sufficient overburden pressure is available for effective primary migration, the oil is trapped within sand lenses associated with organic rich source beds then it is known as long distance migration.

Some key elements in petroleum migration, accumulation and entrapment are capillary pressure, buoyancy, dissolved gas, tilted contacts, stratigraphic barriers and vertical migration. The basic requirement for the migration of oil in patches within a water-wet reservoir is, the capillary pressure of oil water interface should exceed the displacement pressure of the larger openings between the pores. Petroleum moves along with water and the rate and the direction of water movement are determined by the magnitude of fluid potential difference. The fluid potential may change with geological time due to changes in overburden load, erosion if any structure and geochemistry of water. A petroleum trap is also an area of low potential energy towards which fluids move from an area of higher energy. Similarly the fold not only forms a structural high it forms an area of lower fluid pressure due to reduced overburden over the crest. Further, at the crest, rocks may fracture which also reduce the pressure.

If sufficient concentration of oil and gas has reached by accumulation, the upward movement due to buoyancy will start. When oil and gas reach an anticlinal area further movement towards lower limb is resisted by downward flow of water, due to which it leaves behind oil and gas at the point of lowest potential energy and it becomes a structural trap.

In case of stratigraphic trap, the permeability reduces in up dip direction. The oil and gas migrate up dip by their buoyancy till buoyancy or capillary pressure no longer exceeds the displacement pressure of the fine grain rock.

The mobility of oil is greatest when its viscosity is the lowest and its buoyancy is the highest that is at its saturation pressure. At saturation pressure gas dissolved in oil is up to maximum extent possible beyond which gas is called free gas and form gas cap in the area of entrapment.

In a series of anticlines, the first one falling in the pathway is filled first once the closure is full of oil and gas (up to spill point) the moving water takes oil and gas into the next up-dip anticline. This will continue as long as the supply of oil and gas is there. If there is insufficient supply highest anticline may remain to be filled with water, whereas lower ones may have oil and gas. This is known as Gussov's theory of accumulation.

For vertical migration of hydrocarbons, the primary requisite is the upper layers should have higher fluid potential than the lower ones, then only oil/gas migrate across the bedding planes. In normal circumstances, it is not possible because due to more overburden over the lower beds, its fluid potential is ought to be more. However, in areas of faulting, the fault planes may become passages for fluid escape from lower sequences to the higher one due to which the top layers may be charged with more fluids making higher fluid potential in upper layers. In multi-layered sequences thin clay beds constitute as physical barrier for vertical migration. During compaction because of low permeability in way beds the fluids cannot be escaped there by higher fluid pressures will be generated forming a pressure seal for vertical migration of hydrocarbons. Sometimes upward migration can be zig-zag through the continuity of sand facies in different areal positions.

Time of formation of trap is another important factor. It is also known that, in the dynamic earth, due to tectonic and structural changes trap can be formed any time. As and when a trap is formed in a hydrocarbon potential basin, with mobility of hydrocarbons, these can be filled in. The clues to understand time of formation of trap can be geological.

Sometimes good structures are found to be devoid of hydrocarbons in petroliferous basin which indicates that either the structure has formed after the migration has been completed or the structure is not falling within the pathway of migration.

Plate Tectonics and Oil Exploration

Basins are created globally by plate tectonic concepts. In the case of continental subduction deep linear basins formed in the area. In case of continental collisions the orogeny generate synclinal mega basins within the zone. Thirdly when continent brake in to small masses, on both sides of such landmasses, ocean dipping low lying basins are formed called peri-cartonic Basins.

Newly developing ocean basins, formed by plate tectonics and continental rifting, provide just the right conditions for rapid burial in anoxic waters. Rivers rapidly fill these basins with sediments carrying abundant organic remains. Because the basins have restricted the water circulation, they also have lower oxygen levels than the open ocean.

The same plate tectonics that provides the locations and conditions for anoxic burial is also responsible for the geologic paths that these sedimentary basins subsequently take. Continental drift, subduction and collision with other continents provide the movement from swamps, river deltas and mild climates where most organics are deposited.

Plate tectonics is also responsible for creating the requisite pressure and temperature conditions that slowly matures the organics into oil and gas. This process usually takes millions of years, giving the oil and gas deposits plenty of time to migrate around the globe on the back of plate movements. Alternatively, rifting, collisions between land masses, and other tectonic forces can free the mature oil and gas from deep within sedimentary basins and then trap these organic fluids in reservoirs before they escape to the earth's surface.

Analysis of sedimentary basins in relation to plate-tectonics should identify styles and times of basin evolution and trends that embody favourable combinations of the important attributes for hydrocarbon occurrence: Organic-rich source beds, appropriated heat for thermal maturation of source beds and permeable migration paths for the concentration of hydrocarbons and their entrapment.

The importance in oil and gas exploration is to understand the genetic type of the basin and its tectonic history which determine the provenance and the composition of the sediment fill. The drift or tectonic history of a basin determines the changing position of drainage, provenance and paleo-latitudes which control the depositional environments for deposition in the basin in place and time.

Thermal regime also equally important for petroleum generation and primary migration, which is related to tectonic location and basin evolution. More heat input occurs near the subduction zones along convergent margin as well as in rift basins, whereas low thermal gradients are present in deltas and in pull apart basins. Basins showing a deltaic regime superimposed over earlier rift and pull part basins are found to be rich in hydrocarbons. Hence the idea of thermal and subsidence history of the basin will help in knowing the prospectivity rating of a basin for hydrocarbon exploration. In case of rapid or optimal subsidence of the basin will protect the organic matter to take into reducing environment early. Further, the faster subsidence will also facilitate the providing requisite temperature regime for early maturation.

In case of pericratonic basins associated with divergent margins most of requisite conditions for deposition of source, thermal regime for maturation and sufficient overburden necessary for expulsion and primary migration from source and its secondary migration with in the permeable beds till it got entrapped both vertically by regional cap, large size low amplitude anticlines and most favourable stratigraphic traps formed due to changing depositional pattern along the basinal dip direction.

In case of convergent margin areas, the quick subduction of argillaceous section into greater depths quickly may facilitate maturation and the compressional forces will generate high amplitude structural closures for entrapment. In most cases the generated oil and gas may be moved towards thrust areas at shallower depth.

In case of transform faults of global nature, the generation of hydrocarbons and their entrapment is seems to be protected for longer geological periods without destruction by plate movements.

Initiation of seafloor spreading from continental rifting may have special significance for hydrocarbon exploration. Here high temperatures are present concurrently with shallow water deposition of sediments which favor the formation of source and reservoirs for hydrocarbon exploration.

Diverging Boundaries and New Ocean Basins

When a new ocean basin opens at a diverging plate boundary, it will be long and narrow. There will not be established currents that move water sideways or vertically. This means that, the water is pretty stagnant with well-defined vertical layers. The surface waters do not get carried to the bottom, so the bottom waters lack oxygen. This is a perfect environment for organic material to settle and accumulate, and then be buried in the sediment.

As the ocean basin gets wider, this buried organic material is left behind on the edges of the continents. So we should expect that the edges of continents now called passive margins, since they aren't plate boundaries anymore are good places for oil and gas exploration. But it can be an expensive place to look for petroleum because it involves offshore drilling, sometimes in deep water.

Oil Migration in Subduction Zones

Petroleum and natural gas pools are abundant within the thick prisms of sediment deposited along many rifted continental margins. The rifted-margin sediment prisms are drawn down against and beneath the suture belts formed by crustal collisions that assemble composite continents. During such partial subduction, fluid hydrocarbons may be driven updip away from the subduction zones to accumulate in reservoirs along adjacent platform margins and within growing foreland fold-thrust belts.

Collisions and narrowing ocean basins: Ocean basins can also be long and narrow when they are closing up. This happens when continents colloide, and the ocean in between them is squeezed shut. It can also happen when smaller chunks of land, such as fragments of continent (like Indonesia) or volcanic island chains (like Japan or Hawaii) run into a continent at a subduction zone. The ocean floor between the continent and the island chain gets reduced at the subduction zone, and that section of ocean gets thinner and thinner until the island chain collides with the continent.

Either way, whether it is a big collision or a smaller collision when an island chain hits a continent, the narrow bit of ocean between the landmasses will not have much vertical mixing of water, and the bottom will become oxygen-poor. Organic material gets layered into the sediment and becomes petroleum. When the collision happens, the rocks containing petroleum get pushed upward. So the best place to find petroleum in a collision zone is next to the high mountains formed by the collision.

Chapter 4

Hydrocarbons Entrapment and Petroleum System

As it is known, the important pre-requisite for oil and gas occurrence is availability of good source rocks, maturation of source material and entrapment of hydrocarbons. For hydrocarbon exploration the basic necessity is a sedimentary rock with pore space in which fluids that is oil and gas will be stored.

Hydrocarbon Traps

Trap is a configuration of rocks suitable for containing hydrocarbons and sealed on all sides by a relatively impermeable formation through which hydrocarbons will not migrate further. Apart from structural features the hydrocarbon traps that result from changes in rock type or pinch-outs, unconformities, or other sedimentary features such as reefs or buildups. A trap is an essential component of a petroleum system. Now that many of the structural traps have been found out, the attention is to find subtle (stratigraphic) traps in matured basins and opening up challenging frontier areas.

If the droplets of oil encounter an impermeable surface through which they cannot flow, they will continue to flow upward along the underside of this impermeable sealing rock and collect in traps. If no traps are encounter, the oil droplets will migrate all the way to the surface creating an oil seep which are also common in areas having petroleum potential (Davison, I., *et al.*, 2000).

Natural gas, being less dense then it tends to float above water and oil in the subsurface. Sometimes, the natural gas is dissolved in the oil, but it very often forms a separate layer of its own which floats above oil layer. In such a case, a layer of natural gas upon a layer of oil which is in turn underlain by a layer of water (Knebel GM and Rodriguez-Eraso G., 1956).

Accumulation depends on the sealing capacity of the overlying cap rock sequence, which is normally a low permeability in shale. The task in the exploration of hydrocarbons is to find available traps whose cap rocks are intact, and whose reservoir rocks are (i) sufficiently porous to contain commercial quantities of hydrocarbons and (ii) sufficiently permeable to allow them to flow fast enough towards production wells (Friedman, G. M., 1961).

Heat, Pressure and Diagenesis

The predominant dynamic controls on hydrocarbon distribution are heat, diagenesis and fluid pressure. Heat is largely derived from deep in the earth's crust and flows upwards through conductive, sediment-filled basins. This phenomenon is expressed by geothermal gradients, which define the average, linear rate of increase in temperature per kilometre depth of burial. Geothermal gradients can range from over 80°C per kilometre in hot basins to less than 20°C per kilometre in cold basins. The global average for sedimentary, oil-bearing basins is about 30°C per kilometre.

Pressure also increases with depth, hydrostatic pressure is the unbroken column of formation water held in the pores. This implies that if the water can free to rise in a borehole it would reach the surface. The litho-static pressure gradient represents the total load of the overlying water-saturated sediments –the overburden.

These two pressures of pore fluids in a sediment column normally varies. Initially, the pore pressure closely follows the hydrostatic boundary, but at a certain depth in a given basin it swings towards the litho-static limit. In other words, at and beyond the overpressure zone the fluid pressure at a given depth is considerably greater than the hydrostatic pressure (Kinji Magara, 1978). The difference between the normal hydrostatic pressure and that approaching the litho-static limit is of fluid containment, and is related to the overall permeability of a sedimentary section.

The structural relief of the formation may create abnormal or sub-normal formation pressure due to the potentiometric surface which is the height of confined water in the formation rise in the drilled wells in the same aquifer (Figure 4.1).

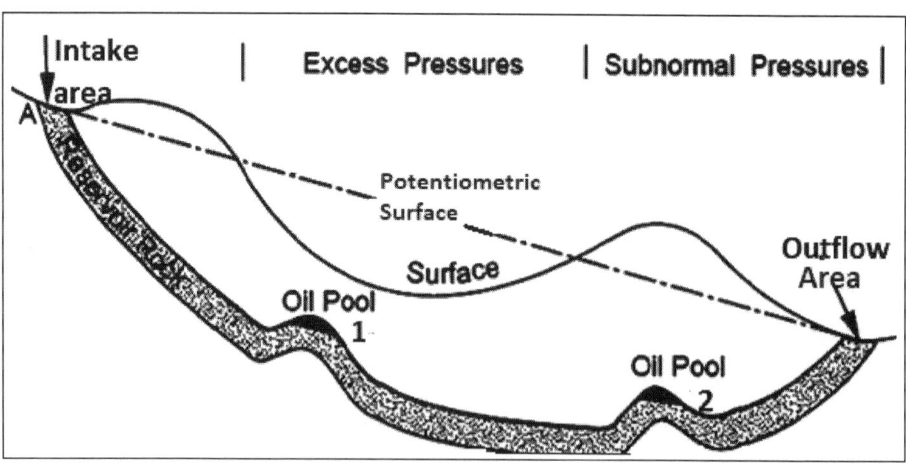

Figure 4.1. Formation Pressures as Factor for Trap.

Both pressure and heat control diagenesis which is the process by which unconsolidated sediments are converted with time and burial into sedimentary rocks through physical consolidation and compaction and the chemical precipitation of new minerals (Rybach, L. 1986).

Cap Rock Integrity and Re-migration

A good cap rock should be as impermeable as possible or have the mechanical strength to withstand fracturing. If the cap rock becomes fractured, some of the escaping hydrocarbons may accumulate in reservoir rocks higher up the previous reservoir, a process known as re-migration. Regarding re-migration, it is fact that hydrocarbons almost lie uppermost in traps, simply because they are more buoyant than water. They are therefore in direct contact with the overlying cap rock shale and will be the first fluids to leave a trap once the cap rock seal has been leaked.

As with geological time the overlying sediment load increases the associated pore pressure may increase which may reach the fracture pressure of the rock there by the generation of fractures which propagate upwards through the cap rock. Initially, the permeability of the fractures will be small. However, as the permeability of the shale matrix decreases over time, the permeability of the fracture will increase. The volume and discharge rate of hydrocarbons leaving the reservoir through fractures will therefore increase. This means that, hydrocarbons probably remain trapped for some periods of geological time before escaping to a lower temperature zone in which they may re-accumulate in more porous and permeable traps. Prospects will therefore become increasingly separated from their primary source rocks, and may no longer depend on primary migration but on vertical migration from one sandstone body to another via lengthy fractures formed in intervening shales.

Uplift and Erosion

In general, sedimentary basin undergoing continual subsidence and sedimentation throughout their geological history. But some basins are subject to one or more periods of tectonic upheaval, resulting in the uplift and possible erosion of significant sediment masses. In general, uplift and erosion will contribute to the partial switching off of thermally-driven, fluid dynamic processes, because the sediments will gradually be moved closer to the surface where the temperatures are lower. In these situations, the properties of the former zonation may be disturbed. The amount of vertical displacement will depend on how much a basin has risen and how much is eroded off the top. The disadvantage is the risk of biodegradation of oils and bacteria in those reservoirs which were relatively hot initially and relatively become cooler later (Figure 4.2).

The movement of oil and gas through the carrier bed to their final positions of entrapment is called secondary migration which begins when the buoyancy of the oil and gas column is sufficiently great to overcome the entry pressure in the carrier bed. As oil and gas migrate along a path, they fill all the traps all along the way until the supplies are exhausted. Traps can be of any size, from a small area to even accommodate many millions or even billions of barrels.

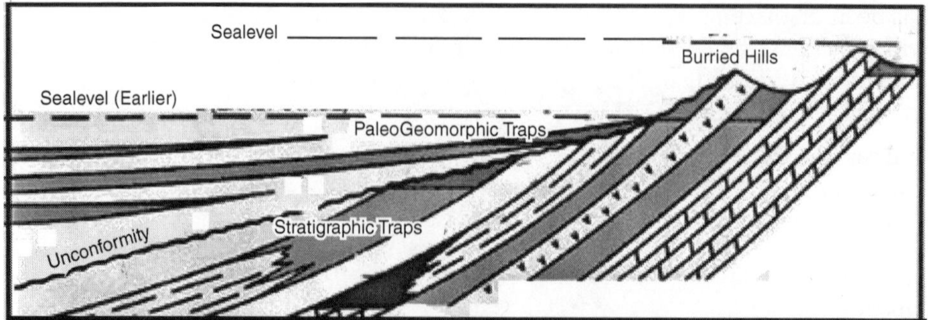

Figure 4.2. Traps Formed Due to Uplift and Erosion.

While oil and gas accumulate in the structural traps, they displace the initial water out of the traps. Under such conditions, in accordance with the law of gravity, gas at the highest part of the trap to from a gas cap above an oil column, and water always occurs below throughout the aquifer. The fluids are bounded, of course, by gas-oil and oil-water interfaces.

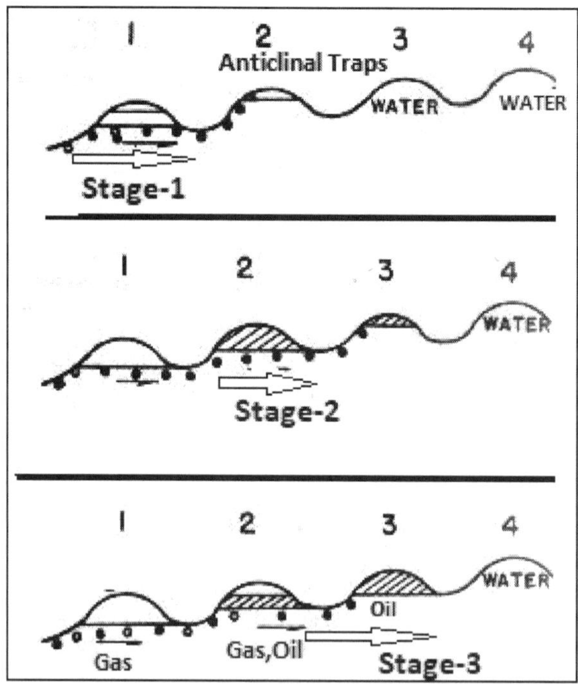

Figure 4.3. Differential Entrapment of Fluids in Trap.

In case the supplies of oil and gas are adequate, these fluids will continues to accumulate until the oil-water interface reaches the spill point. Thereafter, the trap will no longer retain oil, and any additional oil entering it will spill up-dip. Gas will continue to enter the trap, displacing more and more oil until finally all the oil has been displaced and flushed or spilled to further up-dip. In any additional gas will now spill up-dip, and oil will continue to by-pass this gas-filled trap. This is known as differential entrapment (Figure 4.3).

When the movable quantities of oil and gas are exhausted, secondary migration will cease, the reservoir fluids will come to rest, and along the path of migration the following general sequence will rest. The deepest (down-dip) traps will be filled to the spill point with gas. These will be followed by one or more traps containing oil and having a gas cap, and then by one or more traps filled with oil only. All remaining up-dip traps will be filled with water. The number of traps that will be filled with gas only or with oil only will depend on the size of the individual traps and on the amount of oil and gas in mobility till entrapped.

Geologic Structures Become Hydrocarbon Traps

The sedimentary layers in general to be deposited more or less horizontally over large areas. Layers of sedimentary rocks must somehow become bent to provide the geologic structures that may become traps. These geologic structures that are created can become hydrocarbon traps where they must be in the migratory path direction of hydrocarbons with in the permeable rock. The most common and simplest type of structural trap is the anticline which is a structure formed when the layers of rock have been bucked upward (Figure 4.4).

Stratigraphic traps are those formed by changes in the characteristics of the rock formation such as a loss of permeability or porosity or discontinuity of a layer. In oil exploration which the Geologist and Geophysicist may locate which are potential hydrocarbon traps and it is termed prospect evaluation.

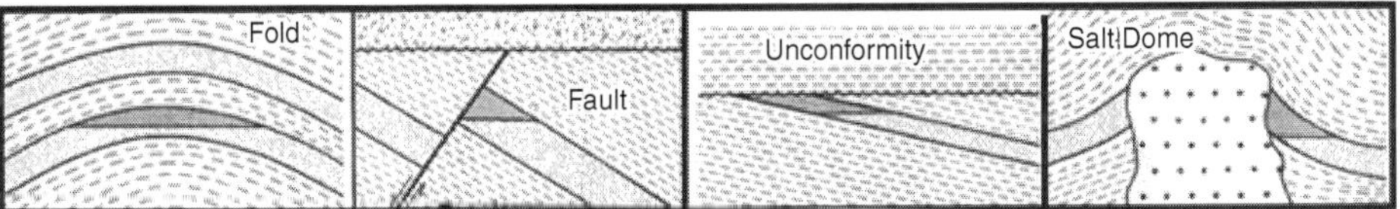

Figure 4.4. Types of Entrapment.

Structural Traps

Faults are fractures in the strata along which the sides have moved relative to each other. The surface against which the movement occurs is termed as fault plane. The block located above the fault plane is known as hanging wall and the block below the fault plane is called foot-wall (Figure 4.5).

A normal fault is the one in which hanging wall has gone down relative to the foot wall. A reverse fault is one in which the hanging wall has gone up relative to the foot wall. Hence it is obvious that vertical faults have neither foot-wall nor hanging wall. Similarly fault without any movement or fault without throw can be known as a joint. Any movement of fault plane is called strike of the fault and the angle between horizontal surface and the fault plane is known as dip of the fault. The hade is the complement (90° – dip angle) of the dip (Sherbon Hills, E., 1963).

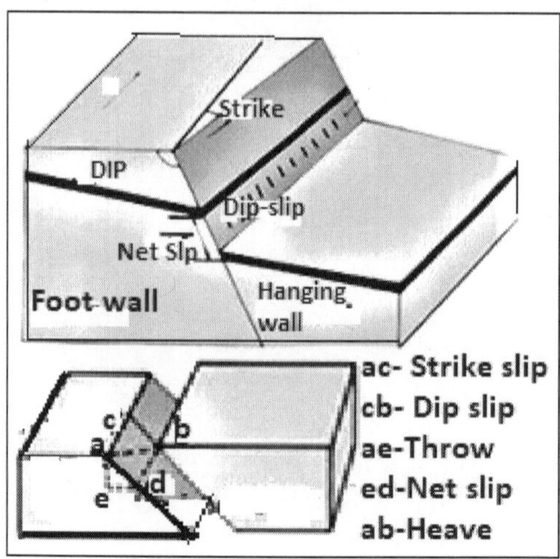

Figure 4.5. Fault Elements.

The term slip is used to indicate relative displacement of formerly adjacent point on opposite sides of the fault and it is measured on the fault surface. The strike slip is the net movement (slip) parallel to the strike of the fault, whereas dip slip is measured parallel to the dip of the fault plane. The net slip is the total displacement which is the distance measured on the fault surface between two formerly adjacent points situated on opposite walls of the fault. The vertical component of the net slip is called throw of the fault and horizontal component is called heave.

The movement along fault may be translational or rotational. In translational movement the strata which is faulted remains to be parallel even before and after the faulting. In case of rotational movement, the bedding surfaces are parallel before the fault and align oblique after the faulting. There can be no evidence in faulting to know which block has movement or which is not moved.

Based on angle of dip of the fault plane, they are termed high angle faults if dip of the fault plane is more than 45° and low angle faults if it is less than 45°.

Criteria to Identify Faults

i) Discontinuity of structures is an indication for the presence of a fault only if other depositional related reasons are not known as discontinuity. ii) Repetition and omission of strata may also indicate presence of a fault.

Characteristics of Fault Planes

Slickensides are polished and striated surfaces that result from the friction along a fault plane. These striations are parallel to the direction of movement and indicate last movement of a fault. Sometimes rocks get pulverized along the fault plane. If grain size of such pulverized material if coarse it is called fault breccia, if fine gravel like clay, it is termed fault gouge. Since fault zones are extreme fractures, solutions may move which may react with rocks by replacing with fine grained quartz, called silicification (Marland P. Billings, 1979).

A scarp is a relatively steep straight slope, if it is crated due to faulting it is called fault scarp. In case rocks on the upstream block are eroded and deposited on the down thrown side, the height of the fault scrap will generally decrease.

Classification-Faults

Based on Movement

A strike slip fault is one in which the net slip is parallel to the strike for the fault. There is no dip slip component (Figure 4.6). A dip slip fault is the one in which the net slip is equal to dip slip and there is no strike slip component. A diagonal slip fault is one in which the net slip is diagonally up or down the fault plane. There are both strike and dip slip components.

A strike fault is the one whose strike is essentially parallel to the strike of the adjacent strata. Similar dip fault strikes parallel to the direction of dip of the beds.

Based on Fault Pattern

If strikes of a series of faults are parallel, that set of faults are called parallel faults. En-echelon faults are relatively short faults that overlap each other.

Peripheral faults are circular or accurate faults that bound a circular area. Radial faults belong to a system of faults that radiate from a point step faults are the faults which form a step like pattern due to down throw on the same direction. Hinge fault is a pivoted or rotational fault. In this one point is pivoted and is rotated from that point. The amount of slip is different from point to point.

Transverse fault strikes perpendicular or diagonal to the regional strike whereas longitudinal fault strikes parallel to the regional strike. A wrench fault is essentially a strike slip fault with lesser dip slip component. However, the strike of the fault can be transverse to the regional strike.

A normal fault if it is of regional nature it can also be called as gravity fault where hanging wall moves down relative to the foot wall. They indicate crustal widening. A graben is a down faulted block, generally linear, bounded by gravity fault. Foot-walls of graben are called horsts. In case of half graben one of the bounding faults of the graben is very steep whereas the other is gentler with series of faults.

The low angle reverse fault is also called thrust fault on which hanging wall has moved up relative to the foot well. Thrust faults indicate shortening of the crust and implies compression. The term up-thrust may be used for high angle thrust faults in which the hanging wall is the active element.

In case of horizontal or gently dipping strata a fracture may follow a bedding plane for long distance and overlying strata may slide and thrusting may occur. Such bedding plane surfaces are called de-colloment planes and these type of thrusts are called bedding plane thrusts.

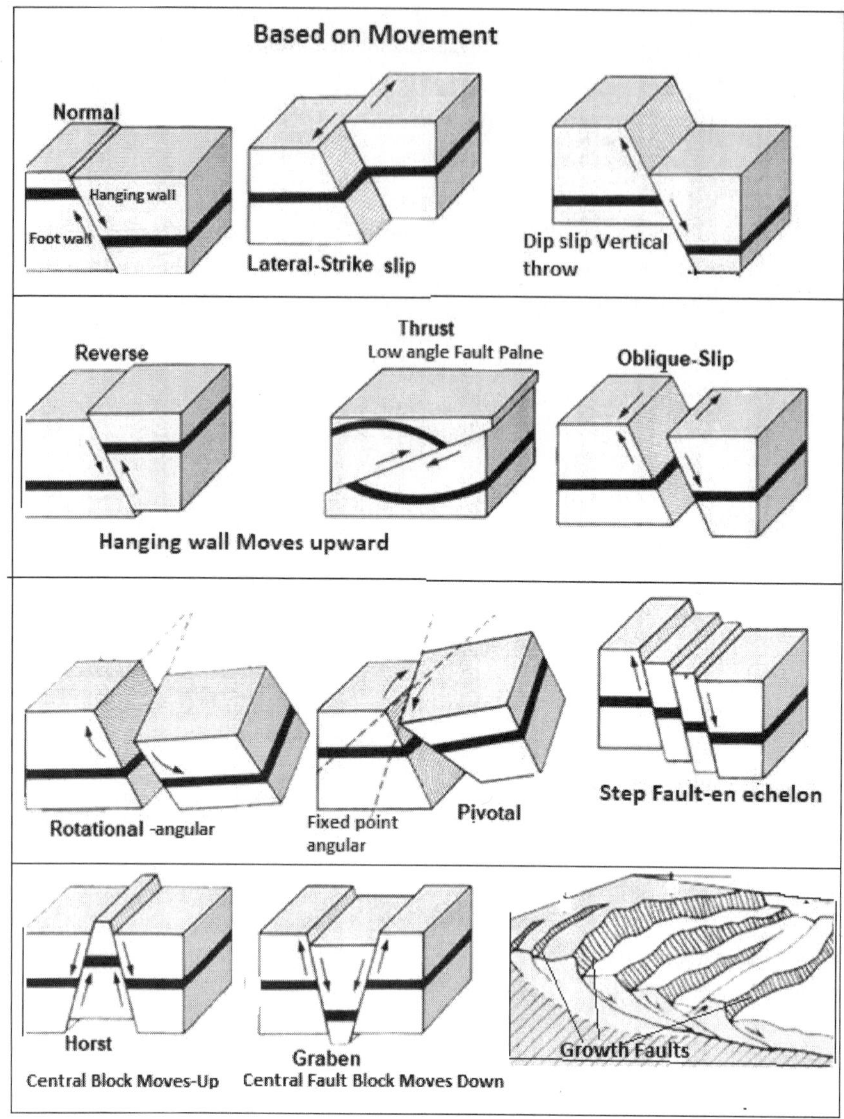

Figure 4.6. Fault Patterns.

When bedding plane thrust assumes a step like appearance in vertical section, through several glide horizons may prevail during upward rise it may be referred as step-bedding plane thrust. In asymmetrical folding, the steep limb weakens by creation of fractures, which due to stresses in the strata sheared through thrust faults. Thrusts generated by breaking of fold are called break-thrusts. In case of recumbent folds, the overturned limb become weak through bending and stretching. If this continued a stretch thrust develops which moves the strata.

Strike slip faults occur in fold belts may cut across fold axes diagonally and faults cut across regional strike perpendicularly, such faults are referred as tear faults.

When sedimentation continued in open marine basins due to overload there could be subsidence of the basin. If weak fracture developed in the strata, because of this continued subsidence the throw of the fault will grow continuously. Such faults are called Growth Faults. The foot wall towards basin margin act as buttress due to which less competent beds may fold against growth fault which are called roll-over structures.

Folds

Folds are undulations in strata which are displayed in layered sequences. Without considering age of the strata, a convex upward structure is called Anti-form and concave downward in syn-form. Anticline is an anti-form with closure towards younger beds and syncline is the closure towards older beds. A limb of a fold is down dip portion from the maximum curvature. Steep limb is called forelimb and gentler one is back limb. Axial line is an imaginary line connecting points of maximum curvature in any one bedding surface of a fold. The plane containing axial line is called axial surface. The axial line of an anticline is called crestal line or crest and for syncline it is trough line (Figure 4.7).

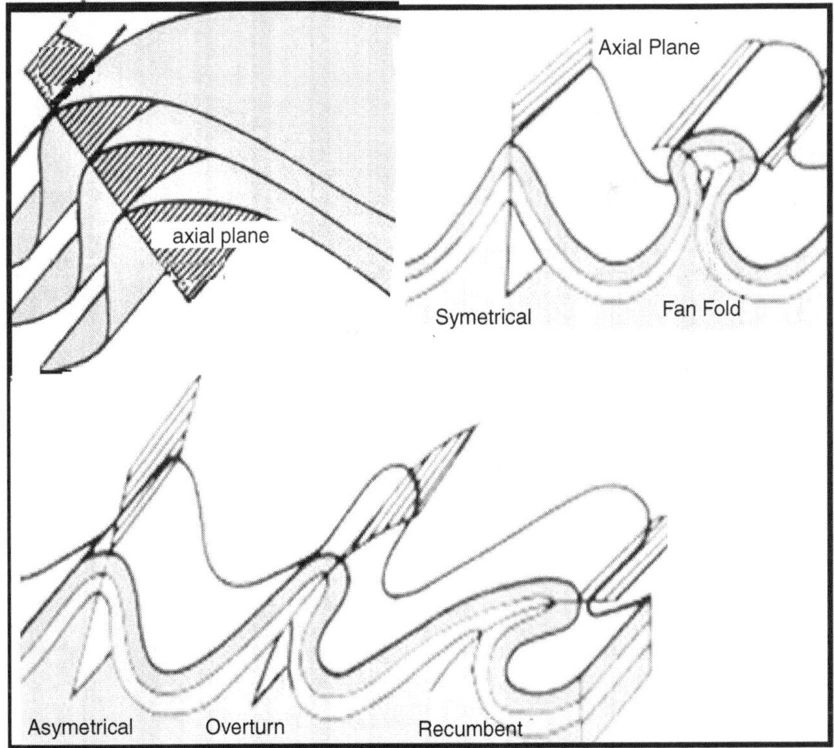

Figure 4.7. Isogonal or Parallel Fold-Types of Folds Based on Geometry.

Classification-Folds

Based on Geometry

If both the limbs of a fold are equidistant from the axial surface it is called symmetrical fold. In asymmetrical fold both limbs have different degrees of dip. If axial surface is vertical it is called upright fold if not it is an inclined fold. In case both limbs are dipping in same direction it is called isoclinal or overturned fold. An isoclinal fold with horizontal axial surface is termed recumbent fold.

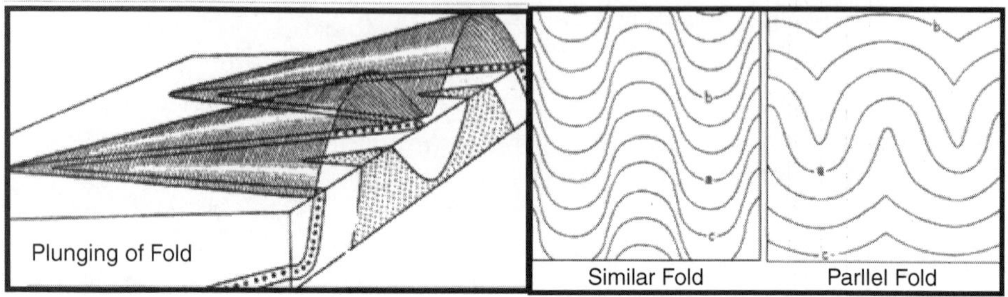

Figure 4.8. Plunging Folds.

In case of open fold the angle between two limbs is more than 90°. In closed or tight fold the angle between two limbs in less than a right angle. If the crest or trough in a fold is more sharp angular, it is called chevron folds. A fanfold is the one in which both limbs are overturned. In an anticlinal fan fold two limbs dip towards each other where as in synclinal fanfold they dip away from one another.

If axis of a fold is at an angle in the vertical plane the fold is said to be plunging in the direction of the angle. This plunging can be in all directions or in one direction. The doubly plunging anticline look like a mountain range (arch) and all sides plunging anticline is called a dome. The angle of plunge is called pitch or rake (Aschoff, J. L. and Giles, K. A. 2005).

In a series of folds if plunges of all are parallel to one another it is called cylindrical folds. A conical fold is the one in which the plunges diverge from a vertex (Figure 4.8).

Based on Morphology

The morphological classification of folds is based on changes in shape of folds in depth and on the patterns formed by folds in planar view. When folds are mapped in these types may not be observed and better known in depth sections. If thickness of a bed is constant all along the fold it is called concentric fold.

These are also called as parallel or competent folds. The size of the anticline decreases down ward while size of the adjoining syncline decreases upward in case of concentric folds. This type of fold is common when intensity of deformation is low and/or when beds are thick and competent.

Similar folds are those in which the strata are bent into similar curves of the same size. The anticlines and synclines do not decrease or increase in size downward.

But the similarity is only maintained by thinning in the axial areas. Such thickness changes are more prevalent in competent strata (Figure 4.9).

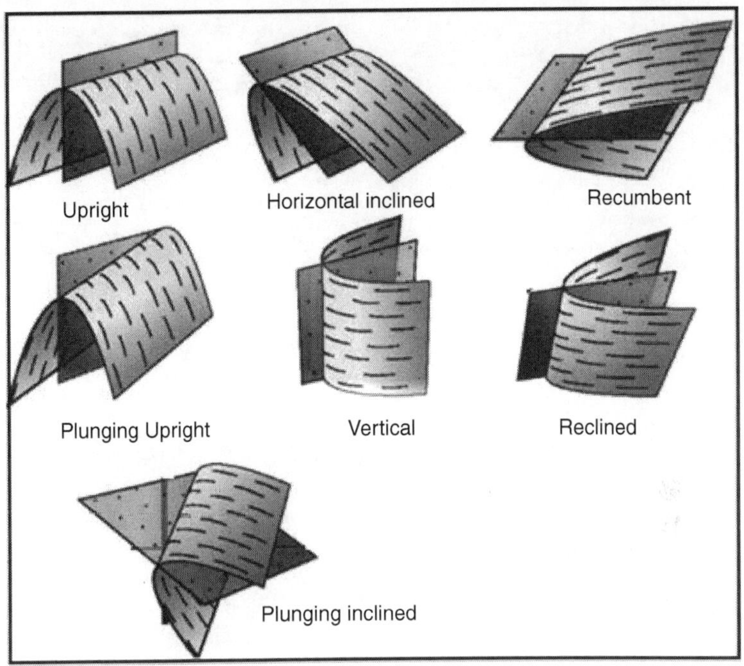

Figure 4.9. Types of Folds-Based on Morphology.

If limits of the folds are concentric but thickening occurs in the axial areas, this disharmonic of fold intensity is known as disharmonic folding and is a very common feature in folds which involve strata of different competencies. Culminations are structurally higher portion along strike in a fold belt remain whereas depressions are structurally lower portions.

An anticlinorium is a composite of anticlines of large dimensions consisting of many small anticlines which together form a folded mountain range. In normal anticlinorium, the axial surfaces of small folds coverage downward, if they diverge it is called an abnormal anticlinorium. Similiarly synclinorium is a composite of many synclines which together form a gently folded elongate deformed basin.

A linear zone of overlapping folds is called en-enchelon folds whose lengths are extensive but overlap with one another in an oblique manner.

Based on Genesis and Internal Kinematics

If sliding of beds happen due to compressive forces, the competent beds move past each other, the most competent shale slide upward. Due to slides drag folds may be formed in the beds. The large scale folding generated due to slide movement are called flexure folds.

In case closely spaced fractures, the shearing occurs along the fracture may displace/slipping of the bedding surfaces. Due to differential movement along the fracture plane a fold pattern can be seen which is called Shear fold. If less competent beds are associated with higher pressures and temperatures the rocks behave like viscous liquids with plastic flow. In such cases, fold structures are developed in incompetent beds due to compressive forces. These are called Flow folds.

Based on External Kinematics

Sometimes fold structures are generated by regional tectonic movements like vertical uplift, regional compressive forces in mobile belts, gravity gliding, block uplift and differential compaction. Due to compressive forces when geosynclinal sediments get compressed rise as mountain chains because of vertical movements but sometimes gravity acts to prevent the central portion from rising too higher in the fold belts. The driving force for the creation of folds is related to the push created by the gliding sediments over the pre-existing sediments.

In mobile belts the initial stretching will generate block faults due to which blocks involve vertical movements. Later due to compressive forces, the sedimentary strata within each block get crumpled and folded with in the blocks. Sometimes regional decoupling due to mega faults release the stresses in both horizontal and vertical directions. These stresses will generate fold structures within the strata.

If two adjacent areas differ in density of sediments the differential pressures created will push the sediments of the adjoining areas. This is the principle for origin of salt dames and shale diapirs. Once such vertical movements occur the drag of sediments will generate fold structures in the strata. The drag folds develop when beds slide past each other especially when an incompetent stratum lies between two competent strata. The upper beds slide away from synclinal areas and drag folds are symmetrically related to the major fold (Gerald M Friedman and John E Sanders, 1978).

Combination Traps

Sometimes due to geological conditions both structure and impermeable seal by pinch out of carrying bed may happen. In such cases it is termed as combination trap. With geological time due diagenesis sometimes porosity may enhance like in dolomitization or loss of porosity can arrest mobility of hydrocarbons within (Figure 4-10).

Hydrodynamic Traps

Due to buoyancy and gravity of moving fluids with in the permeable carrying bed without following general principles of physical phenomenon a different way of entrapment may be possible which are called hydrodynamic traps (Figure 4.10 & 11).

Petroleum Resources Classification System and Definitions

Definitions

The "Hydrocarbon Resources" have been defined as all quantities of petroleum which are estimated to be accumulated. Normally these are calculated on basin-

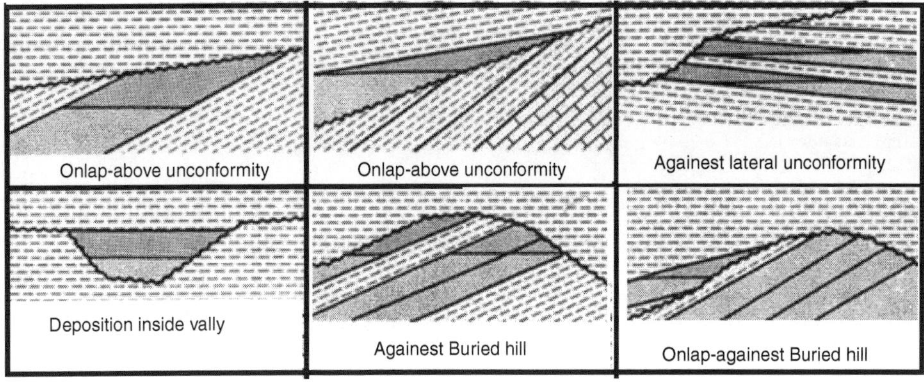

Figure 4.10. Types of Entrapment-Depositional Methods.

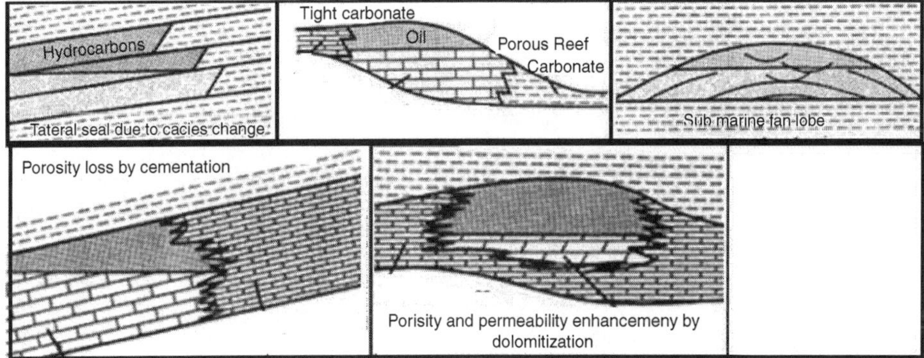

Figure 4.11. Types of Entrapment-Facies Change.

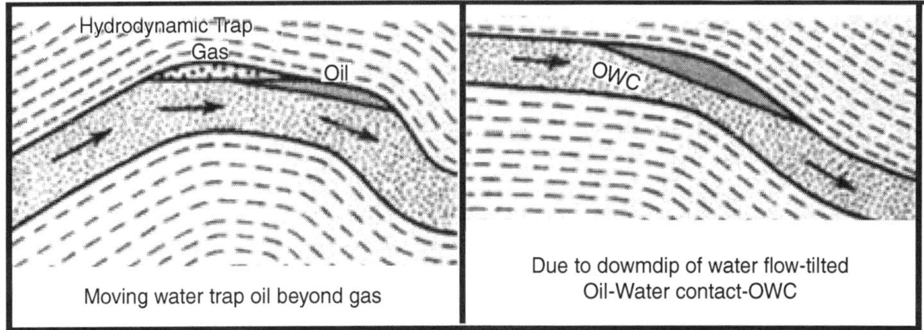

Figure 4.12. Hydrodynamic Traps.

wise, based on quantum of source rock, its expelled and accumulated percentage. So it is merely, a vision in the initial stages of exploration. The reserves constitute a known part of resources, being those quantities that are discovered, recoverable, commercial and remaining. Prospective Resources are those quantities of petroleum which are estimated, on a given date which are undiscovered accumulations.

Total petroleum-initially-in-place is that quantity of petroleum which is estimated at a given time to exist originally in naturally occurring accumulations. Estimated Ultimate Recovery is defined as those quantities of petroleum which are estimated, on a given date, to be potentially recoverable from an accumulation, plus those quantities already produced there from. So the total petroleum-initially-in-place may be subdivided into Discovered Petroleum-initially-in-place and Undiscovered Petroleum-initially-in-place. A portion of those quantities classified as Unrecoverable may become recoverable in future as commercial circumstances change, technological developments occur, or additional data are acquired.

Reserves are defined as those quantities of petroleum which are anticipated to be commercially recovered from known accumulations from a given date forward. Estimated recoverable quantities from known accumulations which do not fulfil the requirement of commerciality should be classified as Contingent Resources. These may include accumulations for which there is currently no viable market, or where commercial recovery is dependent on the development of new technology, or where evaluation of the accumulation is still at an early stage.

The proved reserves will be limited to those quantities that are commercial under current economic conditions, while probable and possible reserves may be based on future economic conditions. The range of uncertainty can be reflected in estimates for Proved Reserves (1P), Proved plus Probable Reserves (2P) and Proved plus Probable plus Possible Reserves (3P) scenarios.

If probabilistic methods are used, the uncertainty distribution can also be called as: most likely/mode, median/P50 or mean. For undiscovered accumulations (Prospective Resources) the range will, in general, be substantially greater than the ranges for discovered accumulations.

In all cases, however, the actual range will be dependent on the amount and quality of data (both technical and commercial) which is available for that accumulation. As more data become available for a specific accumulation (*e.g.* additional wells, reservoir performance data) the range of uncertainty can be reduced.

In general, there should be at least a 90 per cent probability that, assuming the accumulation is developed, the quantities actually recovered will equal or exceed the Low Estimate. In addition, an equivalent probability value of 10 per cent should, in general, be used for the High Estimate.

Proved Reserves

Proved reserves are those quantities of petroleum which, by analysis of geological and engineering data, can be estimated with reasonable certainty to be commercially recoverable, from a given date forward, from known reservoirs and under current economic conditions, operating methods, and government regulations. Proved reserves can be categorized as developed or undeveloped. If deterministic methods are used, the term reasonable certainty is intended to express a high degree of confidence that the quantities will be recovered. If probabilistic methods are used, there should be at least a 90 per cent probability that the quantities actually recovered will equal or exceed the estimate.

In general, reserves are considered proved if the commercial producibility of the reservoir is supported by actual production or formation tests. In this context, the term proved refers to the actual quantities of petroleum reserves and not just the productivity of the well or reservoir. In certain cases, proved reserves may be assigned on the basis of well logs and/or core analysis that indicate the subject reservoir is hydrocarbon bearing and is analogous to reservoirs in the same area that are producing or have demonstrated the ability to produce on formation tests.

The area of the reservoir considered as proved includes, the area delineated by drilling and defined by fluid contacts, if any, and the undrilled portions of the reservoir that can reasonably be judged as commercially productive on the basis of available geological and engineering data. In the absence of data on fluid contacts, the lowest known occurrence of hydrocarbons controls the proved limit unless otherwise indicated by definitive geological, engineering or performance data.

Reserves in undeveloped locations may be classified as proved undeveloped provided (1) the locations are direct offsets to wells that have indicated commercial production in the objective formation, (2) it is reasonably certain such locations are within the known proved productive limits of the objective formation, (3) the locations conform to existing well spacing regulations where applicable, and (4) it is reasonably certain the locations will be developed.

Reserves to be recovered by improved recovery methods that have yet to be established through commercially successful applications are included in the proved classification only after a favourable production response from the subject reservoir and it is reasonably certain the project will proceed.

Unproved Reserves

Unproved reserves are based on geologic and/or engineering data similar to that used in estimates of proved reserves; but technical, contractual, economic, or regulatory uncertainties involved to classify them as proved. Unproved reserves may be further classified as probable reserves and possible reserves. Unproved reserves may be estimated assuming future economic conditions different from those prevailing at the time of the estimate. The effect of possible future improvements in economic conditions and technological developments can be expressed by allocating appropriate quantities of reserves to the probable and possible classifications.

Probable Reserves

Probable reserves are those unproved reserves which analysis of geological and engineering data suggests are more likely than not to be recoverable. When probabilistic methods are used, there should be at least a 50 per cent probability that the quantities actually recovered will equal or exceed the sum of estimated proved plus probable reserves.

In general, probable reserves may include: reserves in an area of the formation that appears to be separated from the proved area by faulting and the geologic interpretation indicates the subject area is structurally higher than the proved area and incremental reserves in proved reservoirs where an alternative interpretation of performance or volumetric data indicates more reserves than can be classified as proved.

Possible Reserves

Possible reserves are those unproved reserves which analysis of geological and engineering data suggests are less likely to be recoverable than probable reserves. When probabilistic methods are used, there should be at least a 10 per cent probability that the quantities actually recovered will equal or exceed the sum of estimated proved plus probable plus possible reserves.

In general, possible reserves may include: reserves which, based on geological interpretations, could possibly exist beyond areas classified as probable and reserves in an area of the formation that appears to be separated from the proved area by faulting and geological interpretation indicates the subject area is structurally lower than the proved area.

Reserve Status Categories

Developed: Developed reserves are expected to be recovered from existing wells including reserves behind pipe. Improved recovery reserves are considered developed only after the necessary equipment has been installed, or when the costs to do so are relatively minor. Developed reserves may be sub-categorized as producing or non-producing.

Producing: Reserves subcategorized as producing are expected to be recovered from completion intervals which are open and producing at the time of the estimate. Improved recovery reserves are considered producing only after the improved recovery project is in operation.

Non-Producing: Reserves subcategorized as non-producing include shut-in reserves. Shut-in reserves are expected to be recovered from the completion intervals which are open at the time of the estimate but which have not started producing or wells not capable of production for mechanical reasons.

Petroleum System

Petroleum system is a concept which is the study of oil and gas formation along with its associated processes such as oil origin, occurrence, migration and their accumulation (Welte DH,1972). The occurrence of hydrocarbons involves processes such as trap formation, generation-migration-accumulation of hydrocarbons and key elements which encompass source rock, reservoir rock, and seal or cap rock and this unified system is called Petroleum System (Figure 4.13) (Bird, K.J., 1994).

These elements and processes must occur in the proper order for the organic matter in a source rock to be converted into petroleum and then to be stored and preserved. If a single element or process is missing or occurs out of the required sequence, a prospect loses viability to be within the specified system (Allen, and Allen, J. R., 1992).

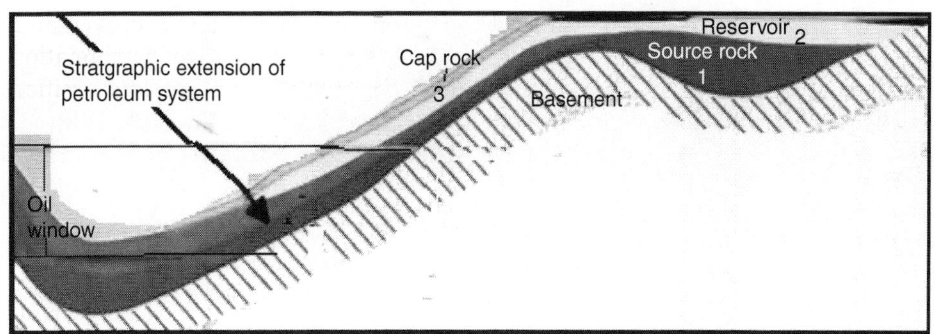

Figure 4.13. Petroleum System.

Naming the Petroleum System

The name of a petroleum system is compound name which has different parts showing the name of the source rock, the part showing the name of the main reservoir rock and another one showing the regional cap rock.

The petroleum system mapping is a way of portraying all the elements of the system basing on geographic (spatial), temporal and stratigraphic extents. The geographic extent of a system is the span of all geographic area in which the active source is found. The burial history is explained by the stratigraphic extent which is traced based on the lithological units of the system. Mapping and studying a system helps explorationists to predict if the trap of interest has oil or not. It also helps them locating the most likely accumulations of the petroleum province (region). The best mapping and study of petroleum system is done by linking all elements such as source rock, reservoir rock, seal and overburden rock to the processes of petroleum geology, generation-migration-accumulation.

A prospect is a potential trap to be drilled to check if it contains sufficient oil for commercial purposes. A play is one or more geological related prospects while a complementary prospect is dependent on a known system.

The first is a petroleum charge (fluids), the second is the trap that encompasses reservoir, and the entire trap geometry and seal rock. The third is the timing by which they knew if the trap was formed after or before oil movement.

Source Rock Hydrocarbon Generation: The generation of oil and gas depends on, presence of organic matter rich enough to yield hydrocarbons, adequate temperature, and sufficient time to bring the source rock to maturity. Pressure and the presence of bacteria and catalysts also affect generation.

Migration: The movement of newly generated hydrocarbons out of their source rock is called expulsion. The further movement of the hydrocarbons into reservoir rock in a hydrocarbon trap or other area of accumulation is secondary migration. It occurs from a structurally low area to a higher area in the subsurface because of the relative buoyancy of hydrocarbons in comparison to the surrounding rock. Migration can be local or can occur along distances in large sedimentary basins, and is critical to the formation of a viable petroleum system.

Accumulation: The phase in the development of a petroleum system during which hydrocarbons migrate into and remain trapped in a reservoir. Sedimentary rocks are the most common reservoir rocks because they have more porosity than most igneous and metamorphic rocks and they form under temperature conditions at which hydrocarbons can be preserved. A reservoir is a critical component of a complete petroleum system.

Seal (cap rock): An impermeable rock that acts as a barrier to further migration of hydrocarbon liquids. Rocks that forms a barrier or cap above and around reservoir rock forming a trap such that fluids cannot migrate beyond the reservoir. A seal is a critical component of a complete petroleum system.

Basin Modelling

It is computer based method to examine the dynamics of sedimentary basins and their associated fluids to determine if past conditions were suitable for hydrocarbons to fill potential reservoirs and be preserved there by understanding the operating petroleum system. Seismic interpretation can delineate closed structures and identify potential subsurface traps, but it does not reliably predict trap content (Poelchau, *et al.*, 1997).

This modelling brings together several dynamic processes, including sediment deposition, faulting, burial, kerogen maturation kinetics and multiphase fluid flow (Weeks LG-1952). These processes may be examined at several levels, and complexity typically increases with spatial dimensionality; the simplest, 1D modelling, examines burial history at a point location. Two-dimensional (2D) modelling, either in map or cross section, can be used to reconstruct oil and gas generation, migration and accumulation along a cross section. Three-dimensional (3D) modelling reconstructs petroleum systems at reservoir and basin scales and has the ability to display the output in 1D, 2D or 3D, and through time. If the time dimension is included it is called 4D.

An important concept is the time of generation, migration and accumulation of most of the hydrocarbons in a petroleum system. The critical moment is the transformation ratio, which is the relative conversion of source-rock organic matter to hydrocarbons. The absolute age of each layer in the basin and petroleum system model is an important parameter for determining the timing of the processes that generate, move and trap petroleum. Age information may be available from bio-stratigraphic determinations. Identification of the lithology and depositional environment of each stratigraphic unit is crucial because such as porosity and permeability, of coarse-grained sediments helps identify their potential as reservoir or carrier rocks that facilitate migration of petroleum from source rock to reservoir.

Characterizing the source-rock depositional environment helps predict the probable petroleum product generated through kerogen maturation. Source-rock properties are needed as inputs to simulate the reactions that govern the degradation of organic material to produce hydrocarbons. These essential properties are the total organic carbon (TOC) measured by combustion of rock samples and the hydrogen index (HI) obtained through pyrolysis of rock samples for petroleum generation potential.

Also required are kinetic parameters for the thermal conversion of the source-rock kerogen to petroleum. Another measure of kerogen maturity is vitrinite reflectance. Simulation of the burial history can be used to predict the expected vitrinite reflectance at any depth or time in the model. Calibration entails adjusting the model so that the simulated vitrinite reflectance matches that measured in samples at varying depths in the well. Several other physical properties must also be specified for each layer.

Porosity and permeability in reservoir and carrier layers are important for fluid-flow computations and reservoir volumetric estimates. Permeability of source rocks affects the efficiency with which generated hydrocarbons can be expelled. Heat capacity and thermal conductivity, usually inferred from lithology and mineralogy, are needed for the thermal calculations that model kerogen maturation and petroleum generation. In addition, density and compressibility data are required inputs to model compaction and burial.

The burial history of basin sediments contains information about burial depth and preservation of organic material, which are related to the temperatures and pressures the sediments were exposed to and the durations of exposure. Temperature is the primary variable in conversion of kerogen to petroleum, and pressure is important for migration of fluids. Key inputs for building a burial history include sedimentation rate, compaction, uplift, erosion and depositional environment. The thermal history of a basin is linked to the basin subsidence, uplift and heat flow. The estimation of petroleum potential of a basin requires reconstruction of the temperature over geologic time and across the basin.

Therefore, in addition to model properties, some specific past conditions must be evaluated. These conditions, treated as boundary conditions by the software, include paleo-bathymetry, which determines the location and type of deposition. Other boundary conditions are sediment/water interface temperatures throughout geologic time which, along with paleo -heat-flow estimates, are required to calculate the temperature history of the basin.

Deposition–Layers are created on the upper surface during sedimentation or are removed during erosion. Depositional thickness, which may have been greater than current thickness, can be calculated by several methods: porosity-controlled back-stripping starting with present-day thickness and estimation from sedimentation rate coupled with depositional environment.

Pressure calculation and compaction–The pressure calculation driven by changes in overburden weight caused by sedimentation. Compaction causes changes in many rock properties, including porosity, and to a lesser extent, density, elastic moduli, conductivity and heat capacity.

Heat-flow analysis–The goal of heat-flow analysis is temperature calculation, a pre-requisite for determining geochemical reaction rates. Heat conduction and convection from below as well as heat generation by emissions from naturally occurring radioactive minerals must be considered. Thermal boundary conditions with inflow of heat at the base of sediments must also be formulated. Contemporary heat flow was calculated by multiplying the geothermal gradient by the thermal conductivity. The resulting map of surface heat flow was an input to the software, which delivered source rock maturity values that matched available maturity measurements (Seifert, W.K., *et al.*, 1980).

Fluid analysis– Fluid-flow models deal with fluid phases that are typically liquid, vapour and supercritical or under-saturated phases. The fluid-analysis step examines temperature- and pressure-dependent dissolution of hydrocarbon components in the fluid phases to determine fluid properties, such as density and viscosity, for input to fluid-flow calculations. These properties are also essential for subsequent migration and calculation of reservoir volumetric.

Fluid-flow calculations–Darcy flow describes multicomponent three-phase flow based on the relative-permeability and capillary pressure concept. With this method, migration velocities and accumulation saturations are calculated in one step. In a hybrid method, flow-path analysis in high permeability zones may be combined with Darcy flow in low-permeability regions. Alternatively, migration and accumulation can be modelled by invasion percolation in the software.

Reservoir volumetrics–The height of a petroleum accumulation is limited by the capillary entry pressure of the overlying seal and the spill point at the base of the structure. Loss at the spill point and leakage through the seal reduce the trapped volume. Other processes, such as secondary cracking or biodegradation, can also impact the quality and quantity of accumulated petroleum.

Risk–The impact of uncertain data can be studied by multiple simulation runs with varying model parameters. Assignment of varying parameters and the corresponding impact on the model can be performed with statistical methods, such as Monte Carlo simulations. These simulations do not provide a unique answer, but rather a range of possible outcomes with estimates of uncertainty. Increased computing power combined with multiple simulations allows to compare the effects of various scenarios and identify which variables exert the most control on the computed results. Final outcomes are mainly scenario probabilities and confidence level in percentiles.

Prospect Evaluation

Once a structure has been identified all required petroleum system data in the area needs to be understood to call it a prospect to drill further. A well can be a geologic success by finding hydrocarbons (Welte DH and Yukler MA, 1981). With a fair geological interpretation, it is better to seek input from other disciplines.

As part of its risk management strategy, the ultimate test for a prospect comes when the geologist can convince partners to drill the best of the known prospects first. This process often yields information or ideas that strengthen the prospect, as well as giving management confidence that industry views the prospect as worthy of drilling.

Chapter 5

Surface Geophysical Methods: Oil Exploration

The study of the earth using physical measurements at the surface is called geophysics. Geology involves the study of the earth by direct observations of rocks, whereas geophysics, on the other involves the study of interior layers of the earth from direct view by measuring their physical properties with appropriate instruments, usually on the surface. It also includes interpretation of the measurements to obtain useful formation on the structure and composition of the concealed zones.

The existence of the structural features in the earth's crust, have been determined by observations upon seismic waves from earthquakes, as well as by measurements of the earth's gravitational, magnetic properties. Many of the tools and techniques developed for such studies have been used in exploration of hydrocarbons. Throughout its history, the tools and techniques of exploration geophysics have been improved along with technology growth both in performance and economy.

The introduction of computer technology and its improvements with time have brought about new capabilities in the recording and processing of all kinds of geophysical data, making it possible to extract useful quality information.

The new developments have been necessitated where exploration is to be carried out in offshore. Moreover, unique types of noise in such areas often cause interference with desired geophysical information, and special techniques must be developed to suppress such interference.

The geophysical techniques most widely employed for exploration work are the seismic, gravity, magnetic, and electrical methods. Less common methods involve the measurement of radioactivity and temperature at or near the earth's surface and in the air (Dobrin, M. B., and Savit, C. H., 1988).

Gravity Method

In this method minute variations in the pull of gravity from rocks within the first few miles of the earth's surface can be measured. Different types of rocks have different densities, and the denser rocks have the greater gravitational attraction. If the higher-density formations are in a structural high, such as an anticline, the earth's gravitational field will be greater over the axis of the structure than along its flanks. A salt dome, on the other hand, which is generally less dense than the rocks into which it is intruded, can be detected from the low value of gravity recorded above it compared with that measured on either side. This method is useful wherever the formations have densities which are appreciably different from those of surrounding formations (Al-Sadi, H. N., and Baban, N., 2014).

The gravitation law states that the mutual attractive force between two point masses is proportional to one over the square of the distance between them. $F= G m_1 m_2/r^2$

The constant of proportionality is usually specified as G, the gravitational constant, F is the force of attraction, and r is the distance between the two masses, m_1 and m_2.

While making measurements of the earth's gravity, the gravitational acceleration, g is measured. The gravitational acceleration is the time rate of change of a body's speed under the influence of the gravitational force. Combining Newton's second law with law of mutual attraction, the gravitational acceleration on the mass m_2 can be shown $g = Gm_1/r^2$

The shape of the gravity curve indicates the variation in gravitational acceleration which is not dependent on the absolute densities of the rocks. It is only dependent on the density contrast between the body and the surrounding matter. The gravitational acceleration is measured with a gravimeter in milli-gals, and it is always acting downward.

To account for variations in the observed gravitational acceleration which are related to elevation variations, so another correction is known as the Free-Air Correction. In applying this correction, the observed gravity values are mathematically converted to ones that look like all were recorded at the same elevation, thus further isolating the geological component of the gravitational field.

If an observation point is at a higher elevation than the datum, there is excess mass below the observation point. So the gravity reading is larger due to the excess mass, and it has to subtract a factor to move the observation point back to the datum. The sign of this correction is opposite to that used for the elevation correction.

The gravity anomalies observed on gravimeter are directly under the instrument and rather close to the same. Thus, the flat slab assumption can describe much of the gravity anomalies associated with excess mass and elevation. Corrections based on the slab approximation are referred to as the Bouguer Slab Correction (Figure 5.1).

In case of Cauvery Basin NNE-SSW trending basement ridges separate Basement lows sub-basins. Also see details of the basin architecture in the chapter-10,

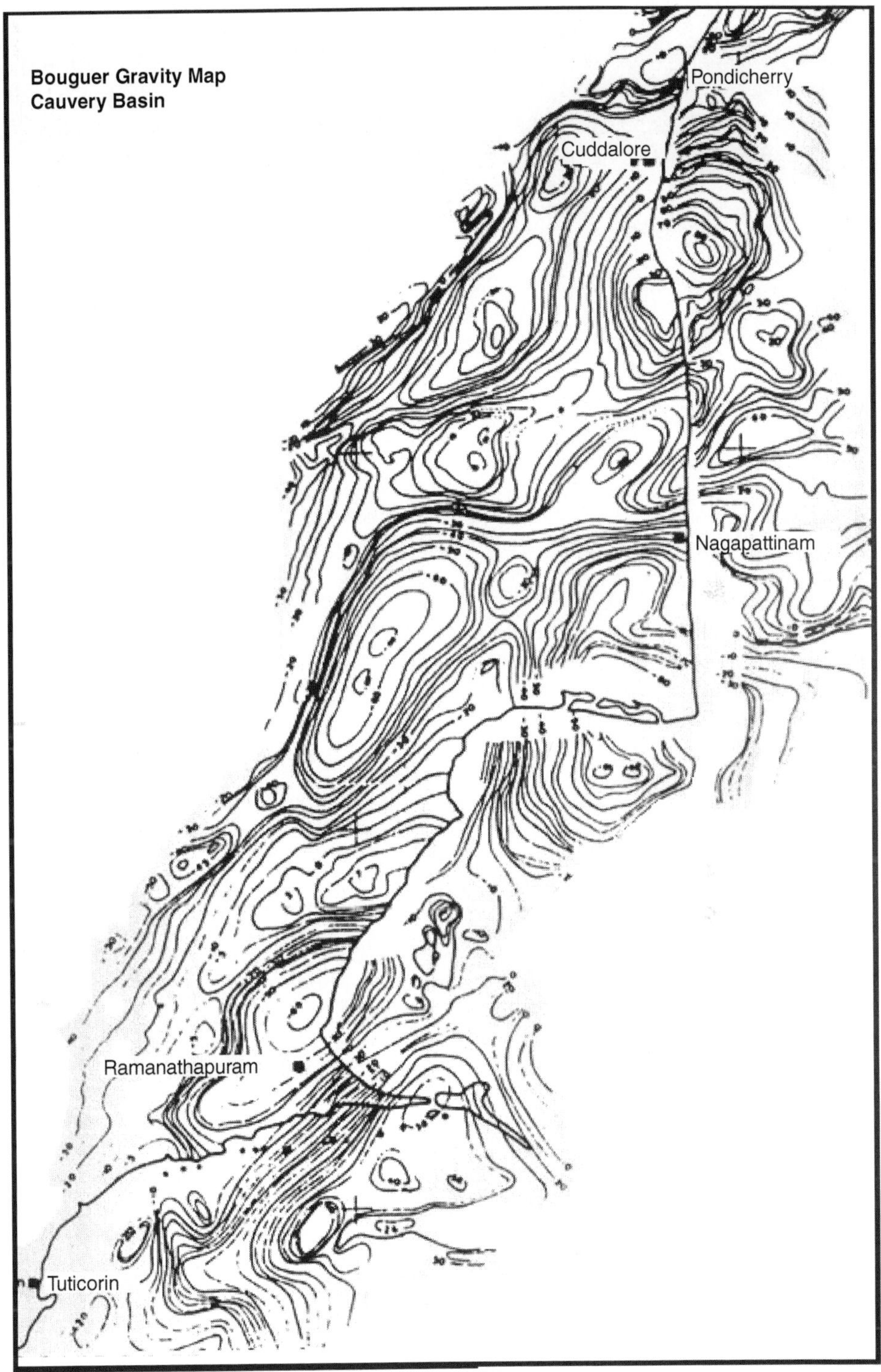

Figure 5.1. Bouguer Gravity Map.

the petroliferous basins). In case of Himalayan basins the northerly dipping of the Indian plate generated deep sedimentary basin all along.

In case of undulating density contrast body say basement, there are variable contributions to the observed gravitational acceleration. The first is caused by large-scale geologic structure that is not of interest. The gravitational acceleration produced by these large-scale features is referred to as the Regional Gravity Anomaly. The second contribution is caused by smaller-scale structure for which is the objective of the survey. That portion of the observed gravitational acceleration associated with these structures is referred to as the Local or the Residual Gravity Anomaly.

The Regional Gravity Anomaly is a slowly varying function of position along the profile line. This feature is a characteristic of all large-scale sources. Local Gravity Anomalies are defined as those that change value rapidly along the profile line. The sources for these anomalies must be small in spatial extent and close to the surface, large or small is defined with respect to the length of the gravity profile. Because Regional Anomalies vary slowly along a particular profile and Local Anomalies vary more rapidly, any method that can identify and isolate slowly varying portions of the gravity field can be used to separate Regional and Local Gravity Anomalies.

The estimates of the regional gravity anomaly determined from an independent data set from gravity survey is conducted within the larger area. The long-wavelength gravity field varies around the survey and then remove its contribution from the local data. After that, it is called Residual Gravity data.

A gravimeter is simply a very precise weighing machine used to find the weight of a certain lump of metal or other material at a series of stations distributed over the area being surveyed. Since the weight of an object is its response to the Earth's gravitational attraction, this weight will be slightly affected by the nature of subsurface materials at the place of measurement. It will be slightly larger, for example, at stations where the subsurface material is of higher density or where dense material comes closer to the surface. The changes in weight are so small that the weighing machine must be capable of detecting changes.

Gravimeters must not only resolve small variations in the gravity field, but must also remain stable over a large range of values and environmental conditions. It is therefore required that practical instruments must be highly precise, portable, robust and simple to use. All gravity meters have inherent limitations, however, and the particular equipment deployed must be consistent with survey objectives. There are clear distinctions between making absolute measurements of gravity and measuring relative gravity, or variation from place to place.

Absolute gravimeters time the free fall of a body in a vacuum, using lasers and optical interferometry, to obtain accuracies of better than 0.01 mGals under favourable conditions. Whereas, absolute gravimeters are transportable, they are still quite bulky and time consuming to set up and read. Normally one absolute reading can be made in a single working day.

Relative gravimeters are devices which can only measure differences in gravity from station to station. When gravity changes, the force on the proof mass will likewise change, and this will be reflected in a change in the length of the supporting spring. The position of the proof mass is sensed by one means or another, and the amount of external force required to bring it back to a standard position provides a measure of the gravity value at the station, relative to other stations. Instruments used for standard surveys are essentially mechanical devices, using optical or electronic means for determination of the proof mass position and manual or electrostatic restoration of the mass to its null position.

Microgravity instruments incorporate extensive electronics in their design for sensing of the proof mass position and its restoration to the null position with great precision. The gravity signal is recorded and processed in the solid state memory by software. In these instruments, corrections for tilt errors, for long term drift, for the temperature of the sensor and for earth tides are applied in real time (Avasthi D.N. and D. Das.1979).

Indian Scenario

During initial geophysical surveys by ONGC in the **Himalayan terrain** in India encountered difficulties in survey due to extremely rugged topography along the river beds which cut through the hills adding to the ruggedness of the topography. The gravity surveys are affected with the extreme variation in the surface topography as also by the isostatic imbalance of the region (Avasthi, D. N., *et al.*, 1969).

In 1982 ONGC acquired seismic and gravity-magnetic data in the **Andaman** area. The scheme of lines selected for the analysis of gravity-magnetic data, Gravity and along E-W profiles, normal to general geological strike, were subjected to spectrum analysis and 2D forward modelling techniques. Regional was fixed through harmonic analysis of gravity data along an E-W profile. Apart from basement mapping, the study could identify zones of shallow magnetism within sediments, zone of unconsolidated sediments and change of basement character towards eastern margin (Singh, T.P., *et al.*, 1991).

The **Cambay basin** is a deep graben in which the Tertiary sediments directly overlie the Deccan Trap lava flows. The gravity survey in this part of the basin has brought out a belt of positive anomalies of high amplitude and large areal extent, in the central part of the basin. During initial interpretation of the data it was believed to probably due to "Sial-Sima" up warping.

The Tertiary sedimentary thickness is about 3 km. at the centre of the basin and no related anomalies with structures are observed. In view of these considerations, these anomalies are considered as part of the regional gravity, and may be originating due to the variation of the thickness of the traps with 2.4 km thick in the central part. The crystalline basement lie below the Mesozoics. This basin appears to have an opening towards the negative low observed in the Bouguer map east of Ahmedabad. Residual anomalies are observed to coincide with the structural features such as Nawagam, Sanand and Mehsana. It may be possible to separate out residuals of small magnitude. The disappearance of the high amplitude positive anomalies north of the Mehsana structure can thus be attributed to the thinning out of the Traps in that direction.

The gravity maps of India, published by the Survey of India, were in use by the geoscientists till 1978. Later, NGRI published gravity maps of India with updated information by taking more gravity measurements over the country. The measurements have been brought to a common datum of the Indian national base at Dehradun with absolute value of gravity as 979.064 gal., which has been tied to the international gravity net.

The Bouguer Gravity Anomaly Map was presented with gravity contours at 10 milligals interval and the anomaly ranges being colour-coded at 50 milligals interval. The isostatic Gravity Anomaly Map was prepared on the with a sea-level crustal thickness of 30 km and crustal and sub-crustal density values as 2.67 gm/cc and 3.27gms/cc respectively. The isostatic anomalies are computed for 3100 stations well distributed over the country. The anomalies are contoured at an interval of 10 milligals. The map is characterised by both positive and negative values.

Positive anomalies are observed over northwest, central and eastern India and the Himalayas, attaining the maximum value of +80 milligals over Shillong, whereas, almost entire South India, alluvial tracts of Indo-Gangetic plains and east of the Aravallis depict negative anomalies with the largest negative value contour of 120 milligals near the Nepal border.

Magnetic Method

Magnetic surveying aims to investigate the subsurface geology by measuring the strength or intensity of the Earth's magnetic field. Sedimentary rocks generally have a very small susceptibility compared with igneous or metamorphic rocks, which tend to have a higher magnetite content, and most magnetic surveys are designed to detect to magnetic minerals directly. The magnetic method was initially used for petroleum exploration in areas where the oil-bearing sedimentary layers appeared to be controlled by topographic features such as ridges or faults, on the basement surface.

Lateral variation in magnetic susceptibility give rise to spatial variations in the magnetic field. It is expressed in so called magnetic anomalies, meaning the deviations from the Earth's magnetic field. The unit of measurement is the tesla (T) which is volts/m^2. In magnetic surveying the nanotesla is used (1nT = 10^{-9} T).

Interpretation of the data is complicated by the fact that intra-basement susceptibility changes have more significant effect on the observed magnetic field than natural relief on the basement surface itself. Interpretation of magnetic data is subject to the same uncertainty as gravity work.

Unlike the gravitational force, which is always attractive, the magnetic force can be either attractive or repulsive. Thus, mathematically, monopoles can assume either positive or negative values. While measuring single magnetic point sources (monopoles) can never be found alone in the magnetic case. Rather, monopoles always occur in pairs. A pair of magnetic monopoles, referred to as a dipole, always consists of one positive monopole and one negative monopole.

A properly reduced magnetic field, however, can have as its origin at least two possible sources. It can be produced via an induced magnetization, or it can be produced via a remnant magnetization. For any given set of field observations, both mechanisms probably contribute to the observed field. It is difficult, however, to distinguish between these possible mechanisms from the field observations alone. The magnetic field is highly time dependent.

The magnetic force experienced between two magnetic monopoles is given by

$$F_m = \frac{1}{\mu}\frac{p_1 p_2}{r^2}$$

where "m" is a constant of proportionality known as the magnetic permeability, p_1 and p_2 are the strengths of the two magnetic monopoles, and r is the distance between the two poles.

The intensity of magnetization, I, is related to the strength of the inducing magnetic field, H, through a constant of proportionality, k, known as the magnetic susceptibility (I=kH). The magnetic susceptibility is a unit-less constant that is determined by the physical properties of the magnetic material. It can take on either positive or negative values.

Positive values imply that the induced magnetic field, I, is in the same direction as the inducing field, H. Negative values imply that the induced magnetic field is in the opposite direction as the inducing field. In magnetic prospecting, the susceptibility is the fundamental material property whose spatial distribution we are attempting to determine. In this sense, magnetic susceptibility is analogous to density in gravity surveying. This wide range in susceptibilities implies that spatial variations observed in the magnetic field may be related to geologic structure.

If the magnetic material has relatively large susceptibilities, or if the inducing field is strong, the magnetic material will retain a portion of its induced magnetization even after the induced field disappears. This remaining magnetization is called Remnant Magnetization.

The Earth's core is the source of magnetism, rocks exist near the Earth's surface that are below their Curie temperature and as such can exhibit induced as well as remnant magnetization. Therefore, if the magnetic field is measured along the surface of the earth magnetization due to both the main and induced fields are recorded. The strength of the magnetic field at the poles is about 60,000 nT.

The induced field is the one of interest because it relates to the existence of rocks of high or low magnetic susceptibility near the instrument. If measurements are taken near rocks of high magnetic susceptibility it will record magnetic field strengths that are larger than the measurements at a great distance from rocks of high magnetic susceptibility. Hence, like gravity the potentially can locate subsurface rocks having high magnetic susceptibilities, by mapping variations in the strength of the magnetic field at the Earth's surface.

Magnetometers are instruments that are capable of measuring the strength, or a component of the strength, of the magnetic field. For land-based magnetic surveys, the most commonly used magnetometer is the proton precession magnetometer which measures the total size of the Earth's magnetic field. These types of measurements are usually referred to as total field measurements. The instrument's sensor orientation need only be set to a high angle with respect to the Earth's magnetic field. No precise levelling or orientation is needed.

Magnetic observations are routinely collected using any one of three different field operational strategies, airborne, ship borne and ground survey.

The problem of interpretation of magnetic data becomes complicated due to the fact that the observed anomalies are greatly influenced by a number of variable factors, such as, the inclination of the normal field, the orientation of the causative structure with respect to the magnetic meridian. A mathematical transformation of the observed anomalies to make into ones, as if obtained when the normal magnetic field were vertical at every point of observation (Baranov, 1957). The great advantage of this method is that it brings the anomalies back to the apex of the causative sources, by removing the effect of inclination of the normal field, thus making the qualitative as well as quantitative interpretations easier and reliable.

Magnetic data has traditionally been considered very useful in the determination of basement depths and it was an established practice to conduct Gravity-Magnetic surveys for reconnaissance. The detailed seismic surveys were considered justified only when interesting gravity or magnetic anomalies were observed to help the oil exploration and also when sufficiently thick sediments were indicated by the magnetic surveys.

Electro-magnetic Surveys

The principle is, rocks saturated with hydrocarbons, typically show higher resistivity than rocks saturated with water. SBL records the presence of electrically resistive bodies in the earth. The technique is able to detect commercial levels of hydrocarbon saturation, effectively discriminating between these and low-saturation (non-commercial) fizz gas.

SBL is a marine electro-magnetic method that has the ability to map the subsurface resistivity remotely from the seafloor. The basis of SBL is the use of a mobile horizontal electric dipole (HED) source transmitting a low frequency electro-magnetic signal and an array of seafloor electric field receivers.

Indian Scenario

In early 1988 ONGC carried out aeromagnetic survey over **Tripura, Mizoram, Manipur** and **Cachar** in the north-eastern region of India with an objective of finding the configuration and composition of basement (Mehra, G., *et al.*, 1989).

The study revealed a maximum sedimentary thickness of 14 km in Mizoram and thinning of sediments towards east, north and northwest in the area; a generalised orthogonal relationship between the outcrops (striking N-S) and the basement features; existence of a 'box' type intra-basement feature in Haflong-Disang thrust area; an E-W trending basement arch across Tripura and Mizoram; an interesting block with detail magnetic at the border of Tripura and Mizoram; and thin skin tectonics for the area in general. The surveys were also carried out in Cauvery basin (Figures 5.2 and 5.3).

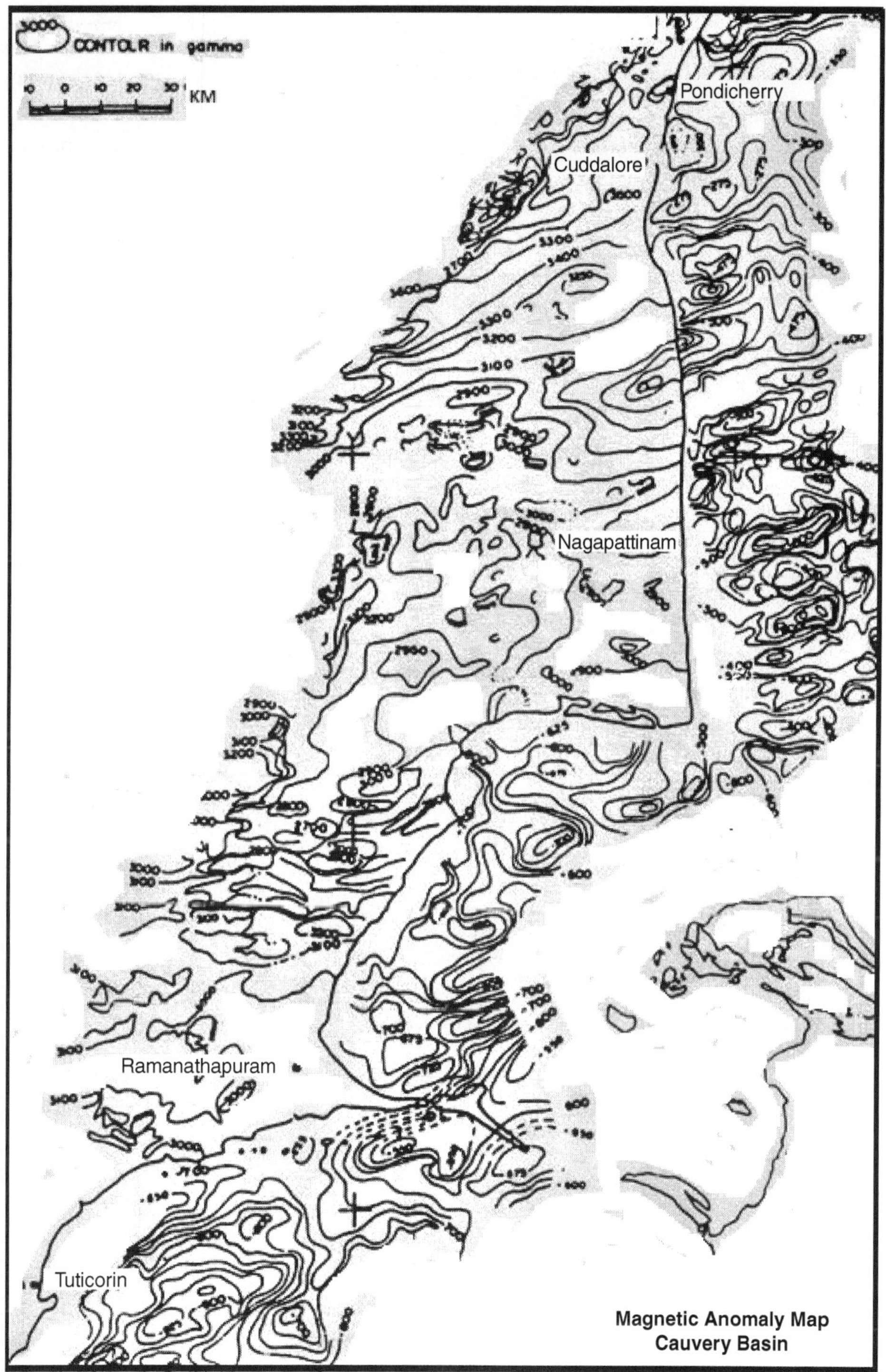

Figure 5.2. Magnetic Anomaly Map (After Mehra G., *et al.,* 1989).

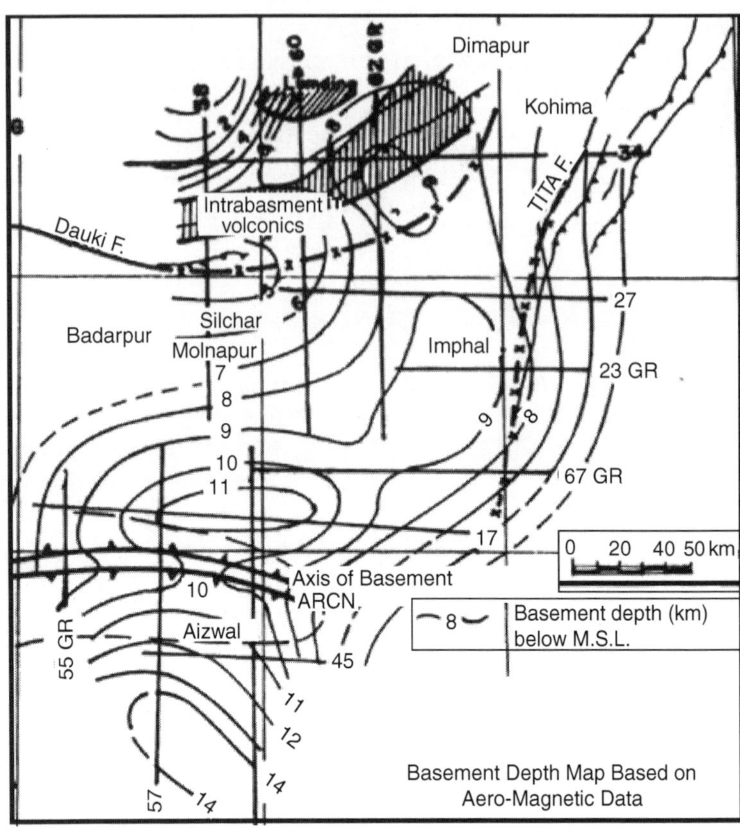

Figure 5.3. Basement-Aeromagnetic-Slcher Area.

The emphasis is on multi-sensor surveys in offshore by the ONGC, wherever possible for economic data acquisition coupled with composite interpretation. This gravity, magnetic and seismic-data were acquired simultaneously along the survey lines by Anveshak, the data acquisition vassal of ONGC. Thus seismic data were available along with the magnetic data. During processing, magnetic data requires observations from a magnetic observatory in the vicinity of the surveyed area. Though absolute susceptibilities (or Magnetizations) cannot be determined, variations in the "Effective Susceptibility" can be determined for prismatic blocks within basement. The studies of the data have been able to detect transition from granitic basement to one overlain by basaltic traps. It should be possible to study trap thickness because higher effective susceptibility values are obtained.

The primary advantages of the gravity and magnetic methods are that, they are faster and cheaper than the seismic method. However, they do not provide the detailed desired information about the subsurface.

Seismic

Of all the geophysical methods, seismic methods particularly seismic reflection method is extensively used for mapping subsurface geology, structures, faults, intrusive bodies or for that matter any geological feature. Seismic surveys use vibration induced by an explosive charge to provide a picture of subsurface rock formations, often as deep as 10 Kms below the earth surface whether onland or offshore. These waves travel downward into the earth's surface and reflect off at various boundaries between different rock strata. In onland, the waves are generated by small explosive charges in the ground or by vibrator trucks, are detected by electronic devices called geophones which receive the reflected energy waves and the data are recorded on magnetic tape which will be further processed to get a final cross sectional view of the Earth's surface. In case of offshore, boats tow cables embedded with hydrophone in the water and instead of dynamite as source pressure pulses are generated by compressed air are used. The waves thus formed travel within water and penetrate subsurface and after reflection from various geological strata return to the cables containing hydrophones where recorded by recording instruments. Further processing of the data almost remains same as in onland data.

Primarily there are two types of seismic surveys *viz.*, Refraction and Reflection surveys. Refraction surveys are not in general widely used because of the limitation in the depth of investigation and also only where the seismic velocities of layers increase with depth which normally do not happen which is the reason why, in the following sections, more elaborate description is given for the reflection methods.

Principle of Seismic Wave Propagation

When the Earth layers are distorted at some point, the energy imparted into the Earth by the source with the distortion can be the energy is transmitted in the form of elastic waves (Figure 5.4). A wave is a disturbance that propagates through, or on the surface of, a medium. Elastic waves satisfy this condition and also propagate through the medium without causing deformation in the medium. Waves that propagate through the earth as elastic waves are referred to as seismic waves. There are two broad categories of seismic waves: body waves and surface waves (Figure 5.5).

Body waves are elastic waves that propagate through the Earth's interior. In reflection and refraction surveys, body waves are the source of information used to image the Earth's interior. Body waves can be further subdivided into two classes of waves: P waves and S waves (Al-Sadi, H. N., 1973).

P waves are called primary waves, because they propagate through the medium faster than the other wave types. In P waves, particles constituting the medium are displaced in the same direction that the wave propagates in the radial direction. Thus, material is being extended and compressed as P waves propagate through the medium. P waves are similar to sound waves propagating through the air.

S waves are called secondary waves, because they propagate through the medium slower than P waves. In S waves, particles constituting the medium are displaced in a direction that is perpendicular to the direction that the wave is propagating. In this example, as the wave propagates radially, the medium is being deformed along spherical surfaces.

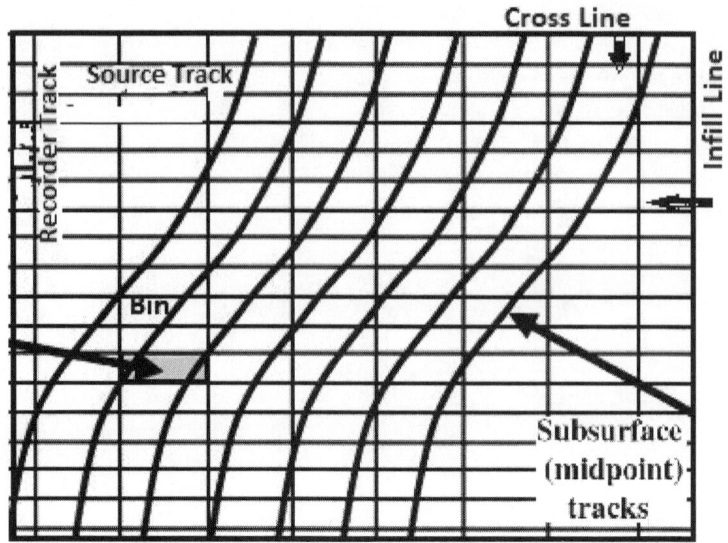

Figure 5.4. Shooting Method of Seismic.

Surface waves are waves that propagate along the Earth's surface. Their amplitude at the surface of the Earth can be very large, but this amplitude decays exponentially with depth. Surface waves propagate at speeds that are slower than S waves, are less efficiently generated by buried sources, and have amplitudes that decay with distance from the source more slowly than is observed for body waves.

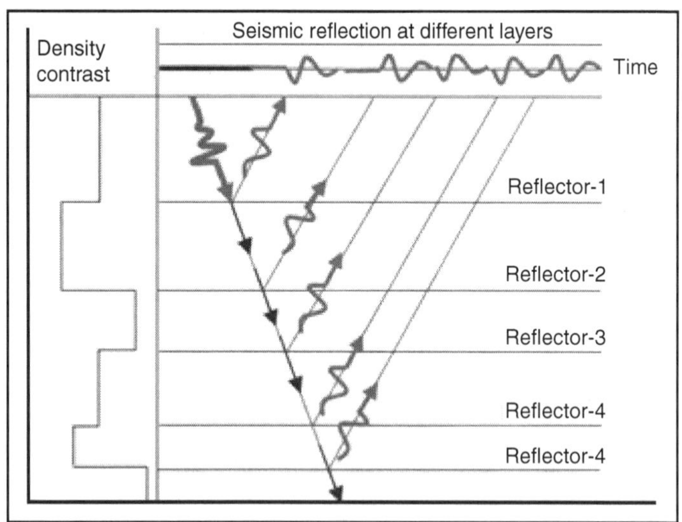

Figure 5.5. Seismic Wave Model.

For exploration surveys, surface waves are a form of **noise** that need to be suppressed. For reflection surveys in particular, suppression of surface wave energy becomes particularly important, because the amplitudes of surface waves generated from shallowly buried sources are often observed to be larger than the amplitudes of the body waves which are important to record and interpret. Surface waves are never the first arrival. The main sub-types of surface waves are Rayleigh waves and Love waves.

Coherent noise is a seismic event characterized by a distinct apparent velocity and well-defined onset. In reflection seismology, coherent noise which appear on shot records, are source-generated seismic events. They are made up mainly of surface waves (ground roll) and air-waves which are of fairly narrow bandwidth with low frequency range. Frequency content of this type of noise is typically below 20 Hz.

The **incoherent noise** consists of seismic events with unpredictable amplitude and onset. This type of noise, which is basically of random nature, forms the seismic-energy background of any seismic shot-record. In seismology, it is called incoherent background noise or ambient noise. In addition to the random nature, the incoherent noise is characterized by a broad amplitude spectrum that covers a wide range of frequencies compared with the nearly limited bandwidth of reflection signals or coherent noises (Figure 5.6).

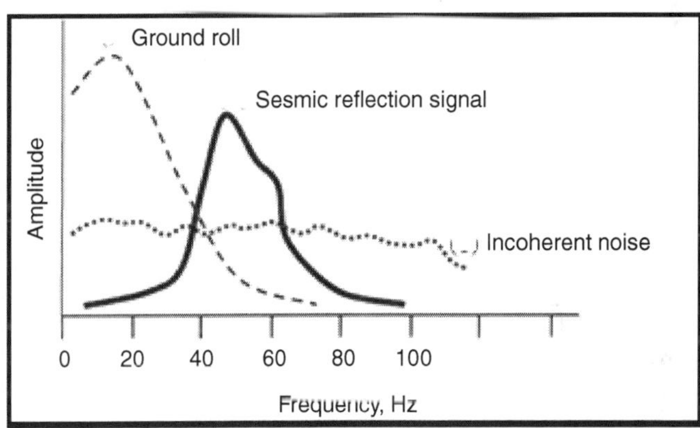

Figure 5.6. Noise Spectrum.

Because of seismic noise which are unavoidable seismic events which get recorded alongside the objective signal, the signal-to-noise-ratio (S/N) becomes an important parameter in signal detection studies. The S/N ratio is used as measure for the signal quality-level.

Shot generated noise and the cultural noise is a major problem in seismic survey because they are time and space coincident energy with reflected arrivals. High dynamic range recording systems are capable of recording weak signals of the order of few nano volts, but the presence of high amplitude ambient noise obscure the signal amplitude which sometimes becomes irretrievable (Shanmugasamy, S., *et al.*, 2006).

Mute is the process of excluding parts of the traces that contain only noise or more noise than signal. The two types of mute used are, front-end mute and surgical mute. The nearer traces are not so affected. When the data are stacked, the far traces are muted down to a time at which reflections are free of refractions is called Front-end mute. Surgical Mute is the muting may be over a certain time interval to keep ground roll, airwave, or noise patterns out of the stack.

Some ground rolls and air blasts are difficult to remove with conventional array based shot records due to the events being spatially aliegned. In the event of using longer array base, high frequency signal is severely affected at far offsets for shallow and dipping reflectors. However, very close spatial sampling and post-acquisition pre-processing with Group Array Formation (GAF) and Group Array Stack (GAS) can result in noise attenuation considerably with increased signal to noise ratio.

Concept of Wave Path and Wave Front

Ray-paths are nothing more than lines that show the direction that the seismic wave is propagating. For any given wave, there are infinite set of ray-paths that could be used. For instance, a valid ray-path could be any radial line drawn from the source (Figure 5.7).

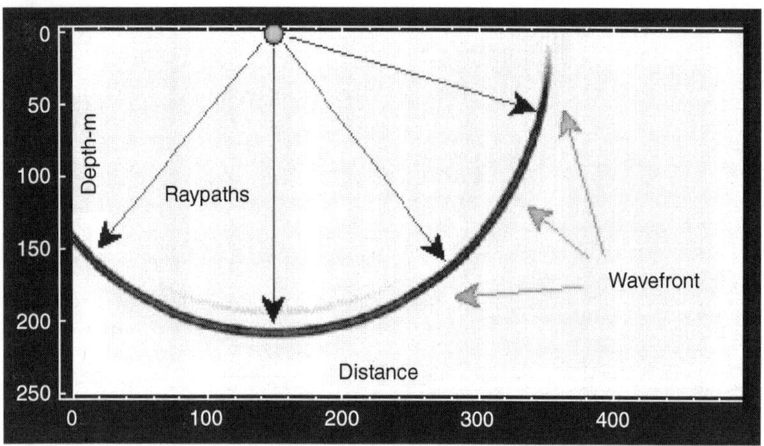

Figure 5.7. Seismic Wave Raypath-Wave Front.

Wave-front connect positions of the seismic wave with same phase at the same time. The wave-fronts are spherical in shape. This surface would connect all portions of the wave that have the largest possible negative or positive amplitude at some particular time.

The change in curvature of the wave front as the wave passes through the interface implies that the ray-paths describing the direction of propagation of the wave change direction through the boundary.

Initially, wave with large in amplitude arrive and that is the direct wave. Then, sometime later, the smaller amplitude reflected wave. The time difference between the observations of these two arrivals is dependent on the distance from the source, the speed of wave propagation in the layer, and the depth to the boundary. Thus, by observing this time difference it is possible to learn something about the subsurface structure.

The velocities of P and S waves through the media are given by the expressions shown:

$$Vp = \sqrt{4/3\mu + k/p} \qquad Vs = \sqrt{\mu / p}$$

Where *Vp* and *Vs* are the P and S wave velocities of the medium, *r* is the density of the medium, and *m* and *k* are referred to as the shear and bulk moduli of the media. Taken together, *m* and *k* are also known as elastic parameters. Any change in rock or soil property that causes *r*, *m*, or *k* to change will cause seismic wave speed to change.

Bulk Modulus is known as the incompressibility of the medium. The shear modulus describes how difficult it is to deform a cube of the material under an applied shearing force.

If a low-velocity layer overlies a higher velocity half-space then in addition to the direct and reflected arrivals, another one may also be observed called as head wave. In refraction seismic surveying the earliest times of arrival of the seismic waves are measured at various distances from the source. For the layer over a half space model, this earliest arriving energy could be associated with either the direct wave or the head wave.

Instruments that are capable of recording ground motion are called geophones which are capable of recording the ground motion produced by the seismic waves.

The first arrival could be the direct arrival or the head wave. Usually, will be recorded the ground motion at a number of different receivers and plot this motion as a function of time and as a function of distance from the source. Plots of the times of arrivals of the various recorded waves versus offset from the source are called travel-time curves.

An interface is that boundary-surface separating two different media. As far as the changes regarding spectral structure and propagation direction of seismic waves are concerned, two media are considered to be different if both of the wave propagation velocity and the medium bulk density are different. Since velocity is function of elastic coefficients, it can be said that density and elastic properties are the factors which control the specific characters of the media. In general, if a reflection event to occur, acoustic impedance (multiplication of velocity with density) contrast is regarded to be the main responsible phenomena.

In passing through an inhomogeneous environment a moving seismic wave would suffer from a number of changes whenever it meets an interface across which there is change in the properties of the medium (Al-Chalabi, M., 1979). In particular, changes in energy content, waveform (spectral structure), propagation velocity, direction of motion, and new wave generation. These changes, are generated at the interface planes defining the layer bounding surfaces (Figure 5.8).

Seismic Refraction Method

Refraction is particularly suitable where the structure of a high-velocity layer, such as the basement or the top of a limestone layer, is a target of geological interest.

If the problem is to determine the depth and shape of a sedimentary basin by mapping the basement surface, and if the sedimentary rocks have a consistently lower velocity than the basement formations, refraction can be more useful. Since velocities in salt and evaporates are often greater than in surrounding formations, refraction has been useful in mapping diapiric features such as salt domes.

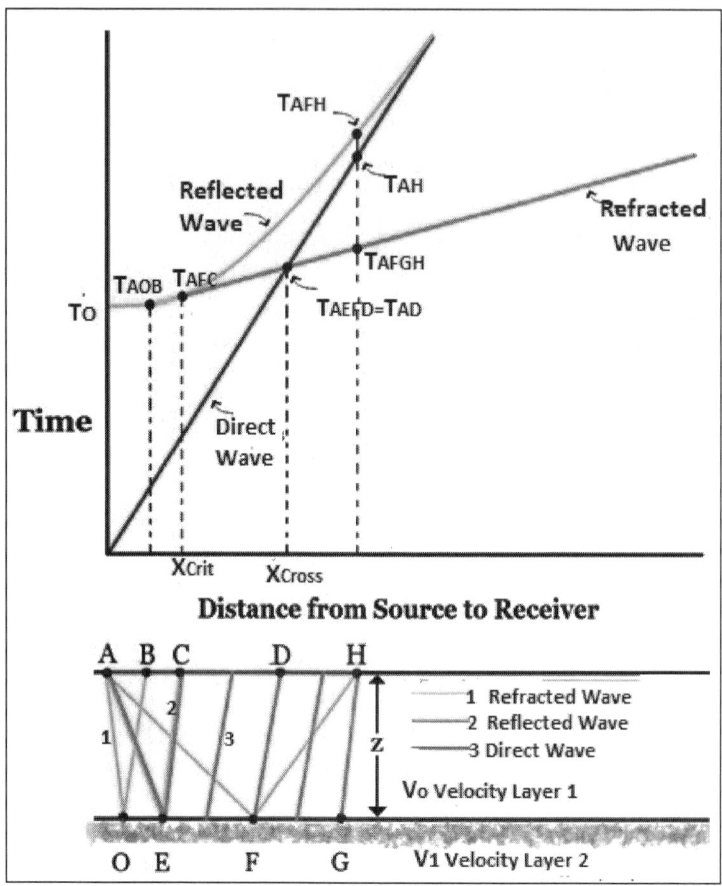

Figure 5.8. Basic Seismic Picture–Direct Wave Reflection and Refraction.

Refraction seismology experiments are based on the times of arrival of the initial ground movement generated by a source recorded at a variety of distances. Later arriving seismic events in the recorded ground motion are discarded in the refraction surveys. Thus, the data set derived from refraction experiments consists of a series of time versus distance data. These are then interpreted in terms of depth to subsurface interfaces and the speeds at which motion travels through the subsurface within each layer (Banerjee, P and K.P. Singh 2006). These speeds are controlled by a set of physical constants, called elastic parameters that describe the material.

The Delay-Time Concept

The travel-times are based on the assumptions that, the interfaces are regular planar surfaces, which may not always be the case. By application of these functions on the more realistic cases, where refracting surfaces are irregular and dipping surfaces, inaccurate results may be obtained. The concept of delay-time, serves as a tool for more accurate interpretation of refraction survey-data where the waves are refracted from irregular and dipping refractors (Gardner et al., 1974).

The delay-time (dTx) is defined to be the difference between the actual refraction travel-time (Ta), corrected for the depth and velocity variations of the weathering zone, and travel time required to traverse the horizontal distance between the source and the receiver at the highest velocity encountered on the ray-path (Sheriff, 2002). The delay time (dTx) can be mathematically expressed as: dTx = Ta-x/V2. Where x distance between source and receiver and refractor velocity is V2. (Figure 5.9).

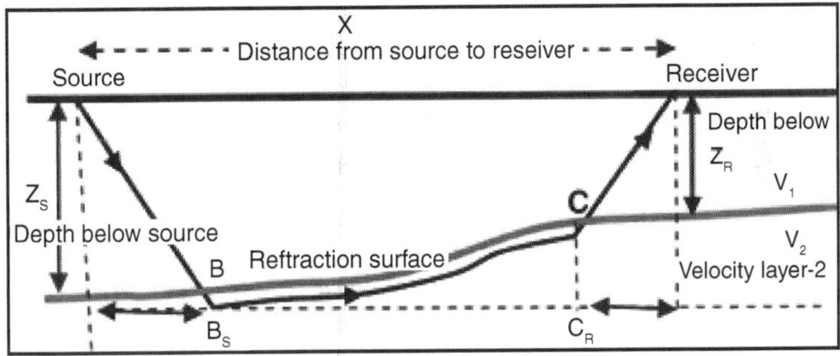

Figure 5.9. Seismic Refraction Method-Principle.

Thus, refraction surveys are more useful in determining the variation of weathered layer thickness and velocity which ultimately very essential for estimating the near surface static corrections while processing the seismic data. Also, this data will be useful for planning the optimum shot hole depths for the survey.

Seismic Reflection Method

Acquisition

In this method, the structure of subsurface formations is imaged by measuring the time required for a seismic wave generated in the earth by a near-surface explosion of dynamite or mechanical impact to return to the surface after reflection from interfaces between formations having different physical properties. The reflections from these geological interfaces are recorded by detecting instruments responsive to ground motion. They are laid along the spread measuring sometimes up to several kilometres on the ground at distance. Depths to reflecting interfaces can be determined from the travel times using velocity information that can be obtained either from the reflected signals themselves or from measurements in wells.

Reflections from depths 10,000 meters can be observed from an explosion, so that in most areas geologic structure can be determined throughout the sedimentary section. The reflection method comes closer than any other prospecting technique to providing a structural picture of the subsurface (Barnes, A., 1993).

The method makes it possible to produce structural maps of any geologic horizons in the form of reflection events, but the horizons themselves cannot be identified without independent geological information (Figure 5.10).

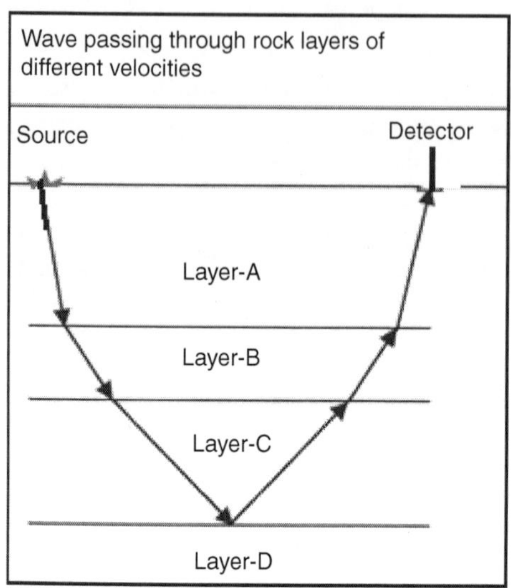

Figure 5.10. Wave Motion in Multi-Layer Sase.

This information provides indication of lithology and one can locate and map such features as anticlines, faults, salt domes, and reefs. Major convergences caused by depositional thinning can be detected from reflection sections, but the resolution of the method is important for finding stratigraphic traps such as pinch outs or facies changes.

Reflection energy arriving on ground movement that has been reflected off the subsurface interfaces. In addition to examining the times of arrival of these, reflection seismic processing extracts information about the subsurface, from the amplitude and shape of the ground motion. Subsurface structures can be complex in shape but can be interpreted in terms of boundaries separating material with differing elastic parameters (Pramanik, A.G, *et al.*, 2003).

Advantages of Seismic reflection methods are to detect both lateral and depth variations, produce images of structural features below and can be used to delineate stratigraphic and depositional features. Response to seismic wave propagation is dependent on rock density and a variety of physical parameters of the rock. Thus, any mechanism for changing these parameters like, porosity, permeability and compaction can be identified.

Seismic Energy Source

Seismic Source is nothing but dynamite explosive source generating seismic energy into the ground. Impact Sources (which are not explosive in nature) that generate seismic energy by impacting the surface of the Earth are known as Vibroseis. Air Gun sources generate seismic energy by transferring the kinetic energy of a moving object into seismic energy (offshore survey). Explosive sources can impart a large amount of seismic energy into the ground (Dave Ridyard., *et al.*, 2006) These sources can vary in size and type from small blasting caps and shotgun shells to larger, two-phase explosives. All explosive sources are triggered remotely by a devise known as a blasting box in a land survey.

Dynamite Source

This is one of the most common types of seismic source used in land seismic surveying. Normally, certain dynamite charge is exploded inside a drilled hole at a depth ranging from few meters to several tens of meters. Typically, the drilled hole is of diameter of about 10 cm and depth of about 5–40 m depending on the nature of the soil cover. Ideally the charge depth should be placed deeper than the base of the weathered layer.

The experience in the seismic surveys carried out in the tough terrain in the ***Himalayan foot hills*** are affected by the near surface corrections in the arrival times arising due to the sudden variations in the topography. One of the major operational difficulty for seismic surveys in these areas is the shot hole drilling. Truck mounted drilling rigs cannot be taken at all places easily due to difficulties in road approach and in drilling a hole in boulder beds restrict the number of shot-points which can be made available for the surveys.

The Weight Dropping method (also called Geograph or Thumper) involves dropping a weight of about 3 tons from a height of about 3 m on to the ground.

Vibroseis method creates mechanical energy which is continuously vibrating for certain time duration. A mechanical, truck mounted vibrator is hydraulically driven, produce electronically controlled vibration. The generated energy is conveyed to the ground by a metal pad attached beneath the truck.

The **Air-Gun** is a widely used energy source in marine seismic surveying. It generates energy by discharging highly compressed air into the water. A variation of this method is the (Unipulse), where the air continues flowing into the generated bubble for some time after the initial discharge.

New land and marine acquisition equipment makes it possible to emit and receive low frequencies (below 10 Hz). This includes: for sources, heavy vibrators and high-pressure air gun clusters; for receivers, digital accelerometers and deep towed solid streamers. The primary interest of recording low frequencies is to improve vertical resolution of surface seismic by widening the frequency spectrum. Enrichment of the reflected signal by lower frequencies is also critical for exploration of deep targets, particularly those located below high velocity basaltic or salt layers if any.

Seismic Energy Receivers

Geophones are devices capable of measuring ground motion generated by the seismic source and these convert the ground motion into electrical signals (voltages) that are recorded by a separate device.

Hydrophone: An underwater microphone designed to monitor underwater pressure pulse response; a measuring instrument referred to as a hydrophone. A typical hydrophone works by transducing (converting) a sound wave into an electrical voltage that is then transmitted in response to a pressure change in the surrounding environment

The hydrophone sensitivity meter (HSM) is used to check the relative sensitivity of hydrophones in a marine streamer section and their polarity. The HSM is loudspeaker sound source in an enclosure which concentrates acoustic vibrations sufficiently to drive a hydrophone directly under it to a detectable output level. Evaluation of individual hydrophones of an active section of marine steamer is based on amplitude comparison between hydrophones. Also, the polarity of the hydrophone can be measured by this device.

In offshore exploration for petroleum, a survey ship pulls 48 or more active sections through the water at constant velocity. Each active section is 50 meters long (Iain Buchan and Andrew Long., 2006). All the active sections are joined together end to end to form the marine streamer. The active section consists of wired hydrophones, spaces and guide ropes which are mounted inside the walled flexible PVC jacket. This PVC jacket is filled with a fluid having low specific gravity so that the streamer is approximately neutrally buoyant in sea water. The streamer is normally maintained at constant depth by dynamic depth controllers.

Seismic Recording System

Recording System is actually consisting of a number of components such that the ground motion is detected by a number of geophones. In addition to recording ground motion, this system must also control the synchronization of the source. It consists of data storage system with numerous electrical connections to the geophones and the source and usually a device to select subsets of the installed geophones to record. Digital sensors meant for acquiring full wave seismic data has the potential to better image reservoir with broader bandwidth and high resolution over the data with analogue sensors.

As a study in India, in Kadi area of **Cambay Basin** to make a comparative study of the digital sensors (Vectorseis) over analogue sensors (SM-24) a 2D seismic profile was shot to record data simultaneously with digital and analogue sensors. All the acquisition parameter were kept same to acquire the data so as to have a meaningful analysis. The wider bandwidth of signal recorded by digital sensors including the very low frequency signals of the range 1-8 hz, were preserved throughout the processing flow, which helped improving the resolution (Jain, M K., *et al.*, 2006).

To attenuate low frequency shot generated noise, instead of frequency filter, f-k filter was used. Use of an identical processing flow, with carefully preserving the signal bandwidth at every stage of processing, the processed output clearly indicates better resolution in the case of digital sensors over that of analogue sensors. The effect is good in the shallower part of the data. Therefore, digital sensors appear to be a better alternative in delivering high quality seismic images with improved resolution,

At surface locations, three separate seismic arrivals can be observed, the direct, reflected, and critically refracted (head wave) arrivals. Only the direct arrival and the head wave are observed as first arrivals.

All parameters of an advancing seismic wave (waveform, speed, and travel-path geometry) may change during the wave propagation. Form and magnitude of these changes depend on the physical properties. The generated seismic energy moves away from the source zone in a form of a wave motion propagation. Under these conditions (seismic energy source within an elastic medium), the seismic wave spreads out from the source zone in every possible direction. A travel ray-path, in a particular medium, is defined once the locations of both of the source-point and detector-point are defined.

A seismic wave spreads out into the three dimensional space of the host medium. If the medium is homogeneous, the seismic energy would advance in every possible direction with constant velocity (Figure 5.11). This means that after any given travel-time the energy would have reached points of equal distances from the source. These points fall on a spherical surface which is marking the wave-front. For a harmonic seismic wave, the wave front is defined to be the locus of points having the same phase of particle vibration. At any point in the wave-field, the line which is perpendicular to the wave front at a certain instant represents a ray.

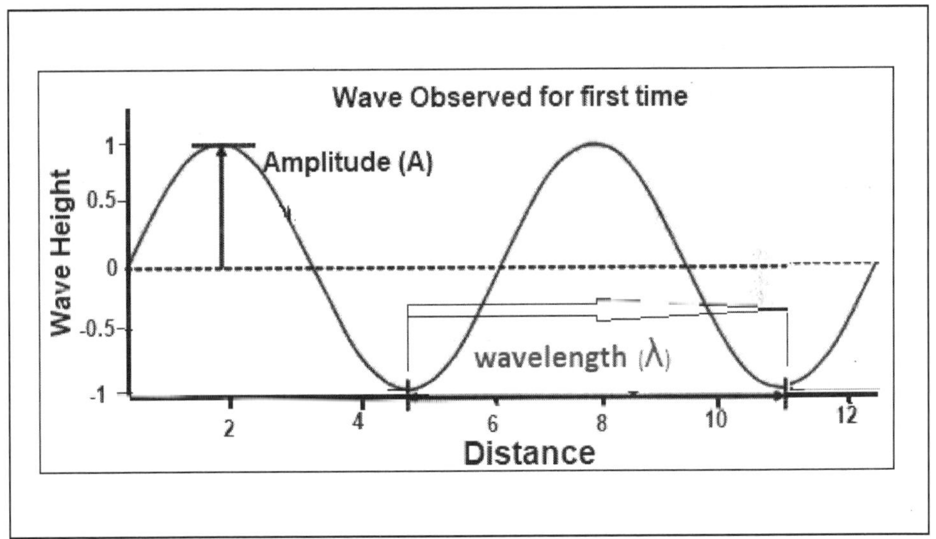

Figure 5.11. Wave Shape.

Recording

The analogue recording system carries out the recording in two stages: the amplification process followed by the recording process. In the first stage, the seismic signal, received from the geophone group, is amplified so that it will be at a level that suits the following recording stage. The seismic signal may be too small for the recorder to detect, or too large for the recording unit to handle without being saturated. In order to cope with this wide range of amplitude variation, amplification factor is made to be variable. This is done by using a special electronic unit called the automatic gain control (AGC), which makes the amplifier gain to vary with the amplitude level of the incoming signal. With this unit the too-small amplitude is increased and the too-large amplitude is reduced. This operation results in compressing the signal amplitude-range to that value which is matching the dynamic range of the recording unit. The technique of amplitude matching with the recorder, for undistorted output, is an essential feature of all seismic recording systems. A modification of the AGC is the programmed gain control (PGC) with which the gain is not tied up to the input amplitude-variation. This is usually made available as an option, applied when it is required.

Next to the amplification part is the magnetic recording system which produces a permanent record of the data stored on magnetic tapes. This is accomplished either directly or after being frequency-modulated (FM). A carrier frequency (typically, 4 kHz) is modulated by the signal received from the amplification system. In this way it can produce a recorded signal of S/N ratio up to 60 db, and frequency response which is practically flat over 3–200 Hz (Evenden, *et al.*, 1971). In digital form, the seismic signal can be subjected to various mathematical analyses using more sophisticated digital software. After being digitized, each sample-value is represented by a binary word which is recorded on the magnetic tape by magnetization status of the cells.

Seismic Survey Design

Seismic wave transmission surveys make use of seismic waves which are directly propagating from sources to receivers. The normal direct wave, which is always recognized on seismic reflection and refraction-shooting records, is a typical transmission method used to calculate the velocity of the surface layer. In exploration seismology, there are several transmission-based methods which are applied for specific exploration purposes. Up-hole surveying, well velocity surveying, vertical seismic profiling (VSP) fan shooting, and seismic transmission tomography, are typical examples of exploration practices employing the seismic wave transmission phenomenon.

In the **up-hole surveying** the source points are placed at different depths inside the drill-hole, while the detector is placed on the surface, close to the head of the bore-hole. In **Well Velocity Surveying** the normal survey procedure starts with placing the detector at a certain depth inside the deep well drilled, while the shot is on the surface. The source-to-detector direct-arrival is recorded (Wuenschel, P. C., 1976).

In seismic **Fan Shooting,** the detection points (receivers) are laid out in fan-like arrays at distances from a common source point (Figure 5.12). More than one fan-spread is usually applied in the one survey. Those ray-paths penetrating the anomalous body will show abnormal travel-times compared with the others which are missing that body. In case of a high-velocity body, a time lead of the arrival time of the received refracted wave will be observed. Thus, by plotting the arrival times against the source-receiver distance, the abnormal values compared with a "normal curve", delineation of the anomalous body is determined (Parr, J. O., Jr., and Mayne, W. H., 1955).

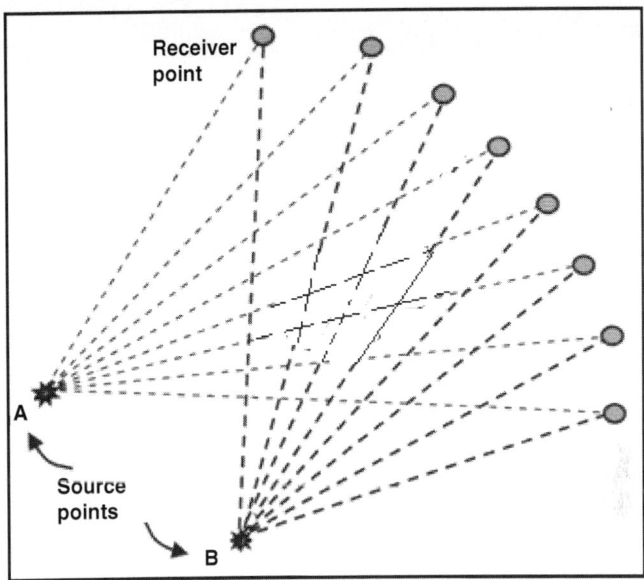

Figure 5.12. Seismic Fan Shooting.

Seismic Reflection Shooting (Onland)

The shooting spread used in seismic surveying is defined as the geometrical relationship between the location of the source-point and that of the receiver-array. This shot-receiver configuration is kept unchanged during surveying of the seismic line (Figure 5.13).

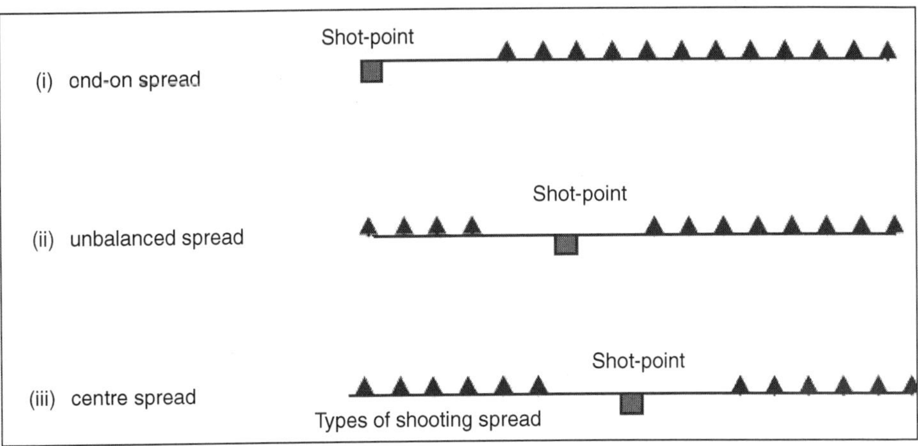

Figure 5.13. Types of Shooting.

The Fold Coverage

The Common Depth Point (CDP) is a subsurface point located vertically below the source-receiver midpoint. This means that each receiver-point of a spread will receive a seismic arrival reflected from a CDP. In fact, shooting of one spread, the reflector will be sampled by a number of reflection points (CDPs) equal to the number of the spread active receivers. From the geometry of the spread ray-path, the CDP spacing will be half that of the receiver points of the spread. Thus, the subsurface coverage (CDP-line) covered from shooting one shot will be about half the spread length (Figure 5.14).

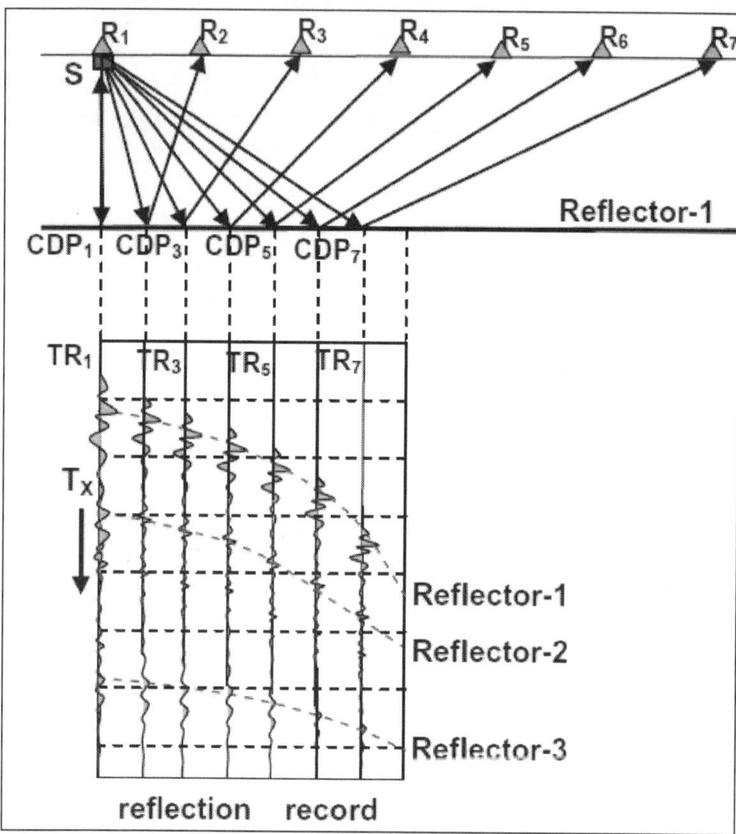

Figure 5.14. Seismic CDP Technique.

Experimental Shooting

The source parameters depend on the type of the source used to generate the seismic energy. For the dynamite source, there are three main parameters, charge depth, charge weight, and shot-hole pattern. The optimum charge depth and charge weight are determined through direct experimental shooting. For the charge depth, several trial shots are conducted such that the charge weight is fixed while varying the depth. The same approach is followed in optimizing the charge weight that is by fixing the depth and varying the weight of the charge.

In **land acquisition** a seismic line is rarely shot in a straight line because of the presence of natural and manmade obstructions such as lakes, buildings and roads. The shot points and the receivers may be arranged in many ways. Many groups of geophones are commonly used on a line with shot points at the end or in the middle of the receiver array. The shot points are gradually moved along a line of geophones (Figure 5.15). The variations in ground elevation in land acquisition causes sound waves to reach the recording geophones with different travel time. The Earth's near surface layer may also vary greatly in composition, from soft alluvial sediments to hard rocks. This means that the velocity of sound waves transmitted through this surface layer may be highly variable. Static corrections a bulk time shift applied to a seismic trace are typically used in seismic processing to compensate for these differences.

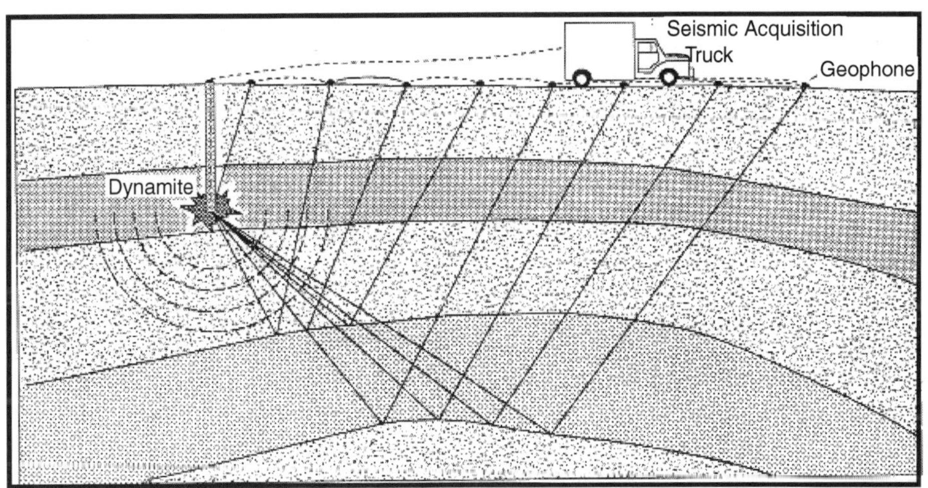

Figure 5.15. Land Seismic Acquisition Pattern.

The most commonly applied shooting spread, used in land 3D surveying, is the rectangular-shaped spread, normally referred to as the spread template. It consists of a number of parallel receiver lines with a shot-line perpendicular (or inclined) to the shot-line. The shot-line is inclined with respect to the receiver lines, usually at an angle of 45°.

Swath Shooting Technique: The swath is defined to be a strip of the survey area of width equal to the width of the used spread that moves along the defined strip in a roll-along movement similar to the "shift-and-shoot" technique used in conventional 2D profiling surveying.

3D surveying of the totally inaccessible area can be carried out using a technique called "Loop Shooting". This technique involves laying out source-points and receiver-points around the inaccessible area, forming a loop that encloses the desired area.

Indian Scenario

Acquiring seismic reflection data in **Himalayan Foot Hill** is one of the challenging jobs for surface geophysicist due to its complex tectonic setting. Rough topographic condition with steep hill, thick thorny shrubs, pine trees and other logistic hindrances are the major disadvantage for better data qualities. It is very difficult to choose a medium for blasting the energy source below the Low Velocity Layer (LVL).

Frequent Uphole survey and experimentation for charge and depth are essential to find a shooting medium. In the complex geological setting of Himalayan Foot Hill, the interference waves coming from different direction to survey line which causes low S/N ratio. Dip angle stratum divert the acoustic wave and ultimately reduce the foldage. So 100 foldge had been adopted to have a sufficient number of fold. To acquire good data, it does not only depend upon the shooting medium but also depend upon the condition of receivers, source and system perturbations (Dasgupta,S, *et al.*, 2006). Uphole study is one way to understanding the complicated structure of limited portion. Seismic noises can be reduced by keeping the energy source in proper depth using optimum charges. But it also depends upon good coupling of geophones, the perturbations occurred in geophone's parameters, like sensitivity, natural frequency, distortion, tolerances and tilting. Controlling the profile at the time of shooting for local noise can reduce the unwanted noises.

The entire area is mountainous and hilly with deep gorges culminating into several streams and rivers in north-eastern India, ***Mizoram***. Elevations of these hill ranges vary from 200m along river valleys to over 1600m along some of the hills. The inclination of hills somewhere is 60-80 degrees. The area is dominated by north-south trending structural systems, the prominent features are the northwest-southeast trending faults, which cut across the entire area. Such a topographical statistics coupled with complex geology poses special challenges and force to do crooked line survey. The CMP gathers are characterized by variable fold and uneven offset distribution during crooked line survey. Straight line binning at the time of crooked line processing is preferred. The static correction utilizes the floating datum concept due to highly undulating topography, while cross-dip move out correction (CDMO) comes into picture due to irregular acquisition geometry. Generally, in thrust-belt area high impedance bodies are very close to surface and become major obstacle to the propagation of seismic energy. Incoherent scattering further complicate energy penetration problem. Wide offset acquisition geometry offers better solution to such type of problem which widens the possibility of undershooting localized complexities as well as gaining the advantage of improved S/N ratio. Wide offset geometry provides enhanced velocity discrimination and larger amplitude at far offsets. Building velocity model in such type of area is an important and critical step during seismic data processing (Shree Prakash Singh, *et al.*, 2006).

Seismic Reflection Shooting (Offshore)

Marine seismic acquisition is generally accomplished using large ships with one or multiple air-gun arrays for sources. Air-guns are deployed behind the seismic vessel and generate a seismic signal by forcing highly pressurised air into the water at a given interval. Receivers are towed behind the ship in one or several long streamer(s) that are several kilometres long. Marine receivers are composed of piezoelectric hydrophones, which respond to changes in water pressure. In marine acquisition, seismic vessels sail along predetermined patterns of parallel circuits. The length of straight segments is calculated from fold plots, and must include additional length run in and run out to allow the cable to straighten after each turn.

3D Survey

The 3D data acquisition systems capable of simultaneously recording more independent data channels have brought back the possibility of point source point detector mode of data acquisition. With such a large channel capacity, this system's approach features long spreads for deep data and simultaneously short channel spacing and short off set for the shallow data. Point source and point detector mode of recording minimizes the smear of ground arrays and reduces the blurring of the data. The benefit from higher fold and longer spreads are given below:

- ☆ Stacked section signal to noise ratio is improved by CDP stack, assuming random noise.
- ☆ Accuracy of static and dynamic velocity corrections obtained from seismic date is improved as fold increases, further enhancing the stacked sections improves signal to noise ratio.
- ☆ Velocity precision, based on normal movement, increases with spread length. Longer spreads provide more accurate stacking velocities.
- ☆ Longer spreads provide greater velocity resolution between primary and multiple energy, particularly for deep events.
- ☆ Short group length permitting high fold for early arrival wide-angle energy optimizes resolution of near surface events.
- ☆ Seismic model studies requiring highly accurate seismic data, benefit directly from both increased multiplicity and more-precise velocity data.

While 2-D techniques can accurately represent only those structures with no cross-line complexities, 3-D techniques can accurately represent complex three-dimensional structures. With 3-D migration, subsurface features, even smaller or subtle structures in geologically complicated formations, can be defined more accurately with respect to depth, orientation and lateral extent. In addition, CDP equivalent cross sections can be reconstructed at any orientation with the 3-D survey to obtain coverage in rugged or difficult access areas.

Due to the introduction of the digital data recording and processing, the main development-trends were in increasing recording channels and in more advanced magnetic tape technology. These developments have led to more effective exploration techniques including the introduction of seismic three-dimensional (3D) surveying. The sub-surface geology targeted by seismic exploration is, in essence, three-dimensional in nature.

In the conventional 2D shooting, the seismic field is created as three-dimensional wave-field, but detected by two-dimensional array of detection-points. Thus, the 2D surveying is concerned with a limited portion of the reflected 3D wave-energy. The rest of the reflected energy, with all the useful information it is carrying, is left out.

In the conventional 2D seismic surveys, where the shot-point and receiver-points are co-linear, the resulting stack section consists of a series of seismic traces each of which belongs to a CMP location. The stack traces of a stack section are uniformly spaced along the distance. Thus the seismic amplitude in the produced section is function of both trace-position on the seismic line and the two-way reflection time. In the case of 3D seismic trace for each of CMPs, forming a data volume in which the seismic amplitude is represented as a function of its position in space, defined by the three dimensions.

Due to the wide spacing of the 2D lines, construction of the isochron contour maps are based on interpolation of the reflection times measured on the survey lines. For this reason, structural details are generally not much realistic by the 2D data.

In 2D marine data acquisition a single streamer is deployed, whereas in 3D acquisition multiple streamers are towed behind the boat. 3D acquired data can be processed in a more consistent manner and be further corrected using modern visualisation tools. In 2D acquisition, the data collection occurs along a line of receivers. The resultant image represents only a section below the line. 2D data can often be distorted with diffractions and events produced from offline geologic structures.

A truly representative image of the subsurface is only obtained when the entire wave-field is sampled. A 3D seismic survey where the vessel is towing many streamers (up to 20) at the same time with multiple arrays of airguns is more capable of accurately imaging reflected waves because it utilises multiple points of observations. After seismic processing the data can then be represented as 3D volume images of the subsurface.

In 2D processing, traces are collected into CMP gathers, while in 3D, traces are collected into common-cell gathers (binning). To perform 3D binning, a grid is first superimposed on the survey area. This grid consists of cells with dimensions of half the receiver group spacing in the in line direction, equivalent to the CMP spacing in 2D processing, and the line spacing in the cross line direction. In reality, midpoint distributions within a cell are not necessarily uniform since cable shape varies from shot to shot and line to line. Such side drift of the cables is called feathering.

Information such as faulting and fracturing, bedding plane direction, the presence of pore fluids, complex geological structure, and detailed stratigraphy are now commonly interpreted from 3D seismic data sets (Brown, A. R., 2011).

Another advanced technique of marine seismic is ocean bottom seismic (OBS) acquisition. It gives the possibility of direct measurement of S-wave data in addition to P-wave data by using ocean bottom cables that have three component geophones (3C) and a hydrophone in addition (thus 4C in total). Most designs can be classified as either patch or swath. In swath designs, the source lines are parallel to receiver lines, while in patch designs, source lines are perpendicular to receiver lines. Patch design produces seismic data of a relatively wide range of azimuths, whereas the swath design produces data of a limited or narrow range. Moreover, the swath design offers a more uniform sampling of offsets with better near-offset coverage.

Broadband Marine Acquisition

A new method was developed for broadband marine acquisition and processing. A 3D shallow towed streamer spread is deployed, designed to optimize the mid- and high-frequency parts of the bandwidth. In addition, data are simultaneously acquired from a small number of deeper towed streamers. The depth of these deeper streamers is optimized for the low frequencies such that the combined overall bandwidth is enhanced. Because the deep streamers only provide the low-frequency part of the bandwidth, one can more sparsely sample these data enabling efficient acquisition scenarios as fewer streamers are required. In addition, inversion for acoustic impedance, imaging, and velocity model building, also benefit from the broadband result. Data acquired in this way are more robust to poor weather conditions than conventionally acquired data.

Indian Scenario

Using 3D seismic data of **Heera–Panna-Bassin area of western offshore,** seismic impedance volume generated through model based post-stack stratigraphic inversion, seismic attributes and attributes derived from logs, quality of limestone and porosity, were analysed with geo-statistical cross-plots method to understand the inter-relationship between litho-facies and seismic facies. Moderate to high amplitude and moderate impedances have been found to be associated with carbonates having good porosity. Various facies assemblages representing carbonate build-ups, off-build-ups at slope of big build-ups and wedges of debris flows (talus) were identified with consideration of structural configurations, attributes and impedance, and depositional geometries. Porosity and impedance have inverse relationship (negative correlation coefficient). Though low gamma high impedance carbonates are clustered with low porosity

The first 3D data was acquired in **Geleki area**, upper Assam, India, during early nineties, with DFS-V which were having limited channel capacity. Due to limitation in offset and signal to noise ratio 3D seismic data could image the sub surface with limited clarity and some data acquisition foot prints were remaining.

The discovery of more hydrocarbons in Gandhar and adjoining fields in the Jambusar-Broach and Narmada Tectonic Blocks of **Cambay Basin** has resulted in acquisition and processing of many 3D volumes since 1994. However, due to different acquisition and processing parameters used by different times, it was not possible to evolve a composite picture of the subsurface, mapping of various sand geometries and evolving a viable composite depositional model, although geological models based on 2D seismic data and available well data were also used.

So, much area with many differently acquired 3D data merging was attempted as pre-stack stage and reprocessed whole data as a single volume. Due to complex topographical and geological set up and acquisition related problems arising out of varying near surface and industrialization in the area, the data were affected by noise, ground rolls, multiples, spikes and other coherent and non-coherent noises. These problems neet to be addressed by using various data preconditioning algorithms, manual and auto editing of shot data which result in improvement in signal to noise ratio. This preconditioned data was used for further processing after merging at pre stack stage. This methodology has helped in bringing out a 3D seismic volume in northern part of the area which is being used to understand complex geology of the area.

Seismic Velocities

The fact that the end product of any seismic survey is a time-image and not depth image of the subsurface geology. Thus, to convert the time-image data into depth domain, velocity must be made available. In seismic reflection surveying, velocity computations, aim at finding the seismic velocity (average- or interval-velocity) expressed as function of depth.

Five velocity-types are in common use in seismic exploration (Faust, L. Y., 1951).

(i) Instantaneous Velocity (V)
(ii) Interval Velocity (IV)
(iii) Average Velocity (AV)
(iv) Root Mean Square (RMS) Velocity (RV)
(v) Stacking Velocity (SV)

At any point of the travel-path, the velocity can be measured or computed at that point. This is the **instantaneous velocity** which can be defined to be the velocity of wave motion at a given point in a medium traversed by an advancing seismic wave. ($v = dx/dt$) x is the distance.

The **interval velocity** (IV) is defined to be the average velocity over an interval of the travel-path. It is usually measured or computed for individual geological layers. Thus, the interval velocity, of a geological formation of thickness (Δh) and interval transit time (Δt), is given by $IV = \Delta h/\Delta t$.

The **average velocity** (AV) is obtained from dividing the total distance travelled by the wave by the time spent in covering that distance. In seismic work, the average velocity is computed for a set of layers, usually starting from the datum plane down to the required reflector level. For a vertical travel-path, the average velocity of a geological section made up of (n) layers will be given by dividing the total thickness hn, $AVn = hn/tn$.

Root mean square (RMS) velocity is defined as the square root of the average of the squares of the weighted interval-velocities, where the weighting factors are the layer thicknesses or the interval transit times. In seismic exploration, the RMS-velocity, like other types of velocities, is usually computed for vertical ray paths of waves traversing a multi-layer (Figure 5.16). The root mean square velocity (RMS) can be calculated from the interval velocity data (IV), down to the nth layer, by use of the formula: $RVn = [(\Sigma IV2n.\ \Delta tn)/\Sigma \Delta tn]^{1/2}$

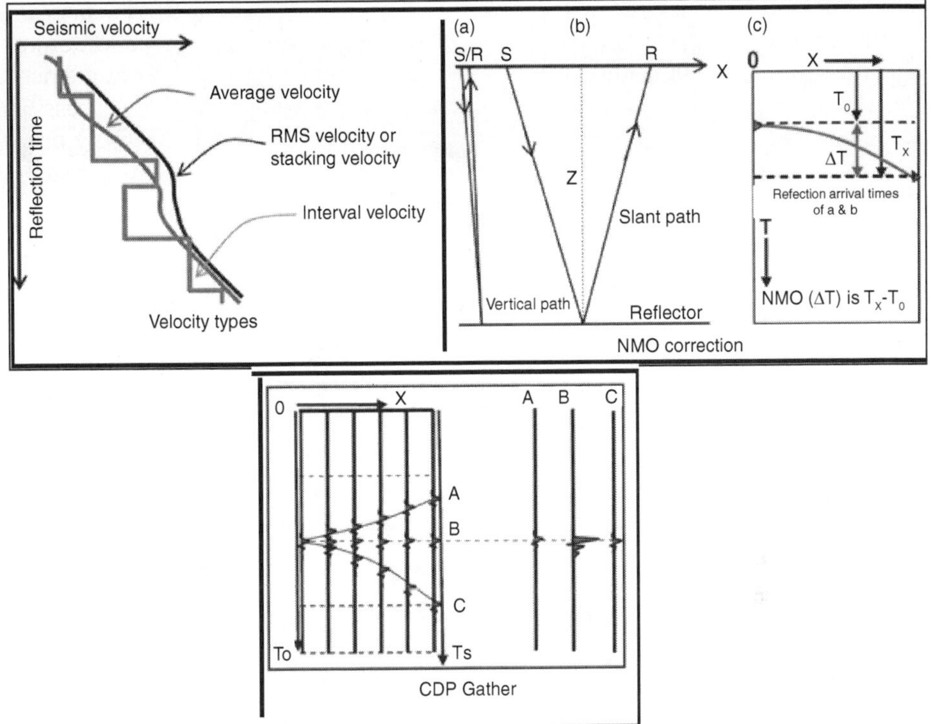

Figure 5.16. Velocity Types - NMO correction- CDP Gather Based on Stack velocity.

Stacking velocity is the main velocity variable that enters the Normal Move Out (NMO)-correction formula. It is applied to remove the time-contribution of the receiver-to-source distance (called, offset) from the total reflection travel-time. Stacking velocity is determined by velocity analysis technique whereby the time contribution of offset is removed before stacking of the CMP gather traces (Wyllie, M. R. J., *et al.*, 1958).

The process of velocity analysis normally follows a trial-and-error approach. A set of trial velocity functions are applied in NMO-correcting the traces of a CMP-gather. The criterion used in recognizing the correct velocity is the S/N ratio of the stacked NMO-corrected reflection signal. The end- result of the analysis is a stacking velocity for each reflection event, plotted as function of reflection two-way vertical time. The stack trace is obtained from summing (stacking) the CMP-gather traces after being NMO-corrected. The correct stacking velocity is the velocity which, on NMO correction, makes all reflection events in the gather in phase, and when these are stacked, a strong reflection signal is obtained.

This is one of the fundamental processing steps (called NMO correction) which bring about coherency of the reflection arrivals which, on stacking, give enhanced reflection event. It is called stacking velocity because of its role in enhancing the stacked reflection signal. Since its direct role is in the NMO correction, the term (NMO velocity) will be more appropriate than the commonly applied term, stacking velocity. Role of the stacking velocity in enhancing the reflection signal (Robinson, E. A., 1983 and 1988). In parallel reflectors with parallel velocity layering, the stacking velocity is equal to RMS velocity.

The Normal Move-Out (NMO) is defined as the difference (ΔT) between reflection travel-time (Tx) and the two-way vertical travel-time (T0). For a single horizontal reflector found at the base of a homogenous layer (of constant velocity), the NMO parameter (ΔT).

By use of certain source-receiver layout, it is possible to shoot a number of shots such that the reflection points resulting from this number of shots coincide on each other that is one point-location will serve as common point to all of the implemented shots (Mayne, W. H., 1967). In this type of repeated shooting-spread, the reflection point becomes known as common depth point (CDP), common reflection point (CRP), or common mid-point (CMP). For a horizontal plane reflector, these points coincide at a point which is located vertically below source-to-receiver midpoint. In the case of dipping reflector, however, the reflection point gets shifted up-dip by a distance depending on the offset as well as on the dip angle of the reflector.

Basic Seismic Data Processing

The main object of seismic data processing is to transform the information of the field record in to useful geological interpretation by eliminating various noises and keep the useful reflection information preserved.

In general various processing stages include:

Demultiplexing

The trace data is recorded in multiplexed format and need to demultiplexed. In multiplexed format, the data are recorded in sequential manner the first sample of the Ist channel is recorded,then first sample of the 2nd channel, first sample of the 3rd channel and so on for all the channels are recorded. Then second sample of all the channels are recorded. This process continues for all the channels. This process is called multiplexed format. Demultiplexing the data means arranging in channel sequence wise the first sample of the first channel, second sample of the first channel, *etc.*

Spherical Divergence Correction

Seismic wave propagation in the subsurface geology involves loss of energy while travelling deep within the media. Main causes are due to spherical divergence, inelastic attenuation, conversion to heat, back scattering, *etc.* As a result, shallow reflectors will have more energy and deeper less energy. Correction to this is applied to compensate the loss of energy.

Deconvolution

Shorter wavelengths (high frequencies) are lost during the propagation of seismic energy due to low velocity or under compact sediments in the near surface below which is referred as Earth is acting as high cut filter. The process of high cut filtering can be compensated by using another filter whose response is opposite to

Earth's response. This process of an inverse filter is called as deconvolution. It tries to recover poorly represented frequencies in the input trace data.

Band-pass Filter

In order to limit the frequencies to seismic range, some filters are designed to pass frequencies containing reflection energy and to reject those frequencies containing mostly noise and no apparent reflection.

Amplitude Normalisation

For computational purposes entire range of seismic amplitudes are preserved by applying amplitude scalar such as the RMS (root mean square). This does not mean that relative amplitudes are lost in the process of normalisation.

Static Corrections

Ideally seismic data needs to be acquired from flat surface to represent true subsurface geology. If the observation points are situated at different altitudes or if the near surface conditions (geological conditions) are dissimilar at source and receiver positions, some corrections are applied to the observations before further analysis. These corrections are known as static corrections which are more significant on the data collected on irregular topography and in areas where fluctuating weathered layer conditions exist. The static corrections are computed at every source receiver positions with respect to a reference datum such as mean sea level. These corrections static because they do not change with time. The static corrections are negligible in the case of data acquired offshore.

CDP Sorting, Velocity Analysis and Normal Move Out Correction

In CDP sorting, all the traces which are reflected from Common Mid-Point (CDP) are gathered and recorded in a single record in the increasing order of offset distances between source and receiver positions. After sorting, each reflection event on the time record is represented by corresponding hyperbolic curve. Each hyperbolic curve is associated with a particular velocity.

Seismic waves have a velocity that is very much dependent on the medium or geology. Thus, velocity changes laterally as well as vertically as the seismic wave travels in to the earth mainly because inhomogeneity. So velocity analysis must be done continuously estimated form place to place in the area. Velocity analysis refers to estimation of velocity that brings best alignment of non-zero offset time to its zero offset time.

A correction termed as Normal Moveout Correction removes the difference in the travel caused due to difference in the offsets and make the primary reflection event of all the traces in a CDP gather coincides with zero offset events.

Stacking

Stacking is summation of two or more traces in to one trace. All the recorded from a common depth point are gathered and grouped as a single record. By applying appropriate NMO correction, all the traces can be summed and normalised to give a stacked CDP trace. Stacking enhances the in-phase primary reflections and reduces the out of phase events ie noise.

Dip Move out Correction

NMO correction is successful only in the case of plane and parallel layers without any dip. NMO correction fails in the case of dipping beds. DMO correction is used in the case of dipping layers for converting non-zero offsets to zero-offsets.

Random and Coherent Noise Attenuation

Noise removal from the seismic data is the main problem in the signal conditioning. There are two types of noise one is random and the other is coherent. Though CDP stacking results in the suppression of random noise to some extent, their explicit estimation is necessary for proper removal. The estimation of random noise is done in Frequency-space domain (F-X). Each seismic trace in T-X domain is Fourier transformed to get corresponding trace in F-X domain. Signal energies are predictable and can be removed through predictive deconvolution. Thus, deconvolution in F-X domain removes signal content of the trace leaving behind the noise content. Total trace field minus the estimated noise gives the signal field. Inverse Fourier transform of this returns in T-X domain.

The surface and near surface waves appear as coherent noise in seismic records and are characterised by their trend, frequency content, *etc.* The main cause for such noise is due to charges fired at shallow depths, lot of wind activity, ground roll, back scatter energy and reflection multiples. Coherent noise is attenuated though multichannel filters implemented in XT, FK, and Tou-P domains; Attenuation of multiples is a challenging in the seismic data processing. Wave equation-based methods are used for removing multiples.

Deconvolution after Stack

NMO correction and stacking act as a high cut filter resulting in loss of high frequencies. The post stack deconvolution though it milder is applied to enhance the resolution aspects in the seismic section.

Time Variant Filter

Earth consists of sets of layers or strata that are distinctly characterised by certain band of frequencies and different parts of seismic section has to be subjected to different sets of band pass filters in time variant manner. The filters are designed through visual inspection of the stacked section by passing through various sets of band pass filters.

Migration

Migration of seismic data moves dipping events to their correct positions, collapses diffractions, increases spatial resolution and is probably the most important of all processing stages. Migration theory has been long established but restricted computer power has driven the industry to a bewildering array of ingenious methods to perform and enhance the accuracy of migration.

The migration techniques have developed through many phases. Seismic migration can be implemented either before CMP stacking (Pre-stack migration, PRSTM) or after the stacking process (Post-stack migration, POSTM). Migration effects in collapsing diffraction arrivals and enhancing fault resolution (Kandpal,M.C., *et al.*, 2006). Development trend of migration techniques seems to be towards building more sophisticated algorithms for computing pre-stack depth migration, based on wave equation modelling (Gazdag, J., 1978).

Effectively, migration is an inversion process transforming seismic data from time domain to depth domain. The input to migration is the seismic record which can be non-zero offset (pre-stack, CMP-gathers) or zero-offset traces (stacked CMP-gathers). The amplitude in the input set of traces, can be considered as a two-dimensional function of distance (x) and time (t) as seen by receivers on the adopted datum plane, that is at zero-depth level (z = 0). After migration, the time image is transformed into the corresponding true image position and true geometrical shape. Migration can be considered as re-mapping of the seismic amplitude as function of distance (x) and depth (z). The migration output is the true geological depth-picture seen.

The most widely used method of imaging "post-stack" does not hold good when reflections and diffractions with conflicting dips are present and when there is a lateral velocity variation. Under such circumstances, the common midpoint (CMP) stacking technique may miss the primary events. The major drawback of conventional normal move-out (NMO) and CMP stacking technique is that it enhance the signal to noise ratio of stacked section with a particular dip and deteriorates the data quality with a different dip.

As a result, technique of dip move-out (DMO) is being used which enables the computation of zero offset seismogram from NMO corrected seismogram. But this technique fails to improve the data quality in regions of severe velocity variations and presence of steeply dipping events. The most direct solution to such problem is to migrate the data before stack or in other words pre-stack migration. Pre-stack migration followed by normal stack yields better results and may be the right solution in such cases (Mitra, PP. 1988).

Seismic Trace Display

The seismic trace is the recorded amplitude variation as function of reflection travel-time. The trace contains all the outputs of an active seismic channel during the recording time. It is, in fact, a record of the whole sequence of the reflection arrivals as well as other arrivals such as direct waves, refracted head waves in addition to the various types of seismic and none-seismic noises.

There are three common modes of trace-display. These are:

(i) Wiggly-trace, the amplitude variation is represented by a wiggly line recorded (usual).

(ii) Variable-area trace, the peak-parts of the wiggly trace are blacked in.

(iii) Variable-density trace, the amplitude variation is represented by an appropriate shade-intensity.

The amplitude of a reflected seismic signal normally decreases with the increase of the distance between source and receiver. This decrease is related to the dependence of reflectivity on the angle at which the seismic wave strikes the interface, spreading, absorption, near surface effects, multiples, geophone planting, geophone arrays and instrumentation. In certain depositional environments, the amplitude variation can also be an important clue to the lithology or to fluid content there in (Figure 5.17).

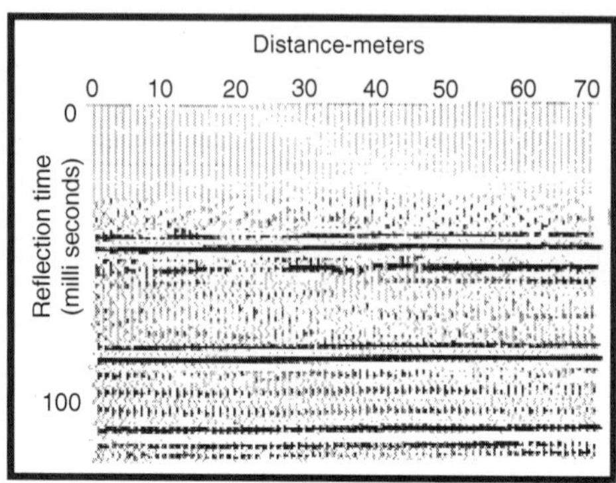

Figure 5.17. Normal Processed Reflection Seismic Section.

Time-to-Depth Conversion

Record sections are conventionally plotted versus time. Average velocity functions are used to convert the time axis to a depth axis. On a depth section, usually made after a migration, the wavelengths of the plotted seismic events lengthen with increasing depth. This effect is caused in part by the increase of average velocity with depth. Thus, at greater times, equal time intervals correspond to greater depth intervals (Figure 5.18).

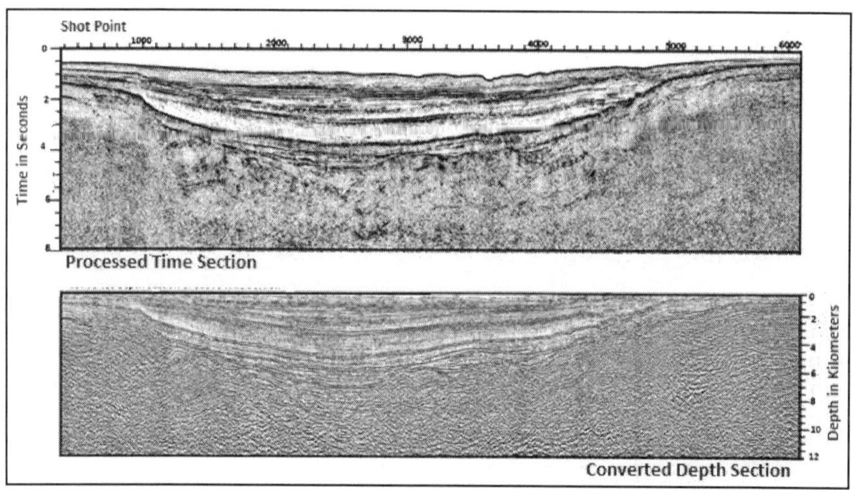

Figure 5.18. Time to Depth Conversion of Seismic Section.

In **Upper Assam** with in the OIL area, most of the identified oil traps are found to be horst graben related closures from the deep Eocene reservoirs of about 3500-4500 m depth. The conventional 2D/3D basic processing which normally includes Dip Move Out and Post stack time migration may not be enough to improve the image. So full Pre-stack time migration (PSTM) was carried out and is very successful when the sedimentary structure is relatively flat, as it improves the reflection quality as diffractions which interfere with primary reflections are collapsed before stack since the lateral resolution is increased as the PSTM process shrinks the Fresnel zone very effectively. Even the 3D PSTM process has shown considerable improvement in data quality both laterally and vertically (Krishna, C.V.G., *et al.*, 2006).

Vertical resolution can be defined as the minimum vertical distance between two interfaces needed to give rise to a single reflection that can be observed on a seismic section. Resolution depends on wavelength of signal at depth in question, which depends on frequency and velocity. Wavelength (λ)= Velocity (v) x period (T) = Velocity (v)/Frequency (f). The higher the frequency of the waveform, the greater or better will be the vertical resolution. Because velocity increases with depth and frequency decreases with depth resolution goes down. Typically tops and bottoms of beds resolved by TWT of $\lambda/2$. Once the bed is about $\lambda/30$ thick or less, reflections from the top and base effectively cancel and there is no detectable seismic response.

Lateral resolution is determined by the radius of the Fresnel Zone, which itself depends on the wavelength of the acoustic pulse and the depth of the reflector. Fresnel zone are typically hundreds of meters.

Thin-bed effect or tuning reflectors that are spaced more closely than one quarter thick ($\lambda/4$) of the wavelength have responses that begin to add constructively to produce a reflection with high amplitude.

The 2D seismic data were acquired with an objective to map the structural features of northern plunge of **Gojalia** anticline and examined the lateral continuity of the pay sands (Bhumija Agarwal,et.,al., 2006). Overall improvement in imaging the crest of Gojalia anticline started with the early detailed evaluation of the earlier 2D data which necessitated the re-acquisition of 2D seismic. Time processing showed tremendous improvement in the data quality over the earlier 2D series. Prestack Time migration showed another level of improvement especially close to the fault and subtle stratigraphy features which were less clear on the POSTM section. These 2D data POSTM will provide more accurate maps giving more confidence/ risk assessment of the area yielding more accurate estimates of the reserves and lower risk.

Seismic tomography is a type of inverse modelling or inversion. Seismic Tomography (meaning - cross-section drawing). In the seismic field, tomography found its way as an effective exploration tool under the name of Seismic Tomography (Figure 5.19). This technique has been applied successfully in imaging seismic velocity variation and producing subsurface geological models.

A 3D tomographic amplitude inversion approach may be used for the estimation of absorption model. The method to mitigate the frequency dependent dissipation effects caused by transmission anomalies and can be considered as one of the processes for amplitude preserving processing that is important for AVO analysis when transmission anomalies are present (Bhuin. N.C., 2003).

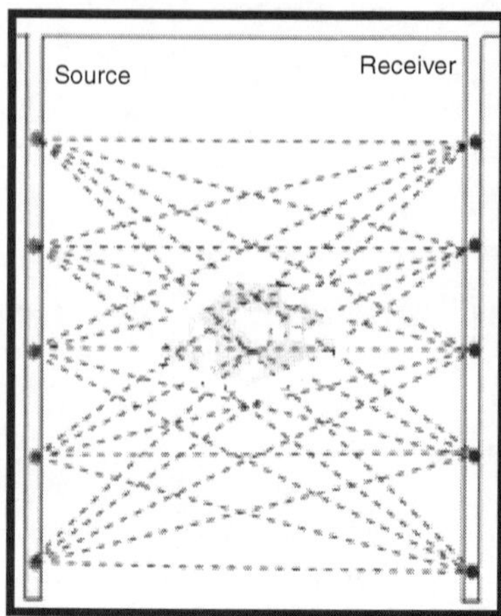

Figure 5.19. Seismic Tomography.

Vertical Seismic Profiling (VSP)

Like well velocity surveying, it is done by lowering a detector into a bore-hole, and taking readings at much closer point-locations (typically at 25 m spacing). The main differences between VSP and well-velocity shooting, is in the detector spacing and in recoding duration-time. Conventional well-velocity survey is concerned with measuring the travel-time of the source-to-detector direct wave. In the case of VSP shooting, recording time is extended to time which is long enough to be able to record arrivals of wave energy reflected from deep interfaces. There are two types of VSP field set-ups; these are the (zero-offset VSP), where a single source is placed at a short distance from the well-head, and the (offset-VSP) where the source is placed at long distance (of about one-to-two kilometres) from the well-head. The zero-offset VSP, which is the more commonly applied type, can help in identifying and removing multiple reflections, whereas the offset- VSP can help in obtaining information about structural and stratigraphic features existing in the areas around the well-location (Galperin, E. I., 1973).

Normally for low dip stratigraphic setup, extent of subsurface coverage obtained from Offset VSP surveys and are governed by the drilled depth (DD) of borehole. In an offset VSP survey, normally to avoid generation and recording of refracted (or mode converted) seismic waves, the energy source is deployed upto 2/3 rd of DD, accordingly subsurface coverage available from offset VSP survey is limited.

A new and innovative technique called as ***'Extended Offset VSP Survey'*** was implemented in a well in **Cambay Basin**, India, wherein the deployment of energy source is stretched much beyond the traditional limit. Consequently a much larger subsurface coverage at the zone of interest was acquired which provides very accurate and high resolution subsurface image (Singh, N. P., *et al.*, 2006). In order to establish the extent of reservoir and to accurate drilling of wells, more and more offset and Extended Offset VSP surveys can be carried out (Hardage, B. A., 1983).

Seismic Inversion and AVO

Seismic Velocity Inversion

The amplitude of a reflection signal (A) is dependent on the reflection coefficient (R) which is, in turn, depending on the acoustic impedance (Z) which is equal to density (q) multiplied by velocity (V). This relationship (Z = qV) shows that velocity can be determined from amplitude measurements, given the seismic impedance and density data.

This is an inverse problem in which the acoustic impedance (expressed by velocity) can be obtained from amplitude data. If the velocity (V1) of the surface layer of a layered medium is known then using the inversion formula, V2 = [(1 + A)/(1 – A)].V1. The velocity (V3) of the neighbouring deeper layer is computed. By repeating the computation, velocities of the rest of layers are sequentially determined. Normally this approach is applied in transforming a seismic stack section into acoustic impedance section, or into what is called pseudo-impedance when density is ignored.

There are two types of modelling: forward and inverse. ***Forward*** modelling makes use of earth models, seismic data acquisition parameters, and theoretical models of physical processes to generate synthetic data that match, within some maximum error, actual seismic data. ***Inversion*** is similar to forward modelling in that it uses data acquisition parameters and theoretical models of physical processes to produce synthetic data. However, the objective of inversion is to produce a model of the subsurface that predicts the observed data.

S-waves amplitude is considered as an effective indicator for studying fracture developments, in particular. For example, abnormally low amplitudes indicate presence of rock fracturing, Poisson's ratio = VS/VP. This is an important relationship, since it is possible to relate the velocity ratio (VS/VP) to the elastic constants of the medium, which means that it furnishes an effective tool for stratigraphic changes as porosity, fracturing developments, sedimentary facies, and fluid contents.

The **AVO** (Amplitude variation with offset), or AVA (Amplitude variation with angle of incidence) is used to express variation of reflection amplitude as function of offset (or function of angle of incidence).

AVO is considered to be one of the effective tools used in direct hydrocarbon detection in exploration. It can be used to explore reservoir and subsurface geological conditions, since it can give indications on lithological changes, reservoir conditions, and presence of high-pressure gas zones. Another significant application of the concept is determination of boundary outlines and areal extensions of hydrocarbon reservoirs.

It is found to be dependent on Poisson's ratio as well as on impedance contrast across the reflection interface. AVO has the same information contained in a combined P- and S-waves data. Reflection coefficient variation with angle of incidence is showing irregular variation with increasing angle of incidence (or with increasing offset). Depending on the geological environment (distribution of the acoustic impedance on both sides of the interface) the reflection coefficient can vary from large-negative to large-positive values. This behaviour can therefore be used as an indicator to predict lithological changes or type of fluid deposits.

For AVO analysis, the reflection amplitude (measured from shot records) must be true amplitude, unmodified by application of processing steps of types which indorse amplitude changes such as severe filtering or equalization. Since AVO analyses are made on gather traces (non-stacked traces) noise effect is large and, therefore, certain data-pre-conditioning processing is needed to remove the incurred distortions. Amplitude variation due to varying conditions near the source- and receiver points must be corrected for variation in seismic reflection amplitude with change in distance between shot point and receiver that indicates differences in lithology and fluid content in rocks above and below the reflector. AVO analysis is a technique by which geophysicists attempt to determine thickness, porosity, density, velocity, lithology and fluid content of rocks. Successful AVO analysis requires special processing of seismic data and seismic modelling to determine rock properties with a known fluid content.

It got limitation to distinguish a gas-filled reservoir from a reservoir having only partial gas saturation. However, AVO analysis using source-generated or mode-converted shear wave energy allows differentiation of degrees of gas saturation. AVO analysis is more successful in less compacted strata. A scheme was developed to estimate the dip of arbitrarily dipping fractures from the analysis of AVO along and perpendicular to the strike of the dipping fractures. Results from simulated examples show that the method yields satisfactory values of the dip estimates for dilute fractures either fully or partially saturated with gas (Ranjit K. Shaw, *et al.*, 2006).

AVO-study of **Dabka Field, Cambay Basin**, India: -The various types of attribute stacks like P-wave, S-wave, Poisson's ratio,, *etc.* have been studied for oil-gas/gas saturated zones and analysed with *in-situ* elastic/petrophysical parameters, lithological parameters, thickness and depth variations within the different pay-units encountered in wells along profiles. AVO attribute stacks analysis for oil-gas/gas saturated zones the gas-saturated zone close to 1340 msec level in well is very clearly demarcated as prominent AVO values on various attribute stacks. The oil-gas saturated zone in nearby well is not seen to be associated with so prominent amplitude variation characteristics on the attribute stacks as seen in case of gas saturated zone in well. The shale intercalations in these two nearby wells in the oil-bearing intervals may possibly the reason.

The amplitude variations with changing offset are analysed over some known oil and gas fields of **Dabka-Gajera-Jambusar area, Cambay Basin**, Gujarat. Both qualitative angle stack as well as 'AVO' attribute stacks studies are carried out a long seismic profile consisting of oil/gas/water wells (Bhuin, N. C., 1995). The oil/gas saturated zones are identified to be associated with significant increase of amplitude values with offset. The continuity/limit of these oil/gas saturated zones within the reservoir/formation are demarcated fairly on the basis of the gradual increase and then decrease in the magnitude of amplitude level. The water saturated zones do not show any 'AVO' response and are identified clearly with respect to the oil/gas saturated zones.

Interpretation

Structural Interpretation

Reflection times of the reflection horizon are read by a process called reflection picking. The process is carried out for the same horizon in all of the sections in the area. Care is needed to be taken when the reflectors are affected by faults. A common difficulty, met with, is the mistie-phenomenon. In the process of tying horizons picked at intersecting sections, reflection times may not coincide. This problem, which may be due to anisotropy or other causes, needs special treatment to make the horizons at the intersections, to tie perfectly at intersections (Das, A. K., *et.al.*, 2006).

The picked reflection values are posted onto the seismic lines at regular intervals, and then a contour map is drawn to produce the isochron map which expresses the structural behaviour of the reflector. Another set of contour maps are also drawn for the iso-time intervals (time-interval between two adjacent reflectors). The produced time-interval contour maps correspond to the depth interval (or isopach) contour map. Interval changes may indicate true thickness changes or velocity changes, which both can give significant geologic information.

The normal procedure is to compute a smoothed contour map of the average velocity functions (derived from the stacking velocity) and get it calibrated with the well velocity functions. Based on the smoothed average velocity map and the isochron map (corrected for mis-ties), the depth contour map is computed.

Seismics for Reservoir Monitoring

3-D seismic has emerged as an important tool in understanding the reservoir structure, petro-physics and fluids-saturation status. In contrast, surface-seismic measurements are laterally continuous but are of lower vertical resolution and provide acoustic rather than petro-physical information. Apart from structural, stratigraphic and depositional understanding of subsurface information using seismic data, in case good quality high resolution seismic information available even net pay map can be made.

Fluid Saturation if in reservoir light oil or gas is present in the porous sandstone reservoirs, the fluid effects can also be interpreted from 3-D seismic data. But the task is to distinguish the fluid effects from other effects on the seismic signal, like varying lithology, fractures, noise (Chen, Q. and Sidney, S., 1997).

Studies have shown that both the ratio of p-wave to s-wave velocities and the acoustic impedance show changes when oil is replaced by brine as the pore fluid. Reservoir simulation results and fluid substitution studies will be used to predict seismic properties at different geographical locations and production stages and to provide realistic input to seismic modelling.

Seismic-amplitude modelling will then be run to learn how different production stages relate to the corresponding seismic data. Rock property relationships and geo-statistical techniques will be used to generate rock-property and saturation maps. These maps should show un-swept and partially-swept areas and enable a clear understanding of the vertical and lateral flooding patterns.

The high-resolution data obtained from cores and logs provide very detailed vertical information about the reservoir but limitations in knowing the lateral variations due to sparse sampling. On the other hand, seismic surveys provide lower vertical resolution but very dense horizontally sampled data in comparison to wells. Thus, to analyse the lateral variations of reservoir properties, away from the boreholes and in between the well regions, it becomes necessary to integrate reservoir properties derived from cores and logs to seismic attributes.

In order to echo the presence of finer formation features in seismic domain, a high-resolution data that is more sensitive to the formation rather than to the borehole condition is needed. In the new method, a high-resolution resistivity curve that has been derived from an image log is used as the input log for making the synthetic seismogram. The resistivity log was converted into an equivalent acoustic log.

After the sonic log calibration, an acoustic impedance log with a closer resolution to the input log can be generated. Then, a synthetic seismogram of comparable resolution to the original image log was generated from the acoustic impedance log. It could also reproduce most of the reservoir architectural elements and barriers even in bad borehole sections.

In **Mumbai High field**, reservoir heterogeneity has complicated the understanding of the water front within the reservoirs making it difficult to locate bypassed oil and optimally distribute water injection for efficient flooding. Reservoir heterogeneity also arises from the variation in porosity/permeability in the carbonate layers affected by changes in clay content and digenesis. The reservoir analysis was based on pre-stack inversion and simulation modelling. This shows that, clay content within reservoir has strong effect on moduli-porosity trends both for bulk and shear data. However, saturation effects are found within range of model uncertainty and clay content is more significant than saturation effects. (Narendra. K. Verma., *et al.*, 2006).

In **Upper Assam**, OIL has carried out Multi-component seismic survey in its fields for having superior subsurface image and improved reservoir model in terms of lithologic information, types of fluid present and presence of anisotropy and fractures, synthetic seismograms were generated to see the responses of S-waves from the prospect horizons and to decide the survey design parameters. From wells, which had P- and S-wave sonic logs available, were used to generate PP and PS synthetic seismograms. The PS synthetic seismogram from a well show an amplitude increase with offset at the top of reservoir (Barail) sand. The **Jorajan** and **Moran** regions also have significant S-wave responses, but are relatively deeper and therefore more challenging (Mandal, K L, *et al.*, 2010).

4D

The goal of any time lapse 3D called 4D seismic survey is to detect the changes in the reservoir parameters like pressure, fluid saturation, movement of fluid contacts, temperature due to production effects on the reservoir. One of the key factors in minimising differences between successive 4D surveys is the ability to record traces with identical source-receiver azimuths for both base and monitor surveys. Fundamental limitations for the utilization of towed streamer seismic data in reservoir characterization are the limitation in bandwidth, and the poor repeatability of the seismic signal for time-lapse (4D) applications. The combination of improved amplitude and phase control, coupled with finer and more precise positioning has been seen to have a significant impact in static and dynamic seismic imaging and hence improved reliability in reservoir characterization and time-lapse interpretation.

A successful 4D imaging requires that the non-repeatable noise due to acquisition or processing between successive seismic surveys should be minimized to bring out 4D signal unambiguously. The accurate matching of different datasets is a key issue in a time-lapse study to extract useful 4D signal.

Tidal effects and changes in water layer properties during acquisition of base and monitor surveys may have a dominating and detrimental effect on the quality of the 4D data set. They can be measured accurately and compensated for in the processing. Weather-related noise can be avoided with solid streamer technology, while high frequency noise is handled by a dense hydrophone spacing. On-board seismic processing allows for a pseudo-real-time quality control of the match between base and monitor seismic data (Nimisha Vedanti, *et al.*, 2006).

Indian Scenario

In **Cambay Basin** the first 3D survey was conducted prior to the commencement of *in-situ* combustion process in the northern part of Balol Field and was meant to serve as base 3D survey. The second 3D survey after 12 months was a monitor survey for studying the changes in the reservoir due to this thermal process. The matching procedure begins with identically processed PSTM gathers of the two datasets as inputs that undergo a step-by-step processing flow for gain correction, frequency balance, differential statics, phase adjustment and finally designing moving window match filter. The matched datasets were then differenced to derive 4D signal. The final difference volume clearly shows high amplitude 4D anomalies around injector wells due to combustion/air injection.

A 4D study of the **Balol Field** in Cambay Basin, shows that repeated seismic data can be used to observe changes in gas saturation. Two 4D analysis methods have been considered, first the conventional method by observing amplitude changes at selected interfaces, such as the top reservoir interface, and secondly to observe shifts in critical offsets for PP and PS reflections (Asit Kumar., *et al.*, 2006). Both methods show good potential for technical success. Typical near offset relative amplitude changes for PP-data are predicted to and the PP critical offset changes are found to be satisfactory.

In western offshore area, in the **Neelam field** the limestone reservoir was karstified due to weathering and hence the well placement was challenging. Instead of acquiring new seismic data, interpretative image processing technique for data conditioning and extraction of advanced seismic attributes was tried to understand the geometry of karst and fault network in 3D, with advancement of interactive interpretative processing for image enhancement to extract and interpretation of such subtle feature from seismic data. Well trajectory can be shown on depth which was designed to avoid karst and fault zone. This also helped to see no mud loss was reported during drilling the well. Well became an oil producer with these successful efforts in real time (Nautiyal, R.,A. Deb, A.M. Tripath. 2010). This technology is becoming importance and called reservoir geophysics.

Seismic Attributes

The aim of seismic attribute analysis is to improve the spatial prediction of structural and stratigraphic features, as well as petro-physical and geo-mechanical rock properties throughout the reservoir

Seismic attributes are sensitive to lateral changes in geology so are utilized to assist in mapping facies, reservoir properties and faults.

- ☆ Complex trace attributes help to interpret lateral continuity of stratigraphic events, bedding geometry, and the presence of hydrocarbon.
- ☆ Geometric attributes are used to characterize stratigraphic features and for fault/fracture detection.
- ☆ Time-Frequency attributes of Spectral Decomposition increase the seismic resolution to highlight lateral variability of depositional features and faults.

There are wide variety of attributes such as:

Mean Amplitude

A post-stack attribute that computes the arithmetic mean of the amplitudes of a trace within a specified window. This can be used to observe the trace bias which could indicate the presence of a bright spot.

Average Energy

A post-stack attribute that computes the sum of the squared amplitudes divided by the number of samples within the specified window used. This provides a measure of reflectivity and allows one to map direct hydrocarbon indicators within a zone of interest.

RMS (Root Mean Square) Amplitude

A post-stack attribute that computes the square root of the sum of squared amplitudes divided by the number of samples within the specified window used. With this root mean square amplitude, one can measure reflectivity in order to map direct hydrocarbon indicators in a zone of interest. However, RMS is sensitive to noise as it squares every value within the window. Very often, this attribute helps to identify geological features such as channels, levee complexes, high impedance layers sometimes corroborate the reservoir geometries,, *etc.*

Maximum Magnitude

A post-stack attribute that computes the maximum value of the absolute value of the amplitudes within a window. This can be used to map the strongest direct hydrocarbon indicator within a zone of interest.

AVO Attributes

AVO (amplitude versus offset) attributes are pre-stack attributes that have as the basis for their computation, the variation in amplitude of a seismic reflection with varying offset. These attributes include: AVO intercept, AVO gradient, intercept multiplied by gradient, far minus near, fluid factor,, *etc.*

Time/Horizon Attributes

Coherence

A post-stack attribute that measures the continuity between seismic traces in a specified window along a picked horizon. It can be used to map the lateral extent of a formation. It can also be used to see faults, channels or other discontinuous features.

Although it should be used along a specified horizon, many software packages compute this attribute along arbitrary time-slices.

Dip

A post-stack attribute that computes, for each trace, the best fit plane (3D) or line (2D) between its immediate neighbour traces on a horizon and outputs the magnitude of dip (gradient) of said plane or line measured in degrees. This can be used to create a pseudo paleo-geologic map on a horizon slice.

Azimuth

A post-stack attribute that computes, for each trace, the best fit plane (3D) between its immediate neighbour traces on a horizon and outputs the direction of maximum slope (dip direction) measured in degrees, clockwise from north. This is not to be confused with the geological concept of azimuth, which is equivalent to strike and is measured 90° counter clockwise from the dip direction.

Curvature

A group of post-stack attributes that are computed from the curvature of a specified horizon. These attributes include: magnitude or direction of maximum curvature, magnitude or direction of minimum curvature, magnitude of curvature along the horizon's azimuth (dip) direction, magnitude of curvature along the horizon's strike direction, magnitude of curvature of a contour line along a horizon.

Frequency Attributes

These attributes involve separating and classifying seismic events within each trace based on their frequency content. The application of these attributes is commonly called spectral decomposition. The starting point of spectral decomposition is to decompose each 1D trace from the time domain into its corresponding 2D representation in the time-frequency domain by means of any method of time-frequency decomposition such as: short-time Fourier transform, continuous wavelet transform, Wigner-Ville distribution, matching pursuit, among many others. Once each trace has been transformed into the time-frequency domain, a band-pass filter can be applied to view the amplitudes of seismic data at any frequency or range of frequencies. Technically, each individual frequency or band of frequencies could be considered an attribute. The seismic data are usually filtered at various frequency ranges in order to show certain geological patterns that may not be obvious in the other frequency bands. There is an inverse relationship between the thickness of a rock layer and the corresponding peak frequency of its seismic reflection. That is, thinner rock layers are much more apparent at higher frequencies and thicker rock layers are much more apparent at lower frequencies. This can be used to qualitatively identify thinning or thickening of a rock unit in different directions. Spectral decomposition has also been widely used as a direct hydrocarbon indicator.

Seismic Stratigraphy

In stratigraphic interpretation of seismic data, using the seismic attribute the seismic reflection wavelet. Typical seismic attributes are wavelet amplitude, frequency, and phase. Usually, marine seismic data give better resolved stratigraphic changes than land seismic survey data. Extraction of stratigraphic information from seismic data depends on a set of principles and techniques of geological interpretation of geophysical data. The main objective of stratigraphic interpretation is determination of the types of stratigraphic features appearing on seismic sections, like facies changes, sand lenses, reefs, and unconformities (Cross, T.A. and Lessenger, M.A., 1988). Primary seismic reflections are generated at physical surfaces of rocks consisting mainly of strata or bedding surfaces and unconformities with velocity and density contrasts. Since geological stratal surfaces do have time depositional concept, the seismic reflections generated by them also can have chronological aspect and hence seismic reflection surfaces are chrono-stratigraphic depositional and structural patterns (Embry, A.F., 2009). A depositional sequence has chrono-stratigraphic significance because all the rocks of the sequence were deposited during the interval of geological time defined by the geological ages of the sequence boundaries even though they are conformable surfaces (Figure 5.20).

Primary seismic reflections follow chrono-stratigraphic called time-stratigraphic correlation patterns rather than time-transgressive litho-stratigraphic units. In other words, seismic reflectors in many cases are time lines. They cut across major lithological boundaries. The impedance contrast correspond with the layering in the subsurface, reflections reflect this layering. It is essential to have knowledge of geological models, their depositional processes, environments and the associated litho-facies to locate oil and gas accumulations in any basin. Now, geological models and depositional systems can also be interpreted using more reliable seismic data.

An important task in petroleum exploration is the preparation of stratigraphic cross-sections on which correlations are made and facies relationships understood. The success of a play involving the delineation of a stratigraphic trap depends on the reliability of the correlations and the subsequent facies analysis within the framework. Each system tract can be seen as an approximate time stratigraphic unit that contains a variety of facies from nonmarine to deep marine. These will occur in a predictable lateral and vertical order within the systems tract.

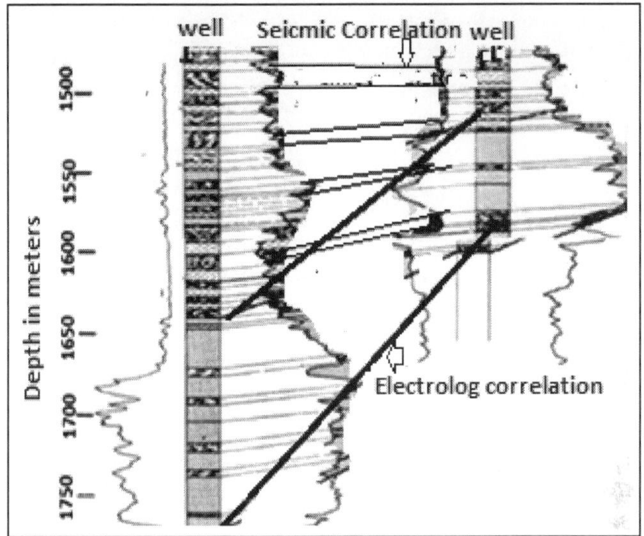

Figure 5.20. Seismic Reflection a Chrono-event and Electolog a Litho-facies.

Sequence analysis also helps to predict how and where porous strata pinch out laterally. Within a regressive systems tract (RST), a shoreline sandstone unit sometimes disappears landward due to truncation by the sequence bounding unconformity and pinches out basinward due to facies change. Impermeable shelf strata of the overlying transgressive systems tract (TST) of the next sequence can seal such strata. In other cases a shoreline sandstone will pinchout landward due to facies change to impermeable coastal plain facies that can also provide a top seal (Vail, P.R., *et al.*, 1987).

Seismic Concepts

The stratigraphic interpretation of a seismic section can be attempted after complete processing to make reliable meaningful cross section which can depict some data in the basin depositional magnitude. Then identify and mark reflection terminations or unconformities, onlap, downlap, truncation. Seismic surfaces on the basis of reflection terminations can also be identifiable. Seismic surface is a line on a seismic section where reflections terminate in a consistent manner. Then mark the individual seismic surface based on its type, like, sequence boundary, transgressive surface, maximum flooding surface (MFS) or age. Identify sequence boundaries which are commonly marked by truncation or onlap, whereas maximum-flooding surfaces are commonly marked by downlap.

Mapping sequence units on the basis of thickness, geometry, orientation, or other features to see how each sequence relates to neighbour sequences. Identify seismic-facies for each sequence. The depositional environments and litho-facies need to be identified. Seismic calibration with wells improve understanding of geologic details. Seismic reflection feature and its geological interpretation can be known. Like amplitude can indicate density contrast of depositional beds and its fluid content. Frequency can indicate bed spacing and its fluid content. Reflector continuity may help to know depositional nature and reflectors configuration can aid to know depositional style of the beds and area of deposition if geological model is visualised.

Seismic Sequence Analysis

Seismic sequence which is a depositional sequence is a stratigraphic unit composed of a relatively conformable succession of genetically related strata and bounded at its top and base by unconformities or their correlative depositional events. Seismic sequence analysis involves identification of major reflection zones that can be delineated by recognizing surfaces of discontinuity (Haq, B.U., 1991).

Identification of depositional sequences, bounded above and below by unconformities, from seismic reflection terminations. Base discordant are known for "on laps and down laps". Top discordant indicate "Top laps and truncations".

Seismic Facies Analysis

Seismic facies analysis is the way of examining smaller reflection units within the sequences that may be the seismic response to litho-facies. Seismic facies are packages of reflectors with a set of seismic characteristics differing from adjacent units. Each facies parameter has geological significance like, the divergent reflections configurations indicate progressive tilting of depositional surface with a balanced sediment influx.

Interpretation of depositional processes, environmental settings, and lithology by mapping sequence-facies parameters, such as: External forms; wedge, bank, mound, fan. Internal reflection configurations; parallel, reflection continuity and amplitude, Interval velocity.

Progradations indicate large influx with low subsidence rate. The reflection terminations at the sequence boundary like onlap, downlap and configuration of reflections within the sequence like parallel, sheet sand, divergent as wedge, sigmoidal as lenses and oblique pattern as mound, can be interpretable by introducing depositional nature to the corresponding nature of time related reflection nature within. This is called geological interpretation of geophysical data.

Reflection continuity and amplitude dependent on continuity and type of strata and the litho-facies as, continuous reflections indicates wide spread, uniformly stratified deposits and chaotic reflections may suggest unstratified slumps or tectonically deformed strata. High amplitudes may indicate shelfal sands/silts/coal/limestone.

Relative Sea Level Change Analysis

Relative sea level changes are important as they control sedimentation pattern, and are caused due to global and eustastic changes, local tectonics, subsidence, and sedimentation rate.

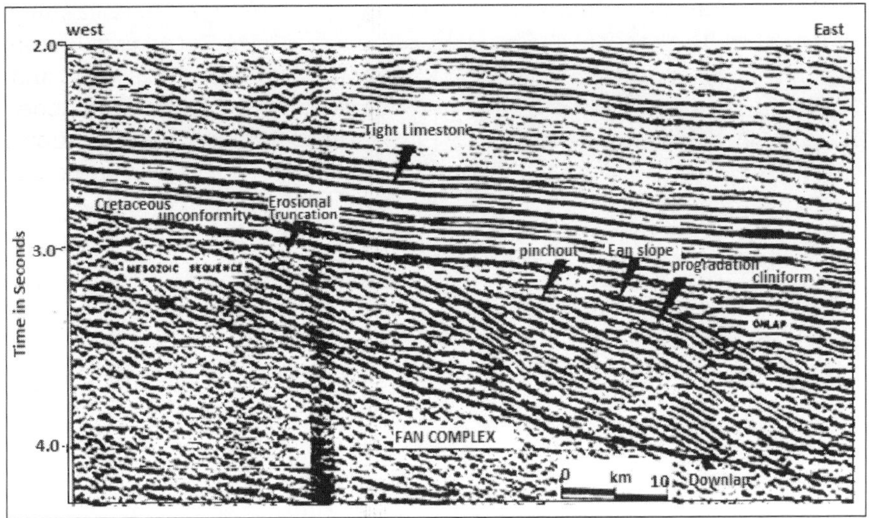

Figure 5.21. Seismic Stratigraphic-Seismic Facies.

Sedimentation pattern and shoreline shifts associated with relative sea level changes have distinct geometries that can be recognized from seismic. Relative rise/ fall and standstills of sea level can be judged from seismic patterns of coastal onlap and top lap. Important to understand transgression/regression of sea in terms of shoreline shifts and not in the context of sea level rise/fall (Figure 5.21).

Recognition of Delta System from Seismic

Due to discharge of large amount of sediments the deltas are commonly characterised by prograding clinoform sequences, such as:

Shingled progradations indicate deltaic facies prograding in shallow waters on shelf, the gently dipping parallel oblique reflections indicating low terrigenous influx. Apparent top laps suggest still-stand sea level.

Oblique progradations are good indicators of deltas on shelf and shore margin. Highly dipping progradational clinoforms suggest high influx in high energy with little or no subsidence. Delta on shelf edge evince large scale oblique progradations by out building into deep water slope system.

Sigmoid progradations show 'S' shaped clinoforms with up building (aggradation) suggesting sea level rise and low energy setting. Possibility of coarser clastics at top set is less, especially at shelf margins. Large scale sigmoid patterns on shelf margins may be a shelf edge progradation with fine clastics (Mitchum, R.M., jr., 1985).

Examples from Indian Basins

The seismic stratigraphic studies carried out in the **Krishna Godavari Basin** east coast of India. Till end the early Cretaceous, the basin was intra-cratonic rift graben and after the continental split due to south-easterly tilt, there after it became a peri-cratonic basin developed well defined shelf-slope system. During that period, a major part of the study area was in the slope regime resulting in deposition of thick basin shales. However, during Paleocene, with further fall in sea level, shelf edge steadily prograded towards southeast exposing a substantial part of the shelf area. South-easterly flowing streams constituted a network of delta system which prograded basin ward. Moderately thick limestones, sandstones and shales are the litho-types in the shelf regime during this period, which grade into thick monotonous shales. Deposition of thick sand bodies in the slope regime due to turbidity currents (?) is envisaged (Narayana Rao., G., 1994).

The offshore seismic data quality is good. More syn-sedimentary subsidence (divergent pattern) is seen in Godavari delta region, whereas erosional truncations and top-laps are features in Krishna river area. Along the slopes, based on depositional pattern both down-laps and top-laps are observed.

In the Seismic facies analyses carried out in the Mahim area of **Mumbai Offshore Basin** and in the north of North Bassein Oilfield revealed conspicuous unconformities at few horizons at the basement level and above. The presence of unconformities has been concluded on the basis of onlaps noticed at these levels. These unconformities have helped to divide the sedimentary section between basement and reflector above into two seismic sequences. Seismic stratigraphic study has enabled to predict carbonate facies in Mahim area which seems to be correlatable to the carbonate section of Bassein area. The zones depicted by "mounded" type of facies are quite significant since the structural trends are situated within this type of facies.

The seismic data of **Heera-Panna-Bassein of Mumbai Offshore Basin**, India, have been evaluated for identification of prospects within Middle to Upper Eocene Sequence. Further, massive to laminar carbonate sections without hydrocarbons were encountered in some wells, which were drilled on the basis of 2-D seismic data interpretation. Subsequently 3-D data were acquired and interpreted by applying concepts of seismic stratigraphy, seismic attribute analysis and 3D visualization techniques. On the basis of seismic reflection configuration and interpretation of electrolog facies, the Bassein Formation has been divided, seismically, into four units which were deposited under transgressive to highstand regime including late highstand sea level fall. Seismic reflection patterns within units were delineated in terms of associated lithological assemblage and porosity and trap development. Carbonate and mixed-carbonate-silicilastics were deposited within these units. Carbonate buildups, off-buildups surrounding the big buildups and wedges of debris flows are identified as potential reservoirs.

More quantities of the seismic data, both 2D and 3D have been acquired in the **Mahandi Basin** located along East coast of India. The post-trappean succession of Mahanadi offshore is divisible into six depositional sequences, separated by unconformities and their correlative conformities. Identification of system tracts within the sequences based on bounding surfaces, position within sequence, and stacking patterns has facilitated identification of hydrocarbon plays in the basin. Based on this study several new plays have been identified *viz.* Oligocene low stand deposits, Middle Miocene basin-floor fans and force regressive packages and

Upper Miocene - lower Pliocene shingled turbidites and channel-levee complexes (Pinakidhar Mahapatra,2006).

Seismic Pit-Falls

In interpretation more value is being attributed to amplitude, but what can be inferred from the amplitude and for what it is not, one should know well. In case of structural interpretation based on seismic data, picking the right reflector and its continuity through faults is more necessary. In short, one has to improve quality control in interpretation. Where does the fault lie in mapping is more important than putting more faults and making interpretation simple. Software cannot interpret, it facilitates to show ideas to the interpreter.

Seismic attributes are routinely used to accelerate and quantify the interpretation of tectonic features in 3D seismic data. Coherence (or variance) cubes delineate the edges of mega-blocks and faulted strata, curvature delineates folds and flexures, while spectral components delineate lateral changes in thickness and lithology. Seismic attributes are at their best in extracting subtle and easy to overlook features on high-quality seismic data.

However, seismic attributes can also mislead otherwise subtle effects such as acquisition footprint and velocity pull-up/push-down, as well as small processing and velocity errors in seismic imaging. In terms of attributes, structural pitfalls fall into two general categories: false structures due to seismic noise and processing errors including velocity pull-up/push-down due to lateral variations in the overburden and errors made in attribute computation by not accounting for structural dip (Kurt J. Marfurt and Tiago M. Alves., 2015).

Historical Growth of Seismic Technology in India

Oil was found in India by surface seepage in Assam in 1867 and shallow well drilling started at Digboi. The surface surveys were carried out by foreign oil companies who own those assets. Torsion balance which was successfully adapted for geophysical surveys of oil was used at Bordubi (Assam) by a geophysical team in 1925. In Assam, successful seismic survey was carried out in Naharkatiya during 1937.

After Independence, Geological Survey of India (GSI) carried out geophysical survey in Cambay area during 1948. After the formation of the National oil Company ONGC in 1956 and country-wide oil exploration activity was initiated. The first seismic crew conducted 2D single fold seismic data acquisition using 24-channel with analogue equipment, in the foothills of Jwalamukhi and Janouri area in Himachal Pradesh in late fifties with the help of Russians. Later on analogue magnetic recording equipment were introduced with the advantage of reproducing the data at any time. Transistorized seismographs PT-100 were introduced in late sixties. The geophones of Russian make 17 Hz, 4 Hz sensors for refraction surveys, 10 Hz sensors, *etc.* were used with imported reversible cables. Topographic surveys, an integral part of the seismic data acquisition for precise positioning on ground used to be carried by Prismatic Compass, Dumpy Level and Plane Tabling. Computation and interpretation of seismic data were done manually by the field crews and based on the study, time structure maps were generated.

Offshore seismic survey started in Western continental margin in 1963, by hiring the vessel S. S. Mahindra. PT-100 PMR-7 seismograph with floating cables and marsh geophones were used for first offshore surveys. Navigation was through shore based. During this period explosives were used as energy source on land as well as in offshore.

For processing of analog data, an analog processor GEODATA was introduced. This was basically a plotter of the seismic traces with static corrections. Static corrections were applied by shifting levels. Computer Services Division (CSD) was established in 1969 with the responsibility to process the seismic data with Honey Well-400, a second-generation computer was commissioned. Interpretation of these plots was done manually by picking time values and drawing time structure maps.

Introduction of techniques like CDP, Vibroseis and Marine surveys along with digital instruments with increased channel capacity (24 to 48 channel) was during early seventies. That time roll along switch and blasters were indigenously designed to meet the requirement of CDP surveys. By this time almost all the geophones were replaced by SM-4.

First fully equipped offshore geophysical vessel M V Anveshak was commissioned in 1975 for conducting 2D CDP seismic surveys with digital recording equipment DFS IV having dual tape transport and 48 channels, neutrally buoyant streamer with hydrophones. Streamer could be kept at any depth up to 20 m in the water. Bolt airgun arrays were used for energy source. Navigation system was a transit satellite based system. Further, the contract Soviet survey vessel "Academic Arkhangelsokey" and vessels from CGG and Western Geophysical were also engaged for offshore data acquisition.

On land seismic crews were then equipped with DFS-III and SN-328 systems (Channel capacity of 24 to 48) with Binary Gain Amplifiers that exhibited higher dynamic range from 60 db to 84 db and better S/N ratio. At the same time SN-338 seismograph was introduced featuring low level multiplexing and with IFP having modular expansion capacity of 24 to 72 channels.

Subsequently, DFS-IV seismograph with channel capacity up to 120 and with IFP having 84 db dynamic range was also commissioned. Rota-long switch and remote blasting systems were introduced with these systems.

Digital era in seismic recording and processing started worldwide in early 1970s and ONGC almost concurrently adopted it. With need for processing of larger volume of data IBM-370/145 computer was introduced with MVS operating system having 1 MB memory and array processor in 1975 in the seismic data processing. Synthetic seismograms were generated on the computer and used to calibrate the seismic reflections on the paper sections.

The digital seismographs like DFS-V from Texas Instruments and SN 358 from with 15-bit recording were introduced in 1984 and the GEOCOR IV, a sign-bit recording seismograph with 1024 channels capacity and line telemetry was in 1985. These digital seismographs provided 96 channels expandable up to 240 channels with better dynamic range up to 84 db and also portability as the size of the modules was very small. This helped in conducting seismic surveys in logistically challenging areas like Schuppen Belt in Nagaland, Cachar in Assam. Calcutta city was also surveyed with MDS-10 seismograph (which is generic to SN-358 and DFS V systems) using vibrators as energy source during 1982-83.

The first 3D campaign was done with GEOCOR IV seismograph in 1985-86 using vibrators as energy source in Balol field of Cambay basin. Subsequently, 3D surveys were also carried by merging capabilities of two DFS V seismographs. VSP surveys were first attempted in 1985 using DSS-10 instrument with PERKINLELMER and MicroVax2 MDS-10 Seismic Data Recording Unit system for in field processing.

Second offshore survey vessel M V Sagar Sandhani with a helipad was acquired in 1986. The vessel was equipped with DFS V seismograph having 120-channel, PRAKLA streamer and Bolt air gun arrays to perform the current state of art offshore 2-D and 3-D seismic surveys concurrently with gravity and magnetic surveys and it also had onboard seismic data processing system ND 570 for quality control and basic processing including 3D binning. First offshore 3D campaign was acquired by Sagar Sandhani under the technical collaboration with GSI, Singapore. To meet more demand of data processing, re-processing and even special processing demand, VAX -11/780 system having 2MB with VMS operating system and Phoenix application software was introduced in 1981 which later changed to IBM3083-JX3 with 4 array processors and terminals having main memory of 32 MB was commissioned at GEOPIC. During 1987 when one crystal Interactive Interpretation Workstation (IIWS) was inducted and three more IIWSs from GEOQUEST were inducted in 1989 at GEOPIC., Dehradun.

Later, in order to diversify the work-load, TATA-ELXSI M-6400 computer with DIGICON application software equipped with Explorer UWS was installed at **Mumbai**. At the same time during 1987-88, ND-570 computer with SINTRON operating system and GECO application software with Charisma-1 IIWS was commissioned at **Chennai**. EC-1061 with MVS operating system and DSS-3 application software with INGOS-IM IIWS were installed at **Calcutta** and **Baroda**. SIDIS-2000 IIWSs with better visualization tools and advanced state of the art application software from GSl were introduced in 1992 in Baroda, Calcutta and Jorhat, which could cater to the need of data analysis by integrating various geo-scientific data.

The system SN-388 with the capacity of recording 8000 channels, 24-bit SIGMA DELTA technology with dynamic range of 110 db and ensure highest fidelity of the signal recording and have inbuilt QC module procured such seismograph during 1996 and mostly engaged in acquiring 3D campaigns. Geophones have been replaced by low distortion SM-24 type. For VSP surveys with slim tools GEOLOCK-2S were introduced and walk away and off set VSP surveys were conducted successfully. Conventional/optical topographic surveying also being replaced by electronic surveying techniques *viz.* GPS since 1996. Acquisition of seismic data in western offshore through Ocean Bottom Cable (OBC) with dual sensors in 1997.

The GEOPIC is equipped with SGI Origin-2000 server with 20 nodes having Paradigm application software for processing. This system is also capable of processing seismic data in pre-stacked and depth domain. IRS Ahmedabad, KDMIPE Dehradun, and RCC, Calcutta are equipped with Sun workstation with PROMAX application software.

For retrieval of valuable data on old magnetic media or on hard copies, the data archival systems were introduced during 1996. These systems transcribe into high-density media *viz.* DLT, 3490, 3590, *etc.* and could also convert paper sections into digital form for further use on workstations or processing. Such archival systems are available at Mumbai, Jorhat, Chennai, Calcutta and GEOPIC Dehradun. GEOPIC also has robotic tape library for handling huge data media. All the work centers were equipped with the powerful desktop Sun workstations using state of the art application software from Landmark (Sharma N.C,2002).

Chapter 6

Drilling Engineering

After visible seepages of oil were exploited, then the idea of making a hole in to the earth layers to locate the rocks which hold oil and gas was started which is called drilling. It is the only direct method to locate hydrocarbons and produce them.

Before drilling, all possible information to have successful location for accumulated oil and gas below need to be identified. At first, geologists will estimate thickness of rocks favourable to generate and hold oil and gas below with an estimated depth. Then the geophysicists will do all surveys to exactly know the depth of such rocks below. Both will map such rock layers and narrow down the most favourable site called structure. At that stage it is only called as "Prospect" to be drilled. After all discussions on pros and cons, the best prospect in the area will be firmed up for drilling with specified target depth.

However, the assumptions of geologists will be broader than that of geophysicist who calculate based on measured data. But the limitations in data acquisition and processing with limited information may have some uncertainties. In all such cases drilling will generate data for fine tuning the data required for explorationists. Based on the reliability of scientific data in any area, the well which will be drilled is categorized.

Categories of Wells

As first step, in a sedimentary basin margin area, **shallow bore-holes** will be drilled to understand the subsurface continuity of the known outcrop (exposed) sediments. Normally these wells will be drilled to 300 to 500 meters.

Though some information is available up to deep target depth where oil and gas can be found, to get more lithological information from both drill cuttings and cores drilling has to be carried out under supervision of geologist. It is called **wild-cat well**.

Once some knowledge of source rock and oil and gas holding permeable rocks nearby are known, an **exploratory well** to locate producible hydrocarbons is planned. Prior to plan such well the complete sub-surface information has to be known by using seismic time or depth maps. Once velocity data were obtained from drilled well more reliable information may help in the area, for geophysicists.

After establishing the commercially producible oil and gas, in order to confirm the limits of subsurface oil pool **exploratory step-out wells** will be planned to the depth range of known oil bearing zone.

With complete knowledge of oil pool extensions in order to exploit or produce effectively more hydrocarbons, **Development wells** will be drilled. At this stage, more reservoir knowledge like porosity permeability quality of oil and gas help in planning the well locations for production. In the process of production to maintain the pressure of the reservoir for effective production, **Injection wells** will be drilled to pump external fluid in to the reservoir.

Well Planning

To optimise the design of a well it is desirable to have an accurate information as much possible about the sub-surface. Therefore, a number of people who know more about the data will have to provide information prior to the design of the well trajectory and before drilling rig with suitable equipment can be selected.

Geologists and seismic interpreters will predict type and depth of the different rock formations to be encountered during drilling. They will advise the drilling engineer where the objective zones will be drilled through along with the target of the well.

Earth scientists will provide a prognosis of pressures along the planned well trajectory. The accuracy of the parameters used in the well planning process will depend on the knowledge about the area. It is important that, the uncertainties are clearly spelled out and preferably quantified. Potential risks and problems expected should be discussed and incorporated in the designing the well trajectory.

The drilling engineer will be able to design the well planning which includes, the setting depth and ratings for the various casing strings, mud weights and mud types required during drilling, and select an appropriate rig and related accessories and consumable like drill bits.

All the information is documented in a comprehensive well drilling programme is called Geo-Technical Order (GTO) in picture mode and possibly can be made as report, incorporating all details both scientific and engineering. The well-bore diagram shows the hole and casing sizes needed to drill the well to its desired depth. So the target depth is very important to decide about the type of rig which can drill to such depth. Further, the location of the site whether it is onland or offshore will decide nature of rig.

The type of rig is known by the ultimate lifting capacity of tubulars to be lowered while drilling and it is known by horse-power capacity of the engineering equipment. To drill shorter depths less capacity rigs like 760 HP, *etc.*, if deeper depths are to be drilled with 2000 HP capacity rigs are in usage.

When oil well drilling started in India in Assam at very shallow depths of the order of few hundred meters very low capacity rigs like that of water well drilling were used. Once the Indian government started its oil exploring national oil company called ONGC, for shallow target wells "Cardwell " rigs and for deeper objectives the rigs called "3D" rigs provided by the then USSR were in practice. These deeper rigs can drill up to 4500 meters in Assam and shallow ones in Gujarat areas based on projected depth of the wells. As the activity increased and old rigs may need to be replaced, new type got manufactured by BHEL in India like, three types of Electrical Rigs and they are of 760 HP, 1360HP and 2000HP capacity rigs. The lesser one can drill up to 3000m, next one up to 5000m and the higher capacity one can drill to 6500m.

Rig Types

Land Rigs

Before rig equipment is brought in, the land must be cleared and graded. Access roads, where possible, must be prepared. For jungle locations, preparing the drill site is more difficult, for access roads do not usually exist and cannot be made. The choice of a drilling location is therefore often dependent on its proximity to reach the site if the distances are too complicated for the construction of approach road, helicopters can airlift everything to the drill site.

The most common arrangement for a land drilling rig is the cantilever mast (often called a jack-knife derrick) which is assembled on the ground, then raised to the vertical position using power from the draw-works (hoisting system). These structures are made up of pre-fabricated sections which are fastened together by large pins. First, the engine and derrick substructures are placed in proper position and pinned together by the drilling crew, then the draw-works and engines are made operational (Figure 6.1). The derrick sections are then laid out horizontally, pinned together, and the mast is raised as a unit by the drill line, traveling block and draw-works.

Offshore Rigs

Barge

The barge is a shallow draft, flat-bottom vessel equipped as an offshore drilling unit, used primarily in swampy areas. This type of vessel can be found operating in the swamps of river deltas, in coastal areas, and in shallow lakes suitable for less than 10 meters of water level. It can be towed to the location and then ballasted to rest on the bottom. In Indian areas of drilling operations such rigs are not used yet.

Jack-Up

This mobile drilling rig is designed to operate in shallow water, generally less than 100 meter deep. Jack-up rigs are very stable drilling platforms because they rest on the seabed and are not subjected to the heaving motion of the sea. They have a barge-like hull which may be ship-shaped, triangular, rectangular, or irregularly shaped, supported on a number of lattice or tubular legs. When the rig is under tow

Figure 6.1. Onland-Rig.

to a drilling location the legs are raised, projecting only a few feet below the deck, and the structure is like a floating box; hence it can be towed only in good seas and at low speed. Upon reaching, its location the legs are lowered by electric or hydraulic jacks until they rest on the sea bed and the deck is level, some 20 meters or more above the waves. Most jack-ups have three legs, some have four or five (Figure 6.2). The legs are either vertical or slightly tilted for better stability. In one design, they are fixed to a large steel mat, which gives it the name of "mat-supported jack-up". The chief disadvantage of the jack-up is its vulnerability when being jacked up or relocated, but as a class, they are cheaper than other mobile rigs.

Semi-Submersible Rig

These are floating drilling rigs consisting of hulls which carry a number of vertical stabilizing columns and support a deck fitted with a derrick and associated drilling equipment. Semi-submersible drilling rigs differ principally in their displacement, hull configuration, and the number of stabilizing columns. The modern types have a rectangular deck, a few are cruciform shaped, others pentagon shaped, while some of the smaller rigs have a triangular deck. The most usual hull arrangement consists of a pair of parallel rectangular pontoons which may be blunt or rounded and house thrusters for position-keeping or self-propulsion, although some have individual pontoons at the foot of each stabilizing column or pair of columns. Eight columns (four stabilizers and four intermediate columns) is a common arrangement as are three and six columns, and both hulls and columns are used for ballasting as well as for storing supplies.

Figure 6.2. Offshore Rig Types-Jack-up Rig and Semi-Submersible.

The semi-submersible is very stable because its centre of gravity is low in the water. It can operate in deeper waters than a jack-up rig. Operational depth is limited principally by the mooring equipment and by riser handling problems, so most semi-submersibles have a limit of about 200 meters. However, some have a capability of drilling in 500 meters of water with the aid of "dynamic positioning". This is a method of maintaining the position of a vessel with respect to a point on the seabed by activating on- board propulsion units in response to signals received from a position-error detector. This method of automatic stationing is often computer controlled.

Drillship

The drillships offer greater mobility than either jack-up or semi-submersible rigs, but are not as stable while drilling. Their main advantage is an ability to drill in almost any depth of water. Many are anchor-moored, but modern ships are fitted with dynamic positioning equipment which enables them to keep on-station above the borehole with more storage capacity (Figure 6.3). A feature of drill-ship is with automated station-keeping facilities and controllable pitch propellers. The drilling slot (moon pool) on a drillship is through the centre of gravity, and the derrick mounted above it.

Figure 6.3. Drillship.

Drill Site Preparation

Once the objectives of the well are clear, decisions are made where to site the drilling location relative to the subsurface target and which type of rig to use. If no prior drilling activities have been not known in the area, usually and environmental impact assessment (EIA) will be carried out and necessary approval need to be obtained from local authorities.

On-land Sites

A site survey will be carried out, from which a number of parameters can be established, like soil strength at the planned location, possible access routes, surface restrictions like built-up areas, lakes, the general topography, possible water supplies/source. The survey will allow the adequate preparation of the location.

The size of the rig site will depend on operational requirements and it will be determined by: the type of derrick or mast, it must be possible to rig up on site, the layout of the drilling, equipment, the size of the waste pit, the amount of storage space required for consumables. A land rig can weigh over 200 tons and is transported in smaller loads to be assembled at the site.

Prior to moving the rig and all auxiliary equipment, the site will to be cleared of vegetation and levelled. To protect against possible spills of hydrocarbons or chemicals the surface area of a location should be coated with plastic lining and a closed draining system installed. Site management should ensure that any pollutant is trapped and properly disposed of. If drilling and service personnel require accommodation at the well site a camp may need to be constructed/arranged. For safety reasons the camp will be located at a distance from the drilling rig and consist of various types of portable cabins. For the camp, waste pits will be required, access roads, parking space and drinking water supplies.

Offshore Sites

The survey requirements will depend on rig type and the extent of the planned development as a single exploration well or drilling jacket installation. A typical survey area is general a 4 km square area centred on the planned location. Surveys may include sea-bed survey: Employing high resolution echo-sounding and side scan sonar imaging, an accurate picture of the sea bottom.

The technique allows the interpreter to recognise features such as pipelines, reefs, and wreckage. Particularly if a jack-up rig is considered, an accurate map of these obstructions is required to position the jack-up legs safely. Such a survey will sometimes reveal crater-like structures, which are common in areas. These are the result of gas escape from deeper strata to the surface and could indicate danger from shallow gas accumulations. Shallow seismic surveys are carried out to provide maximum resolution within the near surface sedimentary layers. The objective is to detect indications of shallow gas pockets close to the surface and may enter the borehole if penetrated by the drill bit, resulting in to dangerous situation.

Soil coring is planned where structures require soil support, *e.g.*, drilling jackets or jack-up rigs, the load bearing capacity has to be evaluated. Usually a series of shallow cores are taken to obtain a sample of the sediment layers for investigation in a laboratory. Particularly for jack-up rigs, site surveys may have to be carried out prior to each re-employment to ensure that the rig is positioned away from the previously formed 'footprints' left by the jack-up legs on a previous job.

Positioning the Rig

The drilling location is generally indicated by a survey vessel, using a marker buoy, prior to the arrival of the drilling vessel. The rig is towed onto the location and anchor-handling tugs are used to drop the anchors in a pre-designed pattern. The anchors are tensioned to ensure that they are securely set into the seabed, then slacked off and adjusted to obtain the final position and heading of the rig. The drilling rig may be held in position over the well by using anchors or by using dynamic positioning techniques. The condition of the seabed directly beneath the rig will generally have been checked by a seabed survey before the rig arrived on location, but a final check is generally made with an ROV (Remotely Operated Vessel) prior to running the equipment.

Drilling Rig-Components

A drilling rig essentially comprises a derrick, the draw-works with its drilling line, crown block and travelling block, and a drilling fluid circulation system including the standpipe, rotary hose, drilling fluid pits and pumps. When drilling is progressing, rotation is applied to the drill-string, and hence drill bit, which is suspended from a kelly or top-drive beneath the travelling block. Drilling fluids are conveyed through the drill-string via the bit into the borehole. The principal rig components can be broken down into hoisting system; rotation system and associated rig floor tools, circulation system, motion compensation system, blowout prevention system (Figure 6.4).

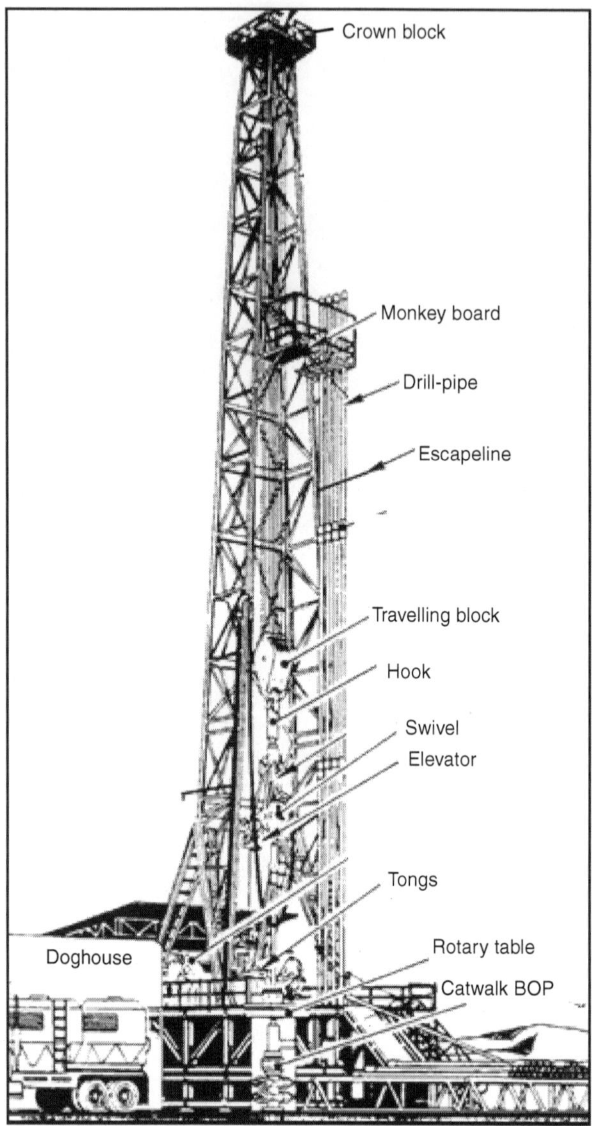

Figure 6.4. Land Drilling Rig-parts.

Hoisting System

The derrick or mast supports the hook and elevators by means of the travelling block, drilling line, crown block and draw-works. The draw-works, which hoists the hook within the derrick, is powered by prime movers which are large capacity diesel engines supplying electric generators (Figures 6.5).

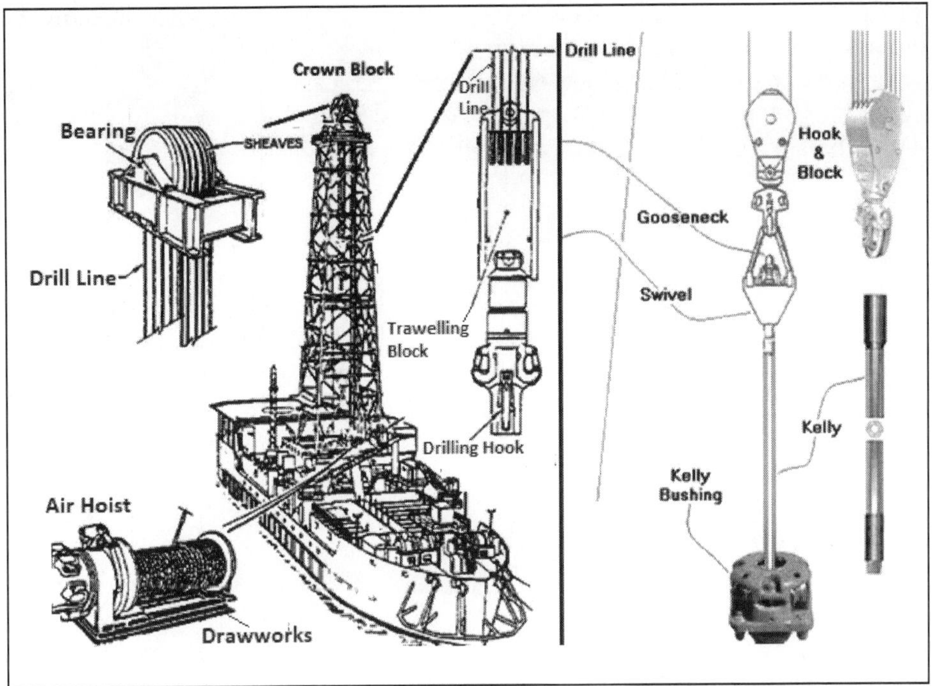

Figure 6.5. Rig Components-Crown Block-Hoisting System.

Derrick or Mast

Whenever the drill-string is suspended by the travelling block and drill line, the entire load rests on the derrick. The standard pyramid derrick is a structure with four supporting legs resting on a square base. In comparison, a mast is much more slender and may be thought of as sitting on one side of the rig floor or work place. The derrick is erected on a sub-structure which supports the rig floor and rotary table and provides work area or the equipment on the rig floor. The derrick and its sub-structure support the weight of the drill-stem at all times, whenever it is suspended from the crown block or resting in the rotary table.

The height of the derrick does not affect its load-bearing capacity, but it is a factor in the length of the sections of drill-pipe that can be removed. The drill-string must be removed from the borehole from time to time, and the length of each drill-pipe section to be removed is limited by the height of the derrick. This is because the crown block must be sufficiently elevated above the rig floor to permit the withdrawal and temporary storage on the derrick the drill-string when it is pulled from the borehole to change the bit or for other reasons.

Travelling Block, Crown Block, Drill Line and Hook

The travelling block, crown block and drill line are used to connect the supporting derrick with the load of drill-pipe to be lowered into or taken-out from the borehole. During drilling operations, this load usually consists of the weight of the drill-pipe, drill collars and drill bit. The drill line passes from the draw-works to the top of the derrick. From here it is sheaved between the crown block and travelling block to give an eight, ten or twelve-line suspension. It is then clamped to the rig floor by the deadline anchor. Suspended from the travelling block, on standard drilling systems, is the hook which, when drilling, carries the swivel and kelly and when tripping lifts the drill-string with the elevators. The elevators are attached to the hook using the links called bail.

Draw-Works

The draw-works is a mechanism commonly known as a hoist. The main purpose of the draw-works is to lift the drill-string out of and to lower it back into the borehole. The drill line is reeled (spooled) on a drum in the draw-works. When engaged, the drum turns and either reels in the drill line to raise the travelling block, or lets out the drill line to lower it. Because the drill-stem is attached to the block, it is raised or lowered.

One outstanding feature of the draw-works is the brake system which enables the driller to easily control a load of drill-pipe or casing. On most rigs, there are at least two brake systems. One brake is a mechanical friction device and can bring the load to a complete stop. The other brake is hydraulic or electric; it can control the speed of the descent of a loaded travelling block, but is not capable of bringing it to a complete stop. It is used to reduce the wear on the primary friction system.

An integral part of the draw-works is the gear (transmission) system. This gives the driller a wide choice of speeds for hoisting the drill-string. The draw-works also has a "drive sprocket" that drives the rotary table by means of a heavy-duty chain. In some cases, however, the rotary table is driven by an independent engine or electric motor.

Another feature of the draw-works are the two cat-heads. The make-up cat-head, on the driller's side, is used to spin up and tighten the drill-pipe joints. The other, located opposite the driller's position on the draw-works is the breakout cat-head. It is used to loosen the drill-pipe when the drill-pipe is withdrawn from the borehole. An independent air hoist is used on many rigs for handling light loads around the rig-floor.

Rotation System

Rigs may be operating a conventional lower rotating (kelly) system or more usually a top drive rotating system.

Lower Rotating (Kelly) System

The kelly is a pipe which fits tightly into the kelly bushing which is fitted into the rotary table. The upper end of the drill-string terminates where the top-most length of drill-pipe screws onto a device called a saver sub. This sub is used to save wear and tear on the threads of the kelly. It is approximately 14 meter long, square or hexagonal on the outside and hollow throughout to provide a passage way for the drilling fluid. Its outer surfaces engage in the kelly bushing. At the top of the kelly there is usually a 'kelly cock valve' – a safety valve which can be closed to prevent fluid back pressure from damaging the swivel, a rotary hose and other surface equipment. The swivel, which is attached to the hook, does not rotate, but supports the kelly which rotates. Drilling fluid is introduced into the drill-string through a gooseneck connection on the swivel via the rotary hose. The rotary table serves two main functions: to rotate the drill-string and to hold slips which support the weight of the drill-string when it is not supported by the elevators or hook and kelly.

The master bushing transmits rotary motion from the rotary table to the kelly drive bushing and the kelly. It is also the connecting link between the rotary table and the slips, which support the pipe during trips. The kelly drive bushing, also called the rotary kelly bushing (RKB), engages the master bushing using four pins which engage four corresponding openings in the master bushing. Rollers within the bushing permit the kelly to move freely upwards or downwards when the rotary is turning or stationary. When the kelly is disconnected and is set back, the drive bushing lifts and is set back with it.

Top Drive (Power Swivel) Rotary System

The drill-string rotation is by a drive motor or power swivel attached to the travelling block. With such a system the kelly and kelly bushing are not required, and the master bushing and rotary table serve only as conduits for the drill-string to be raised or lowered into the borehole. Because the kelly is not required, stands of drill pipe are drilled instead of single joints. As a result, breaks in circulation when drilling occur every 30m rather than 10 m, and for shorter periods of time as the top drive is simply pulled above the next stand of drill pipe to be run and screwed straight into the top. Use of a top drive also allows rotation and circulation to be maintained whilst pulling out of hole for a stand of drill pipe at a time, known as back reaming, making hole problems much easier to handle.

Rotary torque is applied to the drill-string using an electric motor and a simple step-type transmission via a power sub made up on top of the drill string. The power sub is hollow allowing normal drilling fluid circulation from the rotary hose.

Both power swivel and travelling block have guide dollies attached which run on guide rails in the derrick. This allows movement of the top drive system upwards and downwards during drilling and tripping.

Suspended below the power swivel is a pipe handling unit which consists of an elevator system and torque wrench which can be operated hydraulically to make and break connections. The elevators can be moved, via a link mechanism, allowing pipe to be moved towards the "mouse-hole" or "monkey board". A hydraulically actuated valve is also included to be used as a kelly cock.

Rig Floor Tools

The "**slips**" are wedge-shaped steel dies fitted in a frame with handles, which are placed between the drill-pipe and the sides of the master bushing in the rotary table when making a connection or tripping. Their purpose is to support the drill-string and hold it suspended in the borehole. Due to their wedged shape, the heavier the dill-string, the tighter it is gripped. If the pipe has no external upset above the slips, any failure or slippage may result in the string falling down hole. For back up in this case a device known as a dog collar is securely fastened around the pipe just above the slips. Slips for tubing and casing are usually hydraulically activated.

Elevators for drill pipe and collars fasten around, but do not grip, the pipe. They are attached to the bottom of the top drive or suspended from the hook and have an internal profile which matches the external profile of the box connection on the pipe. The pipe is supported within this profile allowing it to be quickly lifted for tripping in or out of hole or for moving around the drill floor. Elevators may be either manually or hydraulically operated. Tubing and casing elevators may have moving internal parts similar to drill pipe slips which physically grip the external surfaces.

Tongs are a type of wrench used for tightening and loosening drill-pipe and drill collars. Two sets of tongs are used, one to hold the drill-string and the other to tighten the joint. They are called the 'make-up' and 'breakout' tongs. **Spinning Wrench** (Pipe spinner) is a pneumatically-powered wrench used for rapidly spinning the drill-pipe or collars when breaking out or making up the pipe. When using a top drive, this tool may be rarely used.

An **iron roughneck** takes the place of both the tongs and pipe spinner for handling of drill pipe or collars. This machine runs on rails attached to the rig floor and is easily set aside when not in use. At the top a set of rollers closes around and spins the pipe in or out, whilst underneath a pair of jaws grip and torque the pipe for either making or breaking connections. This allows precise control of torque, minimal damage to threads and reduced drill crew fatigue as manual handling of tongs and spinners is not required. For casing or tubing similar dedicated devices known as power tongs are used.

Type of Bits

The drill bit is the critical item for drilling. To select the proper bit, some information must be known about the nature of the rocks to be drilled.

Harder formations are generally drilled at more weight on bit (WOB) than softer formations, and therefore require stronger cone shells and bearings. Since harder bits have shorter teeth, it is possible to increase the cone shell thickness and bearing journal diameter in order to meet the higher strength requirements. The thicker cone shells are more resistant to fatigue failure while the greater bearing journal surface area is capable of supporting higher loads.

Bit tooth metallurgy is varied according to formation hardness. Different approaches are used on steel teeth and tungsten carbide insert bits. Steel teeth cutters are toughened by a carburizing that is hardening process. Wear resistant hard facing is then applied to one or more sides of the teeth on soft and medium bits.

Fluid erosion leads to cutter reduction and/or loss of cone shell material. The loss of cone shell material on tungsten carbide insert bits can lead to a loss of inserts due to the reduced support and grip of the cone shell material. Erosion can be caused by, abrasive formation contacting the cone shell between the cutters, caused by tracking, off-centre wear, or excessive WOB. Abrasive formation cuttings eroding the cone shell due to inadequate hydraulics. Excessive hydraulics resulting in high velocity fluid erosion.

Junk damage can be detected by marks on any part of the bit. Junk damage can lead to broken teeth, damaged shirttail, and shortened bit runs and therefore can become a problem.

There are two main types of bits used for rotary drilling. There are several variations within these types, primarily based on the cutting structure used for drilling the rock. The types are: Roller Cone Bits and Fixed Cutter Bits. The other drilling tools used are Reamers and Hole Openers.

Roller Cone Bits, commonly called tri-cone bits, are the most common bits used in which,the cutting structures are located on three rolling cones attached to the bit body. A variety of types are available depending upon any specific conditions involved. Two main categories of tri-cone bits are milled-tooth and insert bits.

Milled-Tooth Bits (Figure 6.6): These bits have steel teeth which have been milled on the cones. The teeth vary in size and shape, depending on the formation they are expected to drill. Long, slender teeth are used in soft formations and short, broad teeth are used in harder formations.

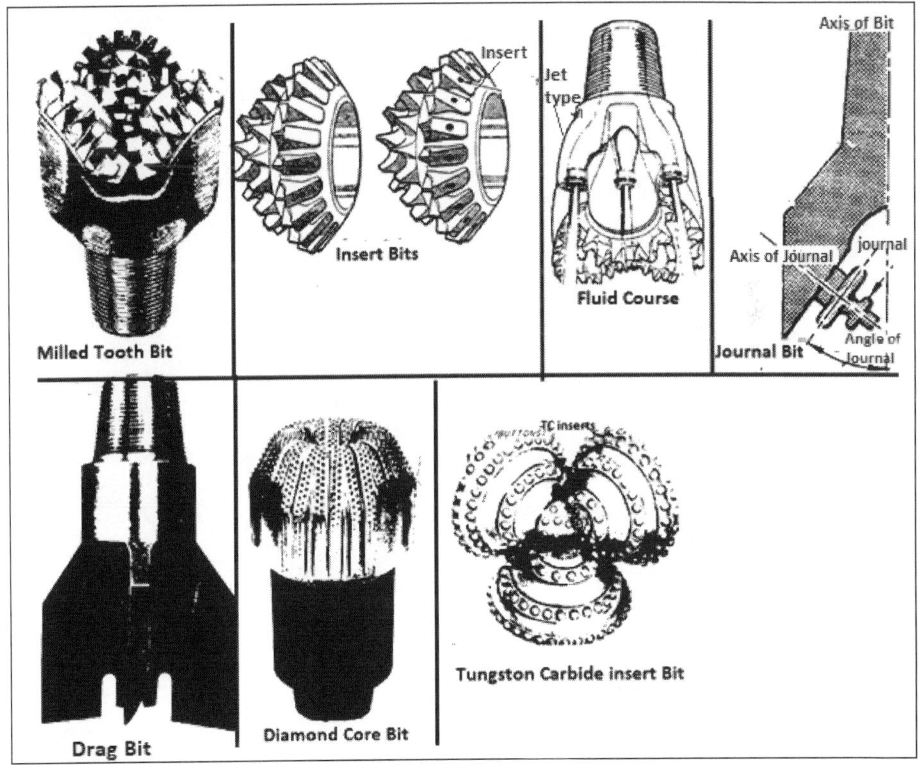

Figure 6.6.Types of Drill Bits.

Insert Bits: These bits do not have steel teeth instead, tungsten carbide inserts (teeth) are pressed into the cones. These are very much harder and last longer when drilling hard formations. The inserts can come in a variety of shapes, from long chisel shapes for firm formations to short round buttons for hard, brittle formations.

Fluid Courses: These allow for the drilling fluid to leave the bit and carry away the cuttings. The earliest arrangement was a water course, where the fluid passed through holes drilled in the center of the bit body. As wells were drilled deeper, more hydraulic force was necessary to carry the cuttings from the bottom of the hole. "Jet nozzles" were introduced and have been used.

The outer rows of teeth receive the greatest wear. This is because, if the bit drilled an in-gauge hole, then the outer rows are always in contact with the formations. To compensate for this, one method is to "hard-face" the bit's teeth with an outer layer of tungsten carbide, another is to "T- shape" a bit tooth to give it a much larger surface. Another method is to press tungsten carbide inserts into the outer row teeth.

The journal is the metal shaft upon which the cones are set. They are manufactured for the insertion of bearings, which allow the cones to rotate. The angle, from horizontal, will dictate which type of formation the bit will be used.

Two types of bearings are used in tri-cone bits. Roller/ball bearings are used when bit life is not a problem. Journal/friction bearings, a specialized metal ring, enhance the drilling performance because they will last longer. To withstand the pressures and temperatures of drilling, the bearings have to be lubricated.

Two types of lubrication are used: Non-sealed bearings: This type of lubrication uses the drilling fluid to cool the bearings. However, because of the solid content of the drilling fluid, the bearings do not have a particularly long life.

Sealed bearings: This lubrication system uses a graphite-type lubricant sealed in a reservoir to cool the bearings. Pressure changes cause the lubricant to be pumped around the bearings.

An off-centre alignment of the cones makes the teeth scrape and gouge the formation as the cones rotate. The amount of scraping depends on the offset.

Fixed Cutter Bits have no moving parts. The bit body and cutting structures rotate as one maning there are no cones. The main categories of fixed cutter bits are drag bits, diamond bits, PDC (Polycrystalline Diamond Compact) bits, and TSP (Thermally Stable PDC) bits.

Drag Bit was the earliest type. The cutting structure was sharpened steel. As such, it could only be used in soft formations, and is used where known thick clay (Girujan clay Upper Assam Basin) has to be drilled-in.

Diamond Bits use natural diamond as the cutting structure. They are usually slightly smaller than tri-cone bits to prevent diamond damage while being run into the borehole. The design of diamond bits varies greatly in the shape of the head, the size and setting of the diamonds. The main advantages of diamond bits is that, with no bearings, they can be run for long periods of time and they can drill almost any formation and used to cut a "core" which is the cut piece of the formation.

PDC Bits: The cutting structure of these bits is composed of diamond dust/crystals bonded to a tungsten carbide stud. These studs are then either pressed or moulded into the bit body. They are used in soft to medium-hard formations. A TSP cutter is composed of diamond dust/crystals as a PDC bit. The difference is that the cutters are used by themselves and not bonded to a stud.

Reamers/Hole Openers

These are tools that are run immediately above a bit to maintain or enlarge the hole size. Their cutting action is by rotating cones built out from the central stem (Figure 6.7). Both types perform similar functions, but the cones of an under-reamer are built into collapsible arms which are held out while drilling by the pressure of the drilling fluid being circulated down through the centre of the tool. In this way, it may be run in or pulled out through a smaller section of the borehole. After drilling any major clay section, since shales will heave and reduce the dried dia reamers are used to stabilise the drilled hole before further drilling. After drilling major clay section by drag it normally there reamers are used before well is stabilized for further drilling.

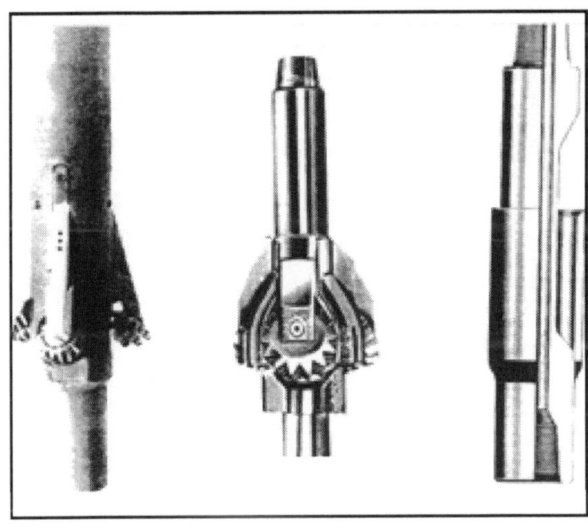

Figure 6.7. Drill Reamers.

Drill collars are tubulars which have a much larger outer diameter and generally smaller inner diameter. The function of drill collars is to provide enough weight on bit for efficient drilling, to keep the drill string in tension, thereby reducing bending stresses and failures due to fatigue and to provide stiffness in the BHA (Bottom Hole Assembly) for directional control. A typical drill string would consist of bigger size drill collars below whose length depend on the amount of weight on bit to be provided as (L = WOB/w) where "w"is the weight per foot of the drill collor. The drill pipe is fixed above the drill collars.

Drill-pipe is the major component of the drill string. It generally constitutes major portion of the entire length of the drill string. Drill pipe is a seamless pipe with threaded connections, known as tool joints. Heavy wall drill pipe has a greater wall thickness than ordinary drill pipe and is often used at the base of the drill pipe where stress concentration is more.

Drill pipe is used to transmit power by rotary motion from ground level to a drilling bit at the bottom of the hole and to convey flushing media to the cutting face of the tool. Length of Drill pipes of longer are of 12 meters to 14 meters. The basic requirements for designing a drill string are the burst, collapse and tensile strength of the drill string components must not be exceeded. The bending stresses within the drill string must be minimised.

When laying down the drill string, specific operations should include:

- Th tool joints and drill string are to be washed internally and externally with clear fresh water. This will remove any salt or other corrosive agent which might bring about more rapid deterioration.
- A rust preventive compound is to be applied to the threads and shoulders, particularly if drill string is to be stored for any length of time.
- Check the thread protectors are installed tightly on boxes and pins.

A rubber O-ring at the base of the pin of drill-coller, immediately adjacent to the face of the pin, called a mud groove. The selection of the proper O-ring may fill this groove and offer contact with the counter bore of the box tool joint, when the box and pin are screwed together. When pressure is applied, the O-ring will be compressed and will move to impede the flow of fluid across the faces of the tool joint.

Drilling System

Whether onshore or offshore drilling is carried out, will be the rotary rig and the basic functions carried out during rotary drilling operations: - Torque is transmitted from a power source at the surface through a drill string to the drill bit, drilling fluid is pumped from a storage unit down the drill string and up through the annulus. This fluid will bring the cuttings created by the bit action to the surface, hence clean the hole, cool the bit and lubricate the drill string. Subsurface pressures are controlled by the weight of the drilling fluid and by large valve assemblies at the surface.

Stabilisers are added to the drill string at intervals to hold, increase or decrease the hole angle. Up to this it is called the bottom hole assembly (BHA) and is screwed together to the drill-pipe. The drill string is connected to the kelly saver sub. A saver sub is basically a short piece of connecting pipe with threads on both ends. The kelly is hung from the travelling block which does not rotate, a bearing is required between the block and the kelly. This bearing is called a swivel.

After some hours of drilling, depending on the formation and bit type, the teeth will become dull and the bearings wear out. Then the bit needs to be changed and for that the whole drill-string has to be lifted to the surface or for some times to log the well.

In order to "pull out of hole", hoisting equipment is required. On a rotary rig this consists of the hook which is connected to the travelling block. The latter is moved up and down via a steel cable ("block line') which is spooled through the crown block on to a drum ("draw works'). The draw works, fitted with a large brake, move the whole drill string up and down as needed. To change the bit or drilling assembly, the drill string may have to be brought to surface. It is normal practice to pull stands consisting of 30m speeds up the operation and allows better reaming of the hole is known as top drive. Sections of drill string and rack them in the mast rather than disconnecting all the segments. The procedure of pulling out of hole ('POOH') and running in again is called a round trip.

If the hydrostatic head of the drilling fluid is insufficient to control subsurface pressure, formation fluids may flow into the well. To maintain well control, back pressure is applied by routing the returns through adjustable chokes until the well flow condition is corrected. The chokes are connected to the blowout preventer stack through arrangement of valves, fittings, and lines which provide alternative flow routes or permit the flow to be halted entirely. This equipment assemblage is called **"choke manifold**."

While drilling of wellbore fluids have a hydrostatic pressure which is greater than the pressure found in the formation fluids said to be of overbalance. Circulation Pressures is resistance to flow can be considered as friction. Friction acts in the opposite direction of flow. Because of the inherent resistance of liquids to flow, force must be applied to circulate drilling fluids around the well. The pressure caused by resistance to flow in the annulus is applied in the annulus, and the sum of all the annulus friction is focused at the bottom of the hole is called ***ECD*** (Equivalent Circulating Density) of mud. ECD is slightly more than that of the mud weight used.

Most drilling fluids are usually made up using water and are called water based muds. Another frequently employed system is based on oil, oil based mud (OBM).The advantage of OBM is better lubrication of the drill string, compatibility with clay or salt formations which facilitates higher rate of penetration. Diesel fuel is usually used for the preparation of OBM.

Now synthetic drilling fluids are introduced which has a major impact on the evaluation of logging operations and the potential damage to well productivity caused by mud invasion into the formation.

For high angle wells, in general, cuttings will fall to the low side of the hole and away from the primary fluid flow at the top of the hole. To ensure good hole cleaning, it is critical that flow rates and pipe RPM are maintained at high levels throughout the drilling process. The fluctuations are worse because of the longer fluid travel distances (Measured Depth), and the relatively shallow vertical depths.

The driller controls the drilling console on the rig floor from where instrumentation will enable them to monitor and control the drilling parameters; hook-load, torque in drill string, weight on bit (WOB), rotary speed (RPM), pump pressure and rate, rate of penetration (ROP in meter/hour), mud weight in and out of the hole, volume of mud in the tanks.

Sub Sea Drilling

When drilling from a Jackup, the drilling operations are very similar to platform based or land-based operations with a conductor being driven into the seabed and conventional wellhead and surface BOP stack equipment being used. However, since the Jackup will have to move off location when the drilling operation is complete the casing strings must be physically supported at the seabed and it must be possible to remotely disconnect the casing strings between the seabed and surface when the operation is complete. Seabed support for such wells is provided by a Mud-line suspension (MLS) system. The MLS system is a series of full bore housings and hangers run with the casing strings. When drilling with an MLS system the casing strings are temporarily extended back from the mudline to surface and the conventional wellhead and BOP stack is nippled up on top of these extension strings just beneath the rigfloor. The MLS system only provides physical support for the casing strings. All annulus sealing and monitoring functions are provided by the wellhead at surface. When the well has been drilled and the well is either suspended for later completion or it may be completed immediately and a subsea Xmas tree installed on the wellhead.

Drilling Techniques

If an attempt to drill one straight hole from the derrick to all the way down to the depth there is possibility of either with the hole collapsing or the loss of drilling fluid into formations. Hence, from time to time, the borehole needs to be stabilised and the drilling progress needs to be safeguarded. This is done by lining the well with steel pipe (casing) which is cemented in place. In this manner the well is drilled like a telescope to the planned target depth.

The diameters of the 'telescope joints' will start usually with a 23" (conductor), then 18 5/8 " (surface casing), 13 3/8" (intermediate casing), 9 5/8" casing and/or 5 1/2 "casing based on target depth. In general to a maximum of 1,500 m will be in open condition in between next stage of drilling hole for safety reasons. Another option of 7" 'liner' over some section may help to reach the desired section (Figure 6.8).

A liner is a casing string which is clamped with a packer into the bottom part of the previous casing; it does not extend all the way to the surface. Casing joints are available in different grades, depending on the expected loads to which the string will be exposed during running, and the lifetime of the well.

The main criteria for casing selection are, collapse load which originates from the hydrostatic pressure of drilling fluid, cement slurry outside the casing. This is the internal pressure the casing will be exposed to during operations. Next is tension load, caused by the string weight during running in; it will be highest at the top joints.

Top Hole Drilling: - For the very first section of the borehole a base from which to commence the drilling is required. In a land location this will be a cemented "cellar" in which a conductor pipe will be piled prior to the rig moving in. The cellar will accommodate the 'christmas tree' which is a set of valves, once the well has been completed and the rig has moved off the location.

Once the drill bit has drilled below the conductor the well is said to have been spudded. As a result of the bigger hole diameter and the rapid penetration rate, vast

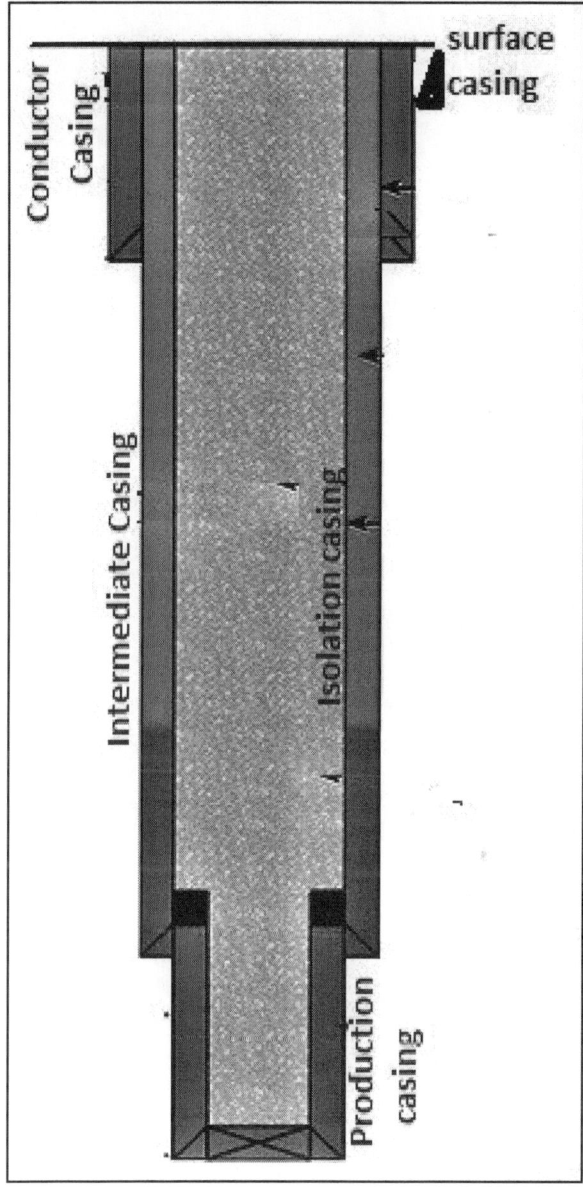

Figure 6.8. Well Bore Diagram.

quantities of drilled formation will have to be treated and removed from the mud circulation system. Often the rate of penetration will be reduced to allow adequate removal of cuttings and conditioning of mud. In some cases first drilling a hole with smaller diameter bit and later re-drilling the section to the required size using a hole-opener. The hole openers are also run if the hole has to be logged because most logging tools are not designed for all diameters.

Intermediate and reservoir section between the top hole and the reservoir section in most cases an intermediate section will need to be drilled. This section consists of more consolidated rocks than the top hole. The deviation angle is often increased in this interval to reach the subsurface target and eventually a casing is set prior to entering the reservoir sequence. An intermediate casing is usually set above the desired productive zone in order to keep it against one casing.

In exploration wells, the information from logging and testing may not be sufficient to fully evaluate the formation if the hole is not on gauge. If needed, for early confirmative results for hydrocarbons open-hole drill stem testing can be conducted based on logging tools.

Leak off Test

A leak off test is made to determine the pressure at which a formation will begin to leak off. Leak off tests are usually run after drilling a short distance below all the casing shoes after cementation and before starting the next stage of drilling.

A leak off test is performed by pumping drilling fluid into the wellbore at a slow rate or in increments of volume, with blowout preventers closed. The resulting pressures are to be carefully plotted versus the volume pumped. The pressure at which the plotted curve begins to flatten, *i.e.,* when the pressure increases a smaller amount for a volume pumped, is the surface leak off pressure. Pumping should be stopped immediately. The surface leakoff pressure plus the hydrostatic pressure of the drilling fluid at the shoe is the formation leak off pressure. This leak off test will indicate up to what desired mud weight next phase of openhole can be drllied safely.

Circulation System

Mud Tanks and Pits

Mud consists of a continuous liquid phase of water in which clay materials are suspended. A number of reactive and non-reactive solids are added to obtain special properties. A water-based mud system is a three-component system consisting of water which may be fresh water or salt water. Seawater is commonly used in offshore drilling. Reactive solids or clay is the basic material of mud is commonly referred to as "gel". It affects the viscosity, gel strength and water loss. Common clays are with Bentonite for fresh water muds and Attapulgite for saltwater muds.

A drilling rig usually has two ***mud pumps,*** and these are the heart of a fluid-circulating system. Their function is to circulate the drilling fluid under pressure from a surface pit called mud pit, through the drill-string, to the bit, return it up the annulus, and back to the pit.

The volume of fluid pumped per minute is determined by multiplying the volume per complete cycle by the number of strokes/cycles per minute. Strokes per minute (spm), on a duplex pump actually means cycles per minute, because the double acting, twin cylinders stroke four times during each cycle.

In addition to the mud pumps, the surface portion of the fluid circulating system consists of high pressure piping from the pumps to the standpipe. The standpipe is firmly clamped to the derrick and topped with a goose-neck fitting. The standpipe anchors one end of the rotary hose and keeps the hose clear of the rig floor when the drillstring is lowered during drilling. The other end of the rotary hose is connected to a gooseneck on the swivel.

The **mud pits** are usually a series of large steel tanks, all interconnected and fitted with agitators to maintain solids in suspension. Some pits will be used for circulating whilst others will be used for storing and mixing of drilling muds.

The out-coming mud from the well will contain suspended solids, perhaps some gas and other contaminents. These must be removed before the mud is recycled. The mud passes over a **shale shaker**, which is basically a series of vibrating screens of various mesh sizes. Settling tanks allow further settling of residues before the mud is passed back to the mud pits. Fine solids and gases can be further removed by other auxiliary components if required such as de-sanders, de-silters, de-gassers or centrifuges.

Having been cleaned, the mud is transferred into mud tanks, large treatment and storage units. From there a powerful pump brings the mud up through a pipe (stand pipe) and through a hose connected to the swivel (rotary hose) forcing it down the hole inside the drill string.

Originally, "mud" was made from clay mixed with water, a simple system. In order to effectively lift the cuttings out of the hole a certain viscosity needs to be achieved, yet the fluid must remain pumpable. If the mud circulation stops, for instance to change the bit. It has to be stable under high temperatures and pressures as well as at surface conditions. Cuttings are flushed from the bottom of the borehole up to the surface and separated from the drilling fluid at the shale shakers, thus cleaning the bottom of the hole and providing the logging and to geologist with samples at the surface.

Drilling fluids have to be capable of carrying heavy clay called barites in order to control required formation pressures. They have to be compatible with the formations being drilled,, *e.g.*, they should prevent the swelling of formation clay and not permanently damage the zone.

The shale shaker takes out large cuttings and the tanks should have adequate volume for settling out sand and releasing contained gas from the mud. Particular care should be taken in placing the suction inlet of the pump in the pit or tank. The inlet should be far enough off bottom so that flow will not be restricted and sand will not be drawn into the pump (Figure 6.9). There should be no loops or sharp bends in the suction line that would restrict the flow of fluid going to the mud pump.

Drilling Fluid-Engineering

Now much varied types of mud are made due to the varied hole conditions. Factors such as depth, types of formations, local structural conditions are considered into the choice of a particular mud. The functions and corresponding properties of a drilling mud are to:

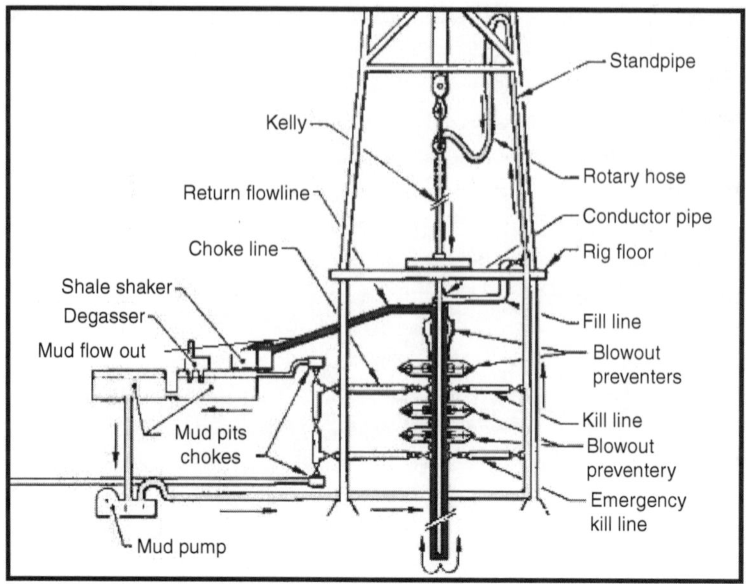

Figure 6.9. Mud Circulating System

- ☆ Control subsurface pressures and prevent caving (mud density)
- ☆ Remove cuttings from the borehole (viscosity)
- ☆ Suspend cuttings when circulation stops (gel strength)
- ☆ Cool and lubricate the bit and drillstring (additive content)
- ☆ Wall the borehole with an impermeable filter cake (water loss)
- ☆ Release the cuttings at the surface (viscosity/gel strength)
- ☆ Help support the weight of the drillstring/casing (density)

Controlling Subsurface Pressures: The pressure of the mud column at the bottom of the borehole is a function of the mud density and column height. This pressure must be adequate at all times to prevent the flow of formation fluids into the borehole. In case, mud density fall below that which is necessary to hold back formation pressures, then formation fluids can enter the well which is called "kick". If this condition is allowed to continue unchecked for even a short period of time, the mud density may be reduced so severely that uncontrolled flow will result termed "blowout".

On the other hand, the excessive mud weights result in low rates of penetration and in the fracturing of weak formations, and may cause the loss of drilling mud into them (lost circulation). Density is also important in preventing unconsolidated formations from caving into the borehole. The most important factors involved are also the speed at which the mud travels up the borehole (annular velocity), the viscosity and gel strength of the drilling mud.

Gel strength is the ability of the drilling fluid to develop a gel as soon as it stops moving. Its purpose is to suspend the cuttings and mud solids, while they are in the borehole and not permit them to settle around the bit when circulation is stopped.

The pressure of the drilling fluid exerted against the walls of the borehole helps prevent the caving of unconsolidated formations of the borehole with a thin, tough filter cake called "Mud cake".

Sodium carboxy-methyl cellulose (CMC) can be polymerized into different grades depending on the desired viscosity. Lost Circulation Material (LCM) is added to the mud system in order to bridge-over or plug the point of mud loss. It is available in many types to suit particular circulation loss which may also include mica flakes.

Motion Compensation System-Offshore

It is used on floating rigs, semi-submersibles and drill-ships. There are four basic motion compensation components:

The ***drill-string compensator system*** is designed to nullify the effects of rig heave on the drill-string or other hook-supported equipment. It is mounted between the hook and travelling block. The compensator is connected to rig-floor mounted air pressure tanks via a hose loop and standpipe, and is controlled and monitored from the driller's control console. While drilling, the drill-string compensator controls the weight-on-bit. As the rig heaves upward, the compensator cylinders are retracted, and the hook moves downwards to maintain the selected loads. Actually, the hook remains fixed relative to the seabed; the rig and compensator moves, producing relative motion between the hook and rig. The motion of the drill-string is relative to the rotary table.

The functions of the **Marine riser and Guideline Tensioners** are to provide a path for the returning drilling fluid from the borehole to the surface and to guide the drill-string and other tools to the wellhead on the sea floor. Riser tensioners provide tension to the marine riser through a system of cables joined via sheaves to a series of pneumatic cylinders. Its purpose is to maintain the riser tension at all times regardless of rig heave. Guideline tensioners operate on the same principle as the riser tensioners. Their purpose is to maintain the correct tension in the guidelines between the rig and the guide-base, located on the sea floor, regardless of rig heave.

The **telescopic joint** is used at the top of the marine riser and is used to compensate for the vertical motion of the rig. All floating drilling rigs will have some lateral movement relative to the wellhead whether they are anchored or dynamically positioned. To take account of this a **ball or flex joint** is installed in the riser system above the BOP stack. Ball joints usually allow up to 10 degrees lateral movement from vertical. An anti-rotational pin between the ball and socket prevents the upper member from turning in relation to the lower part of the joint.

Blowout Prevention (BOP) System

When noticeable amounts of formation fluids enter the bore hole, the event is known as a 'kick'. An uncontrolled flow of formation fluids is a 'blowout'

The blowout preventers are a series of powerful sealing elements designed to close off the annular space between the pipe and the hole through which the mud normally returns to the surface. By closing off this route, the well can be 'shut in' and the mud and/or formation fluids are forced to flow through a controllable choke, or adjustable valve. This choke allows the drilling crew to control the pressure that reaches the surface and to follow the necessary steps for 'killing' the well, *i.e.*, restoring a balanced system. The BOP stack consists of a number of different blowout preventers. Their arrangement is decided by the degree of protection deemed necessary, and the size and type of drillpipe in the borehole.

The positioning of the BOP stack is dependent on rig type. On floating offshore rigs, the BOP stack sits on the seabed. On jack-ups and platforms it rests on the 'spider deck' below the rig floor, and on land rigs, it is located in a dug out chamber under the rig floor, called the 'cellar'.

Blowout preventers may be classified as:

Annular Preventer: This consists of an annular rubber sealing element which, when pressure is applied, closes around the drill-pipe or kelly. Since pressure can be applied progressively, the annular preventer can be made to close on any size or type of drill-pipe. A slight relaxation of pressure may allow a small leakage of fluid and permit the pipe to be rotated within the annular preventer or pulled through slowly.

Ram preventers use pairs of hydraulic rams and solid ram blocks to close around the pipe or shut off the hole. Ram preventers may be classified into three types depending on the blocks used internally: **pipe rams** have a rubber face moulded to fit around a certain size of drill-pipe. If more than one size drillpipe is in use, there must be one set of pipe rams for each size of pipe; **blind rams** completely close off the borehole. As such, they are used only when there is no drillpipe in the borehole; **shear rams** have specially designed cutting structures, which when closed on drill-pipe, will cut through the drill-pipe and completely close off the borehole. The hydraulic pressure used to close the blowout preventers is supplied by 'accumulators', containing high pressure nitrogen. In an emergency, the ram preventers may be manually closed (but only on land rigs and jack-ups).

Bell Nipple is a length of casing attached to the top of the BOP to extend the annulus to just below the rig floor. The top of this pipe is belled-out (nippled) to prevent tools from 'hanging-up' when lowered through the rotary table. On one side of the conductor pipe is attached the drilling fluid return line (flow-line), through which the drilling fluid is carried to the surface cleaning equipment.

Choke and Kill Lines: After the BOP is closed, high-pressure fluid can be released at carefully controlled rates by use of a hydraulically controlled valve. The choke line carries these high pressure fluids to a reserve pit where they are discarded, or through a separator where the fluid is degassed and saved.

The kill line is used to introduce heavy drilling fluid to the well through a check-valve on the BOP in order to control high formation pressures. It fills the borehole from the top, instead of using the drill-pipe to fill the borehole from the bottom. It is also used to fill the annulus when round trip in progress.

Wellhead is the casing attachment to the B.O.P. or production (Christmas) tree. It is a permanent fixture, bolted or welded to the conductor pipe or surface casing. Wellheads installed on land rigs are located in the cellar, on the spider deck on jack-ups and fixed platforms, and on the seabed when used with barges, semi-submersibles and drillships.

The wellhead is the starting point for most blowout-preventer assemblies, and the vital link between the casing and B.O.P. stack. The conductor pipe is not usually provided with a pressure connection to connect the two. The surface casing is nearly always welded to the wellhead, and subsequent casing strings are inserted inside the wellhead housing and supported in a casing hanger.

Corrosion service: Carbon dioxide (CO_2) or hydrogen sulphide (H_2S) in formation fluids will cause rapid corrosion of standard carbon steel and special steel may be required.

Buckling resistance: The load exerted on the casing if under compression.

Drilling and Completing Well

After the well is spudded, routine drilling consists of continuously drilling increments of drill-pipe, making connections and continuing until it is time to change the drill bit. The bit must be changed when it is worn or when a formation is encountered for which the particular bit being used is not suitable. Changing the bit is accomplished in an operation called "tripping". A round-trip includes coming out of the hole, changing the bit, and going back into the hole.

When the Kelly has drilled all the way down, it is pulled out and a new length of drillpipe is added. In the kelly is near the "kelly down" position, where another joint of pipe is made-up and tightened with the tongs. The new length of pipe has been placed in the "mouse-hole", ready to be connected. The mouse-hole is used to store the next measured joint of pipe until it is required for threading into the drill-string.

The difference with "top drive" systems is that the kelly is not used and stands of pipe are drilled instead of joints. The next stands to be drilled may be placed in the mousehole or may be left standing in the monkey-boards. The kelly, rotary bushing and swivel are placed in the rathole during the trip with the kelly out of the way, the elevators are latched around the pipe just below the tool joint's box end.

When making a trip, drill-pipe is handled in stands of three joints each. Pipe is removed from the borehole and set back on the rig floor. The tool joint's box end provides a shoulder for the elevators to pull against. The pipe is then pulled from the borehole, and after being secured in the rotary table (using the slips), the connection is loosened with the break-out tongs. A spinning wrench (power tongs) are commonly used to unscrew the pipe at this time.

The drill collars and bit are the last to come out of the borehole. The master bushing may have to be removed to allow the large diameter collars to pass through the rotary table. When the bit appears, the master bushing is replaced and a "bit breaker" is placed in the rotary table. Using the break-out tongs, the bit is loosened and removed from the bit sub.

Casing and Cementing

In order to protect the drilled borehole from collapse it is protected with steel pipe called casing periodically. Casing in a well serves number of functions:

- ☆ Prevents caving of the borehole.
- ☆ Provides a means of containing well (formation) pressures by preventing fracturing of upper, weaker zones.
- ☆ Provides a means for attaching surface equipment (blowout preventers and production tree).
- ☆ Confines production to the wellbore.
- ☆ Allows segregation of formations behind the pipe and thereby prevents inter-formational flow, and permits production from a specific zone.
- ☆ Permits installation of artificial lift equipment for producing the well.
- ☆ Provides a borehole of a known diameter for further operations.

Lowering Casing String

The thread protector from field end of casing until ready to stab. If necessary, apply thread compound over entire surface of threads just before stabbing. The brush or utensil used in applying thread compound should be kept free of foreign matter and the compound and the compound should never be thinned. In stabbing, lower casing carefully to avoid injuring threads. Stab vertically, preferably with assistance of a man on the stabbing board. If the casing stand tilts to one side after stabbing, lift up, clean, and correct any damaged thread with three-cornered file, then carefully remove any filings and reapply compound over the thread surface. After stabbing, the casing should be rotated very slowly at first to insure that threads are engaging properly and not cross-threading.

Casing strings should be picked up and lowered carefully, and care exercised in setting slips to avoid shock loads. Dropping a string even a short distance may loosen couplings at the bottom of the string.

Liner String: Unlike casing, which is run from the surface to a given depth and overlaps the previous casing, a liner is run only from the bottom of the previous string to the bottom of the borehole. There is a minimal overlap with the previous casing, and the liner is suspended in this overlap by means of a liner-hanger. They are often cemented in place. Any type of casing can be used as a liner.

With the depth of wells increasing to more than 5000 metres, lowering of liner is normally become part of the well programme. Liners permit deeper drilling by providing protective sealing against lost circulation zones, heaving shales and geo-pressured formations. Cementation of liner is a technique due to short annular clearance between holewall and the liner. As these liners don't extend upto surface, hydraulics is improved for further drilling as larger size pipes can be used above drilling liners.

Production Liners set through producing zones are called production liners. Stub Liner extends from top of a liner to a point inside another casing or liner. It is primarily used to provide protection against a damaged casing or liner.

Liner Overlap: Liners are extended from 20 mts to 100 mts inside the available bottom most casing. This overlap places the liner hangers above any weak or suspect casing. Longer overlaps are used when high pressures are isolated behind liners so that more cement volumes can be provided between casing-liner annulus.

Liner Hanger is installed at the top of liner. Hangers can be mechanically or hydraulically operated. In both types slips wedge against the casing wall.

Tie-back casing is used to extend a liner to the well head. This is lowered at the end of drilling operations and assures a fresh unworn casing for producing operations.

Providing an elective cement seal at top and bottom of liner is most important function of cementation. For a drilling liner these seals help in further drilling. For production liner, the seal between top and between formation and liner is the basic requirement.To develop these seals, the slurry should be displaced effectively and the properties of cement slurry should be such as to provide these seals with minimum of Wait on Cement (WOC) time.

Casing Accessories for Cementation

Once installed, the casing is cemented into place to provide additional support and a pressure-tight seal.

Guide Shoe is attached to the first length of casing to be lowered into the hole. It is aluminum with a hole in the center and rounded, to guide the casing into the borehole, around obstructions.

Float Collars permit the casing to literally float into the borehole, by virtue of being partially empty. It is a back pressure valve which is closed by the outside fluid column, thereby preventing entry of the fluid as the casing is lowered into the hole. It also serves as a check-valve in the casing string, to prevent back flow of cement after being pumped outside the string. This is important because the density of the cement slurry is always greater than drilling mud. This back pressure valve serves to prevent a blowout through the casing, if a kick should occur during casing operations.

A float collar also serves as a "stop" for the two plugs when cement is displaced. This action allows a quantity of slurry to stay inside the casing at the casing shoe, so that the operator has reasonable assurance of there being adequate cement outside the casing at that point.

Centralizers and Wall Scratchers: These are attached to the casing, to ensure a reasonably uniform distribution of cement around the pipe and to obtain a competent seal all the way around the casing and with the adjoining formation. Centralizers hold the casing away from the borehole wall and therefore also serve to prevent differential sticking. Wall scratchers are mechanical wall cleaning devices, attached to the casing, that abrade the hole when worked by reciprocating or rotating. This helps to provide a more suitable surface for the cement to bond to borelole wall (Figure 6.10).

Cementing Head provides the union for connecting the cementing lines from the cementing pump to the casing. This type of head makes it possible to circulate the mud in a normal manner, release the bottom plug, mix and pump the cement and pump it down, release the top plug, and displace the cement without making or breaking any connections.

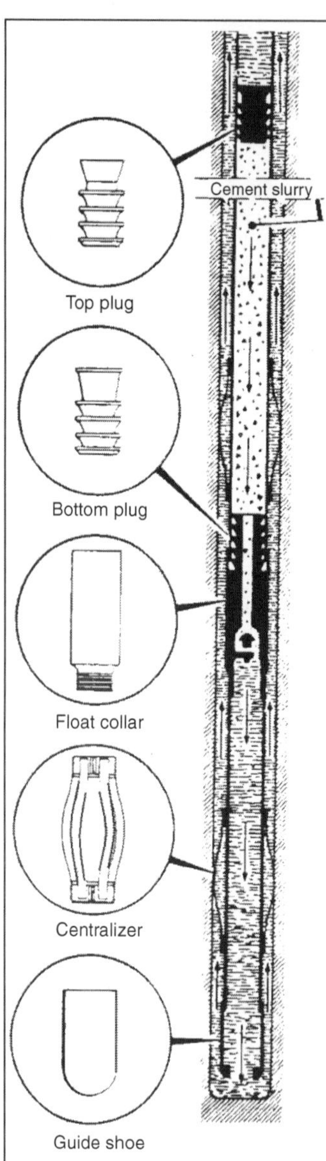

Figure 6.10. Cementing Accessories.

Oil well cementing is the process of mixing and displacing a cement slurry down the casing and up the annular space behind the casing where it is allowed to set, thus bonding the pipe to the formation.

Cementing procedures are classified as primary or secondary. Primary cementing is performed immediately after the casing is run into the borehole. Its objective is to obtain an effective zonal separation and help protect the casing. Cementing also helps in:

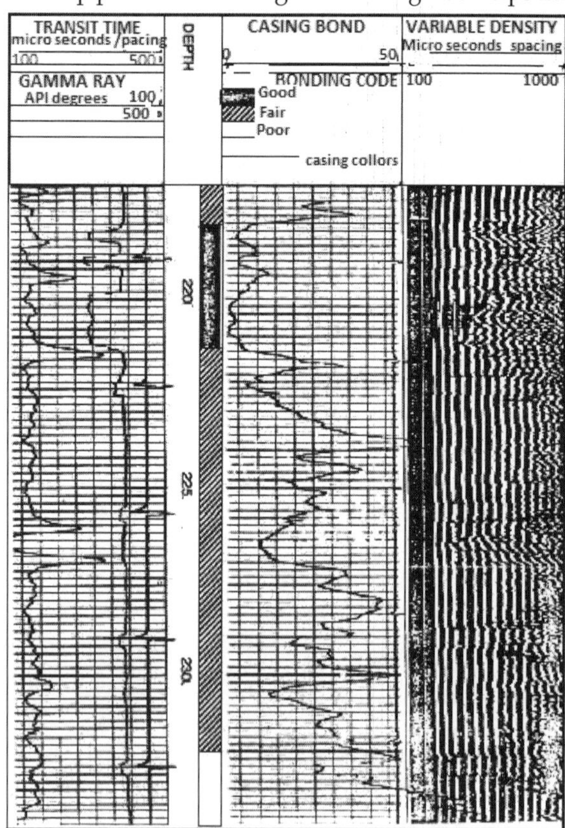

Figure 6.11. Cement Bond Log (CBL).

☆ Bonds the casing to the formation.
☆ Protects the producing formations.
☆ Seals off troublesome zones (*i.e.* lost circulation zones).
☆ Provides support for the casing.
☆ Prevents casing corrosion.

Once the casing is set and circulation has been assured, is to pump "spacer" liquid ahead of the bottom plug, which is immediately followed by the cement. The spacer serves as a flushing agent and provides a spacer between the mud and cement. It also assists in the removal of wall cake and flushes the mud ahead of the cement, thereby lessening contamination.

Cement plugs consist of an aluminum body encased in molded rubber. Two plugs are usually contained in the cementing head to facilitate the operations. When the bottom plug reaches the float collar, the diaphragm in the plug ruptures to permit the cement to proceed down the casing and up the annulus. The top plug, which is solidly constructed, is released when all the cement has been pumped. It is dropped on top of the cement, followed by drilling mud, to displace the cement from the casing. This plug causes a complete shut-off when it reaches the float collar.

Pumping is stopped as soon as there is a positive indication (pressure increase) that the top plug has reached the float collar. To ensure good cement circulation and drilling mud displacement, movement of the casing, either by reciprocation or rotation, may be continued throughout the pumping and displacement operations.

The first property of a cement slurry that should be considered is that which is commonly referred to as the pumping time or thickening time. A cementing slurry must remain fluid for a sufficient length of time to allow it to be pumped down the casing and up the annular space behind the pipe. A suitable cement should possess an adequate safety factor in case of unavoidable shut down while pumping the cement slurry. Secondly, the cement, after having been properly placed in the well, must set in a reasonable period of time and should develop sufficient strength to allow continuation of normal drilling operations.

After the cementation ensure CBL- VDL/CET logs taken and interpreted properly. The cement bond logging tools have become the standard method of cement evaluation jobs since they not only detect the top of cement, but also indicate how good the cement bond is (Figure 6.11).

Secondary cementation is the cementing operation performed after the primary cementation, to repair some segments in bore hole having poor cement in annulus. This is also known as remedial cementing. The cement squeezing and cement plugging are the main two types of secondary cementing processes.

Purpose of Squeeze Cementing

1. To repair a primary cement job that failed due to cement by passing the mud (channeling) or insufficient cement height in the annulus.
2. To eliminate water intrusion from above, below or within the hydrocarbons producing zone.

3. To reduce the producing gas oil ratio by isolating gas zones from adjacent oil intervals.
4. To repair casing leak due to corrosion or split pipe.
5. Plug all or part of one or more zone in a multizone injection well so as to direct the injection into the desired intervals.
6. Plug and abandon a depleted or watered out producing zone.

Then **Hermetical test** is conducted after the last cementation was done successfully or not. It is a pressure holding test of annulus. To detect leakage in casing after cementation the casing is tested under water for about 30 minutes at a pressure of above 2000psi.if the casing holds this pressure, the casing is said to be hermetically sealed. If the casing was found not holding pressure of water during hermetically tested or if interlinking of different formation zones is found or any other similar defect was found the well is declared as sick well. The sickness of a well can be directly linked to the failure of primary cementation or casing failure.

Completing the well as Producer

The drilling rig is used to run and cement production casing. The blowout preventers are removed and a production wellhead is attached to the top of the casing. The production well head seals the tops of the various casing strings in the well, provides a place to suspend and seal production tubing as needed, and provides the valves that control flow out of the well.

Modern Types of Drilling

Normally first few wells in a structure will be drilled vertical. Once available reservoirs, their areal extent, thickness are known, in order to enhance production more directional wells are drilled to save rig time by maintaining the well spacing with in the reservoir for drainage requirement. Some times to increase production horizontal wells will be drilled.

In case of offshore fields more wells from the same platform will be drilled to connect to process platforms. So such wells have to be directional wells. If the angle of direction is more it is called Extended Reach Drilling (ERD).

Directional Drilling

For some reasons it may not be possible to drill a vertical well. There may be constraints because of the surface location. In the subsurface if multiple targets are palnned, the shape of the structure, faults directional well can be planned to reach the desired target in the subsurface at the reservoir/target level.

In case of onland locations if the identified sub-surface location found to be in an inhabited area. In such cases well can be directional from shallow levels and may reach the desired point in the structure below. In case of some coastal basins, the promising structure may be located very near to the coast and in very shallow waters. In such cases first availability of such suited rigs and its cost of operation matters. In such cases directional drilling can be planned from nearest onland to reach the bottom at the warranted location on the offshore structure.

Directional drilling requires equipment to deflect the drill bit from the vertical and subsequently to steer it along the planned trajectory. Initially, angle has to be gradually built up in the desired direction. The kick off point is the depth at which the deviation commences. To build angle several techniques are employed. It is important that the hole angle is not increased or decreased rapidly creating 'dog legs' which will result in excessive torque and drag. If dog legs occur, the string may develop keyseats in the borehole wall and eventually get stuck. Until down-hole motors became available a 'whipstock' which is a slightly asymmetric steel joint, was inserted in the drill string. The assembly is oriented down-hole and a 'rathole' is drilled which is then enlarged to full bore hole size.

If a shallow kick off in soft formation is required a technique using jet bit deflection or 'badgering' is employed. A rock bit is fitted with two small and one large jet. With the bit on bottom and oriented in the desired direction the string is kept stationary and mud is pumped through the nozzles. This causes asymmetric erosion of the borehole beneath the larger jet. Once sufficient hole has been jetted, the drill bit will be rotated again and the new course followed. This process will be repeated until the planned deviation is reached.

Another technology in operations is a mud motor or a mud turbine are mostly used for directional drilling. Rotary drilling may be carried out between mud motor/ turbine drilling, *i.e.*, the use of these is often restricted to a certain interval only. A mud motor is a positive displacement hydraulic motor, driven by the circulated drilling fluid. As fluid is pumped through the tool it flows through those cavities creating a rotational force. To deflect the bit in the desired direction a 'bent sub'(with typically to 2°bend) which can be oriented from surface is inserted in the string just above the motor. An alternative type of down-hole mud motor is the mud turbine, (multistage axial flow turbine) which directly drives the bit. The tool consists of an upper section containing the turbine blades and lower section with bearings. As mud is pumped through the upper section the blades are turned. Turbines are designed to rotate at higher speed than the displacement motor. The higher rotation speed requires diamond or composite bits.

Mud motors in general have several advantages over the rotational directional drilling methods. They provide a full gauge hole, dog legs are more likely to be smooth and the penetration rate is usually high. Torque is transmitted more effectively to the bit since the whole drill string is not rotated. This is particularly advantageous in long reach, highly deviated wells. Downhole motors may cause problems if losses occur and lost circulation material needs to be pumped, as this may plug up the motor. Wells may be drilled at a constant angle to the target or dropped off to a lower angle through the reservoir section. To build, maintain or drop the deviation angle stabilisers are run in the bottom-hole assembly. A change in deviation used to require a round trip to change the position of those stabilisers in the bottom-hole assembly.

High deviation angles (above 60 degrees) may cause excessive drag or torque while drilling, and will also make it difficult to later service the well with standard wire-line tools. Not only can the inclination (relative to vertical) of a well be changed, but also the azimuth at which it is being drilled. The azimuth refers to the compass bearing relative to magnetic north. Obviously it is essential to know at all times where the bit is. Therefore a number of directional surveying techniques are employed to monitor the progress of the hole. Essentially compass readings and inclination measurements are taken at intervals and transmitted to the drill floor.

Horizontal Drilling

Given the lateral distribution of reservoir rock or reservoir fluids, a horizontal well may provide the optimum trajectory. In order to have more produciability of hydrocarbons from a single well in order to have maximum exposure with in the reservoir, horizontal wells may be drilled.

The build-up rate of angle is the main distinction from a drilling point of view. Medium radius wells are preferred since they can be drilled, logged and completed with fairly standard equipment. The horizontal drilling target can be controlled within a vertical window of about 2m. To target the horizontal section with the required accuracy a pilot hole may be drilled first for depth control. Accurate directional surveys are critical for the optimum positioning of the well. The success of horizontal wells was largely dependent on the development of tools which relay the subsurface position of the drill bit in real time to the drill floor.

Measurement while drilling (MWD) is done by the insertion of logging equipment into the drill string close to the bit. Data transmission is via a mud pressure pulse which is translated through a decoder into an electric signal. Initially providing only directional data, the tools have been improved to the point where petro-physical data gathering (gamma-ray, resistivity, density and porosity) can be carried out while drilling. It also helps in real time monitoring to see well is positioned within the reservoir thickness.

The electrical resistivity measurements enable the driller to steer the bit above a hydrocarbon water contact, a technique used, for example to produce thin oil rims. These techniques, known as geo-steering, are being applied to field development optimisation.

Limitations of Horizontal Wells

The major advantage of a horizontal well is a large reservoir contact area. Now-a-days horizontal well of 1000- to 1500-meter-long can be drilled. The major disadvantage is that, only one pay zone can be drained per horizontal well. Recently horizontal wells have been also used to drain multiple pay zones. This can be accomplished by drilling a "staircase" type well where long horizontal portions are drilled in more than one layer.

Extended Reach Drilling (ERD)

An extended reach well is defined whose horizontal portion is of at least twice the vertical depth. ERD wells are technically more difficult to drill and because of the degree of engineering required. Extended reach drilling (ERD) will be considered, where surface restrictions exist, where marginal accumulations are located quite distant away from the existing platforms.

The high deviation (up to 85°) and the long horizontal displacement expose the drill string to extreme drag and torque. The hole cleaning (cutting removal) and cementing of casing is more difficult due to the increased effect of gravity forces compared to low angle wells. A thorough understanding of the dynamics behind good hole cleaning is critical to the success of ER wells. For high angle wells, in general, cuttings will fall to the low side of the hole and away from the primary fluid flow at the top of the hole. This makes removing the cuttings from the hole difficult and requires special techniques for various well inclinations. The drilling parameters, BHA and bit designs, mud rheology and hole condition will all have a significant effect on the rig system's ability to effectively clean the hole. To ensure good hole cleaning, it is critical that flow rates and pipe RPM are maintained at high levels throughout the drilling process. Extended reach wells inherently have higher ECD's (Equivalent Circulating Density) fluctuations than conventional wells.

The fluctuations are worse because of the longer fluid travel distances (Measured Depth), and the relatively shallow vertical depths. Furthermore, the drilling parameters must usually be more aggressive to maintain hole cleaning, whilst the mud system can allow less scope for compromise. Temperature and pressure variations (and their effect on mud properties) are also more extreme in these wells.

Well control is generally more challenging in an ERD environment. Not only is detecting a kick often more difficult, but the ability to manage and kill the well control problem is more difficult.

Slim-Hole Drilling

The slim hole is a well in which 90 per cent or more of the length has a diameter of 7 inches or less. The oil industry has been developing rigs, drill string components and logging tools that will allow smaller diameter holes and completions. The cost of slim-hole drilling is less. A slim-hole rig weighs about one fifth of a conventional rig and its small size can open new frontiers by making exploration easy and economical. The maximum depth drillable with slim-hole configurations is one limitation of the technology. They are ideally suited for remote locations since they can be transported by helicopter.

Coiled Tubing Drilling (CTD)

A special version of slim hole drilling currently emerging as a viable alternative is coiled tubing drilling (CTD). Whilst standard drilling operations are carried out using joints of drill pipe, coiled tubing drilling employs a seamless tubular made of high-grade steel. The diameter varies between 1¾" and 3 ½". Rather than being segmented, the drill string is reeled onto a large diameter drum. The advantages of CTD are, nearly no pipe handling, better well control allows at-balance or even underbalanced drilling, resulting in higher penetration rates and reduced potential for formation damage.

Under-Balanced Drilling (UBD)

It is the process of drilling into a formation in which the formation pressure, or pore pressure, is greater than the pressure exerted by the drilling fluid or gas. UBD is an excellent method to drill under pressured depleted formations and formation containing mud filtrate sensitive materials. Underbalanced drilling is used to avoid or limit lost circulation and as a method to protect reservoirs, prevent differential sticking, and increase the drilling rate. Drilling with a hydrostatic pressure that is less than formation pressure may create a condition similar to a "kick," or well-control condition. Controlling these pressures and maintaining a safe environment require special surface pressure control equipment. The type of equipment required depends on primarily the lithology, permeability, and pressure of the formations that are to be drilled. The variety of "under-balanced fluids" include, dry air, natural gas and foam.

The drilling fluid in this case must remove the cuttings from the bottom of the hole and carry them to the surface. At the surface, there must be adequate equipment to remove the solids from the fluid before it is circulated back down the drill string. In cases where the fluid is not recirculated, such as air drilling and some foam drilling, provisions must be made for handling the solids and liquids.

Decreased bottom hole pressure typically causes higher penetration rates. However, higher penetration rates can increase the circulating bottom hole pressure and bring the well back to overbalanced conditions. Moreover, due to the annular fluid segregation, there is an increased risk that the wellbore will be packed-off resulting in stuck pipe. Large cutting volumes generated by high penetration rates are also difficult to remove. Therefore, penetration rates should be carefully adjusted to ensure sufficient hole cleaning.

When drilling with gas, the gas is usually flared. The intense heat at the end of the line prevents catching representative samples. The heat destroys some of the samples and any hydrocarbons associated with them. Foam in the hole is an emulsion of air or gas in water, but at the flowline a proper foam breaks to a mixture of droplets of water in an air stream. Most foam drilling is done with disposable foam.

During under balance drilling well control equipment is handled at surface rather than down hole. Under balance drilling results in less formation damage due to mud or mud filtrate, faster drilling rates, extends bit life, minimizes the need for costly stimulation.

Indian Scenario

In **Kalol** area of Ahmedabad area to minimize formation damage, in Mumbai offshore wells which are subjected to severe loss situation in limestone and in Gandhar field where permeability ranges more and skin effect is a concern, it can enhance the productivity of wells where around 150-200 m of productive section was drilled with reduced mud weight of 1.15 gm/cc. and slug removal.

Possible Complications during Drilling

Drilling equipment and drilling activities have to be carried out in complex and often hostile environments. Surface and subsurface conditions may force the drilling rig and crew to operate close to their limits and constant vigil on operating parameters. Even small delay or over-look may cause serious troubles in drilling operations. Sometimes due to non-routine or unexpected operating conditions some drilling problems can occur.

Well Bore Instability/Stuck Ripe

Wellbore instability is usually more critical in deviated wells due to the increased wellbore angle, increased hole exposure time and increased ECD fluctuations and effects. The drill-string and BHA will be on the low side of the hole and at least partially buried in cuttings. The effects of differential sticking are even more important for wells where torque or drag or buckling problems exist, since even minor differential sticking increases friction considerably.

Differential Sticking

Differential sticking is caused by a difference between hydrostatic head of the mud and pore pressure. When the drilling combination is not moving more mudcake can be formed by this pressure differential. Very high pulling forces are required to free the pipe. A thin impermeable filtercake is essential to prevent differential sticking. Once the pipe is stuck rapid action is required as on the stuck point the filtercake dehydrates and sticking increases with time. Priority is to reduce the pressure differential as far as safely possible. Pills of lubricating oil are often used in an attempt to reduce friction losses. Furthermore, mixtures of lubricating oil and surfactants (pipe lax b-free) are applied to "dissolve" the mud-cake.

Stuck Pipe

This term describes a situation whereby the drill string cannot be moved up or down or rotated. The pipe can become stuck as a result of mechanical problems during the drilling process itself or because of the physical and chemical parameters of the formation being drilled. Most common reasons for stuck pipe are:

- ☆ Excessive pressure differentials between the borehole and the formation. For instance, if the pressure of the mud column is much higher than the formation pressure the drill pipe may dragged against the borehole wall (differential sticking). This often happens when the pipe is stationary for some time. Prevention methods include: reduced mud weights, adding of friction reducing components to the mud, continuous rotation/moving of string, adding of centralisers or use of spiral drill collars to minimise contact area between string and formation, low fluid loss mud system.
- ☆ Some clay minerals may absorb some of the water contained in the drilling mud. This will cause the clays to swell and eventually reduce the borehole size to the point where the drill pipe becomes stuck. As prevention, mud additives which prevent clay swelling like potassium salt can be used.

☆ Unstable formations or a badly worn drill bit may result in under-gauged holes.

In order to prevent stuck pipe sometimes high mud weights may help delay deformation of the bore hole. If the well trajectory shows a severe dogleg, that is, sudden change in angle or direction the movement of the string may result in a groove cut into the borehole wall by the drill pipe. Eventually the pipe will become stuck, a process termed key seating. To avoid dog-legs frequent reaming, insertion of stabilisers on top of drill collars, insertion of key seat wipers in the string may help.

In many cases, the point at which the pipe is stuck can be determined by means of a 'free point indicator tool', a special electrical strain gauge device run on wire-line inside the drill pipe will measure axial and angular deformation. An initial estimate of where the string is stuck can be calculated by applying a pull on the drill string in excess of the drill string weight and measuring the observed stretch in the pipe.

This information may be used to decide where to 'back off' the string if the deeper part cannot be recovered. If the string indeed cannot be retrieved by over-pull, an explosive or chemical charge is lowered inside the pipe to the top of the stuck interval and the pipe above the stuck point is recovered after severing the string. The tubulars left in the well called 'fish' and the attempt to retrieve the same is called "fishing"

Fishing

Fishing describes the retrieval of a fallen or left-over object (rig floor tools, logging tools, MWD tools, *etc,*) in the borehole. Fishing operations will be required if the object is expected to hamper further drilling progress either by jamming the string or damaging the drill bit. This 'junk' often consists of small non-drillable objects,, *e.g.,* bit nozzles, rock bit cones, or broken off parts of equipment or left over string because of stuckpipe. Some logging tools will have radioactive sources which may need to be recovered or isolated for safety and legal reasons.

However, prior to commencing fishing operations, a cost-benefit assessment need to be made to establish that the time and equipment attributable to the fishing job is justified by the value of the fish or the cost of side-tracking the hole.

The drill collars which are off bottom can fall to the bottom of the hole after they are washed over. The other is that the top of the collars, if free, can lean over in an enlarged hole and make it impossible to get over them with tools. The wash-over back-off tool connect to the fish prior to washing over. If the fish is not free after the first wash-over, a string shot back-off may be performed to remove all free pipe.

Fishing small Objects

Hand wrenches, bolts, drill bit cones, clamps, hammers, pieces of steel are examples of small objects. These may be successfully fished in open hole with magnets, Jet charged cutters, Junk baskets and mill magnets.

If the object is lying on the bottom of the hole and not covered with cavings or cannot be removed by circulation, magnets may be successful. Magnets may be run in conjunction with rotary shoes and junk baskets. Magnets may be run on wire line or on pipe. Wire line runs are faster while pipe runs allow circulation.

Jet charges cut steel with a high velocity hot jet gas stream. The purpose of the jet charge is to fragmentize larger objects which will not fit in a junk basket or cannot be held by a magnet. The operation of the jet charged cutter is to run it to the top of the object and fire into it. Several runs may be required to cut up larger objects.

Many types of junk baskets exist. The three categories are the plain, the reverse circulating, and the pump junk baskets. The operation of the plain junk basket is to run into the hole, circulate, cut a short core to push the object into the barrel of the junk basket, grasp the core and/or object with the fingers or grapple of the junk basket. The pipe is usually chained out of the hole (not rotated) to prevent the dropping of a partially held object.

Reverse circulating junk baskets are the most popular; however, they sometimes plug preventing circulation. The operation is similar to the plain junk basket with the exception that a jet fluid cross above the barrel diverts the mud stream out of the pipe, down the annulus, and then up through the barrel of the junk basket. In practice the objects are swept into the barrel of the basket by the mud.

Larger objects may be cut up with a rotary shoe mill. Mills are usually run below either a junk basket or magnet. Their purpose is to dislodge or fragmentize an object. Full gauge complete bottom-hole coverage mills these tools permit faster recovery because collars or pipe may be washed over and pulled in one trip. A junk sub is usually run above the mill to catch the small pieces which are lifted off bottom by mud circulation.

Case Study-Borholla – India

In ***Borholla*** area the well was drilled down to the basement, but the hole was lost following total loss of criculation which resulted in drilling pipes getting stuck. As the mud used was of s.p.gr. 1.28.-1.30, it appears very probable that pipes were stuck on account of differential pressure through the very long sand sections in which formation pressure are close to hydrostatic. Stuck point indicator actually showed that pipes were free only inside the protection casing. When drilling with high weight of mud, the sand content in the mud was 3.5 per cent and it appears that settling down of this sand when circulation was stopped, must also have aided a lot in sticking the pipes. Hole was recovered only after a prolonged fishing operation.

As the second fishing job, following the total loss of circulation after reaching total depth was unsuccessful, a side track was drilled from just below the shoe of the casing. So it was realised that, chances of successfully completing the well will also depend on maintaining a low solids in mud. (Khanna, S.N., 1976).

Crooked Holes

The planned vertical wells are never vertical. Every well deviates, in some cases it is intense and in others it can be less. Deviation of wells occurs when the bit meets the formations of different hardness at an angle, depending upon the dip angles of formations. Loss of stability of the Bottom Hole Assembly (BHA) under the effect of axial and centrifugal forces is the main reason of deviation of wells.

In the process of drilling, if the bit weight and rotary speed exceeds the critical values, the bottom hole assembly loses its stability which leads to hole deviation. Where the formations are horizontal, the hole can deviate in any direction. In order to prevent deviation of planned vertical wells, a bottom hole assembly with not more than two stabilisers but without near bit reamer can be used.

If with no weight on the bit, the only force acting on the bit is the result of the weight of the string on the bit. When weight is applied, there is another force on the bit which tends to direct the hole away from vertical. The resultant of these two forces may be the deciding factor for deviation. Dog legs in hole may be problematic as sticking can occur by sloughing or heaving of the hole and by pulling the large OD drill collars into a key seat while pulling the drill stem out of the hole. Logging tools and wire lines can become stuck in key seats.

The lateral force of the drill pipe rotating against the casing in the dog-leg or dragging through it while tripping, can cause a hole to wear through the casing.

The lowering casing through a dog-leg can be a very serious problem. Even if running the casing to bottom through the dog-leg is successful, the casing could be severely damaged creating troubles during cementation. The dog-leg will force the casing over tightly against the wall of the hole, preventing a good cement bond because no cement can circulate between the wall of the hole and the casing.

A packed hole assembly to overcome crooked hole problems and the pendulum is used only as a corrective measure to reduce angle when the maximum permissible deviation has been reached. The packed hole assembly is sometimes referred to as the "gun barrel" approach because a series of stabilizers is used in the hole already drilled to guide the bit straight ahead. The object should be to select a bottom hole assembly to be run above the bit with the necessary stiffness and wall contact tools to force the bit to drill in the general direction of the hole already drilled. If the proper selection of drill collars and bottom hole tools is made, only gradual changes in hole angle will develop.

The condition of a hole is linked with the rate of drilling (Belov, V. P. and R. D. Ghildiyal, 1964). The dominating factor in the caving in drilling soft formations is the quality of mud and regime of circulation in the hole, end especially in the bottom hole zone. In jet bit drilling, the action of the jets upon the walls of a well has to be taken into account.

Lost Circulation

During drilling operations sometimes large volumes of drilling mud are lost into a formation. In this case normal mud circulation is no longer possible and the fluid level inside the borehole will drop, creating a potentially dangerous situation. A larger overbalance would encourage excessive loss of mud into the formation, slow down drilling and potentially cause differential sticking.

The formations in which lost circulation can be a problem are more porous, coarse or vuggy formation which does not allow the build-up of an effective mudcake, the limestone formation which has been eroded resulting into cave, a fractured interval and low strength formation in which open fractures are initiated by higher mud pressure.

The consequences of lost circulation are dependent on the severity of the losses, *i.e.*, how quickly mud is lost and whether the formation pressures in the open hole section are hydrostatic or above hydrostatic, *i.e.*, over-pressured. Mud losses lead to a potentially hazardous situation. Moderate losses may be controlled by adding 'lost circulation material' (LCM) to the mud system, such as mica flakes or coconut chippings. The LCM will plug the porous interval by forming a sealing layer around the borehole preventing further mud invasion. However, LCM may also plug elements of the mud circulation system, *e.g.*, bit nozzles and shale shaker screens and may effect productivity or injectivity of the objective intervals.

In severe cases the losses can be controlled by squeezing cement slurry into the trouble horizon. If sudden total losses occur in a hydrostatically pressured interval,, *e.g.*, in a cavernous limestone, the decision may actually be taken to drill ahead without drilling mud but using large quantities of surface water to cool the bit. The fluid level in the annulus will usually stabilise at a certain depth.

As an example, while drilling Bhimanapalli well in which the formation is cavernous limestone, in early days of exploration in Krishna-Godavari Basin, was encounterd with heavy mud loss which could be controlled by cooling the bit with fluids at some depth.

In the event of a sudden loss of mud in an interval containing overpressures the mud column in the annulus will drop, thereby reducing the hydrostatic head acting on the formation to the point where formation pressure exceeds mud pressure. Formation fluids (oil, gas or water) can now enter the borehole and travel upwards. The last line of defence left is the blowout preventer.

While drilling through a shale into an over-pressured formation, the mud weight must be increased to prevent influx. If this increased mud weight could cause large losses in shallower, normally pressured formations, it is necessary to isolate the normally pressured formation behind casing before drilling into the over-pressured formation.

The prediction of overpressures is therefore important in well design. Considerable effort will be made to predict the onset of overpressures ahead of the drill bit. The most reliable indications are gas readings, porosity - depth trends, rate of penetration and shale density measurements. If a situation arises whereby formation fluid or gas enters the bore hole the driller will notice an increase in the total volume of mud. Other indications such as a sudden increase in penetration rate and a decrease in pump pressure may also indicate an influx. Much depends on a quick response of the driller to close in the well before substantial volumes of formation fluid have entered the borehole. Once the BOP is closed, the new mud gradient required to restore balance to the system can be calculated. The heavier mud is then circulated in through the kill line and the lighter mud and influx is circulated out through the choke line. Once overbalance is restored, the BOP can be opened again and drilling operations continue.

Well Activity in Entering High Pressure Zones

Rate of Penetration (ROP) or the drilling rate (metres per hour) is a function of weight on bit (WOB), revolutions per minute of rotary (RPM) density of drilling fluid, bit type and size, and finally on lithology. With depth, the drilling rate decreases due to compaction. But in under-compacted sequences it will increase lower penetration rate is often observed in overlying pressure seal sequence. In case of abnormally pressured formations below, when other variables remain same the drilling rate is directly proportional to the quantum of abnormal formation pressures. Further, balancing of formation pressure the careful increase of mud weight is needed. If not, formation fludies start entering into mud. Initially, it is called influx, if could not carefully balance the mud weight it may be a "kick". If it is severe it can be blow out. So for safe drilling well control at right time is very essential.

"Kick"-Kill Objective

After a kick has been stopped by well closure, it should be circulated to the surface at constant bottom-hole pressure to avoid both further influx of formation fluids and excessive borehole pressures. Also, drilling fluid density should be increased to re-establish primary well control. A drilling fluid of required density may be pumped while circulating out the kick (Wait and Weight Method), or the kick may be pumped out and then drilling fluid of required density circulated (Drillers Method).

Abandoning a Well

In case, a well is not at all hydrocarbon bearing, or producer of non-commercial oil and gas or most of oil or gas which can be produced have been completely done such wells need to be abandoned. As per the local government rules no well can be left as it is, keeping in mind the safety of area, people and environment.

To plug and abandon a well, the drilling rig pumps several cement plugs through the drill pipe. The cement plugs are used to isolate and seal unprofitable hydrocarbon zones from non-hydrocarbon-bearing zones and to seal freshwater zones from saltwater-bearing zones. The intervals between cement plugs are left full of drilling mud. At this time, it might be possible to cut off and recover some of the intermediate casing string for use in other wells. The surface casing string is always left in place and sealed at the bottom and top by either cement plugs or a combination of mechanical and cement plugs. The surface casing will be cut off below the ground level or mud line and a cap placed on the stub.

The first step when starting a well abandonment operation is removing existing tools. This can be done using an existing drilling or conventional work over rig with the capacity to pull out of hole all downhole equipment previously used by the operator, such as production tubing, downhole pumps and packers. After the removal operation, the wellbore needs to be cleaned from fill, scale and other debris.

In the offshore, all strings of casing are cut below the seabed, and that all structures above this point should be recovered. Also any debris lying on the seabed, within permissiable radius of the drilling location should be removed.

Geo-scientific Data Collection while Drilling

Well-site Geological Work

The well-site geologist should be completely familiar with all aspects of the drilling prognosis also called as Geo-Technical Order (GTO). Particular attention should be paid to any sections which may require geological decisions. These include: Determination of Primary and Secondary Objectives. Determination of Casing Points, Idea of Formation-pressures with Intervals of Over-pressured sections if any, indication of Lost Circulation Zones, selection of type Logging Run Intervals required for better evaluation of the well data, indicate possible productive zones and intervals where conventional cores need to be taken.

It is also need to be known which interval the cutting sample required to be collected and based on nature of the well from which depth drilled formations need to be described based on cutting samples collected and described. Most commonly, a log-type format is used with cuttings descriptions and drilling parameters displayed graphically on a litholog/striplog.

When washing cuttings samples for lithologic examination, the amount of washing should be the minimum necessary to remove the drilling fluid, so that all components are easily recognized, but not enough so that soft sample materials are washed away.

Mud-Logging Unit: In the unit desired equipment is available to know or calculate parametrs like, Cutting, Geochemistry, Lithology, Correlation, Density, Calcimetry, Hydrocarbons, Shale Factor Hole Stability, Bit Condition Hydrocarbons Total gas, Chromatograph, Gas Ratios, Connection gases, Trip gases, Oil shows Gases CO_2, H_2S Engineering Dxc, Torque, Drill Rate, Formation Pressures

MWD/FE-MWD data: Directional Borehole Trajectory (MWD), Dogleg Severity Gamma Ray Lithology Determination, Shale Content, Resistivity Correlation, Hydrocarbon Evaluation, Pressure Indication, Sw Estimations Density Lithology, Correlation, Pressure Indication, Gas/Oil Contact.

Oil Shows: Evaluation of oil in the drill cuttings (and drilling fluid) begins with inspection of those samples under the microscope and inspection under the ultraviolet-light. Tests and visual inspection should be performed on the mud, unwashed and washed bulk cuttings, as well as individual grains.

Gas Shows

Background Gas: This is the gas recorded while drilling through a consistent lithology. Often it will remain constant, however, in over-pressured formations this value may show considerable variation. This is the gas baseline which is plotted on the strip-log and mud log.

Real Gas Show: This is a gas reading that varies in magnitude or composition from the established background. It is an observed response on the gas detector and requires interpretation as to the cause. Not all gas peaks are from drilled formation, some may occur as post-drilling peaks.

Connection Gases: Gas peaks produced by a combination of near-balance/ under-balanced drilling and the removal of the ECD by stopping the pumps to make a connection. They are often an early indicator of drilling overpressured formations.

Trip Gases: Gas peaks recorded after circulation has been stopped for a considerable time for either a bit trip or a wiper trip. As with connection gases, substantial trip gases can indicate a near balance between the mud hydrostatic pressure and the formation pressure.

Examining Cuttings: - The rock type can be identified from the cuttings. A rough idea of the porosity of a rock can be determined by viewing cuttings under a microscope. If a rock contains oil, trace amounts of oil will coat the cuttings even after they have been circulated in drilling fluid and brought to the surface.

Oil is a polarizing compound. It will have a fluorescent shine when viewed under ultraviolet light. The oil stain on cuttings can be confirmed by flushing the oil off the cuttings with a solvent. The streaming solvent will also show fluorescence under the UV-light. In this way, an oil stain can be differentiated from other rock mineral that might also fluorescence.

Coring-Analysis

Taking the rock sample as it is without powdering it is called coring. It is very necessary in case of assessing reservoir properties without ambiguity. In case of exploratory wells, in case good oil and gas indications are observed normally conventional core will be taken by lowering core-bit. It gives 8 metre of 4 ½″ dia actual sub-surface rock. Sometimes if due to bad-hole conditions core bit not advisable to lower, then electro-logging is planned and Side Wall Cores (SWC) at suitable depth will be taken.

Side-Wall Core (SWC)

These SWC samples help to identify a rock's type and also can be used to determine reservoir fluid contents. In addition, SWCs provide information when not evident on the logs. The percussion-type guns are accurately positioned by using an SP or gamma ray measurement. A surface-controlled power charge fires a hollow-core barrel into the formation at each predetermined sample depth. Each core barrel uses retaining wires attached to the gun for retrieval of core and sample. The core barrels are fired one shot at a time, and the samples are retrieved once the core gun is brought back to surface. Core-barrel designs are available for various types of formation consolidation.

Core Preservation is a method when core is recovered, it must be laid out, measured, described, and prepared forlabortory studies. For each core, preservation and packaging should be designed for the special core analyses based on type of rock like, composition, degree of consolidation, and other such factors.

Measurement While Drilling (MWD) Log

Gamma rays in the formation are emitted mainly by radioactive isotopes of Potassium, Thorium and Uranium. These elements occur primarily in shales, and so the GR log is a good shale indicator. Since any change in lithology must be known as quickly as possible the GR sensor should be placed as near the bit as possible, below the directional sensors. Running an MWD GR log has the added problems of rigging up a depth tracking system. Since the logging speed is different than that when the same log is recorded after the drilling there may be some differences in comparision of same logs of both episodes and also the resolution of the two logs will be different.

DST

The drill-stem test is a method of sampling the fluid in a subsurface formation during the course of drilling operations. Most modern drill-stem tests, however, yield three types of hydrologic data, a sample of the formation fluid, the undisturbed formation pressure, and a co-efficient of permeability for the stratigraphic interval tested (Figure 6.12).

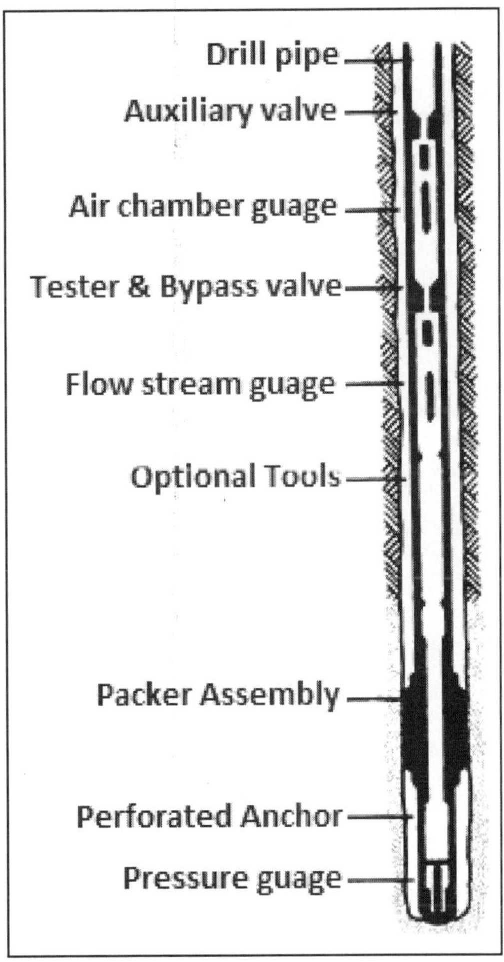

Figure 6.12. Drill Stem Testing (DST) Tool.

During the drill-stem test the stratigraphic interval of interest is isolated in the hole by the use of packers attached to the drill string and is allowed to yield fluid into the drilling pipe under the influence of the formation head. In the usual relatively shallow test the drill pipe initially is completely empty and open to atmospheric pressure. By opening the tester valve in the test string, the operator allows the formation to yield fluid into the drill column for a relatively short period of time.

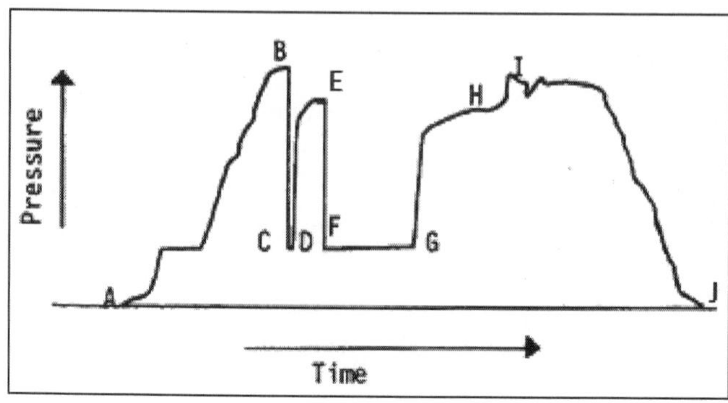

Figure 6.13. DST Chart.

The period usually ranges from 30 minutes to 2 hours. Following the production period the tester value is closed, shutting in the formation and causing the formation pressure to recover. Pressures are recorded throughout the test by a recording gauge contained within the drill string near the bottom. The test can be subdivided into five phases, each of which is represented by a segment of the pressure graph (Figure 6.13):

1. The initial phase of the test, represented by segment AB on the lower pressure curve, involves lowering the drill string into place in the well. The pressure gauge records the increasing pressure of the drilling fluid as the pipe is lowered into the hole. The very "noisy" appearance of the pressure curve is due to the stepwise procedure of adding joints of drill pipe used in making up the drill string.
2. Once the pipe is in the desired position, the packer is expanded, or "set," isolating the interval of interest, as well as the gauge, from the column of mud above the packer. The tester valve is momentarily opened (point B on the pressure graph), allowing the formation to produce for a few minutes reducing the pressure in the isolated zone to very nearly atmospheric (point C), and then closed. The pressure builds up again (segment CD of the curve) to some point (D) which asymptotically approaches the undisturbed formation pressure. The initial shut-in period commonly ranges from 15 to 45 minutes.
3. After the initial production and shut-in period the tester valve is again opened and the formation fluid flows into the drill pipe. Within the isolated interval the pressure reduces to essentially atmospheric (point E on the pressure graph), but as the column of produced fluid builds up in

the drill pipe, the pressure increases (segment EF of the pressure graph). Commonly this production period lasts from 30 minutes to 2 hours.

4. After the production period the tester valve is again closed (point F), isolating the formation and allowing the pressure to recover. During this final shut-in (recovery) period the pressure gauge records the pressure build-up in the isolated section of well (segment FG on the pressure curve). The final shut-in time is often made equal to the time of production.

5 After the final shut-in period the packer is unseated (point G), and the pressure returns to that produced by the column of drilling mud in the well (point H), the so-called hydrostatic mud pressure. The drill pipe is then pulled out of the hole. The pressure-recording gauge records the decrease in pressure as the pipe is removed from the hole (segment IJ of the pressure curve).

The column of produced formation fluid remains in the drill pipe until the pipe reaches the surface. This affords a sample of fluid produced from the formation tested. From the measurement of the length of drill pipe filled with fluid the volume produced can be calculated.

MDT (Modular Formation Dynamic Tester)

In the earlier wire-line formation testers there was one objective specifically to collect reservoir fluid samples, but it could only collect one sample per trip in the well. This tool was replaced first by the Formation Interval Tester (FIT) and then by the Repeat Formation Tester (RFT) tool. The development of the Modular Dynamics Formation Tester (MDT) tool offers significant improvements in pressure measurement. The collection of condensates and critical fluids at the surface. In this, one of the down-hole sampling operations can be made out quickly and efficiently with very small pressure drawdowns. This MDT tool was used to determine lateral hydraulic continuity and can run in a horizontal well system. One of the most important improvements offered by the new tool is the ability to control a multitude of tool functions from the surface.

Conventional tools can collect up to two samples with each run into the borehole. The quality of these samples is often impaired by the presence of mud filtrate associated with invasion during drilling. In the MDT tool up to 12 sample chamber modules can be connected to the tool. However, weight limitations determined by well conditions and cable strength restrict/decide the number. The multi-sample module contains each with a 450cc capacity, and so can provide additional fluid samples during a single trip.

The surface unit can use the resistivity cell on the probe module, or the Optical Fluid Analysis module, to identity the nature of fluids like, mud filtrate, oil, water and gas before taking samples. The resistivity cell often has difficulties in identifying fluids when a well has been drilled in oil-based muds and may, in some cases, be unable to differentiate oil from gas. The optical fluid analyzer has been designed to cope in these circumstances, identifying mud filtrate, oil, water and gas quickly and accurately. Mud filtrate if so in the sampler can be displaced by the pump out module, a miniature down-hole pump which pushes unwanted fluids into the bore-hole before sampling begins.

The flow control module takes litre of formation fluid into a chamber, displacing a piston in the process. During the test, flow rates are monitored. Acquired flow rate and pressure data from the observation probes can be analysed to yield formation properties.

Mini Drill Stem Tests: Defining pressure points for fractured, vuggy, very tight or highly-laminated formations has often presented problems for wire-line formation testers. The dual packer module available with the MDT tool provides a much bigger flow area. The radius of investigation may reach tens of feet in a test completed within a few minutes. The Dual Packer Module helps to overcome the testing problems encountered in highly fractured reservoirs.

High Pressure and High Temperature Wells

The Pore pressure, also called as Formation Pressure is defined as the pressure acting on the fluids in the pore spaces of the rock. Depending upon the magnitude of the pore pressure, it can be classified into three types, normal, abnormal and subnormal, based on whether the pore pressure is equal to, greater than or less than the hydrostatic pressure at the given depth.

Due to geological reasons, the formation pressure may be abnormal, but in the case of producing reservoirs in case reservoir pressure in not maintained properly if the pore space is filled with air or gas, it may be sub-normal.

As for high pressure, that definition is met when the maximum anticipated pore pressure of the porous formation to be drilled exceeds a hydrostatic gradient of 0.8 psi/ft, or the well requiring pressure control equipment has a rated working pressure in excess of 10,000 psi. In other words the indication of high pressure can also be expressed in relation to equipment rating which may help to caution the work force at such drilling location.

The high temperature (HT) is where the undisturbed bottom hole temperature is more than 300°F or 150°C.

Technical concepts in HPHT drilling are significant to design the tools for the key planning processes. Formation pressure prediction, fracture pressure determination, casing setting depth, drilling fluid's rheological properties, hydraulics, bit selection and cementing program should be all highlighted more carefully when drilling through the HPHTwells.

To maintain the safety during drilling, the fluid pressure exerted by the cement, drilling fluids must not exceed the fracture pressure of the formation at any depth. The fracture pressure is the upper safe pressure limitation of the formation before it may be fractured. The hydrostatic pressures of the fluids in the wellbore, along with the friction pressures (ECD) created by the fluids' movement, cannot exceed the fracture pressure. If the formation is fractured it can no longer be controlled easily, and lost circulation happens.

The fracture pressure of a rock is dependent on its compaction. In case of less compacted formations normally at shallow depths the fracture pressure is lesser and it increases with depth and related compaction. It is very essential parameter while drilling, though it can be predictable based on rock strength, drillers believe direct measurement. As mentioned earlier, after lowering each casing and drilling few meters after that,"leak-off test" is being conducted in order to know the fracture pressure at that depth. It will help in calculating maximum mud-weight up to which next phase of drilling can be done safely.

Abnormal pore pressure is developed as a result of a combination of geological, and tectonic processes. If sedimentation rate in a basin faster and also associated with syn-sedimentary subsidence then sediments will not have required time to get compacted by the over burden load of sediments, due to which the sedimentary sequence is less compacted in the level below. Because of this, some quantity of depositional fluids may be locked with in the pores which exert pressure over and above the overburden pressures. For this, an impervious bed above will act as barrier which is not allowing the transmission of the rock pressure to upper formations. This situation will generate abnormal formation pressures within such formations. In the sedements deposited off Godavari Delta in the Krishna Godavari basin indicated much abnormal pressures in different formations due to under-compaction related to the sedimentation (?).

In the case of post-depositional tectonic changes some deep beds may be thrusted over the normally compacted younger beds. Further such over thrusted formations may also have sealed within an impervious bed which are not allowing the transmission of fluid pressured within the adjacent formations. So the depth of such bottom beds become shallower and if formation pressure are calculated based on present overburden pressure, they are treated as tectonically abnormal pressured formations. In the north-eastern fold belt of Assam-Arakan region abnormal pressured related to thrust belts are observed.

Similarly in case of deep dipping beds, the actual formation pressure in the same bed which is located shallower levels will be of abnormal based on relatively lesser over burden available.

The piezometric fluid level in the rugged topographic area may also cause abnormal pressure because hydrostatic pressure is even fluid connective level and topographic level is uneven over burden rock pressure.

Determination of pore pressure can be done in two ways, pre-drilling and post drilling. Pre-drilling techniques consists based on the quality of available seismic data and any drilling mud-parameter information available in the area.

Methods used during drilling include, accurate monitoring of important drilling/mud parameters like rate of penetration, weight on bit, drag, torque, flow rate, pit volume, mud weight in and out, flow line temperature, conductivity/resisitivity of mud using offline/online geologging units. Plotting and interpretation of 'd' exponent and Sigma log (Figure 6.14) and timely measuring and plotting of shale density will help to know the zone of transition.

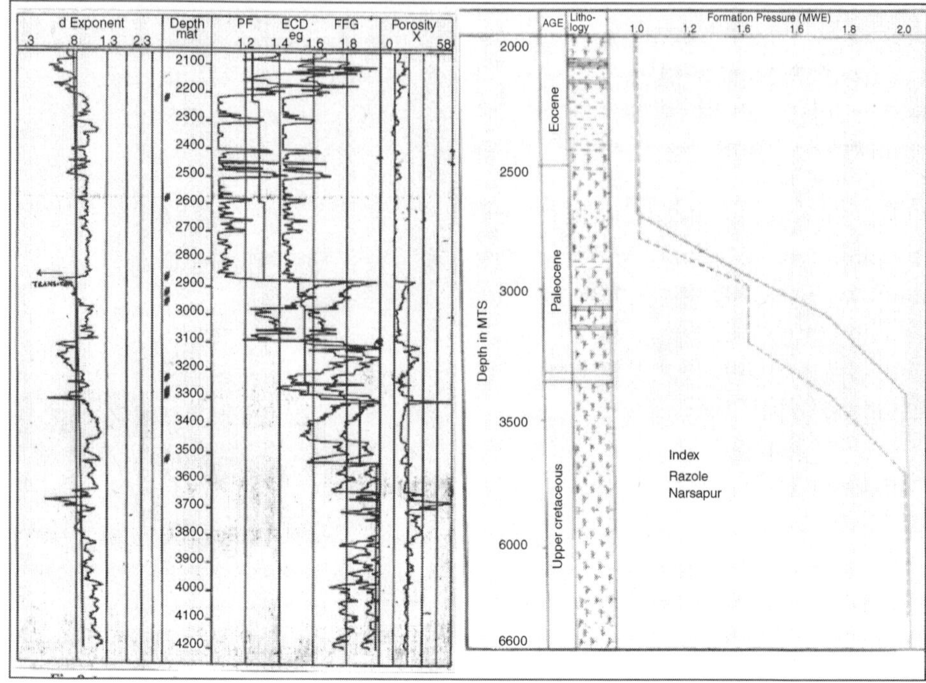

Figure 6.14. d-exponent Plot-well Drilled in Krishna-Godavari Basin.

The post drilling techniques can be classified into three categories. Firstly, the mud logging methods which includes measurements of drilling parameters and evaluation of drill cuttings gas levels at surface. Secondly, Measurement While Drilling, logging while drilling and wire line logging methods and lastly, the direct methods which includes DST, production tests and RFT.

'd' Exponent is the normalized penetration rate computed during drilling and such techniques were used in Krishna-Godavari basin and is found to be more accurate in predicting and demarcating the pressure regimes. Hence, the corrected 'd' exponent (dcs) plotted in various wells drilled in Krishna-Godavari basin has been analysed and interpreted to classify into different pressure regimes. In normal compacted sediments having normal formation pressures, the 'dcs' log motif follows a general compaction trend. In over pressured formations, the 'dcs' values are reduced and the curve shifts from normal trend towards left. The separation between the normal compaction trends to the over-pressured trend is related to the formation pressures gradient.

Indian Scenario

Krishna Godavari Basin

The exploratory well in the basin called Narasapur#1 was started in 1977. Though well was planned to drill upto 4,500 meters, but due to unexpected abnormal formation pressures the well could not be drilled through the target. The maximum mud weight used was 1.98 specific gravity during initial phase most of the wells drilled beyond 4000 meters in on-land Godavari delta area are found to be having abnormal pressures. Some times to control kick heavy mud of 2.08 mud specific gravity was used.

When *Ravva* area in shallow waters off Godavari was taken up for drilling at depth below 2500 m in Miocene section is found to be having high formation pressures (Gudemane.,S M, 1989). The genesis of abnormal pressures was analysed and found to be due to under-compaction of sediments related to fast subsidence (Rao, G N., 1994). So the level of abnormal zone is found to be shallowing up in depth in the basin (refer KG Basin in Chapter-10). While drilling at Amalapuram #3 well, high temperature regime was encountered below 4000 metres reaching more than 150°C.

The *Nandigama South* well #2 with a target depth of 4,600m was drilled vertically to a depth of 4,359m., with maximum 1.95 specific grvity of mud and drilling was terminated due to non-availability of kick-tolerance. Pore pressure analysis from "d- exponent" plot show normal pressure from surface to 3,930m, transitional pressure regime from 3,930m to 4150m and an abnormal pressure regime from 4,150m below. In *Kottalanka #2* well was drilled with mud of 1.98 specific gravity and kill mud weight of 2.02 to depth of 5,462 metres.

Assam-Arakan Basin

In the fold belt zone of north eastern India thrusting is the main tectonic regime due to which older section were thrusted up which created abnormal formation pressures in the section due to sealed nature of section initially. Silcher well was drilled in Cachar fold belt in Assam during 1976 and Rokhia well drilled in Tripura fold belt in 1982 were found to be of high pressures up to double the hydrostatic pressure. Mud weight of 2.10 was used to control the activity. (Refer Basin AAFB-Chapter-10).

West Bengal Basin

The well Bodra #2 was drilled which was located in the deep basinal area and within Miocene section. From 3,600 meters on wards the formations are more than hydrostatic. Mud weight of more than 1.80 was used to control the kick activity. Since the formation is Miocene and less compacted and so its fracture pressure also found to be very close to desired mud pressure.

The role of hydraulic factors on the ROP while drilling with oil base mud in the exploratory well Bodra # 2 in the depth interval 3,760-4,270 m in which the ROP at every 5 meters was compared with the calculated pressure drop across the bit nozzles, nozzle velocity and bit hydraulic horse-power. The weight on bit was through out was 8 to 10 Tons and other mechanical parameters were essentially held constant.

Indian Offshore Drilling-Bombay High

The first exploratory well in Bombay High was drilled by 'Sagar Samrat' in early 1974 and oil and gas was struck in Miocene Limestone. Then few exploratory-cum-assessment wells have been drilled to delineate the structure and assess the potentiality of the field.

Some of these exploratory-cum assessment wells were vertical wells and were completed on mud lines suspension. Other wells were drilled by a floater Rigs and were completed subsea. All these wells were plugged back after testing.

To start with, 'Sagar Samrat' drilled two wells on one platform and three wells on another one located in the northern part of the field. All these wells were completed on mud line suspension and were disconnected, to be subsequently tied back and completed after the erection of the platforms on these locations. After few platforms were erected old ones connected and more wells have been directly drilled from the platform (Mittal S.C, 1978).

Right from exploratory stage the casing policy followed for the drilling of the wells was 30" X 20" X 13 3/8" X 9 5/8". 30" drive pipe is driven to a depth of 15/160 Mts. 20" casing is lowered to a depth of 250 meters. The 13 3/8" casing is lowered to a depth of about 1000 mts. The 9-5/8" production casing was lowered to target depth and coveres all hydrocarbon bearing layers.

The wells are deviated with 17 ½" bit about 20/25 metres below the shoe of 20" casing. The kicking off is done with a down hole motor (Dyna Drill or Turbine) and kick-sub. After attaining the required azimuth, angle building-up assembly is lowered and the desired maximum angle is attained. There after the angle is maintained by stiff assembly till target depth so as to achieve the planned horizontal drift.

The formations in Bombay High field being very soft have posed a problem in maintaining the planned direction and the angle, since the bit has a tendency to move towards right. Also the angle keeps on falling or increasing which requires a very strict control in holding the angle. Various bottom hole assemblies have been tried to hold the angle but still frequent trips have become necessary to maintain the required direction and angle. In some cases Dyna Drill had to be re-run to correct the azimuth. To meet the situation casing policy was changed and 30" conductor pipe has been eliminated and in its place 20" drive pipe was driven through the sea-bed. The drilling through 20" drive pipe was started with 17.1/2" bit and the well was deviated in the planned direction. Thereafter the main pay could be drilled through by 8 ½" bit and 7" production liner. This casing policy helped in controlling mud loss and improved the cementation of the casing in the pay zone.

From the reservoir point of view a spacing of 2 km is required between two wells and it will be preferred to have a spacing of 2.5 km. This being the case, for the 2 Km spacing a horizontal drift of 1.4 km is required if the wells are considered to be at the four corners of a square. Because of desired deviation is more from a single platform not more than four wells can be drilled from one platform (Sarin.,LN, 1995).

Indian Scenerio

As of now the deepest well drilled in India is in Jwalamukhi to 6,750m in 1994, in Krishna Godavari Basin to 5,462 in Kottalanka in 2001 and in West Bengal Jaguli#1 to 5655m.

In the deep waters of 3,600 meters, gas producing field was identified in Krishna-Godavari offshore called UD field.

Chapter 7

Well-Logging in Oil and Gas Exploration

Introduction

Well logs are used in geological studies, they present the results of instrumental records, but these measurements are directly against the desired rock sample in the bore-hole. So the term borehole geophysics is used for such measurements (Figure 7.1).

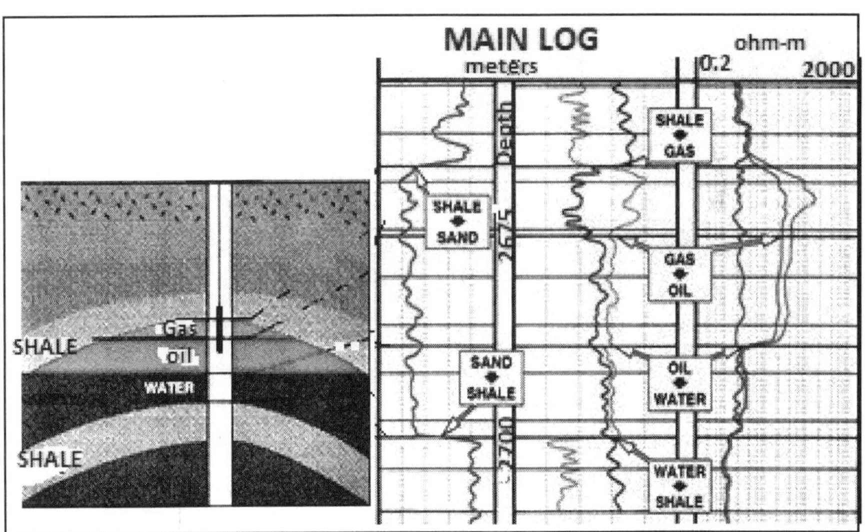

Figure 7.1. Scheme of Logging Showing Subsurface Zone and Log.

It is not possible for to physically view inside well bore the hydrocarbon bearing formations. So facilitate to know the fluid content in the formations geophysics has developing many methods to understand the subsurface with technology developments. All logs are based on the principle of measuring the contrast of measurable character of rocks. It means if there is any contrast in physical properties of subsurface formations it can differentiate the fluid content within it.

Electrical log is the record of rock properties measured using electrical recording systems to know the nature of rock and its fluids. These are recorded by placing the measuring device against the well wall in a drilled well. Electrical methods makes use of a large variety of liquids, each based on some different electrical property or characteristic of the rock. The resistivity method is to record information on rock bodies having anomalous electric conductivity (Bateman, R. M., 1985). The induced-polarization method, employed in the exploration will give diagnostic readings where ionic exchanges take place on the grains. Each type of log needs certain speed of movement for sensor reaction and recording of the parameter. If it is more speed erroneous graph will be the possibility. In general all logs are recorded while pulling out from the bottom, except temperature log which is recorded while lowering the tool.

The log, a single graph of the electrical resistivity or any physical parameter of the rock formations drilled, was recorded by the down-hole measuring instrument, called "sonde" which can be stopped at periodic intervals in the well where measurements were made. The type of sensor depends upon the nature of the measurement. Resistivity sensors use electrodes or coils; acoustic sensors use transducers; radioactivity sensors use detectors sensitive to radioactivity.

The sonde housing may be constructed of steel and/or fiberglass. The other component of the down-hole tool is the cartridge. The cartridge contains the electronics that power the sensors, process the resulting measurement signals, and transmit the signals up the cable to the truck. The cartridge may be a separate component screwed to the sonde to form the total tool, or it may be combined with the sensors into a single tool. That depends upon how much space the sensors and electronics require and the sensor requirements. The cartridge housing is usually made of steel. In recent adavancement in technology the sondes and cartridges of several tools can be connected to form one tool and many measurements and logs can be recorded in one run from the borehole.

Among rock properties being logged with such instruments are, electrical resistivity, self-potential, gamma-ray generation (both natural and in response to neutron bombardment), density, magnetic susceptibility, and acoustic velocity.

Once an exploratory well was drilled, one has to determine whether hydrocarbons exist in the area, and if it exist, would it be available to produce economically? So the well logs are being used to know various reservoir parameters like, Depth, Thickness, Lithology, Porosity, areal extent, Fluid content, Dip of the individual beds.

Logs also help to know the quality of cementation, possible leakage, gas and water coning and channeling. The logs also help in geological correlation and fracture identification. Logging can be done in different phases of drilling a well:

- ☆ Logging while drilling (LWD): Logging a well while drilling.

- ✰ Measuring while drilling (MWD): Measuring of parameters during drilling
- ✰ Open hole logging: Logging in a well after drilling but before casing.
- ✰ Cased hole logging: Logging in a well after casing.
- ✰ Production logging: Logging well during production.

But based on process, different types of logging tools are being used. With advancement of technology the tools, quality, precision and nature of parameters are being developed. In such improvements, in Measurement While Drilling (MWD) logging, the sensors are integrated into the drill string and the measurements are made while the well is being drilled.

History of Electro-Log Technology

The first electrical log was recorded in 1927 in a well. This log, a single graph of the electrical resistivity of the rock formations cut was recorded by the "station" method. The down hole measurement instrument, the sonde was stopped at periodic intervals in the borehole, measurements were made, and the calculated resistivity was hand plotted on a graph. In 1929, electrical resistivity logging was introduced on commercial basis. In 1931 the spontaneous potential (SP) measurement was included with the resistivity curve on the electrical log. In the same year, the Schlumberger brothers, found a method of continuous recording and the first pen recorder was developed.

Introduction of various logs as mentioned year wise were: in 1938 Gamma ray log and first Neutron log, in 1947 Resistivity dip meter, Induction log, in 1949 Latero log, in 1957 Sonic and Densitylogs, in 1965, commercial digital recording of log data, in 1988, LWD tool, in 1991, Formation Micro Imager tool (FMI) and in 2013, Microscopic High definition Imaging While Drilling.

Well Logging Technology – Indian Scenario – Logging Services

Oil industry, in India, has been progressing in a gradual manner right from 'Digboi' days. Only Schlumberger had the monopoly of providing logging services in India. This company in logging, has a business policy of selling services only and not the logging equipment.

There are few logging companies, who sell their equipment. ONGC and OIL purchased logging equipment and established their own in-house logging facilities including several wireline logging units and MWD units. Training was obtained from the suppliers and the operations have been going on with in-house efforts. Due to fast changing technology coupled with procedure followed for procurement of equipment sometims tools becomes obsolete. Thus, there is a distinct technology lag in India so far as the logging equipment is concerned.

One of the ways to combat technological set-back may be, to obtain the equipment on lease instead of purchasing it. In case of leasing, operators can utilize their own manpower to operate. At the end of the lease period, the equipment may be returned and state-of-art equipment may be obtained again. During early days of logging, geoscientists use to assume all high resistivity was an indicator of presence of hydrocarbons. After wards, it was proved that current flow is also controlled by porosity.

Types of Logs-Open Hole

The **spontaneous potential** (SP) method is used to detect the presence of certain minerals which react with electrolytes in the earth in such a way as to generate electrochemical potentials (Figure 7.2). A rock body oxidized to on its top than along its bottom will give rise to such potentials, which are detectable with electrodes at the surface. The SP curve is a recording versus depth of the difference between the electrical potential of a movable electrode in the borehole and the electrical potential of a fixed surface electrode.

Againest shales, the SP curve usually defines a more or- less straight line on the log, called the shale-base line. Opposite permeable formations, the curve shows variations from the shale base line; in thick beds, these deflections tend to reach an essentially constant deflection defining a sand line. The deflection may be either to the left (negative) or to the right (positive), depending primarily on the relative salinities-of the formation water and of the mud filtrate.

If the formation water salinity is greater than the mud filtrate salinity, the deflection is to the left. For the reversed salinity contrast, the deflection is to the right. The position of the shale baseline on the log has no useful meaning for interpretation purposes. The SP sensitivity scale is chosen and the shale baseline position is set by the recorder such that, the curve deflections remain in the SP track.

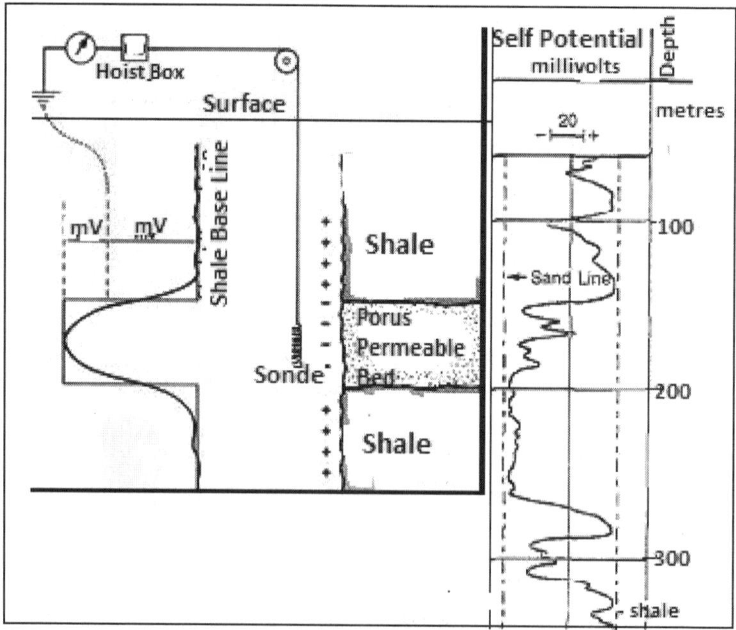

Figure 7.2. Self Potential Log.

The SP tool consists of an electrode (lead) mounted on an electrically isolated bridle on the down hole tool. A 1.5 volt battery is included in the circuit to give a bucking current to bring the SP to the scale. The tool's galvanometer records only change in potential.

The SP log is measured in millivolts (mV). An SP curve cannot be recorded in holes filled with non-conductive muds because such muds do not provide electrical continuity between the SP electrode and the formation. If the resistivity of the mud filtrate and formation water are about equal, the SP deflections will be small and the curve will be rather featureless. The deflections on the SP curve result from electric currents flowing in the mud in the borehole.

The SP currents are caused by electromotive forces in the formation, which are of electrochemical and/or electro-kinetic origin. The slope of the SP curve at any level is proportional to the intensity of the SP currents in the borehole mud at deflection opposite a permeable bed depend on several factors. The factors, which affect the distribution of the SP current lines and the potential drops taking place in each of the media through which the SP current flows.

The shale baseline (from which the SP deflections are measured) is usually fairly well defined on the SP log. However, in some wells, baseline shifts are observed. A baseline shift occurs whenever formation waters of different salinities are separated by a shale bed that is not a perfect cationic membrane. Large shifts make definition of the shale baseline and determination of the SSP value quite difficult.

Factors influencing amplitude and shape of SP curve:

Bed thickness: In thick and permeable beds with high Rmf (resitivity of mud filtrate)/Rw (water) contrasts, the SP curve will display a deflection with a large slope where the amplitude will remain steady at an SSP (static spontaneous potential) value. In thin beds the full SP deflection is never developed. If Rmf more than Rw, the amplitude of the SP curve will be large and negative and if Rmf is equal to Rw, there will be no SP deflection. If Rmf is less than Rw, the amplitude of the SP curve will be positive.

Bed resistivity: As Rt increases, the amplitude of the SP curve decreases and boundary deflection become more rounded and spread marking it difficult to determine bed boundaries.

Permeability: Weighted drilling mud creates a pressure differential between the borehole and reservoir rock. High permeability zones and differential pressure will cause a mud-cake to build rapidly effectively sealing the zone. Very little invasion will be present. If the bed is thick and free of shale the measured SP will be equal to SSP. Low permeability zone and pressure differential conditions will also build mud-cake, but at a much slower rate. Invasion can become quite deep in low permeability zones. As hole size increases, the SP amplitude will decrease and boundary slope will decrease. A look at the SP over a certain series of beds in sand-shale sequences will show water salinity changes. The SP is used to calculate formation water resistivity using the relationship between resistivity and ionic activity.

It is considered the volume of shale Vsh in a water wet, shaly sandstone can be simply calculated using the SP as:

Volume of shale (Vsh) = 1-PSP/SSP,

where,

PSP=pseudo-static spontaneous potential= the SP read in the water-bearing shaly sand zone, SSP=static spontaneous potential = maximum SP value in a clean sand zone.

This simply assumes that the SP deflection between the shale base line (100 per cent shale) and the static SP in clean sand (0 per cent shale) is proportional to the shale volume. All deflections on the SP indicate a permeable bed. But, the amount of deflection does not indicate the amount of permeability a good permeable bed will give the same value as a normal permeable bed.

The SP is one of the first logs to permit correlation in sand-shale sequences because of its typical log shapes. This shape is related to shale abundance. As shaliness is related to grain size, the SP is a good facies indicator. The SP log is useful for correlation should made between sand with similar salinity values.

The hydrocarbon detection is facilitated by SP behavior against oil bearing part. SP log has four known components namely Electro-filtration, Electro-chemical, Electrode and Redox potential. Since hydrocarbons are related to reducing environment, Redox mode is ideal for hydrocarbon detection. Spontaneous potential (SP) log is well known to respond to oil water contact (OWC) and in certain cases to GOC also. This phenomenon is due to the Redox component of SP, Conventional SP log using stainless steel electrodes shows OWC but subdued.

In ***Lakwa*** field of Upper Assam, to test Redox SP mode, gold plated component was used as gold electrode because geometry of electrode is not critical. It was observed that Redox mode was better in picking OWC than conventional SP.

Caliper Log

It is a tool for measuring the diameter and hence the shape of a borehole. It uses a tool which has extendable arms. The arms can move in and out as the tool is withdrawn from the borehole, and the movement is converted into an electrical signal by a potentiometer. The arms are linked to the cursor of a variable resistance. Lateral movement of the arms is translated into movement of the cursor along the resistance and hence variations in electrical output which translated into diameter variations after a simple calibration. Four arm caliper gives more information than simple caliper tool. Data from a four arm caliper enables the shape of a hole can be seen to be on gauge and round or oval and wash out.

The uses of caliper log are for lithological assessment, Indicator of good permeability and porosity zones due to development of mud-cake in association with gamma ray log. Thickness of mud cake formed and volume of the borehole can also be calculated using caliper log (Figure 7.3).

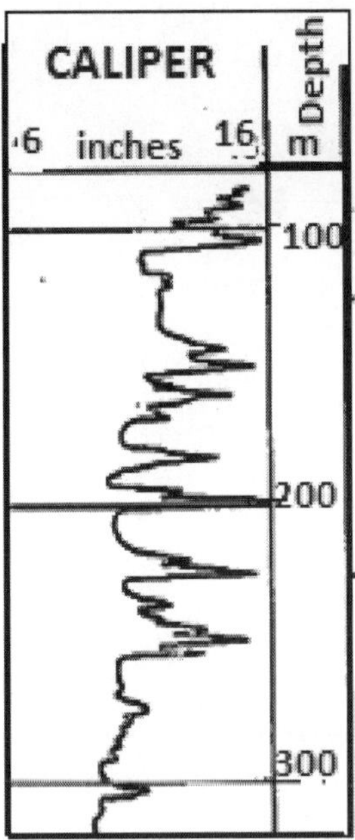

Figure 7.3. Caliper Log.

Gamma Ray Log

The Gamma Ray (GR) and *Neutron* tools represent the use of radioactive properties in well logging.They are able to log formations through steel casing, as well as in air- or gas-filled holes or in oilbased muds. The GR log indicates the natural radioactivity of the formations. Nearly all rocks exhibit some natural radioactivity and the amount depends on the concentration of potassium, thorium, and uranium.

Gamma rays are bursts of high-energy electromagnetic waves that are emitted spontaneously by some radioactive elements. Nearly all the gamma radiation encountered in the earth is emitted by the radioactive potassium and by the radioactive elements of the uranium and thorium. Each of these elements emits gamma rays; the number and energies of which are distinctive of each element.

There are two types of GR logs. One, the standard GR log, measures only the total radioactivity. The NGS (natural gamma ray spectrometry) log, measures the total radioactivity and the concentrations of potassium, thorium, and uranium producing the radioactivity separately.

In sedimentary formations the log normally reflects the shale content of the formations. This is because the radioactive elements tend to concentrate in clays and shales. Clean formations usually have a very low level of radioactivity, unless radioactive contaminant such as volcanic ash or granite wash is present or the formation waters contain dissolved radioactive salts. The GR log can be recorded in cased wells, which makes it very useful as a correlation curve in completion and work-over operations. It is frequently used to complement the SP log and as a substitute for the SP curve in wells drilled with salt mud, air, or oil-based muds.

One of the characteristics of gamma ray is that when they pass through any material their energy is progressively absorbed. This effect is known as Compton scattering. Natural gamma rays are gradually absorbed and their energies reduced as they pass through the formation. The rate of absorption depends on the density of the formation; more energy loss in high dense formation. In borehole logging, Compton scattering is due to the collision between gamma ray and electrons in the formation.

Method: The simple gamma ray tool is a sensitive gamma ray detector consists of scintillation counter and a photo multiplier. The scintillation counter is typically a sodium chloride crystal. When gamma rays pass through the crystal, they cause flash. The photon energy (light) strikes a photo-sensitive surface or cathode causing electron emission. The electrons so produced are accelerated to anode which impact, release additional electrons that are directed to another anode. Those anodes are supplied with higher voltages by an internal or external resistor chain. The high flow of electrons to be measured and recorded as an indication of the incident gamma ray radiation. The electronic signal results a voltage signal which is directly proportional to the photon energy.

The basic construction of a spectral gamma ray tool is a large scintillation crystal is used to identify the gamma radiations in several, predefined energy bins or windows. These windows are designed to separate the distinctive energy peaks of the individual radioactive elements. The relative contributions of the three radioisotopes are therefore measured by the tool. The log results from the spectral tool are the elemental abundances of thorium, uranium and potassium (Figure 7.4).

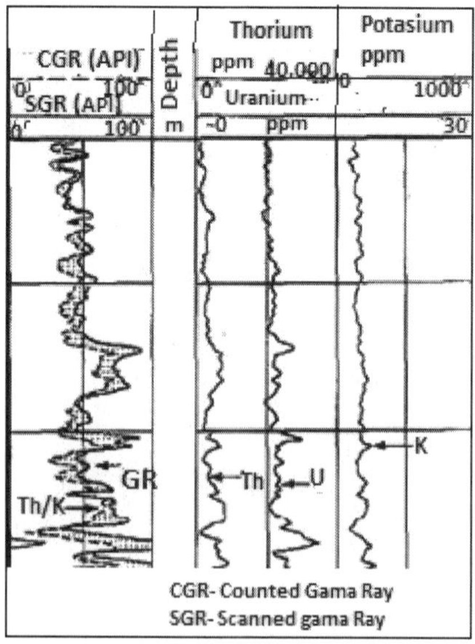

Figure 7.4. Spectral Gamma Ray Log.

The unit for radioactive logging is the API (American Petroleum institute) unit. The simple gamma ray log is usually recorded in left track along with the caliper. Spectral gamma ray log is the measurement of individual radioactive contribution of the three radio-isotopes. The individual channels of the spectral tool can be empirically calibrated. The log usually recorded in right tracks on arithmetic scales. Thorium and uranium are given in ppm while potassium is given in per cent.

The maximum depth of investigation for the tool depends upon formation and mud density. For average values of drilling mud and formation density, approximately 50 per cent of the gamma ray signal comes from within 18 cm of the borehole wall, and can be increasing to 75 per cent from within 30 cm. This will decrease for denser formations of the same radioactivity and increase for less dense formations of the same radioactivity.

The gamma ray log is useful for discrimination of lithology. Its main use is the discrimination of shale by their high radioactivity. Note that shale, organic rich shale and volcanic ash show the highest gamma ray values. Halite, anhydrite, coal, clean sandstones, dolomite and limestone have low gamma ray values. If sandstone contain feldspars, micas, glauconite, or heavy minerals, will give the sandstone higher gamma ray values. The gamma ray log values can also be used to calculate the shaliness or shale volume of the rock.

The gamma ray log is one of the most frequently used logs for correlation. Facies identification in detrital sediments (sand-shale) has been developed using gamma ray log shapes. It depends on the relationship between grain size and shale content. The shaliness does not change suddenly, but occurs gradually with depth. Such gradual changes are indicative of the litho-facies and the depositional environment of the rock. The Spectral gamma ray log, can also be used to calculate the volume of radioactive minerals.

Neutron Log

Neutron logs are for delineation formation-porosity. It is based on response primarily to the amount of hydrogen in the formation. In formations, whose pores are filled with water or oil, the neutron log reflects the amount of liquid-filled porosity. Gas zones can often be identified by comparing the neutron log with another porosity log or a core analysis. A combination of the neutron log with one or more other porosity logs yields even more accurate porosity values and lithology identification and evaluation of shale content (Figure 7.5). This enhances lithological interpretations and well-to-well stratigraphic correlations and also as a porosity indicator.

With the introduction of the SNP (sidewall neutron porosity) tool and the CNL (compensated neutron tool) the neutron gained acceptance as a porosity measurement. The Dual Porosity neutron tool combines those two neutron measurements into a single tool.

Method

Neutrons are electrically neutral particles, each having a mass almost identical to the mass of a hydrogen atom. High-energy (fast) neutrons are continuously emitted

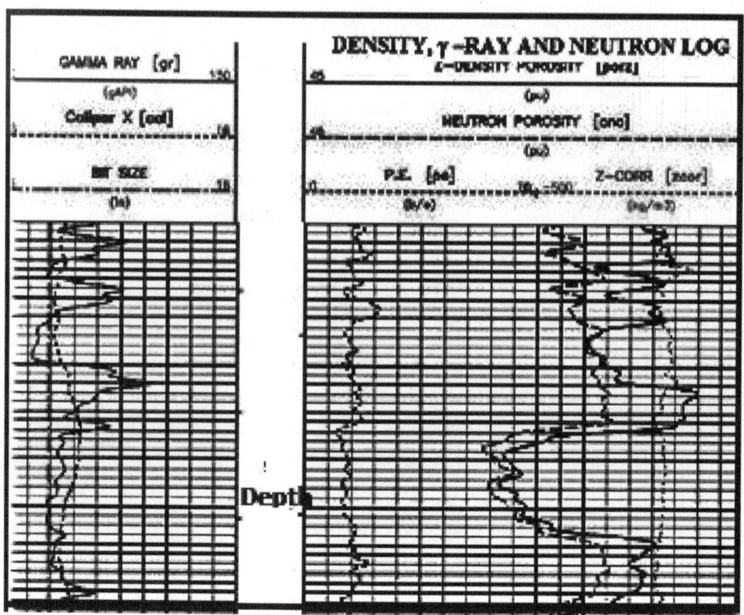

Figure 7.5. Neutron-Density–Gamma Log.

from a radioactive source in the sonde. These neutrons collide with nuclei of the formation materials. With each collision, the neutron loses some of its energy. The amount of energy lost per collision depends on the relative mass of the nucleus with which the neutron collides. The greater energy loss occurs when the neutron strikes a nucleus of practically equal mass, *i.e.*, a hydrogen nucleus. Collisions with heavy nuclei do not slow the neutron very much. Thus, the slowing of neutrons depends largely on the amount of hydrogen in the formation. Within a few microseconds the neutrons have been slowed by successive collisions to thermal velocities. They then diffuse randomly, without losing more energy, until they are captured by the nuclei of atoms such as chlorine, hydrogen, or silicon. The capturing nucleus becomes intensely excited and emits a high-energy gamma ray of capture. Depending on the type of neutron tool, either these capture gamma rays or the neutrons themselves are counted by the detector in the sonde.

When the hydrogen concentration of the material surrounding the neutron source is large, most of the neutrons are slowed and captured within a short distance of the source. On the contrary, if the hydrogen concentration is small, the neutrons travel farther from the source before being captured. Accordingly, the counting rate at the detector increases for decreased hydrogen concentration and vice versa.

The Gamma Neutron Tools were non-directional devices that employed a single detector sensitive to both high-energy capture gamma rays and thermal neutrons. Although the GNT tools responded primarily to porosity, their readings were greatly influenced by fluid salinity, temperature, pressure, borehole size, standoff, mudcake, mud weight, and, in cased holes, by the casing and cement.

The **SNP** (sidewall neutron porosity) tool the neutron source and detector are mounted on a skid, which is applied to the borehole wall. The neutron detector is a proportional counter, shielded so that only neutrons having energies above epithermal are detected. It has advantages, because it is a sidewall device, borehole effects are minimized. The SNP equipment is designed for operation only in open holes. The minimum hole diameter in which the tool can be used is 5 inches. A caliper curve is recorded simultaneously with the SNP neutron data.

More recent tools include the **CNL** (compensated neutron) and the Dual- Energy Neutron Log (DNL). The CSU program automatically provides the corrections necessary in liquid-filled holes for mud weight, salinity, temperature, and hole-size variations (Figure 7.6). Correction for mudcake is done manually using a nomogram. The CNL log and the DNL log are recorded in linear porosity units for a particular matrix.

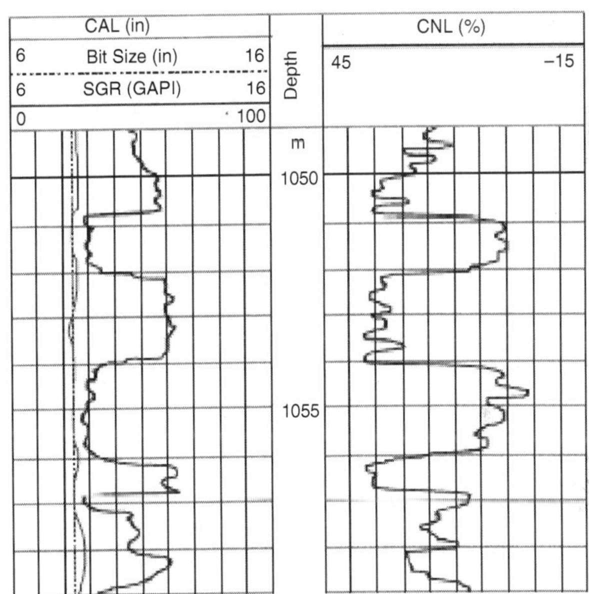

Figure 7.6. Compensated Neutron (CNL) Log.

When a CNL tool is run in combination with another porosity tool, all curves can be recorded on the same porosity scale. This overlay presentation permits visual qualitative interpretation of porosity and lithology or the presence of gas. Since oil and water contain practically the same amount of hydrogen per unit volume, the responses reflect the liquid-filled porosity in clean formations. However, the tools respond to all the hydrogen atoms in the formation, including those chemically combined in formation matrix minerals. Thus, the neutron reading depends mostly on the hydrogen index of the formation. The hydrogen index is proportional to the quantity of hydrogen per unit volume, with the hydrogen index of fresh water at surface conditions taken as unity. Liquid hydrocarbons have hydrogen indexes close to that of water. Gas, however, usually has a considerably lower hydrogen concentration that varies with temperature and pressure. Therefore, when gas is present near enough to the borehole to be within the tool's zone of investigation. This characteristic allows the neutron log to be used with other porosity logs to detect gas zones and identify gas/liquid contacts.

In the producing field the Gas Water Contact (GWC) or Oil Water Contct (OWC) shifts upwards with time and quantum of production from the reservoir. Sometimes, new gas cap forms depending on pressure decline. The cased hole Neutron tool is simple and provides gas fluid contacts in particular with saline formation water environment due to low hydrogen index. Periodic monitoring of old wells with cased hole neutron log will indicate the shift of gas contacts which is essential to know the status of the well as well as the field.

There are also tools to monitor fluid contacts in the reservoir with time like Reservoir saturation tool (RST) and Pulse Neutron Logging Tool (PNLT).

Nuclear Magnetic Resonance Imaging (NMR)

NMR logging tools to the magnetic resonant frequency of hydrogen, the signal is maximized and can be measured. The quantities measured are signal amplitude and decay. NMR signal amplitude is proportional to the number of hydrogen nuclei present and is calibrated to give porosity, free from radioactive sources and free from lithology effects.

Many nuclei have a magnetic moment which behave like spinning bar magnets. These spinning magnetic nuclei can interact with externally applied magnetic field. The decay of the NMR signal during each measurement cycle, called the relaxation time. Relaxation times depend on pore sizes. Small pores shorten relaxation time, the shortest time corresponding to clay bound and capillary bound water. Large pores allow long relaxation times and contain the most readily producible fluids.

Relaxation times of a vuggy carbonate might in rock pore space has shorter relaxation times. Relaxation times for sandstone typically range from 10m sec for small pores to 500m sec for larger pores. Relaxation times and their distributions may be interpreted to petrophysical parameters such as permeability, producible porosity and irreducible water saturation and hydrocarbon identification.

Density and Neutron porosity measurements require environmental corrections and are influenced by lithology and formation fluid. The porosity derived is total porosity, which consists of producible fluids, capillary-bound water and clay-bound water. NMR porosity includes only capillary-bound water and producible fluids and is not influenced by lithology.

For example, in ***Gandhar Field*** of Cambay Basin, in a well the porosity evaluation was uncertain through N-D combination as it is highly shaly reservoir with poor producibility. NMR log successfully yields matrix independent (not affected by complex lithology) porosity with producible component. In ***Limbodra*** well, the N-D separation shows good clean reservoir, however, on account of bad hole it appears uncertain producibility. NMR logging shows extremely poor producibility (Prasad, K K, 2006).

In ***Krishna-Godavari Basin*** offshore (Vashista), an attempt has been made to identify Clay Mineralogy, to determine depositional environment using open-hole NGS (Natural Gamma Ray Spectroscopy), ECS (Elemental Capture Spectroscopy), Litho Density and Neutron logs. In this field the reservoir sands are mostly channel sands as evident from log motifs. Smectite-Mixed clay mineralogy is identified from the cross plot studies. Depositional environment is of non-marine nature with development of stacked channel sands. The framed petro-physical model was adopted in the log data processing. Log motifts depict fining upward nature. Sonic log against these sands read 140-160μs/ft which indicates that these sands are unconsolidated. This finding corroborates with the laboratory studies as well. These sands are compositionally quartz rich "wackes" unconsolidated with variable amounts of feldspar, fine mica and glauconite, *etc.* Thorium-Uranium ratio cross plot has generated for these wells to understand the depositional environment. The Petro physical (Multi-mineral) model has been framed comprising Quartz as rock matrix, The Th/U ratios indicating that, the depositional environment is of non-marine nature and so the reservoir sands were deposited in oxic environment (Rao., BA, 2010).

In another case in the deep offshore of **Krishna-Godavari Basin,** an examination of an improved method of estimation of water saturation in virgin zone using NMR data. The reservoir is a gas bearing with no Gas-Water-Contact penetrated in the well. The litho-facies of the section are finely laminated sandstone with inter-laminations of silty sandstone, clayey silt and shale. Conventional approaches to water saturation estimation using NMR data in the zone of investigation do not necessarily give water saturation in virgin zone. The method of computing reliable water saturation was used by integrating NMR with other measurements. Wherever, lamination with shale or other factors make Sw from conventional petro-physics to be on lower side, this method become as an alternative for realistic hydrocarbon volumetrics.

In the ***Ankleshwar field,*** Cambay Basin, the log data were examined, where the reservoir is a highly complicated detrital formation. NMR data has made it possible to quantify the results and correlate petrophysical information provided by core analysis where conventional radioactive tools are not of much help. The NMR allow the neutron-density, sonic porosities to be redundant in such type of formations. All the conventional GR, SP, Density, Neutron data add to the complexity of log data interpretation. The clear cut delineation with NMR and Resistivity information from a reservoir to non-reservoir in the same lithological unit makes NMR data superior to other conventional logs (Raghunath, H.K. and J. Abraham,2006).

Density Log

The Density log is a continuous record of formation's bulk density. This is the overall density of a rock including solid matrix and the fluid enclosed in the pores. The formation density tools are induced radiation tools (Figure 7.7). They bombard the formation with radiation and measure how much radiation returns to a sensor.

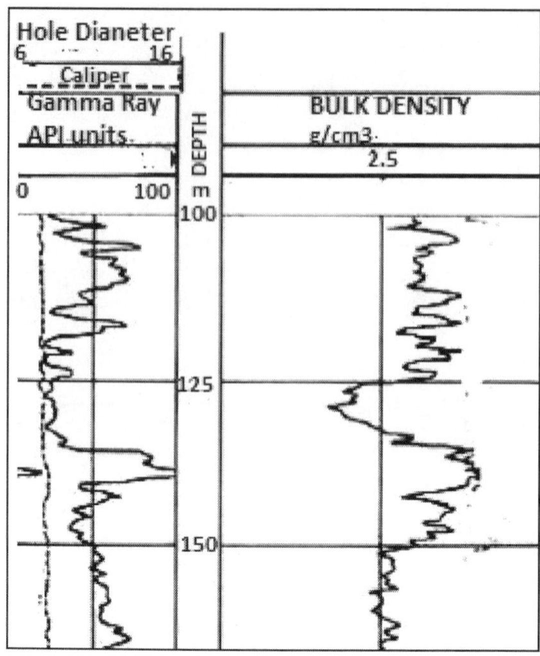

Figure 7.7. Density log-Correction

Method

The density log recorded with two detector tool called compensated formation density (**FDC**) log. Compensated formation density tools have one focused radiation source, one short spacing detector from the source, and one long spacing detector from the source. The source and both detectors are heavily shielded to ensure that the radiation only goes into the mudcake and formation, and that detected gamma rays only come from the mudcake or formation (Alger, R P., *et al.*, 1963).

The photo-electric factor log or the litho-density log (**LDL**) is a continuous record of the effective photo-electric absorption cross section index of a formation. The photo-electric absorption index is strongly dependent on the average atomic number of the constituents of the formation. It provids an improved bulk density measurement and a lithology-sensitive photo-electric absorption cross section measurement. The use of litho-density log is ineffective when barite based mud is used in wells, since the photo-electric absorption index for barite is much more that of most common minerals.

A neutron and density log combination provides a more accurate porosity and a value of minimum gas saturation. The determining porosity is one of the most important uses of neutron logs. Corrections for lithology and borehole parameters are necessary for accurate porosity determinations (Figure 7.8).

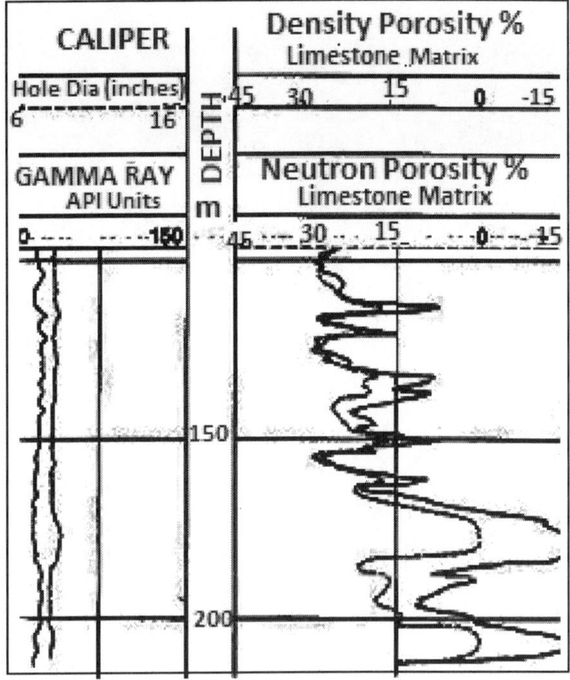

Figure 7.8. FDC-CNL Log.

The compensation features of the CNL and Dual Porosity tools greatly reduce the effects of borehole parameters, and the tools are designed for combination with other open-hole or cased-hole tools. The neutron-density combination is best in clean formations since the responses to gas are in opposite directions. In shaly formations, the neutron-sonic combination is an effective gas detector.

Azimuthal Density Neutron (ADN)

Azimuthal Density Neutron (ADN) tool measures borehole compensated formation density, neutron porosity and photo-electric factor with improved measurement accuracies. These measurements are acquired in four quadrants along with the average measurement around the drain-hole.

A two-coil magnetometer is used to define quadrants around the drain-hole. One of these quadrants is always oriented by gravity in non-vertical wells. Therefore, bottom, top, left and right quadrants are defined. This orientation is made with the direction and inclination information from tool. The position is keyed to the gravitometer sensor, which allows the detector position in the quadrant to be known precisely at any instant of time. In addition, an RPM curve is derived from the magnetometer sensors and used to determine whether the BHA is sliding or rotating. The radioactive logging sources used within the ADN tool are wire-line-retrievable in case the BHA is stuck in the hole.

Sonic Log

The logging cables were used to lower geophones into wells to measure long-interval acoustic travel times from sound sources at the surface. It is considered as a reliable porosity log; its measurement responds primarily to porosity and is essentially independent of saturation. Subsequent improvements in sonic logging included the (BHC) borehole compensated sonic, the (LSS) long-spaced sonic, and the Array-Sonic tools. The latter tools permit the recording of the entire sonic wave-train.

The Sonic tool consists of a transmitter that emits a sound-pulse and a receiver that picks up and records the pulse as it passes the receiver. The sonic log is simply a recording versus depth of the time,(Δt), required for a sound wave to traverse 1 ft of formation. Known as the interval transit time, transit time, Δt is the reciprocal of the velocity of the sound wave. The interval transit time for a given formation depends upon its lithology and porosity. This dependence upon porosity, when the lithology is known, makes the sonic log very useful as a porosity log. Integrated sonic transit times are also helpful in interpreting seismic records. The sonic log can be run simultaneously with many other services (Figure 7.9).

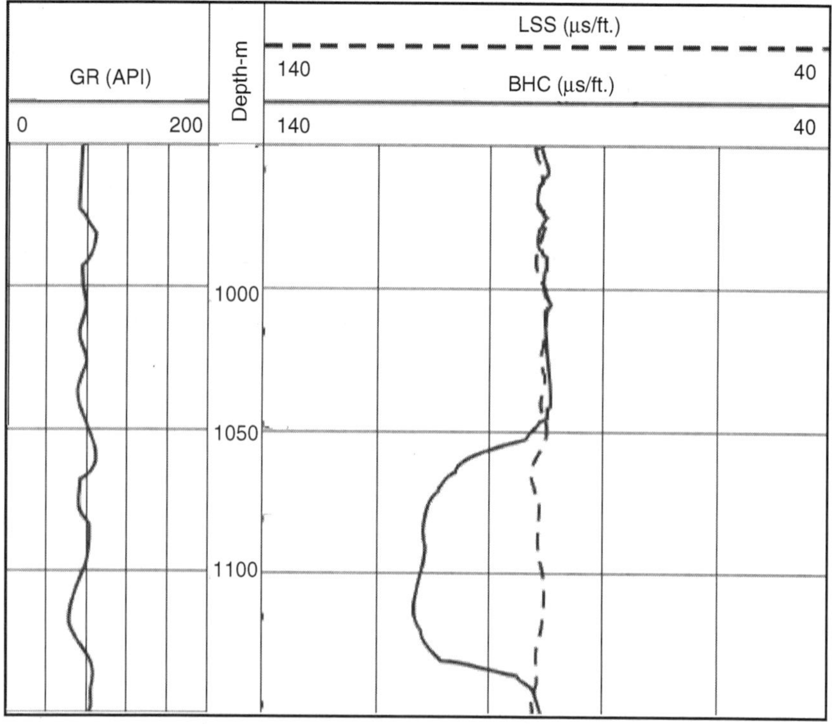

Figure 7.9. Sonic Log.

The velocity of the compressional wave through a rock formation is influenced both rock matrix and fluid filled pore spaces. A relationship defining a uniform intergranular porosity is in terms of the total formation velocity, rock matrix velocity and fluid velocity.

The relationship is expressed to;

$\varphi_{sonic} = (\Delta t_{log} - \Delta t_{ma}) / (\Delta t_f - \Delta t_{ma})$

φ_{sonic} = sonic derived from porosity.

Δt_{log} = interval transit time of formation.

Δt_{ma} = interval transit time of the matrix.

Δt_f = interval transit time of the fluid in the well (fresh mud = 189; salt mud = 185).

Method

The sound emanated from the transmitter impinges on the bore-hole wall. This establishes compressional and shear waves within the formation, surface waves along the borehole wall, and guided waves within the fluid column. In the case of well logging, the borehole wall, formation bedding, borehole rugosity, and fractures can all represent significant acoustic discontinuities. Therefore, the phenomena of wave refraction, reflection, and conversion lead to the presence of many acoustic waves in the borehole when a sonic log is being run. The acoustic energy arrivals are recorded by the receivers of a sonic logging tool. These waveforms were recorded with an array of eight receivers located at different distances from the transmitter. The first arrival or compressional wave is one that has traveled from the transmitter to the formation as a fluid pressure wave, has been refracted at the borehole wall, has traveled within the formation at the compressional wave velocity of the formation, and has traveled back to the receiver as a fluid pressure wave.

The shear wave is one that has traveled from the transmitter to the formation as a fluid pressure wave, has traveled within the formation at the shear wave velocity of the formation, and has traveled back to the receiver as a fluid pressure wave. The mud wave (not strongly evident in these wavetrains) is one that has traveled directly from transmitter to receiver in the mud column at the compressional wave velocity of the borehole fluid. These values are read on alternate pairs of receivers. The values from the two sets of receivers are averaged automatically by a computer at the surface for borehole compensation. The computer also integrates the transmitted readings to obtain total travel times.

Sometimes the first arrival, although strong enough to trigger the receiver nearer the transmitter, may be too weak by the time it reaches the far receiver to trigger it. Instead, the far receiver may be triggered by a different, later arrival in the sonic wave train, and the travel time measured on this pulse cycle will then be too large. When this occurs, the sonic curve shows a very abrupt and large cycle skipping. Such skipping is more likely to occur when the signal is strongly attenuated by unconsolidated formations, formation fractures, gas saturation, aerated muds, or rugose or enlarged borehole sections.

It is known that in some shales a lateral velocity gradient exists. Sound waves travel at lower speeds near the borehole; at some greater distance from the borehole, they propagate at the true speed of sound in the shale. Similar variations may exist in the radial velocity profile in some unconsolidated rocks. In large-diameter boreholes, it is possible to have a mud wave arrival at the near receiver before the formation signal. This problem is particularly prevalent at shallower depths where sonic logs are often run for seismic calibration. In all these cases, a sonic tool with long spacing is required to provide a correct measurement of the velocity in the non-altered zone. When the receivers are far enough from the transmitter, the first arrival is not the refracted ray traveling just inside the borehole wall but a wave penetrating beyond the borehole into the faster non-altered zone.

LSS (Long spaced sonic) tool, with different transmitter-receiver spacing measure the interval transit time of the formation at a much greater depth into the formation than the usual BHC sonic tool. This tool is more likely to yield a measurement free from the effects of formation alteration, relaxation damage (from drilling process), and enlarged borehole. These more accurate measurements are always desirable when the sonic data are to be used for seismic purposes. The Array-Sonic tool contains an array of eight wideband piezoelectric receivers. The receivers are spaced 6 inch a part with the closest receiver 8 ft from the upper transmitter.

The acoustic log can be used to determine porosity in consolidated formations, it is also valuable in other applications, such as:

- ☆ Indicating lithology (using the ratio of compressional velocity over shear velocity).
- ☆ Detecting fractures and evaluating secondary porosity.
- ☆ Evaluating cement bonds between casing and formation.
- ☆ Detecting over-pressure.
- ☆ Determining mechanical properties (in combination with the density log).
- ☆ Determining acoustic impedance (in combination with the density log).

Fracture Recognition

The density tool records the bulk density of the formation. The porosity derived from this will include all pores and fractures whether they are connected or not. The sonic tool can be used to measure the porosity of the formation. The sonic tool is not sensitive to fracture porosity. Hence, the difference between the porosities derived from these two measurements can be used as an indicator of the extent of fracturing in a reservoir interval.

Electromagnetic Propagation Tool (EPT)

The Electromagnetic Propagation Tool (EPT) transmits electromagnetic wave into the formation and measures its propagation, attenuation rate between a pair of receivers 4 cm apart. The antennae are mounted in a pad which is rigidly fixed to the tool body, and contact with the borehole wall is achieved by means of a powered mechanical back-up arm. The two transmitters are alternately energized to allow similar to that in the sonic tool. The EPT being a mandrel type tool, the bore-hole size has no significant effect on the measurement as long as good pad contact is maintained. A vertical resolution of 2 or 6 inches may help in very good bed-boundary definition. This technique reduces the effects of tool-tilt, varying mud cake thickness and slight instrumentation imbalance.

The EPT measures the travel time and attenuation of an electromagnetic wave. These parameters are related to the dielectric permittivity, electromagnetic waves are not plane waves, owing partly to the closeness of the transmitter to the receiver.

Because in most cases, rocks are composed of non-magnetic minerals, the dielectric permittivity, measured at very high frequency, is function of the water-filled porosity, and independent of the salinity (Calvat, T.J, *et al.*, 1977).

Dipole shear Sonic Imager (DSI)

The DSI tool is a multi-receiver tool with a linear array of eight receiver stations, a monopole transmitter and two dipole transmitters. The receiver array provides more spatial samples of the propagating wave field for full wave form analysis. Compressional or P-waves of rock compression and expansion. Here small particle vibrations occur in the same direction the wave is travelling. Shear or S-waves of shearing action as would occur when striking the rock column from the side. In this case, particle motion is perpendicular to the direction of wave propagation. Slowness time coherence (STC) processing is a full waveform analysis technique aimed at finding all propagating waves in the composite waveform.

USIT (Ultra Sonic Imager Tool)

The Ultrasonic Imager tool (USIT) uses a single transducer mounted on an Ultrasonic Rotating Sub (USRS) on the bottom of the tool. The transmitter emits ultrasonic pulses between 200 and 700 kHz and measures the received ultrasonic waveforms reflected from the internal and external casing interfaces. The rate of decay of the waveforms received indicates the quality of the cement bond at the cement/casing interface, and the resonant frequency of the casing provides the casing wall thickness required for pipe inspection. Because the transducer is mounted on the rotating sub, the entire circumference of the casing is scanned. This 360° data coverage enables the evaluation of the quality of the cement bond as well as the determination of the internal and external casing condition. The very high angular and vertical resolutions can detect narrow channels, Cement bond, thickness, internal and external radii, and self-explanatory maps are generated in real time at the well site.

Ultra Sonic Imaging Tool in combination with CBL has been used in few wells to evaluate cement behind casing in the **Assam Oil fields**. The importance of cement evaluation is very critical in these reservoirs as they are thin sandstone beds separated by thin shale/clay beds. Sometimes months of rig time is wasted without any injectivity resulting in futile cement squeeze jobs due to false prediction of poor isolation on CBL. USIT helped in determining the nature of cement behind the casing in the object of interest. It further helped in deciding the feasibility, validity and necessity of a cement squeeze.

Resistivity Logging

The resistivity is the most fundamental of all measurements in logging. All rock materials possess some amount of resistance which is inherent to the flow of an electrical current. Resistivity (R) is the physical measurement of resistance and is defined as the reciprocal of material's electrical conductivity (C).

Rock matrix, oil, and gas are electrical insulators and therefore their resistivities are to be infinite. Water will conduct electricity depending upon its salinity. This implies that any current flow through a formation is taking place in the formation water and not hydrocarbons or the rock matrix. Salt water, with high concentrations of dissolved solids will conduct electricity much more readily than fresh water.

Therefore, salt water has a much lower resistivity than fresh water. In most instances, the water present in a formation at depth will be moderately saline. Water-bearing zones, therefore, have higher conductivity–or lower resistivity–than hydrocarbon-bearing zones.

Because oil and gas will not conduct electrical current, it is difficult to distinguish them from rock matrix on the basis of resistivity. These fluids do, however, fill the pore space of a formation, leaving less room for conductive formation water. The electrical current that does flow through a hydrocarbon bearing formation is forced to take a more tortuous path. The overall effect of the presence of hydrocarbons is an increase in resistivity.

Resistivity logging is the measurement of resistivity of the formations with respect to depth. Resistance (R) is the force which opposes flow of electrical current through the substance. Measuring unit is- ohm. The Resistivity (ρ) is the resistance measured between opposite faces of a unit cube of the substance at specified temperature, so unit is-ohm-m. (Archie, G. E., 1942).

The basic principle of a conventional electric log is that a current is passed through a formation via some electrodes and the voltage between other electrodes is measured. From this reading, resistivity is determined. The conventional electric log can only be run in a conductive mud. But usually resistivity is not directly measured; voltage is measured with the current held constant. From measured voltage and constant current, resistance is computed by ohms law. Then that resistance is converted into resistivity of the formation.

Factors influencing resistivity are:

- ☆ Matrix material – High resistivity
- ☆ Formation water – Low/High resistivity
- ☆ Oil – High resistivity
- ☆ Gas – High resistivity
- ☆ Water-based mud filtrate – Low resistivity
- ☆ Oil-based mud filtrate – High resistivity
- ☆ Salinity is more – Resistivity low
- ☆ Water content is more – Resistivity low

- ☆ Temperature is more – Low resistivity
- ☆ Clay content is more – Resistivity is low
- ☆ Metallic content is more – Resistivity is low (because, resistivity is the inverse of conductivity).

Bore Hole Invasion

Normally, the drilling mud pressure is greater than the fluid formation pressure. The penetration of drilling mud into the formation it is called invasion, if so, there is a build-up of mud particles on the inner wall of the borehole, and this is called the mud cake. The remaining liquid part of the drilling mud enters into the formation, pushing the reservoir fluids. This part of the drilling mud is called the mud filtrate. The zone where the mud filtrate has replaced the reservoir fluids is called the flushed zone and there is a zone further into the rock where the replacement of reservoir fluids with mud filtrate is incomplete, which is called the transition zone. The only reservoir fluids occupy the uninvaded zone further deep into the formation.

The depth of invasion is related to the permeability of the rock. The mud filtrate diffuses into the formation water in the rock at a rate that depends upon the permeability of the rock. The invasion is limited by the rate of filtration, and the build-up of mud cake.

The replacement of oil by the water-based mud filtrate is by pressure driven displacement. In water-bearing formations the mud filtrate replaces all of the formation water close to the borehole and this decrease with depth of invasion. In oil-bearing formations the mud filtrate replaces all the formation water and most of the oil close to the borehole wall, again decreasing with distance into the formation (Figure 7.10).

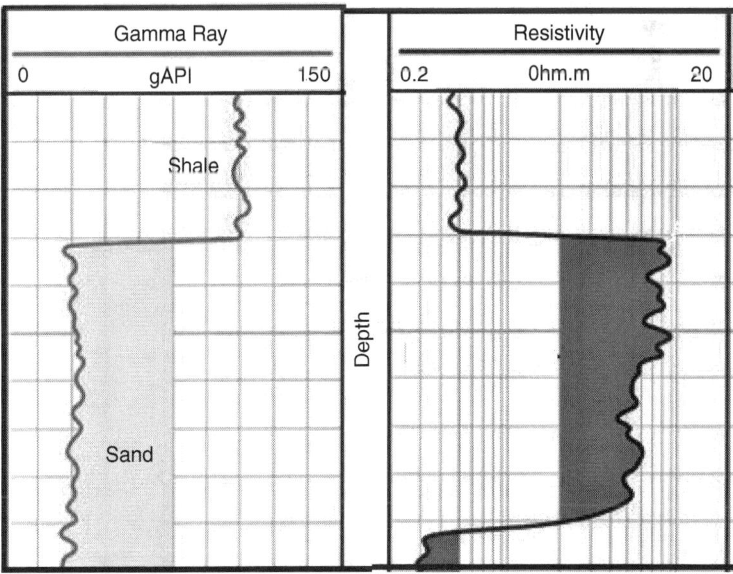

Figure 7.10. Resistivity-Log.

In normal tool, surface electrode passes a constant current to an electrode down hole on the tool. The potential difference is measured between another electrode on the tool and one at the surface. The spacing of a short normal is usually 16 inches and the long normal is 64 inches. Short normal is used for picking bed boundaries and records resistivity of the invaded zone. Long normal is used for finding true resistivity in thick beds.

In lateral tool, constant current is passed from a surface electrode to one on the cable. On the tool there are two potential electrodes. The spacing is defined as the distance between the electrode on the cable and the midpoint between the two potential electrodes. This distance is usually 18′ 8″. The best interpretation using this tool is in thick, homogeneous sediments. It runs approximately a three foot average on the resistivity and has a depth of investigation around 19′.The lateral logs are not suitable for identifying bed boundaries but generally detect thin resistive beds better than normal tools.

If the material surrounding the electrode system is not uniform, these logs read only an apparent resistivity. Other factors which affect the log readings are, hole diameter, mud resistivity, bed thickness, resistivity of surrounding bed, resistivity of invaded zone, true resistivity of zone and diameter of invaded zone.

In unfocussed resistivity logs the direction of survey current is not controlled. This current take the path of least resistance, favouring conductive mud and conductive beds over the resistive beds at the level of tool. Further, very thin beds less than measured spacing values are not dependable and for corrections need to be applied only with the help of charts.

The **Microlog** tool, is a miniature linear array of three electrodes embedded in the face of an insulating pad, which is applied to the borehole wall. A borehole caliper is provided by the arm carrying the electrode pad and an opposite backup arm. The Microlog recording is also useful to delineate permeable beds, and other microresistivity devices help establish the resistivity profile from the invaded zone near the borehole to the non-invaded formation.

Latero Log

It utilizes a multiple electrode array to force survey current to travel laterally across the mud into the adjacent formation. The advantages of this technique are the ability to operate in highly saline mud, while providing excellent bed definition, independent of neighboring bed resistivity. In this all the electrodes are maintained essentially at the same potential and send current out in the same phase.

The Dual Latero log (DLL) was originally developed to give two resistivity readings, one deep (LLD) and one shallow (LLS). These two readings help indicate the invasion profile (Figure 7.11).

Dual Latero log is the MSFL, SFL which provides a better picture of the invasion profile. The dual latero log system uses both small and large guard electrodes. The sonde have both deep and shallow system working simultaneously. The systems using the same electrodes operate at different frequencies and have different depths of investigation (Figure 7.12). The deep tool needs long guards and remote current

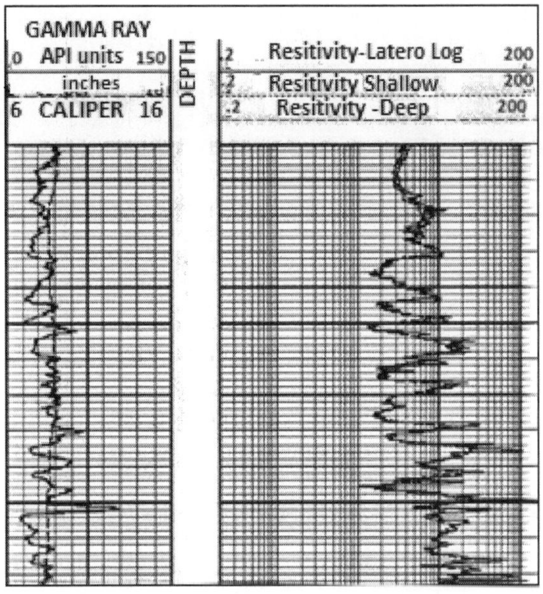

Figure 7.11. Deep Latero Log (DLL).

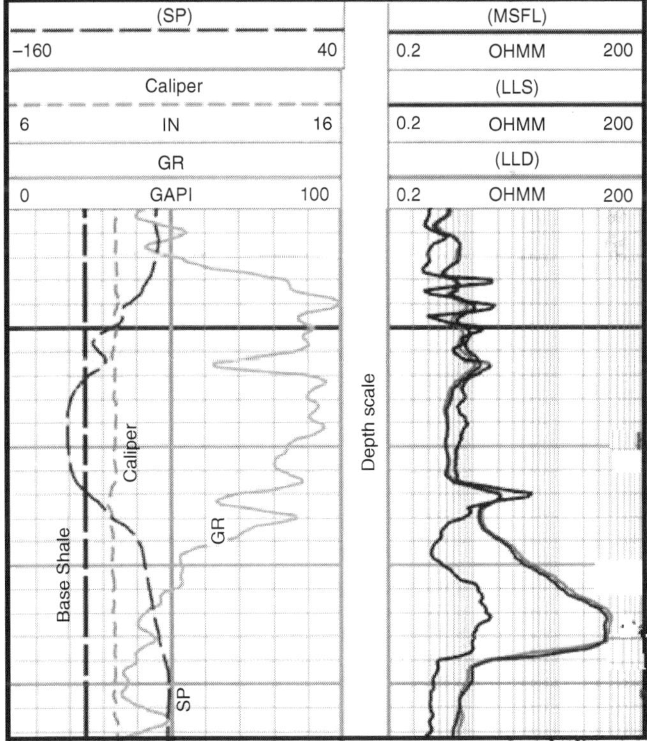

Figure 7.12. Micro-Sperically Focussed Log-LLD Log.

return, while for the shallow it is necessary for the guards to be short and the return relatively near area.

The **Micro-laterolog** is a padded tool was developed for salt muds. The Micro Proximity log and Micro log have followed the laterolog tool, the first focused deep investigating resistivity device, was introduced. The Proximity log tool design is slightly different, but the principle is similar to that of the micro laterolog. It looks a little deeper into the formation (4-6 inches) and is less affected by the mud cake. A mud cake thickness affects the micro laterolog however, much thicker mud cake has very little effect on the proximity log.

It uses a focusing system to constrain the surveying current (emitted from a central electrode) to substantially a horizontal disc for some distance from the sonde. These are pad devices which have very short spaced electrodes in a pad which is pressed up against the borehole wall. There are two types of contact logs, non-focused and focused. The non-focused contact log the accuracy of the measurement was questionable. The focused contact log gives a better determination of porosity, more detailed bed definition, and a more reliable Rxo reading.

The non-focussed logs fall under different names depending upon the logging company, The purpose is to read the resistivity in the flushed zone (Rxo), obtain a detailed bed definition and show where zones of permeability exist from mud-cake build-up. If Rxo could be determined, Rmf was known, and the Archie or Humble relationship assumed, porosity could be calculated. This is good only if Rxo represents a zone saturated with complete mud filtrate. In a hydrocarbon bearing zone, residual saturations within the formation result in an error in computed porosity.

Focused resistivity logs are well adapted for investigation of thin beds drilled with low-resistivity muds. The laterolog device quickly supplamented conventional resistivity logs in salt muds and highly resistive formations. MSFL is used to determine Rxo, Exact thickness of formation beds.

Rxo can be used with Archi's equation to calculate saturation of the flushed zone $Sxo = \sqrt{(a/\varnothing m) * (Rxo/Rmf)}$.

Each latero log is affected differently by the borehole mud, the invaded zone, and adjacent beds. Consequently, there are correction charts for these effects depending upon the nature of tool used. It works best in muds of low resistivity (*i.e.*, salty). It gives true resistivity when the invasion is shallow.

Azimuthal Resistivity Imager (ARI)

The ART Azimuthal Resistivity imager, a new laterolog tool, which makes directional deep measurements around the using 12 azimuthal electrodes provides deep oriented resistivity measurements while retaining the standard deep and shallow readings. A very shallow auxiliary measurement is incorporated to fully correct the azimuthal resistivity for borehole effect. To overcome electro-chemical effects across the electrode/mud interface, the azimuthal array is implemented in a monitored laterolog configurations. The mud in front of the azimuthal current electrodes is effectively equipotential. The 12 azimuthal currents and the mean potential of the electrodes relative to the cable armour are measured.

The azimuthal resistivity measurements are sensitive to tool eccentering in the borehole and to irregular borehole shape. To correct these effects, a simultaneous auxiliary measurement is made with the array at a frequency, which is sufficiently high to avoid interference with the monitor loops.

The orientation of the ART tool is measured with a GPIT (General Purpose Inclinometer Tool), the device used to orient many Dipmeter and imaging logs. The tool can be logged with different speeds for greater accuracy. In addition, the conventional deep and shallow laterolog measurements (LLD and LLS) are available.

Compensated Dual Resistivity (CDR) Tool

CDR tool is an electromagnetic propagation tool. It responds to conductivity and in turn measures the resistivity of the formation. The tool can be operated in water and oil base muds and provides two depths of investigation. The transmitter of the tool propagates electromagnetic wave and measures the phase shift and the attenuation of the wave between two receivers. The phase shift is transformed into a shallow resistivity and the attenuation is transformed into a deep resistivity.

Induction Log

A new logging technique tool was introduced to measure resistivity in oil base muds. The induction tool has several transmitters and receiver coils which are wound coaxially on a supporting insulated sonde. The transmitter creates an alternating current of constant intensity and high frequency which induces eddy currents into the formation. These currents create their own magnetic field which induces a current back in the receiver that is proportional to the formation conductivity. This is known as skin effect; however, modern logs automatically correct for this. The size of the borehole can affect the induction log's response. The bore hole effect increases as the hole diameter increases (Doll,H.G; 1949).

The induction log was a five-coil induction device was combined with the SP curve which was replaced by with coil array with deeper investigation. The DIL (dual induction log,) consists of deep induction, medium induction, and shallow resistivity measurements. The shallow resistivity-measuring device is now a focused resistivity device provides improved thin-bed response, deeper depth of investigation, and greater dynamic resistivity range.

Multicomponent Induction Log

A new resistivity measurement is now available to the industry to overcome the insensitivity of the conventional resistivity measurements in a laminated hydrocarbon-bearing sand-shale sequence. In those formations, the vertical resolution of the conventional induction tool is insufficient to resolve individual layers. Instead, the traditional measurement is highly biased by the conductive shales and insensitive to the resistive hydrocarbon bearing sands.

The multicomponent induction tool measures multiple magnetic components, generated by three mutually orthogonal transmitter – receiver coil configurations at multiple frequencies (Kriegshauser, *et al.*, 2000). Inversion of appropriate selection of these measurements yields two resistivities, one perpendicular to the bedding plane, called Rv and the other parallel to the bedding plane called Rh. The vertical resistivity measurement is very sensitive to the hydrocarbon bearing sands and used as a key for more reliable formation evaluation.

Indian Scnerio

In ***Cambay Basin***, India, the conventional Induction log behaves anomalously at each interface of a high-conductivity contrast. The deep and medium induction curves show a very high value while crossing the bed and an unrealistic value in the form of *peaks* appear which is not seen on the shallow latero log. Addition of the auxiliary coils in the Dual Induction Tool, having spacing less than the main-coil span to achieve better focusing results, can add anomalous character to the curve shape while crossing the interface of highly conductive bed. Dual Latero Log (DLL) and ***High Resolution Induction Log*** (HRI) were recorded after *horns* were observed on the Induction Log. HRI tool is comparatively more suitable for highly conductive formation because of its improved tool design and measurement principle.

Dipmeter

The three-arm dipmeter device, with an associated photoclinometer which also contained a fluxgate compass, was permitted both the direction and angle of the formation dip to be determined. Initially arms contained an SP sensor which were later replaced by short resistivity devices. This made dip measurements possible in wells where the SP had little correlatable detail. Now four-arm dipmeter tool records more micro-resistivity curves simultaneously, and a triaxial accelerometer and magnetometers provide highly accurate information on tool deviation and azimuth. (Allaud, L.A. and Ringot, J.1969).

The dipmeter tools can detect the very thin events that are related to sedimentary features. Dips are computed using computers at many more levels, and computations are made by correlating the dipmeter curves over shorter intervals. These short-interval correlations reveal the fine structure of current bedding and other sedimentation-related dips (Figure 7.13). When long-interval correlations are made, this detailed information is averaged out, and essentially what remains is the structural dip.

The geologic definition provided by short-interval correlations has been further enhanced by the use of the HDT (high-resolution dipmeter) and the other Dual Dipmeter (SHDT) tools (Campbell, R.L., Jr 1968). The Dual Dipmeter (SHDT) tool was designed to maintain pad contact also in a deviated rugose hole, Data are also recorded for an automatic correction of variations in sonde velocity. Plotting symbol indicates the direction of the dip. The displacement of the symbol from the left edge of the plot represents magnitude of dip angle. Vertically, the symbols are plotted versus depth.

The HDT dipmeter tool has four pad-type arms. Four microresistivity (or microconductivity) curves, spaced at 90° intervals around the borehole, are recorded.

Although three curves are enough to define a dipping plane, the fourth curve provides valuable redundancy in the dip computation, particularly in rugose, irregular, or deviated boreholes in which one pad may make good contact with the formation. Opposite pairs of pads are independently linked to provide a dual caliper measurement of borehole diameter. One pad contains an additional electrode, vertically displaced from the dipmeter micro-resistivity measurement electrode.

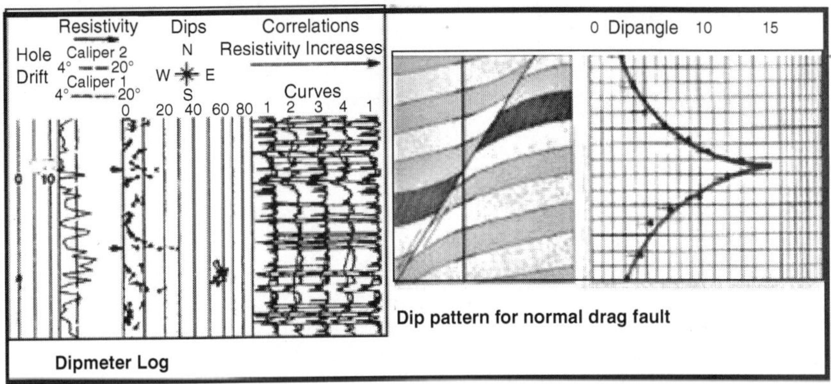

Figure 7.13. Dipmeter-log with Dip Patterns (After Chattopadhyay T. and D. K. Ghosh, 2006).

The main advantage of Dipmeter log is it provids complete borehole image with continuous directional survey and fracture identification and orientation. Dipmeter logging defines the sub-surface sedimentary structures with depositional dip patterns along with tectonic or structural dip.

The ***Baramura field*** is an asymmetric doubly plunging anticline trending roughly NNE-SSW. The eastern flank of Baramura structure was affected by a thrust dipping westward. It is not possible to characterize structural deformations fully from dip data. Moreover, steeply dipping events such as faults are not picked up by computed dips due to the different response of the tool on the opposite sides of the borehole. In a well, dipmeter log was recorded using four electrode dipmeter. This necessitates the critical examination of raw dipmeter curves and other log curves along with computed dip data. A visual inspection of formation micro-resistivity image log will be of great help in these situations. The dipmeter log of a well in Baramura field processed and in which, a minor normal drag fault has been identified and its orientation has been determined (Chattopadhyay T. and D. K. Ghosh,2006).

The Cretaceous sandstone formation in a well located in the Tranquebar sub-basin of ***Cauvery Basin,*** imaging clearly show debris flow, grain flow, shale-sand laminations, slumps and turbidites on Borehole image log. Some of these features can be interpreted from Dipmeter also (Figure 7.14). Two objects were tested in well, out of which one flowed hydrocarbons and the other water. Borehole image logs against these two objects show the differences in the reservoir depositional facies.

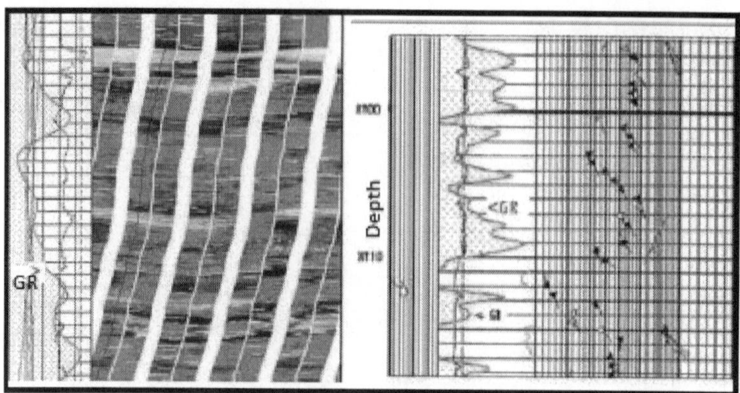

Figure 7.14. Bore Imager Log Showing Depositional Dips (After Subba Rao, V V, *et al.*, 2006).

Another useful presentation of dipmeter micro-resistivity measurements is called the **FIL** (***Fracture Identification Log***) and the simplest and most effective methods for fracture detection. When mud filtrate invades a fracture system, it usually causes a lower dipmeter micro-resistivity reading from the pad positioned opposite the fracture. A comparison of measurements from adjacent pads indicates fractures. If no differences exist, the probability of fractures is low and if large differences exist, the probability is high.

In India, the ***Cambay basin*** was on production since more than five decades from various stratigraphic formations of clastic sedimentation. The production has been on decline and presently the exploration is focusing more on the tight silty formations. Fractures do play an important role in such reservoirs to provide the secondary porosity and flow conduit. Borehole images by virtue of their high resolution on borehole images with the shear splitting was used for identification of fractures. In OBM (Oil Base Micro Imager) all fractures, open/closed with secondary mineral deposit seems resistive as the open fractures get filled with resistive mud. Orientation of the fractures found following the the regional structural trend.

Bassein Field

An integrated approach has been adopted to evaluate fractures based on close examination of conventional core analysis data, drilling history and qualitative and semi quantitative study of well logs. Fracture porosity and Fracture Intensity Indices (FII) have been computed to evolve Fracture Proximity Index (FPI). These studies reveal the presence of fractures at discrete intervals of *Bassein Formation* in the axial part of the structure. The high vertical permeabilities observed in the fractured reservoirs of wells may cause severe coning problems in the axial part of the structure, whereas the near absence or minimal fracture intensity in the western flank wells suggests that this part of the structure is ideal for exploitation of thin oil column of the field.

Caliper, FDC- CNL, Sonic, VDL, MSFL, PB and Resistivity logs have been critically evaluated. Resistivity Ratio Plots of LLD/LLS Vs depth were made to evolve fracture identification. A fracture filled with conductive water/mud filtrate and parallel or sub-parallel to bore hole axis offers a path of least resistance to the current (Rao.VK, 1989).

Formation Micro Imager (FMI)

Full bore Formation Micro Imager (FMI) tool is the electrical imaging device. The FMI images overcome the inadequacies of traditional dipmeter data as the images produce 'core like' borehole wall picture, which allows identification of micro geological features with good vertical resolution and lateral continuity. The image analysis aimed to evaluate sedimentary features, textural characteristics and to recognize depositional environments (Figure 7.15).

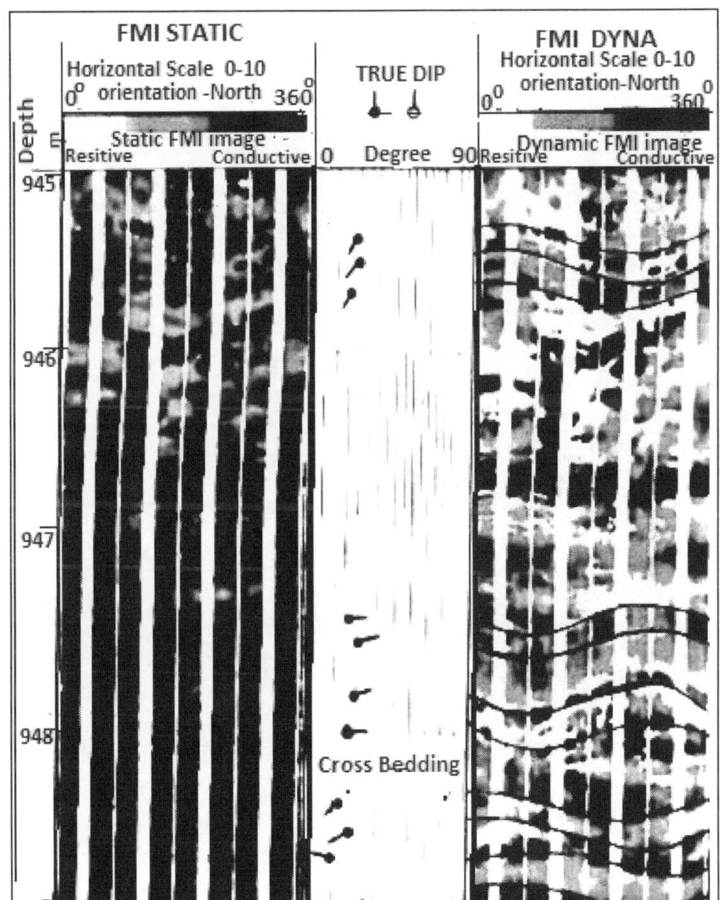

Figure 7.15. FMI-KG Image Tool.

The tool has four arms carrying eight pads – four fixed pads and four folding flaps. Each pad and flap have 24 button electrodes arranged in two rows, the upper row displaced horizontally by 0.1 inch with respect to the lower row and vertically by 0.3 inch. In 8½ inch borehole, about 80 per cent of the borehole wall surface is covered. Depth of investigation varies according to the resistivity of the formation. Against resistivity beds the tool reads deeper than against conductive beds. The mode of operation provides wide coverage of the bore hole and high resolution image. the bore hole coverage depends on the bore hole size; the larger bore hole, the more the arm expand and the grater separation between the pads.

Extended Range Micro Imaging (XRMI)

The high resolution electrical borehole imaging tool has shown method to study geology of an area in terms of depositional environment. Due to its high borehole wall coverage and very high vertical resolution, fine scale sedimentary features can be picked up more confidently. This information, combined with gamma-ray and SP log signature, can be used to reconstruct depositional environment with certain degree of confidence.

North Chandmari is a hydrocarbon bearing structure discovered in Upper Assam Basin. A thick column of oil bearing sand was identified within the Langpar formation of Lower Eocene age. Detailed study of borehole image log combined with GR log and drill cuttings evidences has been carried out in order to understand reservoir facies and sand geometry for development planning. Analysis indicates that, the Langpar formation was deposited in a fluvio-deltaic environment. The direction of extension of the bars in the NS direction as well as direction of distributary channel development and delta progradation towards east of the deltaic sequence is interpreted with fair degree of confidence (Bharali, B, *et al.*, 2006).

Side Wall Cores

The recovery of real rock samples and formation fluid samples are being done using a hollow, cylindrical "bullet" shot into the formation and retrieved by pulling it out. With time, this technique has undergone improvement and now for very hard rocks, down-hole mechanical coring tools exist that actually drill out the rock samples.

The **formation tester** is to recover a sample of the formation fluids and the pore pressure was measured during the sampling process. The **FIT** (formation interval tester) and the RFT (repeat formation tester) are further developments with time. The older tools could make only one pressure measurement and recover only one fluid sample per trip into the well; the **RFT** tool can make an unlimited number of pressure measurements and recover two fluid samples per trip. To handle those formations in which the formation water is fresh, or varies in salinity, or in which the salinity is unknown, dielectric measurements have been developed.

The **EPT** (electromagnetic propagation) log the **DPT** (deep propagation) log are also available.

Other logging measurements include nuclear magnetic resonance, nuclear spectrometry (both natural and induced).

Real Time Logging-While-Drilling (MWD and LWD)

These tools are part of the bottom-hole assembly during drilling, near the bit and can have sensors for resistivity, gamma radiation, density, down-hole torque, *etc.* similar to wire-line logging tools. MWD (Measurement While Drilling) and LWD (Logging While Drilling) have basically the same meaning, the parameters are measured down-hole while drilling.

Two concepts of MWD tools are commonly available: memory tools are MWD collars that records the parameters in a solid state memory. The data can be played back every time the collars are pulled, *i.e.,* usually on a bit trip and the other tools have the capability to transmit the data to surface while drilling.

MWD tools provide the display similar to wireline logs, which help for correlation with offset wells (Figure 7.16). But, since the sensors are several feet above the drilling bit. Thin layers were penetrated by the time data analyses through log-plots.

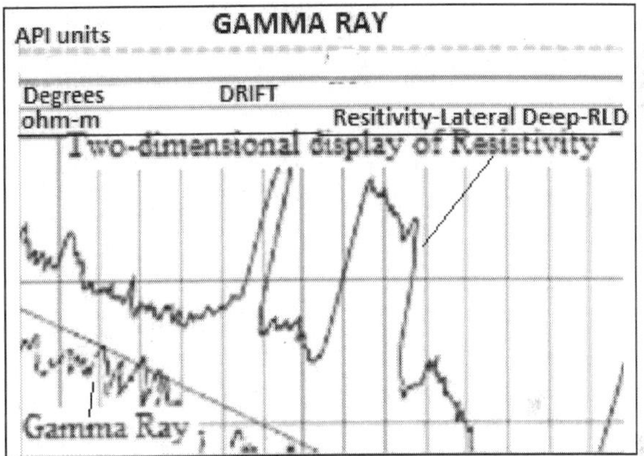

Figure 7.16. Deviated Horizontal Well-Corrected to Azimuth (After A D Mathur, *et al.*, 2006).

The MWD tool in the BHA is a restriction for the mud flow and may limit some operations. The MWD tool may plug if coarse LCM material is pumped through it. It also may pose a problem if cement needs to be pumped through it. It is not preferred if the section to be drilled is prone for lost circulation.

The data quality of MWD logs depends amongst other parameters on the ROP. MWD data are transmitted or recorded in regular time intervals. Therefore, in times of fast drilling, less measurement are taken per depth increment. The MWD data transmission through the mud stream is a low frequency signal. The pulsations may coincide with the pulsation from the mud pumps and if the two frequencies come near to each other, the interference need to be avoided. This effect is particularly strong on radioactive tools (neutron, density) which require a longer time to take a reading. Though corrections can be applied in some situation real, fine resolution data may be lost.

Geo-Steering Tool (GST)

Geo-steering Tool (GST) consists of instrumentation package between the power and the motor which consists of a 4 ft. oil filled unit that carries the motor drive shaft through its centre. It contains the electronics and power supply that is required to make measurements of Inclination, Gamma Ray, Bit Resistivity and motor RPM.

The data from the Geosteering Tool (GST) is transmitted by magnetic waves to a receiver sub that is placed on the MWD tool. The GST data is then pulsed up to the surface and used in conjunction with data from other tools in the bottom hole assembly, such as ADN, Resistivty and Weight on Bit Sub. At the Surface Acquisition Unit, the data are displayed versus trajectory which when viewed with azimuthal density sensors with respect to tool face orientation.

One horizontal well was drilled in ***Mumbai High*** South field to tap un-drained oil in a sub-layer of L-III reservoir. On drilling the well with LWD, heavy mud losses were encountered and LCM pills were placed to control mud losses. With great difficulty only 101 mts. of 6″ drain hole could be drilled and drilling had to be terminated because of severe uncontrolled mud losses. Real time analysis of 6″ hole LWD logs revealed that the well is having good reservoir facies so it was completed barefoot.

Vertical Seismic Profile (VSP)

Vertical seismic profiling is a technique of simultaneously recording the up-going and down-going wave-train. By recording a sufficient number of fairly regularly spaced levels in the well, the up-going and down-going wave fields are separated by computer processing. This analysis permits the detailed study of the change of the seismic wave-train with depth. The acoustic properties of the earth layers can then be directly linked to and interpreted in terms of the subsurface lithology. The use of down-hole sensors reduces the signal distortion caused by the low-velocity shallow layers since the signal passes only once through the surface layers. The total wavefield recorded at the detector in the borehole consists of signals arriving from above the tool.

In this technique, an air gun, vibroseis, or other seismic source on the surface, generates the input signal that is detected by a down-hole geophone. As the sound energy travels only once through the weathered surface layers, the resultant profile has much better resolution than the surface seismic around the borehole, and, in favorable cases, can identify reflectors far below the total depth of the well. Unlike many wire-line services, open-hole and cased hole seismic results are similar since the casing typically does not affect the seismic signal. In fact, cased holes eliminate some of the open-hole operational problems associated with poor hole conditions and highly deviated wells.

The ***equipment*** consists of a downhole Tool (WST), the Seismic Acquisition Tool (SAT), and the Downhole Seismic Array tool (DSA). The WST tool has four uni-axially stacked geophones that are primarily sensitive to movement in the vertical direction. The SAT tool has three mutually orthogonal geophones (which may also be gimbal-mounted for use in deviated wells) for three-dimensional operation. This arrangement provides a system of reference where each arriving ray can be represented by a vector. Among other applications, the ability to record and process signals in three axes allows the recording and interpretation of shear waves and long-offset VSP surveys. The DSA tool uses eight sensor packages (shuttles) which are positioned along an insulated multi-conductor bridle cable at intervals of up to 50 ft. The sensor package contains a vertical geophone for signal acquisition, a magnetic clamping device to secure the package to the casing, a shaker element to generate mechanical vibrations for reference, and electronic circuitry to transmit.

At each depth, the interval velocity of the formations between the source and the borehole geophone is measured with an air gun source, the hydrophone monitors the signature and timing of the source signal, and the down-hole geophone records the direct and reflected arrivals. Transit time is measured from the first break of the hydrophone (surface) recording to the last break of the geophone (down-hole) recording. Several shots are usually made at the same level and stacked in order to improve the signal-to-noise ratio. If the hole is deviated or if there is a significant source offset, the transit times obtained must be converted to True Vertical Depth (TVD) transit times. Correction to the seismic reference datum (SRD) is also necessary if the source is above or below the seismic datum (Figure 7.17).

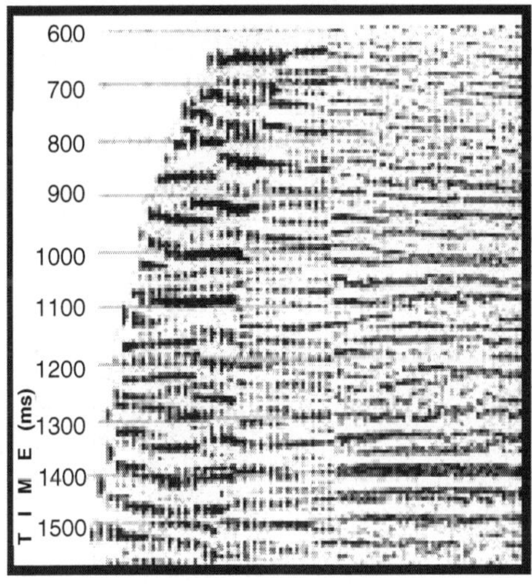

Figure 7.17. Offset-Vertical Seismic Profile.

Offset Vertical Seismic Profile

The normal VSP survey in a vertical borehole with horizontal bedding gives very limited lateral information. However, with dipping reflectors, a normal VSP survey can provide some information on the updip features, however, offers the possibility of large lateral coverage. Lateral coverage of up to one-half of the source offset distance can be achieved in the direction of the source. Profiling of a feature can be done by using a fixed offset source position some distance from the well and moving the geophone(s) in the well, or by having the geophone(s) fixed and moving the source.

A new technique ***'Extended Offset VSP Survey'*** was implemented in a well in ***Cambay Basin***, India, wherein the deployment of energy source is stretched much beyond. Consequently a much larger subsurface coverage at the zone of interest was acquired despite logistic constraints. The technique provides very accurate and high resolution subsurface image. Seismic attribute analysis can be carried out for VSP CDP stacks.

Walkaway Surveys

A walkaway survey provides a two-dimensional seismic picture of the formations on either side and below a well. In walk-away surveys always employ a multiple source and single receiver arrangement. A typical offshore walkaway survey would be carried out by a boat with the energy source, moving at a constant speed and direction along a 3 km line that passes close to the well. Seismic shots would be generated with an air gun at 30 m intervals along the line and a down-hole geophone would monitor the arrivals. Each pass of the boat would generate 100 seismic wave traces at each geophone position. Accurate navigation fixes are obtained for each shot position along the survey line by placing navigation equipment on both the rig and boat. The seismic wave pattern created by this multi-source/single receiver arrangement is particularly useful in investigating complex formations.

A software is available for the calculation of a full response synthetic VSP as well as of many primaries synthetic VSP using only a sonic log. The structural information obtained by mapping the VSP data has a much higher vertical and horizontal resolution than that of the conventional surface seismic data.

With the objective of delineation of pay sands in ***South Kadi*** field in Cambay Basin, Multi offset VSP surveys were carried out in three wells. The pay sand extension could be clearly brought out and also helped in optimally locating a water injector wells in the reservoir. Continuous updating/refinement of the model was carried out with each drilled well data being incorporated.

Types of Cased Hole Logs

Casing Collar Locator

The CCL is based on the principle that a changing magnetic flux within a coil generates a voltage across the terminals of that coil. The magnetic field of a CCL magnet is affected by any magnetically susceptible material close to it such as casing in the borehole. A collar or joint in the casing changes the magnetic flux field including the flux passing through the coil ends adjacent to the magnet or magnets. Thus an electric voltage is generated in the CCL by the discontinuity (collar or joint) in the casing. It is customary to run Collar Locator (CCL) or Tubing collar (TCL) and Gamma Ray along with CBL tool, will help for depth control during perforation job (Figure 7.18).

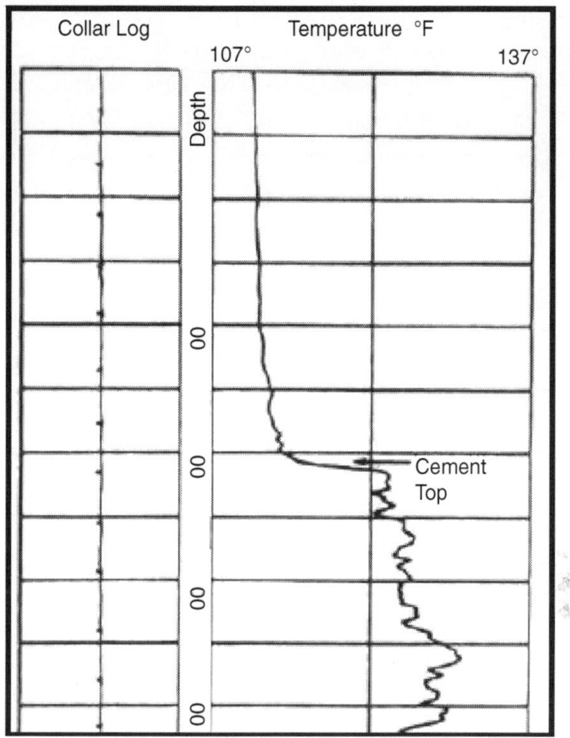

Figure 7.18. Casing Collar Locator (CCL)-Cement Top.

Cement Bond Log – Variable Density Log

Cement evaluation is meant that, the determination of bond quality between casing and cement and that from cement and formation. The cement should be well bonded to both the casing pipe and the formation so that fluids cannot migrate easily from one formation to another.

The purpose of the CBL-VDL log is to determining the quality of bonding of annular cement (Pipe to cement and cement to formation) against the depth intervals, to determine the cement top, cement coverage and compressive strength for fracturing, to evaluate the right position for the hydraulic seal within the casing, to determine the internal and external damage or deformation of casing, corrosion detection, and measurements of ID of casing and thickness.

Several other tools namely, RBT (Radial Bond Tool), USIT (Ultra Sonic Imager Tool) SASL, CAST-V (Circumferential acoustic scanning tool), SBT (Segmented Bond Tool) CBIL (Circumferential borehole imaging tool) of some of the trade names of different logging companies which are also used for cement bond evaluation.

The CBL measurement is based on the sound wave propagation through casing. The transmitter emits sound waves that travel in all possible directions and the different components of the signal are received. The receivers record the composite signal comprised of casing, formation, cement and fluid arrivals. In general the casing has the highest sonic velocity so this signal arrives first followed by the formation signals. In very hard formations with high sonic velocity (called 'fast' formations) the formation signals can arrive first and interfere with the casing returns.

The curve shows how much acoustic signal reaches a receiver and is an important indicator of cement bond. This curve shows the amount of time it takes an acoustic signal to travel between the source and a receiver. It is used as log quality indicator. For free pipe of a given size and weight, the travel time between points is very predictable.

When casing is held by cement, a major part of the energy is transmitted into the cement and attenuated by it depending on its compressive strength. As a result, the amplitude of the first arrival is very low which indicates good bond. In case of contaminated cement; the more the free pipe or contaminated cement in an interval, the poorer the bond where the amplitude is less attenuated. Quantitatively, the compressive strength against the zone of interest is measured with respect to the well cemented bond zone of the CBL-log. So the causes for increase in CBL-amplitude are: Poor cement bond, Channelling, Micro-annulus.

The pipe signals, formation signals and fluid signals are usually easy to recognize on the **VDL**. If these signals can be identified a practical determination for the presence or absence of cement can be made. It provides the qualitative analysis of the bond between the cement and the formation. It is also a qualitative indicator of the cement to casing bond.

The waves of interest in VDL measurement are the ones that travel through the cement-formation interface with formation speed and get recorded at another receiver placed at 5 feet distance from the transmitter to improve their separation from the casing arrivals. Unlike CBL, full waves are recorded in either wiggle -trace mode or variable density mode.

In case of very good bond among casing, cement and formation, most of the energy travels through cement–formation interface. As a result, the casing arrivals are very feeble but the formation arrivals are quite strong. The bands corresponding to the formation arrivals are not straight but wavy because the acoustic velocity of the formation varies with depth. The bands are somewhat mirror image of sonic transit time because they are presented in left to right increasing time scale whereas transit time is presented in right to left increasing scale. The strong formation arrivals, thus, indicate good bond between cement and casing as well as cement and formation. In case, there is a good bond between cement and casing but poor bond between cement and formation, both casing arrivals and formation arrivals are very feeble (Figure 7.19).

Cased Hole Formation Resistivity Tool (CHFRT)

The formation resistivity measurements through steel cased hole can also be recorded through CHFRT log. This tool has two major advantages over nuclear measurements which are deeper depth of investigation up to 11metres and wide dynamic range of measurement. The CHFRT is capable of measuring formation resistivity behind casing close to open hole resistivity in low porosity and salinity reservoirs. The major limitations of this tool are the measurements are affected by multiple casing, casing collars, scaling, thick and resistive cement behind the casing and casing sizes beyond 9 5/8".

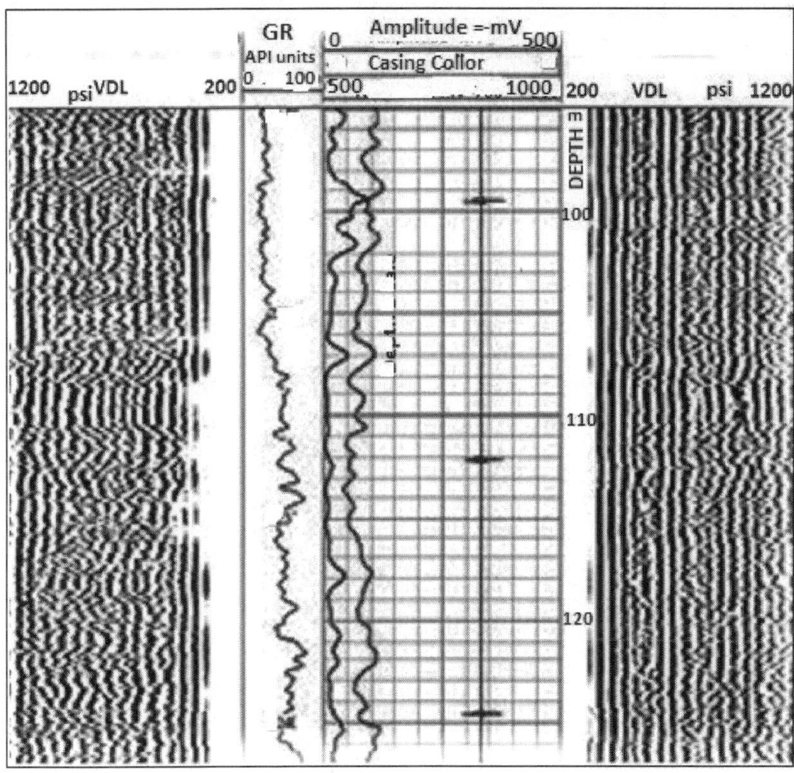

Figure 7.19. CBL-Variable Density Log (VDL).

The measurements are not affected by caving of the formation and larger borehole diameters. The better measurements depend on the thickness and resistivity of cement and the thickness of the steel casing. The vertical resolution of CHFRT is around 1.2 m as it is the function of the spacing between the measuring electrodes of the tool.

In some complicated wells in ***Mumbai High field***, the formation evaluation behind casing was achieved through pulsed neutron measurements. Some recent developments have made analysis behind casing using resistivity based conventional methods. Formation evaluation behind casing is carried out using cased-hole gamma ray/spectral gamma ray, cased-hole neutron and cased hole resistivity. In an exploratory well, where open hole logs could not be recorded due to bad hole. Sonic log was required for geophysical applications. Cased hole dipole sonic was recorded and compressional and shear transit times were extracted. Synthetic seismogram generated with the extracted compressional ΔT and constant density.

The well which was drilled in 1988 open hole logs could be recorded. Cased hole neutron and gamma ray logs were recorded. Initially the well was producing 1800 bopd without any water cut. By the middle of 2001, the oil rate came down below 200 bopd and water cut increased to 70 per cent. Cased hole formation evaluation was carried out using formation resistivity, neutron and gamma ray. Based on the analysis work-over job was carried out and oil production was doubled in the well.

Temperature Log

The temperature measurement in a well is used for locating cement top behind the casing and locating the depth of lost circulation. The temperature survey is always recorded slowly while going down in the hole. The logging speed is kept low to allow the thermometer to attain sufficient time to reach the temperature of the surrounding medium. Continuous logs can observe intervals of raised (or lowered) temperature caused by the influx of hotter (or colder) fluids into the borehole.

Temperature measurements were earlier made with a thermometer whose sensing element was temperature sensitive resistance. If constant current is passed through this resistance the voltage drop will be proportional to the resistance and may be recorded as temperature.

Distributed Temperature Sensing (DTS)

In this, the technique in use is an optical fiber sensor in temperature logging. The key technique is optical time domain reflectometry (OTDR). A laser pulse is launched into an optical fiber. As the pulse passes through the fiber, energy is lost owing to scattering. The intensity of the back-scattered light decays exponentially with time, given uniform losses within the fiber. Therefore, knowing the speed of the light in the fiber, it is possible to convert this intensity against distance.

With a difference to conventional wire-line logging, where the sonde is moved along the section of the borehole to be measured, the DTS cable remains in place during the measurement of the temperature profiles. For long-term monitoring or in cases when full access to the interior of the borehole is needed, the sensor cables are installed behind the borehole casing. Because it is not necessary to move point sensor during measurement, the DTS technique enables us to make simultaneous monitoring of the temperature profile of the well at a time interval of a few to several minutes.

Perforation

To establish fluid communication between cased well and formation for production/injection. It uses high explosives with shaped charges like: Intiator/ detonator, detonating chord, shaped charges.

The two types of perforations are, over balanced method (+ve head) and under balanced method (-ve head). The positive head is achieved by keeping higher density mud in the borehole as compared to formation pressure. The advantage is it is easy to handle the well during perforation. The negative head is achieved by keeping lower density borehole fluid compared to formation. The advantage of the well is it can be straightway connected to the production pipeline. However caution is exercised to avoid blowouts. There is a system of *through tubing,* under balance perforation, using gun with big shot charges, having penetration power of a casing gun.

Bridge plugs are mainly used for isolation of zones in casing. This prevent the movement of fluid from either direction.

Back off services: Sometimes the drill string or tubing gets stuck in the bore hole. To release it a shock at a joint just above free point is given, when detonating cord is fired it release the string.

Puncture job: It is a perforation operation with specially design small charge so as to have big holes with lesser depth of penetration.

Production Logging

It is the measurement of fluid parameters on a zone-by-zone basis to yield information about the type and movement of fluids within and near the wellbore. Production logging is aimed primarily for knowing the performance of production wells. It provides information, depthwise where fluids such as water, oil and gas are entering a well and gives an indication about the situation of the perforations.

In production wells, the reservoir zone may be open-hole, perforated casing or gravel pack. There may be single or multiple zones and single or multiple tubings. The log is normally run in dynamic conditions, the well is flowing mixtures of liquids and gases.

It is also to measure flow rate downhole with a stationary production logging tool at or near the top of the reservoir. It involves measurements of flow, density and readings are used in quantitative production logging analysis. Temperature and pressure data have normally been used in a qualitative way to compute *in-situ* flow properties and locate zones of entry of fluid into a well. These tools consists of a number of sensors which make the measurements inside the well. The main types are: Flow rate (fluid velocity) measurements, wellbore temperature, Fluid density measurement (gradiomanometer), Radioactive tracer tool.

Spinner flow meter is a device for measuring *in-situ* the velocity of fluid flow in a production or injection well, based on the speed of rotation of a spinner. The spinner can be helical, that is, longer than it is wide, which is similar to a fan blade turned by the flowing fluid. The speed of rotation of the spinner is measured and related to the effective velocity of the fluid. The spinner revolutions generate electrical currents or pulses that are measured by the surface equipment and converted into spinner revolutions per second.

The most common device uses a small spinner, about 3.8 cm in diameter, allowing the logging tool to pass through the tubing and other restrictions before reaching the reservoir interval. The small spinner captures only part of the fluid flow in the casing. Other devices have been designed to capture full-bore spinner and various types such as the packer flow meter and the basket flow meter.

Fluid Density Tool, the **gradiomanometer** tool uses the pressure differential between two pressure sensors spaced a known distance apart to infer the density of the fluid between the sensors. The device uses a bellows system which will compress with pressure.

The ***radioactive tracer ejector tool*** carries a small quantity of radioactive solution into the well where it may be selectively released into the flow stream. Fluid movements in the well are then traced by means of one or more gamma ray detectors mounted above and/or below the surface- controlled ejector or "squirt gun". Radioactive tracer log is a record of the presence of tracer material placed in or around the borehole to measure fluid movement in injection wells. Radioactive tracer methods are sometimes used to establish flow profiles in water injection wells. The tracer solution must be density balanced with and soluble in the well fluid. The technique for recording is the tracer-loss measurement, in which a tracer material is added to the completion fluid and its progress monitored with a gamma ray tool.

In ***Panidighi field*** of Upper Assam, India, PLT data recorded to identify the source of water while flowing. It was observed from the PLT profiles that the water is not coming from the objects but it is entering the well bore through the squeezed perforated interval from high pressure water bearing sand lying below. Similarly, in northen part of ***Lakwa field***, good quality of PLT data was obtained and the source of water was identified within the reservoir and so the bean size was reduced and well was on nominal water cut.

Data Processing

The main processing task is to remove depth discrepancies between the different logging runs. Such discrepancies are caused by cable stretch, incomplete heave compensation, and by tides. The natural gamma ray log (SGR and HSGR) is generally used to match between the logging runs, as this log is recorded on all tool strings. One gamma log is chosen as the reference, on the basis of the length of the logged interval and data quality. The other gamma logs are matched to the reference using an automatic routine; the match of each log is checked to make sure distinctive peaks and troughs line up, and the match is adjusted. The resulting depth shifts are then applied to the other logs on the tool strings.

Signal processing can be performed at least three levels: down-hole in the tool, up-hole in the truck, and at a central computing center. Where the processing is done depends on the importance of processed result in relation to operational time involved and also technological considerations. Where it seems desirable, the logging tool is designed so that the data are processed down-hole and the processed signal is transmitted to the surface. The Log Interpretation Standard (LIS) is a standard for the exchange of well log information. Not only well log information, but also mud logging data are exchanged and stored on LIS format tapes.

Formation Evaluation

Formation evaluation can be defined as the practice of determining both the physical and chemical properties of rocks and the fluids they contain. The objective of formation evaluation is to locate, define, and produce from a given reservoir by drilling as few wells as possible. To this end, oil companies utilize a variety of formation evaluation methods (Dewan, J. T., 1983).

Well logs of deviated wells are generally displayed against measured depth along vertical axis. ***TVD log*** is the display of logs against True Vertical Depth. It is the log corrected for vertical shortening and thus gives the true depth of various zones identified by log motifs. However, TVD log does not take in to account the horizontal drift. This creates problems in log correlation of dipping beds. Two-dimensional displays of well logs resolve such problems and helps in proper correlation of logs. In conventional display of logs for every depth point there is a single value of each log in the recorded section. Two-dimensional display of logs involves TVD, Horizontal drift and the value of the log recorded.

The neutron porosity and resistivity alone provide no useful information about the productive capabilities of a reservoir. However, by using these data to calculate porosity and water saturation, some of the important questions regarding nature of lithology can be answered.

The objectives of reservoir evaluation by using well logs include: Computation of porosity (total, primary, secondary, effective). Computation of water saturation, Calculation of pay thickness and determination of the productivity of the well (Asquith, G. B., 1982).

This involves: determination of mobility, calculation of permeability, determination of formation pressure, determination of the lithology, facies and depositional environment to get an idea of reservoir geometry as well as petro-physical properties of the reservoir. Exact nature of the principal minerals and, if possible, of the accessory ones, since any error in mineral type can lead to significant errors in calculation of porosity and saturation.

Porosity determination using different logging devices relies on the knowledge of the rock type. In the case of the density tool, the density of the rock matrix must be known. The matrix travel time must be known to determine porosity from sonic log. In order to get porosity from neutron log the matrix setting for the neutron tool must correspond to the rock type. Determining these parameters is not much of a problem if one has good geological knowledge of the formation. If one is not sure about lithology, the graphical methods can be used.

The three types of porosity logs are available (sonic, density, neutron), are recorded in different units (microseconds per foot, grams per cubic centimeter, and neutron counts or percentage porosity unit). These logs can be compared if calibrated on a common reference scale. The most widely used scale for this purpose is defined in terms of equivalent units of limestone percentage porosity. The density scale may be transformed by setting the grain density of calcite (2.71) to zero porosity, the fluid density (1.00 for fresh water) to 100 per cent porosity and interpolating intermediate values.

By an analogous procedure the transit time of the sonic log can be converted to the limestone porosity equivalents by setting the matrix transit times of calcite and pore fluid as the two porosity extremes for interpolation. Limestone porosity scale is the most common reference of the neutron logs and hence requires no scaling.

An overlay of any combination of the three porosity will give immediate indications of the lithology of logged units by virtue of the different responses of matrix minerals to the individual porosity logs. In log interpretation the main focus is to identify the formations susceptible of containing displaceable fluid and the nature of the fluid content in the pores.

Another method of obtaining a single value for porosity from density porosity and neutron porosity data is through the use of the ***cross-plot porosity*** equation. The value obtained from this equation likewise may be assumed to represent the actual formation porosity, regardless of which matrix value was used during logging.

Cross-Plot Porosity $\Phi_{XPLOT} = \sqrt{\frac{\Phi_D^{\ 2} + \Phi_N^{\ 2}}{2}}$

The presence of gas in a formation has a profound effect on the neutron porosity measurement. Because the tool measures hydrogen index, the low hydrogen density of gas (compared to hydrogen density of liquids) causes the neutron tool to underestimate porosity. Gas also affects the density measurement, causing density porosity to be overestimated.

The empirical relationships between resistivities, fluid types and porosity.

F (formation factor) **=Ro/Rw** Where:

- ☆ Ro is bulk resistivity if pore space is filled 100 per cent with brine (connate water).
- ☆ Rw is resistivity of the connate water itself.

This relation applies best for clean (clay free) rocks of constant porosity when Rw < 1 Ohm-m, at 25°C. The Ro is not really measurable in the field. But the relation can be made useful because the resistivity ratio (*i.e.* F) changes consistently as porosity changes.

Also the Formation factor and porosity are usually related: -

F=a/m

a is a constant between 0.6 and 1.5. It is often left at 1.

m is the cementation exponent

The cementation exponent is usually 1.5-1.8 in sandstones and 2.0 in limestones, dolomites and tight consolidated sandstones.

The **water saturation (Sw)** and formation resistivity were also found to be related. Water saturation is the percentage of the pore space filled with water, which is of course important when evaluating a potential hydrocarbon reservoir. Sw=1 means all fluids are water, and Sw=0.1 means 10 per cent of fluids are water, implying 90 per cent are non-conductive or oil/gas.

The relation is Sw = (R0/RT)1/n where,

n is a saturation exponent, usually close to 2.0.

Ro = rock resistivity with 100 per cent brine for fluid,

Rt = true resistivity (with hydrocarbons and water in pore space).

Archie's Equation (combined)

Sw=nv′aRw/O^m Rt

where

Sw= Water saturation

n= Saturation exponent (usually near 2)

a=empirical constant usually near unity

Rw= resitivity at formation water –m

O^m = Porosity fraction, m cementation exponent (usually near 2)

Rt=True formation resitivity –m

The above formula is being used to estimate water saturation in formation, and indirectly oil saturation as well. The a, m,n and Rw are constants over a zone, which means to estimate Sw we need to measure Rt and Phi. Logging techniques to estimate Rt have been developed, however, to estimate porosity we have to use indirect methods, such as density, sonic and Neutron logs.

This relation is commonly used for oil/gas reservoir characterization. Three parameters must be measured: (i) porosity, (ii) resistivity of the undisturbed formation (Rt), and (iii) resistivity of connate water (Rw). Rw can be obtained from laboratory analysis of samples extracted from the formation rocks or from the SP log.

So obtaining accurate values for water saturation is the primary goal of open-hole log analysis. With knowledge of water saturation, it is possible to determine what percentage of porosity is filled with a fluid other than water (*i.e.*, hydrocarbons) and therefore, hydrocarbons inplace.

The basis for log analysis is to compare the measured resistivity of a formation with the calculated resistivity of that formation assuming its porosity is 100 per cent water-filled. The resistivity of a rock at 100 per cent water saturation is referred to as wet resistivity (Ro). If, for a given porosity, the measured resistivity is significantly higher than the wet resistivity, then the presence of hydrocarbons is indicated.

Log analysis calculations require values of resistivity, in particular mud filtrate resistivity (**Rmf**) and formation water resistivity (**Rw**). A single measured or calculated value of Rmf and/or Rw may need to be applied over a wide range of depths. Resistivity decreases with increasing temperature, hence, requires resistivities be corrected for the appropriate temperatures at depth. So stabilized bottom-hole temperature and geothermal gradient of the area are needed to calculate exact temperature at the desired reservoir depth. The value of Rmf may be obtained from the mud resistivity (Rm) through the cross-plot, in case not readily available.

Determining an accurate value of formation water resistivity (Rw) from logs is often quite difficult. A zone that is assumed to be 100 per cent water saturated may not be in actuality. The presence of hydrocarbons may suppress any SP deflections, resulting in erroneous calculations. Furthermore, in a slightly shaly formation, clay minerals may result in abnormally low resistivities.

Determination of Rw

The relation between the SP and formation water and mud filtrate is known as

SP= -K*log (Rmf/Rw)

SP measured in millivolts and K is dependent on temperature

K= (T+505)/8

where T is the formation temperature in degrees Fahrenheit

But this equation does not explain the true electrochemical behaviour of the salt solutions. The actual development of SP log is controlled by relative activity of formation water and mud filtrate.

So the equation is written as SP= -K*log (Aw/Amf)

Aw and Amf are activities of formation water and mud filtrate.

The interpretation of logs in Tipam Sandstones of ***Lakwa and Lakhmani*** was a problem because of the very fresh formation waters and varying amounts of clay minerals. Also values of formation water resistivity (Rw), even from samples collected during testing, do not give a consistent value it was observed for TS-I and TS-2 sands of Lakwa that the water saturations computed from the Archie's equation were quite high 40 to 60 per cent for even those zones which produced clean oil. In Lakhmani, the problem got still more aggravated, in that it was not possible even to distinguish the oil water contacts from the resistivity logs.

Permeability

The most commonly used permeability indicators are the Micro Electric (or Microlog) and the Spontaneous Potential (SP) tools. The Microlog indicates permeability when there is separation between the Micro-normal (or Normal) and Micro-inverse (or Lateral) curves. The Micro-normal curve will read a higher resistivity than the Micro-inverse curve because of the effects of mudcake (**Rmc**) on the resistivity measurements. Mudcake can only be present opposite a permeable formation, therefore the presence of this separation is used as a qualitative indicator of permeability.

The productive horizons of ***Rudrasagar field*** of Upper Assam, India show well marked transition zone in different parts of the field. The transition zone is not uniform throughout, but is marked by the presence of hard streaks or shales. Under such circumstances, it was not possible to estimate permeability values objectively for all wells. The data have been compared with the permeability data obtained from cores and production tests by plotting. There is a wide scatter on the plot of permeability from cores vs. permeability and it is not possible to draw an average line giving the trend of relationship. Even though a fair amount of scatter, it is possible to draw an average line giving a general trend of relationship. An idea of the order of magnitude of permabilities can be had from this average line. Higher permeability values from cores are possibly due to the fact that they represent air permeability values which are usually higher than liquid permeability values. Furthermore, the absence of the overburden pressure during laboratory measurement of core permeability could have been responsible for the generally high values. In fact, there is wide discordance between permeability values obtained from core and production data.

Estimation of Shale Volume

For estimation of volume of shale (**Vsh**), gamma ray and spontaneous potential (SP) are common methods. Other measurements can be used under special conditions to estimate shale volume, such as the resistivity in very high resistivity formations, the compensated neutron in very low porosity formations (Asquith, G. B., 1985).

Shale Volume from Gamma Ray

vgr = (Grlog – Grmin)/(Grmax – Grmin)

where,

Grlog – gamma ray from log.

Grmin – Minimum gamma ray value from the log on sand portion.

Grmax – Maximum gamma ray value from the log on shale portion.

Shale Volume from SP

vsp = (SSP – PSP)/(SSP)

where,

PSP – pseudo static potential, *i.e.*, the SP read in the water bearing shaly sand zone.

SSP – static spontaneous potential, *i.e.*, maximum SP value in a clean sand zone.

Shale Volume from Resistivity

VshRt = a0.59 + 1.44a – 1.29 a^2

where,

a = (Rlim   Rt)RSh/(Rlim – RSh)Rt. – Rlim – max resistivity in the entire zone

Rt – Laterologdeep (LLD) value at each depth.

Rsh – Maximum resistivity value in the shale portion.

Shale Volume from Density

VND = (ØN – (ØDØN)max – ØD)/(ØNSh(ØDØN)max – (ØDSh)

ØN- Neutron value at each depth from log on curve.

ØD - Porosity value from Density log.

ØNSh - Maximum value at shale portion from log on TNPH curve.

ØDsh - Maximum value at shale portion from density curve.

ØD = (pb – plog)/(pb – pf)

where,

ØD Porosity log from density log.

pb Bulk density of the formation.

plog density value at each depth from log.

pf density of the formation fluid.

Matrix Values- pma (sandstone) = 2.65gm/cc, pma (lime stone) = 2.71gm/cc, pfluid = 1.0gm/cc.

In ***Lakwa and Lakhmani*** fields of Upper Assam, India, no relationship of cation exchange capacity with weight percentages of the finer fractions comprising the rock samples could be established. The electrical core measurements and their plots indicate that, there is a general increase of formation resistivity factor with shaliness. Average Qv values for TS-2 reservoirs of Lakhmani are around 0.1 which happens to be half of the corresponding value reported for Lakwa field. This is also reverse to the general perception that Lakhmani reservoirs are low resistive as compared to Lakwa. For Geleki well, fairly good relation was noticed between the cation exchange capacity per unit pore volume and the shale percentage of particles less than 5 microns in diameter (Koitharaj, *et al.*, 1971).

In ***Lakwa Field***, the TS2, the largest and most extensive pay sand and highly permeable except its shaly silty top portion which is less permeable and fining upward. While the highly permeable lower portion produces with high influx and generally high water cut, the less permeable top silty portion gives low influx and generally negligible water cut (Figure 7.20). The overlying subsidiary of TS2, designated as TS2S. In a well after 20 years of production, water-cut was started and even after some efforts to shift the producing interval to higher level became futile. After the layer transfer in the top silty portion, total liquid rate reduced and the water cut also reduced. The above understanding will be helpful in better exploitation of TS2 reservoir (Ashok Kumar, *et al.*, 2010).

In eastern ***Cambay Basin*** in the ***Gamij*** and **Padra** fields, the Deecan Trap basaltic rocks form the technical basement and is proven hydrocarbon producer. The Padra field showed moderately weathered basalt as the reservoir rock, and identification and evaluation of oil bearing zones in the basement section using well logs. The same method to the analysis of basaltic basement in the Gamij field. The analysis of well logs of the basaltic basement section of Gamij field by a method based on cross plot, seems to be much oil left in the basement section a well below 1240m (Ashok Kumar, 2003).

In ***Ankleshwar*** field of Gujarat, India, silty/shaly zones within Hazad and Ardol members were found to be oil producer. So the main objective was to identify the prospective zones within the silty reservoirs by integrating the log, testing and other geological information using reverse modeling. For this purpose, different log overlays mainly SP–Resistivity, Resistivity–GR and Density–GR have been used. Out of these combination first two overlays along with other log information were more useful for identification of reservoirs, whereas the overlay of Density Vs GR

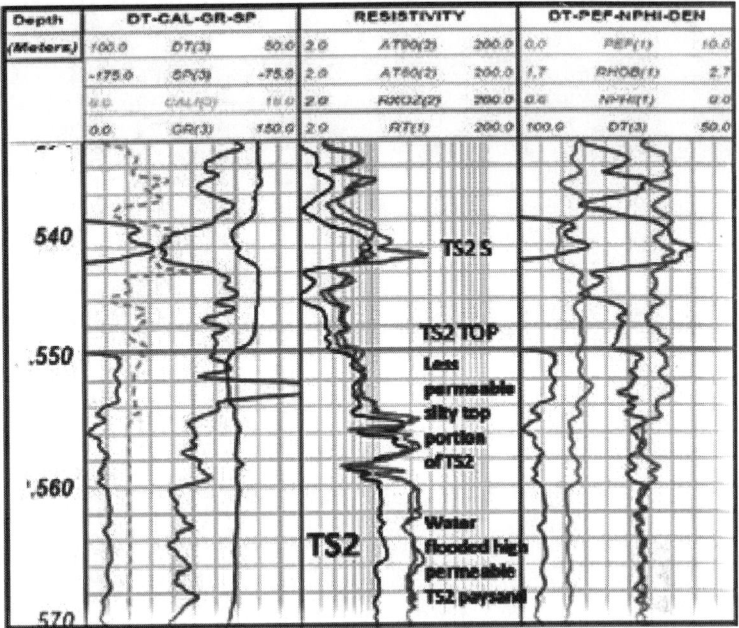

Figure 7.20. Processed Log in Major Lakwa Reservoir (After Ashok Kumar, *et al.*, 2010).

(in reverse scale) found to be good. After analyzing the criteria for identifying from hydrocarbon point of view.

- The zone should have significant SP development and moderately clean (preferably GR<30 API).
- Presence of mud-cake from caliper log is also desirable which indicates the presence of porous/permeable zone.
- Resistivity should be > 2 ohm-m, Density should be < 2.3gms/cc and Neutron should have a tendency to match with density.

The ***Demulgaon*** field near to Lakwa field was put to oil production from Barail Main Sand (BMS) of Oligocene since 1985. BMS is a layered sandstone reservoir with the top two layers L-1 and L-II being important from hydrocarbon point of view. The most prospective top most layer L-I is further sub divided into L-IA and L-IB. The Layer L-IA comprises of mainly silty sandstone whereas L-IB is the cleaner most sandstone layer of BMS. Water injection in BMS was started in 1993. To identify the source of high water cut, the open hole well log responses of wells drilled before and after water injection were studied and integrated with the available core studies. The study concludes that, the preferential breakthrough of injected water through high permeability streak identified in layer L-IB of BMS may be the cause of excessively high water cut and hence decline in oil production.

The evaluation of low resistivity reservoirs developed within ***Kamalapuram field*** of Cauvery Basin, India, is warranted because the conventional interpretation of the pay sands are characterized with low resistivity, erratic GR response and higher Neutron porosity. The SP response is also not consistent with production profiles of wells. The analysis of various logs and lithological cross-plots suggest sand-silt-shale composition of these reservoirs. The computed shale volumes are higher. The computed effective porosities are found to be lower, because of higher shale volumes. The multi-mineral model with quartz, silt and clay as the major constituents of the formation will be most suitable for evaluating such reservoirs. The computed results are consistent with production data.

In the gas reservoir of Upper and Middle Bhuban formations of ***Baramura field*** of Arakan Basin, India, instead of conventional 'a', 'm' and 'n' parameters being used, the laboratory determined values were found to be more suited. The data were validated against water bearing zones. The hydrocarbon saturations for Upper and Middle Bhuban formations have increased by about 5 to 10 per cent as compared with the earlier estimated values using conventional parameters.

Encouraging log responses in Bokabil sands in ***Agartala Dome*** has necessitated exploring shallow prospects also in Tripura basin. The Neutron-Density overlay technique is used to identify gas sand. In shaly sand reservoirs, this technique fails to detect gas. Sonic travel time is around 100 m sec/ft, whereas as density is 2.5 gm/cc against shale. It may be due to the presence of some clay mineral like Kaolinite and illite and heavy mineral like mica. Due to these reasons, the density and neutron porosity are comparatively higher in shaly sand. Hence, Neutron-Density cross-over is not observed clearly against shaly sand reservoirs. Neutron porosity (∅N) – Sonic travel time (DT) overlay technique has been used successfully to detect gas and found useful to detect gas in shaly sand reservoirs encountered in Bokabil and Bhuban formations.

Identification of Over-Pressured Zones

The bulk density of shale increases with compaction, the departure from this trend is observed in over-pressured zones in which shale density decreases with increasing depth. A high-density zone, top barrier, usually lies at the top of the interval of decreased density. Density logs run at intervals during the drilling of a well can be used to predict abnormally pressured zones so that precautions can be taken to eliminate possible hazards (Anderson, R.A., Ingram, J.S., and Zanier, A.M. 1972). Variations in sonic travel time and formation resistivity are good indicators for identifying the high pressure zones. The plot of sonic travel time on lograthimic scale against depth provides most reliable results.

In order to know the beginning of high-pressure zones in Wamaj low and Tankari low of Ahmedabad and Baroach blocs of ***Cambay Basin***, India, deep wells data were studied. The top of high-pressure zone lies just above the Cambay Shale in all the wells and maximum pressure is observed within Cambay Shale. The maximum pressure computed is in accordance with mud weight used. After reaching the maximum pressure, the formations in both the blocks follow the normal compaction profile but pressure never returns to hydrostatic level.

The bulk-density of shales measured during drilling for pressure prediction purposes in well Bodra-2 situated in the deep basinal part of ***Bengal Basin***, were utilized for checking the accuracy of FDC log as the measurements were available below 2700 m only. The availability of shale cutting was restricted to 2700-3760 m in water base mud and 3760-4172m in oil base mud.

The sonic values are converted to bulk density values with the help of cross plot relationship. These derived bulk density values are plotted along with the corrected FDC values to make up the curve. The bulk - density versus depth curve, thus obtained, has been extrapolated from 400 m upward to surface. Formation pressure, overburden pressure and fracture pressure data have been plotted versus depth for well Bodra-2 in terms of gradients equivalent to mud weight (MWE).

It can be seen that the formation and fracture pressures are very close to each other near the bottom of the well. The proximity of the overburden stress gradient with formation and fracture pressures indicates that most of the overburden is resting on the fluid and the stress on the matrix is tending towards zero. The well has been drilled in the lower section by under balancing mud to the extent that ECD covers up the formation pressure because there is very less difference between formation and fracture pressure. If increase mud loss was observed due to fracturing.

Depositional Model Using Electrical-Logs

The ***Gandhar field*** located in South Cambay Basin represents sequences of mouth-bars and deltaic distributary channels. Some of these sands have become prolific hydrocarbon producing in spite of low resistivities. The formation water salinities also vary widely, both laterally and vertically. The water salinities vary from 1 gm/litre to 35gm/litre (*i.e.*, 1.5 Ohm m to 0.06 Ohm m). It is found to increase from bottom to top and laterally it increases from SSE (which is input direction) to NNE in the field.

The standard logs have been analyzed along with processed dipmeter logs used for stratigraphic interpretation. The open hole logs of well located on the crestal part cover the sand in which the maximum resistivity is around 3 ohm-mt with the average value of resistivity is around 1.7 Ohm-mt. The Gamma Ray log indicates the sand is cleaner and gross coarsening up sequence upwards culminating in the transgressive Kanwa Shale. The Density-Neutron log do not indicate clean reservoir sands or the presence of hydrocarbons besides a shaly sand section 4 m thickness. The processed results show movable hydrocarbons against the shale layer. Anomalously low resistivity sand units which are likely to produce hydrocarbons have been identified on log evaluation. One reason could be a localized sea water influx possible along a creek.

Corrections for Perfectness (Cross-plots)

Factors affecting log responses may include hole size, mud weight, bed thickness, depth of invasion, and other properties of the logging environment and formation. Many of these effects have strong impacts on analysis and evaluation and must be corrected for prior to evaluating the formation. Many cross-plots are available for these corrections. Sometimes the correction obtained can be so less or insignificant. In case of modern tools the software within the tool also taking care for required application of the correction.

Resistivity tool responses generally must be corrected for a variety of conditions; namely, borehole effects, thin beds, and invasion. Microresistivity measurements must be corrected for the presence of mudcake. Just as neutron corrections were particular to tool type, resistivity corrections also are particular to tool type.

Some of the most important corrections include those required for neutron porosity measurements. Neutron porosity is affected by a number of factors, most importantly temperature. If these effects are left uncorrected, then neutron porosity will be in error and the Archie equation will not yield accurate results.

Gamma Ray value also need correction for two separate effects, the distance between the tool and the borehole wall and the density of the mud filling the borehole. Both of these conditions affect the amount of natural radiation that is measured at the gamma ray detector.

In relatively thin beds the deflection of the SP curve is suppressed because of the influence of underlying and overlying formations. This is particularly true in sand/shale sequences where sandstones are overlain and underlain by shales.

After applying necessary corrections based on other log values, readily usable computer processed logs are available on need based called Para-logs

The objective of the **Faciolog** program is to segment a well, or part of a well, into log-consistent layers;, *i.e.,* layers where all levels of the layer have similar log responses. The number of levels, or clusters, is then reduced until each cluster can be related to a meaningful geological facies. The environmentally corrected logs are transferred into principal component logs.

Electrofacies

The term electrofacies was defined as the set of log responses which characterizes a bed and permits to be identifiable from others." Electrofacies can usually be assigned to one or more litho-facies as log responses are measurements of the physical properties of rocks (Serra and Abbott, 1985). Identifying facies may help in petroleum prospecting and reservoir characterization. Traditionally facies have been identified manually with the aid of graphical techniques like cross-plotting from wire-line logs and thereby correlating their behaviour to cores. Many mathematical models have been introduced to automate the task of facies identification.

The main idea of the present comparison was to predict electrofacies accurately which has very significant bearing on other reservoir parameter calculations like permeability, lithology and depositional facies. (Bishnu Kumar and Mahendra Kishore,2006).

In a geological set-up where correlatable markers do not exist, the lateral facies variation and extent of the reservoir become difficult to decipher. The electrofacies analysis in such cases happens to be an effective instrument for understanding the depositional model. The environment of deposition, the litho-facies encountered while drilling and electrofacies analysis using syndip and dip-meter data provided a clue to sand geometry (Figure 7.21).

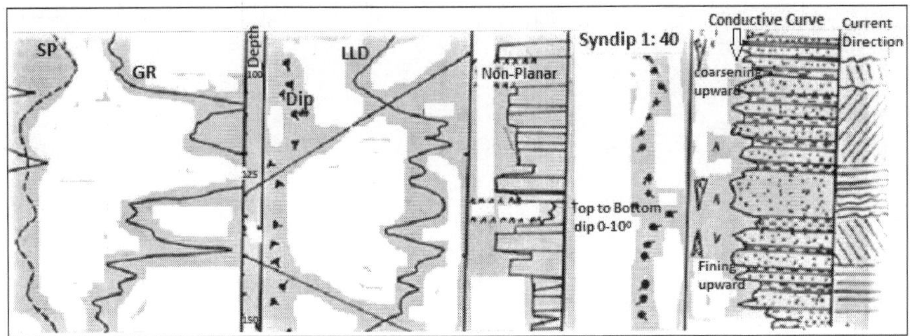

Figure 7.21. Micro-electrofacies Analysis-K G Basin.

In a well in KG Basin at 2,540 m sand bed of 50m thick was encountered and to understand depositional style the complete elctrofacies analysis was carried out, which indicated very thin sand beds with both coarsening and fining upward channel and sand bars (Figure 7.22). Similarly, in wells drilled near Kaza –Kaikaluru Horst area, certain interval of high gamma zone was encountered. The neutron gamma spectral log showed that high gamma was due to thorium. In the Easternghat terrain some places are known to have radioactive black sands in the beaches (Rao.G N., 1993). This zone possibly similar in the subsurface at that point of geological time (Figures 7.23).

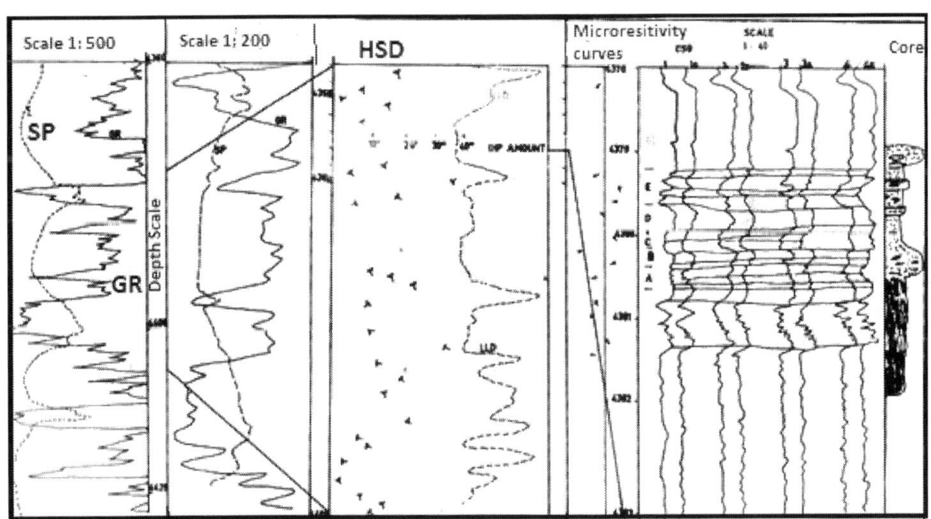

Figure 7.22. Sedimentary Pattern from Dipmeter Data and Core-K G Basin.

Electrofacies analysis was carried out in G-1 and G-2 area in ***Godavari offshore*** to understand the depositional model for further exploration. Identified eight electrofacies in G-1 area and six in G-2 area. These electrofacies could be linked to five despositonal sequences which were deposited under two drainage system. The sands are found to be lenticular in nature and they thin out in the landward as well as in the basin ward direction (Bakshi., S. S.1992).

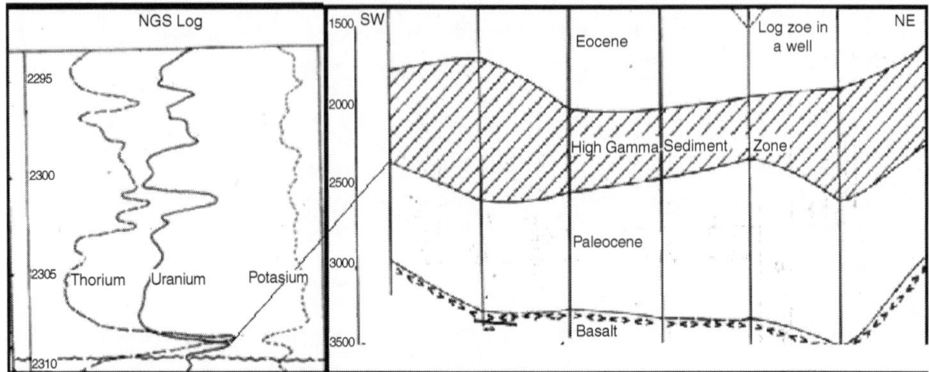

Figure 7.23. NGS-Spectral data to know nature of deposition-K G Basin

Chapter 8

Reservoir Engineering

The lithological unit which produces the hydrocarbons in commercial quantities analyzed after the discovery is called a reservoir. A reservoir is formed of one (or more) subsurface rock formations containing liquid and/or gaseous hydrocarbons, which have porosity and permeability located in a large structural or stratigraphic area, to produce sufficiently more recoverable oil and gas. In general, sedimentary rocks like sandstones and limestone do have depositional porosity which can be enhanced by fracturing or reduced by compaction. Sometimes, even hard rocks with fracture porosity may hold hydrocarbons.

A reservoir life begins with exploration that leads to discovery, which is followed by delineation of the reservoir, development of the field, production by primary, secondary and tertiary means and finally abandonment. In the process of going through these various steps, it is absolutely necessary that sound technical and economic decisions are taken at appropriate stage to achieve economic recovery of oil and gas from the reservoir.

The goal of reservoir engineering starting with the discovery of a reservoir, is to set up a development project that attempts to optimize the hydrocarbon recovery as part of an overall economic policy. Reservoir specialists thus continue to study the reservoir throughout the life of the field to derive the information required for optimal production from the reservoir. So the reservoir engineer applies scientific principles to the drainage problems arising during the development and production of oil and gas reservoirs so as to obtain an economic recovery. The scope of reservoir engineering includes the analysis of optimum ways for recovering oil and natural gas, and the study and implementation of enhanced recovery techniques for increasing the recovery above that which can be expected from the use of conventional and available technology (Amyx, J., Bass, D., and Whitney, R., 1960).

So the reservoir engineer must accordingly estimate both inplace and efficiently recoverable reserves, with the aim of optimal profitability for a given reservoir. One has to estimate even the well production potential which include initial productivity and later changes if any.

The data for the reservoir engineer are subsurface geology, applied mathematics, and the basic laws of physics and chemistry governing the behavior of liquid and vapor phases of crude oil, natural gas, and water in reservoir rock. This aspect also include numerical reservoir modeling, production forecasting, well testing, well drilling and work-over planning, economic modeling, and PVT analysis of reservoir fluids.

The image of the reservoir develops when the nature, boundaries, internal heterogeneities, distribution and volumes of the fluids contained in the reservoir are known. The methods used are partly included under the term of reservoir geology, and are based on petroleum geology and geophysics.

- ☆ *Direct analysis*: core analysis, and PVT analysis of fluid (Pressure, Volume, Temperature). These measurements are performed in the laboratory.
- ☆ *Indirect analysis*: well logging, in which some logs are recorded during drilling and some during production, these wire-line logs help to obtain vital data on the lithology, porosity and fluid saturation (Dake, L.P., 1977).

Structural Position within the Reservoir

The gravity segregation causes a natural stratification of reservoir fluids, with gas on top of oil and oil over water. In the absence of inter-connected porosity though the gas, oil, and water will form distinct layers.

The upper level is mainly oil; the lower level is all water, while the middle level shows ever-increasing concentrations of water with depth. The two dimensional plan view along with sectional view of the oil accumulation have to be understood with three dimensional extension or existence of the reservoir (Figure 8.1).

Tilted OWC

If the water in a reservoir rock is in motion, the oil water contact assumes the same dip as that of the water's movement. Hydrodynamic traps of various shapes and sizes, depending on the characteristics of the porous-permeable rock in which the reservoir is located, may thus be generated. A tilted oil-water contact could be caused by a change in permeability across the reservoir coupled with water dynamics.

Physical Characteristics of Reservoirs

Porosity

It is defined as presence of openings in rock. The total (absolute) porosity ***m*** is defined as the ratio (**m**=Vpor/Vo) between the total pore space volume Vpore and apparent volume of the rock (Vo). Porosity is expressed as a percentage that is mX100.

To a large extent the properties of porous media are governed by the size of the pore channels which are classified as capillary and sub-capillary (Figure 8.2). Capillary when diameter ranges appreciable, if less than that, it is sub-capillary.

Fluids cannot move in sub-capillary pores due to small distance between the sides of the channels or pores.

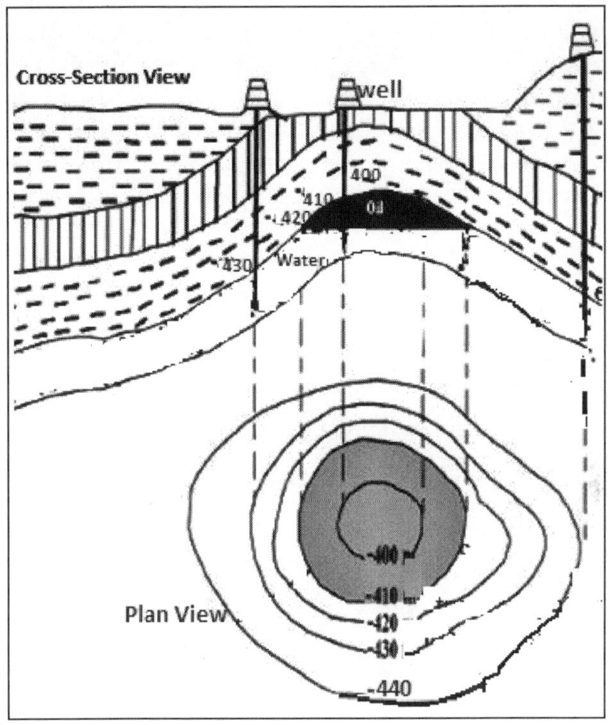

Figure 8.1. Section and Plan View of Oil Field.

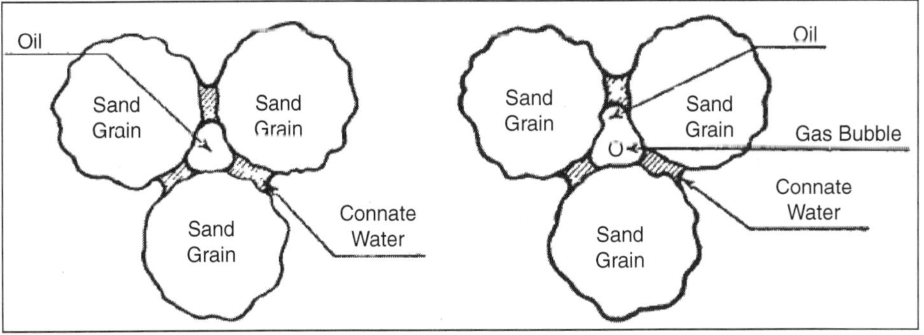

Figure 8.2. Grain Porosity.

More over fluids do not move through pores which do not have inter-connection. So in reality, not all pores permit the movement of fluids. The effective volume of pores of the rock through which fluids can move to the apparent volume. The dynamic porosity factor is defined as the ratio of the volume of a moving fluid in the rock to the apparent volume. So the dynamic porosity factor, does not take in to account either of the in-effective pores or volume of effective pores occupies by static films of fluid adhered to the surface of rock particles or of the volume of fluid retained by capillary forces.

Hence, two types of porosity can exist in the rock, total and effective. Total porosity contains all the pore spaces whether they are connected or isolated, while effective porosity refers only to the interconnected pore spaces. This value is of interest in reservoir engineering; however, in many reservoirs the difference between the two porosities is negligible. The difference may be significant in highly vuggy or fractured reservoirs, where some vugs or fractures may be isolated (Figure 8.3).

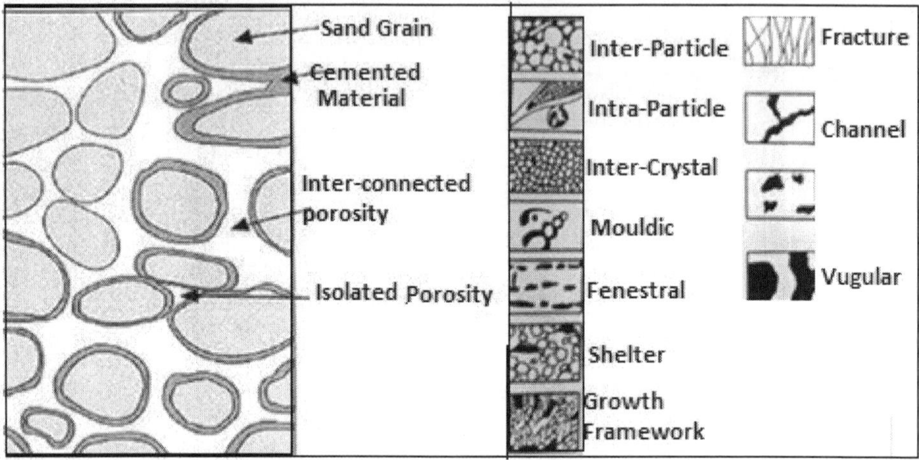

Figure 8.3. Porosity Types.

Various methods exist for measuring porosity. Some are based on measurements *of* a rock sample's bulk volume and solid volume, and obtain the pore volume by subtracting the solid from the bulk volume.

Other methods are based on measuring the pore volume directly in addition to the bulk volume. Such methods utilize gas expansion, fluid saturation, or mercury injection. Porosity measured by these techniques is the effective porosity. Porosity in sandstones vary from 3 to 30 percent, whereas in Limestone and dolomites 1 to 33 per cent.

Permeability

The permeability of rock is defined as their ability to permit the movement of oil and gas and water through them. If sufficient pressure is applied, fluids can be forced to move through all bodies. The permeability of a porous medium may vary depending on the substance conditions that are moving through it and on the nature of the movement (Figure 8.4). Permeability is a measure of the ability of porous rock to transmit fluid (Berg, R.R., 1970).

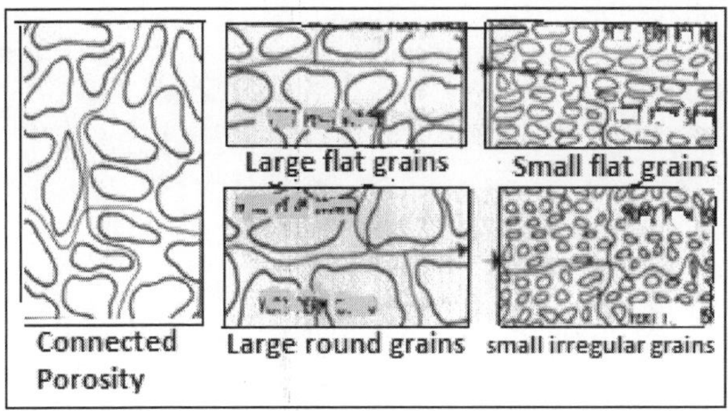

Figure 8.4. Permeability Types.

Absolute permeability is defined as the permeability of a porous medium when any single phase (gas or homogeneous liquid) is moving through it. The absolute permeability factor depends only on the physical properties of the rock. It has been established that, when a fluid moves through a porous medium, permeability depends on the physico-chemical properties of the fluid. That is why permeability to gas is usually accepted as the absolute permeability of a rock. Absolute permeability is calculated by Darcy's law using laboratory-measured data.

Unit of Permeability

The permeability of one **Darcy** may be defined as that permeability which will allow the flow of one cubic centimetre per second of a fluid of viscosity one centipoise through a rock sample of one square centimetre in cross-sectional area under a pressure gradient of one atmosphere per centimetre.

A permeability of one darcy is a large value, and hence normal usage of the unit is millidarcy (one thousandth of a darcy) to describe the permeability of most reservoirs. In some reservoirs the permeability may be as low as a fraction of a millidarcy, while in others it may be several milli-darcies. The well-flow rate is directly proportional to permeability. Thus, wells with very low permeabilities are normally marginally productive, and may require remedial action to improve their production.

Effective permeability is defined as permeability of a rock to a given fluid when a multi phase system is present which are moving through the pores. Effective permeability not only depends on physical proprieties of the rock but also on the extent of its saturation with oil, gas or water and on their physical and chemical properties. The effective permeability to oil, gas and water is represented by symbols ko, kg, and kw respectively (Chilingar, G.V., 1964).

Relative permeability of a porous medium is the ratio between effective and absolute permeability. The relative permeability is defined as the ratio of effective permeability of a fluid at a given saturation to the effective permeability of that fluid at hundred percent saturation or absolute permeability of the formation. The relative permeability for oil, water and gas is expressed by symbols kro, krw and krg respectively. The relative permeability curves are normally established through experiments in lab on cores of the formation. The other ways of calculating relative permeability are using capillary pressure data and from field performance data. Relative permeability is a function of saturation. A critical saturation value must occur before relative permeability exceeds zero, that is to say, before the fluid starts to flow.

All of these equations assume that only one fluid saturates the porous media; thus **k** is the absolute permeability. However, as mentioned previously, water is always present. Furthermore, in oil reservoirs, oil, gas, and water exist together. In such cases, one must use the effective permeability to the phase of interest in place of the specific permeability. Normally, one replaces the effective permeability by keff = krk where: kr = the relative permeability (Doyen, P., 1988).

Permeability of rocks can be determined quantitatively using Darcy's linear law of filtration according to which the rate of filtration of fluid through a porous medium is proportional to pressure differential and inversely proportional to viscosity (Ahmed, U., *et al.*, 1989).

Darcy's Law

Darcy's law is an empirical relation that describes the fluid flow in porous media as a function of pressure gradient and the viscosity of the fluid. It is basically an extension of the principles of fluid dynamics to flow of fluids in porous media and represents the equation of fluid motion. The rate of flow through the sand bed was proportional to the pressure head above the bed and to the cross-sectional area of the filter, and inversely proportional to the viscosity of the water and the thickness of the bed. The constant in Darcy's equation was written as a ratio of permeability to viscosity.

This equation, stated in differential form is

$$V = - K\, dp/\mu dl$$

where,

V = apparent fluid flowing velocity- cm/sec.

K = proportionality constant, or permeability- darcies.

μ = viscosity of the flowing fluid-centipoise

dp = pressure drop per unit length, atmospheres/cm

The velocity 'V', in the above equation is not the actual velocity of the flowing fluid, but is the apparent velocity determined by dividing the flow rate by the cross-sectional area (A) across which fluid is flowing.

Substituting, Q/A in place of 'V' in the above equation and solving for q results in

$$Q = - KA\, dp/\mu\, dt$$

Note that the negative sign in front of the equation is needed to obtain a positive q, since dp/dt is negative.

Wettability

It is effect of energy of adhesion. Wettability is the result of combination of all the active surface forces prevailing between the fluids and the rock surface in the reservoir. This property governs the tendency of a particular fluid to adhere to the porous rock in preference to the other fluid. The tendency of liquid to spread over the surface of a solid is an indication of the 'Wetting' characteristics of the liquid for the solid. It can be measured the contact angle between the liquid and solid surfaces. This angle is always measured through the liquid to the solid, is called the contact angle 'Ø'. In general in oil-water-rock system, the solid is said to be non-wet by water when the contact angle 'Ø' is greater than 90°, and water-wet when the angle is less than 90°.

The wetting phase fluid will preferentially cover the entire solid surface of the reservoir rock and will be held in the smaller pore spaces of the rock because of the action of capillarity. On the other hand, the non-wetting phase will tend to be expelled from contact with the surface of the rock. Thus at small saturations, the non-wetting phase will tend to collect in the larger openings of the reservoir rock.

In such cases it becomes important that which fluid wets the grain surface as relative strategy for production and recovery optimization will depend. In general smaller the grain water will adhere to it and bigger the grain oil may be adhered to. This percentage of wettability is calculated using well logs and called Sw (water saturation with in oil column).

The capillary pressure/saturation relationship depends on the interaction of wettability, pore structure, initial saturation, and saturation history. No simple relationship exists that relates the capillary pressure determined at two different wettabilities. Therefore, the most accurate measurements are made with cores that have native reservoir wettability.

Capillary Pressure

Capillary pressure in porous media is defined simply as the pressure difference existing across the interface separating two immiscible fluids, one of which wets the surfaces of the rock in preference to the other (Figure 8.5). The pressure required to displace the wetting phase from the core is exactly equal to the capillary forces holding the remaining water within the core after equilibrium has been reached, the pressure data can be plotted as capillary pressure data. The water saturation reaches some minimum saturation which is equivalent to the connate water saturation (Dumore, J.M., 1974).

Using the capillary pressure equation for a single capillary tube

$$Pc = h \times \Delta P \times g = 2\gamma \cos \varnothing\ r$$

where,

Pc = capillary pressure

h = height above the free liquid surface

ΔP = difference in density of the two fields

g = acceleration due to gravity

γ = interfacial tension between the fluids

$\varnothing$ = contact angle

r = radius of capillary

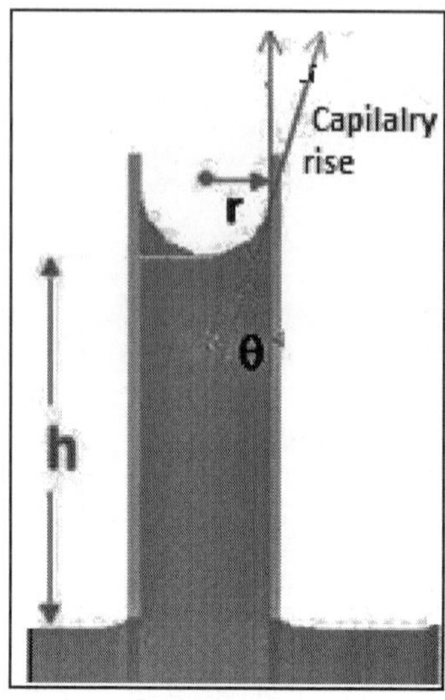

Figure 8.5. Capillary Pressurs.

The capillary pressure can be explained as the combined effect of surface tension, interfacial liquid tensions, pore size and shape, and the wetting properties of the reservoir rock. The force of capillary pressure is not used directly as a means of reservoir control or evaluation but as an aid in explaining the distribution of connate water through a reservoir and as a tool in detailed reservoir studies. There is no such thing as an oil-water contact, but there is a gradual transition from 100 per cent water lower level to maximum oil saturation above. If the oil-water contact is defined as the lower point in the reservoir, which will produce 100 per cent oil, transition zone in between produces oil with water if perforated.

Changes in pore size and changes in reservoir fluid densities will alter the shape of the capillary pressure curve and thickness of transition zone. If all other factors remain unchanged, a low API gravity oil reservoir with an oil-water contact will have a longer transition zone than a high API gravity oil reservoir. So the density of oil and size of the pore may control the thickness of the transition zone.

The pore size can often be related approximately to permeability and hence high permeability reservoirs will have shorter transition zones than low permeability reservoirs. It is the pressure differential between mobile and interlaced fluid phases within the pores of the reservoir. It is the suction capacity of the rock for a fluid phase that will wet the rock.

Formation Pressure

The liquids and gases in reservoir are under the so called formation pressure, which is the pressure exerted by the overlying rocks of the reservoir at given depth. So depth of the oil or gas pool decides the formation pressure. So if the oil reservoir stratum dips at steep angle, the pressure exerted in different parts of the reservoir rock may vary considerably, more towards flanks and less at upper part of the reservoir. Similar is the case in sub-surface reservoirs located under higher topographic terrains.

When formation pressure is measured by lowering pressure gauges to a desired depth in a drilled well it called Bottom Hole Pressure. Under the formation conditions density of oil is influenced by pressure, dissolved gas and temperature. Volume of liquid will increase when gas is dissolved in it. Gas Oil Ratio (GOR) is the good indication of the overall composition of the fluid. Production with GOR greater than 10,000 SCF/bbl is commonly called lean or dry gas. The GOR and consequently API gravity of the liquid will vary with pressure and temperature.

Connate Water

Connate water is water trapped in the interstices of sedimentary rocks at the time of their deposition. This water was displaced by oil and gas migrating into the trap. The water displacement was not 100 per cent because of the natural capillary forces present. The reservoir, at the time of discovery, is normally believed to be in equilibrium in that sufficient time has elapsed for all forces to come to a balance.

Connate water can be defined as the irreducible water saturation. It is a function of permeability in that a low permeability rock will have higher connate water saturation than a high permeability rock, which is a direct function of the capillary pressure forces. The water-oil contact, that is, the point of contact between oil and water levels, is not sharp, due to capillary action. In a tight reservoir, division is broad. The thickness of the transition zone depends upon the individual fluid, porosity and permeability.

Connate water values are determined in the laboratory, using core plugs. If core plugs are not available it is possible to estimate water saturations by using electric logs. The electric log method is reasonably accurate in areas of good control for data calibration but sometimes can be very misleading in wildcat areas (Hagoort, J., 1974).

Residual Oil Saturation

It is oil saturation that cannot be produced from an oil reservoir by gas or water displacement. It is usually considered the immobile oil saturation. It is based on result from a laboratory test. A core sample is taken from a reservoir is first saturated with oil and then displaced (pushed out) with water or gas. The test is ended when no more oil can be produced from this core. The oil that cannot be pushed out and remains in the core is termed "residual oil." The residual oil to gas displacement may be different from the residual oil from water displacement.

Water Saturation

Saturation is a measure of the relative volume of each fluid in the pores. Thus the oil saturation is defined as the ratio of the volume of the oil in a porous rock to the pore volume of the same rock. It is expressed in fraction or in percent. Water is always present in all reservoirs, and its saturation is always greater than zero. In contrast, the oil saturation is zero in gas reservoirs, and the gas saturation is zero in oil reservoirs when the pressure is above the bubble-point.

The water saturation is normally obtained from log data. The oil or gas saturation is then calculated by subtracting the water saturation from unity (in two phase reservoirs). Sometimes the fluid content and saturations are measured directly in the laboratory on fresh core samples. In case cores are obtained from a well in which oil-base mud is used considerable care need to be taken during the coring operation. Oil or gas saturations are needed to volumetrically calculate the initial oil or gas in place.

During hydrocarbon accumulation in the reservoir, water saturation can be reduced to some small value, typically 5-40 per cent, after which no more water can escape from the pores. This occurs when water saturation becomes immobile, at the irreducible water saturation like, Swi, Swc, Swir.

Swir – irreducible water saturation, below which water cannot flow.

Swc – connate water saturation existing on discovery of the reservoir. It may or may not be irreducible.

Swi – may mean irreducible, connate, or interstitial, Swi may also mean initial or original, which truly means the water saturation on discovery, but it may or may not be irreducible.

Density differences between gas and oil, as well as between oil and water result in normal reservoir situations in which oil floats on water. If there is a free gas phase, the gas floats on the oil. Keep in mind that there will be some water saturation (at least the irreducible water saturation) throughout the reservoir, even in the pores are at the very top of the reservoir.

Irreducible Water Saturation

Irreducible water saturation (sometimes called *critical water saturation*) defines the maximum water saturation that a formation with a given permeability and porosity can retain without producing. This water, although present, is held in place by capillary forces and will not flow.

The critical water value should be compared to the reservoir's in-place water saturation calculated from downhole electric logs. If the in-place water saturation does not exceed the critical value, then the well will produce only hydrocarbons. These saturation comparisons are particularly important in low permeability reservoirs, where critical water saturation can exceed 60 per cent while still can produce only hydrocarbons.

Reservoir Pressure and Temperature

With changes in pressure and/or temperature, the volumes of the rocks get affected and the fluids that contained in the rocks vary. Much of the petroleum engineering has to do with relations between pressure, volume and temperature, called PVT. Both Pressure and Temperature influence volumes, and therefore the relative buoyancies, of the fluids, but pressure has the greater influence, especially on natural gas. The main effect of increasing temperature is a decrease in the viscosity of the liquids; this causes them to flow more readily.

Reservoir Pressure

The fluids confined in the pores of the reservoir rock occur under a certain degree of pressure, generally called reservoir pressure, measured in Psia. The hydrostatic pressure gradient for a fresh-water (sp. gr. – 1.0) system is 0.433 Psi/ft. Reservoir pressure is the original or virgin pressure, before commencement of any production. When a well is shut-in, after some production, is called Static Bottom Hole Pressure (SBHP). Similarly a pressure recorded under stabilized flowing conditions is flowing B.H.P.

Where static B.H.P. decline per unit of petroleum produced from a pool is rapid, the volume of the reservoir is small and if the decline is slow, the reservoir may be extensive. An early understanding of such pressure behavior is of great importance for estimating available energy and consequently its productive possibilities (Braj Nandan,1971). The main sources of reservoir pressure are due to interconnected pores of nearly every potential reservoir rock that are filled with water and the over burden pressure of rock (1Psi/ft).

Depending upon initial reservoir pressure oil reservoirs can be sub-classified into the following categories:

- ☆ ***Under-saturated*** oil reservoir. If the initial reservoir pressure "pi" is greater than the bubble-point pressure (pb) of the reservoir fluid, the reservoir is called an under-saturated oil reservoir.
- ☆ ***Saturated*** oil reservoir. When the initial reservoir pressure is equal to the bubble-point pressure of the reservoir fluid, the reservoir is called a saturated oil reservoir.

If the initial reservoir pressure is below the bubble point pressure of the reservoir fluid, the reservoir is termed a gas-cap or two-phase reservoir, in which the gas or vapor phase is underlain by an oil phase. The appropriate quality line gives the ratio of the gas-cap volume to reservoir oil volume.

Bubble point pressure is the pressure at which the first bubble of gas appears when the pressure decreases. From the thermodynamic point of view it is the point where a liquid is in equilibrium with gas whose phase volume is zero.

Dew point pressure: during isothermal expansion the fluid in a liquid phase expands until gas appears at equilibrium pressure the volume of liquid decreases and gas volume increases at constant density. After the last droplet of liquid disappeared the gas expands. In summary, the sequence is liquid to gas when pressure decreases.

For condensate systems at high pressures a liquid deposit forms, this opposite sequence is called retrograde and the pressure at which liquid forms is the dew point pressure. In the normal pressure at which liquid disappears in pressure decrease, case is called direct dew point

Temperature

The temperature underground normally increases with depth below surface, and the rate of increase with depth is called the geothermal gradient. Average geothermal gradient is 3°C/100m. Unlike reservoir pressures, which generally decline with oil and gas production, the reservoir temperature in a well remains nearly constant.

The important effect of increase in temperature results in increase in the viscosity of gas (at low pressures), decrease in the viscosity of oil and water, increased the volume of gas, oil and rock, can increase fluid pressure where they are confined, causes decrease in the solubility of gas in oil and causes an increase in solubility of salts in water. Different hydrocarbon system would have a different phase diagram, the general configuration is similar.

Reservoir Mechanics

It includes a study of those physical and chemical changes in the reservoir and the reservoir fluids that take place during dynamic period when a petroleum reservoir is under production.

Inter-phase Phenomena

When fluids are contained in the pores of the rocks, the interfaces between fluids, and between fluids and solids, are the loci of many physical and chemical relations that explain what happens in the reservoir with the production of a pool. Surface tension, interfacial tension, adsorption and wettability are some of the examples of phenomenon that occur at the interfaces.

Surface tension: All the molecules in a liquid, except those at the surface, are acted by molecular attraction or pull from all sides. But those at surface, which are in contact with air, gas or the liquid vapour, are partially surrounded by other molecules of liquid, and are attracted only towards the body of the liquid. The tension at a liquid surface in contact with either air or its own vapour is commonly called surface tension and is expressed as the force in dynes necessary to extend the contracted surface, a distance of one centimeter (dynes/cm).

Interfacial Tension: If the interface is between two immiscible liquids or between a liquid and a solid, the term interfacial tension (γ) is used for the force tending to reduce the area of contact. Many reservoir phenomena are dependent on the interfacial tensions between the reservoir liquids and between the reservoir solids and the liquids. The concentration of surface active agents, either in water or in oil, lowers the interfacial tension.

Some factors that may affect the interfacial tension between crude oil and reservoir waters:

- ☆ Increase in temperature decreases the interfacial tension.
- ☆ Increase in pressure decreases the interfacial tension.
- ☆ More gas in solution above the bubble point pressure, the lower the interfacial tension, the more gas in solution below bubble point, the higher the interfacial tension.
- ☆ Viscosity – a decrease in viscosity difference between oil and water results in a decrease in interfacial tension.
- ☆ Specific Gravity – a decrease in the difference in specific gravity between oil and water generally results in a decrease in interfacial tension.

Physical and Chemical Properties of Petroleum

In order to understand and to predict the behavior of oil and gas reservoirs, knowledge of the physical and chemical characteristics of the fluids involved will help. It is preferable that these properties be measured on representative samples obtained under reservoir conditions at the time of discovery.

Viscosity is the property of a liquid that describes its resistance to flow. It is the opposite of fluidity. As a rule, the viscosity of a liquid increases with an increase in pressure, and decreases with an increase in temperature. In general, reservoir conditions will be at a higher pressure and temperature than stock tank conditions and dissolved gas will be present.

The effect of the gas is to reduce the density and is the most critical of the three effects. There could be a ten-fold variation between reservoir oil viscosity and stock tank oil. When the saturation pressure is passed, further reduction of pressure, causes rapid changes, gas escapes from solution the liquid shrinks and becomes denser and more viscous.

Relative volume factor: As gas dissolves in crude oil, there is an increase in the total liquid volume. The change in oil volume, upon liberation of gas, must be described in some way. The generally preferred basis is stock tank oil.

At a given reservoir pressure, the oil is capable of holding in solution only a given amount of gas. When this condition exists, the oil is said to be saturated. Any excess gas beyond that necessary to saturate the oil at the existing pressure, will be trapped in the formation as free gas, generally at or near the crest of the structure in a "gas cap." Oil is not necessarily saturated with dissolved gas at the available reservoir pressure even though surplus gas is available in the reservoir. The oil in this case would be "under-saturated." When a higher pressure exists than that at which the oil is saturated, subsequent reduction of pressure by production will not release any gas from solution in the oil, until a pressure is reached below that at which the oil is saturated. This is the saturation pressure or bubble point of the oil.

The bubble point signifies a pressure at which gas is released from solution and is reflected by the formation of small bubbles of free gas in the oil. The importance of the bubble point on the relative permeability characteristics of the reservoir rock justifies a short review of permeability at this point. It can be seen that the inter-granular space occupied by the oil without any gas present is greater and offers a clear unobstructed channel to flow. However, when the first bubble of oil breaks out of solution, it will occupy a position in the center of the inter-granular space and will act as an obstruction to the flow of oil. This hindrance increases as the gas bubble grows in size. When the gas bubble becomes large, the ability for flow oil reduces because it is much easier for the gas to move through the reservoir rock.

Solution gas-oil ratio: This is the amount of gas dissolved in reservoir oil at any pressure below saturation pressure and is described as a given number of cubic feet per barrel of oil. It is an important parameter to be determined for estimation of solution gas reserves of an oil reservoir. The gas in solution normally refers to the total amount for gas that has been liberated bringing the oil under reservoir pressure and temperature conditions to stock-tank conditions and is reported in cubic meters of gas produced per ton of stock-tank oil. This parameter can, therefore, be determined from the surface measurements through separators, of the volume of gas produced from a measured quantity of stock-tank oil. This can also be determined in the laboratory from the bottom hole fluid samples (PVT samples) either by flash or differential liberation process.

API gravity: API gravity is the relationship between specific gravity of water and crude oil at the same temperature base. It has been adopted as a standard by the American Petroleum Institute (141.5/specific gravity at 60° F -131-5). Under API scale, for example, water would have a gravity of 10° API. The higher the API gravity of oil, the greater the light gasoline content.

Reservoir Engineering Data Sources

Geologic and Seismic Interpretations

Reservoir geology helps the engineer to understand the external geometry of the reservoir as well as its internal architecture. Examples of the types of information it provides include, the reservoir extent and its closure, flow barriers, such as faults or pinch-outs, fluid contacts, *i.e.*, oil-water, oil-gas, and gas-water interfaces, aquifer size, lithology variations, continuity of the reservoir in the areal as well as in the vertical direction.

Calculations from Well Logs

Logging provides *in-situ* information about the rock and its content from the immediate vicinity of the wellbore (Coates, G.R. and J.L. Dumanior, 1973). There are many types of logs, information from which may include: location of the productive stratum and its boundaries, continuity of rock strata between adjacent wells, net pay thickness, oil, gas, and water saturations, porosity of the reservoir rock, other miscellaneous information, such as the condition of the hole, the temperature gradient in the wellbore, and the condition of the cement in a cased hole.

Calculations from Well Tests

Well tests measure the pressure response of the well to short-term flow periods and the subsequent pressure buildup performance after shut-in. Some mathematical models can be used to determine the reservoir characteristics responsible for a particular pressure-flow rate behavior. In particular, permeability, the presence of nearby fault boundaries, or fluid contacts may be determined from an analysis of the well test data. Keep in mind that reservoir rock characteristics as determined from well tests are averaged values over the area of the reservoir that is contacted during the test.

Core Analyses

Cores provide petro-physical data essential to reservoir engineering. Basic core data, such as permeability, porosity, and fluid saturations help the engineer decide whether or not to complete the well and where to complete it. Special core analyses also help in evaluating reservoir performance, estimating hydrocarbons in place and reserves, evaluating the feasibility of Enhanced Oil Recovery projects, and providing input data for reservoir simulation studies.

A second type of data used in reservoir engineering concerns the properties of the reservoir fluids and how they react to changes in pressure and temperature. Expressing the original hydrocarbons in place in surface volumes requires such data. Quantitative calculation of recoverable reserves requires estimates or laboratory determinations of formation volume factor, gas-oil ratio, and oil and gas compressibility, all as a function of pressure. Determining production rates of oil or gas requires knowledge of their respective viscosities at reservoir conditions. Any assessment of the practicality of EOR methods requires an understanding of the effects of the particular method employed on the behavior of the oil in the reservoir (*i.e.*, oil viscosity reduction in a steam flood). Reservoir fluid data is generally determined from a laboratory analysis performed on a carefully obtained representative sample of the original reservoir fluid under subsurface conditions. Where sampling is impossible, empirical correlations are available to estimate oil, gas, and water properties.

Production Data

This is another important type of data used in reservoir engineering calculations. By production data, which mean a careful accounting of the volumes of produced oil, gas, and water, as functions of time. Pressure as a function of time is also extremely important. The decline curve analysis and the material balance equation of oil or gas reservoirs require accurate production data in order to be of any value as predictive techniques.

The accuracy of production accounting can vary from field to field, particularly in large offshore developments where isolated wells and "satellite platforms" preclude the individual measurement of well production volumes on a regular basis. In such situations, individual well production is allocated from a total field production volume based on monthly well tests. In areas with high water-production rates the accuracy of measured water cuts also becomes a factor.

Reservoir Fluids

Phase is defined as any homogeneous and physically distinct part of a system which is separated from other parts of the system by definite bounding surfaces.

For example, ice, water and water vapor are three phases: each is physically distinct and homogeneous and there are definite boundaries. The water system consists of one component H_2O each of the phases in equilibrium, *i.e.*, solid liquid and vapor may be regarded as belong made up of this component only.

The sequence is liquid to gas when pressure decreases can seen in relation to temperature (Figure 8.6).

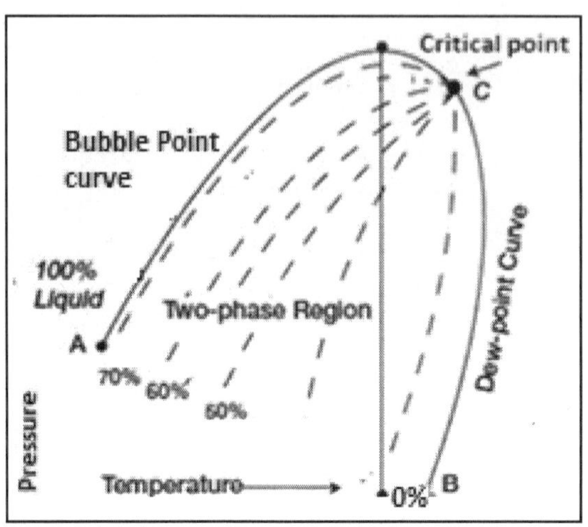

Figure 8.6. Pressure-Temparture Related Phase Diagram-Crude Oil.

Classification of Reservoirs and Reservoir Fluids

Petroleum reservoirs are broadly classified as oil or gas reservoirs. These broad classifications are further subdivided depending on the composition of the reservoir hydrocarbon mixture, initial reservoir pressure and temperature and pressure and temperature of the surface production. The conditions under which these phases exist are a matter of considerable practical importance. The experimental or the mathematical determinations of these conditions are conveniently expressed in different types of diagrams commonly called phase diagrams.

Cricondentherm (Tct) is defined as the maximum temperature above which liquid cannot be formed regardless of pressure. The corresponding pressure is termed the Cricondentherm pressure Tct.

Cricondenbar (Tcb) is the maximum pressure above which no gas can be formed regardless of temperature. The corresponding temperature is called the Cricondenbar temperature Tcb.

Critical point for a multicomponent mixture is referred to as the state of pressure and temperature at which all intensive properties of the gas and liquid phases are equal. At the critical point, the corresponding pressure and temperature are called the critical pressure pc and critical temperature Tc of the mixture.

Phase envelope (two-phase region)–area enclosed by the bubble-point curve and the dew-point curve, wherein gas and liquid coexist in equilibrium, is identified as the phase envelope of the hydrocarbon system.

Quality lines within the phase diagram are shown as dashed lines. They describe the pressure and temperature conditions for equal volumes of liquids. Note that the quality lines converge at the critical point.

Bubble-point curve is defined as the line separating the liquid-phase region from the two-phase region.

Dew-point curve is defined as the line separating the vapor-phase region from the two-phase region.

In general, reservoirs are conveniently classified on the basis of the location of the point representing the initial reservoir pressure (pi) and temperature with respect to the pressure-temperature diagram of the reservoir fluid. Accordingly, reservoirs can be classified into basically two types.

Oil reservoirs–If the reservoir temperature T is less than the critical temperature Tc of the reservoir fluid, the reservoir is classified as an oil reservoir.

Gas reservoirs–If the reservoir temperature is greater than the critical temperature of the hydrocarbon fluid, the reservoir is considered a gas reservoir.

Crude oils cover a wide range in physical properties and chemical compositions, and it is often important to be able to group them into broad categories of related oils. In general, crude oils are commonly classified into the following types:

- ☆ Ordinary black oil
- ☆ Low-shrinkage crude oil
- ☆ High-shrinkage (volatile) crude oil
- ☆ Near-critical crude oil

The above classifications are essentially based upon the properties exhibited by the crude oil, including physical properties, composition, gas-oil ratio, appearance, and pressure-temperature phase diagrams.

Ordinary black oil. The liquid shrinkage curve approximates a straight line except at very low pressures. When produced, ordinary black oils usually yield gas-oil ratios between 200–700 scf/STB and oil gravities of 15 to 40 API. The stock tank oil is usually brown to dark green in color.

Low-shrinkage oil. The quality lines that are closely spaced near the dew-point curve. The other associated properties of this type of crude oil are:

- ☆ Oil formation volume factor less than 1.2 bbl/STB
- ☆ Gas-oil ratio less than 200 scf/STB
- ☆ Oil gravity less than 35° API
- ☆ Black or deeply colored Substantial liquid recovery at separator conditions

Volatile crude oil. This type of crude oil is commonly characterized by a high liquid shrinkage immediately below the bubble-point. The other characteristic properties of this oil include: Oil formation volume factor less than 2 bbl/STB

- Gas-oil ratios between 2,000–3,200 scf/STB
- Oil gravities between 45–55° API
- Lower liquid recovery of separator conditions
- Greenish to orange in color

Another characteristic of volatile oil reservoirs is that the API gravity of the stock-tank liquid will increase in the later life of the reservoirs.

Near-critical crude oil. If the reservoir temperature T is near the critical temperature Tc of the hydrocarbon system, the hydrocarbon mixture is identified as a near-critical crude oil. Because all the quality lines converge at the critical point, an isothermal pressure drop may shrink the crude oil from 100 per cent of the hydrocarbon pore volume at the bubble-point to 55 per cent or less at a pressure 10 to 50 psi below the bubble point. The near-critical crude oil is characterized by a high GOR in excess of 3,000 scf/STB with an oil formation volume factor of 2.0 bbl/STB or higher.

Gas Reservoirs

In general, if the reservoir temperature is above the critical temperature of the hydrocarbon system, the reservoir is classified as a natural gas reservoir (Braj Nandan,1980). On the basis of their phase diagrams and the prevailing reservoir conditions, natural gases can be classified into four categories:

- Retrograde gas-condensate
- Near-critical gas-condensate
- Wet gas
- Dry gas

Retrograde gas-condensate reservoir. If the reservoir temperature T lies between the critical temperature Tc and cricondentherm Tct of the reservoir fluid, the reservoir is classified as a retrograde gas-condensate reservoir. This category of gas reservoir is a unique type of hydrocarbon accumulation in that the special thermodynamic behavior of the reservoir fluid is the controlling factor in the development and the depletion process of the reservoir. When the pressure is decreased on these mixtures, instead of expanding (if a gas) or vaporizing (if a liquid) as might be expected, they vaporize instead of condensing.

Consider that the initial condition of a retrograde gas reservoir the reservoir pressure is above the upper dew-point pressure, the hydrocarbon system exists as a single phase (vapor phase) in the reservoir. As the reservoir pressure declines isothermally during production from the initial pressure to the upper dew-point pressure, the attraction between the molecules of the light and heavy components causes them to move apart. As this occurs, attraction between the heavy component molecules becomes more effective; thus, liquid begins to condense. This retrograde condensation process continues with decreasing pressure until the liquid dropout reaches its maximum point. Further reduction in pressure permits the heavy molecules to commence the normal vaporization process.

This is the process whereby fewer gas molecules strike the liquid surface and causes more molecules to leave than enter the liquid phase. The vaporization process continues until the reservoir pressure reaches the lower dew-point pressure. This means that all the liquid that formed must vaporize because the system is essentially all vapors at the lower dew point.

In most gas-condensate reservoirs, the condensed liquid volume seldom exceeds more than 15 per cent –19 per cent of the pore volume. This liquid saturation is not large enough to allow any liquid flow. It should be recognized, however, that around the wellbore where the pressure drop is high, enough liquid dropout might accumulate to give two-phase flow of gas and retrograde liquid. The associated physical characteristics of this category are:

- ☆ Gas-oil ratios between 8,000 to 70,000 scf/STB. Generally, the gas-oil ratio for a condensate system increases with time due to the liquid dropout and the loss of heavy components in the liquid.
- ☆ Condensate gravity above 50° API.
- ☆ Stock-tank liquid is usually clear or slightly colored.

Oils have been observed with heavier concentrations as low as 10 per cent and condensates as high as 15.5 per cent. These cases are rare, however, and usually have very high tank liquid gravities (Katz, D. L. and Firoozabadi, A.1978).

Near-critical gas-condensate reservoir. If the reservoir temperature is near the critical temperature, the hydrocarbon mixture is classified as a near-critical gas-condensate. The volumetric behavior of this category of natural gas is described through the isothermal pressure. At the point where the liquid ceases to build up and begins to shrink again, the reservoir goes from the retrograde region to a normal vaporization region.

Wet-gas reservoir. A typical phase diagram of a wet gas the reservoir temperature is above the cricondentherm of the hydrocarbon mixture. Because the reservoir temperature exceeds the cricondentherm of the hydrocarbon system, the reservoir fluid will always remain in the vapor phase region as the reservoir is depleted isothermally. As the produced gas flows to the surface, however, the pressure and temperature of the gas will decline. If the gas enters the two-phase region, a liquid phase will condense out of the gas and be produced from the surface separators. This is caused by a sufficient decrease in the kinetic energy of heavy molecules with temperature drop and their subsequent change to liquid through the attractive forces between molecules.

Wet-gas reservoirs are characterized by the following properties:

- ☆ Gas oil ratios between 60,000 to 100,000 scf/STB
- ☆ Stock-tank oil gravity above 60° API
- ☆ Liquid is water-white in color

- ☆ Separator conditions, *i.e.*, separator pressure and temperature, lie within the two-phase region.

Dry-gas reservoir. The hydrocarbon mixture exists as a gas both in the reservoir and in the surface facilities. The only liquid associated with the gas from a dry-gas reservoir is water. Usually a system having a gas-oil ratio greater than 100,000 scf/STB is considered to be a dry gas. It can be observed that hydrocarbon mixtures may exist in either the gaseous or liquid state, depending on the reservoir and operating conditions to which they are subjected.

Properties of Natural Gases

A gas is defined as a homogeneous fluid of low viscosity and density that has no definite volume but expands to completely fill the vessel in which it is placed. Generally, the natural gas is a mixture of hydrocarbon and non-hydrocarbon gases. The hydrocarbon gases that are normally found in a natural gas are methanes, ethanes, propanes, butanes, pentanes, and small amounts of hexanes and heavier.

The non-hydrocarbon gases include carbon dioxide, hydrogen sulfide, and nitrogen. Knowledge of pressure-volume-temperature (PVT) relationships and other physical and chemical properties of gases is essential for solving problems in natural gas reservoir engineering.

These properties include:

- ☆ Apparent molecular weight, Ma
- ☆ Specific gravity, gg
- ☆ Compressibility factor, z
- ☆ Density, rg
- ☆ Specific volume, v
- ☆ Isothermal gas compressibility coefficient, cg
- ☆ Gas formation volume factor, Bg
- ☆ Gas expansion factor, Eg
- ☆ Viscosity, mg

The above gas properties may be obtained from direct laboratory measurements or by prediction from generalized mathematical expressions.

Behaviour of Ideal Gases

The kinetic theory of gases postulates that gases are composed of a very large number of molecules. For an ideal gas, the volume of these molecules is insignificant compared with the total volume occupied by the gas. This relationship for perfect gases is called the **ideal gas law** and is expressed mathematically as:

$pV = nRT$

where,

p = absolute pressure, psia

V = volume, ft^3

T = absolute temperature, °R

n = number of moles of gas, lb-mole

R = the universal gas constant which, for the above units, has the value 10.730 psia ft3/lb-mole °R

The number of pound-moles of gas, *i.e.*, n, is defined as the weight of the gas (m) divided by the molecular weight M, or: n = m/M

Combining the Equations gives:

$$pV = \left(\frac{m}{M}\right)RT$$

Since the density is defined as the mass per unit volume of the substance, equation can be rearranged to estimate the gas density at any pressure and temperature:

$$\rho_g = \frac{m}{V} = \frac{pM}{RT}$$

where,

ρ_g = density of the gas, lb/ft^3

The constant in the equation is called the ideal gas constant and is designated by R; thus the ideal gas equation becomes Pv= RT, where the pressure and temperature are absolute values. The values of the ideal gas constant (R).

The individual gas constant (R) may be obtained by dividing the universal gas constant (Ro) by the molecular weight (MW) of the gas, R=Ro/MW. The units of R must always be consistent with the units of pressure, temperature, and volume used in the gas equation.

No real gases follow the ideal gas law or equation completely.

At temperatures near gases boiling point, increases in pressure will cause condensation to take place and drastic decreases in volume. At very high pressures, the inter-molecular forces of a gas are significant. However, most gases are in approximate agreement at pressures and temperatures above their boiling point. The compression process using ideal gas considerations results in work performed on the system and is essentially the area under a P-V curve.

Specific Gravity (Relative Density)

The specific gravity of a gas, γ, is the ratio of the density of the gas at standard pressure and temperature to the density of air at the same standard pressure and temperature. The standard temperature is usually 60°F, and the standard pressure is usually 14.696 psia.

The ideal gas laws can be used to show that the specific gravity (ratio of densities) is also equal to the ratio of the molecular weights. By convention, specific gravities of all gases at all pressures are usually set equal to the ratio of the molecular weight of the gas to that of air. Although specific gravity is commonly used, it can also be called"relative density."

Specific Gravity of Gas Mixtures

The specific gravity (γ_g) of a gas mixture is the ratio of the density of the gas mixture to that of air. It is measured easily at the wellhead in the field and therefore is used as an indication of the composition of the gas. The specific gravity of gas is proportional to its molecular weight (M_g) if it is measured at low pressures where gas behavior approaches ideality. Specific gravity also has been used to correlate other physical properties of natural gases. To do this, it is necessary to assume that the analyses of gases vary regularly with their gravities.

Gas Specific Gravity

The ratio of density of any gas at standard conditions (14.7 psia and 60°F) to the density of air at standard conditions; based on the ideal gas law ($pV = nRT$), gas gravity is also equal to the gas molecular weight divided by air molecular weight (M_{air} = 28.97). Denoted mathematically as γ_g (where air = 1).

The specific gravity of a gaseous mixture is the ratio of its apparent molecular weight to that of air. Higher the specific gravity the more wet the gas. A dry gas of practically pure methane, for example, will have a value of 0.58 whereas a wet gas high in pentanes will be in the order of 0.8 to 0.9

Real Gases

At low pressures and relatively high temperatures, the volume of most gases is so large that the volume of the molecules themselves may be neglected. Also, the distance between molecules is so great that the presence of even fairly strong attractive or repulsive forces is not sufficient to affect the behavior in the gas state. However, as the pressure is increased, the total volume occupied by the gas becomes small enough that the volume of the molecules themselves is appreciable and must be considered. Also, under these conditions, the distance between the molecules is decreased to the point at which the attractive or repulsive forces between the molecules become important. This behavior negates the assumptions required for ideal gas behavior, and serious errors are observed when comparing experimental volumes to those calculated with the ideal gas law. Consequently, a real gas law was formulated (in terms of a correction to the ideal gas law) by use of a proportionality term.

Real Gas Law

The volume of a real gas is usually less than what the volume of an ideal gas would be at the same temperature and pressure; hence, a real gas is said to be super compressible. The ratio of the real volume to the ideal volume, which is a measure of the amount that the gas deviates from perfect behavior, is called the super compressibility factor, sometimes shortened to the compressibility factor. It is also called the gas deviation factor and given the symbol z.

The gas deviation factor is by definition the ratio of the volume actually occupied by a gas at a given pressure and temperature to the volume it would occupy if it behaved ideally, or:

$$z = \frac{\text{Actual volume of gas at specified } T \text{ and } p}{\text{Ideal volume of gas at same } T \text{ and } p}$$

Note that the numerator and denominator of above *Equation* refer to the same mass. (This equation for the z factor is also used for liquids.) Thus, the real gas equation of state is written:

$$pV = znRT$$

The gas deviation factor, z, is close to 1 at low pressures and high temperatures, which means that the gas behaves as an ideal gas at these conditions. At standard or atmospheric conditions, the gas z factor is always approximately 1. As the pressure increases, the z factor first decreases to a minimum, which is approximately 0.27 for the critical temperature and critical pressure. For temperatures of 1.5 times the critical temperature, the minimum z factor is approximately 0.77, and for temperatures of twice the critical temperature, the minimum z factor is 0.937.

At high pressures, the z factor increases above 1, where the gas is no longer super compressible. At these conditions, the specific volume of the gas is becoming so small, and the distance between molecules is much smaller, so that the density is more strongly affected by the volume occupied by the individual molecules. Hence, the z factor continues to increase above unity as the pressure increases.

The gas deviation factor, z, is determined by measuring the volume of a sample of the natural gas at a specific pressure and temperature, then measuring the volume of the same quantity of gas at atmospheric pressure and at a temperature sufficiently high so that the hydrocarbon mixture is in the vapor phase.

Real Gas Pseudo-Potential

In the analysis of gas reservoirs, well-test analysis, gas flow in pipes, and other calculations can be made more accurate by the use of the real gas pseudo-potential. This is because the z factor and viscosity that appear in such equations along with pressure terms are dependent on pressure. Consequently, the integral of pressure divided by the z factor and viscosity is defined as a separate parameter called the real gas pseudo-potential.

Gas Formation Volume Factor

Gas formation volume factor, which is known as Bg, is equal to:

Bg = volume at downhole T and P/volume at 60°F and 14.7 psia

The weight of any volume of gas can be determined by multiplying the volume of gas and air. The density of air at standard conditions is 0.001223 gr/cc or 0.0762 lb/ft^3. The density of gas at any temperature and pressure can be found from the gas formation volume factor, Bg.

1/Bg = rgwf/rasc

Gas volume at reservoir conditions divided by gas volume at standard conditions. This factor is used to convert surface measured volumes to reservoir conditions, just as oil formation volume factors are used to convert surface measured oil volumes to reservoir volumes.

Gas Density

To calculate the natural gas density one should start with the real gas law. The gas law as applied to the behavior of natural gas is most commonly stated as:

$$PV = znRT$$

where,

P = pressure, **V** = volume, **n** = number of mols, **R** = gas constant, **T** = absolute temperature and **z** = deviation factor (compressibility factor).

Compressibility Factor

The accurate estimation of gas compressibility factor (z) is very important in estimation of free gas reserves. As the gas deviation factor varies usually between 0.70 to 1.20, this has got a great bearing on assessed gas reserves.

The most convenient method of computing the PVT relationship of natural gas, is by use of the compressibility factor, which represents the deviation of the gas in question from the ideal gas laws.

The compressibility factor (Z) is an empirical factor, determined experimentally, which makes the above equation true at a particular temperature and pressure. For a perfect gas, Z is equal to one. For an imperfect gas (hydrocarbon gases), Z is greater or less than one, depending on the pressure and temperature (Hall, K.R. and Yarborough, L., 1974).

At the same reduced temperature and pressure, all hydrocarbon gases have the same value of Z. The reduced temperature and pressures are finding as:

$$Pr = P/Pc \quad \text{and} \quad Tr = T/Tc$$

where,

Pc is the critical pressure and

Tc is the critical temperature

Value for the critical temperature and pressure of several hydrocarbon gases should be tabulated in a table to calculate Tr and Pr, all temperature and pressures must be in absolute unit (°R or °K and psia or atm).

The compressibility factor (Z) may be determined, experimentally by PVT analysis of a gas sample, from an analysis of the gas expressed in more percent or volume percent, and also from the specific gravity of the gas.

Highly Compressive Reservoirs

For some high-pressure gas reservoirs (*e.g.*, geo-pressured or abnormally pressured reservoirs), the combined rock and water compressibility can result in a nonlinear p/z plot. Ignoring this effect can lead to large overestimates of the OGIP.

Gas and gas-condensate reservoirs are compositionally distinguished from oil reservoirs by the predominance of lighter hydrocarbons, such as methane (60 to 95 percent) and ethane (4 to 8 percent). Three types of oil and gas reservoirs are identified based on their hydrocarbon fluid composition, oil (with dissolved gas), gas, and gas-condensate.

In addition, a two-phase region exists between the bubble point and dew point lines; that is, oil reservoirs with a gas cap. Depending on the reservoir pressure and temperature and the fluid composition, some gas reservoirs may produce only gas, whereas others produce mostly gas with some condensate.

In contrast, gas-condensate reservoirs will always produce gas mixed with condensate, with the quantity of condensate varying from one reservoir to another. Gas condensate is a low-density mixture of liquid hydrocarbons that are present as gaseous components in the natural gas. The reservoir management of gas-condensate reservoirs becomes challenging when the condensate begins to build up near the bottom of the wellbore, affecting well productivity due to the relative permeability effect of a two-phase flow (Bhushan,Y A.K. Das and B.L. Lohar, 2003).

Reservoir Drive Mechanisms

The recovery of oil by any of the natural drive mechanisms is called primary recovery. During primary recovery the natural energy of the reservoir is used to move hydrocarbons towards the production wells.

Solution Gas Drive

For an under-saturated reservoir no free gas exists until the reservoir pressure falls below the bubble-point. In this regime reservoir drive energy is provided only by the bulk expansion of the reservoir rock and liquids (water and oil).

If a reservoir contains oil initially above its bubble point then, as production continues, the production from the reservoir of the produced oil will be compensated for by an expansion of the oil left in place within the reservoir. This will lead to a reduction in pressure and eventually the pressure within the reservoir will drop below the bubble point. Gas will then come out of solution and any subsequent production of fluids will lead to an expansion of both the oil and gas phases within the reservoir.

The gas will come out of solution as dispersed bubbles throughout the reservoir wherever the pressure is below the bubble point but will be concentrated in areas of low pressure such as the near wellbore area around production wells. The gas which has come out of solution will build up in the reservoir until its saturation allows it to produce more easily.

As the gas saturation increases, free gas will be produced in increasing quantities associated with the produced oil. Further, the gas may migrate to above the oil in the reservoir and form a free gas cap if the vertical permeability permits and sufficient time is allowed for gravity segregation. The produced GOR increase once the bubble point is reached due to the liberation of gas. The other effect will be a reduction in the oil production rate because as the gas comes out of solution from the oil, the viscosity and density of the oil phase increases and its formation volume factor decreases, meaning, less shrinkage will occur with production. In addition, as the gas saturation in the pore space increases, the relative permeability to oil will decline. Later the observed production GOR will steadily increase due to increased gas saturation and mobility.

There is usually no production of water during oil recovery unless the reservoir pressure drops sufficiently for the connate water to expand sufficiently to be mobile. Even in this scenario little water is produced.

Gas Cap Drive

A gas cap drive reservoir usually benefits to some extent from solution gas drive, but derives its main source of reservoir energy from the expansion of the gas cap already existing above the reservoir. The presence of the expanding gas cap limits the pressure decrease experienced by the reservoir during production. The actual rate of pressure decrease is related to the size of the gas cap (Figure 8.7).

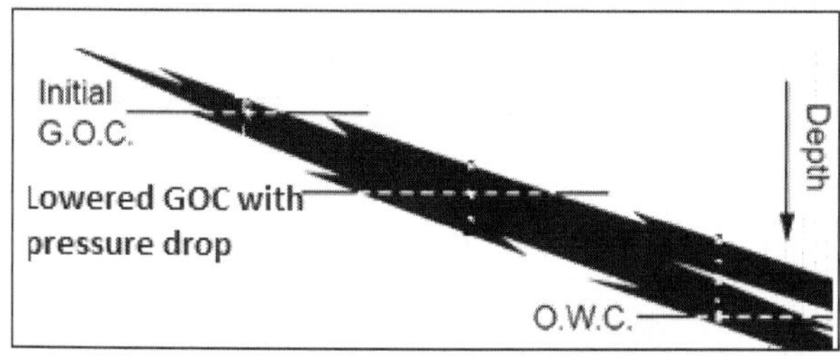

Figure 8.7. Contacts-Gas Cap Diagram.

The GOR rises slowly in the early stages of production from such a reservoir because the pressure of the gas cap prevents gas from coming out of solution in the oil and water. As production continues, the gas cap expands pushing the gas-oil contact (GOC) downwards. Eventually the GOC will reach the production wells and the GOR will increase by large amounts. The slower reduction in pressure experienced by gas cap reservoirs compared to solution drive reservoirs results in the oil production rates being much higher throughout the life of the reservoir, and needing artificial lift much later than for solution drive reservoirs. Gas cap reservoirs produce very little or no water.

The recovery from the gas cap reservoirs is better than the solution drive reservoirs. The recovery efficiency depends on the size of the gas cap, which is a measure of how much latent energy there is available to drive production, and how the reservoir is managed considering economic considerations.

Water Drive

Many oil reservoirs are bounded by their peripheries by water bearing rocks called aquifers. The aquifer itself may be entirely bounded by impermeable rock so that the reservoir and aquifer together form a closed (volumetric) unit. In case, the reservoir is outcropped at one or more places where it may be replenished by surface water. It is common to speak of edge water or bottom water in discussing water influx into a reservoir (Figure 8.8). Bottom water occurs directly beneath the oil and edge water occurs off the flanks of the structure at the edge of the oil.

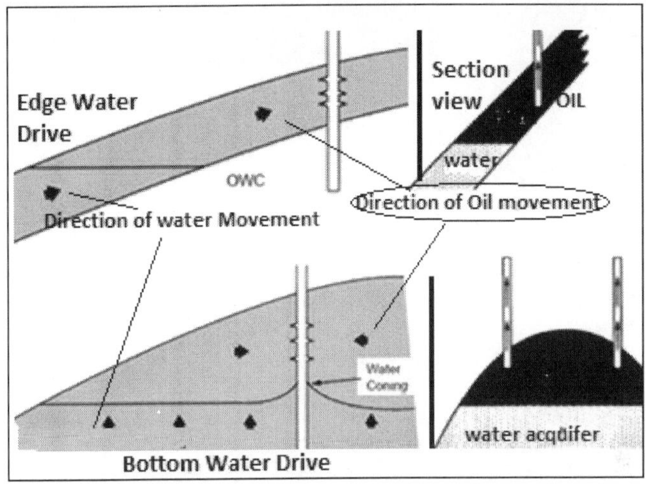

Figure 8.8. Edge water-Bottom water Drive drive

Regardless of the source of water, the water drive is the result of water moving into the pore spaces originally occupied by oil, replacing the oil and displacing it to the producing wells. Water production is indicative of a water drive (Agarwal, R.G, *et al.*, 1965).

The reservoir pressure decline is usually very gradual. The reason for the small decline in reservoir pressure is that oil and gas withdrawals from the reservoir are replaced almost volume for volume by water encroaching into the oil zone. Variable rates of pressure decline, depending principally upon the amount of gas conservation. Where the gas is conserved and reservoir pressure is maintained, the reservoir would be operating under combined gas-cap drive and gravity-drainage mechanisms. Therefore, for the reservoir to be operating solely as a result of gravity drainage, the reservoir would show a rapid pressure decline. This would require the up structure migration of the evolved gas where it later would be produced from structurally high wells.

Since the more production of oil is needed, it is extremely important to reduce water and gas production to an absolute minimum. This can usually be accomplished by shutting in wells producing large quantities of these fluids and, where possible, transferring their allowable to other wells producing with lower water-oil or gas-oil ratios.

The drive energy is provided by an aquifer that interfaces with the oil in the reservoir at the oil-water contact (OWC). As production continues, and oil is extracted from the reservoir, the aquifer expands into the reservoir displacing the oil. Clearly, for most reservoirs, solution gas drive will also be taking place, and there may also be a gas cap contributing to the primary recovery. Two types of water drive are commonly recognized. The pressure profile of a water driven reservoir depends upon, the size of the aquifer, the permeability of the aquifer and finally the production rate.

If the production rate is low, and the size and permeability of the aquifer is high, then the reservoir pressure will remain high because all produced oil is replaced efficiently with water. If the production rate is too high then the extracted oil may not be able to be replaced by water at par, especially if the aquifer is small or low permeability and hence the reservoir pressure will fall.

The GOR remains very constant in a strongly water driven reservoir, as the pressure decrease is small and constant, whereas if the pressure decrease is higher, the GOR increases due to gas exsolving from the oil and water in the reservoir. Likewise the oil production from a strongly water driven reservoir remains fairly constant until water breakthrough occurs. When water breakthrough does occur the well can either be shut-down, or assisted using gas lift.

The recovery from water driven reservoirs is usually good, although the exact figure depends on the strength of the aquifer and the efficiency with which the water displaces the oil in the reservoir, which depends on reservoir structure, production well placing, oil viscosity, and production rate. If the ratio of water to oil viscosity is large, or the production rate is high then water coning can occur which leaves oil behind in the reservoir.

Gravity Drainage

The mechanism of gravity drainage occurs in petroleum reservoirs as a result of differences in densities of the reservoir fluids. The fluids have separated as a result of the gravitational forces acting on them. The fluids in petroleum reservoirs have all been subjected to the forces of gravity, as evidenced by the relative positions of the fluids, *i.e.*, gas on top, oil underlying the gas, and water underlying oil. The density differences between oil and gas and water result in their natural segregation in the reservoir. This process can be used as a drive mechanism. The best conditions for gravity drainage are, thick oil zones and more vertical permeabilities.

Due to the long periods of time involved in the petroleum accumulation-and-migration process, it is generally assumed that the reservoir fluids are in equilibrium. If the reservoir fluids are in equilibrium, then the gas-oil and oil water contacts should be essentially horizontal. Gravity segregation of fluids is probably present to some degree in all petroleum reservoirs, but it may contribute substantially to oil production in some reservoirs. Low gas-oil ratio from structurally low wells. This is caused by migration of the evolved gas up structure due to gravitational segregation of the fluids. On the other hand, the structurally high wells will experience an increasing gas-oil ratio as a result of the upward migration of the gas released from the crude oil.

Formation of a secondary gas cap in reservoirs that initially were under saturated. Obviously the gravity-drainage mechanism does not become operative until reservoir pressure has declined below the saturation pressure, since above the saturation pressure there will be no free gas in the reservoir.

Ultimate recovery from gravity-drainage reservoirs will vary widely, due primarily to the extent of depletion by gravity drainage alone. Recovery will be high where gravity drainage is good, or where producing rates are optimal to the gravitational forces. In reservoirs where depletion drive also plays an important role in the oil recovery process, the ultimate recovery will be less.

In operating a gravity-drainage reservoir, it is essential that the oil saturation in the vicinity of the well bore must be maintained as high as possible. There are two basic reasons for this requirement: a high oil saturation means a higher oil flow rate, a high oil saturation means a lower gas flow rate.

If the evolved gas migrates up structure instead of toward the well bore, then high oil saturation in the vicinity of the well bore can be maintained. In order to take maximum advantage of the gravity-drainage-producing mechanism, wells should be located as structurally low leaving some area at the top.

Factors that affect ultimate recovery from gravity-drainage reservoirs are, permeability in the dip direction of the reservoir, reservoir producing rates, oil viscosity, relative permeability characteristics.

The rate of production by gravity drainage is very low compared with the other drive mechanisms. However, it is extremely efficient over long periods and can give good recoveries. Consequently, it is often used in addition to the other drive mechanisms.

Combination or Mixed Drive

In practice a reservoir usually incorporates at least two main drive mechanisms. The management of the reservoir for different drive mechanisms can be diametrically opposed, because low level of perforation for gas cap reservoirs compared with higher perforation for water drive reservoirs. If both occur as in, a compromise must be sought, and this compromise must take into account the strength of each drive present, the size of the gas cap, and the size/permeability of the aquifer. It is the job of the reservoir manager to identify the strengths of the drives as early as possible in the life of the reservoir to optimise the reservoir performance.

Formation Volume Factor (FVF)

The ratio between the volume of a liquid under formation conditions and the volume of the same liquid after degassing that is at surface is defined as volume ratio and also called *Formation Volume Factor* (FVF).

The Oil Formation Volume Factor (FVF) parameter is a very important fluid property in reservoir engineering computations. Ideally, this property should be obtained from actual measurements. In circumustance this can also be empirically derived.

Gas Oil Ratio (GOR)

It is the good indication of the overall composition of the fluid. Production with

GOR greater than 10,000 SCF/bbl -dry gas The GOR and consequently API gravity will vary with pressure and temperature.

An accurate knowledge of Pressure-Volume-Temperature (PVT) properties is essential in reservoir and production engineering calculations. Estimation of reserves, determination of oil reservoir performance, recovery efficiency, production optimization and design of production systems are some of the areas which require precise determination of a fluid's physical properties at different conditions of pressure and temperature. Ideally, the physical properties of the reservoir fluids are determined experimentally in the laboratory.

For saturated systems, gas is liberated as pressure is reduced below the bubble point. This results in a corresponding shrinkage in oil volume. Solution GOR accounts for the largest change in FVF. Increases in temperature, crude oil gravity, and gas gravity provide a small increase in FVF.

Shrinkage Factor (SF)

Crude oil in the reservoir almost always mixed with lighter hydrocarbons such as methane, ethane,, *etc.*, that are released when oil is flashed at lower pressure. As the gas evolves, the liquid volume decreases; this phenomenon is know as shrinkage. Gas produced is called solution or associated or separated gas. Shrinkage is expressed in terms of volume of stock tank oil per volume of reservoir fluid.

Lab Measurement of Crude Oil Properties (PVT)

In the PVT lab the idea is to simulate what takes place for fluids in reservoir conditions. Main aspect of PVT studies to analyze how gas evolves from oil when pressure falls below bubble point. Two methods are followed called flash liberation and Differential liberation (Anil Kumar, *et al.*, 2005). Flow in the reservoir conditions are best simulated by differential Liberation and flow up of the well through the separator best simulated by series of flash liberation called flash separation.

Flash Liberation

- ☆ No gas removed from the PVT cell – gas remains in equilibrium with oil.
- ☆ Overall hydrocarbon composition remains same (components are re-distributed between the gas and liquid phase).
- ☆ Experiment starts with pressure far above bubble point pressure and then is gradually reduced.

Differential Liberation

- ☆ At each stage of depletion gas is physically removed from contact with oil by injecting Hg at constant pressure.
- ☆ There is continual change of fluid composition in cell.
- ☆ Remaining hydrocarbons in cell progressively become heavier (mol. wt. continually increases).
- ☆ Experiment starts at bubble point.

Well Testing

The aim of well testing is to characterize a reservoir to get information about a well and about a reservoir. One major purpose of well testing is to determine the ability of a formation to produce reservoir fluids. Further, it is important to determine a well's productivity (Earlougher, R.C. Jr: 1977).

Well testing is the ultimate method to flow oil and gas to surface, from the known formations. As we know, while drilling, porous reservoir formations can be first identified by rate of penetration, followed by examination of cutting samples after lag-time at surface.

Geologist at site will report oil gas indications in the drilled formations using lithological description, florescence, mud parameters, gas chromatographic observations. This will be followed by, electro-log data which precisely locate such favorable zones to more accuracy. Sometimes while logging, if log interpretation is encouraging, RFT (Repeat Formation Test) will be conducted to check the sample. It is also to be remembered, in case good permeable formations, infiltration of mud in to formation may leave lot of filterate in formation which may be first sample in to RFT chamber. So to say, sometimes RFT may not give real oil sample and also flow potential of oil and gas if any, cannot be completely assessed by RFT.

Well testing is the method to see flow of oil and gas to surface and also collect data for completely assessing potential of hydrocarbons in the reservoir. The testing of potential zones can be done during drilling and after drilling also. The need when to carry out test is based on necessity how fast one can establish proved reserves in an exploratory area. In case of testing during drilling it is called Open Hole Drill Stem Test (OHDST) which is based on well condition. After drilling and casing if DST is conducted it is called Cased Hole DST. Different types of tests are designed to provide reservoir characteristics as per the nature of the test.

Open Hole DST

It is one of the initial methods of formation testing during drilling stage based on initial indications of any oil and gas while drilling and also positive indications of hydrocarbons in electro-log data. For planning Open hole DST in a well, it is essential to have gauged hole with normal dia, which can be known by caliper log. It also good to have same in a compacted lithology section such that packer may not give way during testing. It provides much needed information which are essential in evaluating the hydrocarbon bearing formations:

- ☆ Nature and type of fluid in the formation
- ☆ Initial reservoir pressure
- ☆ Formation permeability
- ☆ Extent of well bore damage
- ☆ Flow rates with certain draw down/productivity index
- ☆ Extent of depletion with production
- ☆ Indication of reservoir anomaly-fault/barrier, if any
- ☆ Radius of drainage during testing

Procedure to be adopted during DST is initial flow period for 5 mts, initial shut-in for 30 mts, final fow period of about 1 to 2 hours and final shut-in for about 1.5 time of initial flow period.

All pressure data recoded against time on a chart which will be interpreted for desired information (Figure 8.9).

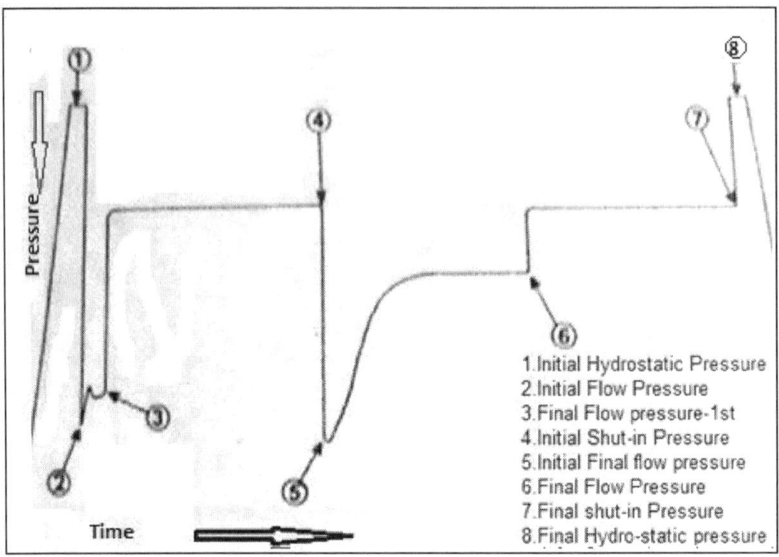

Figure 8.9. Open Hole Drill Stem Test (OHDST)-Chart.

The quantitative evaluation of permeability and well productivity is carried out from the pressure build up curve generated from initial and final shut-in curves. The increase in pressure with time is recorded. The entire period of shut-in time is break down in to small increments and corresponding pressures are read and pressure versus time on semi log paper and plotted for evaluation of reservoir parameters. This plot is called Horner Plot.

Cased Hole DST

It is useful in wells where number of pay zones are encountered during drilling and to ascertain each zone may take if normal conventional time followed one after another will be more and it is necessary to ascertain economic life of each zone.

Formation Pressure Measurement

This class of application uses the direct static formation pressure measurement. It includes:

- ☆ Static pressure measurement and depletion determination
- ☆ Determination of the inflow performance and productivity index (PI) of the reservoir and, in gas wells, the absolute open flow (AOF) potential of the reservoir.

- ☆ Determination of reservoir fluid density from gradients.
- ☆ Determination of reservoir fluid contacts.
- ☆ Identification of reservoir vertical permeability barriers.
- ☆ Identification of vertical flow through layered sequences in developed reservoirs.
- ☆ Numerical reservoir simulation applications.

Permeability and Skin

The pressure and associated measurements are interpreted to yield reservoir dynamic parameters relevant to fluid flow, such as formation permeability and any occurrence of skin that would impair the flow. The measurements will help determine:

- ☆ Reservoir permeability both vertical and lateral
- ☆ Well deliverability
- ☆ Damaged or stimulated well condition, if so
- ☆ Efficiency of stimulation treatments.

Formation Fluid Characterization

The essence of formation testing is flowing the well, which presents the unique opportunity to recover samples of the reservoir fluid. It enables:

- ☆ Collecting representative reservoir samples.
- ☆ Characterizing the fluid composition, its phase behavior and its pressure-volume-temperature (PVT) properties.

Reservoir Characterization

The pressure response during a well test provides the characteristic signature of reservoir fluid flow events that will be interpreted in terms of boundaries, heterogeneities and reservoir volume. It enables:

- ☆ Determining the total reservoir pore volume connected to the tested well.
- ☆ Determining the average reservoir pressure.
- ☆ Determining reservoir boundary conditions such as impermeable barriers and constant pressure conditions.
- ☆ Characterizing reservoir heterogeneities such as layered systems and natural fractures.
- ☆ Quantifying vertical and horizontal reservoir communications.

Wire-line Testing

Wire-line tests are performed mostly in open hole using a cable-operated formation tester and sampling tool anchored at depth while reservoir communication is established through one or more pressure and sampling probes.

Production or Injection Test with Completion String in Place

Production and injection well tests, performed using production logging tools, are conducted to obtain pressure and optional flow measurements. Well tests are conducted at all stages in the life of a reservoir: exploration, development, production and injection. At each of these stages, tests are performed with set objectives.

Exploration and Appraisal Well Tests

At the exploration stage, tests are performed both with an open-hole wire-line tester, to measure pressures and collect fluid samples, and with the drill-string, to simulate production at the time of the completion. A wire-line tester is used to measure the static pressures of all the permeable layers of interest. Pressure versus depth plots help establish the formation fluid gradients and identify the fluid contacts in the reservoir (Bourdet, D, *et al.*, 1984).

The MDT tool can also be used with a packer module to perform a smaller-scale production test. A conventional DST is necessary because at the exploration stage the operator needs to know whether the discovery is of commercial value. The objectives are to establish a production rate and the volume of the reservoir. If production rates are low, the operator needs to know if it is because of poor reservoir deliverability or because of a high skin factor. In the latter case, flow rates could be increased if the skin were removed, and the well would have better potential.

If the volume of hydrocarbons in place, inferred from the initial test, was too small, the discovery would lose its economic potential even if production rates were high. A primary reason for testing exploration wells is to recover a fluid sample. This is the best stage at which to characterize the reservoir fluid, because the reservoir is in a virgin state and no fluid has been produced. In later stages of a reservoir's life, the fluid composition changes and the surface fluid composition may not exactly match the down-hole fluid composition. Because the fluid sampling program in an exploration well is particularly important, it will feature both a wire-line sampling to carry out PVT studies.

Development Well Tests

During the development phase, the test objectives are different from those of the exploration and appraisal stages. Presumably the reservoir deliverability has been assessed and the reservoir fluid has been characterized. It is done for better understand the hydraulic communications at this stage which mainly consists of openhole wireline pressure testing.

The emphasis is on static reservoir pressures, which are used to confirm fluid contacts and fluid density gradients. On that basis, the different hydraulic compartments of the reservoir will be determined and linked to validate the model. Often, field production has already started while additional development wells are being drilled.

In the new wells, pressure gradients already may reflect the influence of the production on the reservoir pressure. On such wells, the reservoir simulator may be used to predict vertical pressure profiles to be confirmed by the wireline tester measurements. Any differences would be used to refine the geological model and introduce suitable compartments in the dynamic model. Another reason for testing development wells is to prepare them for stimulation operations, which may be necessary to produce them economically.

Production and Injection Well Tests

In the production phase, the objectives of testing shift from flow evaluation to reservoir monitoring, data collection for history matching of reservoir simulators, and productivity tests to assess the need for stimulation. Well tests are performed to check for skin resulting from migration of fines to the near-well bore region, and to assess the need for acidizing to remove those fines.

Partial completion effects are diagnosed and rectified by re-perforating or by extending perforation intervals. Gravel-packed wells are tested to evaluate the gravel pack skin and decide on any subsequent treatment. Most of these tests are performed using a pressure gauge suspended from an electric cable or a slickline, or using a production logging tool that will provide complementary measurements such as downhole flow rate.

The complexity of tests will depend on the well and may vary from a simple buildup test to a series of step-rate tests intended to analyze the dynamic performance of complete multilayer systems. Cased hole wireline testing also may be performed to obtain formation pressures in layered sequences, collect samples of bypassed oil detected by saturation measurements and, in some cases, measure permeability.

Productivity Well Tests versus Descriptive Reservoir Tests

The interpretation of well tests is one of the most widespread sources of dynamic reservoir data. Tests on oil and gas wells are performed at various stages of drilling, completion and production. The test objectives range from simple measurements of reservoir pressure to characterization of complex reservoir features. Well tests can be classified either as single-well productivity tests or descriptive reservoir tests.

Single-well productivity tests are conducted to:

- ☆ Determine well deliverability.
- ☆ Characterize formation damage and other sources of skin.
- ☆ Identify produced fluids and determine their respective volume ratios.
- ☆ Measure reservoir pressure and temperature.
- ☆ Obtain representative fluid samples suitable for PVT analysis.
- ☆ Evaluate completion efficiency.
- ☆ Evaluate workover or stimulation treatments.

Descriptive reservoir tests are conducted to:

- ☆ Assess reservoir extent and geometry.
- ☆ Determine hydraulic communication between wells.
- ☆ Characterize reservoir heterogeneities.
- ☆ Evaluate reservoir parameters.

Reservoir Pressure

The primary data obtained from reservoir testing are continuous measurements of reservoir pressure versus time; flow rate is typically controlled to follow a planned schedule. In effect, formation pressure is probably the single most important measurement in the production history, dynamics and economics of the field. Pressure measurements are used in volumetric calculations (reserves), dynamic reservoir property determinations (permeability), reservoir characterization (compartmentalization), fluid characterization (phase behavior, fluid properties) and well completion design (lifting systems).

Pressure also provides information on the evolution of reservoir energy and fluid contacts with time, and it is an essential input to reservoir simulation models. Further, repeated and comprehensive use of pressure measurements is fundamental to the success of production optimization programs. Such programs address not only the well itself but the whole reservoir and even the surface facilities. In addition to well testing applications, reservoir pressure testing is used in:

- ✰ Determining rock stresses and hydraulic fracturing characteristics.
- ✰ Optimizing well flow characteristics and control of hydraulic losses.
- ✰ Optimizing surface production systems.
- ✰ Determining material balance.
- ✰ Characterizing reservoir energy.

Transient Well Testing

It is one of practical method to characterize reservoir properties or the ability of the formation to produce fluid. It has advantages compare to other technique in determining reservoir properties. One of them is that, well testing covers wider area to interpret, even more it can reach up to its boundary. In addition, many types of tests are also available depending on specific parameter to be analyzed.

The well test concept is basically sending a signal to the reservoir, then receiving its response from the formation, at the end the permeability can be obtained from its decline. Response which is received at the wellbore is used to evaluate the near wellbore properties. If response from a boundary is reached, so the distance is possible to estimate from the time delay. The whole process of well test requires specific set-up. The standard well test set-up consists of surface rates, well-head pressure (WHP), bottom-hole pressure (BHP) including bottom hole temperature (BHT), and acquisition interpretation.

Transient well testing applies the inverse solution of indirect measurement where input and output are known from the test, while the system is going to predict or estimate from interpretation, reservoir characteristics, where output represents pressure responses, and input shows a change of rate.

As part of field data, well test data contributes in production analysis model (*i.e.* well test models, material balance models, and decline curve analysis). Those data are collected become reservoir information that furthermore designed to be a predictive model. This model allows engineers to simulate production forecast and run various scenarios with different production strategies. Finally, economic study and decision making for the field development is conducted by considering many aspects.

Well test investigates a much larger volume of the reservoir compare to cores and logs. Approximate depth of investigation of coring is only 10 cm and logging is 50 cm, while well testing may reach 50-500 meters of investigation. Its larger area allows the estimations of reservoir permeability, porosity, skin, average pressure, fracture length, heterogeneities, drainage are, shape, open-flow potential, distances to the boundaries, and some others conditions.

Types of Transient Tests

Well tests, if properly designed, can be used to estimate the following parameters:

- ☆ Flow conductance, kh/μ
- ☆ Skin factor, s
- ☆ Non-Darcy coefficient, D (multirate tests)
- ☆ Storativity, ϕc_t
- ☆ (interference tests)
- ☆ Fractured reservoir parameters λ and ω
- ☆ Fractured well parameters x_f and $k_f w_f$
- ☆ Drainage area
- ☆ Distance to faults
- ☆ Drainage shape

Most well tests consist of changing the rate, and observing the change in pressure caused by this change in rate, for this the basic requirements for a pressure transient test one need to be able to measure, time, rate, pressure and control rate. To control rate at the sand face (that is, at the perforations, or bottom hole) but in most instances it can only control at the surface. An important exception is when bottom hole shut-in devices are used. Bottom hole shut-in devices are generally only used on drill stem tests.

Similarly, rate and pressure are best measured at the sand face. This can be accomplished using down-hole meters. It is usually difficult to measure rate at sand face, and in practice it is rarely done. It is somewhat easier to measure pressure near the sand face, and because surface pressure data are very difficult to interpret, most well tests use bottom-hole gauges. Generally, the gauges are lowered into the well via wire-line

Certain types of tests are dedicated to specific stage of reservoir discovery, development, and production. In exploration and appraisal wells, Drill stem tests (DSTs) and wire-line formation test are normally run. During primary, secondary, and enhanced recovery stages, the conventional transient well tests (*i.e.*, drawdown, buildup, interference, and pulse tests) are run. Step-rate, injectivity, falloff, interference, and pulse tests are executed during secondary and enhanced recovery stages. Some tests are implemented throughout the life of reservoir, such as multilayer and vertical permeability tests.

Each type of test has various reservoir properties which can be obtained. Some tests interpret the same properties, but the level of accuracy might be different. For instance, permeability estimated from buildup test gives higher level of accuracy than drawdown test; otherwise skin calculation from buildup test deliver lower accuracy than drawdown test.

By analyzing the measured pressure response in an "active" or "observation" well caused by a change in flow rate, it is theoretically possible to estimate various reservoir and well parameters: -

- ✰ Reservoir "transmissivity"/"permeability thickness (kh)"
- ✰ Reservoir "storativity" or "porosity thickness (Ôh)"
- ✰ Well "skin" factor and wellbore storage

Old analysis involves plotting the data on semi-log plots to estimate the transmissivity from the slope and then using log-log plots to "match" type curves to check whether the measured response is consistent with a "fracture" or "homogeneous" reservoir. Now possible to use well test analysis software to analyze the data, which provides a lot more options in terms of reservoir properties.

With the advancement in computer aided interpretation tools and introduction of dimensionless variables of pressure, time and area, the scope of pressure transient studies has increased further for estimating reservoir heterogeneity, fault, facies change, pinch out, constant pressure boundaries (aquifer or gas cap), *etc.* which are essential features required to build a geological model of the reservoir.

Build-Up Test

The primary purpose of performing a build-up test is to determine the wellbore damage (skin) and the reservoir permeability. However, during the course of a build-up, it is possible to encounter reservoir boundaries. If all the reservoir's boundaries are contacted during the build-up, the size of the reservoir can also be determined. If the well has been pressure tested before, subsequent testing allows relative material balance calculations (decline curve analysis), as well as the determination of the drive mechanism for the reservoir.

A build-up test is one of the simplest tests to perform. The biggest advantage in performing a build-up is that it is a constant rate test. Simply install the equipment while the well is flowing, then shut the well in upstream of the choke. Nevertheless, if the test objectives are to determine skin and the rock has permeability greater than 2 milli-darcies, most build-up tests can be limited to 2 days. After a build-up it is advisable to perform a single rate drawdown, or a Modified Isochronal test.

Flow Regimes Categories

At different times, fluid flows in the reservoir with different ways generally based on the shape and size of the reservoir. Flow behavior classification is studied in terms of pressure rate of change with respect to time. Three main flow regimes are steady-state flow, pseudo steady state flow and transient state flow.

Steady State Flow

In steady state flow, there is no pressure change anywhere with time. It occurs during the late time region when the reservoir has gas cap or aquifer support. This condition is also called constant pressure boundary which pressure maintenance might apply in the producing formation.

Pseudo Steady State Flow

This flow regime also occurs in late-time region, but it forms when there is no flow in the reservoir outer boundaries. No flow boundaries can be caused by the effect of nearby producing or presence of sealing faults. It is a closed system or acts as a tank where a constant rate production results constant pressure drop for each unit of time.

Transient State Flow

When the pressure/rate changes with time due to well geometry and the reservoir properties (*i.e.* permeability and heterogeneity), it indicates that transient (unsteady state) flow occurs. It is observed before boundary effects are reached or also called infinite acting time period. Higher compressibility of the fluid leads the more pronounced the unsteady state effect of the reservoir fluid.

A step change in the production rate of a well causes a pressure disturbance that propagates radially outward into the reservoir at a velocity determined by the hydraulic diffusivity of the total reservoir and fluid system based on the reservoir porosity compressibility of the total reservoir.

The velocity of propagation is independent of the magnitude of the change causing the disturbance. There are two basic types of pressure transient tests:

- ✰ Pressure drawdown tests are performed after the well has been shut in for a period sufficient to establish static pressure conditions. The well is opened and produced at a steady flow while the pressure (and optionally the rate) change is observed at the sandface.
- ✰ Pressure buildup tests are performed after the well has flowed for a period sufficient to establish a radial flow regime. The well is closed while the pressure (and optionally the rate) change is observed at the sandface. The "open"and "close"cycles can be generated at the surface by changing the choke or closing the master valve, or downhole by using a downhole shut-in valve or a wireline tester flow control schedule.

Basic Interpretations

The rock properties within the homogeneous formation do not change. This ideal condition will not actually occur in real reservoir, but only close enough to be assumed as homogeneous reservoir. Sequences of fluid regimes during well test is started from wellbore storage, near wellbore conditions, and then ended by late time boundary effects.

Transitional characteristic between flow regimes is described from recorded pressure analysis. Using log-log plot, a complete production test can be diagnosed where all flow regimes can be distinguished from one single plot. Therefore, well test interpretations mostly prefer using log-log scale method. How far the pressure response travels in the reservoir is important to measure, or if specific radius of investigation is already set, then how much time the test is performed can be also estimated. Three types of tests, drawdown test, buildup test, and interference test, use log-log or semilog scale as analysis method to recognize different flow regimes and predict several reservoir properties. In the field, buildup is more common than drawdown.

Radius of Investigation

Quantitatively and qualitatively, radius of investigation has great significance both in planning and analyzing a well test. It describes the distance (from the tested wellbore) of the transient pressure into the formation if there is an unstable pressure caused by production or closure of a well. It will show that this distance has a correlation with physical properties of the rock and fluid and also depends on the length of time of well testing. The radius of investigation is a guide to plan a well test.

Drawdown

Drawdown test is ideally performed when the pressure is equalized throughout the formation. Basically, this test measures bottom hole pressure during a period of constant production rate. The equipment is first set into the well, and then begins a constant flow rate. The consideration for having this test is simply when there are some uncertainties in buildup interpretations. Therefore, analysis from drawdown test can be used for comparative analysis. Drawdown Pressure -Δp- differential of the pressure drop between the reservoir (ps) and the flowing bottom hole pressure (pwf).

Interference Test

Interference test is part of multiple-well test which involves more than one well during the test. Predicting inter-well reservoir properties and interpreting communication between wells are the basic purposes of this test. As a multiple-well test, interference test can be applied not only between near-by wells, but also between different sets of perforations in one wellbore. Near-by wells test is used to determine areal reservoir properties or horizontal properties, while different perforation sets test is used to determine vertical properties.

Productivity Index

A commonly used measure of the ability of the well to produce is the *Productivity Index*. Defined by the symbol J, the productivity index is the ratio of the total liquid flow rate to the pressure drawdown. For a water-free oil production, the productivity index is given by:

$$J = \frac{Q_o}{p_r - p_{wf}} = \frac{Q_o}{\Delta p}$$

where,

- ☆ J = productivity index, (STB/day/psi)
- ☆ Qo = oil flow rate, STB/day
- ☆ Pr = volumetric average drainage area pressure (static pressure)
- ☆ pwf = bottom-hole flowing pressure
- ☆ Δp = drawdown, (psi)

The productivity index is generally measured during a production test on the well. The well is shut-in until the static reservoir pressure is reached. The well is then allowed to produce at a constant flow rate of Q and a stabilized bottom-hole flow pressure of pwf. Since a stabilized pressure at surface does not necessarily indicate a stabilized pwf, the bottom-hole flowing pressure should be recorded continuously from the time the well is to flow.

Inflow Performance Relationship (IPR)

It is the behavior of the well's flowing pressure and production rate, which is an important tool in understanding the reservoir/well behavior and quantifying the production rate. The IPR is often required for designing well completion, optimizing well production, nodal analysis calculations, and designing artificial lift.

The Vogel IPR curve was originally developed for oil wells producing from solution gas drive reservoirs. As the Static Reservoir Pressure (P_r) falls below the Bubble Point Pressure (P_b), gas starts to come out of solution in the reservoir. This free gas reduces the relative permeability and inflow capacity for oil. Thus, the unit inflow rate of oil per unit pressure drop in the Flowing Bottom-Hole Pressure (P_{wf}) reduces as the drawdown ($P_r - P_{wf}$) increases.

Although the Vogel IPR was developed for oil wells producing from solution gas drive reservoirs, it has been found to work reasonably well for gas wells. Vogel's IPR equation is:

$$Q_w/Q_{max} = 1.0 - 0.2\,(P_{wf}/P_r) - 0.8\,(P_{wf}/P_r)^2$$

where,

Q_w – The oil production rate at any Flowing Bottom-Hole Pressure P_{wf}

Q_{max} – The maximum possible oil production rate at $P_w = 0.0$

P_{wf} – The Flowing Bottom-Hole Pressure, expressed in Psig.

P_r – The value of the Static Reservoir Pressure at the time of test, in Psig.

Values of Q_w and P_{wf} can be measured by performing a well test (Q_{wf}) and associated flowing bottom-hole pressure measurement. The value of P_r can be measured with a static bottom-hole pressure measurement.

The absolute open flow potential (AOFP) of a well is the maximum flow rate that the well can produce against zero sand face back pressure and FBHP is equal to atmospheric pressure. Deliverability of the well depends upon the following factors like level of water contact, performance of sand production, consolidated grains and other reservoir parameters. Deliverability of the well is generally taken 1/4th of AOFP.

Gas Well Testing

Gas well test analysis is conducted, to obtain reservoir parameters, to determine whether all the drilled length of gas well is also a producing zone, to estimate skin factor or drilling and completion related damage to well. Based upon magnitude of the damage a decision regarding well stimulation can be made. Throughout the life of a gas well, from exploration to abandonment, enough well test data are collected to describe well condition and behavior.

Efficient Gas Well Test Analysis Programs

Initial bottom hole pressure measurements should be made at a selected "Key Gas Well" periodically. The key to determining whether or not a well is a good candidate for stimulation treatment is diagnosing the well to find the cause for its low productivity. Buildup, drawdown, or drill-stem tests, core analyses, offset well data, and other information can be used to accomplish this. After diagnosis, the optimum well stimulation treatment, either small or massive hydraulic fracturing, can be designed for the well.

Isochronal Tests for Gas Wells

The isochronal test is a series of single-point tests developed to estimate stabilized deliverability characteristics without actually flowing the well for the time required to achieve stabilized conditions at each different rate. The isochronal test is conducted by alternately producing the well then shutting it in and allowing it to build to the average reservoir pressure before the beginning of the next production period.

Pressures are measured at several time increments during each flow period. The times at which the pressures are measured should be the same relative to the beginning of each flow period. Because less time is required to build to essentially initial pressure after short flow periods than to reach stabilized flow at each rate in a flow-after-flow test, the isochronal test is more practical for low-permeability formations. A final stabilized flow point often is obtained at the end of the test (Figure 8.10).

Formula

$$Qmax=Qwf/(1.0-0.2(pwf/Pr)-0.8(Pwf/Pr)^2$$

where,

Pb=Bubble Point Pressure

Pr= Static Bottom hole Pressure

Pwf =Flowing Bottom hole Pressure on well to test

Pw= Flowing Bottom hole Pressure

Q max= Maxumum Prosuction rate

Qwf= Production rate on well Test

Qw= Production Rate.

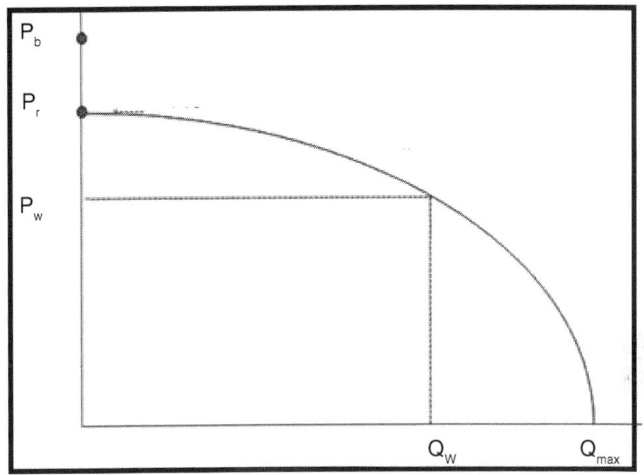

Figure 8.10. Isochronal Test.

The isochronal test is based on the principle that the radius of drainage established during each flow period depends only on the length of time for which the well is flowed and not the flow rate. Consequently, the pressures measured at the same time periods during each different rate correspond to the same transient radius of drainage. Under these conditions, isochronal test data can be analyzed using the same theory as a flow-after-flow test, even though stabilized flow is not attained. In theory, a stabilized deliverability curve can be obtained from transient data if a single, stabilized rate and the corresponding BHP have been measured and are available.

Modified Isochronal Test Procedure and Analysis

The modified isochronal test is conducted like an isochronal test, in tight reservoirs, except the shut-in periods are of equal duration. The shut-in periods should equal or exceed the length of the flow periods. Because the well does not build up to average reservoir pressure after each flow period, the shut-in sand face pressures recorded immediately before each flow period rather than the average reservoir pressure are used in the test analysis.

As a result, the modified isochronal test is less accurate than the isochronal test. As the duration of the shut-in period increases, the accuracy of the modified isochronal test also increases. Again, a final stabilized flow point usually is obtained at the end of the test but is not required for analyzing the test data.

The well does not build up to the average reservoir pressure during shut-in; the analysis techniques for the modified isochronal tests are derived intuitively. Recall the transient flow equation, expressed in terms of the reservoir pressure at the start of flow, on which isochronal testing is based.

Volume of Hydrocarbons

The volume of hydrocarbons in a reservoir can be calculated: *directly* by volumetric methods, *indirectly* by material balance methods.

Volumetrics provide a ***static*** measure of oil or gas in place. Accuracy of volumetrics depends on data for:

- Porosity
- Net thickness
- Areal extent
- Hydrocarbon saturations

Material balance methods provide a ***dynamic*** measure of hydrocarbon volumes. Accuracy depends on quality of data for:

- Pressure surveys
- Temperature surveys
- Analysis of recovered fluids. Normally material balance methods increase in accuracy as the reservoir is produced.

Volumetric Analysis

Also known as the ***geologist's method*** because it is based on geological maps, core logs and analysis of wireline logs. Isopach maps are used to compute the bulk volume of the reservoir (V).

Gas-in-place Volume

The ultimate goal of any exploration and production effort is to increase hydrocarbon reserves while ensuring their efficient and economical recovery. Gaining detailed knowledge of reservoir geometry, rock properties, and productive capacity is the first step toward reservoir optimization.

Net pay is the part of a reservoir from which hydrocarbons can be produced at economic rates, given a specific production method. The distinction between gross and net pay is made by applying cut-off values in the petro-physical analysis. Net pay cut-offs are used to identify values below which the reservoir is effectively non-productive. In general, the cut-off values are determined based on the relationship between porosity, permeability, and water saturation from core data and capillary pressure data.

Data required to estimates of the gas-in-place volume are:

- Reservoir pressure and temperature, or depth to calculate the required parameters,

- ☆ Compositions of oil and gas or their gravities and molecular weights,
- ☆ Gravities and production rates of separator condensate and gas,
- ☆ Rock porosity,
- ☆ Gas or interstitial water saturation, and
- ☆ Area and thickness.

Volumetric estimates based on cores, well logs, fluid analyses, and geological estimates of reservoir size provide a "rock-based" estimate of gas in place, while material-balance relationships provide a "fluids-based" or "pressure-based" estimate. These two types of estimates are essentially independent (except for the use of consistent values of B_{gi}). Thus, when the two estimates are comparable, there is greater certainty in the OGIP estimate.

It is now a common practice to develop geologic and simulation models of a reservoir to determine reserves and depletion strategies and to evaluate alternative development scenarios.

The volume of free gas in a gas reservoir or gas initially-in-place is given by:

$$GIIP = Gr = (V.n).(1 - Swi)$$

In terms of standard volumes at STP, the gas volume is:

$$G = (V.n).(1 - Swi)/Bg$$

$$G = (V.n).(1 - Swi).E$$

Where, Bg is the gas formation volume factor and E is the gas expansion factor.

For oil and gas reservoirs below the bubble point, the total Hydro-Carbon in Place Volume is given by the HCPV:

$$HCPV = (V.n).(1 - Swi)$$

The stock tank oil volume can be computed as:

$$STOIP = (V.n).(1 - Swi - Sg)/Bo$$

The standard gas volume at STP is given by:

$$G = (V.n).(1 - Swi - So)/Bg$$

Discusses various aspects of gas reservoir performance, primarily to determine initial gas in place and how much is recoverable. The equations developed can be used to form the basis of forecasting future production rates by capturing the relationship between cumulative fluid production and average reservoir pressure.

There are ***gas condensate reservoirs***, which yield significant amounts of condensates along with the gas production. The reservoirs operating under depletion drive generally show a rising trend in GCR as the reservoir pressure goes below dew point pressure. In a gas condensate reservoir it is generally required to estimate the condensate reserves along with the gas reserves.

Material-Balance Determination of OGIP

Material balance equation (MBE) is a basic reservoir engineering tool for interpreting and predicting reservoir performance. MBE is applied to estimate initial hydrocarbon volumes in place, predict future reservoir performance and predict ultimate hydrocarbon recovery under various types of primary driving mechanism. It provides a relationship between original fluids in place, cumulative fluid production, and average reservoir pressure. For many gas reservoirs, a simple material-balance equation can be derived on the basis of the following assumptions:

MBE calculation is based on changed in reservoir conditions over discrete periods of time during the production history.

Assumptions include: - filled pore volume is constant, gas dissolved in water or liberated from the rock is negligible and reservoir temperature is uniform and constant and also:

- ☆ Volume changes in the reservoir are assumed to occur without any temperature changes. If temperature changes occur, they are usually significantly small to be ignored without significant error.
- ☆ All parts of the reservoir have the same pressure and therefore fluids properties through-out are constant. Minor changes near the wellbore can be ignored.
- ☆ Reservoir volume is assumed to be constant except for those conditions of rock and water expansion or water influx that are specifically considered in the equation.

A simple expression for the oil MBE is the following:

(Cumulative net withdrawal in STB) = (original oil in place in STB) - (oil remaining in place in STB)

The cumulative net withdrawal is the difference between the oil that leaves the reservoir and the oil that enters it. In this basic analysis, there is no oil entering the reservoir since the boundaries are considered impermeable to flow.

Reservoir Characterization

Reservoir characterization is the process of ascertaining petro-physical properties in space. Reservoirs can be extremely heterogeneous. Reservoir heterogeneity is defined as a variation of reservoir properties in space. These randomly distributed heterogeneities can significantly reduce the transmissibility. Proper reservoir characterization is the key to model fluid dynamics more accurately. Integrated reservoir characterization combines all geological, geophysical, and engineering data pertinent to understanding fluid flow and reservoir behavior, and uses the data to produce a 3D-earth model that can be used in all phases of field development. In this way, it differs from traditional reservoir characterization projects, which tend to focus on geologic and engineering analysis of proposed enhanced recovery projects, as well as drilling, production engineering, and reservoir simulation.

Integration of seismic data with well logs and other subsurface information makes it possible to significantly enhance reservoir description and thus improve reservoir economics. Porosity predictions away from drilled wells allow a better characterization of reservoir in economic.

Permeability, one of the key parameters to reservoir characterization, is a function of pore properties such as pore space and its distributions within the matrix. In general logarithm of permeability is linearly proportional to porosity in sandstone and it depends on the type of formation mainly on grain size, sorting, pore size distribution or irreducible water saturations (Beard, D. and P. Weil, 1973). If for various reasons the data needed to estimate these parameters not available, the permeability prediction can be calculated through regression analysis on core measurements. The build-up test data and RFT/SFT measurements also help in knowing permeability. In most of the cases the core data is relatively less and to obtain porosity and permeability relations the empirical relations are used.

In reservoir simulation studies, it is necessary to characterize fine scale heterogeneity within the reservoir. Complex problem matching through reservoir modeling and subsequent prediction of performance depends much on the spatial distribution of petro-physical properties like saturation, porosity and permeability.

Even discontinuous unproductive zones can restrict vertical flow, causing tortuous flow of the fluid some distance horizontally, before moving vertically. Geostatistics determines the spatial variability of properties through certain correlations. Using these, values of a variable at un-sampled locations can be estimated. The uncertainties in estimation can also be defined (Alden, J., *et al.*, 1997).

The goal of geostatistical analysis is precisely to generate maps for porosity models and quantify uncertainty by conditional simulation (Bhattacharya, S., *et al.*, 2003). Probability maps, deviation map, stochastic simulation maps and its average map were generated for analysis of the occurrence of computed porosity for future wells or history matching of seismic based porosity map. New technologies like, seismic trace inversion and porosity modeling have helped the interpretation of seismic data to know the hydrocarbon bearing potential of a reservoir.

The approach for reservoir characterization by quantifying the flow units can be based on petro-physical rock types, storage capacity, flow capacity and reservoir process speed (k/ϕ). Using these parameters, it provides a quantitative approach to transform rock types into flow units. A flow unit is a stratigraphically continuous interval of same reservoir process speed that has honored the geologic framework and maintained the characteristics of rock types. It is the inter-relationship of flow capacity to storage capacity of the unit that relates directly to reservoir flow.

In reservoir characterization, the thin lamination of sand and shale in addition to thick clean sands is an important issue. The thick sands are resolved properly but over thin beds log records average properties of sand and shale. Saturation estimated from the conventional resistivity measurements are dominated by high conductive thin shale and fine grained sands which obscure the presence of more resistive hydrocarbon bearing sands. Direct interpretation of the log reading therefore results in a significant underestimation of hydrocarbon saturation.

In general, reservoir engineers prefer to estimate hydrocarbon saturation using capillary method. Reservoir water saturation decreases with increasing height above the FWL (free water level), where capillary pressure is zero. Minimum water saturation (Swirr) is reached at a great height above the FWL and this water saturation is immobile. Variation in the capillary radius is controlled by the pore geometry, which is a function of rock properties such as permeability and porosity. Hence, pore size distribution has a major influence on the magnitude of the irreducible water saturation and the extent and height of the transition zone (Rajesh Kumar, P. K., *et al.*, 2002).

The distribution of vertical and areal variations of permeability is critical in determining the dynamic flow behavior of a layered heterogeneous reservoir in time and space (Gunter, G.W., *et al.*, 1997). The main constraint to map the variations has been the availability of combined permeability of layers due to multilayered completions besides, the single layered permeability, wherever available, are too meagre to supplement a viable reservoir model (Bloch, S., 1991).

Permeability is directly determined from core plugs but it is difficult to use the plug data alone for field scale mapping as the core plug do not give a continuous variation of permeability with depth. Cores are taken in vertical wells. It was observed from pressure build up studies that, the permeability of a layer shows wide areal variation, even between two wells. Core permeability is known to be less than build up permeability. Availability of electro-logs are possible more area in the field and can be used for estimating as well as mapping the layer-wise permeability. It has been observed that the core derived flow capacities are mostly at the lower side compared to those obtained from pressure build up studies.

Examples

The reservoir characterization, in terms of layer-wise permeability assessment and mapping in partly weathered carbonates, has been carried out in **Bombay High North (**Basu, D.N., *et al.*, 1980). A successful attempt has been made to link the coefficient of permeability, determined from core samples, with the log characteristics through the help of transforms using multiple Regression Technique (Bhandari, L.L. and S.K. Jain,1984,). Having calibrated against the permeability interpreted from pressure build up data, the transforms were further used for estimation of layer-wise permeability (Azim, S.A, *et al.*, 1994). The permeability maps, prepared with these data have brought out vertical and spatial variation of permeability and thereby delineating areas which are poorly drained by early water and gas encroachment. The maps are in good agreement with field observation, thus giving a good confidence level to the transforms (Amitabh Pandey., 2003).

The analysis in exploratory and development wells in **South Heera field** depicts good number of flow units. The core analysis shows poor reservoir characteristics with porosity varying from 7 per cent to 16 per cent and permeability in the range of very less to 10 md., with grain density 2.68 gm/cc to 2.77 gm/cc. The pore throat radius is found to be in the range of 0.05 to 2.8 micron of the pore structure of a well. Different types of petrophysical rock types like packstone, wackstone, mudstone and shale have been identified on the basis of cross plots. The rock types identified have been integrated with the microfacies and conventional core analysis. This has also been observed in SEM microphotographs. It clearly indicates the heterogeneous nature of the reservoir.

Neelam field with two structural culminations which is located in Western offshore of India was proved petroliferous, the reservoir is limestone with shaly intercalations (Dipankar Ghosh., *et al.,* 2005). The field is located approximately 45 kms south west from Mumbai. It was discovered in year 1987. Its initial oil in place is 110 MMT with Ultimate recoverable reserve of 32 MMT. The cumulative production was about 26 MMT.

The production was started from this field in year 1990 with the peak production was 95000 BOPD. Since wells started ceasing and production from field declining, wells had to be put on gas lift during early period. It was also witnessed a very high GOR in wells of northern sector with high water-cut. The tight zone got preferentially fractured due to its less ductility which is believed to be a barrier, is providing sufficient transmissibility to fluid through its fracture network (Chandra Prabha Verma., *et al.,* 2006). The field was developed through years with more than 120 wells which include some water injection wells. The porosity is 26.70 per cent uniform but permeability varies very widely few milli-darcies to hundreds of milli-darcies.

In the **Bassein Gas Field** the Middle Eocene Limestone is a thin (10-15 m) oil column a between thick (~60m) gas cap and bottom water. Saturation at any point in the reservoir depends on the porosity and its height from free water level. Using the Geocellular model the saturation height function approach appeared to be more suitable than the other method as it takes into account the height of data point above the free water level (FWL) along with the petrophysical data (Cherukupalli, P.K., *et al.,* 2003). This has improved the saturation model and the volumetric estimates were in good agreement with production history of the field. The model has clearly characterized the heterogeneity of the formations and modeled the fluid distribution more accurately.

North Kadi oil field in north **Cambay Basin,** has been under production since 1969. The hydrocarbon accumulation in the field is mainly in Middle Miocene sandstone of Kalol Formation having high permeability and strong aquifer support. Earlier, the pays within Kalol Formation were believed to be hydro-dynamically connected and accordingly the fluid contacts (GOC and OWC) were also considered common for all the pay units (Bhandari L.L. and L.R Chowdhary,1975). This necessitated acceptance of a tilted water contact in the southern culmination for inexplicable reasons.

The new well data drilled later gave a new insight when the well showed the presence of gas in deeper zone, and oil in upper zone. A complete review of reservoir characterization was taken up with log correlation along with reservoir pressure data of various pay units. These have led to identification of five pays identified within Kalol Formation. The KS pays are highly permeable (upto 3000 md) and holds most of oil-in-place. The average pay thickness varies from 15m to 4m. The reservoirs within Kalol formation have strong aquifer support (Ghosh B.N.et. al; 2003 and 2005).

South Kadi field of **Cambay Basin,** western India, comprises of multi layered reservoirs of Kadi and Kalol formations. Lower Kadi pays are discrete sand bodies having limited areal extent; occur in the depth range of 1700m to 2200m, with sand thickness varying between 2m to 20m. All these pools are operating under depletion drive mechanism, with initial more hydrostatic pressure. Rapid lateral facies variation of Lower Kadi sands has resulted in high degree of uncertainty in deciphering the dispersal and geometries making the field development a very challenging (Rakesh Dhir, K., *et al.*, 2003). Pressure transient studies and reservoir limit tests of some pools were carried out to decipher pool limit, volume of reserves and different reservoir parameters. Pressure-production history together with geological data has indicated more number of hydro-dynamically independent pools than initially known. Such small pools also added reserves when calculated through Material Balance method.

The hydrocarbons in the **Gandhar field** are contained within the Middle-Upper Eocene Ankleshwar Formation, which reaches a maximum thickness of 450 m. The field contains 13 separate reservoirs, separated by shale intervals. Twelve of the reservoirs occur in the Hazad Member (termed sand bodies as GS-1 to GS-12) and one other in the overlying Ardol Member. The individual reservoirs are of the order of 3-10 m thick. It is contained within a combination trap, formed by east wards structural dip on the NW margin of the Broach depression, combined with stratigraphic wedge-out of deltaic sands towards the northwest. Porosities range from 17-24 per cent and average 19.6 per cent, while permeabilities are commonly 300 mD, ranging from 20-1800 mD. Porosity is impaired by the presence of detrital and diagenetic clay. The API gravity of the oil varies from 34° to 59° with an average pressure of 3,912 psi. Recovery is facilitated by the low fluid viscosity, which is generally in the range 0.15-0.35 cp. The reservoir temperatures are 130 to 140°C. The dew points obtained from experiments were used to interpolate the dew point at 132°C and it was found to be 277.9 kg/cm^2.

Initially the oil was produced by pressure depletion drive, with partial support from aquifer edge-water and gas-cap expansion drive. With drilling of more wells, the gas production started increasing. The Gas Condensate Ratio has increased from a level of about 1700 v/v to about 5000 v/v. The average reservoir pressure has declined from the initial value of 312 kg/cm^2 to about 170 kg/cm^2 slowly. The condensate API gravity is also showing a rising trend from the initial value of 51.8°API.

The decline in reservoir pressure led to a drop in the oil rate from most reservoirs within 2-3 years of coming on stream. Water injection commenced in 1991. Oil production from the field reached a peak of over 40,000 BOPD in 1995. Development wells were drilled in a five-spot pattern.

The **Viraj field** is in Cambay basin which is a small field occupies an area of 4 sq. km. The field which was discovered in 1977 with three oil-bearing zones was put on production in 1980.The estimated oil reserves are 7.88 MMT. About 50 wells were drilled and the field has produced nearly 20 per cent of oil in-place.

The reservoir is mainly sandstone with good permeability. Occurrence of siltstone, often grading into shale and coal gives rise to reservoir heterogeneity. Initial average pressure for the reservoir works out to be 140 kg/cm^2 at -1260m MSL. Initial GOR (Gas-Oil-Ratio) from PVT data is very low at 15v/v with saturation pressure of 65kg/cm^2. With time the average field GOR has marginally increased. The reservoir pressure has not declined (Panda, B.K., *et al.*, 2005). Sand cut problem has been encountered in some wells due to unconsolidated nature of sand. Most of the wells have been on production with successful gravel pack jobs. High water cut is an inherent problem of all wells. Since beginning the reservoir has water cut problem. Many wells have abnormal water cut and conventional water shut off job has shown no effect. The crude from this field being heavy and viscous is prone to foam. Silicone based de-foaming chemical, DM-silicone fluid 10,000 CTSK was found most effective and decrease the foaminess index (Bassi. K.D. and R.K. Gang.2003).

The **Bechraji field** lies in the northern part of of Cambay Basin which was discovered in 1987 and was put on early production. The field has an OOIP of 32.65 MMT but because of high viscosity of crude oil (250cp) and low API gravity (15.6°) the primary recoverable reserve was only 3.41 MMT, *i.e.,* 10.4 per cent of OIIP. The porosity was computed using density-neutron log combination. Water saturation (Sw) was calculated. The geocellular model has effectively defined the heterogeneity in sand – shale - coal intercalated. Some shale layers and coal streaks are developed within the sand unit which are isolated in nature and do not prohibit the reservoir communication in horizontal direction (Dey, A.K., *et al.,* 2005).

The three heavy oil fields ***Lanwa-Balol-Santhal*** **(LBS)** complex comprises a continuous, heavy oil bearing belt on the eastern flank of the Mehsana High of Cambay Basin. It comprises a continuous, 40 km long by 2-3 km wide, heavy-oil belt that has been divided into the Lanwa, Balol and Santhal fields. The **Santhal Field** is one of the fully developed commercially producing heavy oil fields. The reservoir consists of 4 sand layers in the Upper Suraj Pay, KS-1, KS-II and KS-III+LS of Kalol formation, with a good cap of Tarapur shale. The sands have a common oil water contact (OWC) varying between 992m and 998 m from south to north and are supported with active edge water drive. The reservoir pressures are hydrostatic. In the Santhal Field, net pay ranges from 0 to 70 m, but is generally 1-40 m. Sandstone reservoirs in the heavy oil belt have gross thickness of 40-60m.

Most of the wells have shown high water cut from the initial stages of production. The work-over jobs attempted to shut-off water have met with limited or no success especially in the wells located in wedge zone. Above 970 MSL depth clean oil production is obtained for a longer period, Sw at this level is just around the irreducible saturation of 22 to 35 per cent. The high viscosity of the order of 525cp, coupled with active water drive has resulted in adverse mobility ratios for oil. With primary recovery of only 11-15 per cent.

The Middle Eocene pay zone (Kalol Formation) is characterized by unconsolidated sand units with active aquifer as the drive mechanism. Porosities range from 20 to 36 per cent while permeabilities are of the order of 8 to 15 Darcies. Most of the wells have shown high water cut from the initial stages of production. Detailed analysis of regular production performance of wells has indicated that premature water cut is a real time problem due to adverse mobility ratio. The edge water encroachment adds to the problem. Because of high viscous crude oil at water saturation (Sw) as low as 36 per cent, the fractional flow (Fw) of water could be 54.3 per cent and at Sw=40 per cent the Fw can be 76.9 per cent. Quite early during the production, as soon as Sw reaches 46 per cent, the wells flow virtually free water (Gupta. A.,and B.C. Jaiprakash., 1993).

The **Charali Field** is located in the Assam and in all there are five hydrocarbon bearing sands, which were deposited under paralic conditions and the BCS comprises of cyclic sedimentation of coal and shale with various channel sands, whereas the Tipam sands deposited under fluvial conditions. The Barail sand with initial pressure of 330 ksc and was put on production in 1979. The sand is having bottom water and the OWC is at -3288 m. The well has produced clean oil for almost 15 years and aquifer water breakthrough started. The water cut suddenly shot up to 60-70 per cent and the stabilizing at around 40-45 per cent. The reason for this sudden rise in water cut is suspected to be due to channeling. Using geo-statistical methods identified the small scale heterogeneities in form of intervening shales and their extension and continuity. Log correlation was attempted to conceptualize the geological model and layering of all sands. The history match was satisfactory when this model was used.

The pressure transient testing have been utilized in **Suryaraopeta field** in **Krishna Godavari Basin**. On the basis of pressure transient analysis it was revealed that the reservoir is a confined one with area about 27. 6 acres, thus confirming marginal status to the field. Well testing was used as a tool for knowing well productivity and estimating oil and gas in-place through application of material balance method. The width of reservoir body is about 100 metres and length is about 1,200 metres. The well location is at the corner of the channel fill bar as evident from boundary data. However, the structure contour map of the area indicates that the well is located on a culmination, which covers substantial area. This clearly demonstrates that the reservoirs have more stratigraphic control.

G-1 structure in Krishna-Godavari basin situated in water depths ranging between 250-300m. Four hydrocarbon bearing sands have been encountered, of which sand-8 has a maximum hydrocarbon volume (Gupta, S.K., *et al.*, 1998). The major challenge is to design the exploitation strategy which is having large gas cap with 14 m oil column.

The details of 3 phase black oil simulation studies have been carried out for the sands, simulator include the depletion case, in which the production rates assigned are based on test rates and productivities of the well during performance prediction. The life under depletion drive is about 3 years. Exploitation of upper sand through depletion drive one well only. This gives an oil recovery of about 12 per cent. However, pressure maintenance through water injection improves oil recovery substantially to 36 per cent by adding one injector well (Suri. N.K., 2003).

The **Narimanam** structure is located in the Cauvery basin, southern India and found oil and gas in 1985. The field has multiple oil and gas pays of Oligocene and Eocene age. In the Oligocene section several different sands were identified and put on regular production. The estimated recoverable reserves are at 2.8 Mmt of oil with 5.5 Mmt of original oil in place. So far about 72 wells were drilled. Log correlation data indicated the presence of gas cap in some wells and H_2S was also observed in the gas. Whereas PVT analysis indicated under saturated crude in some wells. So on the basis of PVT data together with production behavior of the wells it was concluded that the sands comprises of two different lobes one having gas cap and another having under saturated crude.

Reservoir Management

Petroleum reservoir management is a dynamic process that recognizes the uncertainties in reservoir performance resulting from our inability to fully characterize reservoirs and flow processes. It seeks to mitigate the effects of these uncertainties by optimizing reservoir performance through a systematic application of integrated, multidisciplinary technologies. It approaches reservoir operation and control as a system, rather than as à set of disconnected functions. As such, it is a strategy for applying multiple technologies in an optimal way to achieve synergy.

One of the most important stages of reservoir management is **reservoir description**, this stage is to identify and present a model for describing the reservoir which its behavior and answer be similar to the actual behavior of the reservoir as much as possible. Reservoir description is a dynamic process which is repeated as soon as receiving the new information. Reservoir model is yield from integrating the interpreted models of different types of information and by engineering software. The aim is to define a **reservoir model** which shows the dynamic and static information of the reservoir. As soon as the reservoir model is constructed its coordination and compatibility with available interpretative data and information must be proved. In the proving stage, the response of the reservoir should be computed and evaluated by different simulators such as seismic, well logging, flow and reservoir in order to prove the ability to reproduce the information and data.

After the model was confirmed, the model can be used for predicting the future behavior of the reservoir in different development scenarios. Reservoir production behavior such as pressure, flow rate, fluid saturation percentages are calculated by using the flow adjusted simulators. For predicting the field behavior considering the entire system is necessary and all of reservoir, wells and well head facilities should be studied together.

Reservoir Simulation

The purpose of reservoir **simulation** is estimation of future field performance under a variety of producing schemes. Observation of model performance under different producing conditions then aids in selecting the optimum set of producing conditions for the reservoir (Haldorsen, H.H., 1986). It allows estimation of field performance under various reservoir mechanisms such as natural depletion or water, gas or polymer injection, *etc.*, the effects of well locations and spacing, and the effect of producing rate on recovery (Singh, V.K., *et al.*, 2005).

Reservoir simulation models are used by oil and gas companies in the development of new fields. Also, models are used in developed (Brown) fields where production forecasts are needed to help investment decisions. Improvements in simulation software have reduced the time to develop a model. Also, models can be run on personal computers rather than more expensive workstations.

For new fields, models may help development by identifying the number of wells required, the optimal completion of wells, the present and future needs for artificial lift, and the expected production of oil, water and gas. For developed fields, models may help in improved oil recovery by hydraulic fracturing. Highly deviated or horizontal wells can also be represented.

In reservoir simulators the objective is much the same to simulate the exploitation of a real reservoir without the costs of real life trial and error,, *e.g.*, to test different productions scenarios to find an optimal one before the reservoir is actually put on production (Coats, K.H., Dempsey, J.R. and Henderson, J.H., 1971).

Primarily, the pore system itself and the flow in this system happen on a detail level that is impossible to describe or model. In addition to that, the detail structure of the pore system will obviously be unknown. So the difficulties are, generalization of real micro-scale data from macro scale observations (seismics and well data), upscaling of detail level behaviour to a manageable scale and problem size and some uncertainty in the descriptive model itself (equations for multiphase flow in porous media).

Basic Analysis

If a reservoir is fairly homogeneous, average values of the reservoir properties, such as porosity, are adequate to describe it. The average pressure, time, and production behavior of such a reservoir under a production behavior of such a reservoir under a solution gas drive, are normally calculated by the familiar methods. All of these methods use the material balance equation normally referred to as the MBE.

The Meterial Balance Equation reduces to its simplest form. Such a reservoir model is called the ***tank model***. It is zero dimensional because rock, fluid properties, and pressure values do not vary from point to point. Instead, they are calculated as average values for the whole reservoir. This tank model is the basic building block of reservoir simulators. Now let us consider a reservoir represented by a sand-bar. Let the two halves of the sand-bar vary in lithology. The sand-bar as a whole cannot be represented by average properties, but each half can. Thus, the sandbar consists of two tank units, or cells, as they are normally called. The MBE describes the fluid behavior in each cell as in the previous tank model.

In order to manage and optimize hydrocarbon production, oil companies need to model oil and gas recovery under different production scenarios. These simulators numerically model the subsurface flow of oil, water and gas. A model of the porous reservoir rock is constructed on a three-dimensional mesh using a finite-volume discretization. Within the model, each individual mesh cell models the physics of the flow behavior. Production elements such as producer or injector wells are treated as boundary conditions.

Practical limits on the size of reservoir simulation models are computational expense or capabilities. These constraints may control the size of the reservoir segment being simulated must be reduced or the grid block size increased. Simulating small characteristic segments of a field using a fine grid may be preferable to simulating larger segments using a coarse grid.

Despite the significant advances in computing power in recent years, the average cell size for large reservoirs in the horizontal direction has remained in the order of hundreds of meters. This implies that petro-physical properties, pressures, and saturations are assumed to be constant over significant rock volume (of the order 100m x 100m x 10m). To fully utilize high-resolution seismic data and drilling well measurements at the desired one-metre scale, one would need the capacity to run reservoir simulations on meshes containing billion of cells.

A simulation study requires description of the reservoir's rock and fluid properties, validation of completion and production history, and extensive history matching to validate and modify this input data. When history matching is complete, numerous predictions of field and well performance characteristics are calculated for various development scenarios.

Two types of fluid descriptions are used in reservoir simulation studies: black oil and compositional. The black oil description expresses the fluid's volumetric (formation volume factor and solution gas-oil ratio) and flow (viscosity) characteristics as a function of pressure for three phases: oil, gas, and water.

The term **black-oil model** refers to the fluid in which water is modeled explicitly together with two hydrocarbon components, one (pseudo) oil phase and one (pseudo)gas phase. A black oil simulator does not consider changes in composition of the hydrocarbons as the field is produced. This is in contrast with a compositional formulation, in which each hydrocarbon component (arbitrary number) is handled separately. Black oil fluid descriptions are used to describe most oil and gas fields. Primary depletion, water-flooding, and gas injection can all be simulated with black oil models. Volatile oil reservoirs or gas condensate reservoirs generally require compositional models. The black-oil equations are a set of partial differential equations that describe fluid flow in a petroleum reservoir, constituting the mathematical framework for a black-oil reservoir simulator. Black oil thermal models describe fluid properties as a function of temperature as well as pressure.

The **compositional model**, is a more complex model, where the PVT properties of oil and gas phases have been fitted to an equation of state (EOS), as a mixture of components. The simulator then uses the fitted EOS equation to dynamically track the movement of both phases and components in field.

The simulation model computes the saturation change of three phases (oil, water and gas) and pressure of each phase in each cell at each time step. As a result of declining pressure as in a reservoir depletion study, gas will be liberated from the oil. If pressures increase as a result of water or gas injection, the gas is re-dissolved into the oil phase.

These models may exhibit such complexities as a fluid whose density is linearly proportional to depth or whose phase switches repeatedly between oil and gas. Thermal models, used to simulate steam injection, may use either black oil or compositional fluid descriptions.

Fluid saturations and the produced fractions of oil, gas, and water are determined by capillary pressures and relative permeabilities specified as functions of water saturation. Equilibrium (initial) fluid saturations are directly dependent on capillary pressure, which is itself a function of height above the fluid contacts, the fluid densities, porosities, permeabilities, and the surface chemistry of the fluids.

Once production or injection commences, fluid movement is controlled by the relative permeability of each phase (except at very low velocities where the effects of capillary pressure are important). In a reservoir consisting of two fluid phases, oil-water, gas-oil, or gas-water capillary pressures and relative permeabilities must be specified. For a three-phase system, relative permeabilities and capillary pressures for two of the three possible systems are specified.

Matrix properties describe reservoir characteristics specified at each point of the grid (matrix) overlying the reservoir model. Once a grid has been selected, the average depth, thickness, porosity, and permeability are calculated for each grid block. Digitized structure and isopach maps may be used with mapping and gridding software to calculate average depths and thicknesses. Mapping software is used to convert the digitized contour maps to an interpolated grid of values.

Porosities are calculated for each grid block in a method similar to that described for reservoir depths and thicknesses. First, the porosities calculated at 0.5- to 1-ft intervals from well logs must be averaged for each simulator layer. These layer porosities are then contoured and gridded. More advanced techniques for calculating three-dimensional porosity distributions (such as stochastic and geostatistical methods) are topics of current research. Permeability is often correlated as a function of porosity.

Initializing the Model

The first step in any model study is the calculation of pressures and fluid saturations before the onset of production. At this point in time, the model should be in ***static*** equilibrium. Distinct fluid contacts should appear at the appropriate depths, and the potential (the sum of pressure and the gravity head) should be equal everywhere in the model. A model that exhibits indistinct fluid contacts or uneven potentials (meaning that flow will occur in the next time step without production) has not been initialized properly.

Software Available

BOAST - Black Oil Applied Simulation Tool simulator is a free software package for reservoir simulation. Boast is an IMPES numerical simulator (finite-difference implicit pressure-explicit saturation) which finds the pressure distribution for a given time step first then calculates the saturation distribution for the same time step isothermal.

MRST - The MATLAB Reservoir Simulation Toolbox (MRST) consists of two main parts, a core offering basic functionality and single and two-phase solvers, and a set of add-on modules offering more advanced models, viewers and solvers. MRST is mainly intended as a toolbox for rapid prototyping and demonstration of new simulation methods and modeling concepts on unstructured grids.

The Open Porous Media (OPM) initiative provides a set of open-source tools centered on the simulation of flow and transport of fluids in porous media.

ECLIPSE- In this, simulators include black oil, compositional, thermal finite-volume, and streamline simulation. 1D model is in one direction and used for linear or radial flow of reservoir, 2D model used for both types of flow where as 3D model is three directions like pattern,segment or full field understanding.

The **Petrel** platform enables the selection of more representative geologic models that capture a full range of static and dynamic variability. It also enables the study of dynamic connectivity across different geologic realizations in tandem with volumetric calculations. By visualizing connectivity in 3D as flow paths and 2D as maps, clearer and more certain well locations can be targeted for data acquisition and improved production potential. Furthermore, the Petrel platform allows the automation of uncertainty studies, thus enabling productivity gains. In Static model in Petral one has to define work on geological characteristics to be formulated and fed in to software, where as in dynamic model all pressure and production data both available and logically predictable have to be fed in to.

Case Studies

Reservoir management of **Jotana field** of Cambay Basin, western India, is a challenging task as it include many small reservoirs differentiated by faults and permeability barriers. Continuous monitoring, analysis and evaluation are key to problem identification and immediate corrective measures (Deepti Tiwari, *et al.,* 2003). This field is an elongated NNW-SSE trending structure, Hydrocarbon accumulation was established at 4 levels within depth range at 1000-2000 mts. (Chandra P.K. and L.R. Chowdhary; 1969). *Mandhali* sands of Lower Eocene age host majority of hydrocarbon reserves. Flow unit characterization, on the basis of electro log correlation and consistency of lateral continuity, indicates that 4 main units consist of sub-units separated by fine clastics/shale. In the main producing Mandhali sands porosity ranges between 21-28 per cent and permeability ranges between 100-500 md in the western part near main fault. In Kalol Sands, average porosity is in the range 20-35 per cent and average oil saturation is in the range 59-70 per cent. In *Sobhasan sands*, porosity ranges between 20 to 30 per cent and oil saturation ranges between 50 to 70 per cent. In *Linch sand*, porosity and oil saturations are in the range 18-30 per cent and 50-90 per cent respectively. In general, most of the pools are acting under depletion drive. Average productivity of wells in the field ranges between 2 to 45 $m^3/d/Kg/cm^2$. Initial reservoir pressures in Kalol, Sobhasan and Mandhali sands were hydrostatic but super-hydrostatic in Linch sands. In general, oil properties are good but wax and parafin contents make the oil thick in some wells.

The **Kharsang** field is located in the state of Arunachal Pradesh, which is approximately 60 km east of the Digboi Field. It is producing from the multi-layered fluvial sandstone reservoirs of Upper Miocene age. The reservoirs, occurring at depths ranging from 700 to 1500m, are stacked medium to fine grained sandstones that form 2 to 15 m thick, mostly upward fining successions (Handique, G.K. and Bharali,B. 1984). The shale bands within Girujan reservoir, although only a few meter in thickness act as pressure seals in many cases.

While drilling, indications of high pressures in Kharsang area were obtained mainly in the lower Girujan argillaceous section. The abnormal pressure zone necessitating use of heavy mud (Sp. gr. 2.17) for drilling at a depth of about

2050 m in the eastern part of the structure while the corresponding range was met at a much shallow depth of about 550 m in the updip south-western part of the structure.

Trends show that, wells completed in laterally continuous sandstone reservoirs have low production-decline rates compared to wells completed in laterally discontinuous sandstone reservoirs. Sand has been recognized as a potential problem of the field in number of wells particularly in reservoirs occurring at shallower depths (Debabrata Panda and S. Mahapatra, 2003). Moreover, because of the low solution GOR and with fairly low mobility due to viscous nature of the oil (paraffin content are as high as 15 per cent.), artificial lifting was implemented in many wells from early stages of production. The oil is a viscous medium gravity crude (19- 34 deg API) in a majority of the wells. The gas-oil ratio is typically low (below 75 cu m/kl or 425 scf/bbl), but high in the crestal culminations of certain sand units.

Because of comprehensive reservoir management methods, the production level of the field have increased from about 40 m^3/day to over 160 m^3/day. The reservoir pressure of the pools are typically 10-15 per cent higher than hydrostatic. In pools which are characterized by high connectivity, the reservoir pressure shows very little reduction with production.

Distribution of sub-surface temperature (values of geothermal gradients) is erratic in the Kharsang area down to a depth of about 1830 m, below which the geothermal gradient is in line with that (2.01° C/100 m) encountered in the fore-land part of basin. The reservoir temperature is generally low ranging from 38 - 45°C.

Reservoirs of the gas field located D-6 block **(Dhirubai)** in the **Krishna-Godavari Basin**, offshore East India are sands originally deposited in a deep water channel-levee system with relatively clean channel and laminated levee facies (Bastia, 2003). The water depths range from 600 to 1300 m and the bulk of the gas is at depths of about 2000 m. The channel facies comprise of poorly consolidated clean sands with excellent porosities and darcy-range permeabilities. A well test was conducted over two thinly laminated sand/shale zones with low resistivity and flowed dry gas. The beds are thinner than the vertical resolution of logging devices and so the first step is a high-resolution petro-physical analysis.

The data were integrated to enhance well logs, with a vertical resolution of a few centimeters, and the seismic data, an iterative approach was adopted between which has a vertical resolution of several meters. The petro-physical model at one end and synthetic-to-seismic matching on the other in order to create a rock physics model. The porosity models from the geostatistical inversion were used to calculate hydrocarbon pore volumes (Bastia. R. 2003 and Bastia R., *et al.*, 2005). In this process, it was assumed that the gas saturations are a function of height above the gas water contact. The reservoir properties net pore volume and net-to-gross are upscaled in a straight forward manner by averaging. The volumetric estimates of the geostatistical models greatly reduced the uncertainty gas-in-place volumes. The upscaled net sand distributions with their different gas-in-place scenarios are input to flow simulations in order to determine the optimum development well spacing which is critical in this deep water setting.

Chapter 9

Petroleum Economics: Development of Oil and Gas field

Introduction

If the occurrence is in sufficient quantities and flows to the surface they are termed as oil and gas reserves and the rock holding them is termed as reservoir rock. If these oil and gas reserves can be produced economically then such trap in the subsurface is called Oil or Gas Field.

The development of an oil field include to plan, how many wells are to be drilled and the sequence of their drilling, their location, in order to ensure a maximum recovery of petroleum and gas at minimum expense. The development project is based on the data obtained during exploratory drilling and their exploitation of the first few oil wells, and also on the experience gained in operating oil fields characterized by similar conditions. The important factors which determine whether an oil gas accumulation is of commercial value are controlled by: technical, economic and political.

Technical

These factors mainly the quantum of reserves, include depth and area of occurrence. Depth and area determine the cost involved in drilling and for example on land drilling much cheaper than offshore. Quality of oil also sometimes important because light oils easy to produce where as heavy oils need extra efforts as well as more cost for production and further,the cost of also depends on quality.

Petroleum Economics - Logistics

These factors include market cost of oil and gas which decides whether the field can be put on production or the said reserves can be only future resources. Unlike other products the cost of oil will not be in raising mode always. It depends on the decisions of Organization of Petroleum Exporting Countries (OPEC). Since these countries will supply oil to oil importing nations and by based on demand they can control the selling price of oil.

Further, the quantity of oil and gas reserves will not always decide whether to produce from a field or not it is its monetized value which is important. Similarly the exploration ventures to find new oil also depends on cost of oil rather than its presence. If oil prices are low no company will carry out exploration, they simply produce from the existing fields, and even new finds cannot be put on production. In other words, though, the projected oil prices at the time of preparing plan of production is more than the oil price, the field has to wait for development to reach the cut off price.

It is also important that, in case the field is located nearer to utility place even small quantities can be producible economically, where as in case of offshore fields, the cost of pipe lines based on water depth is an additional cost to develop an oil field. So to say in case of offshore fields the production rate and reserve base should be more than that of the on-land areas.

Even in on-land areas if the fields are located in remote areas the cost of pipe lines will be an additional financial burden. Fiscal terms, means the conditions under which government granted license is also one of the important criteria, it include "profit oil" share to be given to the government once production start and finally the country's tax regime decide pofits in the venture.

International Relations

Though the inter-country relations not belong to critical criteria, but do bear some meaning in case of oil fields to be produced in other countries. Say, the land locked countries may need long pipe lines to bring to shore as well as permissions from the countries falling on the path. It also depends on bilateral relations between the countries. In case sanctions are imposed against some country, it is difficult to carry out production operations there. Such factors bear meaning in developing an oil field.

As first step one should know, in case of new oil find to develop as an economical field is the quantum of in-place reserves in the reservoir and recovery efficiency, indicating how much of such reserves can be produced. It depends on the geological characteristics of the reservoir and technology available while production.

The term oil field development is meant the whole complex of work connected with the drilling of an oil field and withdrawal of oil economically to the surface. The work involves the drilling of number of wells over a reservoir, control of the movement of oil and gas to the wells by proper spacing patterns and operating conditions and the reservoir energy control.

The first step the **estimation of hydrocarbon reserves** is one of the most important criteria to develop a field. But during initial phase of exploration, with minimum data, reliable estimate of reserves are required for decision making.

In case of oil and gas mining/production the drilled well is the only connecting link to the surface and the samples obtained by drilling and the information obtained from borehole are the means by which geography, geometry and lithology of the reservoir can be established.

The first well which flowed oil and gas to the surface with sufficient rate and pressure is called "Discovery well". The conventional core helps in determining rock properties through laboratory techniques. If not, the desired rock properties like porosity and permeability can also be calculated with the help of electrical logs.

So in estimating discovered quantities of oil and gas many uncertainties are inherent. These estimates are based on various judgments of geological data and assumptions involved in estimating desired parameters using indirect data. In the initial stages, the data necessary for estimating reserves will be too meager it is required to be modified as and when new data are obtained.

Basic Geological Data for Reserves Calculation

Geometry of the reservoir, structure for knowing the areal extension, gross thickness of reservoir and any impervious layers with in the reservoir to calculate net pay thickness of the reservoir. These data are usually presented as structure contour maps of the reservoir, Iso-pach maps (thickness) and geological sections, the geometrical data serve as basis for calculating the reserves and establishing the well spacing. Physical characteristics of the rock: Porosity, permeability, elasticity and mechanical composition which are used in various hydro dynamic calculations.

After knowing in-place reserves one should know other characteristics of oil and gas for calculating the recovery factor based on formation pressure and temperature and by examining bottom-hole oil samples. The data are used in hydrodynamic calculations and to know the quality of oil, the saturation pressure is often taken as the limiting condition of well performance, the solubility factor the initial gas saturation of the oil, and the oil to gas viscosity are used for a solution gas drive in a reservoir.

The initial reservoir pressure and permissible pressures in producing wells during the production period determined by the saturation pressure, the minimum free flow pressure, the maximum permissible production from a well is usually determined by the stability of the reservoir rock. Sources of the reservoir energy are established by comparing initial formation pressure with saturation pressure and also by establishing dimensions of the entire drive system.

Volumetric Method for Estimation of Oil and Gas Reserves

One of the earliest methods of estimating in-place hydrocarbon is the volumetric method. The volumetric method requires calculation of the oil and gas in-place from the computed rock volume.

For volumetric estimation, basic requirement is to prepare the structural and iso-pach map of the pay zone on the basis of available sub-surface data obtained from drilling, logging and testing of the wells in the area to obtain the total rock volume of the hydrocarbon reservoir (Axh). Assessment of core analysis data and the electric log interpretation data provide the average value of the other parameters like porosity, connate water saturation (or hydrocarbon saturation). Laboratory analysis determines the specific gravity, formation volume factor, GOR of any oil sample.

While calculating the in-place volume of oil and gas by volumetric method, it is necessary to determine the average values for the different parameters of the reservoir and the reservoir fluids. These parameters, as determined in the early stage of exploration, are likely to change with time as the oil/gas fields are brought into the stage of exploitation and more reliable data are obtained from exploratory and development wells.

The volumetric method requires calculation of the approximate reservoir volume, which is subject to well drainage, and the recovery of the petroleum in place from the computed rock volume. Under performance analysis, mathematical or graphical models are solved for oil and gas in-place and ultimate reserves of hydrocarbons.

For evaluation of recoverable (Ultimate) reserves recovery factor will have to be introduced in the above formulae. As per the present knowledge, the recovery factor could be 25 per cent primary recovery, and additional about 10-25 per cent for secondary recovery.

When oil and gas limits have been adequately delineated and their commercial viability has been ascertained the development phase commences.

Main Steps of an Oil or Gas Field Development

Field Discovery

Oil and gas are trapped in reservoir rocks buried underground (on shore/off shore), to locate hydrocarbon accumulation, geoscientists analyze images of the earth's subsurface produced by seismic data. They build a model of geological layering of the subsurface, and then identify potential reservoirs called 'prospects', exploration wells need to be drilled to check if the identified prospect does indeed contain hydrocarbons.

In multiple zone reservoir, due to presence of separate productive intervals, it is necessary to plan an independent development system for each pay zone and decide the order in which they are to be brought into production or if found feasible to produce several pay zone intervals simultaneously.

Field Evaluations

Once discovery has been confirmed, 3D numerical reservoir simulation models are built: to estimate the initial volume of oil and gas in the reservoir, to simulate the reservoir fluid flow behavior and optimize the field development scenario like, number, type and location of wells, level of field production. Appraisal wells are drilled to improve the field description through further data acquisition. These appraisal wells can be planned to know exact fluid contacts in the periphery of the field area, and pressure and temperature data more precisely and layering if any within the oil bearing zones.

In sandstone or inter-granular reservoirs most production wells can be represented by a radial flow system and it is well-known that large pressure drops sometimes occur in the vicinity of well-bores and the bottom-hole flowing pressure

(BHFP) does not represent the pressure in the drainage area of the well. It is normal practice to shut-in producing wells for a certain periods and determine the static pressure of the oil body at the point, at some time, close to the producing horizon (BHSP). By using the well-established concepts of pressure build-up, the maximum pressure can be obtained. For determining average pressure, corrections are applied to all well measurements to give values a convenient datum level. The datum level may be a mean level for the oil saturated horizons, at the gas-oil level.

Field Development

A field development plan establishes the number of wells to be drilled to reach production objectives, the recovery techniques to be used to extract the fluids within the reservoir, the type and cost of installations, such as platforms, depending on the marine environment (tides, storms, waves, winds, corrosion), the separation systems for gas and fluids, the treatment systems needed to preserve the environment, an offshore production platform.

After obtaining reliable recoverable reserves, the work involved in preparing Field Development Plan (FDP) to know the drainage area around the well bore, that means in how much area one well can be drained. Based on the reservoir characterization, a model can be developed how many wells may be required to drain the entire recoverable reserves. The quantum of recoverable reserves are based on type of recovery like Primary, Secondary and/or enhanced or Tertiary recovery. Accordingly the time span and expenditure at different recovery methods have to be understood with clarity.

Another exercise for the same is by knowing the average rate of production from a well and by knowing the life of the field in years one can calculate how many wells need to produce recoverable oil in a given number of years.

Apart from technical data the life span of a field also decided by the Government regulatory body which indicates the number of years for tax concessions if any, this will indirectly speaks the National Oil Policy of the country where hydrocarbon field has to be developed. Production profile for the field life has to be prepared based on reservoir characterization.

One of the main and most important problems appearing in working out a development scheme consists in the optimal spacing of the producing wells, this determining in the long run the total number of wells in a field. This choice of well spacing is controlled by the geological features of the field and is based on the hydrodynamic calculations.

Theoretical considerations and the experience gained in developing oil fields show that with a close spacing of producing wells, their interaction becomes more pronounced, and average rate of production drops. On the contrary, with an excessively large spacing of wells in an oil field, there may remain undeveloped sections, current oil recovery will be low owing to the insufficient number of wells drilled, and duration of oil field development will increase.

Oil, gas, and water are typically trapped within fine rock pores under high pressure in the oil reservoir. Expansion of pressurized water and gas in solution forces oil out of the rock pores into the well and up to the surface. This is known as the "primary recovery". Oil flowing out of the rock drains energy from the formation; pressure in the reservoir begins to slowly decline; primary drive reduces and the production rate falls. The oil cannot be produced unless pressures within the reservoir are maintained or restored to cause displacement of the oil being held in the rock and to drive it to the wellbore. Installation and implementation of a secondary and enhanced recovery system significantly increase a field's productivity, economics and longetivity.

Development of Multiple Pay Zone in a Field

In case, with in a structure or field area if more than one reservoir of oil and gas are found vertically stacked then such field is known as multi-pay field. In multiple zone reservoir, due to presence of separate productive intervals, it is necessary to plan an independent development system for each pay zone and decide the order in which they are to be brought into production or if found feasible to produce several pay zone intervals simultaneously.

The earlier system consists of producing the bottom most zone, then the well is transferred to the overlying pay zones, one after other. This system makes it possible to reduce the total number of producing wells necessary to ensure exploration and exploitation of all the strata of an oil field. A short coming of this system of development is the long exploitation period of the oil field as a whole.

Simultaneous development of several pay zones of an oil field is also need to be studied. Here, each pay zone is developed with by an individual grid of wells. This system is usually practiced in developing oil fields with large reserves of oil. From a single well, two to three pay zones are exploited through different set of tubing strings, and packers isolating the pay zones, from any communication. Cost of drilling will reduce, but operationally more cumbersome.

Usually medium to large oil producing reservoirs, adopt water injection from the initial stages of field development. Gas injection is carried out, where excess gas is available for injection and where ever it is preferred over water injection or in simultaneous water and gas injection projects.

In some cases, producing wells can be converted to injection wells, while in other cases it may be necessary or desirable to drill new injection wells. Development planning takes place for production and injection wells simultaneously and their sequence of completion is also decided as per plan. One of the basic objectives in a fluid injection project is to contact as much of the reservoir as possible with the displacing phase.

Case Studies

The **Demulgaon field** is a small oil field located in the NE part of India. The structure was identified by conventional 2-D seismic survey in 1967-68 and oil was discovered in 1976 in sands within coal shale sequences of Oligocene age locally known as BCS. Subsequently, oil was found in lower arenaceous unit named as Barail Main Sand (BMS) at a depth of 3650m MSL. Oil production started from this well in 1985. Tipam sands of Miocene age are also oil bearing at places. Demulgaon structure is ENE-WSW trending fault closures having distinct horst feature bounded by two prominent depressions in the north and south. The areal extent of the field is around 8 sq. km. There are normal faults oblique to the general ENE-WSW fault and they have played important role in entrapment of hydrocarbon. The field has passed its peak production in 1991 and is in declining stage with high water cut. The reservoir is operating under active water drive. The reservoir got OIIP around 4 MMt. All the studies have indicated recovery factor of around 40 to 50 per cent. The BMS sand has gross thickness of about 170 to 250 m with average iso-pay thickness of around 30 m. The initial reservoir pressure of 382 ksc has dropped around 320 to 330 (bubble point pressure is 310 ksc). The simulation study was carried out and water cut was considered at 3668m msl as against OSC consideration in the feasibility studies (Talukdar, R., *et al.*, 2003).

The **Jotana field** located in Cambay Basin, is a major fault controlled structure. Intersection of NNW-SSE trending Jotana Main fault with NE-SW aligned transverse faults form two-way closures entrapping hydrocarbons. Multilayered "Mandhali" member, the prolific producer with gross thickness of 50 to 160m is affected by faults varying in throw from 10- 40m. The strati-structural data reveal that reservoir connectivity across faults form a critical aspect of reservoir management. A two-stage analysis has been attempted involving identification of reservoir-juxtaposed areas over the fault surface and mapping structural disposition of lithology offset by the fault plane. Lateral communication across these faults for particular flow units are confirmed from history match of pressure-production data. Fault seal is commonly recognized by differences in fluid contact on either sides of the fault. Much more comprehensive is when differences in pore pressures across the fault are detected. A dynamic reservoir model was conceived by examining the reservoir pressure and fluid histories within the confines of recognized permeability barrier in the form of transverse faults. Faults act as horizontal permeability restrictions and juxtapose different beds. The commencement of water injection through a peripheral well had reactivated an updip producer that had ceased, which revealed that communication across faults do occur. A fault, which is a seal for one flow unit, may be communicating for another flow unit.

In **North Kadi oil field** of Cambay Basin, the Kalol sands of Eocene age are oil producers. Hydrocarbons were entrapped in multiple culminations on the southerly plunge of a palaeo-structural feature 'Mehsana Horst' in highly permeable sandstone reservoir operating under active water drive. These culminations have under-saturated oil characterized by the occurrence of gas cap. In a well-located up-structure in the northern culmination, the gas cap was put on production. Repeated gas compositional analysis has confirmed that the gas cap contains dry gas lacking oil staining. Gas compositional analysis coupled with geological evidence has been used to reconstruct the palaeo-configuration of the field that controlled the distribution of oil and gas. The migration and filling model suggests two different migration phases for oil and gas with gas predating oil emplacement in Miocene for oil and gas accumulation based on tectonic activity within the structure. The two-phase migration theory offers a rational explanation for the existence of gas cap in an undersaturated oil reservoir. North Kadi oil field has gas cap in all the discovered culminations. The oil column resting below gas is underlain by edge water on either flank, the free water level being governed by the steepness of dip (Srivastava. M.C. and V. Pratap.2005).

Production Installations

After the complete understanding of total quantum of oil and/or gas production and duration of production,then comes the stage of planning for required production facilities like oil collecting facilities: GGS (Group Gathering Station) in on-land which will have number of CTFs (Collective Tank Forms) based on per day production of oil. In case of offshore, similar facilities are termed as Process platforms, drill platforms. In some cases, they are called FPSO (Floating Production Storage and Offloading) or TLP (Tension Leg Platform) facilities. What methods of facilities can be chosen will depend on overall economic scenario of the development plan.

Next stage of planning include transporting of oil and gas, called pipe lines to the place of usage, main factor is diameter of the pipe lines, which depend on quantity of fluids to be transported. If the utility place is far away, sometimes road transport may be economical in on-land areas. In case of deep offshore, storage ships will be placed and they carry oil to the shore areas for further transportation.

For Gas fields, technology available for converting gas into liquid form (LNG) which can be easy to transport economically and safely.

Further, after certain production the reservoir pressure drops, if it is not maintained at optimum level, flowing wells may cease to flow. At that stage in-fill wells will be drilled through which water will be injected in to reservoir to maintain reservoir pressure.

Deep-Water Oil and Gas Field Development

The definition of deep-water has been changing with evolving technology. Presently deep-water refers to water depth in excess of 500 meter. Ultra deep-water means water depth in between 1500 and 2500 meters and super deep-water is in excess of 2500 meters.

This has been possible through a series of continuous improvements and technological advance in floating platform design, manifolds and template systems, riser system, mooring systems, drilling and completion techniques.

Giant deep-water oil fields should have sufficient reserves to justify installation of massive structures. Deep-water fields with low reserves of about 150 million bbl or less require smaller, optimized solutions to meet the cost economics. Two approaches possible for deep-water field development based on quantity of reserves and reservoir characteristics, sub-sea well completions tied back to existing nearby facilities or stand-alone floating installations with onboard processing and export facilities.

There are two categories of facilities called TLPs and Floaters. A semi-submersible is a Floating Production System (FPS) that has capabilities for both drilling and production. But the FPSO is the most mobile of all production system as it can easily be shifted to a new location. It is preferred for remote locations as it has integrated production and oil storage capability. Oil is periodically offloaded by shuttle tanker.

Now in deep water field development, geo-engineering studies like, soil investigations including in situ testing, sampling and laboratory testing, instrumentation to measure the in situ pore pressure, foundation of floating structures with emphasis on the successful use of suction pile technology, and finally geo-hazards evaluations (Mokkelbost, K.H. S. Strandvik,1999) which have proved to be extremely important in deep-water field development (Lunne, T. *et al.*, 2001).

Project Cost analysis

Field Development Planning is the detailed multi-deciplinaty process of evaluating multiple development options for a field and selecting the best option based on assessing trade-offs among many factors: -

- ✰ Net present value, typically the key driver of decisions for publicly-traded operators.
- ✰ Oil and gas recovery, which is heavily valued by National Oil Companies and regulators.
- ✰ Operational flexibility and scalability, particularly important factors for deep water developments.
- ✰ Capital versus operating cost profiles.

Investors sell oil company valuations in part on the company's ability to design and execute credible field development plans. As a result, oil and gas company management is under increasing pressure from investors to translate discoveries into cash flow as quickly as possible. On the other hand, growing field complexity in many parts of the world has made field development plans (FDPs) increasingly difficult to formulate and execute. Industry studies have shown that FDPs routinely over-estimate field production rates and ultimate recovery due to causes ranging from poor initial characterization of the reservoir to not fully integrating available development technologies.

An economic assessment is performed taking into account revenue according to production forecasts and the estimated development costs. If the required economic criteria are met, the field is developed and then produced.

This phase includes development drilling, construction of production and pipeline facilities and at the end commencement of production. The drilling and production operation would have yielded considerable geological and reservoir engineering data, and good idea of the operation recovery and drive mechanisms would also have been obtained to give a more reliable estimation of recoverable reserves.

While assessing economic scenarios, prediction of oil and gas price through the production period is a critical, because whole project economics mainly based on this. The input data for this is prediction of oil economists with past market tends which will be quoted and adopted while carrying out project economics. These studies based on past experience coupled with demand versus supply in future scenarios in world. But in the complex oil game all the time predictions for future may not be dependable.

The decision making parameter is IRR (Internal Rate of Return) which is more or less similar to annual profit percentage, if it is positive and more than 10 per cent such projects are considered to be viable. Most of the financial calculations are indicated in NPV (Net Present Value). It indicates the present value of futuristic investments and returns of cash flows. It is simple way to understand what we spend now and what we get if it is today. While assessing futuristic estimates based on the data available the rate of uncertainty one has to be aware.

Payout period is defined as the number of years required to payout the original investment from net income from the investment. The payout method of computing investments is simple, straight forward and usually is a reliable guide. This method does not take into consideration the present value of money invested.

Profit–Investment Ratio (PIR) = Cumulative net income – Original investment/ Original investment

Discount factor is not applied to the net income when determining PIR. All the above factors should be considered in the overall analysis of investment opportunities.

The abandonment cost of the field is also to be included in the economic calculations and of course care has to be taken fiscal regime based on Production Sharing Contract (PSC) made with the local government, which indicate how much profit oil will be given to government with time.

It is to be noted that, in case abandonment cost is not included in project economics, the project may seem viable and it is risky to participate in such projects because State laws are very stringent.

Once Development project is finalized, the initial expenditure to be made for construction of installations, drilling of wells till production starts is termed as CAPEX (Capital Expenditure). Once production starts, the annual expenditure met to maintain the men and machinery is called OPEX (Operating Expenditure).

After all these calculations if development project of a field if found to be viable then that scenario is adopted as final development plan to proceed further. Presently in the market, various expertise are available to apply game-changing technologies across the field development.

Advanced geophysical and logging technologies to help characterize reservoirs and identify potential "sweet spots".Drilling, completion and stimulation technologies to enhance reservoir contact, well productivity and ultimate recovery. Lift and flow assurance technologies to ensure high well deliverability. Implementation of digital oilfield technologies to optimize surface and subsurface assets and enhance production.

Risk management software to help in modeling multiple uncertainties at a field or portfolio level, Well planning and design software,Modeling and simulation software at the well and field levels, Software to enable rapid assessment of field development options and costs for offshore fields, Industry-leading reservoir characterization, drilling, completion and stimulation technologies.

Field Production

The time period over which hydrocarbons may be extracted varies between 15 to 30 years and may be extended up to 50 years or more for giant fields.

The lifetime of a reservoir is composed of different successive phases: a period of production increase, a stabilization phase or 'plateau', injection phases (water, gas or chemical products) to help the hydrocarbon recovery and thus maintain a satisfactory volume of produced resources, the depletion period when hydrocarbon production declines progressively.

Important task in developing oil field is projection of production profile from the field through years, generally during initial phase of production there is steep rise of production because whenever production facilities are completed at each stage of development more wells will be added. There will be moderately long plateau stage, and there will be sharp decline due to drop in reservoir pressure (Figure 9.1).

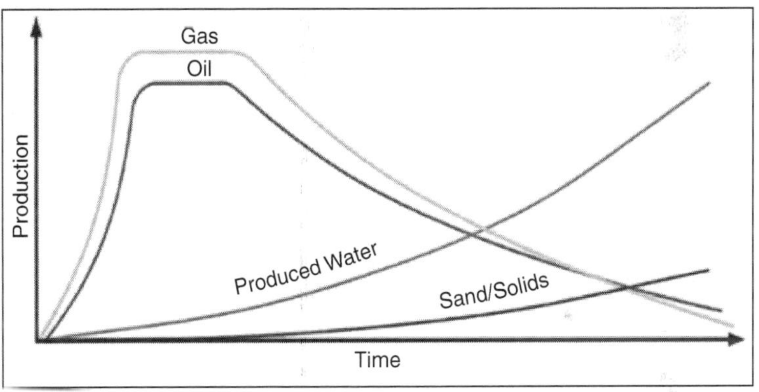

Figure 9.1. Oil Field Production Profile.

Field Abandonment

When the hydrocarbon production rate becomes non economical, the reservoir is abandoned. Abandonment Pressure is the minimum pressure of the reservoir when the wells are abandoned. An acceptable abandonment includes, the plugging of the wellbore and reclamation of the land surface to a stable and productive use.

The purpose of plugging a well is to prevent fluid migration between zones within the wellbore to protect aquifers of useable quality water and to protect other minerals from damage. Well plugging requirements vary with the characteristics of the rock, geologic strata, well design, and reclamation requirements. For wells no longer capable of production, all perforations must be isolated so as not to allow fluid to migrate up hole to the surface. The perforations may be isolated by plugging the well(s) and restoring the site.

A producing well after ceasing the production will be abandoned by placing a cement plug 100m above the top of perforated zone and tested for pressure leaks if any. Similarly another top plug will be placed 100m below the seabed depth inside the well.

Once individual wellbores have been plugged and abandoned, the pipelines, facilities and other structures in the field must be decommissioned and removed. The surface site must be returned to same as previous condition. These operations can be challenging onshore; in offshore environments, especially in deep water, the decommissioning procedures require careful coordination of several specialized crews. Offshore production-platform decommissioning is subject to extensive regulation worldwide. Decisions about when and how to decommission platforms involve complicated issues of environmental protection, safety and cost. Limited availability of heavy-lift equipment necessitates significant advance planning to remove platforms. Operations typically are scheduled to avoid rough weather.

Offshore field abandonment and platform decommissioning encompass abandonment of each well in the field. Permeable subsurface formations are isolated permanently from each other and the surface. Each well is plugged and the casing is cut at some depth below the seabed, as specified by local regulations. Pipelines also must be decommissioned and removed. The pipelines may be reused, sold as scrap or treated as waste. Next, surface facilities and other structures are decommissioned, which may involve partial or complete removal or toppling in place. This may begin with removing the platform deck or topsides, followed by removing the supporting structure, known as the jacket, or the entire structure may be removed in one piece. Depending on the method selected, extensive diving operations may be necessary to cut the structure into pieces.

Finally, the seabed must be made clean. Typically, platform removal is the most expensive part of decommissioning operation because of the expensive lifting equipment that must be mobilized. Ongoing advancements in lifting technology should make platform removal safer, quicker and easier. Most offshore platforms are customized, so decommissioning operations are tailored for the specific configuration and conditions. The removal of fixed steel platforms involves detaching the platform from the seabed. Sometimes oil companies may reuse some of the platforms in another nearby field or sell them at cheaper price.

Now the oil companies cannot walk away leaving the oil field area as it is. Many governments have enacted legislation and introduced codes of practices and fiscal provisions covering the abandonment of offshore installations. Government legislation stipulates the condition that the abandoned field area has to be restored as it was before the oil exploration. It is related to environmental concern.

Stages of Recovery Methods

The **primary recovery** refers to the volume of hydrocarbon produced by the natural energy prevailing in the reservoir. Each reservoir is composed of a unique combination of geometric form, geological rock properties, fluid characteristics, and primary drive mechanism. It is observed that, each drive mechanism has certain typical performance characteristics in terms of ultimate recovery factor; pressure decline rate; gas oil ratio; water production. For a proper understanding of reservoir behavior and predicting future performance, it is necessary to have knowledge of the drive- mechanisms that control the behavior of fluids within reservoirs (Atwood, D.K.1964).

During primary stage, the recovery of oil is by either one or combination of available natural drive mechanisms. Primary recovery was the only method available during the early years of the oil industry. The natural energy or reservoir drive that is used during primary production can be visualized by considering that each unit volume of oil produced must be replaced by some other fluid available in the reservoir.

It is determined by analysis of production data (reservoir pressure and fluid production ratios). The earliest possible determination of the drive mechanism is a primary goal in the early life of the reservoir, as its knowledge can greatly improve the management and recovery of reserves from the reservoir in its middle and later life.

History Matching

History matching is the process of building one or more sets of numerical models (representing a reservoir) which account for observed, measured data. History matching is the act of adjusting a reservoir model until its simulated production response closely reproduces the past behavior of the reservoir. Once a reservoir model has been history matched, it can be used to simulate future reservoir behavior with a higher degree of confidence, particularly if the adjustments are constrained by known geological properties of the reservoir. Based on expected future behavior, production scenarios may be optimized.

During any kind of model calibration process, it is important to note that:

- Reservoir models are models not reality. The model input is uncertain, and the uncertainty is almost always under-estimated. There are inevitable errors and approximations that are found in any model of physical phenomena.
- The history matching process is undertaken for the purpose of decision making. History matching serves no purpose on its own.

A simulation project of a developed field, usually requires "history matching" where historical field production and pressures are compared to calculated values. During this process, significant modifications to the reservoir description may be required to match historical performance. History matching is often a trial-and-error process requiring many model runs before a satisfactory match between the history-matching. In new fields, little data exists and reservoir performance predictions may be made without well tests from discovery and delineation wells, should be used to calibrate the reservoir model.

Predicting Performance

Predictions always begin by calculating the projected performance of the base case. The base case represents the mode of operations where everything remains exactly as it stands on the last day of the history match. In its strictest interpretation, the base case means no new wells, recompletions, work-overs, or changes in a well's artificial lift status. In practice, the base case means these changes usually occur to the extent that they could have been foreseen in plan or by extrapolating current producing wells to injecting or artificial lift wells in the future.

Each prediction, case must specify each well's producing rules. The location of each well to be produced or injected during the prediction must be specified, including the time at which it is considered active. In the simplest case, wells are constrained to produce at a constant rate or at a constant bottom-hole pressure. Economic limits can also be specified. Using simple rules, the simulator output must be run in increments of a year to reflect the field's operating practices (Babu D.K. and Odeh Aziz S., 1989).

Artificial Lift Methods

When pressure in the oil reservoir has fallen to the point, where the well does not produce at the economical rate by natural energy, some methods of artificial lift should be used. Artificial lift help in mature, depleted fields, where production has declined such that the reservoir can no longer produce under its natural energy. But these methods are also used in younger fields to increase production rates and improve project economics. It is used to lower the producing Bottom Hole Pressure (BHP) on the formation to obtain a higher production rate from the well.

Artificial lift can be used to generate flow from a well in which no flow is occurring or used to increase the flow from a well to produce at a higher rate. Most oil wells require artificial lift at some point in the life of the field, and many gas wells benefit from artificial lift to take liquids off the formation so gas can flow at a higher rate.

The most common methods of artificial lift are:

Beam pumping, or the **sucker-rod** lift method, is the oldest and most widely used type of artificial lift for most wells. A sucker-rod pumping system is made up of several components, some of which operate above the ground and other parts of which operate underground, down in the well. The surface-pumping unit, which drives the underground pump, consists of a prime mover which is usually an electric motor.

Gas lift is a method of artificial lift that uses an external source of high-pressure gas for supplementing formation gas to lift the well fluids. The principle of gas lift is that gas injected into the tubing reduces the density of the fluids in the tubing, and the bubbles have a "scrubbing" action on the liquids. Both factors act to lower the flowing bottom-hole pressure (BHP) at the bottom of the tubing. There are two basic types of gas lift in use today–continuous and intermittent flow.

Electrical Submergible Pumping (ESP) system is normally tubing hung from the wellhead with the pump on top and the motor attached below. The down-hole components comprising of: multistage centrifugal pump with a three-phase induction motor. The rest of the system includes a surface control package and a three-phase power cable running down-hole to the motor. Because of the ESP's unique application requirement in deep, relatively small-bore casings.

Plunger lift has become a widely accepted and economical artificial lift alternative, especially in high-gas/liquid-ratio (GLR) gas and oil wells. Plunger lift uses a free piston that travels up and down in the tubing string. It minimizes liquid fallback and uses the well's energy more efficiently.

Progressing cavity pumping (PCP) systems are increasingly used for artificial lift, and have been adapted to a range of challenging lift situations like heavy oil, high sand production, gas wells, directional or horizontal wells.

Secondary Recovery Methods

Secondary recovery refers to the volume of hydrocarbon produced as a result of the addition of energy into the reservoir, such as fluid injection, to complement or increase the original energy within the reservoir.

Fluids are injected into selected wells at or near original pressure levels to achieve maximum recovery efficiency. Two of the more common fluid injection methods are the injection of liquids or gases from extraneous sources into the well bore. This injection increases reservoir pressures and enhances the mobility of unrecoverable oil through the reservoir. Gas injection can also be used to restore reservoir pressure. The installation of a secondary recovery system involves drilling of injection wells and new recovery wells or conversion of production wells to injection wells.

Water Flooding

This method involves the injection of water at the base of a reservoir to maintain the reservoir pressure, and improve the mobility oil (usually with gas and water) towards production wells. The water displaces oil from the pore spaces, but the efficiency of such displacement depends on oil viscosity and rock characteristics. However, it should be noted that the successful outcome of a water flood process depends on designs based on accurate relative permeability data in both directions. Water flooding is the most dominant among enhanced oil recovery techniques. Rock and fluid compatibility in conjunction with permeability around the well bore often comes up in the performance of injector wells.

The oil field that does not have strong natural water drive is considered for water flooding. Water pressure maintenance is a process whereby water is injected into an oil producing reservoir to supplement natural energy indigenous to the reservoir and to improve oil producing characteristics of the field prior to the time that economic productive limits have been reached.

In determining the suitability of a given reservoir to water flooding or pressure maintenance the following factors must be considered:

(i) Reservoir geometry, (ii) Lithology, (iii) Reservoir depth, (iv) Porosity, (v) Permeability Magnitude and degree of variation, (vi) Fluid properties and relative permeability relationships, (vii) Continuity of reservoir rock properties, (viii) Magnitude and distribution of fluid saturation.

Also of interest is the initial saturation of connate water. Connate-water content may be estimated from cores obtained by using oil-base mud, electric-log information, laboratory oil floods or capillary pressure tests. The reasons for success behind water-flooding were identified that, water has viscosity, density, and wetting properties, compared to oil, that affect how efficiently it will displace various oils from reservoir rock.

Most oil reservoirs are heterogeneous rock formations. The primary geological consideration in water-flooding evaluation is to determine the nature and degree of heterogeneities that exist in a particular oil field. Second, the diagenetic history of the reservoir rocks must be ascertained. Then, the structure and faulting of the reservoir must be determined to understand the interconnectivities among the various parts of the reservoir, particularly the injector/producer connectivity. Fracture trends that developed because of regional tectonic stresses on the rock and the effects of burial and uplift on the particular rock layer. Fault trends that affect the connection of one part of an oil reservoir to adjacent areas, either because they are flow barriers or because they are open conduits that allow unlimited flow along the fault plane.

Finally, the water/oil/rock characteristics need to be understood because they control wettability, residual oil saturation to water-flooding, and the oil relative permeability at higher water saturations. Because of these needs, there always should be a developmental geologist on the water-flood-evaluation team (NecMetin Mungan,1965).

The structure of the reservoir and how it affects water-flood performance is another geological consideration. Structure creates dipping beds that dip at various angles. The interplay between the bed angle, gravity, and the oil/brine density difference at reservoir conditions significantly affects the relative vertical and horizontal flow behaviors. Structural considerations also can include whether the oil column has an underlying aquifer or an overlying gas cap, either of which can significantly affect the likelihood of successfully water-flooding the oil column.

The water-flood production-well data, the water and oil rates as a function of time are critical because they are the first data that relate directly to the inter-well connectivity within the reservoir and that validate or cause modification of the geoscientists' concepts of the various levels of reservoir heterogeneities.

After the preliminary reservoir evaluation indicates that water-flooding is likely to be economically justified and that it will increase significantly the volume of oil recovered, the next consideration is to find an acceptable source from which to obtain enough water for the proposed water-flood project. Onshore locations typically obtain injection water from subsurface aquifer intervals or nearby streams or rivers. Near-shore and offshore water-flood projects typically use seawater.

During a water-flood, tracers can be injected to track which injector/producer pairs are well connected and which are poorly connected. Other monitoring techniques include the use of specially drilled observation wells and 4D-seismic interpretations to track the directionality and shape of the higher-pressure water-swept reservoir areas that are centered on the injection wells.

Compatibility of the planned injected water with the reservoir's connate water. Interaction of the injected water with the reservoir rock (clay sensitivities, rock dissolution, or generally weakening the rock framework). Injection-water treatment to remove oxygen, bacteria, and undesirable chemicals. The challenges involved in separating and handling the produced water that has trace oil content, naturally occurring radioactive materials and various scale-forming minerals. Source water/connate water compatibility mainly concerns whether mixing the two waters causes any precipitation of insoluble carbonate or sulfate compounds that might impair reservoir permeability.

The design aspects include: Injection/producer pattern layouts, Injection-water sensitivity studies, Injection wells, injectivity and allocation approaches, including well fracturing, Pilot water-flooding. At offshore locations, the number of well slots on the drilling platforms limits the number of producers and injectors and their layout.

Several aspects of the design and operation of water injection wells are critical to their success. The near-wellbore "skin" (a rock volume of reduced permeability around the wellbore) is not known until an actual well test is conducted. Injection wells can be fractured to eliminate positive skin in the near-wellbore region; however, fracturing must be done carefully to avoid fracturing out of the reservoir interval and into adjacent porous and permeable intervals into which injection water can be lost.

In general the water injection wells are drilled based on several studies made, but in case the earlier production wells where OWC has gone above the producing interval, if that area need an injection well such well can be converted such.

Examples

The laboratory studies for rock and fluid compatibility done in cores of oil field, to carryout Water-flooding in **Karaikal** region of Cambay Basin, India, reduction in permeability retention with tube well water was noticed. This phenomenon of declining permeability at initial stage can be attributed to damage in the well bore resulting from the salinity created in contact with fresh injection water. The differential salinity causes dispersal of fines of clay materials to choke the pore throats around the well bore. The study also revealed that about 40 per cent salinity of the connate water is to be maintained in injection water so as to make it compatible with *in-situ* formation water for the purpose of injection into the formation for pressure maintenance.

The **Lakwa field** with another culmination adjacent to the field called **Lakhmani field**. Oil was established within multiple reservoirs in the bottom even in Tura formation, Barail sands (2 sands), Tipams (6 sands) and in the top sandstone with in Girujan Clay Formation, with bulk production coming from the TS-2 Sand of the Tipam Sandstone.

The TS-2 pool is a major pool and comprises a single 90 m-thick oil column and compartmentalized by many listric faults with throws of up to 20 m. The sand is up to 120 m thick and comprises four, laterally extensive braided-fluvial sandstone subunits that are separated by thin, continuous to semi continuous, floodplain shale layers. The oil has been produced under combined aquifer and solution-gas drive. The porosity of the TS-2 Sand ranges from 22 to 26 per cent. Water-cuts in individual wells were started as early as in early ninetees and the range 30-90 per cent.

In Barail Main Sands were deposited by distributary channel as mouth bars, point bars, channel sands. Thickness map indicates the sand bodies to be oriented in NW-SE direction. Lakwa Barail Sand-2 (LBS-2) is widely distributed in the field. Most of the fault blocks are hydro-dynamically not connected. Most of the reservoir sands in this unit are mouth-bars and point-bars (Arun Garg and Sanjoy Goswami.2003). Due to decline in productivity and fall in reservoir pressure from initial 380ksc to pressure of 122-170ksc, water injection was initiated but desired result could not be achieved and pressure continued to fall below bubble point pressure. A new geological model has been prepared. The prolific producers from this reservoir are generally localized in the eastern part of the area on a mouth-bar. Porosity in these blocks varies from 12 per cent to 18 per cent and oil saturation values from 35 per cent to 56 per cent. The GOR gradually increased with production and has reached to the level of around 1000 v/v. The purpose of water-injection has been defeated as the performance of the wells did not improve substantially.

Gas Injection

This method is similar to water-flooding in principle, and is used to maintain gas cap pressure even if oil displacement is not required. Again accurate relative permeabilities are needed in the design, as well as injector/producer array geometry and crustal stresses. There is an additional complication in that re-injected lean gas may strip light hydrocarbons from the liquid oil phase. Gas injection method can be subdivided into two categories.

The immiscible gas injection is very inefficient fluid for additional oil recovery. The gas is non-wetting to reservoir rocks. The gas will move through the larger spaces of the reservoir rock by passing much of the reservoir oil. Some gas saturation will be present. Thus the initial gas may be displacing gas not oil. Due to low viscosity of gas its mobility is quite high which results in excessive channeling and by passing of the oil.

The displacement of oil by non-aqueous injected solvent, lean hydrocarbon gases, such as CO_2, N_2 or other gases are generally described as miscible fluids. The various conditions of pressure and temperature that are required for miscibility whether on first multiple contact or normally dealt. In multiple contact system residual oil behind displace center may stripped of light and intermediate fraction reducing substantially the residual oil saturation. This is known as vaporization gas drive. Another mechanism called condensing gas drive involves the transfer of intermediate components from the displacing gas to the residual oil. The residual oil becomes of a lower viscosity and has increase oil mobility. These volume effects can be significant even when full miscibility is not attempt. In an ideal process the swelling and the mobilization dispersed the discontinuous residual oil phase.

Tertiary Recovery-(Enhanced Oil Recovery-EOR) Methods

EOR refers to the recovery of oil through the injection of fluids when enough energy not normally present in the reservoir. The injected fluids must accomplish boost the natural energy in the reservoir, Interact with the reservoir rock/oil system to create conditions favorable for residual oil recovery that include reduction of the interfacial tension between the displacing fluid and oil, Increase the capillary number, Reduce capillary forces, Increase the drive water viscosity, Provide mobility-control, Oil swelling, Oil viscosity reduction, Alteration of the reservoir rock wettability.

The ultimate goal of EOR processes is to increase the overall oil displacement efficiency, which is a function of microscopic and macroscopic displacement efficiency. Microscopic efficiency refers to the displacement or mobilization of oil at the pore scale and measures the effectiveness of the displacing fluid in moving the oil at those places in the rock where the displacing fluid contacts the oil. For instance, microscopic efficiency can be increased by reducing capillary forces or interfacial tension between the displacing fluid and oil or by decreasing the oil viscosity.

Macroscopic or volumetric displacement efficiency refers to the effectiveness of the displacing fluid(s) in contacting the reservoir in a volumetric sense. Volumetric displacement efficiency indicates the effectiveness of the displacing fluid in sweeping out the volume of a reservoir, both areally and vertically, as well as how effectively the displacing fluid moves the displaced oil toward production wells.

The overall displacement efficiency of any oil recovery displacement process can be increased by improving the mobility ratio or by increasing the capillary number or both. Mobility ratio is defined as the mobility of the displacing fluid (*i.e.* water) divided by the mobility of the oil.

At the end of the water-flooding process, experience has shown that at these low capillary numbers an important amount of oil is left behind in the reservoir trapped by capillary forces at the pore scale. Thus, if the capillary number is increased through the application of EOR processes, residual oil will be mobilized and recovered. The most practical alternative to significantly increase the capillary number is through the application of surfactants or alkaline flooding (chemical flooding).

Polymer Flooding

Polymer flooding is an augmented water flooding technique mainly used for heterogeneous reservoirs, to retrieve oil after areas in the reservoir with high permeability have been highly water flooded. It consists in mixing long chain polymer molecules with the injected water in order to increase the water viscosity. This method improves the vertical and areal sweep efficiency as a consequence of improving the water/oil Mobility ratio.

Surfactants may be used in conjunction with polymers to decrease the surface tension between the oil and water. This reduces the residual oil saturation and improves the macroscopic efficiency of the process. Primary surfactants usually have co-surfactants, activity boosters, and co-solvents added to them to improve stability of the formulation.

Caustic flooding is the addition of sodium hydroxide to injection water, which help in lowering the surface tension, reversing the rock wettability, emulsification of the oil, mobilization of the oil and helps in drawing the oil out of the rock.

For very large field projects, polymer solutions may be injected over a 1-2 year period of time; after which the project reverts to a normal water-flood. Incremental oil recovery is on the order of 12 per cent of the original oil in place (OOIP).

The emerging technological trends in polymer flooding include the development of temperature and salinity resistant polymers, high-molecular weight polymers, the injection in the reservoir of larger polymer concentrations, and the injection of larger slugs of polymer solutions.

Relative-Permeability-Modification (RPM) water-shutoff treatments, also termed Disproportionate Permebility Reduction (DPR) treatments are water-soluble polymer systems and weak gels that reduce the permeability to water flow to a greater degree than to oil and gas flow; particularly in wells where water and oil are produced from the same zone and the water-bearing cannot easily be isolated.

These systems perform due to adsorption onto the pore walls of the formation flow paths. Several mechanisms are available including changes in porous media wettability, lubrication effects, and segregation of flow pathways, gel dehydration, and gel displacement, among others.

After polymer flooding for two to three years, the oil reservoir will have serious scaling (*i.e.*, materials sticking to metal surfaces - one of the major problems in the development of oil and gas fields). The complex requirements of the project make polymer flooding an expensive option and therefore difficult for small or medium sized fields.

Micellar-polymer flooding is also known as micellar, micro-emulsion, surfactant, low-tension, soluble-oil, and chemical flooding. The differences are in the chemical composition and the volume of the primary slug injected.

Chemical Flooding

In a chemical flood, chemicals are injected with the water flood to improve the displacement efficiency. A chemical solvent is specially developed for adaptation to the specific structural characteristics and physico-chemical properties of a reservoir.

After injecting with water, chemical reactions form new chemical sediment, which can reduce the contradiction between layers, increase volume and amount of water injected. This can improve the degree to which reserves can be recovered, while improving production efficiency.

Chemical flooding is a generic term for injection processes that use special chemicals (*i.e.*, surfactants) dissolved in the injection water that lower the interfacial tension between the oil and water from an original value and is possible to break up the oil into tiny droplets that can be drawn from the rock pores by water. There are two common chemical flooding include polymer flooding and alkaline or caustic flooding.

Sandstone reservoirs rather than carbonate formations that contain anhydride or gypsum, which can consume large amounts of alkaline chemicals. These chemicals are also consumed by clays, minerals, or silica, and the higher the temperature of the reservoir the higher the alkali consumption. Another common problem during caustic flooding is scale formation in the producing wells.

The **Nandej oil field** of Cambay Basin is on production since 1990 and producing mainly from Kalol formation. The north-western part of the field is having better reservoir facies where the permeability varies from 10 md to 42 md. Central part of the field is considered to be the poor reservoir with permeability is 5 md. The southern part of the field is having moderate reservoir permeability to the order of 10 md. These wells had been stimulated with mud acid but the well performance could not be improved. A number of acidization and hydro-fracturing jobs were carried out. New acid system- fluoboric acid is being designed to befit for migratory clays. The problem of fines generation and subsequent damage is not observed in case of fluoboric acid system and permeability is regained (Bahuguna, V.K, *et al.*, 2003).

Alkaline or Caustic Flooding

In this type of chemical flooding, alkaline or caustic solutions are injected into the reservoir. Common caustic chemicals are sodium hydroxide, sodium silicate, or sodium carbonate. These caustic chemicals react with the natural acids (naphtenic acids) present in crude oils to form surfactants *in-situ* (sodium naphthenate) that work in the same way as injected synthetic surfactants (reduction of interfacial tension between oil/water) to move additional amounts of oil to the producing well. These chemicals also react with reservoir rocks to change wettability. Alkaline flooding can be applied to oils in the API gravity range of 13° to 35°, particularly in oils having high content of organic acids. The preferred oil formations for alkaline flooding are alkaline chemicals, polymer as a mobility buffer to displace the primary slug. Modifications of alkaline flooding are the alkali-polymer (AP), alkali-surfactant (AS), and alkali-surfactant-polymer (ASP) processes.

In addition to the beneficial formation of natural surfactants (surfactants *in-situ*) driven by the reaction of alkali with naphthenic acids in the crude oil, the role of the alkali in the AS and ASP processes is to reduce the adsorption of the surfactant during displacement through the formation. The presence of alkali can also alter formation wettability to reach either more water-wet or more oil-wet states.

Non-Hydrocarbon Gas Injection

Miscible flooding is a general term for injection processes that introduce miscible gases into the reservoir. This allows for total displacement efficiency. Gases used include carbondioxide, or Nitrogen. The fluid most commonly used for miscible displacement is carbon dioxide because it reduces the oil viscosity and is less expensive than liquified petroleum gas. Oil displacement by carbon dioxide injection relies on the phase behavior of the mixtures of that gas and the crude, which are strongly dependent on reservoir temperature, pressure and crude oil composition.

Injection gases also include liquefied petroleum gases (LPGs) such as propane, methane under high pressure, methane enriched with light hydrocarbons, nitrogen under high pressure, and carbon dioxide used alone or followed by water.

When a reservoir's pressure is depleted through primary and secondary production, carbon dioxide flooding can be an ideal tertiary recovery method. It's particularly effective in reservoirs deeper than 2,000ft., where CO_2 will be in a supercritical state. Carbon dioxide flooding works on the premise that by injecting CO_2 into the reservoir, it dissolves in oil, the oil swells and the viscosity of any hydrocarbon will be reduced and hence, it will be easier to sweep to the production well.

Normally the CO_2 injection is alternated with more water injection, with the water acting to sweep the oil towards the production zone. In these applications, between one-half and two-thirds of the injected CO_2 returns with the produced oil. This is then usually re-injected into the reservoir to minimize operating costs. But unless natural CO_2 exists in the nearby area, it's generally difficult to collect sufficient amounts of CO_2 for industry use.

Although CO_2 is not truly miscible with the crude oil, CO_2 extracts the light-to-intermediate components from the oil, and, if the pressure is high enough, develops miscibility to displace the crude oil from the reservoir. CO_2 has a greater viscosity under pressure than many other gases and displaces oil at low pressures. CO_2 flooding has been the most widely used EOR recovery method of medium and light oil production in sandstone reservoirs, but safety in handling is more important.

Thermal Recovery Processes

Thermal recovery processes use thermal energy in some form both to increase the reservoir temperature and therefore to decrease the viscosity of the oil, which makes possible the displacement of oil towards the producing wells. Thermal recovery processes are globally the most advanced EOR processes, which are classified in three main techniques:

Steam Drive

Steam drive, also known as steam injection or continuous steam injection, involves generating steam of about 80 per cent quality on the surface and forcing this steam down the injection wells and into the reservoir. When the steam enters the reservoir, it heats up the oil and reduces its viscosity. As the steam flows through the reservoir, it cools down and condenses. The heat from the steam and hot water vaporizes lighter hydrocarbons, or turn them into gases. These gases move ahead of the steam, cool down, and condense back into liquids that dissolve in the oil. In this way, the gases and steam provide additional gas drive. The hot water also moves the thinned oil to production wells, where oil and water are produced.

Cyclic steam injection operation involves only one well that functions either as injection and production well. In this process steam is injected into the reservoir for several days or weeks to heat up the oil. Then, steam injection is stopped and the well is shut in to allow the reservoir soak for several days. In the reservoir, the steam condenses, and a zone of hot water and less viscous oil forms. Later on, the well is brought into production and the hot water and thinned oil flow out. This cyclic process of steam injection, soaking, and production can be repeated until oil recovery stops.

In-situ Combustion

In-situ combustion is a thermal recovery technique in which a part of heavy oil in place is burned to generate heat. This heat brings reduction in viscosity of the crude oil to get improved mobility and hence oil production rate and recovery. It is also called fire flooding, is a process in which an oxygen containing gas is injected into a reservoir where it reacts with the oil contained within the pore space to create a high-temperature self-sustaining combustion front that is propagated through the reservoir.

Ignition may be induced through electrical or gas igniters of may be spontaneous if the crude oil has sufficient reactivity. In most cases the injected gas is air and the fuel consumed by the combustion is a residuum produced by a complex process of cracking, coking, and steam distillation that occurs ahead of the combustion front. The heat from the combustion thins out the oil around it, causes gas to vaporize from it, and vaporizes the water in the reservoir to steam. Steam, hot water, and gas all act to drive oil in front of the fire to production wells. *In-situ* combustion is possible if the crude-oil/rock combination produces enough fuel to sustain the combustion front. *In-situ* combustion field tests have been carried out in reservoir containing API gravities from 9 to 40°API.

The **Balol field** was discovered in 1972 and put on production in 1982. Tertiary recovery techniques (primarily *in-situ* combustion) are forecast to increase the ultimate recovery factor the heavy oil. The major pay zone in the Balol sandstone reservoir. The reservoir zone is 32 m in thickness and top of the reservoir is at depth 996 m. The porosity in the range of 28-30 per cent and its permeability varies from 8-15 milli darcy (Asit kumar et al., 2006). The oil in this reservoir is asphaltic in nature. The primary recovery of viscous and heavy oil from Balol is only 10-12 per cent, hence the pay zone is under thermal EOR process. The *in-situ* combustion process is carried out to improve the recovery of oil from the sandstone reservoir.

Foam Flooding

Foam is a metastable dispersion of a relatively large volume of gas in a continuous liquid phase that constitutes a relatively small volume of the foam. The gas content in classical foam is quite high (often 60 to 97 volume per cent). Bulk foams are formed when gas contacts a liquid containing a surfactant in the presence of mechanical agitation.

In oilfield applications, the use of CO_2 foams has been considered a promising technique for CO_2 mobility control and steam-flooding mobility control. The use of foams for mobility control in surfactant flooding, specifically at high temperatures (due to polymer degradation).

The **Viraj field** is in Cambay basin which is a small field occupies an area of 4 sq. km. Though there are three oil-bearing zones in the field which was discovered in 1977 and put on production in 1980.The estimated oil reserves are 7.88 MMT. The crude from this field being heavy and viscous is prone to foam. Foaming tendency of crude not only decreases the effective efficiency of separator but also reduces storage capacity of storage tanks and pumping efficiency of pumps. The foaminess index for crude oil of Viraj at 30°C is 0.675 minutes. Silicone based defoaming chemical, *i.e.*, DM-silicone fluid 10,000 CTSK was found most effective and decrease the foaminess index from 0.675 minutes to 0.2 minutes at a dose of 2.0 ppm (Bassi. K.D. and R.K. Gang.2003).

Microbial Enhanced Oil Recovery (MEOR)

Microbial enhanced oil recovery (MEOR) is a tertiary oil recovery process where micro-organisms and/or their metabolic by products are utilized for the mobilization of crude oil trapped in mature oil formations. This is categorized into two categories: to alteration of oil/water/rock interfacial properties and changes in flow behavior.

Microbial Enhanced Oil Recovery (MEOR) relies on microbes to ferment hydrocarbons and produce by-products such as bio-surfactants and carbon dioxide that help to displace oil in a similar way than in conventional EOR processes. Bacterial growth occurs at exponential rates, therefore bio-surfactants are rapidly produced. The activity of bio-surfactants compare favorably with the activity of chemically synthesized surfactants. The injection of nutrients such as sugars, nitrates or phosphates stimulates the growth of the microbes and aid their performance. MEOR applications are limited to moderate reservoir temperatures, because high temperatures limit microbial life and the availability of suitable nutrients.

Three approaches have been used to achieve microbial injection. In the first approach, bacterial cultures mixed with a food source (a carbohydrate such as molasses is commonly used) are injected into the oil field. In the second approach, nutrients are injected into the ground to nurture existing microbial bodies. These nutrients cause the bacteria to increase production of the natural surfactants they normally use to metabolism crude oil underground. After the injected nutrients are consumed, the microbes go into near-shutdown mode, their exteriors become hydrophilic, and they migrate to the oil-water interface area where they cause oil droplets to form from the larger oil mass and more likely to migrate to the wellhead. The third approach is used to address the problem associated with the paraffin wax components of the crude oil, which tend to precipitate as the crude flows to the surface.

Microbial injection is part of microbial enhanced oil recovery and is rarely used because of its higher cost and because the developments are not widely accepted. These microbes function either by partially digesting long hydrocarbon molecules and generating bio-surfactants, or by emitting carbon dioxide, which then functions as described in Gas Injection above. MEOR oilfield applications have shown mixed results. On average, MEOR field trials results have been poor.

Therefore, MEOR is potentially a "high risk, high reward" process the reward refers to the fact that the difficulty and the logistical costs of implementing the process would approach those of implementing a water-flood rather than an EOR process. The risk is associated with the many and severe performance constraints that a microbial system would have to satisfy to take advantage of an *in-situ* carbon source.

MEOR process applied effectively mobilized residual crude oil. The MEOR process implemented worked effectively under reservoir conditions and is environmentally friendly. Common operational problems associated with paraffin, emulsion, scale, and corrosion were significantly reduced and oil production decrease was not observed during the MEOR projects time frame. It may take longer duration for genetic reactions within the reservoir, possibly in years.

Injected microbes and nutrients are cheap; easy to handle in the field and independent of oil prices. Economically attractive for mature oil fields before abandonment. Low energy input requirement for microbes to produce MEOR agents.

But, the oxygen needed in aerobic MEOR can act as corrosive agent on non-resistant topside equipment and down-hole piping. Anaerobic MEOR requires large amounts of sugar limiting its applicability in offshore platforms due to logistical problems. Microbial growth is favored when, layer permeability is greater than 50 md; reservoir temperature is less than 80°C, salinity is below 150 g/L and reservoir depth is less than 2400m.

The MEOR job in ONGC was done by INBIGS in the year 1997 in Assam region. Microbes for MEOR were isolated from oil field sources, and acclimatized, tested, bulk cultured and then injected along with nutrients into the selected wells. After successful implementation of MEOR job in low temperature wells of **field–A**, attempt was made in the well having BHT temperature up to 75°C and incremental oil gain was noticed. Most of the wells have produced incremental oil, by gradual adaptation method and job was quite successful (Uniyal, N.P., *et al.*, 2005).

EOR Processes and Technical Maturity

The selection of a suitable EOR process for a specific oil formation requires and integrated study of the reservoir and its characteristics. Some basic questions that the operator must address during the decision making process.

If initial calculations indicate the process may be profitable, then there will be a need for additional laboratory and field measurements followed by more modeling. In many cases, a single well test will be justified to evaluate injectivity, reduction in oil saturation, and other performance indicators that can only be assessed with field tests. When comparing the economics of different processes, many factors must also be taken into account. Chemical methods have the advantage of lower capital costs than miscible gas and thermal methods, and commercial projects can start small and be expanded if successful without the need for expensive infrastructure such as pipelines.

Detailed outlines of EOR screening criteria are to be studied based on matching particular EOR processes to reservoir properties such as API gravity, reservoir depth, oil saturation, rock permeability and porosity distribution, oil viscosity distribution, type of rock formation, and reservoir temperature distribution, among others.

As previously indicated, thermal recovery processes are the most advanced EOR processes and contribute significant amounts of oil to daily production. Most thermal oil production is the result of cyclic steam and steam drive.

The fracture direction during the hydraulic fracturing process cannot be controlled. It can only produce cracks that conform to the *in-situ* fracturing direction or high permeability area – however the fracture direction may not be the one needed. Because the injected water is flowing generally along the direction of primary cracks in high permeability area, if an opening fractures in this direction it may result in large water mix in the reservoir. In such cases any EOR method can be of help in recovery of oil.

Infill recovery is carried out when recovery from the previous three phases have been completed. It involves drilling production wells to produce the left over/ un-drained oil if any, in between in the whole reservoir has been fully recovered.

Bombay High North-Field Development

Mumbai High is an offshore oil field in water depth of 80m in the Arabian sea and 160Km west–north-west of Mumbai city. Discovered in 1974 and put on production in 1976. This giant field holds nearly 1660 MMt of oil and about 420 MMt have been produced (~25 per cent of OIIP). The good part of the main oil bearing reservoirs display premature water breakthrough, gas conning with differential pressure depletion. Major oil production comes from carbonate reservoirs in the field. Mumbai High production built to a plateau, sustained for 4-5 years and then experienced gradual decline. With these reservoir problems the ultimate recovery from the fields is forecasted to be about 30-35 per cent (Narasimham, J.L., *et al.*, 2005).

Having reached to a peak production level of nearly 400,000 barrels of oil per day during 1989-90, the field in subsequent years experienced steep decline in the output. Increase in gas oil ratio (GOR) levels, reduction in reservoir pressures and undesirable flow of water were some of the reasons for decline in the field output. Since then, number of steps were taken up in the field to arrest the rate of decline of oil production. With the implementation of development drilling and treatment of reservoirs, the declining trend was arrested and upward trend in production could be re-established.

Deposited in the mid-Miocene period in a shallow marine environment, this reservoir is essentially composed of limestone layers separated by thin shale/marl or shales/marls sections. The reservoir characterization, in terms of layer-wise permeability assessment and mapping in partly weathered carbonates, has been carried out in ***Bombay High North*** (Basu, D.N., *et al.*, 1980). The L-III reservoir divided into a number of sub-layers. A successful attempt has been made to link the coefficient of permeability, determined from core samples, with the log characteristics through the help of transforms using multiple Regression Technique (Bhandari L. L. and Jain S. K., 1984). The primarily inter-granular porosities range from less than 10 percent to greater than 30 percent, with an average between 15 and 20 percent. Permeabilities range from 0.01 milidarcies (mD) to 1,000 mD with averages in the range of 100 to 250 mD. The crude oil is one of the high quality. For instance, Bombay High crude has more than 60 per cent paraffinic. The field is exploited using the gas lifts method, a major artificial lift method used for oil production.

Having calibrated against the permeability interpreted from pressure build up data, the transforms were further used for estimation of layer-wise permeability (Azim, S.A, *et al.*, 1994). The permeability maps, prepared with these data have brought out vertical and spatial variation of permeability and thereby delineating areas which are poorly drained of affected by early water and gas encroachment. The maps are in good agreement with field observation, thus giving a good confidence level to the transforms (Amitabh Pandey., 2003).

With significant advances in drilling related equipment, LWD technology and realization of higher per well productivity, drilled horizontal and multi-lateral wells (Babu, *et.al.*, 1989, Mukherjee, *et.al.*,1989, Vij et. al.,1998). Now more horizontal/ multi-lateral wells were drilled and some sub-optimally flowing conventional wells were side-tracked and most of them re-completed as horizontal wells successfully (Dikken, B.J., 1990).

Besides these, concept of drilling high angle to increase the length of exposure in the pay zone, well was drilled for at a high angle of 80° which resulted in the pay-zone exposure to almost four times (Drilled 344m of 6″ hole section at 80°). Fast activation using surge plug for higher draw down, has contributed to an increase of production.

The **Heera and South Heera fields** have more than 172 development wells including 18 horizontal and 6 multi-lateral wells have been drilled from 15 platforms. The peak oil production rate from Heera and South field was 88,625 bopd with an average water cut of 22 percent with increase of average water cut of 50 per cent. Few wells are under water injection. The field has produced about 41.03 MMt of oil, *i.e.*, 14.5 per cent of OIIP (Narasimham, J.L., *et al.*, 2005).

The **Neelam field** was discovered in 1987 and was put on production in 1990. The field developed by installing 11 production and 2 injection platforms. In Neelam field, the Bassein limestone constitutes the main pay and reservoirs both oil and gas. It is a highly heterogeneous reservoir, representing the various solution induced features of Karst topography, like vugs, solution channels, roof collapses, *etc.* The peak oil production rate from this field was about 95,000 bopd in 1994. Water injection was also initiated in this field in 1994.

Since June, 2001, the Improved Oil Recovery (IOR) scheme was implemented,. About 112 development wells including 8 horizontal wells have been drilled. About 74 wells are producing an average 23,600 bopd of oil, while 15 wells are injectors, injecting 67,870 bwpd water for voidage compensation and pressure maintenance. Till date, the field has produced about 22.5 MMt of oil, *i.e.*, 21 per cent of OIIP. A large number of wells in Neelam field are cutting water in excess of 80 per cent (Narasimham, J.L., *et al.*, 2005).

The **Panna Field** reservoir is a heterogeneous limestone with in-place volumes of 1 billion barrel of oil and 1.9 TCF of gas. The field is characterized by a 20-meter initial oil column. The field consists of seven wellhead platforms with a total of 65 wells. The field produce approximately 24,000 BOPD and 110 MMSCFD of gas.

Both water and gas starts coning from the beginning with a very low drawdown. To control the same new design was implemented based on the principle of segmented liner, where the entire horizontal section is divided into multiple segments each separated using isolation packers. Well fluid comes inside the tubing through which can be opened or closed as per requirement. At any point of time, contribution from each zone can be determined by carrying out a PLT, and based on the data remedial measure can be taken to isolate the offending segment.

In the **Jotana field**, situated in Cambay Basin, in general, oil properties are good but wax and paraffin contents make the oil thick in some wells. Most of the pools are acting under depletion drive (Deb, G.C., V. Pratap and S.C. Gawali., 2003). It is observed that impact of water injection was not uniform in the whole area due to heterogenities. With water injection, average reservoir pressure increased by 40 kg/cm^2. Intake of injected water by some of the layers only, led to water break-through in some good producers. Enhanced oil recovery techniques MEOR, ASP and chemical flooding are also part of the field.

Identification and Treatment of Sick Wells

In case the optimal and economic flow rate of oil and gas from a flowing well falls less than that, the well is called sick well, which means a possible remedial measure is needed to bring the well back to normalcy. If there is major problem or no technology available to achieve that then normally such wells will be abandoned. Most of the old and matured fields, may result in an increased rate of wells falling sick. These fields are often associated with the problems of declining reservoir pressure, change of fluid characteristics, resulting in wells becoming sick due to poor influx, sand incursion and choking of production tubing, among others. Operator should have an action plan for liquidation of sick wells, application of new technology for revival of sick wells requires a study of cause for becoming sick and latest technological methods which can be implemented economically.

The well can be sick due reservoir performance or due to mechanical or physical damage in or around well bore with time. Problems due to reservoir performance related sickness either due to high water-cut which reduces economics of well flow and water conning around wellbore or sand cut around well bore.

Water Cut

In a producing field water cut will increase with time which may be due to upward movement of OWC, due to which the bottom perforations may be below new OWC and contribute water production. In some cases, if excess drawdown was given water preferentially enters in to perforation forming a conning effect around well bore due to which water production from well increase. Once water enters in to well bore unless its entry is shut-off, clean oil production will be difficult. Once allowable economic threshold of water production is crossed well become sick and water shut-off job has to be carried out.

The required repair job is to first know the latest OWC depth and squeeze cement through bottom perforations up to that level and re-perforate the object within the latest oil zone.

Squeeze cementing forces a cement slurry behind the pipe to repair leaks or shut of fluid loss. Squeeze cementing is normally thought to be a repair step, but is also used to seal off depleted zones or unwanted fluid production.

Sand Cut

In case of relatively less compacted sandstone reservoirs if excess draw down is given along with oil flow sand also collapse and enter into well bore. Based on the quantum of sand incursion, sometimes perforations will be blocked and well bottom will also be filled with sand and also in some cases tubing will be with sand. In order to revive the well first task is to sand cleaning at the bottom of the well. With fair knowledge of reservoir sands screens will be kept against perforations such that collapsed sand cannot make obstacle in the flow path. In case of producing field, hydraulic jet clean inside the well and re-perforation can be a better alternatives adopted.

Cement Bond of Casing - After lowering the casing it will be cemented behind the casing with an idea to isolate oil zone and water zone behind casing within the reservoir, such that the selective perforation in the reservoir if done it can only contribute oil/gas.

USIT log enables measurements for cement and casing evaluation-corrosion and wear. Field results show that channeling, contaminated cement, light cement and gas can be diagnosed and that external hardware such as centralizers can be detected. The corrosion measurements can detect mechanical wear, corrosion and deposits.

If cement bond between casing and cement is good and bond between cement and formations is bad, it may lead to water interface in the reservoir and needs isolation. Further, the common phenomenon is channeling behind casing. To do secondary cementation at desired intervene to arrest such situations is called "block cementation". In this two sets of perforations are made at top and bottom of the required zone. A drillable plug will be set in between these sets of perforations. Cement slurry will be pumped under pressure through bottom perforations which will go behind casing and likely to re-enter casing through top perforation. This ensures behind the casing cement occupy all weak/channel zones. Drill the plug and re-perforate the required interval for production purpose.

Casing Damage -Sometimes the production casing may slip down by parting. Casing also get damaged by corrosion with time based on composition of fluids. The sulphur content in the crude if any, will quickly damage the casing. The casing repair is very costly operation because it requires the high capacity drill-rig again at site. The operation is simple pullout the upper part of the casing where from it can be unscrewed and lower new one and tie back to the casing in the well again.

Choking of Tubing: There is narrow one inch dia coil tubing is available which is normally truck mounted. It will be lowered through regular tubing and clean jobs will be done inside the tubing.

Zone Transfer

In case of multi zone reservoir, generally the lower most zone will be put on production first and once its recoverable reserves are produced and or the zone is ceased and found economically not viable for revival, then next upper zone will be perforated. For this smaller capacity rig called "Work-Over rig" will be used to retrieve earlier tubing and isolate the bottom zone by a plug and perforate desired upper interval. In case of work-over rigs their capacity will be to lift tubing only not the casing.

Chapter 10

Petroliferous Basins of India

Introduction

Generally, sedimentary basins where oil and gas occur are categorized or classified based on their origin in relation to their prospectivity level for hydrocarbon exploration. So, the categorisation of sedimentary basins is a dynamic process in conformity with the geological knowledge, status of exploration, known hydrocarbon accumulation and latest visualization of prospectivity.

The total sedimentary area of the India up to 200 meter isobaths is about 1,784,700 square kilometres. This total area was spread in 26 major Sedimentary Basins (Figure 10.1). Of these three Basins further extends into deep waters with in the Indian Territory. The deep water area covers 1,349,120 sq. km.

Categorization

Based on the exploration carried out so far and status of knowledge in terms of occurrence of hydrocarbons, the sedimentary basins of India are divided into 4 categories. Areal spread of the Basins as per Category wise are shown in Table 10.1.

The exploratory efforts in different basins have resulted in estimation resources of hydrocarbons and success has resulted in establishing the commercially producing basins. The estimated oil and gas resources are about 28,100 MMT of O+OEG. Out of these, 7,000 MMT are in very deep waters in which, the exploration in ultra-deep waters in K G Basin showed positive results. The established in-place reserves are of about 11,100 MMT, hence the remaining 10,000 MMT, excluding ultradeep water area, will remain as yet to find (YTF) hydrocarbon potential of India.

The exploratory and production efforts carried out in the country are compiled from various data available in many places are basin-wise (Table 10.2). Since these efforts are of dynamic in nature with time and some efforts are not known, the statistical information may had its limitations.

Category I

Basins with proved hydrocarbon reserves and where commercial production has already started. These basins are:

1. Cambay
2. Assam shelf and Assam-Arakan Fold Belt
3. Mumbai offshore
4. Cauvery
5. Krishna-Godavari
6. Rajasthan

Category II

Sedimentary basin with proved occurrence of hydrocarbons but from which no commercial production has been obtained yet. These basins are:

1. Andaman – Nicobar
2. Kutch
3. Mahanadi-NEC (North East Coast) Basin

Category III

Basins having hydrocarbon shows that are considered geologically prospective, these basins are:

1. Himalayan Foreland Basin
2. Ganga Basin
3. Vindhyan Basin
4. Saurashtra Basin
5. Kerela Konkan Basin
6. Bengal Basin

Category IV

Basins having uncertain potential which may be prospective by analogy with similar basins in the world. These basins are:

1. Karewa Basin
2. Spiti-Zanskar Basin
3. Satpura–South Rewa–Damodar Basin
4. Chhattisgarh Basin
5. Narmada Basin
6. Deccan Syneclise

7. Bhima-Kaladgi
8. Bastar Basin
9. Pranhita Godavari Basin
10. Cuddapah Basin

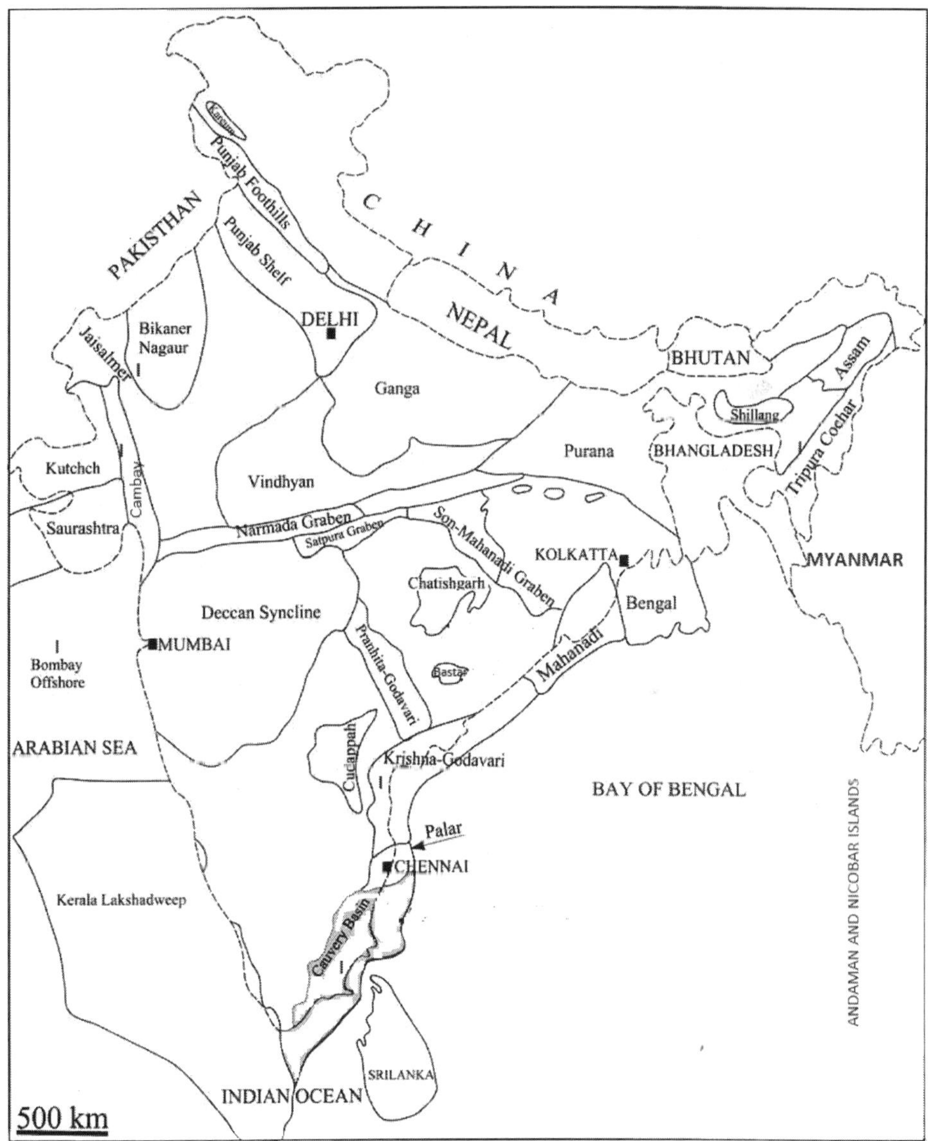

Figure. 10.1. Sedimentary Basins - Category (Modified after Biswas., S K, 2012)

Table 10.1. Area-wise Details of these Categorized Basins

Category	*Basin*	*Basinal Area (Sq. Km.)*		*Total*
		On land	*Offshore*	
I	Cambay	51,000	2,500	53,500
	Assam Shelf	56,000	- - - -	56,000
	Bombay offshore	- - - -	116,000	116,000
	Krishna Godavari	28,000	24,000	52,000
	Cauvery	25,000	30,000	55,000
	Assam-Arakan Fold Belt	60,000	- - - -	60,000
	Rajasthan	126,000	- - - -	126,000
II	Kutch	35,000	13,000	48,000
	Mahanadi-NEC	55,000	14,000	69,000
	Andaman-Nicobar	6,000	41,000	47,000
III	Himalayan Foreland	30,000	- - - -	30,000
	Ganga	186,000	- - - -	186,000
	Vindhyan	162,000	- - - -	162,000
	Saurashtra	52,000	20,000	80,000
	Kerala-Konkan-Lakshadweep	- - - -	94,000	94,000
	Bengal	57,000	32,000	89,000
IV	Karewa	3,700	- - - -	3,700
	Spiti-Zanskar	22,000	- - - -	22,000
	Satpura-South Rewa-Damodar	46,000	- - - -	46,000
	Narmada	17,000	- - - -	17,000
	Decan Syneclise	273,000	- - - -	273,000
	Bhima-Kaladgi	8,500	- - - -	8,500
	Cuddapah	39,000	- - - -	39,000
	Pranhita-Godavari	15,000	- - - -	15,000
	Bastar	5,000	- - - -	5,000
	Chhattisgarh	32,000	- - - -	32,000
	TOTAL	1,390,200	394,500	**1,784, 700**
	DEEP WATERS			1,350,000
	GRAND TOTAL	- - - -	- - - -	3,134, 700

After Biswas, S.K., 2012.

Table 10.2. Basin-wise Exploratary Results - Category I Basins

Basin	*Resources-Mmt*	*In-place Reserve-MMT*		*Ultimate Reserve-MMT*		*Cumulative Prod-MMT*		*Balance Reserv-MMT*		*Wells*
	HC	*Oil*	*Gas*	*Oil*	*Gas*	*Oil*	*Gas*	*Oil*	*Gas*	
Mumbai	9,190	3,680	2,400	810	1,040	485	555	325	485	4,300
Cauvery	700	130	45	45	30	35	13	10	17	800
KG	1,130	123	280	42	200	18	42	24	158	960
Cambay	2,050	1,050	450	350	350	242	240	108	120	7,700
Assam	5,040	1,747	850	510	503	367	180	143	323	4,950
Rajasthan	380 (!)	250	40	120	20	90	5	30	18	560
Total		**7,087**	**4,055**	**1,877**	**2,153**	**1,257**	**1,032**	**620**	**1,121**	
Grand Total	**18,490**	**11,142 (65 per cent)**		**4,030**		**2,289**		**1,741**		**18,770**

Category I

Upper Assam Basin

Introduction

Oil seeps were reported in Assam as early as in the year 1825. The oil exploration in India commenced with the discovery of the Digboi oilfield in Upper Assam based on the surface oil shows in 1867. Experimental bore-well, to assess an oil show at Makum, close to Digboi was drilled up to 102 feet at Nahorpung near Jaipur area of Upper Assam but failed to establish satisfactory production. In the second attempt, oil was struck at 36m in well at Makum near Margherita area.

The first well at Digboi field in Assam in 1889 at depth of about 225 m by AR and T Company was regarded as the first commercially successful oil discovery with flow of 200 gallons per day. By 1893 more wells were drilled at Digboi and produced 757.08 liters/day. AR and T established Assam Oil Company (AOC) in 1899. Between 1922 and 1932 Burma Oil Company (BOC) mapped and drilled 10 structures in the Schuppen Belt with no economic successes. BOC also conducted seismic surveys in the area.

After India became independent, the next major discovery was the Nahorkatiya field, in 1953. In 1959, the Oil India Private Limited Company (OIL) was incorporated and in 1961 it became a joint venture company between the Indian Government and Burmah Oil Company. In 1981, OIL became a wholly-owned Government of India enterprise. The continued exploratory efforts by OIL have resulted in many discovered fields like Dikom, Kathaloni, Tengakhat, Tamulikhat, Baghjan and Panidihing. These oilfields, discovered, are situated mainly in the areas south and southeast of the Brahmaputra River and few in the thrust belts, associated with Naga-Patkai hills in the Naga Schuppen zone, after the Digboi, Kharsang field was discovered. However, the area to the north of the Brahmaputra River up to the Eastern Himalayan foothills has remained poorly explored.

After its formation in 1956, the area south of Moran is under Oil and Natural Gas Commission (ONGC). All the outcrop areas have been covered by detailed geological surveys. The deepest well drilled by the ONGC in the basin has penetrated the lower part of Barail rocks at 4805 m. depth. The deepest oil production is from 4250m in the Eocene. Analysis of the geo-scientific data obtained so far from this basin enabled an understanding of tectono-environmental conditions of deposition of interesting sediments, basin evolution and the geological factors controlling hydrocarbon accumulations. This has also helped in providing leads to look for new pools.

Most of the oil and gas fields, discovered in the Upper Assam shelf, are situated mostly on the southeastern slope of the Brahmaputra arch, and almost all the major oil fields like Nahorkatiya, Lakwa, Lakhmani, Geleki, Dikom, Kathaloni lie in a belt bordering the Naga thrust. In the Dhansiri valley also, oil fields like the Borholla and Khoraghat and Nambar lie in the same belt.

Geological History

The Assam Shelf is comparatively a stable area characterized by epirogenic movements has an extent of about 40,000 sq. km and includes the Shillong Plateau, the Garo, Khasi, Jaintia, Mikir hills and the Upper Assam Valley. Geologically this is the north-eastern part of the Indian Peninsular Shield with the Garo-Rajmahal Hills gap in between.

The Assam-Arakan sedimentary Basin is a shelf-slope-basinal system. The shelf part of the basin spreads over the Brahmaputra valley and the Dhansiri valley, the slope lying between the Mikir hills and the Naga foothills. From the Digboi, the shelf runs westward to the southern slope of the Shillong plateau. The shelf-to-basinal slope, *i.e.*, the hinge zone lies below the Naga schuppen belt (Figure 10.aa.1). The basinal (geosynclinal) part is occupied by the Cachar, Tripura, Mizoram and Manipur fold belts (Mathur L.P. and Evans P. 1964).

The shelf part rests on Precambrian granitic basement, whereas the basinal part lies on transitional to oceanic crust. The area within the Upper Assam shelf, having high petroleum potential, measures approximately 56000 sq. km. and contains about 7000m thick sediments, and the area in the basinal part with moderate to high hydrocarbon potential measures about 60,000 sq. km. and contains more than 10,000m thick sediments.

The Shillong Plateau and Mikir hills Uplift, composed mostly of Precambrian granitic and metamorphic rocks. The southern slope of the Shillong Plateau exposes Gondwana, Cretaceous and Tertiary rocks.

The Indo-Burmese trench system that developed during the oblique subduction of the Indian plate below the Burmese plate became the locus of deposition of Upper Disang shales under deep marine conditions (Mukhopadhyay, M., and S. Dasgupta, 1988).

The development of the frontal foredeep in front of the rising Himalaya, during Mio-Pliocene due to tectonic loading by thrust slices were filled with coarser sediments. During this time, sedimentation in the Surma basin (including Sylhet trough) and the Kohima synclinorium took place in inter-montane basins, depositing the arenaceous sediments.

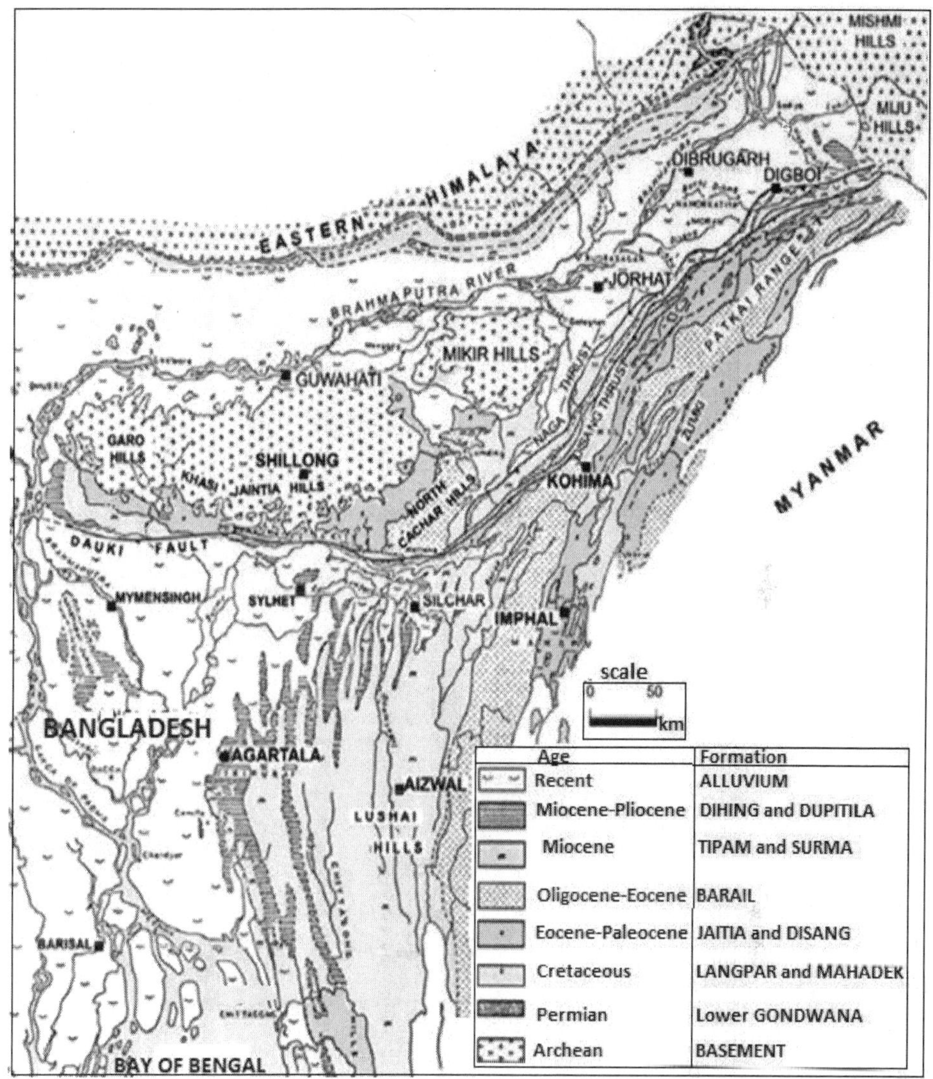

Figure 10.aa.1. Geological Map of North Eastern India (Modified After Mathur L.P. and Evans P. 1964).

During Pleistocene time, there was the last major folding movement and further upliftment of the Barail Range, the Central Disang uplift, the Mishmi Hills and the Himalaya. The Dihing boulder conglomerates, shed by the rising mountains were deposited at the slopes of the rising mountains.

During the deposition of Plio-Pleistocene sediments, the Assam Platform formed part of the frontal deep of the Himalayas and was consequently sunk north west-wards to accommodate a huge thickness of molasses derived from the rising Himalayans. Simultaneously there was intense orogeny in the Naga Mobile Belt (Ranga Rao A. and Samanta M.K.1987).

Tectonic History

The basin evolved through four tectonic phases. The first represents formation of intra-cratonic grabens along Precambrian weak zones during Permo-Carboniferous, when India was part of Gondwanaland. The second phase was the rifting in Early Cretaceous with separation of India and Antarctica, when the sedimentation is largely over older sediments of Gondwana grabens. End of rift is marked by lava flows and discrete sedimentation in Early Albian, when the Indian plate detached completely from Antarctica and entered drift phase. The drift phase from Late Cretaceous to Oligocene represents passive margin setup, which continued until the collision of Indian plate with Eurasian and Burmese plates. The fourth and final phase represents deposition of Foredeep sequence of Miocene to Recent age in the foreland tectonic setting.

During Middle to Late Cretaceous, when the Indian plate was moving northward, a number of horst and graben features developed on the granitic crust in the southern slope of the Shillong Plateau and Dhansiri valley.

The continental collision between Indian plate and Eurasian plate was oblique one with first episode at north western margin of the Indian shield during Early Eocene, whereas north eastern part, that is Assam plains form the southern fringes of Tethys Sea.

The formation of the trench system was, possibly, initiated in the north-eastern part and gradually progressed southward. The closing of the trench system was also initiated in the northeast and then gradually progressed southward (Desikachar S. V., 1971). The Andaman trench, which was receiving mostly argillaceous sediments since Upper Cretaceous to Paleocene, is the southward extension of the Indo – Burmese trench system (Figure 10.aa.2).

Following completion of collision and subduction of the oceanic part of the Indian plate during Late Oligocene (to Early Miocene?) when the continental part of the Indian plate seems to have come close to Tibetan and Myanmar (Burmese) plates, there was up-liftment and erosion all over the shelf and in a major part of the basinal area.

The presence of oceanic crust visualized below this thick pile of sediments may be subjective within the oceanic basins. However, the presence of younger ophiolitic belt at the Indo-Myanmar border, can be considered towards east of continental collision line the sediments are derived from westerly dipping Burmese plate.

This event was followed by a pronounced south to southeastward tilt of the basin, mostly the geosynclinal part, which was, perhaps, caused by subduction related tectonics. These lateral tectonic movements were accompanied by upliftment and total withdrawal of the sea, heralding the onset of continental sedimentation on the Assam Shelf as well as on the earlier basinal area (Mitra, R.N.1971).

Tectonically the area is located between two convergent plate margins and is under the influence of active compression. The NE-SW trending Naga Thrust demarcates the zone of thrust sheets and plains in the area. The zone of intensive thrust area is called Schuppen belt covering between Naga and Diang thrusts. The crustal shortening due to thrusting is estimated to be 150 kilometers.

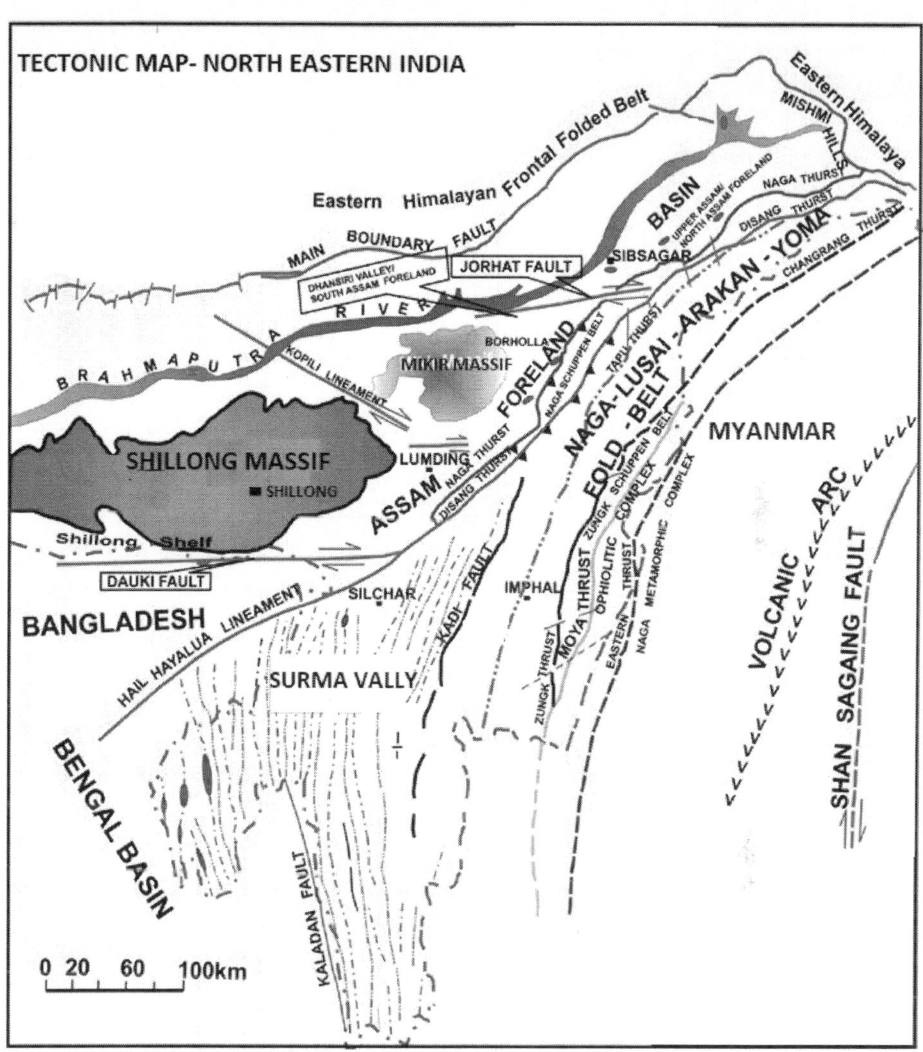

Figure 10.aa.2.Tectonic Setting of North Eastern India (Modified after Roymullik SK, *et al.*, 2009).

Towards south-west the Disang became Haflong fault which merges with Dauki fault there by separating the Tripura-Cachar fold belt towards south of this fault. The Mishmi Hills delineate towards north-east. The subsurface data reveal the presence of broad arch at the basement level in NE-SW alignment which runs a parallel to the present river course of Brahmaputra in the Upper Assam Valley. This area also slopes in all directions and towards northeast the deepest point is in the front of Mishmi Hills. Continued synchronous convergence of both orogens resulted into flexing of the foreland on either side of the narrow cratonic platform and deposition of thick sediment wedges (Sahoo, M. and K.D, Gogoi, 2008).

It is a Peripheral Foreland Basin, with South Assam Shelf as the southern unit. The area is characterized by well-defined northeasterly plunging linear ridge intervened by depressions oriented in NE-SW direction. Within the ridges there are number of horsts and grabens. These ridges and depressions represent the extensions of spurs or linear tectonic elements of Mikir Hills.

Though the general strike of this arch is NE-SW it was dissected by number of ENE-WSW and E-W aligned faults. The other lineament is NW-SE. The Brahmaputra valley is separated by Dhansiri Valley by a major E-W trending fault zone called Jorhat fault.

The post-collision convergence lead to uplift and foreland ward advancement of orogen by progressive footwall deformation. Continued tectonic loading of the foreland results in down flexing, frontal to the advancing orogen, leading to formation of pheripheral foreland basin.

Major tectonic elements of the basin are:

- ☆ Assam Shelf
- ☆ Naga Schuppen belt
- ☆ Assam-Arakan Fold belt

The Bougher anomaly map shows two distinctive trends a East-West trend parallel to the Himalayan structural trend which changes to NE-SW over Upper-Assam and north-South trend parallel to Arakan-Yoma ranges (Figure 10.aa.3).

The Shillong plateau which marks the south western margin of the upper Assam basin, is separated towards south by a faulted monocline called Dauki Fault (Murthy, M.V.N., *et al.*, 1969). This is a tear fault with a probable horizontal displacement of about 250 kilometers. This horizontal movement detached the Shillong plateau from the main Indian shield and the principal movements occurred mostly in the Pliocene (Idem 1977).

The platform conditions were controlled by block movements along the basement faults and the platform merges gradually into geosynclinal belt and is marked by steepening of basement slope towards southeast. Basement highs were already in existence in some places like Haldibari, Rudrasagar, Nahorkatiya and Moran. The difference between shelf and geosyncline is less pronounced and there is predominance of faulting towards the pericraton. The structures in the pericraton were gentle and bounded by basement faults.

Major post-Miocene faulting towards north western edge of platform resulted in strong basement tilt in this direction. Reactivation of some of the criss-cross faults in the basement during this period resulted in formation of structures, like Geleki and Lakwa, in Tipam sediments particularly in the peri-cratonic area. The tectonic movements during this period also resulted in initial formation of the Brahmaputra Arch and the fore-deep in front of the then rising Eastern Himalayas (Desikachar S.V. 1984).

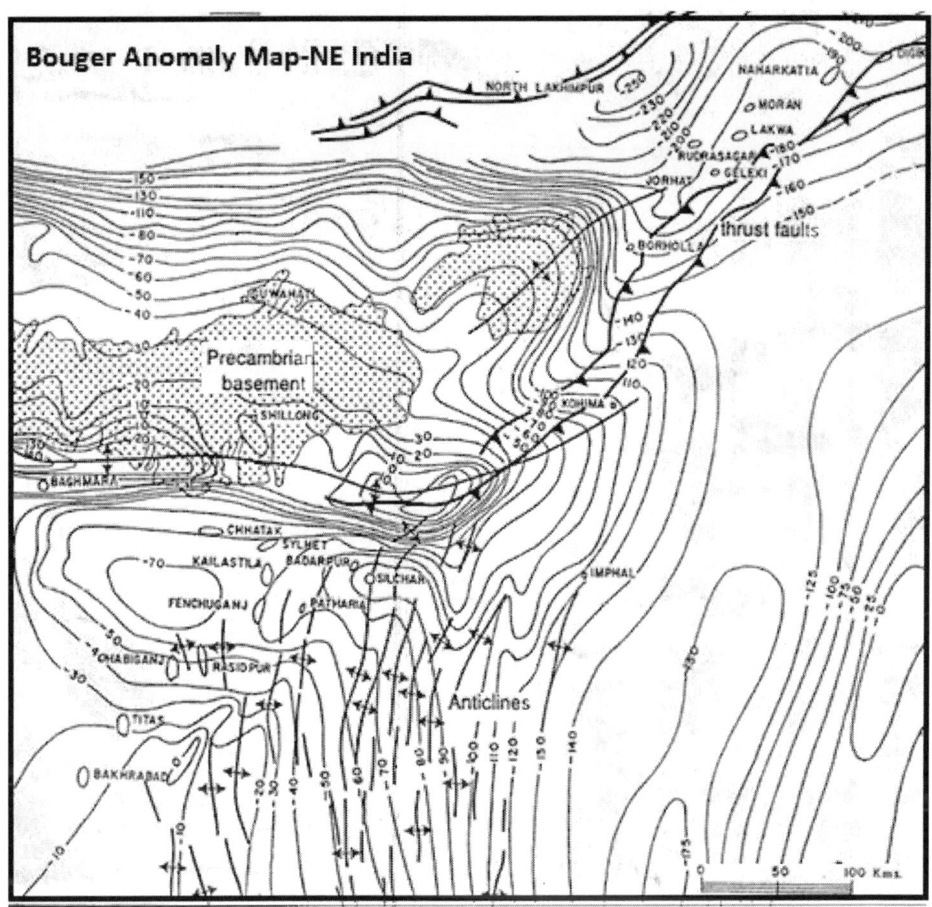

Figure 10.aa.3.Bouger Anomaly Map of North Eastern India.

Most of the faulting activity ceased towards the end of Pleistocene. The major structural configuration of Upper Assam platform is, therefore, an arch parallel to and just a little south of Brahmaputra River which splits the Upper Assam Shelf basin into northern and southern parts. The easternmost uplifts lying in an NW-SE trend mark the limit of the Brahmaputra Ridge, and the Upper Assam shelf.

The southeastern slope of the Upper Assam Shelf, southeast of the Brahmaputra arch, having local structural highs and lows, upto the Naga thrust, and extending 8 to 10 km beneath the Naga schuppen belt. This element contains most of the oil fields of the Upper Assam Shelf.

Upper Assam

Upper Assam Shelf is defined as the alluvium-covered extension of the Shillong and Mikir Massifs to the east –northeast. Generally it conforms to ENE-WSW trend of Indian Peninsula. It is a narrow, about 100 km wide valley bounded by two mobile belts to the north and southeast. Geographically the area can be divided into

Dhansiri valley, south eastern flanks of Mikir Hills, Brahmaputra Valley, plains of Upper Assam, Naga Hills towards south east and North Bank area covering north of Brahmaputra River. Mishmi block with an NW-SE trend lie to the east and the Mikir hills mark the boundary to the west. During the Paleogene, the Assam valley was a platform shelf which originated as a marginal sag grading into a pericratonic rift (Goswami, P and Goswami, P.C, 2008).

A spur of pre-Tertiary sediments associated with gently dipping terrain extend from Shillong plateau to underneath the alluvial cover of Upper Assam plains. Over this spur the eastern Himalayas have been thrust southwards and Naga Hills have thrust northwest wards.

The deep well data indicated depositional lows at Lakhimpur, Nazira, Dimapur, Safrai and near Schuppen belt. The maximum thickness of sediments are of the order of 5000-6000 meters. The sediment fill in the southern parts of Dauki fault can be estimated to the tune of 10,000 m. Upper Assam Shelf extends into the sub-thrust block of Naga Schuppen Belt in southeast and in to the subthrust block below the upthrusted Neogenes in the eastern Himalaya Foothills in the north.

Sadiya and Pasighat Lows are depressions found in the extreme northeast of the Upper Assam valley. The Brahmaputra Ridge, which is a series of domal uplifts takes a down plunge beyond a NW-SE trending hingeline connecting the easternmost uplifts of Laimakuri, Dhauldhoa and Dum Duma. The northern low is named Pasighat Low and the southern one is named Sadiya Low. Thick Paleogene sediments are expected in these lows.

Lower Assam

The post collision convergence lead to uplift and advancement of footwall deformation because of continued tectonic loading of the foreland results in down flexing. The result is preorogenic sediments are overlain by post- orogenic molasses and the uplift and recycling of molasses also occurs as orogen keep advancing. The litho-columns of wells drilled along the Naga Hills located towards NE-SW is shown for a better understanding (Figure 10.aa.4).

Naga Schuppen Belt

It is a narrow linear belt of imbricate thrust slices which follow the eastern and southeastern boundary of the Assam valley along the flank of the Indo-Myanmar range. This belt consists of overthrusts along which Paleogenes of Indo-Myanmar mobile belt have moved northwest wards relative to the buried basement. The Schuppen belt is limited in the southeast by Disang-Haflong thrust (Ganju J.L. Khar B.M and Chaturvedi J.G. 986).

The Schuppen Belt is a typical Neogene molasse basin, related to intense tectonism and uplift along the main axis of the N-S trending Indo-Myanmar area at the close of Oligocene resulting in the development of a narrow, linear foredeep to the west. The Paleogene sediments in the Schuppen belt are enormously thick and compared to the Paleogenes in the Assam platform. Their current proximity both laterally and vertically, indicates substantial horizontal movement or crustal shortening. This is due to the thrust faults and also some strike-slip faults developed in the basin (Kent, W.N., Hickman, R.G., and Dasgupta, U., 2002).

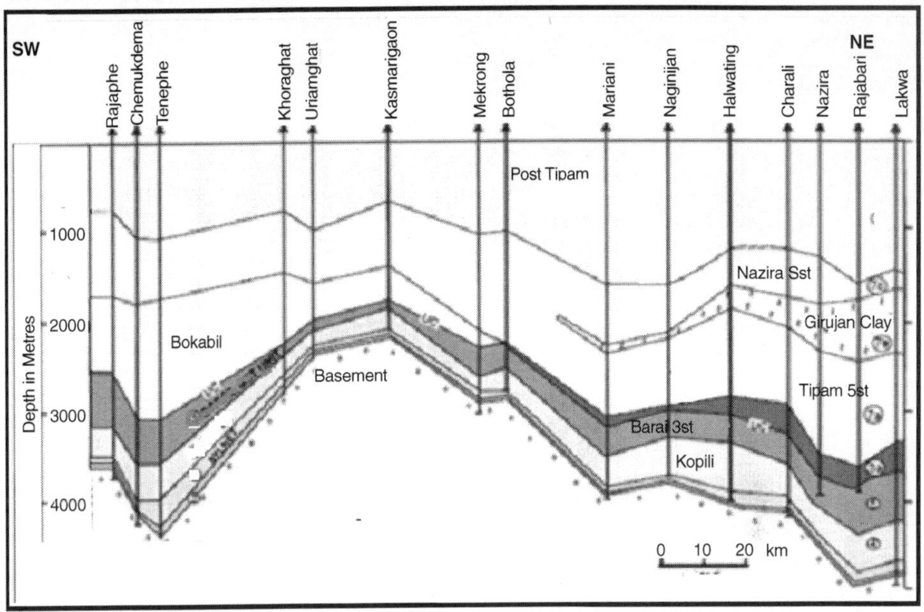

Figure 10.aa.4. Well Section Along Foot-Hills of Naga Thrust (After Saha, D., 2012).

Assam Arakan Fold Belt

The south of Shillong plateau or Dhauki-Haflong fault the deep Neogene basin is termed as Surma basin covering southern parts of Assam. The area of Tripura – Cachar –Mizoram – Kohima is in the Surma basin. Forearc basins are essentially the northward continuation of the Andaman forearc and have since been deformed by continued convergence and have got joined to the orogen.

The E-W trending Dauki fault marks the southern boundary of these platform sediments in the Meghalaya sector, where Sylhet trough of Bengal Deep Basin occurs just south of the fault. Further east, the platform sediments of the Mikir hills are separated from the folded Tertiary rocks of the North Cachar Hills and Naga Hills by the Disang thrust and the Naga thrust. Sedimentation in this belt started in a marginal sag morphing into a divergent margin pericratonic system but overtaken by events of plate convergence in the eastern parts.

South Shillong Front includes the domain of stable shelf sedimentation along the southwestern, southern, and eastern margins of the Shillong plateau (Meghalaya) as well as along the southern and eastern margins of Mikir hills, since Cretaceous times. These are now occurring as outcrops which are largely unaffected by any fold movement. The northwestern boundary is defined by the NE-SW trending Sylhet Fault. The Sylhet Depression separates the Tripura folded belt in south from the Shillong plateau in the north.

In this vast terrain the Neogene sediments occur as westerly convex N-S fold belt for a strike length of about 550 km with a maximum width of 200 km. The eastern limit is generally defined by the N-S trending Kaladan fault, which separates the Neogenes from the folded Paleogenes of the Indo-Myanmar mobile belt.

Sedimentation

The oldest known Tertiary sediments represent coarser classic sandstone whose thickness is found to be more towards north eastern part of the area with less thickness both towards southeast and northwast. The NW-SE trending fault indicate the north westerly sloping limit of Mishmi Hills, which was the lineament since Paleocene.

The early Cretaceous sedimentation which is mainly of continental nature is restricted to NW-SE trending rift grabens. Since late Cretaceous a general NE-SW trending shore has developed extending from south of Shillong plateau to east of Lakwa with drainage pattern from northeast to southwest.

The first stage of development of this platform starts with the deposition of Basal sandstone and Sylhet limestone (**Palaeocene-Eocene**). The sediments of this period suggest shallow marine to lagoonal environment and during periods of least terrigenous supply, carbonate precipitation took place. During Paleocene, there was a marine transgression on the southern edge of the Shillong Plateau, depositing sediments of the Therria Formation consisting of limestone, sandstone and shale. The Lakadang Formation (Early Eocene) comprising limestone and coal bearing sandstones was deposited in shallow marine to lagoonal conditions, while the overlying Tura Sandstone Formation (Early Eocene) was deposited under fluvio-deltaic environment. The Tura Formation is extensively developed in the Upper Assam Shelf.

In the grabens, a sequence of sandstones, shales and subordinate limestone towards top, assigned to the Khasi Group, was deposited in the southern slope of the Shillong Plateau, and a sequence of sandstone and shale, assigned to the Dergaon Group, was deposited in the Dhansiri valley. Presence of pelagic fauna indicates that, these sediments were deposited in shallow shelf to open marine conditions during Maestrichtian to Early Paleocene time. During this time, the basinal area to the east and southeast witnessed deposition of Lower Disang shales.

Intermittent periods of slight emergence in the depositional area during this time are evident from the deposition of coals. During this period, the Assam Shelf was being evolved in a passive margin tectonic setting and under shallow marine to brackish water sedimentation conditions.

Following the deposition of the Tura Sandstone, there was a wide spread marine transgression in which the Sylhet Limestone (**Middle Eocene**) was deposited almost all over the Upper Assam Shelf (Biswas.B,1962). Towards the close of Middle Eocene, limestone deposition ceased because of an increase in the influx of finer clastics in the shelf. These clastics, making the lower part of the Kopili Formation, were deposited in open marine conditions during Late Eocene. Further increase in the clastic influx in the stable shelf during Late Eocene to Early Oligocene resulted in marine regression with the deposition of the upper part of the Kopili Formation, consisting of shales, siltstone and subordinate sandstones, in shallow marine to pro-delta environments.

In the North Bank of the Brahmaputra River, however, environmental conditions were deltaic with the deposition of sandstones with minor shales and siltstones. On the basis of litho facies the paleo-shore line is positioned in the trend Naginijan-Dergaon to Disangmukh and shelf edge is marked towards east of Geleki to Khoraghat trend.

East of the hinge zone, *i.e.,* in the basinal area, Upper Disang shales, which are lateral facies equivalent of the Sylhet and Kopili formations, were deposited in deep water basinal conditions. The Dhekiajuli Formation, consisting of mostly soft sandstones, was deposited at the mountain fronts in the Upper Assam Shelf and in areas now overridden by younger Naga thrust.

These are shelf edge carbonates and eastern limit of which represent the shelf edge. The sediments are also more argillaceous in the east of this trend. North-Western of this trend the sediments are of coarser clastics. These carbonates are inter-bedded with sandstone beds which represent due to coastal transgressive and regressive phase, to sea level fluctuation are resulting in clastics.

During shallowing of the sea in the basinal area, the succeeding sediments of the Barail Group were deposited under environments ranging from moderately deep marine to deltaic. The thickness of Barail sequence indicate prominent depocentre in Brahmaputra valley with depositional slope towards northeast showing maximum thickness more than 900m.

The ***Miocene*** (Tipam) sediments were deposited and thickness of Tipams indicate three different depositional lobes. A prominent Surma basin has taken its shape towards Tenephy, Rajaphe area of Dhansiri Valley. However, the southeastern parts of upper Assam that is Brahmaputra valley, the thinning down of sediments suggest the initiation of compressive forces due to westernly movements of Burmese plate. Discrete lows have been generated along the western limit of abduction zone. These are now known as Borholle, Naginajan, Nazira and Sonari lows the major depocentre for Noegene sedimentation (Figure 10.aa.5). Tipam Sandstone in many of the Dhansiri Valley and Upper Assam wells suggest that a certain fraction of the sediments making the lower part of the Tipam Formation came from the rising Barail Range towards east (Desikachar, S V. 1984). Towards the end of the Tipam Sandstone deposition, there developed a series of N-S to NE-SW trending compressive structures in the basinal area. During the growth of these structures, the Girujan Clay Formation was deposited in the lows. The most prominent structural depression was formed in Kumchai – Manabhum area in front of the Mishmi uplift, where the Girujan Clay Formation attains a thickness of about 2300m. On the northwestern side of Brahmaputra arch another low has became prominent probably due to ongoing subduction of Indian Plate along Himalayan foot hill zone.

Thickness of Girujan Formation show uniform south-easterly slope towards southeast of Gelaki-Lakshmijan area. The important tectonic activity during this period could be reactivation of Mashimi Hills area towards northeast, due to which

Sonari area which has a prominent low during Tipam deposition has became a higher area.

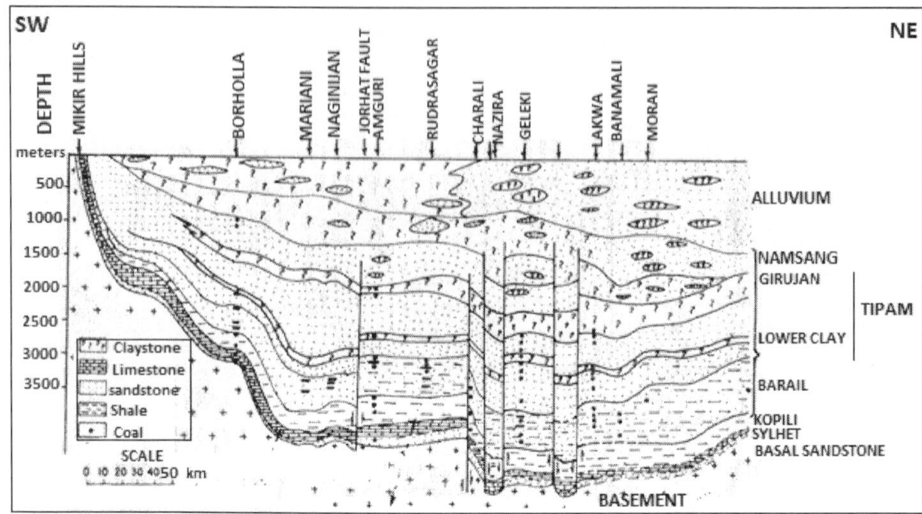

Figure 10.aa.5. Geological Cross Section Across Drilled Wells in Upper Assam (After Desikachar, S V. 1984).

The post-Girujan thickness of sediments have indicated a prominent northeasterly compression restricting the effect of Brahmaputra arch during deposition limiting to northeast and Jorhat Fault. Major depocentres were towards northwest of Brahmaputra area and Lakwa-Sonari-Banamali low towards southeast of the arch.

The Surma basin is continued to be a persistent low and its influence which was seen in southern parts of Dhansiri valley with its southeasternly depositional slope. This could possibly due to lateral movement of Dauki fault located further southwards which shifted the axis of Surma basin towards further south.

Stratigraphy

Sedimentary sequences ranging in age from Late Mesozoic to Cenozoic are exposed in the Assam-Arakan Basin. The sequences can be divided into shelf facies and basinal (geosynclinal) facies. The shelf facies occur in Garo hills, Khasi-Jaintia hills, parts of North Cachar hills and Mikir hills, and below the alluvial cover in Upper Assam, Bengal and Bangladesh. The basinal facies occur in the Patkai range, Naga Hills, parts of North Cachar hills, Manipur, Surma valley, Tripura.

The oldest known sediments in the area are of pre-Cretaceous Lower Gondwanic suit penetrated in a well drilled near Mikir Hills called Barpeta, where 340 meters thick sediments are overlain by dolerite. The Cretaceous rocks are exposed in the Garo-Khasi and Jaintia Hills known as Mahadeo and Longpar formations (Bhandari.L.L, *et al.*, 1973). Pre-Tertiary rocks outcrop in Eastern Himalayas, Mishmi Hills, Shillong Plateau and Mikir Hills and form the floor of the basin.

The Paleogene sequence was deposited on the flanks of the passive continental margin as a fairly simple transgressive (Basal/Tura Sandstone, Sylhet Limestones, Kopili Shale)/regressive (Barails) wedge.

In the subsurface in Dergaon well drilled on the southern banks of Brahmaputra River, about 500 m of thick quartzitic sandstone and argillaceous sequence with Mastrichiean fauna has been encountered (Mohan, M. and Pandey,J.1973). Disang Series and the Naogaon Sandstones occur south-east of the Disang Thrust, the Barail coals and the oil-bearing Tipam sands occur on the north-western side of the Disang Thrust.

In Upper Assam plains, the Tertiaries rest directly over the Pre-Cambrian basement. The Paleogene sequence is divisible into lower Jaintia group and Upper Barail group. The Jaintias include Theria, Tura and Basal sandstones, Sylhet limestone and Kopili sand-shale alternations (Singh,N.P., Boruah,R.M. and Dave,A. 1986).The Barails are composed of sandstone with shale and coal beds.

The overlying Neogene sediments are known to be deposited in fluvial conditions and called Tipams in Upper Assam and called Surmas in southern part which are mainly argillecous and deep basinal equivalents of Tipams and within the Tipam group the thick Girujan clay is a basinal lithomarker. The overlying Moran group comprise mostly arenaceous. Most of the sedimentary sequences are continental to fluvial in nature and hence lack define age determining fauna. Except in the case Middle Eocene Sylhet limestone sequence. Some places Kopili sequence also indicated biostratgraphic age upto Early Oligeocene.

Two Groups are recognized in the Cretaceous, *viz.* the Khasi and Dergaon Groups. The Dergaon Group is present in the Dhansiri Valley and the Mikir Hills, while the Khasi Group, which extends into the Paleocene, is present in the South Shillong Shelf. The Dergaon Group is divided into the Bamangaon, Mikir Trap and Moabund Formations. The Khasi Group is divided into the Mahadek and Langpar Formations.

The Disang Group recognized in the Schuppen and Fold Belts, is divisible into two units and extends from the Late Cretaceous to Late Eocene. The Jaintia Group, present in the shelf areas, is divisible into four formations and is Late Paleocene to Late Eocene in age. It is divided into the Therria, Tura, Sylhet and Kopili Formations. The Therria Formation is present in the Khasi and Jaintia Hills and divided into two members. The Sylhet Formation is divisible into five members in the Khasi and Jaintia Hills only.

The Kopili Formation is divisible into two members, *viz.* Charali and Amguri, in the subsurface of Upper Assam, Dhansiri Valley and in the Mikir Hills.These members are recently recognized.

The Barail Group, present in all the tectonic units of the basin is divided into nine formations, in different tectonic blocks, based upon lithological consideration. In the Fold Belt and the south western part of the Schuppen Belt it is divided into the Laisong, Jenam and Renji Formations. In the north eastern part of the Schuppen Belt, it is divided, into the Naogaon, Baragolai and Tikak Parbat Formations. In the subsurface of Upper Assam and Dhansiri Valley, it is divided into the Disangmukh,

Demulgaon and Rudrasagar Formations. The Disangmukh Formation represents the onset of coarse clastic sedimentation in a deltaic regime in the post-Jaintia period (Murty, K.N,Tongaonkar, C.S and Kaul, S.K. 1979).

The Surmas, are divided into the Bhuban and Bokabil Formations. The Bhuban is further subdivided into three units. The Khoraghat Sandstone, a Member recognized in the Bokabil Formation, in the Dhansiri Valley and Mikir Hills (Mohan, M. and Pandey, J. 1973).

The Tipam Group is subdivided into Jaipur and Girujan Formations in the Schuppen Belt, while in the Folded Belt, the Jaipur and Gobindpur Formation are recognized. In the subsurface of Upper Assam, the Tipam is divided into the Geleki Sandstone, Lakwa Sandstone, Girujan Clay and Nazira Sandstone Formations. The Tipam Group remains undifferentiated in the Dhansiri Valley and South Shillong Plateau (Deshpande, S.V., *et al.*, 1993).

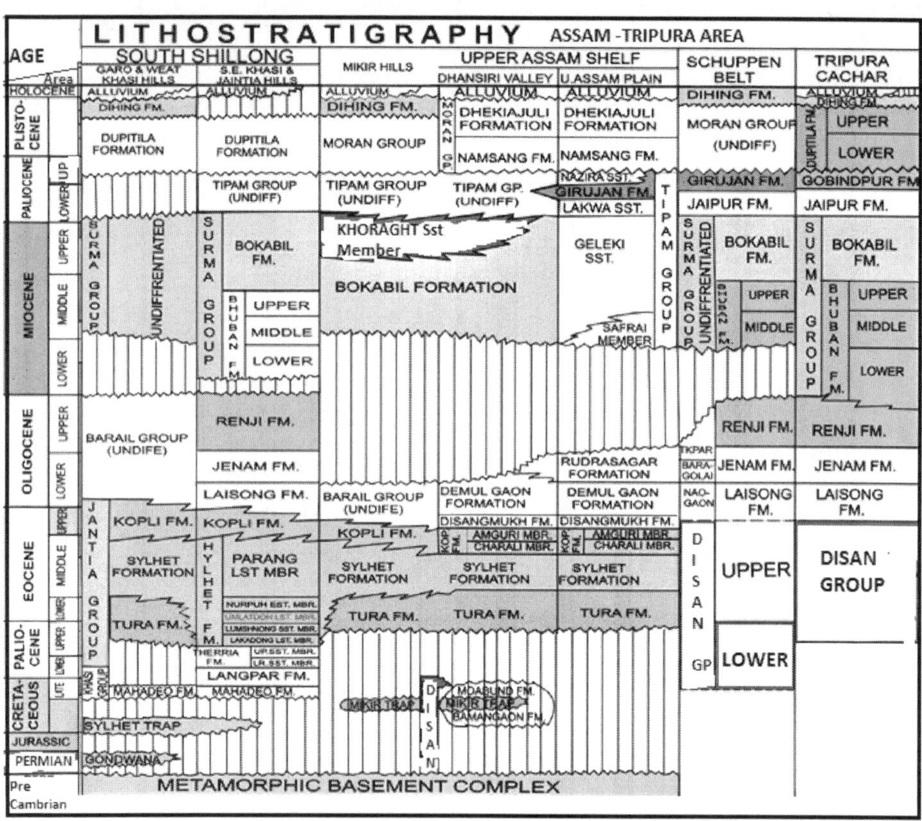

Figure 10.aa.6. Generalized Stratigraphy of Assam-Tripura (Modified after Roymullik SK, *et al.*, 2009).

The Moran Group overlies the Tipam Group and is divided into the Namsang and Dhekiajiuli Formation, in the subsurface of Upper Assam and Dhansiri Valley (Figure 10.aa.6). The Dupitila Formation is the coeval equivalent of the Moran Group in the Fold Belt. The Dihing Formation overlies the Moran Group or the Dupitala Formation in the Schuppen and Fold belts. The Recent Alluvium constitutes the youngest stratigraphic unit in the basin.

Petroleum Geology

Assam-Arakan basin was widely explored specially in the platform part in Upper Assam. This is a proven petroleum province and has generated sizable hydrocarbons from known petroleum generating depressions (Harding, T.P. and J.D. Lowell, 1979).

The initial discovery was from the Barail rocks of Oligocene age. However, Tipam reservoirs (Miocene) became important subsequently, followed by Kopili sandstones (Eocene), Basal sandstone (Paleocene) and the basement.

Most of the oil and gas fields, discovered are in the Upper Assam shelf and situated mostly on the south-eastern slope of the Brahmaputra Arch and also in a belt bordering the Naga thrust (Murty, K.N.1983). In the Dhansiri valley also, oil fields the Digboi and Kharsang lie in the same belt. In the Naga Schuppen belt, oil accumulations occur in that zone of the shelf which is overridden by the Naga thrust (Figure 10.aa.7).

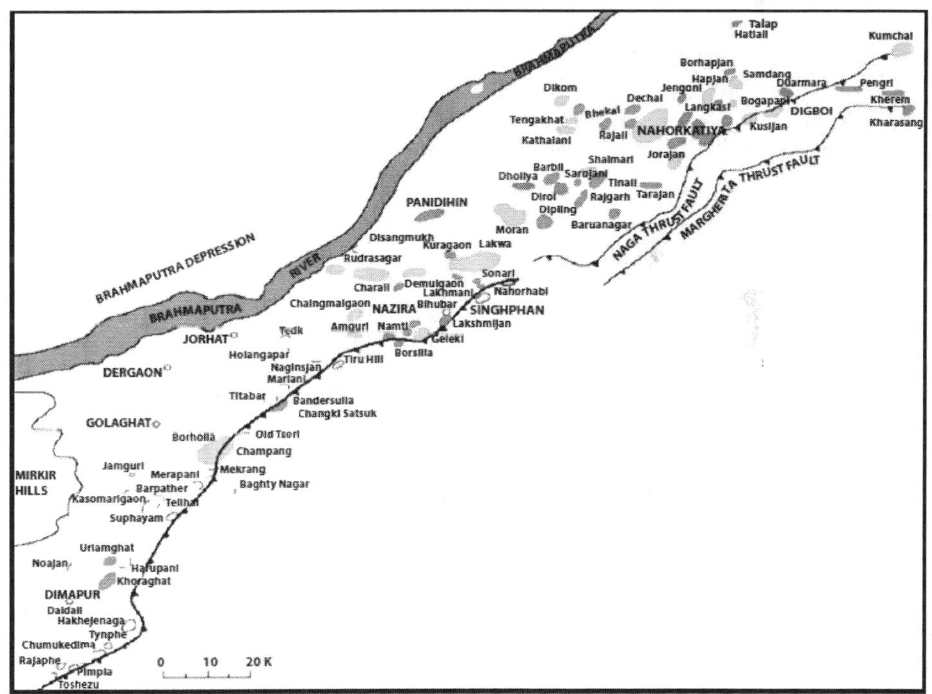

Figure 10.aa.7. Oil and Gas Fields of Upper Assam (After Bhowmik, P K).

The compressive tectonics prevalent on the area had produced good high amplitude structural anomalies. The reservoir facies are available in every potential sequence. Girujan clay and Barail coal shale sequences are major regional caps available for Paleogene and Neogene Petroleum systems prevailing in the basin (Idem, 1983).

Source Rock

The source potential is in Sylhet, Kopili and Barail sequences. The organic rich concentration in Sylhet lie in around Teok well with 0.5 per cent TOC. The rich source facies are seen in Kopili upto 2 per cent TOC in Khoraghat Uriamghat areas and in northeast of Lakwa. The organic carbon concentration in Barail sequence is maximum upto 5 per cent in Gelaki-Charaideo area.

The potential source rocks have been identified as belonging to Eocene with the Kopili Shales of Upper Eocene, as well as the coal-shale alternation forming the top part of the Oligocene Barail Group. In the overall transgressive-regressive sedimentation scenario during Eocene-Oligocene periods, Kopili and Barail formations have good souce potential, whereas Sylhet and Basal sandstone formations have poor rating.

Maturation

The area south of Borholla has became prominent source basin after Barail deposition which might have facilitated maturation of Pre-Barail source sequence. In the case of Upper Assam the lows developed all along the compressive high (Naga Thrust) during Tipam period are potential Kitchens for Barail source Facies (Naidu, B.D., and Panda, B.K., 1997). The maturation of source facies is possibly only towards southeast of Khoraghat-Borholla and Gelekia-Lakwa trend.

Thermal maturity values range from Ro 0.5 to 0.7 per cent in both the Sylhet and Kopili Formations. Maturity is expected to increase with depth to the southeast toward the Naga Thrust fault and in the sub thrust areas.

The oil window and "cooking pots" for the main source rock is to the south east in the down deep portion of the Assam Shelf, *i.e.,* deeper than 3800-4000m and close to the outcropping edge of the Naga Schuppen Belt. Thus the hydrocarbon generating centres are Nazira-Sonari-Safrai lows in the east, Mekrang low in south-west and Dimapur low in south southwest all along Naga thrust.

Migration

Oil generated beneath the thrust sheet has probably migrated to the leading edge of the thrust sheet. Formation salinity map of the area indicates such movement of fluids.Vertical migration through reactivated basement-rooted faults may predominate in near the edge of the Naga thrust fault (Hunt, J. M., 1996).

Bedding and unconformity planes in sediments appear to have provided permeable paths for migration into reservoir space in fractured parts and weathered portions of basement rocks or oil has entered the granite from the sediments that were deposited along the flanks of the buried hill. The updip intersection of the planes of the two major unconformities (basement-sediment and pre-Barail-Barail unconformities) will be of considerable interest in the western part of Dhansiri Valley and Mikir Hills as hydrocarbons which migrated to any of the pre-Barail or pre-

Bokabil reservoirs will be effectively sealed by the overlapping Bokabil sequence. There is also every possibility of migration of hydrocarbons from the younger to older sediments at this unconformity surface.

Regional caps are well developed in the eastern and northeastern part (Girujan clays, Barail coal shales and Kopili shales and in extreme south western part (Bokabil shales). The central part (South of Borholla) there is no regional cap/seal. Girujan Clay also thins out towards west and northern part of North Assam Shelf and does not extend beyond Brahmaputra River in the north and Dikhumukh – Teok in the west and southwest. It is quite thin over Borholla and does not extend south west beyond Borholla. The Kopili shales have played a major role as seals/caps and in preservation of oil deposits occurring below it all over the basin.

Entrapment

The Kopili Shales and the Barail Coal-Shales attained their maturity during Middle and Upper Miocene times. This is the time when the post- Barail uplifts and hiatus have already taken place and the NW-SE/NE-SW trending conjugate strike–slip fault system have been in position due to plate collision. The entrapment would happen in the Eocene/Oligocene reservoirs. The Coal-Shales in Barails can act as an effective cap rock because of its generally widespread nature (Murty K.N. 1970).

The Upper Assam shelf areas had uplifts and differential subsidence events after Oligocene, and synchronizing with migration and entrapment of most of the oil and gas which are sourced in Eocene/Oligocene.

The fractures have affected the Eocene calcareous reservoirs and even Archean basement and through fracturing and differential uplifts, creation of secondary fractures and solution (Basement, Tura, Sylhet) porosities, resulting in accumulations in older reservoirs sourced from younger shales (Kopili). Generally heavy nature of these oils and accompanied by the composition of formation water indicates a complex history of probably more than one phase of migration.

The second stage of migration of oil in this basin seems to have occurred towards the end of Miocene. The tilt of the platform towards northwest disturbed the equilibrium of the fluids as also the pressure and temperature of every reservoir involved. The buoyant oil and gas moved during this time and got trapped in the available structures.

After the post-Girujan uplifts of Late Miocene time, there was a quantum change in the facies distribution in Upper Assam. The Girujan Clays is the regional cap rock in the central and western part of the basin.

Basement highs, fault closures and strati-structural traps are the exploration targets in the shelf part. In the Naga Schuppen belt, structural closures, both in the sub-thrust and supra-thrust blocks are also attractive exploration targets. Large exposed anticlines in Tripura-Manipur fold belt and structural highs in the intervening valleys hold high potential for hydrocarbons in Oligocene and Miocene sediments.

The Late Miocene Uplift is associated with up-thrusting process in the Schuppen Belt and at least some of the source beds would have been raised and inversion structures created along the Naga Thrust belt. This would see reversal of the migration direction back to the southeast. It has been seen that in the Brahmaputra Ridge area, the basement blocks have been going up and down differently. The re-migrated oils are likely to be heavier due to loss of lighter compounds because of exposure if any.

Play Types

Structural Prospects

The Paleogene and Miocene extension which affects both the Lower Paleogene and the Upper Paleogene-Miocene hydrocarbon systems. Some of these paleo-structurs have undergone a mild inversion during Neogene. Strongly tilted Paleogene to Miocene Extensional structures. Late Neogene basement-involved compressional uplifts associated with reverse faulting. Regional Dome with faults and obvious Paleogene extensional structures.

Most of the discoveries are primarily stratigraphic traps with some structural element. These wells are mostly located near the basin edge. Some non-commercial indications of hydrocarbons are obtained from Teok-Amguri area and outcrops in Mikir Hills at Samkher.

Stratigraphic Traps

Carbonates associated with the Sylhet carbonates include shelf margin plays, or carbonate turbidite play in a more basinal facies. Clastic sandstone bodies associated with the deltaic and shallow marine environments of Kopili Formation, Barail Group and Surma Group. It will be necessary to differentiate and map channel sands from flood plain mudstones, meander-belts, strandplain barrier-bar system associated with deltaic systems. The stratigraphic phenomenon occurring within the Coal Shale Unit is a regional feature. The discontinuous sands which have developed on the localized folding may be promissing.

Petroleum Systems

The operating petroleum systems in Upper Assam area based on GME model are mainly three as per stratigraphic envelop of sedimentary sequences.

Sedimentologically, Nahorkatiya Block had been an area of continuous low energy deposition throughout Paleogene. Shales and fine sands predominated even during Middle Eocene when limestone was the dominant lithology in southwest. Paleo-environment wise, the organic rich Kopili and older strata as well as the coal-rich Margherita Syncline indicate a general anoxic environment and a good source rock development in the Nazira-Sonari-Safrai Depression and its likely extension into the Naga Schuppen Belt. In such an overall setup, the likely reason for Nahorkatiya block having larger reserves are availability of comparatively better reservoirs and cap/seal rocks within Eocene and Barails.

Paleogene System

The Kopili Formation is both a significant source bed and a regional seal. It is source for both lower Paleogene and an upper Paleogene hydrocarbon system. On the other hand the Upper Paleogene-Miocene system is fed by the same source beds in the Kopili and the Rudrasagar Formation with its local seals. The Bokabil of the Dhansiri Valley and the Girujan Clay of the Assam Shelf are the most important regional seals.

The stratigraphy of the lower Paleogene hydrocarbon system includes from bottom to top the late Paleogene Therria limestones and sandstones of the Shillong uplifts and the granite wash and overlyin clastic of the Paleogene-Eocene fluvio-deltaic Tura formation.

For working petroleum system it will be required that the hydrocarbons migrate from the overlying Kopili source into the underlying reservoirs. It is likely that hydrocarbon migration within the system occurred well before the late Neogene deformation in the adjacent fold belts and before the uplift of the Shillong plateau. Thus primary traps would be either stratigraphic or paleo-structural.

So far discovered reservoirs of the system produce only in the structurally controlled Disangmukh, Panidihing, Lakwa–Lakhmani, Borholla, Champang, Mekrang and Tengakhat–Dikom fields in Assam Shelf.

Although drilled many exploratory wells for deeper prospects no significant reserves have been established yet except for minor accumulations in Lakwa and Panidihing areas of Upper Assam and Borholla of Dhansiri Valley to the further south. In Lakwa area, one well producing oil from the Upper Tura Formation, whereas in Panidihing#2 well, heavy oil has been reported at 4250-4252m depth from Upper Tura Formation.

Miocene System

In Dhansiri valley system, with the Bokabil as major regional seal (Khoraghat field) and the Kopili as major source bed. The Upper Assam platform system, with the Girujan Clay Formation as main seal and the Rudrasagar and Kopili as main source beds. Only minor shows and the lack of significant accumulations testify the absence of regional seal.

The Miocene–Lower Pliocene interval is sandwiched between two major unconformities. Regionally this interval thickens toward the southeast. The thickening and the basal onlap suggest a foredeep associated clastic wedge. It has no internal source bed.

Oil Fields–Oil India Limited Area

As it was known, the first oil find in India was in this area in 1867 through seep. The continuous efforts since then, resulted in locating many oil and gas fields in the area. The exploration and production efforts since then resulted in drilling wells by different companies.

The oil reserves are about 806 Mmt with 247 MMt recoverable and 86 MMt of balance of reserves. Similarly the GIIP is 344 MMT OEG, 190 recoverable and 109 balance of reserves.

Digboi Field

Located in the north-east of Tipam hills in Dibrugarh district of Upper Assam, Digboi is the oldest oil field of India. The oil bearing strata cover an area of about 13 sq km where oil is available at 400 to 2,000 metre depth (Figure 10.aa.8). Many oil wells have been drilled.

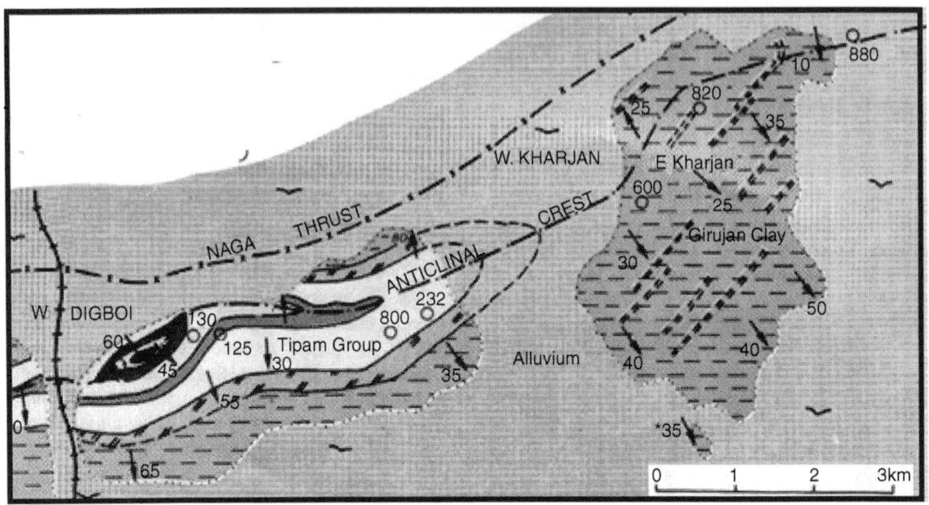

Figure 10.aa.8. Geological Map–Digboi Field-Section.

Naharkatiya Field

The Nahorkatiya Block is a large sized basement block, situated in the eastern part of North Assam Shelf. It is located at a distance of 32 km southwest of Digboi at the left bank of Burhi Dibing River where oil was discovered in 1953. Oil is available at depths varying from 4,000 to 5,000 metres. Out of the 60 successful wells drilled, 56 are producing oil while the remaining 4 are producing natural gas. Tectonically this Block was uplifted probably higher than the western and south western blocks (Azad, *et al.*, 1971). (Fig,10.aa.9).

Moran-Hugrijan Field

It is located about 40 km south-west of Naharkatiya. Oil at Moran-Hugrijan field was discovered in 1953 and production started in 1956. Drilling has proved an oil bearing Barail horizon at a depth of 3,355 metre (Bhattacharya and Chaliha,1973). In one of the recent Moran well the interval 3187-3203m, produced gas @40,000 SCMD through 5mm bean, from Barail Sand.

Kharsang Oil Field

The **Kharsang** field is located in the state of Arunachal Pradesh, which is approximately 60 km east of the Digboi Field. Occurrence of oil seepages in the vicinity of Kharsang area were reported in 1940. It was discovered by Oil India Ltd.

and the first exploratory well was drilled in 1968. A total of 36 wells were drilled out of which 27 are oil wells, 5 as gas wells.

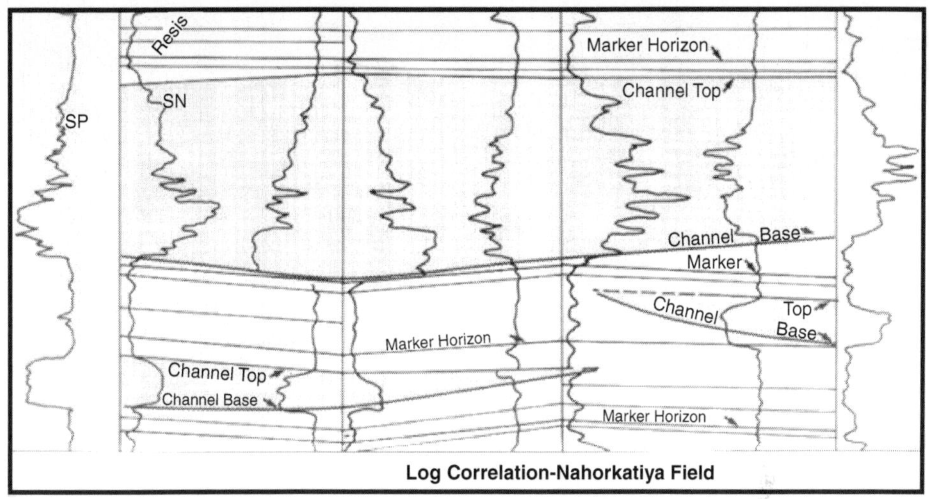

Figure 10.aa.9. Log Correlation–Naharkatiya Field.

It is producing from the multi-layered fluvial sandstone reservoirs of Upper Miocene age. Owing to the presence of Margherita thrust, the Barail Formation of Oligocene age over-lies the Namsang/Girujan formations of Mio-Pliocene age. This change in the dip directions indicates the presence of an angular unconformity between Namsang and Girujan beds. The total Recoverable Reserves are 2.83 MMT, from an estimated 18.83 MMT oil in place in the upper Girujan reservoirs. Gas in-place was similarly estimated as 1167.56 MMm^3.

The reservoirs, occurring at depths ranging from 700 to 1500m, are stacked medium to fine grained sandstones that form 2 to 15 m thick, mostly upward fining successions (Handique, G.K. and Bharali,B. 1984). The shale bands within Girujan reservoir, although only a few meter in thickness act as pressure seals in many cases.

While drilling, indications of high pressures in Kharsang area were obtained mainly in the lower Girujan argillaceous section. The abnormal pressure zone necessitating use of heavy mud (Sp. gr. 2.17) for drilling at a depth of about 2050 m in the eastern part of the structure while the corresponding range was met at a much shallow depth of about 550 m in the updip south-western part of the structure.

Trends show that, wells completed in laterally continuous sandstone reservoirs have low production-decline rates compared to wells completed in laterally discontinuous sandstone reservoirs. Sand has been recognized as a potential problem of the field in number of wells particularly in reservoirs occurring at shallower depths (Debabrata Panda and S. Mahapatra, 2003). Moreover, because of the low solution GOR and with fairly low mobility due to viscous nature of the oil (paraffin content are as high as 15 per cent), artificial lifting was implemented in many wells from early stages of production. The oil is a viscous medium gravity crude (19- 34 deg API) in a majority of the wells. The gas-oil ratio is typically low (below 75 cu m/kl or 425 scf/bbl), but high in the crestal culminations of certain sand units.

Because of comprehensive reservoir management methods, the production level of the field have increased from about 40 m^3/day to over 160 m^3/day. The reservoir pressure of the pools are typically 10-15 per cent higher than hydrostatic. In pools which are characterized by high connectivity, the reservoir pressure shows very little reduction with production.

Distribution of sub-surface temperature (values of geothermal gradients) is erratic in the Kharsang area down to a depth of about 1830 m, below which the geothermal gradient is in line with that (2.01°C/100 m) encountered in the fore-land part of basin. The reservoir temperature is generally low ranging from 38 - 45°C.

Presently the field is being operated by Geo-Enpro Petroleum Ltd. since 1996 with Geo-petrol and Oil India Ltd as partners. (Figure 10.aa.10).

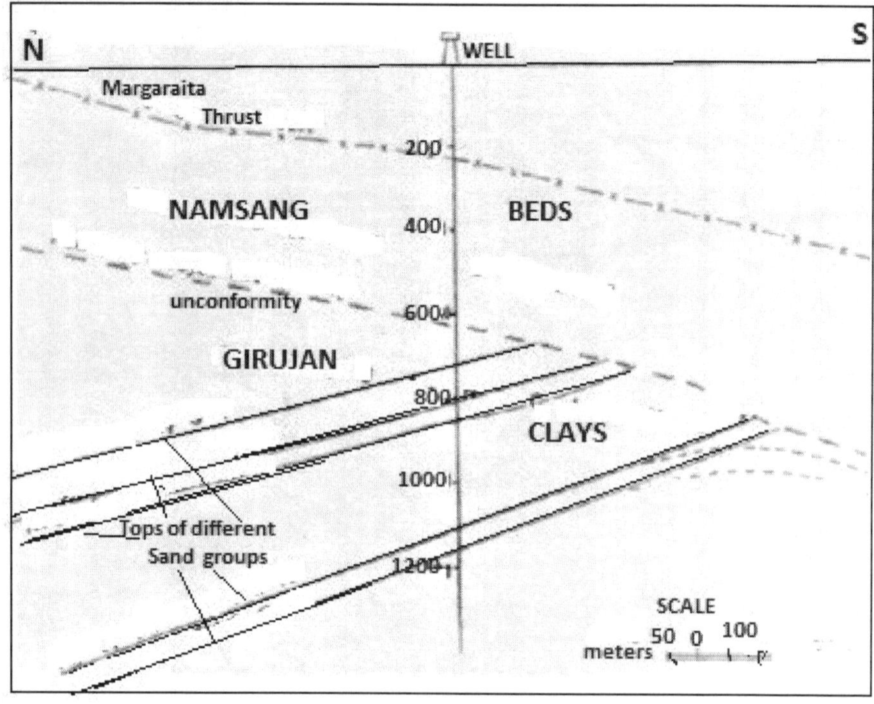

Figure 10.aa.10.Kharsong-N-S Geological Section (After Handique and B Bhrali, 1984).

Samdang

The Samdang well was drilled down to 3620 m to probe the hydrocarbon prospects within the Barail and Tipam Formations and produced oil from sands within Barail Formation.

Baghjan

The areal extent of the structure is about 8 sq. km. In a recent well drilled, discovery of gas was in Narpuh Formation of Eocene in the interval 3809-3821m which produced gas @90,000 SCUMD through 5mm bean. SITP: 4400 psi. The well South Baghjan-1 was drilled down to 4154 m. The well encountered few prospective sands within Palaeocene-Eocene Formations produced oil@ 92 klpd.

Sapkaint

In a well Sapkaint (#2) located in the central part of the Structure with an areal extent of about 3 sq. km., situated near Dumduma, oil was discovered new/un-appraised Tipam sand. The well is producing @ 55 klpd oil in the interval 3237-3241 and 3235-3236 metres through 4mm bean.

Dirok

Dirok structure is bounded between the Kumchai and Margherita thrusts. Discovery was in the Girujan formation of Miocene age. Multiple hydrocarbon sands have been encountered in the wells.

ONGC Area

After its formation ONGC, took up exploration activity in its licensed area located to the South of OIL's licensed area in 1960. Its first efforts were resulted in finding oil and gas in Rudrasagar, Lakwa and Geleki. Since then, the un-tiresome efforts established nearly 60 oil and gas fields in its areas of operations. Of which, about 35 are located in Assam shelf area and 25 are in Assam Arakan fold Belt area. These E and P efforts resulted in drilling of about 4,500 wells by the company. Some of the details of few fields located in shelf are discussed below.

Lakwa–Lakhamni Field

The **Lakwa** field was discovered in 1964 and put on production in 1968. There is another culmination adjacent to this field called **Lakhmani field**. Oil was established within multiple reservoirs in the bottom even in Tura formation, Barail sands (2 sands), Tipams (6 sands) and in the top Sandstone with in Girujan Clay Formation, with bulk production coming from the TS-2 Sand of the Tipam Sandstone. So far about 530 wells were drilled in the field which include both exploratory and production wells in the multi-layered reservoirs. The field produced cumulatively about 25 MMt, out of in-place reserves of the order of 180 MMt(!) from the field.

The TS-2 pool is a major pool and comprises 90 m-thick oil column and compartmentalized by many listric faults with throws of up to 20 m. The sand is up to 120 m thick and comprises four laterally extensive braided-fluvial sandstone sub-units which are separated by thin, shale layers. The oil has been produced under combined aquifer and solution-gas drive. The porosity of the TS-2 Sand ranges from 22 to 26 per cent. Water-cuts in individual wells were started as early as in early ninetees and the range 30-90 per cent.

In Barail Main Sands were deposited by distributary channel as mouth bars, point bars, channel sands. Thickness map indicates the sand bodies to be oriented in NW-SE direction. Lakwa Barail Sand-2 is widely distributed in the field. Most of the fault blocks are hydro-dynamically not connected. Most of the reservoir sands in this unit are mouth-bars and point-bars (Arun Garg and Sanjoy Goswami.2003). Due to decline in productivity and fall in reservoir pressure from initial 380ksc to present day pressure of 122-170ksc, water injection was initiated but desired result could not be achieved and pressure continued to fall below bubble point pressure. The prolific producers from this reservoir are generally localized in the eastern part of the area on a mouth-bar. Porosity in these blocks varies from 12 per cent to 18 per cent and oil saturation values from 35 per cent to 56 per cent. The GOR gradually increased with production and has reached to the level of around 1000 v/v (Figure 10.aa.11).

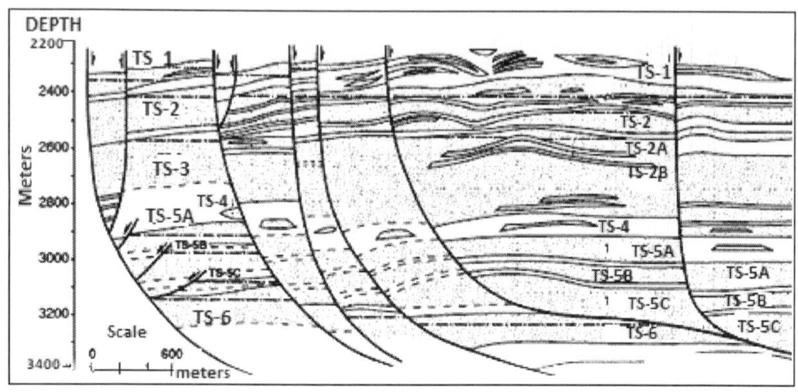

Figure 10.aa.11. Geological Section Lakwa Field.

Amguri

ONGC discovered the field and created facilities but production was stopped in 1994 which was producing @1,600 barrels of oil per day. It was given to Assam Company India Ltd. (ACIL) with the promise to restart production.

Geleki

Geleki with area is about 28 square kilometres was discovered in 1968. The area form the part of North Assam shelf and had been producing hydrocarbons for more than four decades. The main hydrocarbon bearing formations are Tipam Sandstone of Miocene age, Barail of Oligocene age and Kopili of Eocene age containing many sand layer so far about 300 wells were drilled, both exploration and development of multi-layered Reservoirs.

Reservoirs in different stratigraphic levels starting from sand units of Kopili Formation (KS-1 to KS-13) of Eocene age, Barail Group (BMS, BCS sands) of Oligocene age to sands within Tipam Group (TS-6, TS-5 A and B, TS-4B, TS-3, TS-2 and TS-1) of Miocene age have been charged in favourable entrapment conditions (Murty.K.N.*et al.*, 1979).

Among these layers, TS-5B is one of the prolific reservoir in Geleki field. It is about 200m thick comprises predominantly thick sandstone with intervening thin siltstone/claystone/shale layers of about 1 -10 m thick. The sandstone is generally dirty white to grey, moderately hard and compact, sometimes friable, fine to medium grained, sub angular to sub rounded, moderately sorted and occasionally calcareous. Though the TS-2 sand is developed in entire Geleki field, hydrocarbon accumulation is observed only in few blocks (Figures 10.aa.12). TS-2 Sand of Geleki Field is under water injection @ 300m^3/d through five injectors. The initial reservoir pressure was 261 ksc with a saturation pressure of 179 ksc. The reservoir is operating under partial aquifer support. The field has OIIP 127 MMt and Ultimate Reserves of 32 MMt. The field has so far produced about 18 MMt of oil.

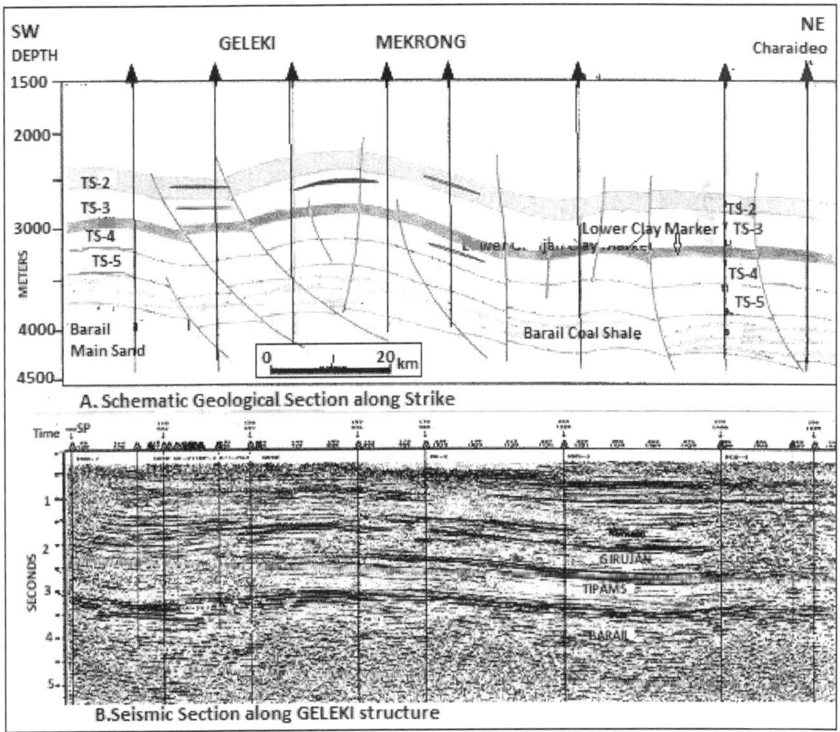

Figure 10.aa.12. Geological-Seismic Section Geleki Field.

Rudrasagar

The field was discovered in 1960 and put on production in 1966. Structurally, it is a broad anticline with its longitudinal axis trending ENE- WSW direction. The field is dissected by fault with considerable displacement on the northern flank of the structure whereas towards southern and western part of structure ENE-WSW and NE-SW trending faults define its limit. Hydrocarbon accumulation is confined mostly to the topmost part of Barail main sand (BMS), which is the main pay in the field (Figures 10.aa.13). The main BMS pool has been dissected by faults into blocks (Murty, K.N. 1968).

The Barail Main sand consists of thick sand unit with thin shale intercalations. The thickness of BMS unit in Rudrasagar structure varies between 150-180m and it is quite extensive except towards northwest where the facies gets deteriorated. The sands are medium to fine grained, subangular to subrounded, poorly consolidated and less compacted. The Log motif exhibits typical bell shape signature against the sand indicating it is deposited in a lower deltaic distributary channel system. The hydrocarbon accumulation is confined to the upper part only (Figure 10aa.13).

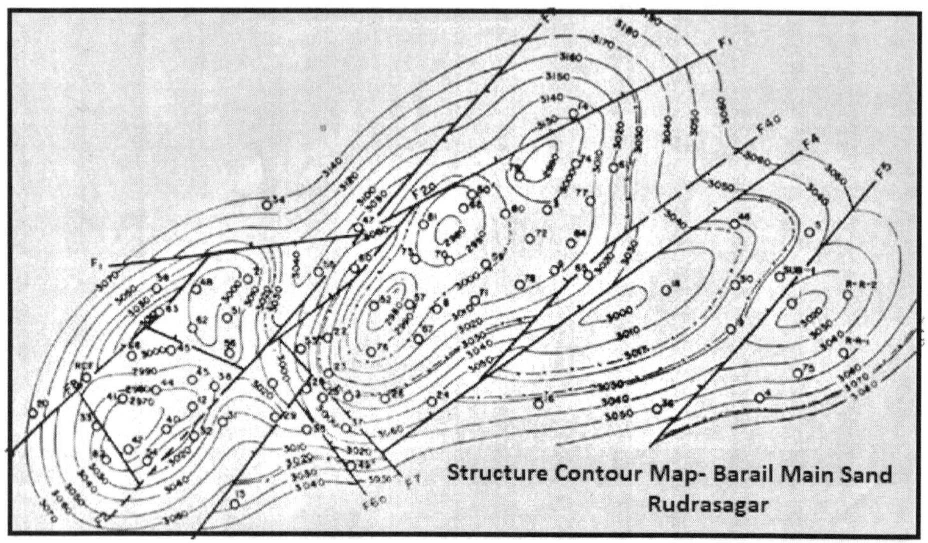

Figure 10.aa.13. Map Barail Main Sand-Rudrasagar.

Tipam sand (TS- IV – Miocene) also have contribution to oil production of the field in the northern part. But these TS-IV sands are discrete because it is a channel sand deposit and thinner as compared to nearby Charali-Changmaigaon area. About 250 wells were drilled in the field for finding new reservoirs as well as developing the known ones.

Borholla-Champang

Borholla and Champang have shown the presence of commercial quantities of hydrocarbons within the fractured basement which is granitic to dioritic in character. The Borholla and Champang fields are parts of the same structure, the Borholla part falling within Assam State in the plains region and the Champang part as an extension of the same structure below the Schuppen belt falling in the Naga hills part of Nagaland beyond Naga thrust (Figure 10.aa.14).

The structure was identified on the basis of seismic data, later drilling has revealed that the structure has been intensely faulted into different blocks. Only certain blocks contain oil and gas within the faulted and few blocks including the highest part have only yielded water. In some blocks, the oil column within the faulted blocks of the basement play type is found to be as much as 90 metres thick. The reservoir has been cored in few wells and fracture indications are also in logs. Those logs show that the fractures are not uniformly present but confined to certain levels. The basal sandstones immediately overlying the basement probably equivalent of the Tura sandstones are also oil bearing where the basement is producing. The wells are being completed in the basement part either by barefoot completion or by the use of slotted casings.

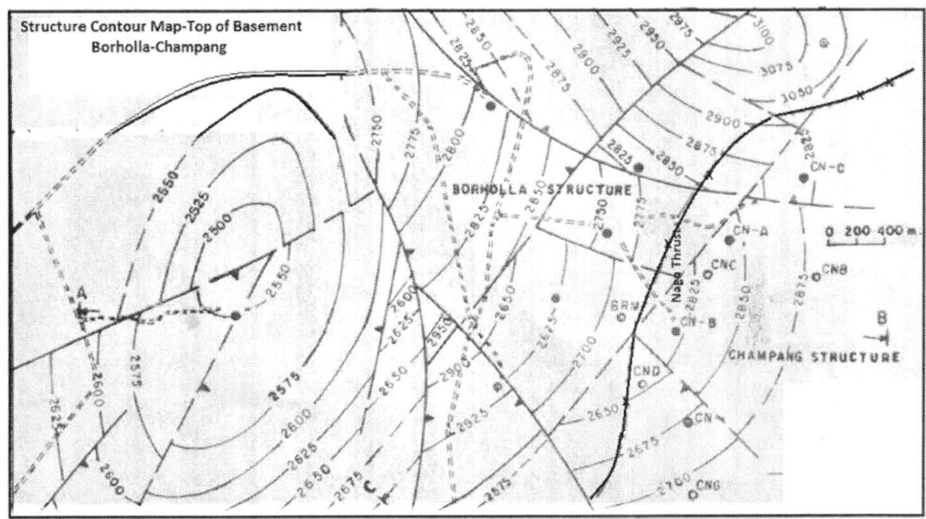

Figure 10.aa.14. Borholla-Champang structure Basement Map.

The character of the oils found in the basement is same that of overlying Sylhet and Kopili reservoir. The oil accumulation in the fractured basement is controlled by the connectivity of fractures as path way from the oil migrated from Eocene source.

Charali

It is located in the Assam in India. The field was discovered in 1974, where hydrocarbon presence was reported from multiple reservoirs within Barail (Oligocene Sands) and Tipams (Miocene Sands). The commercial production from this well started in 1979. The structure is an anticline bounded by faults from all directions. The EW trending reverse fault in the north plays an important role in entrapment of hydrocarbon. The Barail sand has an OIIP of 2.77 MMt in proved category.

In all there are five hydrocarbon bearing sands, which were deposited under paralic conditions and the BCS comprises of cyclic sedimentation of coal and shale with various channel sands, whereas the Tipam sands under fluvial conditions. The Barail sand with initial pressure of 330 ksc and was put on production in 1979. The sand is having bottom water and the OWC is at -3288 m. The well has produced clean oil for almost 15 years and aquifer water breakthrough started. The water cut shot up to 60-70 per cent and the stabilizing at around 40-45 per cent. The reason for this rise in water cut is suspected to be due to channeling. Using geo-statistical methods identified the small scale heterogeneities in form of intervening shales and their extension and continuity. Log correlation was attempted to conceptualize the geological model and layering of all sands. The history match was satisfactory when this model was used.

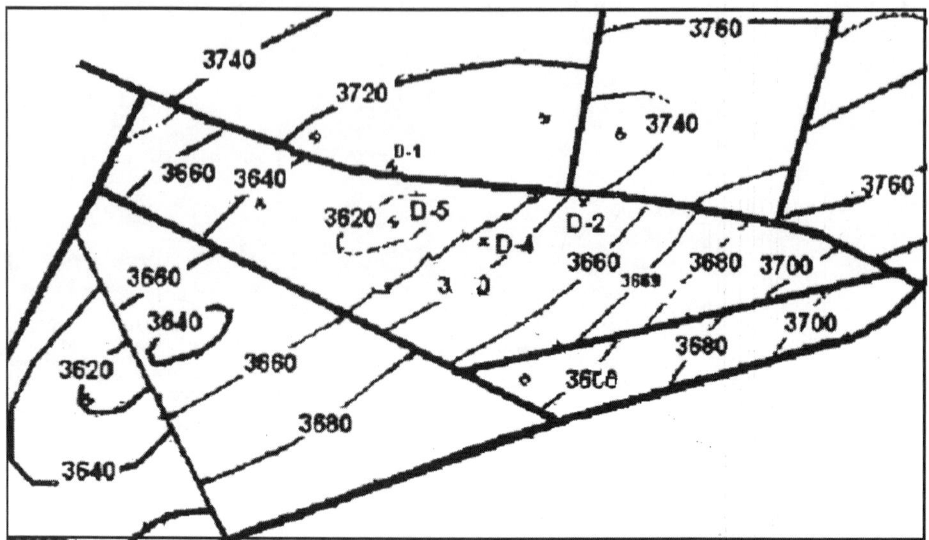

Figure 10.aa.15. Map on Top of BMS-Demulgaon (After Talukdar, R., *et al.*, 2003).

Demulgaon

The structure was identified by conventional 2-D seismic survey in 1967-68 and oil was discovered in 1976 in well D#1 in sands within coal shale sequences of Oligocene age locally known as BCS. Subsequently, oil was found in lower arenaceous unit locally named as Barail Main Sand (BMS) in well D#2. Demulgaon structure is ENE-WSW trending fault closures having distinct horst feature bounded by two prominent depressions in the north and south. The areal extent of the field is around 8 sq. Km. There are normal faults oblique to the general ENE-WSW fault and they have played important role in entrapment of hydrocarbon. The 3D survey was carried out in 1988-89 and its interpretation confirms to the present understanding of structure (Talukdar,R, *et al.*, 2003).

The Barail unit comprising of BCS and BMS was formed in a deltaic environment, with BMS (Barail Main Sand, arenaceous units) in delta front and BCS (Barail-Coal-Shale, argillaceous unit) in delta plain environment. The contact of BMS with overlying BCS is gradational. The BMS sand has gross thickness of about 170 to 250 m with average isopay thickness of around 30 m. So far 39 wells have been drilled out of which 19 are oil wells (Figure 10.aa.15).

It is a small oil field located in the NE part of India and was put on production in 1985. The main producing reservoir is BMS (Oligocene sand) at a depth of 3650m

MSL. The field has passed its peak production in 1991 itself and is currently in declining stage with oil rate of about 200 m^3/d with high water cut. The reservoir is operating under active water drive and current pressures are above saturation pressure. The OIIP are about 4 MMt. The studies have indicated recovery factor of around 40 to 50 per cent. Tipam sands of Miocene age are also oil bearing at places.

It was put to oil production from Barail Main Sand (BMS) of Oligocene since 1985. BMS is a layered sandstone reservoir with the top two layers L-1 and L-II being important from hydrocarbon point of view. Water injection in BMS was started in 1993. To identify the source of high water cut, the open hole well log responses of wells drilled before and after water injection were studied and integrated with the available core studies. One layer L-IB of BMS may be the cause of more water cut and hence decline in oil production.

North Bank of Brahmaputra

Brahmaputra River roughly coinciding with the Brahmaputra Ridge or Arch, divides the Upper Assam Shelf into the North Bank and South Bank areas. Tectonically the North Bank areas fall in the fore-deep part of the Himalayan Orogenic Belt, a situation existing since the post-Barail uplift in this region. Before that the basin developed as a pericratonic basin of the divergent margin type with the Paleogene coast line approximating the Naga Thrust alignment.

The western sector of the North Bank is predominantly sand-rich throughout Tertiary due to proximity to the outcropping Mikir Hills. Additionally the source materials of Eocene and partly Oligocene, matured only in Miocene or later when the Brahmaputra Ridge had already been uplifted and prevented migration of the hydrocarbon charge further north. General absence of cap rocks made accumulations less possible. The large, oval-shaped, N-S trending North Lakhimpur Low in this area, is a "residual low" created after the uplift of the Brahmaputra Ridge belt.

From all considerations, shales are likely to be present in these lows with thickness reaching more than 2000 meters and with good source characters and maturity. Source rock studies from Simen Chapori well, indicate a likely favorable petroleum system further east and northeast. At Eocene level, Shelf edge carbonates, turbidite fans, growth faults are expected. Recent oil-shows in Mechaki well near Sadiya Low drilled by OIL are favorable indicators.

The basement may be as deep as 7500 m but the pools are likely even at 4500-5000 m considering the Mechaki well show at around 4500 m. The Simen Chapori gas show at 4380 m (approx) also has indications of a petroleum system.

Assam Arakan Fold Belt

The arcuate Tripura-Cachar fold belt with convexity towards west is separated in north by the Dauki-Halflong tear fault (Evans, 1934). Eastward it is separated from the structurally compressed Mizo fold belt by Kaladan fault (Jokhan Ram, 1984) while westward it is flanked by Padma-Meghna river alluvium. Towards its west lies the present-day foredeep of Bengal Basin comprising a thick, homoclinal and mainly Cainozoic succession which is covered by the Gangetic Alluvium (Chakraborty A., 1972). Towrds the north, it terminates against Haflong thrust and Hail-Hakalula (HLL) tectonic lineament. Mio-Pliocene Surma and Tipam groups are extensively exposed mainly in the N-S oriented long, narrow and steep anticlines, which form parallel linear ridges separated by wide open synclinal valleys.

Geologically the Tripura-Cachar region is referred to as a 'Frontal Fold belt' of Assam-Arkan Basin which came into existence in an arc-trench convergent plate tectonic setting. Extensive geological mapping was carried out in this region (Figure 10.aa.16). The area comprising narrow box-like anticlines separated by wide, flat synclines and an inner mobile sub-belt of Mizoram and West Manipur consisting of tight, linear folds. Almost all anticlines have been affected by longitudinal faults on one or both flanks and which had either due east or west.

The highly folded and faulted hinterland of the Indo-Burma chain occurs to the east. The sediments exposed in the Tripura-Mizoram-Manipur belt exhibit, from east to west, successive lithological associations from Late Mesozoic pre-flysch and Paleogene flysch and sub-flysch to Neogene molasse and transitional molasse sequences (Bakshi.S.K.1965).

The Assam-Arakan geosyncline, characterized by the deposition of thick sediments. It is flanked on either side by the cratonic continental areas of the Indian Shield and the Shan Massif of Mynmar (Ganguly, 1975). The median zone between these two geosynclins lies along the flanks of the Arakan Yoma-Chin Hills belt of Mynmar and extends southward across the Bay of Bengal through the western Andaman and Nicobar Islands into the outer Indonesian arc system.

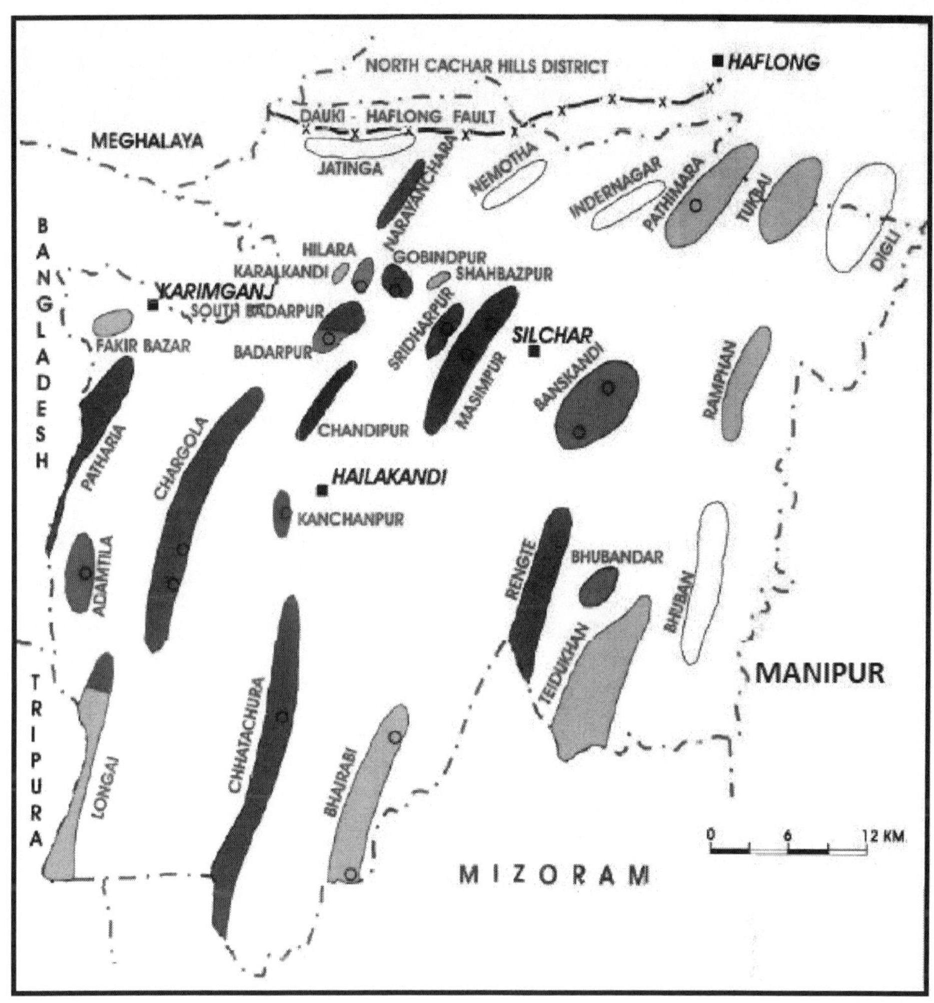

Figure 10.aa.16.Structural Map–Cachar.

The Indo-Burma Chain consists of thick, terrigenous deposits including exotic facies of mainly Paleogene age. These deposits are intensely folded and are bounded in the east by a series of interconnected major faults, uplifting the Mesozoic and pre-Mesozoic sedimentary and metamorphic rocks, associated with ophiolites (Dasgupta., AB, 1977). The fault zone can be linked up with the surface trace of the subduction zone forming the Sunda Trench in the Bay of Bengal, which marks the boundary between the Indian Plate and the Burmese Plate. The compressive orogeny recorded in this region during the Oligocene is believed to have resulted from an oblique collision between the continental plates.

Stratigraphy

In Tripura-Cachar-Mizoram region, the exposers are mainly geosynclinal molasse sediments of Neogene age, comprising of alternating shales, mudstones, siltstones and sandstones in varying proportions (Figure 10.aa.17). This succession has been litho-stratigraphically subdivided from bottom upwards into the standard Surma, Tipam and Dupi Tila Groups following the classification, proposed by Evans (1932) for the Surma Valley. The more argillaceous rocks exposed in the cores of the anticlines are believed to be equivalent to the Surma Group of Mio-

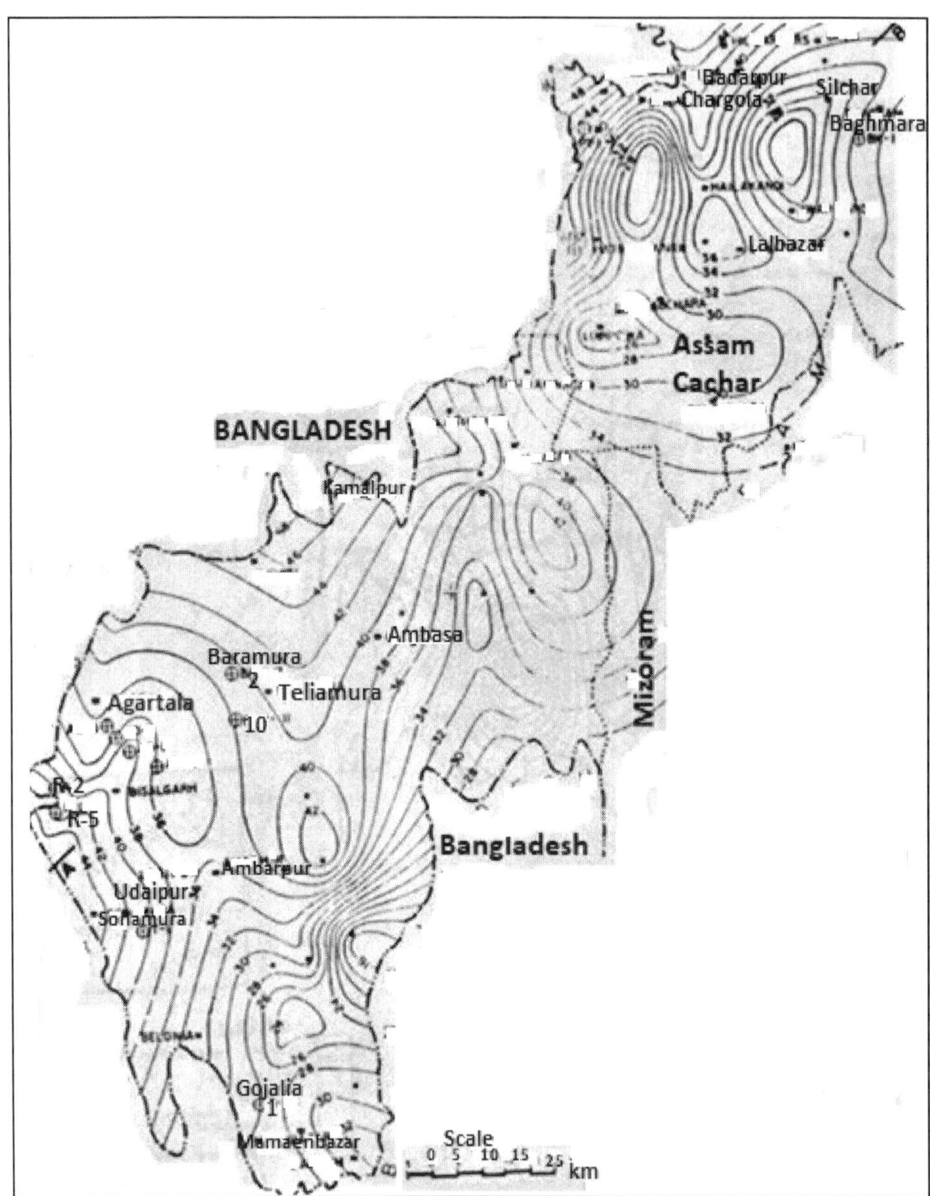

Figure 10.aa.17. Sand Percentage Map–Bokabil-Tripura -Cachar (After Manas K Chakraborty, 1992).

Pliocene age, which is about 4000 m thick in Tripura and about 7000 m thick in Mizoram (Srivastava, 1979). The younger arenaceous beds in the flanks comprise the Tipam Group of probable Upper Pliocene age and are over 2000 m thick (Manas K Chakraborty,1992). The sediments in the synclinal troughs in the western part of Tripura are at places unconformably overlain by semi-consolidated coarse sands and red silts correlatable with Dupi Tila Group of Quaternary age (Figure 10.aa.18).

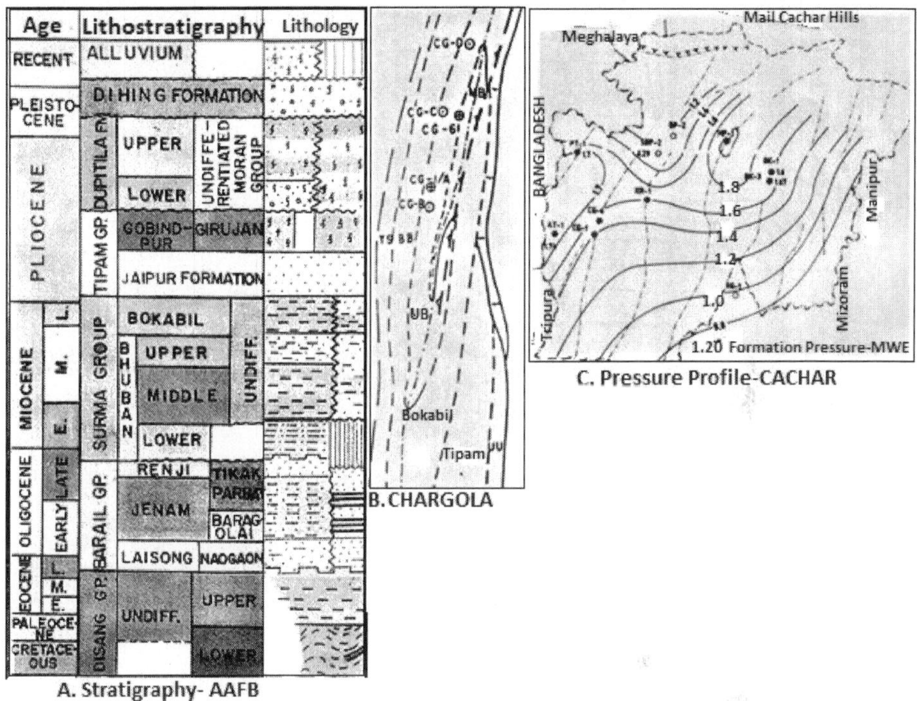

Figure 10.aa.18. Stratigraphic Column- Cargola Structure-Pressure Profile Cachar (After Sudhir Sharma, 1994).

The Surma Group has been further subdivided into a lower, the relatively more arenaceous, Bhuban unit and an upper, the mainly argillaceous, Bokabil unit. The available field, laboratory and well data indicate that, while the younger Plio-Pleistocene sequences of Tipam and Dupi Tila groups were deposited under subaqueous to subaerial, fluviatile to lacustrine conditions, the older Surma Group of Mio-Pliocene age was deposited in an unstable, rapidly subsiding tectonic basin under mainly deltaic conditions, with at least three major intervals representing transgressive marine cycles.

The Lower Bhuban sandstone–shale sequence has either a gradational contact or a minor unconformity at its base and was deposited in a delta front environment. The age ranges from late Oligocene into the early Miocene. In the overlying Middle Bhuban, shales predominate over siltstone and sandstone intervals, suggesting a major marine incursion. There follows an overall regressive regime which during the Middle Miocene is represented by the overlying Upper Bhuban sandstones.

The Bokabil Formation directly conformably overlies Bhuban clastics. Thickness variations of the Bokabil Formation range from 400 m in the Schuppen belt to over 1100 m in the South Shillong Plateau. In part these variations are due to onlap on the basal Miocene foredeep unconformity but also to the transition from a shaly prodelta facies into the sand deltaic deposits of the Tipam Group. Deep drilling has revealed the presence of thick clay sequences with abnormally high fluid pressures in the lower part of the Surma Group.

Tectonics

The east and northeastward drift of the Indian plate and subduction of the leading edges in the trench bordering the Burmese plate generated westward directed compressional forces. The obliquity of the converging Indian plate resulted in early closure of the sea in the north while the oceanic conditions prevailed up to Surma times in the southern part of the basin. While the Indian plate side is represented by imbricate thrusts the Burmese plate side is dominated by imbricate cylindrical folds.

Tripura-Cachar area extended into Upper Assam through the Naga Schuppen Belt. It is a large size tectonic block known as Tripura-Tangail high that came into existence probably during late Eocene as a result of collision – subduction –uplift at the contact between Indian and Burmese plates (Brunnschweiler R.D.1966).

The geophysical and deep drilling data indicate that basement extends under the Schuppean belt upto 8 km. Due to the lack of space at the proximity of Mikir massif, thrust-faulting was seen to accommodate crustal shortening in that region (Khar,B M., 1984). The Halflong-Disang thrust separates the Schuppean belt from the Tripura-Cachar fold belt.

To the west of Tripura, the gravity data indicate the presence of a NE-SW trending subsurface high named as Barisal-Chandpur High to the west of Meghna River. The hinge between the shelf and the mobile zone lies to the west of Sylhet and Faridpur trough.

The Miocene depositional system in the Tripura – Cachar region is characterized by a NE-SW aligned stream flowing over the Basement High flanked by the two remnant oceanic basins, *viz.* the Sylhet Low in the north and the Patuakhali Low in the south in Bangladesh.

Hydrocarbon Prospects

Numerous surface and subsurface manifestations of natural gas are seen in Tripura and Mizoram and of oil in Badarpur in Cachar area. Occurrence of commercial gas fields in the neighboring areas of Bangladesh and a favorable geological history indicate the existence of conditions conducive to petroleum generation in this region. On the basis of geological and geo-chemical studies, it has been opined that, there are good prospects for gas in the relatively shallower horizons and both oil and gas in the deeper horizons in several structures of the region.

The expected hydrocarbon traps include discrete culminations, cross-faulted noses, and subsidiary flexures in synclinal areas, subthrust warps against faults, and various associated stratigraphic traps, as well as combination traps resulting from shale-diapirism.

Considering the fact that, the hydrocarbon migration is essentially short distance, in tens of kilometers, the long N-S anticlines of Tripura and Cachar, are likely to have hydrocarbon charging only in area closest to the Sylhet and Patuakhali Lows.

The petroleum system is largely confined within the Bhubans of Surma Group of Miocene, which is gas-prone with only moderately mature source rocks. The regional cap rocks are expected in Bokabils, and in older Bhubans.

The exploitation of hydrocarbons from different structures of this belt has proved that the hydrocarbon generation potential does exist in this belt. It is known that the main reservoirs are located in the Bhubans. It is also known that the Surma equivalent sediments do not extend beyond the Mizo fold belt implying by that the Surma depocentre lies in the distal part of the subduction complex. Therefore, the sediments might be subjected to sufficient heat flow conditions.

In the sequence of structural deformation, although the folding was initiated by westward compression, the folds were followed by emplacement of longitudinal faults. In such scheme of evolution the faults will act as seals rather than conduits since each fault has limited effect. The Badarpur-Chandipur, Adamtila, Langai-Chargola, Bhairabi-Rengte anticlines offer such locales. The culminations within the broad synclines are neo-tectonics and offer bright prospects for accumulation of hydrocarbons generated in the depressions.

Presently gas is produced from Agartala Dome, Baramura, Gojalia and Rokhia anticlines. Exploration in Cachar in the northern part of the basin did not yield much success excepting two sub economic discoveries of gas in Banaskandi and Adamtila structures.

In terms of Petroleum systems the reservoirs of the Cachar – Tripura gas fields are all within the Bhuban sandstones or the Bokabil Formation. But the data from Khubal and Rokhia wells indicates Lower Bhuban are also have good reservoirs.

Since source rock maturation and trap formation happening simultaneously the geomorphic Highs/Turtle back structure within broad synclines between the steep anticlines are also the likely locales for gas reserves.

Cachar

This region, with its characteristic succession of orogenic ridges and valleys and south Cachar comprising narrow, boxlike, and cuspate anticlines separated by wide, flat synclines, and an inner mobile subbelt of Mizoram and west Manipur consisting of tight, linear, commonly isoclinal anticlines and synclines. Passing from east to west, deformation in this belt becomes progressively younger and less intense. The anticlines are commonly bounded on one or both flanks by longitudinal listric reverse faults. The individual structures are internally segmented by cross faults and oblique faults of multiple alignments, some of which have strike-slip components and have offset anticlinal axes and flank faults.

The expected hydrocarbon traps include discrete culminations and anticlines, cross-faulted noses, subsidiary flexures in synclinal areas, sub-thrust warps against block-bounding faults, and various associated stratigraphic traps, as well as combination traps resulting from clay diapirism. However, on account of logistic constraints at surface, steep dips and high pressures in subsurface conditions, inherent stratigraphic and structural complexities the exploration is facing tough situation. These are also making difficulty in acquiring good quality of seismic data, exploratory operations.

Though in this area exploration for oil and gas started long back, of the nearly 90 wells drilled in the Cachar area only 10 wells are gas producers out of which three wells are in Adamtila field, four from Banskandi structure and one from Bhubandhar field.

ONGC drilled the first well Chargola#1 in 1977 and since then drilled about 82 wells on 19 structures, found six oil-bearing and nine gas-bearing areas, but the flow in the oil fields was ceased (Figure 10.aa.18-B). In these gas fields reserves are about one BCM. About 110,000 m^3 per day of gas is produced from Adamtila and Banaskandi fields.

Abnormal Pressures in Cachar Area

There are nineteen anticlinal structures in Cachar. Exploratory drilling has been carried out on some structures like, Patharia, Adamtilla, Chargola, Kanchanpur, Badarpur, Masimpur, Banskandi and Rengte. The earlier Badarpur oil field of BOC times was re-established opening up new sands in upthrusted Renji and Lower Bhuban sediments (Figure 10.aa.18-C). It is observed during drilling that the lower part of Middle Bhuban Formation and the underlying Lower Bhuban sediments are abnormally pressured. Exception to this are Patharia structure where the Bokabil sediments are abnormally pressured at shallow depths (Chatterjee, 1990), and Rengte structure where the pressures are sub-hydrostatic (Ray, 1994).

The compaction dis-equilibrium and tectonic activity appear to be the two major causes for overpressure generation. The Bhuban Group of sediments having thickness nearly 4000m deposited in a short span of time during Miocene period in fluvio-deltaic conditions, point towards a comparatively high rate of sedimentation. A quick burial along with rate of sedimentation resulted in compaction disequilibrium leading to the initial over-pressuring. Low permeability of Middle Bhuban sediments arrested the expulsion of fluids during subsidence and prevented the pore spaces to attain hydrostatic equilibrium thereby leaving the sediments under-compacted and over-pressured (Sudhir Sharma,1994).

Out the exploratory wells drilled so far by ONGC, abnormal pressures are recorded in wells with drilling complications followed by premature termination. The abnormal pressures generally occurring below 2500 m are found to range in values from 1.6 MWE to 1.76 MWE (60-70 per cent more than hydrostatic).

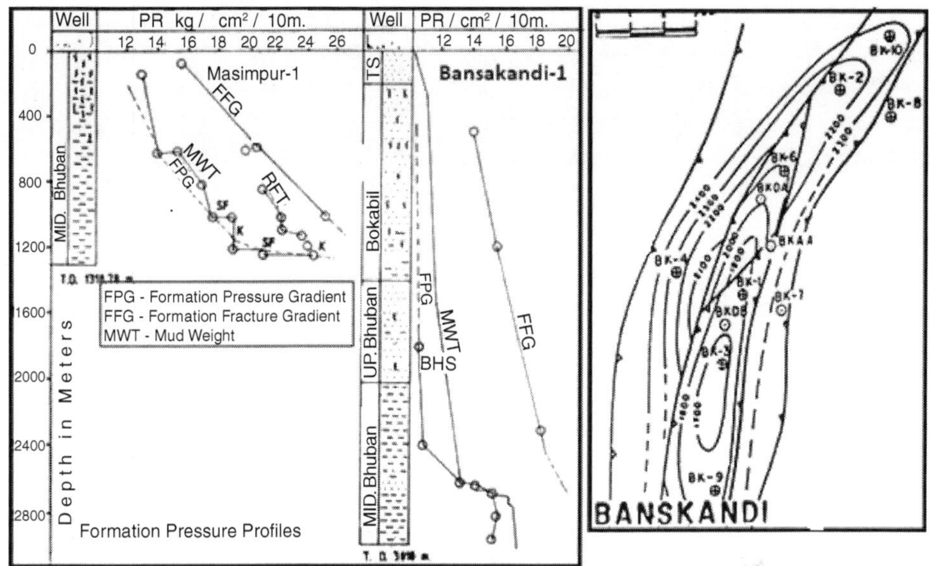

Figure 10.aa.19. Pressure-Profile-Cachar Wells-Structures-Bansakandi (After Sudhir Sharma, 1994).

But in Patharia, where the top of high pressure was met at 640 m in well PT-1 and Masimpur where the uplifted Middle Bhuban sediments have encountered pressures more than 2.4 MWE at shallow depths.

Badarpur

This was an oil field, which was discovered and produced by BOC around 1915 and taken over by ONGC. Thereafter drilled 22 wells in eighties and field was put on oil production in 1987 with very little output. Reservoirs of oil are in Renji formation. Four wells (1,8,11 and 16) produced oil for 15 years with an average rate was 3 – 4 m^3 /day.

Banskandi

Banskandi and Bhubandar are gas fields, the presence of which have been identified mainly based on geomorphic anomalies and seismic surveys (Kamaraju, AVVS., 2010). In the field drilled about 13 exploratory wells Out of these only 3 wells were found gas bearing. They are BK-1,2 and 6. BK-2 and BK-6A are at the north bank of the river Barak but BK-1 is at the south bank (Figures 10.aa.19-A and B). The well BK-2 had sand cut problem during its initial testing.

Well BK-1 is gas bearing in Upper Bhuban Lower Pay horizon and BK-2 is found gas bearing in Upper Bhuban Upper Pay horizon. Though Well BK-6A is gas bearing in both Upper Bhuban Upper Pay and Upper Bhuban Lower Pay, it is completed in Upper Bhuban Lower pay. The Upper Bhuban Lower Pay of BK-1 block, and BK-6 Block, are having separate gas-water contact.

The in-place volume of Upper Bhuban Lower pay for Block of BK-1 is 1054 MMm^3 out of which 727 MMm^3 is placed in proved category (corresponding ultimate reserves 315.0 MMm^3) and 327 MMm^3 is in possible category.

The development well BK-11 was drilled at the south of the river Barak in 1996 and had to be prematurely abandoned due to blowout at very shallow depth-315 m as the well encountered unanticipated shallow gas sand.

Bhubandar

It is located adjacent to the Banskandi field in the south of the river Barak. Total 14 exploratory wells were drilled in this field, of which 10 wells are Exploratory and 4 wells are of Development.

The well BU-2 is completed in sand-I of the Upper Bhuban formation. After the initial testing, the AOFP and well deliverability were computed as 270,000 SCMD and 54,000 SCMD respectively. During gradient study, the SBHP recorded was 187.60 Kg/cm^2 at measured depth of 1700 m and the same extrapolated to mid of sand face works out to be 188.21 Kg/cm^2 and the SBHP gradients are around 0.13 $Kg/cm^2/10m$. During flow after flow test, the gas rate was 109,700 m^3/d through 8 mm bean (correspond FBHP-130.81 Kg/cm^2) and 75,800 m^3/d through 6 mm bean (corresponding FBHP-155.76 Kg/cm^2). The AOFP calculated after the reservoir study was 180,000 SCMD.

Adamtilla

The structure is N-S trending anticline (Figure 10.aa.20-A). Reservoirs are in the depth range of 1800m and two pressure breaks occur in well Adamtila-F at around 2372 m and 2568 m as evident from the plot of pressure predictions parameters. The lower break at 2568 m is a major one being accompanied by severe well activities which necessitated to increase in mud sp.gr, from 1.27 to 1.62 for controlling the activity. The latter break in the Middle Buhuban Formation is characterised by abnormal high pressures and thickness of the transition zone varies from less than 10 m to about 30 m.

The Bokabil top is shallowest in AT-A whereas the upper and middle Bhuban tops are shallowest in AT-F. The well AT-F, though located for downdip at the surface, occupies a structurally higher part of the structure in the subsurface caused by the reserve fault. So the occurrence of abnormal high pressure at a shallower level in Adamtila-F is due to structural uplift caused by reverse faulting.

Nagaland

The Naga thrust is associates with anticlinal folds. The petroleum prospects of Nagaland suggests that the prospects are in closed folds along the Naga thrust where the Girujan clay and/or the Barail coal-shale (Tikak Parbat) major cap rock is preserved, fault closure below the major cap rocks in other parts of the Schuppen belt and also include sub-thurst anticlinal folds.

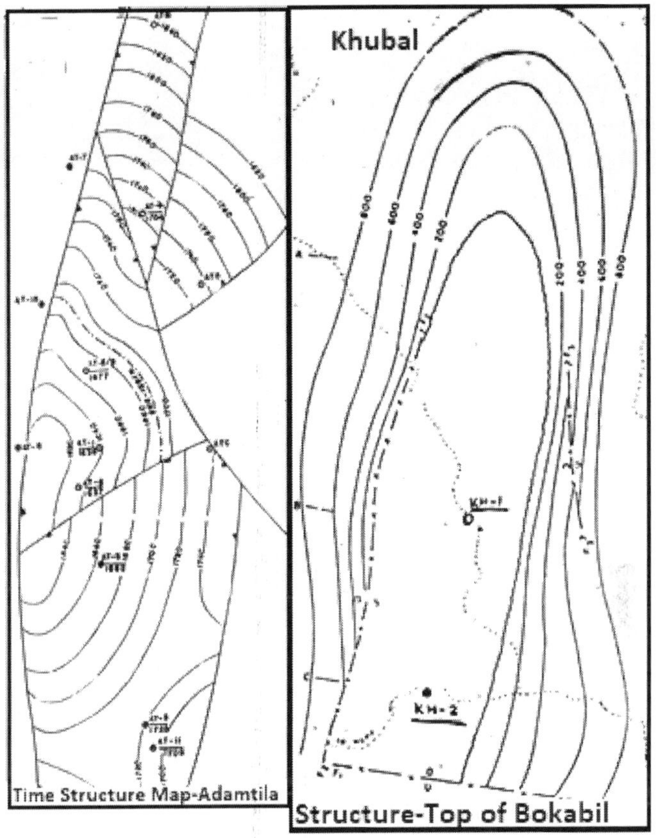

Figure 10.aa.20. Structures–(A) Adamtila- Cachar, (B) Khubal-Tripura.

Meghalaya Mikir Hills and North Cachar Hills

This forms the southern sedimentary fringe of the Archaean craton of Shillong Plateau and its continuation of Mikir Hills. Sediments thicken rapidly from their edge and in North Cachar Hills they are already geosynclinal in character. Numerous oil shows are known from this area. The prospects are in, structural closures in the deeper parts. of the area, as well as combination traps in the Bhubans in the south plunging anticlines along southern border, *viz.* Angartoli, Dewali, and Gopinatkila in Garo Hills and Khasimara in Khasi and Jaintia Hills. However, wells drilled on Baghmara and Gopinathkila did not give any encouraging results.

Mizoram and Arunachal Pradesh

Mizoram has tighter folds with the same general stratigraphy and structural features as are found in Tripura (Jokhan Ram and Venkatraman,1984). In Arunachal Pradesh, the prospects are entirely structural, with having folds in the foothills where rocks are transitional in character between Siwaliks of other parts of the Himalayas and the Tipam-Surma sequence of Upper Assam Valley.

Tripura

Tripura area forms a part of the frontal folded belt on the western flank of Assam Arakan Folded Belt. Tectonically, the folded belt of Tripura-Cachar and Chittagong (Bangladesh) comprises a series of sub-parallel elongated en echelon doubly plunging anticlines separated by wide and flat synclines where turtleback structures have developed at many places (Figure 10.aa.21). The anticlines are usually bounded by longitudinal reverse faults on one or either limbs which disappears towards the plunge of the folds. The young cross faults and oblique faults of multiple alignments with strike slip component offset the longitudinal and the fold axis (Biswas, S.K. and Srivastava, B.P., 1977).

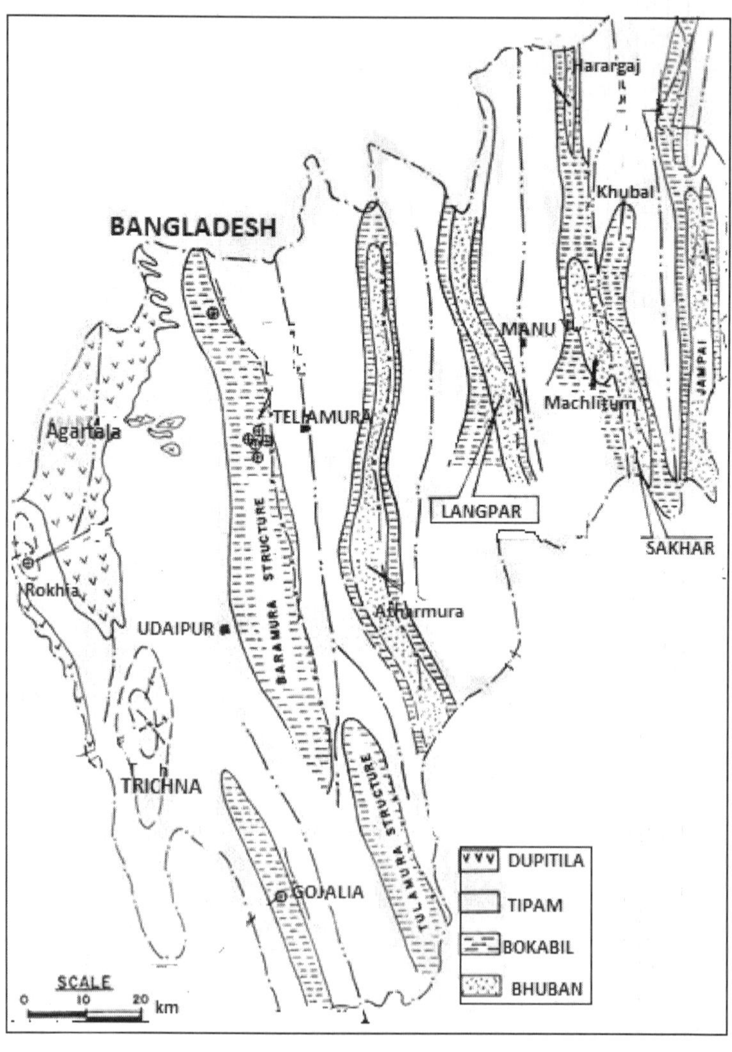

Figure 10.aa.21. Geological Map–Tripura (After Ganguly, S., 1983).

The combined geological unit can be subdivided into four broad tectonic elements; Sylhet Depression to the north, Tripura-Chittagong fold belt to the east and southeast and Patuakhali Depression in the southwest beyond which the basin opens up into the the present Bay of Bengal. The Sylhet Depression and Patuakhali Depression is separated by relatively uplifted Barisal-Chandpur-Tangail-Tripura High (Ganguly, S., 1983).

All the exposed folded structures in Tripura-Cachar area were mapped by geological field parties of ONGC along with photo-geological surveys. The Remote Sensing mapping was used for geological mapping and also number of oil/gas shows were observed.

Exploration and development activities in Tripura were continued since last 45 years. The first discovery was in Baramura#1 in 1972. So far about 120 wells were drilled. The deepest well drilled in 1986 was Rokhia#1 upto 4,600m.

In Tripura, significant gas discoveries were Baramura, Rokhia, Agartala Dome and Gojalia. In 2009, large reserves of gas were discovered in Khubal and in Rokhia in Lower Bhubans. Commercially viable multiple gas pools have been discovered within Bokabil, Upper Bhuban and Middle Bhuban in northern and western part of Tripura.

Plays are essentially structural in nature as Plio-Pleistocne folding has dominated all other trapping mechanism. However, on the folded belt, there are stratigraphic variations which form additional plays in association with the anticlinal folds. One of these variations is found in the presence of channels which are aligned east-west, perpendicular to the folding axis.

Another play is formed by the flank sands which were deposited simultaneously with folding. These sands thin over the crest and are sealed by the shales, thereby forming a combination trap. Since migration is late, after the folding and hence these sands are in good position to hold hydrocarbons.

The total estimated resource of this basin is 1860 mmt (O+OEG) and current in place gas reserve is 20 BCM. This basin remains poorly explored due to difficult logistics and hostile political environment about thirty anticlines have been mapped and semi-detailed geological studies have been carried out by ONGC but seismic surveys were done only in the western part of Tripura-Cachar region.

Khubal

It has struck natural gas in a 2,705-metre-deep well near Pechertal village in Karimganj district on the Assam-Tripura boundary (Figure 10.aa.20.B).

This area forms the western platform of the Arakan geosyncline, and has a thick of Eocene to Pliocene age, with several oil and gas shows and one abandoned oilfield Badarpur. Here the prospects are entirely structural, with large periclines remaining to be tested by drilling. Gas and oil are expected in the Bhuban (Pliocene) and the Barail (U. Eocene-Oligocene) deltaic sequence.

Baramura

The structure is alignied North-South with gently dipping western limb. It was the first structure taken up for drilling in Tripura after lot of detailed geological mapping of the area. The first well drilled was a success and met with commercial gas flow in an Upper-Bhuban sand at a depth of 1900 m. Few more wells have been drilled on the large Baramura structure to delineate this gas pool and to test the possibilities of other sands. About 19 wells were drilled. All reservoirs are in Bhuban formation in the depth range of 2300-3000 meters.

Rokhia/Cenemora

This structure is located very close to Bangladesh boarder and the crestal part is in the Bangladesh territory but both northern and southern limbs are in India. The first well was taken up for drilling in 1984 and well was drilled to 4600 m which is the deepest. Very high formation pressures were encountered and the well was to the depth a drilling challenge and logged (Figures 10.aa.22-A). Commercial production of gas was established. In order to follow safe international ethics, its northern limb was named as Cenemora as separate accumulation, the central part of the anticline is in Bangladesh territory due to which the southern limb is only called Rokhia.

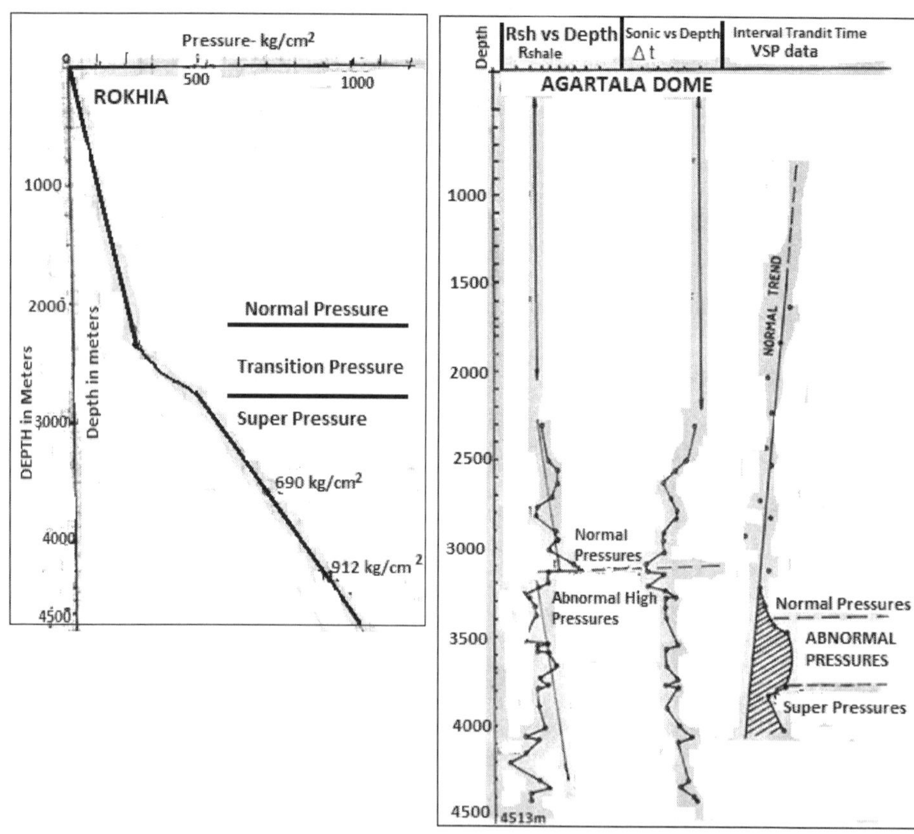

Figure 10.aa.22. Pressure Profile–Rokhia- Agartala Dome.

Agartala Dome

This structure is close to Agartala with less surface manifestation of the structure with clarity through geophysical data (Figures 10.aa.22B). It was taken up for drilling during early 90s and drilled few wells and proved to be commercial gas producer. Abnormal formation pressures were recoded below 3200 m in the structure.

Gojalia

This structure is located at south eastern corner of Tripura. It was taken up for drilling during 1984. Few wells were drilled and proved to be commercial producers of gas.

Trichna

This structure is located towards western boarder of Tripura and was taken up for drilling during late 90s and though the commercial gas is known, due to utility logistics further delineation is deferred.

Amtali

The discovery of gas was established after the drilling of a well up to a depth of 3,515 metres. On testing, one of its sand zone flowed gas at the rate of 113,400 cubic meter per day, through 8 mm bean. The success of this new gas bearing well has opened up the intervening low area between Agartala and Konaban gas drilling structures, and also gives impetus to exploration of more such areas on the flanks of major structures in Tripura fold belt.

Table 10 aa.1. Details of Fields and Prospects Drilled in Upper Assam and Fold Belt

Location	*Field*	*DD m*	*Area Skm*	*Wells*	*Remarks*
UA	**ONGC**				
	Amguri	4221			
	Barsilla	4607	C.22	34	-SF
	Bihubar	4600		12	-SF
	Borholla	3328	32.12	70	
	Changmaigaon	3594	25.00		
	Changpang	3416	12.00	29	
	Charaideo	4805	35.00	1	SF
	Charali	4390	96.64	12	
	Dumulgaon	4546		39	
	East Changmaigaon		35.00		
	East Lakhibari				
	Geleki	4868	35.60	310	
	Golaghat		85.00		
	Kalyanpur		40.00		
	Kasomarigaon		76.00		
	Khoraghat	3339	86.00	29	

Location	Field	DD m	Area Skm	Wells	Remarks
UA	Kurgaon	4805		3	
	Laiplingaon		56.45		
	Lakwa -Lakhmani	5010	172.49	520	
	Langharaj			-	
	Laxmijan	4478	8.9	SF	
	Mekeypore-Santak-Nzr	4379	127.00		
	Mekrang	3037	16.00		
	Nahorhabi	4505	20.00	-	
	Nambar		46.00		
	Namti	4500	35.55		
	Napamua	3750		-	
	North Rudrasagar		149.00		
	Panidihing		34.00		
	Rudrasagar	3916	70.50	255	
	Se Geleki Ext		48.50		
	Sonari	5083	30.00		
	Titabar		10.00		
	Tynphe	4518		-	
	Uriamghat	2596		5	
	West Charali		12.00		
	OIL-Area-Prospects Drilled				
	Moran		489.42		
	DumDuma		194.25		
	Naharkatiya		167.18		
	Hugrijan		725.20		
	Digboi		49.33		
	Ningru		615.67		
	Tinsukia		435.00		
	Dibrugarh		186.00		
	Borhapjan		87.00		
	Dholiya		131.00		
	Chabua		189.00		
	Baghjan		75.00		
	Mechaki		195.00		
	Sapkaint (Murkong- NF)		105.00		
	Mechaki Extension		9.00		
	Small Fields				
	Dipling		10.1		
	Sarojini		8.79		
	Sapekhati		9.28		

Location	Field	DD m	Area Skm	Wells	Remarks
	Duarmara		8.91		
	Jeraipathar		10.1	79	
	ONGC-Dry- Structures Drilled	**Depth m**			
	Baghmara	2541			
	Bhaktinaga	1606			
	Banamali	4208			
	Barpathar	1745			
	Bihubar	4600			
	Changki	2805			
	Chumukidema	4328			
	Dergaon	1814			
	Dikhomukh	3802			
	Disangmukh	3811			
	Gopinathkila	606			
	Halwating	4120			
	Holangapar	3409			
	Hozukenaga	3270			
	Jhanji	3605			
	Jorhat	3250	81.00		
	Kasimara	2777			
	Kasmarigaon	2338			
	Lonepuria	3792			
	Mariani	4092			
	Naojan	1465			
	Nazira	4236			
	Naginijan	4007			
	New Tsori	4193			
	Rajabari	4230			
	Rajmai	4240			
	Rajaphe	3870			
	Teok	3830			
	Tosezhu				
	SF-small field				
	DSF-I				
	Haldibari				
	Madhakali				
	South Rajgarh				
	Umtara				
	Disalijan				

Location	Field	DD m	Area Skm	Wells	Remarks
	Tiphuk				
	Hazarigaon				
AAFB					
	Cachar				
	Adamtilla		152.00	10	
	Badarpur		2.30	14	
	Banamali		50.00	2	
	Banaskandi		15.00	3	
	Bhubandar		6.00	3	
	Hilara		9.6	-	SF
	North Patharia		19.15		
	Patharia				SF
	Kherem		16.45		SF
	Tukbai				
	Tulamura		84.00	DSF-I	
	Tripura				
	Agartala Dome		269.19	7	
	Baramura	3,400	161.00	19	
	Gojalia	3478	271.17	6	
	Konaban		28.0	2	
	Kunjaban		288.00		
	Manikyanagar		109.35	3	
	Sonamura		30.00	2	
	Rokhia	4,600	5.59	12	
	Sundulbari-A Dome		301.00		
	Tichna		195.41		

Table 10.aa.2. Resources vs Reserves

Hydrocarbon Resources – A and AA Basin – (MMt)			
Area	***Resources***		***Inplace***
North Assam Shelf	870	1180	675.11
South Assam Shelf	310		54.8
Cachar	435	1858	3.72
Tripura	588		53.79
Nagaland	555		16.29
Meghalaya	100		
Mizoram	170		
Manipur	10		
Total	**3038**		**803.71**

Cambay Basin

Introduction

The Cambay Basin is a narrow, elongated rift graben, extending from Surat in the south to Sanchor in the north. In the north, the basin narrows down, but tectonically continues beyond Sanchor to pass into the Barmer Basin of Rajasthan. On the southern side, the basin merges with the Mumbai Offshore Basin in the Arabian Sea. The Cambay basin can be geographically divided into northern and southern parts across the Mahisagar River.

This basin extends from Luni River in the north to Tapti River in the south and is about 425 km. long. This basin covers an area of about 53,500 sq. km., the Deccan basalts form the technical basement. The Tertiary sediments record the synrift, post rift features. The sedimentary sequences range in age from Late Mesozoic to Cenozoic. The hydrocarbon entrapment is mainly structural with a few having strati-structural in nature. About 9,000 m thick Tertiary sediments have been estimated in the axial part of the basin.

Cambay basin gave initial success to the ONGC and after continued exploration and development activity for more than 60 years it is at over-matured stage and almost all the possible structures and stratigraphic sequences have already been explored. Less effort has been put to explore the deeper sequences, due to poor quality of seismic data possibly because of basalt layers, greater depths, high pressure and less known quality of reservoir rocks.

Exploration History

ONGC took up its initial oil exploration activity in the basin in 1957 with gravity and magnetic survey. So far about 1,20,100 LKM of 2D seismic and 9,700 sq. km of 3D data were acquired. ONGC drilled some shallow wells to know the subsurface continuity of sediments and then its first deep exploratory well on Lunej structure near Cambay. This turned out to be a discovery well, which produced oil and gas.

The discovery of oil in Ankleshwar structure in 1960 gave boost to the exploration in the Cambay Basin. More than 7,700 wells have been drilled in the basin so far.

Extensive exploration in this basin has resulted in numerous oil and gas discoveries, notably Ankleswar, Gandhar, Nawagam, Dholka, Kalol, Sanand, Jhalora, Viraj, Kadi and S. Kadi, N. Kadi, Jotana, Sobhasan, Balol, Santhal with known 145 large and small-size oil and gas discoveries out of which two, namely Ankleshwar and Gandhar are giant fields (Table.10.ca.1).

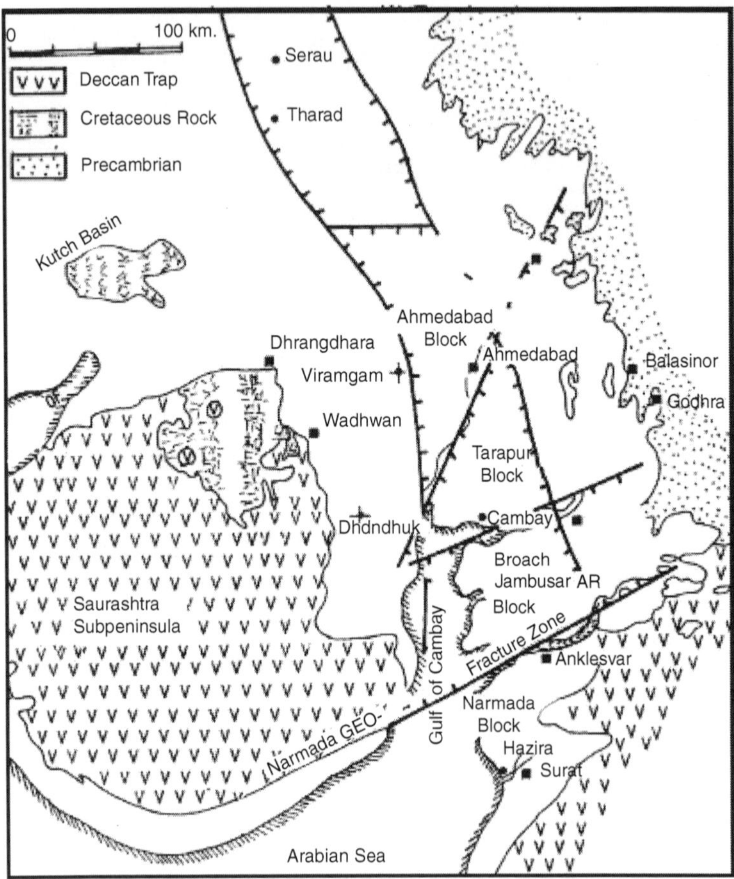

Figure 10.ca.1. Geological and Mega Tectonic Map–Cambay Basin (After Raju, ATR and S Srinivasan, 1993).

With changed government policy to invite foreign participation, some small fields were offered during late ninetees. HOEC, Selan and GSPC took such fields under mining lease of ONGC. The continued efforts of Governemnt through its organization DGH, since 1997 for attracting both foreign and private participation through many policies, the Cambay Basin always attracted most of them.

In Cambay basin where already several oil and gas fields are under exploitation, importance of increasing the percentage of recovery from the aged/brown oil fields, the required efforts by application of secondary and Tertiary recovery methods were already known.

Although the exploration of the basin is in matured stage, there are still chances of discovering small accumulations in subtle traps. The basin could now be rated as low-risk/moderate reward category, which is attractive for small oil companies. Gulf of Cambay still remains to be fully explored. Recent discoveries indicate good prospects of the Gulf, Banner-Sanchor sector which was less explored, turned out to be positive with discoveries of oil gas fields. The future of this basin is still good with almost assured chances of discovering small fields in multiple plays with exploration strategies like "new oil in old fields" and deliberate search for "yet to find oil". This basin is rated as "low risk-low reward".

Geology of the Area

Out crop: The rocks of the Tertiary period, unconformably overlying the Deccan Traps, are exposed on the western flank in Ghogha-Bhaunagar basin in Jhagadia-Tarkesvar area. In the eastern coastal tract of Saurastra, marine rocks occur in a series of isolated outcrops which are of Upper Miocene age.

The 'lateritic rocks', composed of unstratified rocks of red, brown and yellowish brown colour, often considerably mottled and containing tuffaceous material, unconformably overlie the Deccan Traps. They are in turn overlain unconformably by well-bedded, grey to bluish grey plastic gypseous clays with yellow dense marls (Ratanpur Clay). The Bhumbhli Conglomerate, composed of fossiliferous conglomerates, grits argillaceous sandstones and minor clays, overlies unconformably the Ratanpur Clay inturn overlain conformably by the Kuda Sandstone, consisting of grey coloured, argillaceous sandstone with minor sandy clay bands. The Piram bed of a hard and well cemented fossiferous conglomerate, overlie the Kuda sandstone disconformably. The Piram beds are considered to be of uppermost Pliocene age. The alluvium, comprising of dark grey and dark soil with and lateritic soil in its lower part, extends around Deccan Trap exposures.

The Tertiary outcrops in Jhagadia-Tarkesvar area are confined to two small tracts, separated from each other by the alluvium of Kim River (Sudhakar, *et al.*, 1970). The Vagadkhol Formation unconformably overlies the Deccan Traps and is composed of grey clays, tuffaceous sandstones and conglomerates. They are in turn overlain by Nummulities composed of foraminiferal, impure, argillaceous limestone, claysand marls. The limestones and clays are yellowish in colour. An Upper Eocene age is indicated for the Nummulitics' on the basis of their faunal content. These are overlain unconformably by the unfossiliferous Tarkesvar Formation composed of grey and variegated bentonitic clays with thin lateralized bands. The Tarkesvar Formation is overlain conformably by the Baba Guru Formation composed of ferruginous, and current bedded sandstone, conglomerates and grey and white sandstones and clays.

The next unit, the Kand Formation, conformably overlies the Baba Guru Formation and is composed of grey clays with thin bands of fossiliferous marls and limestones, and subordinate sandstones. The Kand Formation is assigned a Burdigalian age on the basis of its fauna. The Jhagadia Formation, conformably overlies the Kand Formation and is composed of white and greyish white, calcareous, micaceous and current bedded sandstone with occasional conglomerates. The aeolian and stream deposits, consisting of dull reddish and greyish sandy clays with kankar, are the youngest formation exposed and constitute the alluvium. During initial days of exploration few structural wells were drilled by ONGC to establish the subsurface continuity of the outcrops.

Tectonics

The Cambay Basin is a narrow half graben trending in NNW-SSE filled with Tertiary sediments. The rifting is due to the extensional tectonics. The evolution of the Cambay basin began following the outpour of Deccan Basalts during late Cretaceous covering large tracts of western and central India (Figure 10.ca.2).

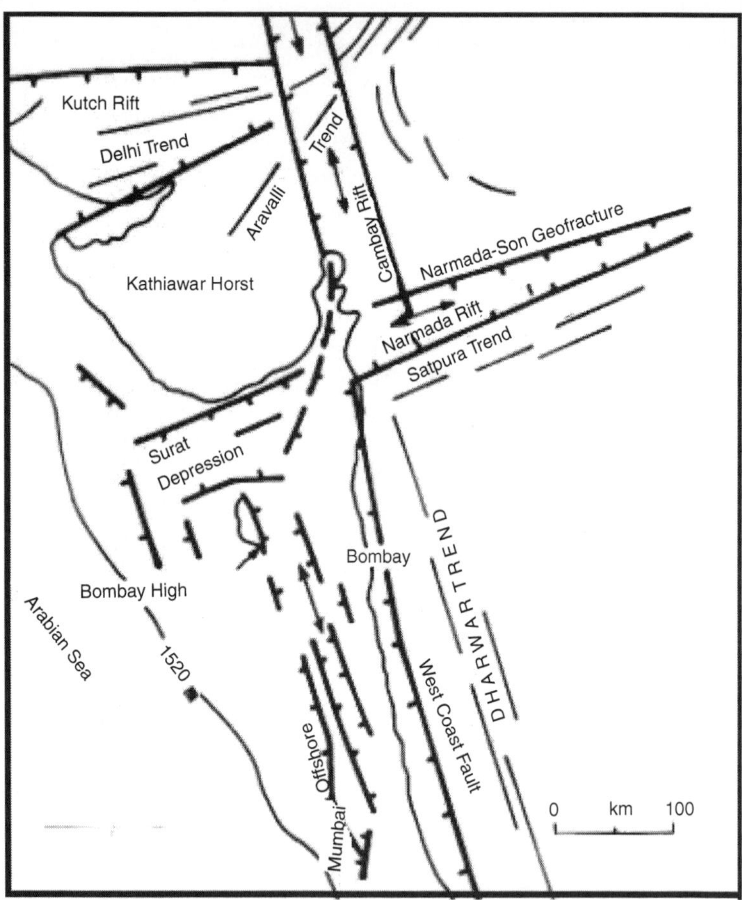

Figure 10.ca.2. Tectonic Elements of Western India (After Madan Mohan, 1995).

Kutch, Cambay and Narmada are three major marginal rift basins (Biswas, 1987) on the western margin platform of the Indian craton which evolved during different phases tectonic history during its break up from Gondwana land, its northward drift and final collision with Eurasia. Kutch and Narmada are Mesozoic basins whereas Cambay is mainly a Tertiary basin, which originated during the Early Cretaceous as a fluvial valley.

The basin, is a narrow elongated (NNW-SSE trending) intra-cratonic rift basin situated between Saurashtra craton to the west, Aravali swell on the northeast and Deccan craton to the southeast. In the south, it extends into Cambay Gulf and further southwards into the offshore. The extensional architecture of the basin is defined by the NNW-SSE-trending limiting faults related to Dharwarian orogeny. In addition, the basin is also transected by NE-SW trending faults related to Aravalli orogeny, and ENE-WSW trend related to Satpura orogeny.

The rift evolved through thermo-tectonic episodes; pre-rift extension. Synrift subsidence and post rift sagging leading to complex sedimentation pattern in the basin controlled by normal listric faulting and transverse transfer zones. Predominantly silici-clastic rift-fill sediments range from Palaeocene to Recent and include several transgressive/regressive sequences (Mathur, L P, *et al.*, 1980).

The Deccan Basalt is the basement for the thick Tertiary sequence. Lower Cretaceous sandstones are present below, overlying the Pre-cambrian gneissic basement. This rift is considered to be the failed arm of a triple junction during rifting of India and Africa in Late Cretaceous.

The dislocation along the eastern fault was more severe than that of the western. The marginal faults are not promiment towards south of Mahisagar River. It may be possible that, the basin was subsiding even during the outpouring of lava flows, accommodating thicker lava sheets in the Cambay graben relative to its margins.

Gravity and magnetic surveys carried out had revealed structures as major highs. The Bouguer anomaly map has helped in identification of the major structural highs and lows in the basin. The magnetic anomaly map also depicts the broad structural configuration of the basin. The structural features were later confirmed by 6 fold CDP seismic survey. All the structures are bounded by major faults on their southern or south-eastern limbs.

The Southern Cambay Basin between Cambay town in the north and Surat in the south, has two distinct units. The northern unit or the Broach Block, is an uplifted unit. Separating these two tectonic units, is a prominent ENE trending fault zone called Narmada Fault. This fault zone has been found to be the westward continuation of the well-known Narmada-Son Lineament.

The basin is further sub-divided into blocks separated by NE- ENE to SW -WSW transfer faults. These are Ahmedabad-Mehsana, Cambay-Tarapur, Jambusar-Broach, and Narmada-Tapti blocks.

Ahmedabad-Mehsana Block

This block lies between the Khari River in the north and the Vatrak River in the south. This is the largest block in the basin, extending from Dholka- Nawagam in the south to Unawa in the north. The Mehsana horst is the prominent feature, which is an extension of the West Patan-Wasna horst (Figure 10.ca,3). The apical part of this horst did not have the sedimentation in Eocene-Miocene. (Kundu and Wani, 1992). In Ahmedabad Sub-Block lies between Kadi in the north and Nawagam in the south, and is traversed by the Sabarmati River. The sedimentary sequence starts with the trap conglomerates succeeded by claystone, trap-wacke, sideritic marls. This constitutes the Lower Cambay shale. This sequence is subareally exposed from time to time, creating local development of reservoir facies with hydrocarbon accumulations in Nawagam.

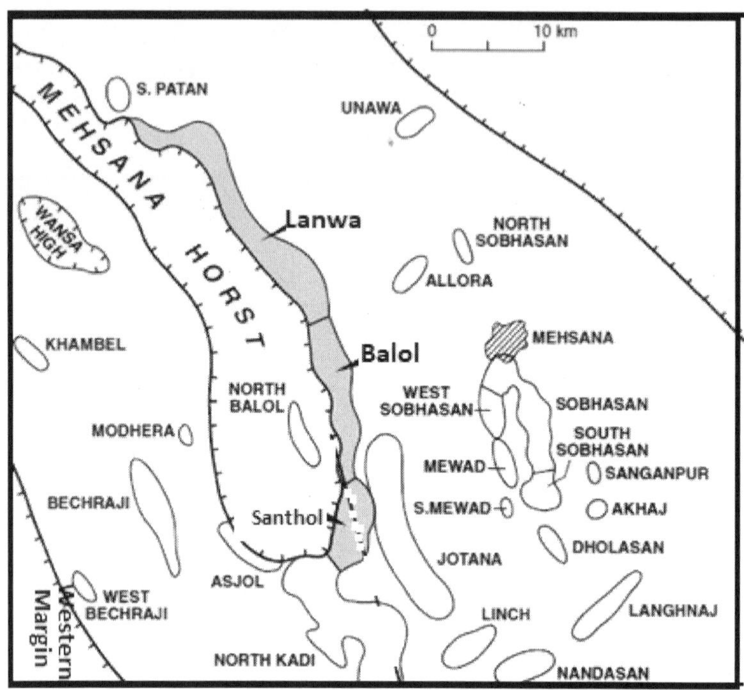

Figure 10.ca.3. Tectonic Elements of North Cambay Basin (After Prabhakar, K.N., *et al.*, 1999).

Broach Depression

This is an oval shaped deep, asymmetrical part lying between the Mahi River in the north and the Narmada in the south and is the deepest depocenter in the Cambay graben. The depression is filled with thick Cretaceous to Quaternary sediments, probably exceeding 9 km and lying over the Archean basement with an intervening thick lava flows of Late Cretaceous to Paleocene age. Very little is known about the pre-Tertiary structures of the basin. During Tertiary, the depocenter shifted from north to south giving it an asymmetric shape. Large number of faults and associated structural highs mark the fringe areas of this depression. The central region of the basin is much less faulted in comparison to the flank areas. The faults in the northeast are oriented in NW-SE direction and the north western flank, they trend NE-SW approximately.

Narmada Geofracture is one of the important tectonic features recorded in the Indian structural trends in approximately ENE to WSW direction for a distance of 1000 km between the northeastern part of the peninsular shield and a point close to the west coast of India. It is also known as "Narmada-Son lineament" which is a well-marked geomorphological by the straight valleys of two large rivers, namely, the Narmada and the Son flowing in opposite directions. The remote sensing data (Agrawal, 1988) clearly indicate the existence of the Narmada lineament and continuity of this fracture zone from the central India, westward into the southern part of the Cambay Basin. Structure mapping of the outcropping belt of the Tertiary, Mesozoic and Deccan Trap, coupled with shallow bore holes along the trend of lineament adjacent to the Cambay Basin (Roy, 1961) have established the continuity of the Narmada-Son geofracture into the basin.

In outcrops, it forms the boundary between sedimentary sequences of widely differing age. In the north of the lineament, no Gondwana rock is present, while to the south the Vindhyan rocks of Precambrian age are absent (Chaubey, 1997). The development of the lineament across the Gulf of Cambay might have affected the oil prospectivity in the region of Gulf of Cambay. The evolution of western marginal basins of India, suggests that the reactivation of the Narmada Fault zone in the Cretaceous that resulted in the opening of the Mesozoic rift basin.

From several deep seismic sounding lines shot across the Narmada-Son lineament it was concluded that the lineament represents a deep fracture extending upto the Moho and the fracture has been intermittently active right from the Precambrian times to present (Kaila, 1986). The position of the geofracture at the Moho level in the Narmada Block is at 7 km south of the Narmada River, which is 3.5 km south of the Narmada Fault zone as mapped at the Middle Eocene level. The fracture zone interpreted from the deep seismic sounding profile and the Narmada Fault zone mapped within the Cenozoic sequence from well and seismic reflection data appear to be same. A series of north-south seismic profiles clearly shows that the fault dies out before it reaches the east coast of the Gulf of Cambay.

The West Coast fault which is a major fault on the eastern margin of the Dahanu depression in the southernmost part of the Cambay graben, apparently has different trend (NW). It appears to continue northeast ward into the onshore area near Surat instead of continuing northward, and coincides with the prominent NE-SW lineaments, Aravalli trend.

Narmada Block

The Narmada block is broadly an uplifted region to the south of the Narmada River containing a number of smaller blocks or slices which have undergone differential vertical movements. Relatively thinner sedimentary cover is present on Deccan Trap in the area. The block's tilt westwards with gradual thickening of sedimentary layers and concurrent changes in the facies of the sediments. To the east, along the fringe of the basin, the Tertiary rocks are exposed. Further to the east of the Tertiary outcrops, the Cretaceous sequences of marine and non-marine facies are exposed as inliers within the Deccan Trap basalt.

The Broach Depression came into existing as a linear structural element trending N-S at the end of the Palaeocene. The area from Tankari to Dahej experienced maximum subsidence during Palaeocene-Eocene time.

To the north of Pakhajan existed a transverse palaeo-ridge between Galodra and Khelod. Subsequent to the deposition of the Olpad Formation over the Trap, the area experienced regional transgerssion when the Cambay Shale was deposited as a marine facies. This palaeo-ridge might have influenced the development of sandstone facies in palaeo shoal environment within the Cambay Shale in Galodra-Khelod area. Another palaeo-ridge, trending almost N-S, existed in the northwestern part between Malpur and Nada. This ridge also perhaps created local palaeo-shoal condition and influenced development of sandstone facies within the Cambay Shale.

Sedimentation

Exploration data indicate a thickness of about 8 km of Tertiary sediments overlying the Deccan volcanics. The sedimentary succession in the basin ranges in age from late Cretaceous pre-rift sediments to Recent. Based on basin formative tectonic events and their impact on sedimentation processes, three distinct sedimentary sequences have been recognized: the pre-rift late Cretaceous sequence, the syn-rift Paleocene sequence and early Eocene to Recent post-rift sequence.

The late Cretaceous sediments were deposited in continental, fluvial setup and comprise sandstones with minor claystone, basal conglomerate and thin inter-beds of shale. The early Paleocene sequence, bounded by unconformities below and above, is restricted only to fault grabens and fault scarps, and is either absent or thinly developed over paleo-highs. These sediments extend as piedmont conglomerate represented by trap-wash, claystone, shale and sand with thin coal beds toward north representing initiation of basin fill.

The late Paleocene witnessed the first marine transgression, followed by regression, during which arenaceous sediments in the northern areas of Mehsana got deposited under tidal regime on the down-thrown blocks. The top of Paleocene marks the end of rifting followed by an unconformity over which early Eocene transgressive shale is observed in the southern part of Cambay Basin. The Eocene to Recent post-rift sequence overlying the Paleocene succession is encountered throughout the Cambay Basin with a minor hiatus. Early Eocene witnessed the deposition of grey shale, silts and coals. The top of early Eocene witnessed regression during which deltaic sands and silt, shale and thick coal bands were deposited (Figure 10.ca.4).

The early/Middle Eocene boundary is marked by a regression in Middle Eocene during which deposition of sandstone, siltstone, shale and coals took place in tidal to estuarine setup. The upper part of Middle Eocene and late Eocene is a mixed facies of laminated sandstone, with limestone nodules and occasional carbonaceous shale. The early Oligocene succession overlying the Eocene strata after a brief hiatus, includes grey shale with claystone, argillaceous sandstone followed by a major regression and unconformity spanning the entire duration of late Oligocene. The Miocene succession in the basin is represented by current bedded sandstone, conglomerates, clays and shale. This is followed by an unconformity above which claystone, sandstone, clay, sand and kankar of Pliocene to Recent age were deposited.

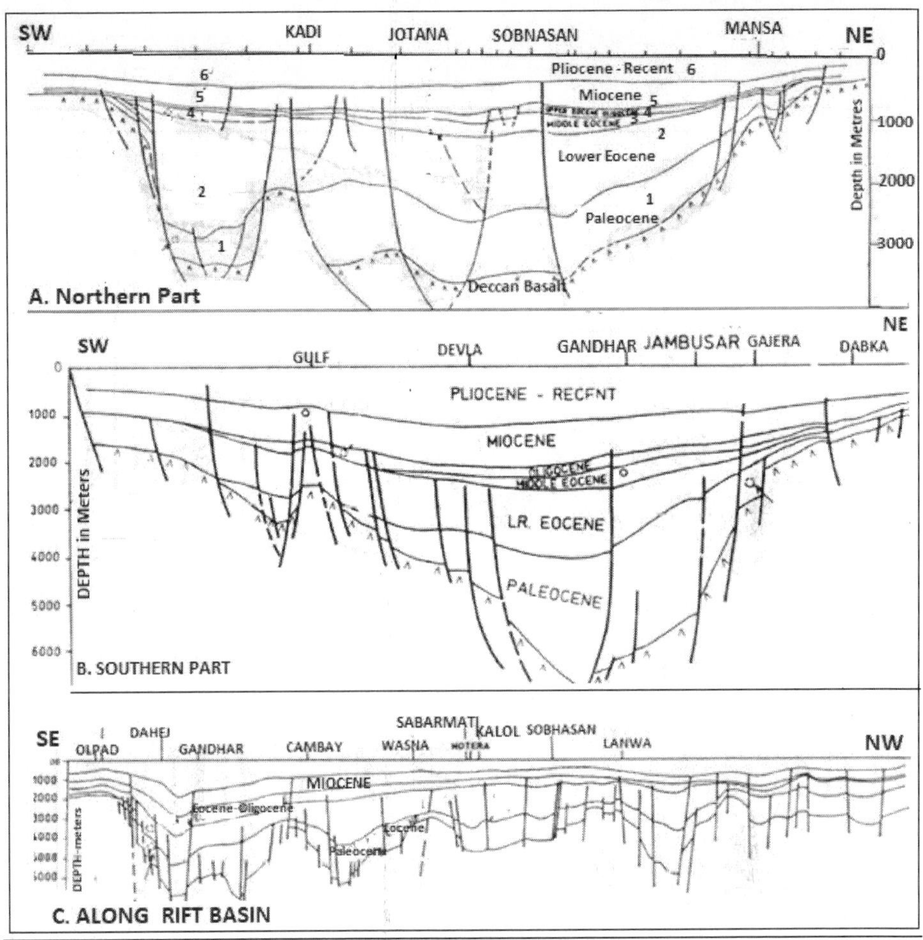

Figure 10.ca.4.Geological Sections–Cambay Basin.

Depositional Style

In the early stages, the basin was differentiated into positive and negative areas by faults cutting across the Deccan Trap. In the fault controlled low areas, thick sedimentary sequence called Olpad Formation, mainly derived from the Trap escarpments was deposited (Figure 10.ca.5-A). The first marine transgression in the basin took place sometime during Upper Palaeocene composed of dark grey to black shales named Cambay Shale.

However, soon after much of the connection with the open sea was lost and euxinic conditions prevailed. Along with the deposition of Cambay Shale, periodic minor fault movements took place which are observed by the presence of silt and shale. During the deposition of the upper part of this sequence, sand coal sequence called Kadi Formation and their probable homotaxial sediments (Surat and West Mehsana formations) were deposited in the north.

Towards the end of early Eocene, the depositional environment changed from marine to paralic in which coarser sediments were deposited in contemporaneous delta complexes. The interlocking of these delta complexes resulted in the formation of the Hajad unit (Figure 10.ca.5-B).

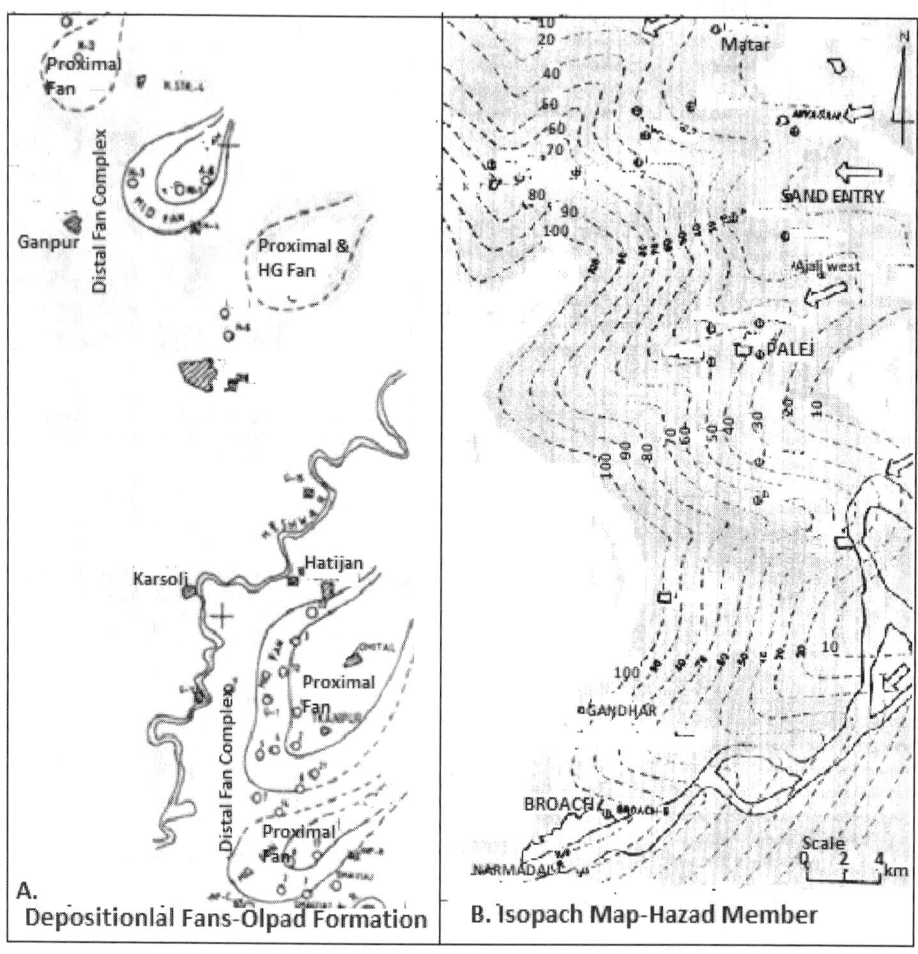

Figure 10.ca.5. Depositional Model–(A) Olpad Formation, (B) Hazad Member-Cambay-Basin (After Dhruvedra Singh, *et al.*, 1999).

The beginning of the Middle Eocene period witnessed retreat of the sea and subsequently the Kalol Formation was deposited as a result of repeated regression (silt-coal sequence) and transgressions (dark grey shales). The subsidence in the basin along the marginal faults was moderated and the resulting deposits are seen overlapping the shelves of the basin. The end of deposition of the Kalol Formation is marked by a major transgression in the basin during Middle Eocene period. The overlying Tarapur Shale and the Ankleshvar Formation were deposited in different parts of the basin under neritic and littoral to shallow marine environments respectively.

Following the deposition of the Hajad Formation, the area experienced widespread transgression in which the Kanwa Shale unit was deposited over a large area. During this transgression, the Tarapur Shale, the lower part of which is known as the Telwa shale unit in the Broach depression, was deposited all over the Cambay basin (Raju, ATR and S. Srinivasan, 1983). Towards the later part of the upper Eocene, the trangressive phase was locally interrupted by transverse influx of coarser sediments supplies by proto-Mahi, proto-Dhadhar and proto-Narmada, drainage systems with Major deltaic depositional lobe associated with more than 40 percent arenaceous facies. This resulted in the deposition of the Ardol Member and the Dhadhar Formation in deltaic environment in the Southern Cambay Basin.

At the end of the Dhadhar Formation, the eastern part of the area experienced a period of erosion and non-deposition during the Upper Oligocene time. This episode was followed by initiation of subsidence in the eastern part of the area around Broach. The succeeding sedimentary sequences were deposited in alternate transgressive and regressive phases all over the basin.

The Broach Depression came into existence as a consequence of continued subsidence along with sedimentation during post-Oligocene period. During this period, the sedimentary sequences in Gandhar area experienced minor up warp. This up warping might have resulted in the formation of Gandhar structural nose.

The end of this period, *i.e.,* late Oligocene, witnessed retreat of the sea and the sediment underwent mild erosion on the edges of the basin (Kundu, J, *et al.,* 1993).

The Miocene period again witnessed marine transgression and the sediments (Kathana Formation) were deposited in a shallow marine to paralic environment. The overlying Miocene sediments (Baba Guru Formation, Kand Formation and Jhagadia Formation) are of unstable platform type, with frequent epiorogenic movements in the basin. The basin was shallow and the environments varied from shallow marine to continental.

During late Miocene-early Pliocene period, the Narmada fault was reactivated resulting in the formation of a depositional sink of the post-Miocene sediments to the immediate north of Narmada River. The existence of marine to brackish conditions until Sub-Recent times in the basin north of Narmada River can be postulated. The present day alluvium of Gujarat was deposited by the major rivers Sabarmati, Mahisagar, Narmada and Tapti. The positive movements in the basin continued till Sub-Recent to Recent period (Figure 10.ca.6).

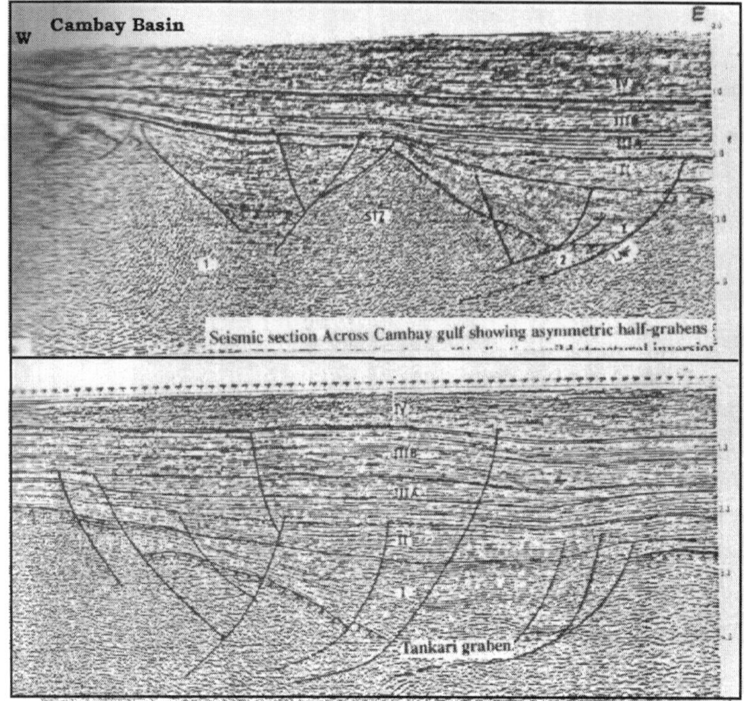

Figure 10.ca.6. Seismic Section–Gulf-of Cambay-South Cambay Basin (After Kundu J, *et al.*, 1993)

Sub-surface Stratigraphic Nomenclature

After detailed study of well lithology and in relation to out crop stratigraphic information coupled with paleontological studies of subsurface lithology, the rock stratigraphic nomenclature was formulated. (Figure 10.ca.7). Based on many micro lithological variations with time and space the litho-stratigraphy was revised by ONGC (Pandey J, *et al.*, 1993).

Further, adding the time concept to known litho-markers with the help of seismic data the lateral variations are shown areally in the Basin for better understanding with time and area (Madan Mohan, 1995),

Petroleum Geology

The Cambay Basin is a proved petroliferous Basin with many discovered oil and gas fields having many reservoir facies deposited in various depositional environments. Since the basin is fault controlled rift basin the understanding both argillaceous and arenaceous facies for exploration and development needed focused attention in relation to depositional input direction (Figure 10.ca.8). About 145 oil and gas fields were identified in the basin. Most of them are either small fields or extension of larger fields in different directions. Most of the reservoirs are very thin clastic beds within depositionally controlled with in major argillceous sequences.

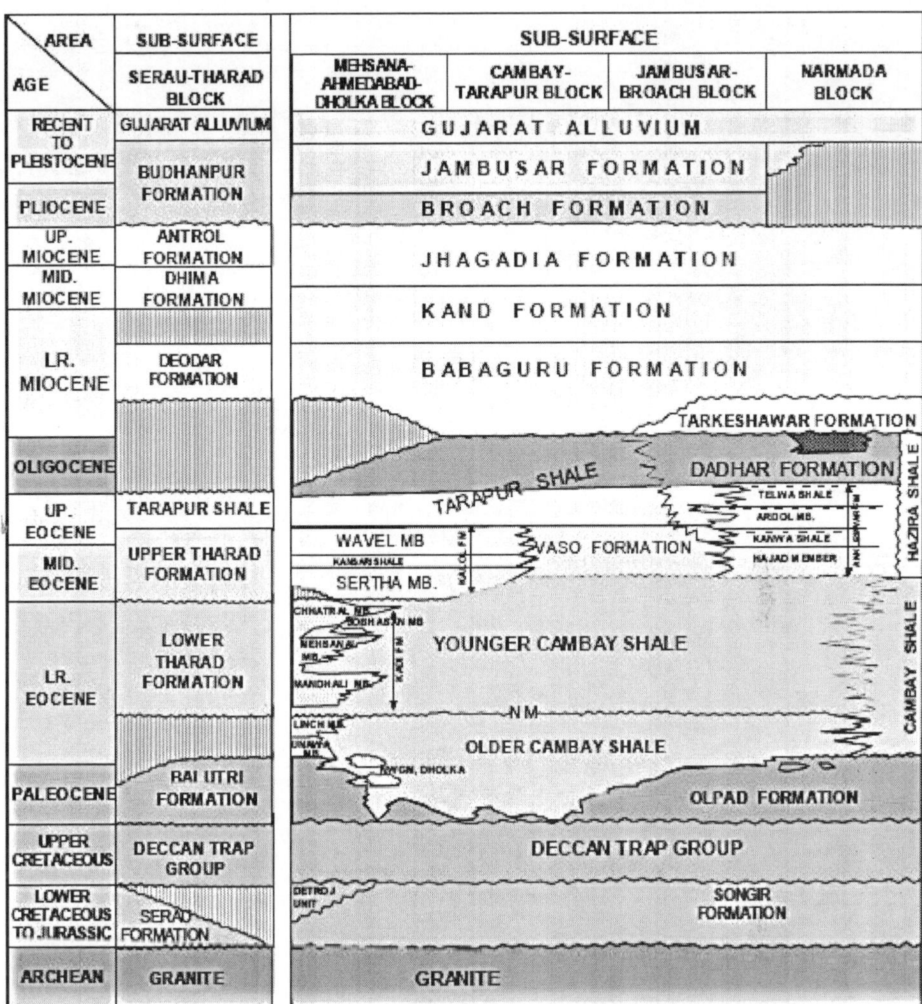

Figure 10.ca.7. General-Sub Surface Stratigraphy–Cambay Basin (After Madan Mohan, 1995).

Source Rock

Thick Cambay Shale has been the main hydrocarbon source rock in the Cambay Basin. In the northern part of the Ahmedabad-Mehsana Block, coal bearing sequence, which is well developed within the deltaic sequence in Kalol, Sobhasan and Mehsana fields, is also inferred to be an important hydrocarbon source rock. The total organic carbon studies suggest that shales of the Ankleshwar/Kalol formations also are organically rich with TOC percentage up to 3.5 in the northern depocenter in Cambay shale and up to 2.5 per cent in the Kalol formation (Figure 10.ca.9-A). The same is true for the Tarapur Shale. Shales within the Miocene section in the Broach depression might have also acted as source rocks (Banerjee A, *et al.*,

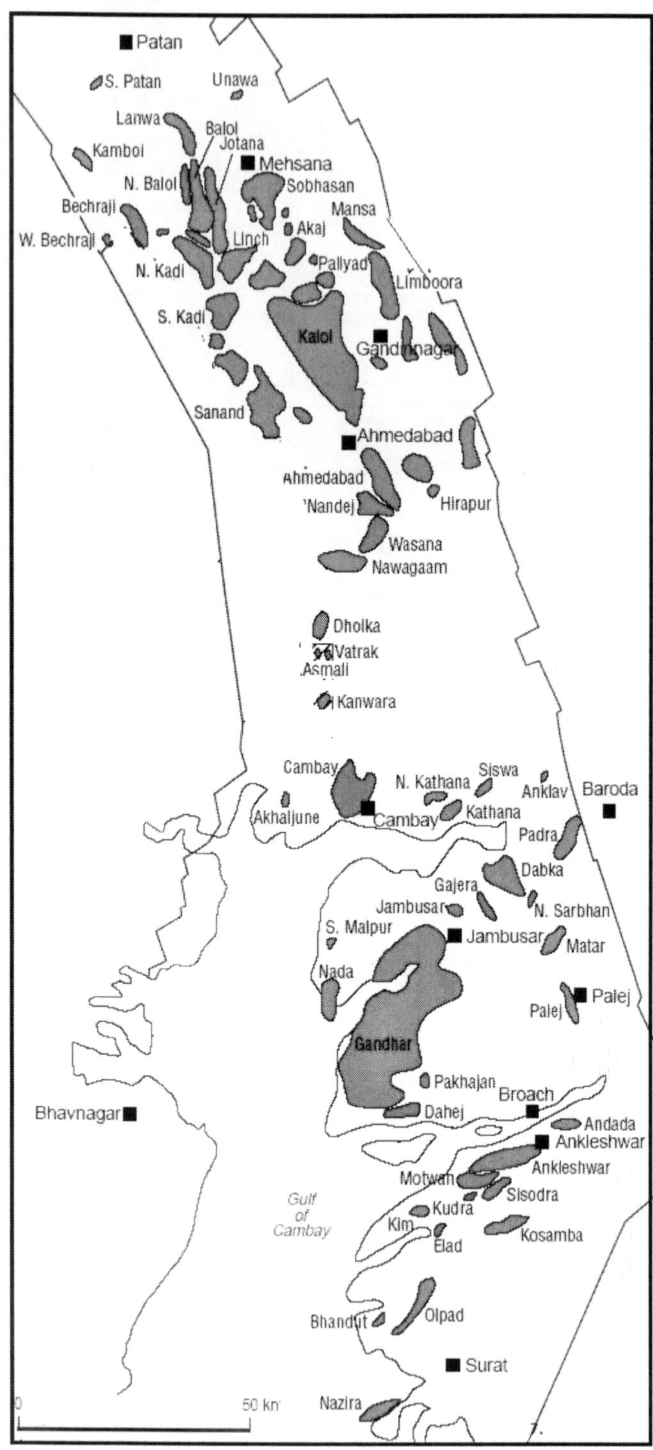

Figure 10.ca.8. Oil and Gas Field–Cambay Basin.

2002). The organic matter is mainly type-II with little type-III (Figure 10.ca.9-B). The type of organic matter is found to be predominantly terrestrial (Type-III) except in Dharnoj-3 where source input is interpreted as marine.

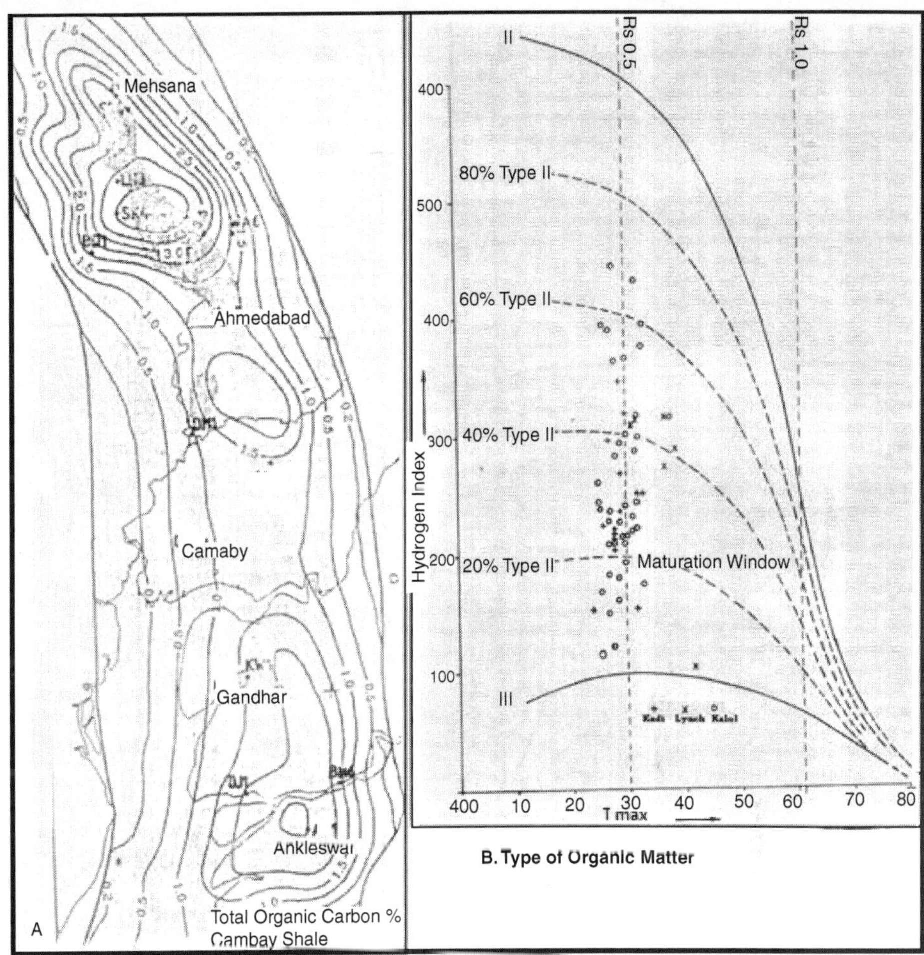

Figure 10.ca.9. A-Total Organic Carbon (TOC) Map of Cambay Shale-B-Type of TOC (After Raju, ATR and S. Srnivasan. 1993).

In the Mehsana-Ahmedabad Block, the Lower Cambay Shale contains mainly Type III and subordinate Type II kerogen and has high oil-generative potential. In the Tertiary sedimentary succession of the Broach and Narmada blocks, principally, three source rock sequences namely (i) The Cambay Shale, (ii) shales within the Hajad/Ankleshear Formation, (iii) the Tarapur Shale and its equivalents *viz.* (a) Kanwa Shale and (b) shales within the Ardol Member and the Dadhar Formation have been identified. Total organic carbon in all these source rocks increase in quality and quantity towards the depocentre.

Maturation and Expulsion

The source rock data indicate that the oil maturation is concentrated in the central part of the block. It also suggests that the direction of hydrocarbon migration is likely from the center to the flanks of the Tankari depression. The Cambay Shale is the main source rock and is found to be thermally well matured and falls within the oil window. In most parts of the area the TOC ranges from 2.25 to 2.0 per cent, average HI from 150 to 100 mgHc/gm TOC and the total hydrocarbon generated and expelled ranges from 0.73 to 1.40 and 0.11 to 0.5 MMt/sq km. Generation and expulsion of hydrocarbons started around 15–10 Ma and 12-8 Ma respectively. The peak of oil generation and migration is understood to have taken place during Early to Middle Miocene.

Hydrocarbons generated in the Cambay Shale migrated to the Hajad Formation through faults or through porous layers acting as carrier beds. The movement of hydrocarbons from the Tarapur Shale and its equivalents is thought to have taken place through the porous layers in the Kathana Formation.

The geochemical studies indicate good to excellent source rock potential of Older Cambay Shale and Olpad Formation.The TOC and VRo values gradually increase towards the basinal part. The TOC values of shales within Older Cambay Shale are 3 to 4 per cent in the area around S. Patan-1 and Dharnoj-3 and 0.5 to 1 per cent in E. Patan-1 and Daswara-12. Similarly the vitrinite reflectance (Ro) values are 0.7 to 0.8 in the basinal part and 0.4 to 0.5 in the basin margin areas. Sediments below 1500m in S.Patan-7 and S.Patan-9 are thermally mature. The upper boundary of the oil window indicating the threshold of liquid hydrocarbon generation (Ro = 0.65) is marked around 1900m. Peak oil generation (Ro-0.8 to 1.0) could be estimated below 2200m whereas the end phase of liquid hydrocarbon generation may be around 3000m.

Migration

The secondary migration of hydrocarbons to the traps is mainly vertical and short distance. However, short distance lateral migration has also played a major role in charging the margin areas. The accumulation of oil and gas in these structures are mainly due to migration of hydrocarbons from the source beds into the adjacent reservoir rocks. After the deposition of sufficient thickness of overburden, the hydrocarbons migrated into the immediately overlying or adjacent porous beds, and then accumulated in the structurally highest part.

It is likely that they must have migrated updip to the margins of the basin. It is envisaged that at least some of this oil might have been trapped in zones of pinch-out and wedge-out of deeper sands.

During the generation of the major qantity of hydrocarbons following the deposition of the Kalol Formation the sedimentary formation were over pressured. Thus, the migration of hydrocarbon in the area is known to have taken place during post-Miocene times.

Reservoir Rock

There are number of the reservoirs within the trap-wacke sequence of the Olpad Formation. These consist of sand size basalt fragments. The associated unconformity also helped in the development of secondary porosity. Besides this, localized sandstone reservoirs within the Cambay Shale as in the Unawa, Linch, Mandhali, Mehsana, Sobhasan, fields,, *etc.* are also present. The most promising hydrocarbon reservoirs in order of importance in northern part of the Cambay Basin are Kalol Sands, Sobhasan Sands, Subsidiary Silts/Sand, Deeper Sands within the Cambay Shale.

The reservoir characters of the Mid Eocene-Oligocene sequences indicate that, the sandstone layers within the Hajad Formation, the Ardol Member and the Dhadhar Formation possess good inter-granular porosity. In general the reservoir sands are very thin but stacked channel sands. In a view sand bodies in certain areas re-crystallization and pore-filling have caused reduction of primary porosity.

The sandstone reservoirs in Kalol Formation of Middle Eocene age in Balol-Santhal fields were formed by carving of sub-aerial erosional valleys on the flanks of a major palaeostructural feature 'Mehsana Horst' along which fluvio-deltaic sands were deposited by south flowing river systems.

Cap Rock

The lithological heterogeneity in the Olpad Formation gave rise to permeability barriers, which facilitated entrapment of hydrocarbons. Transgressive shales within deltaic sequences provided a good cap rock. The formation of traps in the depression is found to be post-Miocene.

Hydrocarbon Plays

Petroleum accumulations are known in all sequences ranging from Palaeocene to Miocene. Major accumulations are known in Middle Eocene structural traps over block uplifts and block edge folds. A few combination traps along block boundaries and over structural noses also occur. Early to Middle Eocene Cambay Shale is the main source rock.The latest estimate of resource is 2,050 MMT and in-place reserve is 1140 MMT.

Structural Highs and fault closures and stratigraphic traps (pinchouts/ wedgeouts, lenticular sands, oolitic sands, weathered trap) in Paleocene to Miocene sequences have been proved as important plays of Cambay Basin.

Mehsana tectonic block is the principal locale of heavy oil accumulation with 150mmt oil in place with a low potential for primary recovery. The heavy oil belt extends from northern part of North Kadi to Lanwa in the north encompassing in between Santhal and Balol fields, an elongate stretch of over 30 km covering 60 sq. km. Besides these, Bechraji field, lying west of Mehsana Horst also hosts heavy oil. Entrapment in this homoclinal structure by way of up-dip pinch out against the palaeohigh resulted in the accumulation.

This oil was biodegraded subsequently resulting in low API gravity. Oil in this faulted reservoir is characterized by high oil density ranging from 0.92 gm/cc

(API 17°) and viscosity 60 cps in Santhal field to 0.98 gm/cc (API 13°) and viscosity 1400 cps in Lanwa field at reservoir conditions. Reservoir permeabilities are of the order of 8-15 Darcies and reservoir energy is in the form of strong aquifer support.

Paleocene – Early Eocene Play

Olpad Formation/Lower Cambay Shale. Sand size basalt fragments and localized sandstone. Unconformities within the Cambay Shale and between the Olpad Formation and the Cambay Shale have played a positive role in the generation of secondary porosities. The Olpad Formation is characterised by the development of piedmont deposits against fault scarps and fan delta complexes.

Middle Eocene Play

Upper Tharad Formation. In southern part, Hazad delta sands of Mid to Late Eocene and in the Northern part the deltaic sequence is made up of alternations of sandstone and shale associated with coal. Plays are also developed in many paleo-delta sequences of Middle Eocene both in northern and southern Cambay and in the Northern Cambay Basin, two delta systems are known.

Late Eocene – Oligocene Play

Tarapur Shale, Dadhar Formation. This sequence is observed to possess good reservoir facies in the entire Gulf of Cambay. North of the Mahi River, a thick deltaic sequence, developed during Oligo–Miocene, has prograded upto south Tapti area.

Miocene Play

Deodar Formation (Lower Miocene), Dhima Formation (Mid-Miocene), Antrol Formation (upper Miocene). The Mahi River delta sequence extends further westward to Cambay area where Miocene rocks are hydrocarbon bearing.

Petroleum Systems

The Cambay shale is the main source rock sequence in the basin along with minor contributions of oil from Kalol, Tarapur and Olpad formation. Regional Cap rock is Tarapur Shale and intervening shales, Clays and Carbonaceous facies within Olpad, Older Cambay Shale, Kadi and Kalol formations also act as local cap rock. Trap is strati-structural. The Olpad, Cambay Shale, Kalol/Hazad, Tarkeshwar sands and Trap wash with thin shale/claystone intercalations, constitute reservoir rocks. The basin has indicated that Cambay Shale has the potential for Shale Oil and Shale Gas.

Two major petroleum systems have been identified in Cambay Basin. In South Cambay Basin, the petroleum system is Palaeocene-Early to Middle Eocene-Middle Eocene and in North Cambay Basin, it is Palaeocene-Middle Eocene. These are also the Cambay-Kadi/Kalol petroleum System in north, and Cambay-Hazad petroleum System in the south.

Description of Fields

North Cambay Area

Khambhat or Lunej

The first well was drilled in 1958 at Lunej near Ahmadabad and confirmed the occurrence of a commercially exploitable oil field. About 62 wells were drilled out of which 19 yielded gas while 3 yielded oil. The annual production is 15 lakh tonnes of oil and 8-10 lakh cubic metres of gas. The total reserves are estimated at 3 MMT.

Ahmedabad

The field, also known as Ramol field is located 11kms south-east of Ahmedabad city, in southern part of Ahmedabad-Mehsana tectonic block. The areal extent of the field is nearly 45sq. km. The first well was drilled in the field in 1965 and subsequent drilling and testing identified the presence of six hydrocarbon bearing zones in the Kalol Formation of Middle Eocene age. These pay zones are: Kalol - III, IV, VII, VIII, IX and X. Pay zones K – III and IV are clubbed into Wavel member, whereas K – VII, VIII, IX and X into Sertha member (Pandey, *et al.*, 1993). Regional Kansari shale separates these two members.

Reservoir units K – IX B and K – IX C are the main producers of the Ahmedabad field. The reservoir unit K – IX A is developed at places towards south. The unit K – VIII is developed to the northern and the central part and is underlain by a coal band which is widely developed in the field. The youngest unit of Sertha Member, K- VII is developed in the northern and the central part having an average thickness of about 2m and a maximum of 4-5m.

Lanwa-Balol-Santhal (LBS)

The LBS Field Complex comprises a continuous, heavy-oil bearing belt on the eastern flank of the Mehsana High. It comprises a continuous, 40 km long by 2-3 km wide, heavy-oil belt that has been divided into the Lanwa, Balol and Santhal fields The LBS Field Complex has a STOIIP of 752 MMBO and reserves recoverable by natural drive of 92 MMBO, representing an average recovery factor of 12 per cent.

The Santhal was discovered in 1971 and brought onstream in 1975.The Lanwa and Balol fields were discovered in 1972 and put on production in 1988 and 1982 respectively. So far in Balol about 120 wells, Lanwa 130 and in Santahl 160 wells were drilled in the field including in situ-combustion purpose wells.

The structure is monocline and dipping to east which is abutting at Mehsana Horst in the west and supported with active edge-water from east direction. The field possess a multi-layered reservoir of late Eocene age. The main producing sands are Upper Suraj Pay, KS-I, KS-II, KS-III and lower sand of Kalol formation. The pay sands are separated from each other by shales (Figure 10.ca.10). The field complex is a classic stratigraphic trap. The heavy oil accumulation is fault-compartmentalized.

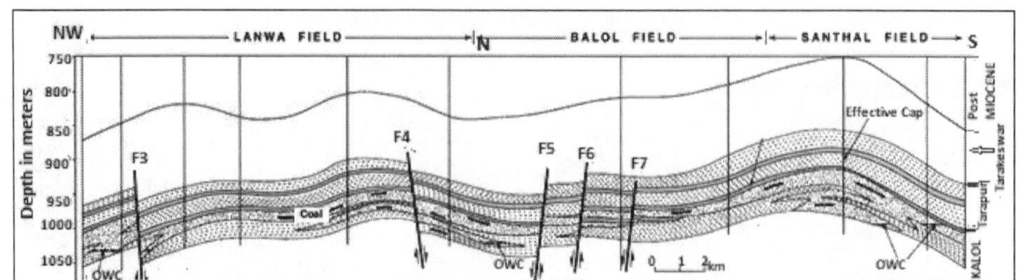

Figure 10.ca.10. Galogical Section along Cambay Basin (Lanwa-Balol-Santhal) (After Srivastava, D.K., *et al.*, 1999).

The accumulation is dip-closed to the east, with minor fault offsets, and stratigraphically trapped to the west by onlap and depositional pinch-out of the Eocene Kalol Formation reservoirs onto the Mehsana High. The accumulation is top-sealed by the Tarapur Shale, which conformably overlies the Kalol Formation. The Kalol Formation reservoirs dip to the east at 4-8°. The western pinch-out line occurs at depths of 840-920 m TVSS in the Santhal Field, 910-940 m TVSS m in the Balol Field and 830-980 m TVSS in the Lanwa Field. The LBS Field Complex has a single OWC, which dips gently to the north. The OWC is at 984-992 m TVSS in Santhal, 992-1020 m TVSS in Balol and 1020-1034 m TVSS in Lanwa. The LBS Field Complex has a maximum hydrocarbon height of approximately 205 m. The northward OWC tilt has been observed. The Balol Field has a net-pay range of 08 m.

The LBS Field Complex contains up to three separate reservoir units. KS-I, which contains about 80 per cent of the STOIIP, is the only unit that occurs throughout the field complex. KS-II is developed in the Balol and Santhal fields and KS-III is developed only in the Santhal Field. Two other minor reservoir layers with patchy oil distribution and minor reservoirs are also present: (1) the uppermost Upper Suraj Pay (USP), which occurs in the Balol Field and (2) the lowermost Lower Sand (LS). Faults divide the field complex into several non-communicating areas with differing original OWCs.

There is a general increase in permeability towards north. This axial drainage system represents fluvio-deltaic deposits. The targeted sands for *in situ* combustion (ISC) are KS-I and KS-II in Santhal field and KS-I in Balol field. Santhal is one of the fully developed commercially producing heavy oil fields of Cambay Basin. The Middle Eocene pay zone (Kalol Formation) is characterised by unconsolidated sand units with active acquifer as the drive mechanism.

The average net pay thickness in the field complex is 2-10 m and in the EOR areas is >5 m. Its porosity ranges from 20 to 36 per cent and permeabilities are average: 57.5 milli darcies. Oil viscosities increase northwards within the field complex, with the highest viscosities (550 cp at reservoir conditions) occurring in the Lanwa Field. The LBS oil is rich in aromatics (25-59 per cent), suggestive of oil degradation caused by water-washing and biodegradation by meteoric water infiltrating into the reservoir from the Mehsana High.

Lanwa

The western limit of the structure lies on the eastern margin of Mehsana Horst and forms a north-south trending homocline with average dip of 7° towards east. Most of the pay sands, developed in the Lanwa field, have shaled out against the Mehsana Horst. The Lanwa Field has a net-pay range of 0-20 m, with the highest values occurring approximately mid-way between the pinch-out line in the west and the OWC in the east.

In Lanwa, hydrocarbon accumulations in Babaguru Formation are identified only in the south-central part of the field where either of the pays or both have developed.BS-I is developed in eleven wells, however, development of better reservoir facies with good hydrocarbon saturation is observed in few wells. Pay sand BS-II is developed in eight wells where the reservoir attributes are good in most of these wells. BS-III is developed only around the two wells.

Balol

Tertiary recovery techniques (primarily *in-situ* combustion) are forecast to increase the ultimate recovery factor the heavy oil. The major pay zone in the Balol sandstone reservoir. The reservoir zone is 32 m in thickness and top of the reservoir is at depth 996 m. The porosity in the range of 28-30 per cent and its permeability varies from 8-15 milli darcy (Asit and Mohan, 2004). The oil in this reservoir is asphaltic in nature. The primary recovery of viscous and heavy oil from Balol is only 10-12 per cent, hence the pay zone is under thermal EOR process. The in-situ combustion process is carried out to improve secondary recovery of oil from the sand stone reservoir.

Santhal

It is located 20 Km S-W of Mehsana town forms a part of heavy oil belt. The field is basically a monocline structure with gentle dip of 1-3° from west to east direction. The reservoir abuts Mehsana horst towards west and is supported by active edge-water from the east. The field is connected to Balol in north and North-Kadi in south without any physical barrier in the reservoirs.

The reservoir consists of 4 sand layers Upper Suraj Pay, KS-1, KS-II and KS-III+LS of Kalol formation, with a good cap of Tarapur shale. The sands have a common oil water contact (OWC) varying between 992m and 998 m from south to north and are supported with active edge water drive. The reservoir pressures are hydrostatic. In the Santhal Field, net pay ranges from 0 to 70 m, but is generally 1-40 m. Sandstone reservoirs in the heavy oil belt have gross thickness of 40-60m. Three major reservoir units are identified *viz.* KS-III, KS-II and KSI in chronological order from bottom. KS-I is widespread and two more reservoir units of limited areal distribution occur stratigraphically above KS-I called Upper Suraj Pay and another unit Lower Sand (LS) below KS-III. The oil in place is about 53 per cent is in KS-I sands, which are medium to coarse grained, loose and are highly porous and permeable. Porosities range from 20 to 36 per cent while permeabilities are of the order of 8 to 15 milli darcies.

Most of the wells have shown high water cut from the initial stages of production. The work-over jobs attempted to shut-off water have met with limited or no success especially in the wells located in wedge zone. Original OWC is fixed at 1010m MSL depth below which Sw is 100 per cent. Producing OWC is at 990m MSL depth, below which Sw is less than 100 per cent but there is no oil production. Above 970 MSL depth clean oil production is obtained for a longer period, Sw at this level is just around the irreducible saturation of 22 to 35 per cent. The high viscosity coupled with active water drive has resulted in adverse mobility ratios for oil. With primary recovery of only 11-15 per cent.

North Kadi

This field is located at a distance of 25 km SW of Mehsana town, (Figure 10.ca.11-A), and is on production since 1969 from sandstone reservoirs of Eocene age. The structure is elongate NW-trending anticlinal structure and covers an area of 35 km^2. This field is with OIIP reserves of about 74 MMt with ultimate reserves of 26 MMt. (Figure 10.ca.11-A)

The reservoir unit is mainly in Middle Miocene sandstone of Kalol Formation having high permeability and strong aquifer support. Earlier, the pays within Kalol Formation were believed to be hydro-dynamically connected (Bhandari L.L. and L.R Chowdhary.1975) (Figure 10.ca.11-B).

Unit KS-I shows more lateral continuity of the three units and KS-III, which is best developed on the eastern flank. The presence of hydrocarbon has been established in six different Kalol sands. Kalol sands have major share in reserves and production. These sands are having permeability of the order of 500 to 3000 md with strong aquifer support. The field is operating under active aquifer drive and less pressure drop was observed after a cumulative production of 17 MMT. Viscosity variation of 20-110 cp is observed from southern to northern part of the field. The top-reservoir surface has three culminations with vertical closures of 10-20 m. The field comprises a 30-40 m thick oil column and a 50 m gas cap. The well spacing of the field was kept at 200m (Optimum drainage radius) keeping the interference to minimum (Figure 10.ca.12). The field achieved plateau production of 1886 tpd during the period 1985-89 which gradually declined.

The new well data drilled later gave a new insight when the well showed the presence of gas in deeper zone, and oil in upper zone. A complete review of reservoir characterization was taken up with log correlation along with reservoir pressure data of various pay units. These have led to identification of five pays identified within Kalol Formation. The KS pays are highly permeable (upto 3000 md) and holds most of oil-in-place. The average pay thickness varies from 15m to 4m. The reservoirs within Kalol formation have strong aquifer support (Ghosh B.N.*et al.*, 2003 and 2005).

South Kadi

This field is located on the Sanand-Jhalora-North Kadi high trend was discovered in 1967 and was put on production in 1968. IOIP of the field has been

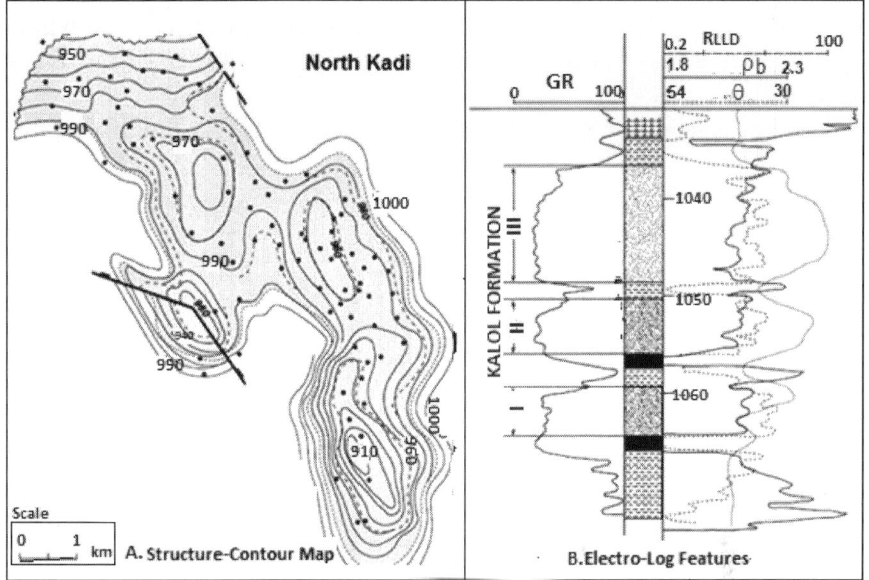

Figure 10.ca.11. (A)-Structure Contour Map – (B)-Electro-log- North Kadi Field.

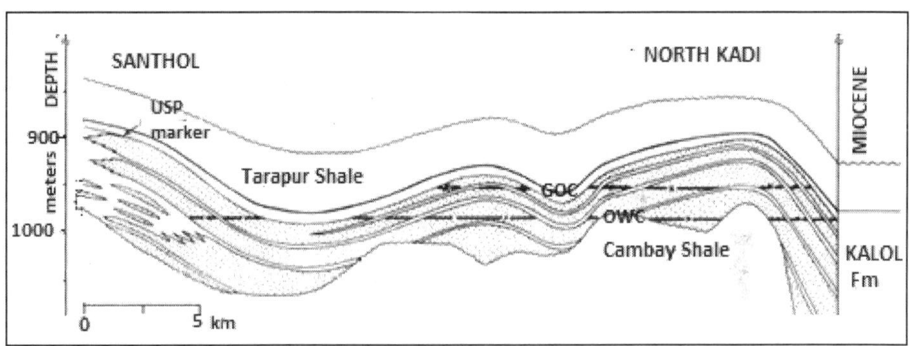

Figure 10.ca.12. Geological Section Across North Kadi Field.

estimated to be about 14 MMt with about 3.5 MMt of recoverable reserves. By now about 94 wells have been drilled

These sands are limited in areal extent and are lenticular. The sands are oil bearing and have got abnormally high pressures. The field comprises of multi layered reservoirs with other than Cambay Shale Formation, sands of Kadi Formation and of Kalol formation.

Lower Kadi pays are discrete sand bodies having limited areal extent; occur in the depth range of 1700m to 2200m, with sand thickness varying between 2m to 20m. All these pools are operating under depletion drive mechanism, with initial super hydrostatic pressure. Pressure transient studies and reservoir limit tests of some pools were carried out to decipher pool limit, volume of reserves and different reservoir parameters. Pressure-production history together with geological data has indicated more number of hydro-dynamically independent pools than initially known. Such small pools also added reserves when calculated through Material Balance method.

Rapid lateral facies variation of Lower Kadi sands has resulted in high degree of uncertainty in the field development a very challenging task with high risk. The multidisciplinary approach has resulted in enhancement of production from the field from 220 m /d to more than 550 m /d.

Jhalora

The field is northern extension of Sanand Anticline. Structurally, it is in the NNW-SSE trending anticline having steeply dipping westerly and easterly flanks with a horst like feature in the left, which has been dissected by a number of faults.

Kalol

This field is located in the Mehsana- Ahmedabad tectonic block over a basement high. It is located about 16 km north of Ahmedabad City covering an area of around 300 sq. km. The Kalol Structure is a doubly plunging anticline trending NW- SE with small culmintions. The first well in Kalol was drilled in 1960 (earstwhile Gandhinagar well) down to depth of 1587 m. It has a STOIIP of about 50 Mmt. The oil production started in 1967 with initial rate of 100 t/d which reached its peak of 1400 t/d. The cumulative total is of 10 MMt (about 20 per cent of STOIIP). After 1986, water encroachment and shutting-in of high water-cut wells resulted in a gradual decline in production. By now about 380 wells had been drilled.

In Kalol field hydrocarbons have been encountered in Olpad, Cambay shale and Kalol formations of Paleogene sequence. The trapping mechanism is strati-structural and stratigraphic. The field is compartmentalized into a number of fault blocks with sub-listric normal faults and cross faults. The sealing or non-sealing nature of faults plays a significant role in fluid distribution, pressure and production behavior of these reservoirs.

The object was in the interval 1421-1452 m which was oil bearing and later in addition to other payzones, were also established in the field. All these payzones were confined to the Kalol Formation, which has an average thickness of about 150 to 200 m (1350 to 1550 m.).

The reservoirs permeability's falling in the range 35mD, and an average porosity of 30 per cent. Initial water saturations average 25 per cent. The oil is heavy, with an API gravity of 21° and an average viscosity of 65 cp at reservoir conditions (163°F and 1523 psi at 1050 m TVSS). Viscosities increase across the field, from 10 cp in the south to 140 cp in the north. The oil has been produced by strong aquifer drive. Units KS-I and KS-II have bottom-water drive, while KS-III has edge-water drive. The predominant drive mechanism in major reservoirs is depletion drive and presently these reservoirs are operating under pressure maintenance. The best oil producers are located mainly on the southeastern flank, where oil viscosities are lowest. The poorest producers are located mainly in the north and west-central areas. Adopted technologies like multilateral/horizontal/drain hole drilling and MEOR to improve recovery.

Wadu

The structure is situated in the northern extension of Kalol field. The main oil procedures are the clastic reservoirs of Kalol Formation which is of Middle Eocene age and its equivalent, however, hydrocarbons have also been produced from Cambay shales, either from fractured shales or intercalated with silt or very fine sand bodies.

Jotana

The field is a fault closure located in the Mehsana-Ahmedabad block. First well came into production in 1980. It was discovered in 1977. It produces from multiple sands in an anticlinal structure. Intersection of NNW-SSE trending Jotana Main fault with NE-SW aligned transverse faults form two-way closures entrapping hydrocarbons. With more than 30 MMt of hydrocarbon reserves about 18 per cent of In-place reserves were produced. Hydrocarbon accumulation has been established at 4 levels from Older Cambay Shale to younger Kalol Formation – Linch, Mandhali, Sobhasan and Kalol at 1000-2000 mts depth.

Mandhali sands of Lower Eocene age host majority of hydrocarbon reserves with gross thickness of 50 to 160m is affected by faults varying in throw from 10-40m. It is divided into many small reservoirs by faults and permeability barriers. In these sands porosity ranges between 21-28 per cent and permeability ranges between 100-500 md in the western part near Main fault. In Kalol Sands, average porosity is in the range 20-35 per cent and average oil saturation is in the range 59-70 per cent.

In Sobhasan sands, porosity ranges between 20 to 30 per cent and oil saturation ranges between 50 to 70 per cent. In Linch sand, porosity and oil saturations are in the range 18-30 per cent and 50-90 per cent respectively. Average productivity of wells in the field ranges between 2 to 45 $m^3/d/Kg/cm^2$. Initial reservoir pressures in Kalol, Sobhasan and Mandhali sands were hydrostatic but super-hydrostatic in Linch sands. Current pressures in Linch are in the range 95-205KSC and 40-136 KSC in other sands. In general, oil properties are good but wax and paraffin contents make the oil thick in some wells. Most of the pools are acting under depletion drive (Deb, G.C., V. Pratap and S.C. Gawali., 2003). Reservoir pressures had dropped drastically in some of the wells, resulting in drop in production and ceasure of wells. Flow unit characterization, on the basis of electro log correlation and lateral continuity, indicates that 4 main units consist of sub-units separated by fine clastics/shale.

It is observed that, impact of water injection was not uniform in the whole area due to heterogeinities. With water injection, average reservoir pressure increased by 40 kg/cm^2. Intake of injected water by some of the layers only, led to water break-through in some good producers. With conversion of some wells into water injectors, change of perforation in existing water injectors were based on the evaluation by reservoir simulation studies (Deepti Tiwari, *et al.*, 2003). Enhanced oil recovery techniques, *e.g.*, MEOR, ASP and chemical flooding are also part of enhancing oil recovery from the field.

Sanand

The field is located in the western part of the northern Cambay Basin. approximately 25 km northwest of Ahmedabad. Sanand structure is an elongated doubly plunging anticline trending NNW-SSE. Discovered in 1962, field has been producing oil and gas since 1969. The northern plunge is gentle whereas the southern plunge is steep. Both limbs of the anticline are faulted by normal longitudinal faults running roughly parallel to the crest of the fold. The eastern portion of the field is further dissected by traverse faults (Figures 10.ca.13 and 14).

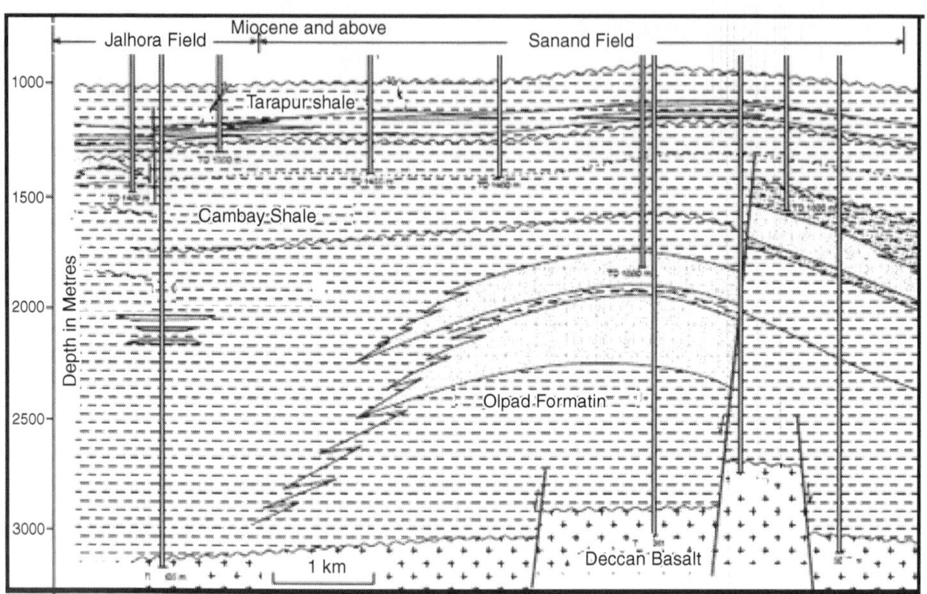

Figure 10.ca.13. Geological Section across Jhalora and Sanand Fields.

It is a multipay field, with major payzone is K-III of the Eocene Kalol Group. Payzone K-III has a STOIIP of 12.5 MMt and ultimate recoverable reserves of 3.5 MMt, representing an approximate recovery factor of 25 per cent. Payzone K-III contains medium-gravity oil and a gas cap. The reservoir sands consists of thin channel sandbody with good reservoir properties with average well-test permeability: 1000 mD. The oil was produced by natural drive, gas-cap expansion and weak aquifer influx.

The main reservoir rock is silt/sand of K-IV unit (Wavel member) of Kalol Formation. Additionally, sands/siltstone of Chhatral and Olpad- formations have also been found to be accumulation of oil and gas in Kalol, Cambay Shale, and Olpad formations. K-III pay zone is the main producer in northern part of the field and Jhalora, whereas in the southern part, K-IV is the main producing layer.

Gamij

This field is situated near the eastern margin of Ahmedabad-Mehsana block of the Basin. It is an elongated field about 25 km long and 5 km wide running North-

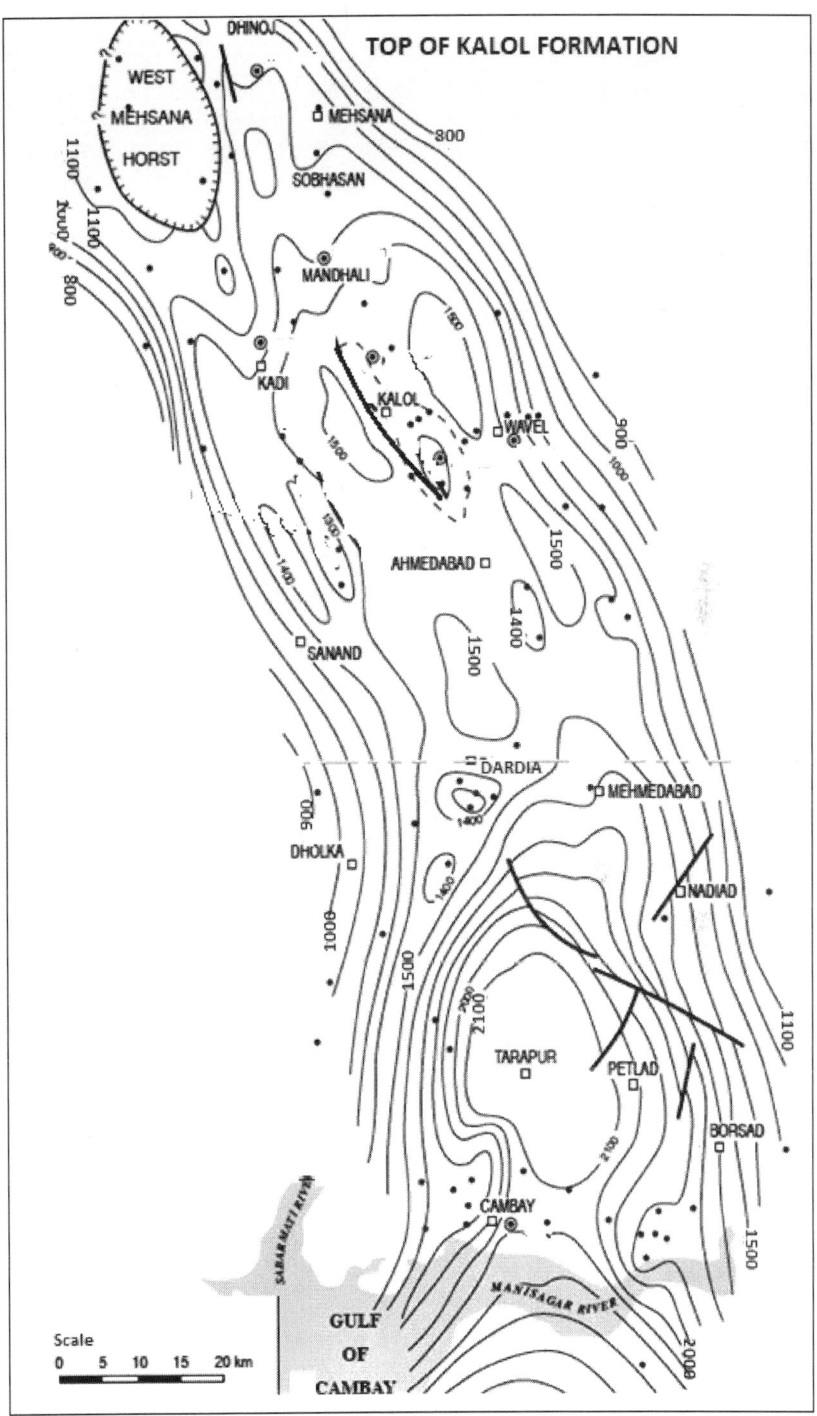

Figure 10.ca.14. Structure Contour Map–Top of Kalol Formation with Sanand Field.

South parallel to the eastern margin fault. The beds are dipping west with about 6-8° slope. There are several North-South longitudinal faults and NE-SW transverse faults dividing the area into several blocks. The field has multiple pays in different formations are in a shaly, silty reservoir, the free fluid porosity may be much less than the effective porosity derived from conventional core and log analysis. The large bound water associated with very fine clay and silt grains is the major cause of low resistivity.

Nawagam

In the Nawagam field, oil is confined to the lower and upper parts of Eocene. Trap derivatives represented by grey and brown ferruginous shales. Overlying this formation there occur grey fissile pyritic shales, sometimes containing thin coal bands alternated with silts and sandstones. Within the lower part of the section, several productive horizons are identified. Upper part of Eocene consist of black carbonaceous sideritic shales with occasional intercalation of limestones. Data of the Nawagam crudes show that the average properties of these oils are similar to those of Kalol and Cambay crude.

Wasna

The field, covering an area of 42 sq. km was discovered in 1969 and put on production since 1985, is located in the south – eastern part of Ahmedabad-Mehsana block of the basin and is flanked by *Nandej* field in the north and Nawagam field in the south-west. The field has an inplace reserve of 22 MMt in proved category yet the recovery is low due to poor reservoir properties. The field was delineated and developed by drilling 85 wells.

The Kalol payzones comprises of K – VIII, K – IX and K – X parasequences/ subunits which are composite bodies having thin sublayers of siltstone, very fine grained sandstone and silty shale and at many places as finely interlayered with finer clastics, *viz.*, shale, sideritic claystone, carbonaceous shale and also coal laminations. The pay zones occurring between 1470 – 1515 m depth in the field belong to Sertha member of Kalol Formation of Upper to Middle Eocene age. The drive mechanism is with solution gas.

The second stage of development and exploitation programme was initiated with workover activities and drilling of 54 new locations for even water injection were made. Due to reactivation job the production of the field gradually increased from 82 m^3/d to 206 m^3/d.

The **Bechraji field** lies in the northern part of Cambay Basin which was discovered in 1987 and was put on early production. The field has an OOIP of 32.65 MMT but because of high viscosity of crude oil (250cp) and low API gravity (15.6° API) the primary recoverable reserve is only 3.41 MMT, *i.e.*, 10.4 per cent of OIIP. (Dey, A.K., *et al.*, 2005).

The **Viraj field** is in Cambay basin which is a small field occupies an area of 4 sq. km. Though there are three oil-bearing zones in the field which was discovered in 1977 and put on production in 1980. The estimated oil reserves are 7.88 MMT. About 50 wells were drilled and the field has produced nearly 20 per cent of oil in-place.

The reservoir is mainly sandstone with good permeability. Occurrence of siltstone, often grading into shale and coal gives rise to reservoir heterogeneity. Initial average pressure for the reservoir works out to be 140 kg/cm^2 at -1260m MSL. Initial GOR (Gas-Oil-Ratio) from PVT data is very low at 15v/v with saturation pressure of 65kg/cm^2. With time the average field GOR has marginally increased. The reservoir pressure has not declined (Panda, B.K., *et al.*, 2005). Sand cut problem has been encountered in some the wells due to unconsolidated nature of sand. Most of the wells have been on production with successful gravel pack jobs. High water cut is an inherent problem of all wells. Since beginning the reservoir has water cut problem. Many wells have abnormal water cut and conventional water shut off job has shown no effect. The crude from this field being heavy and viscous is prone to foam. Silicone based de-foaming chemical, *i.e.*, DM-silicone fluid 10,000 CTSK was found most effective and decrease the foaminess index (Bassi. K.D. and R.K. Gang.2003).

Limbodra

The first well on the structure is situated in the eastern flank of the Cambay basin, has flowed thick oil from a 4 meter thick sand layer. This pay-zone was correlated as horizon-V of Kalol formation and within Tarapur Shale. This shale thickness increase towards western area (Figure 10.ca.15).

Nandej

This field of Cambay Basin is on production since 1990 and producing mainly from Kalol formation. The north-western part of the field is having better reservoir facies where the permeability varies from 10 md to 42 md. Central part of the field is considered to be poor reservoir with permeability of 5 md. The southern part of the field is having moderate reservoir permeability to the order of 10 md. These wells have been stimulated with mud acid but the well performance could not be improved. A number of acidization and hydro-fracturing jobs were carried out. New acid system fluoboric acid is being designed to be fit for clays. The problem of fines generation and subsequent damage is not observed in case of fluoboric acid system and permeability is regained (Bahuguna, V.K, *et al.*, 2003).

Sanganpur

The area of the field is 4.49 sq. km., which was fully covered by 3D seismic data with proved reserves are 0.86 MMt. The year of discovery was 2006 in Madhali Formation of Lower Eocene. Two wells were drilled and both flowed oil and gas. Though the well initially produced oil @ 18 m^3/day for a short period but many activation efforts only flow of oil, water and gas was observed. The gradient survey indicated oil, however, self flow could not be sustained.

Subsequently additional intervals (1893.5-95.5-1873.5-77.5, 1865.5-67.5m) were perforated well became active and the influx study indicated a higher skin. Testing was called off and decided to test Mandhali-II in selected intervals in zone 1770- 1793 mts) which flowed emulsion. Another zone 1277-1278 m was tested dry and influx of water. The cumulative production 2.48 Mt of oil and 1.726 MMSCM of gas till.

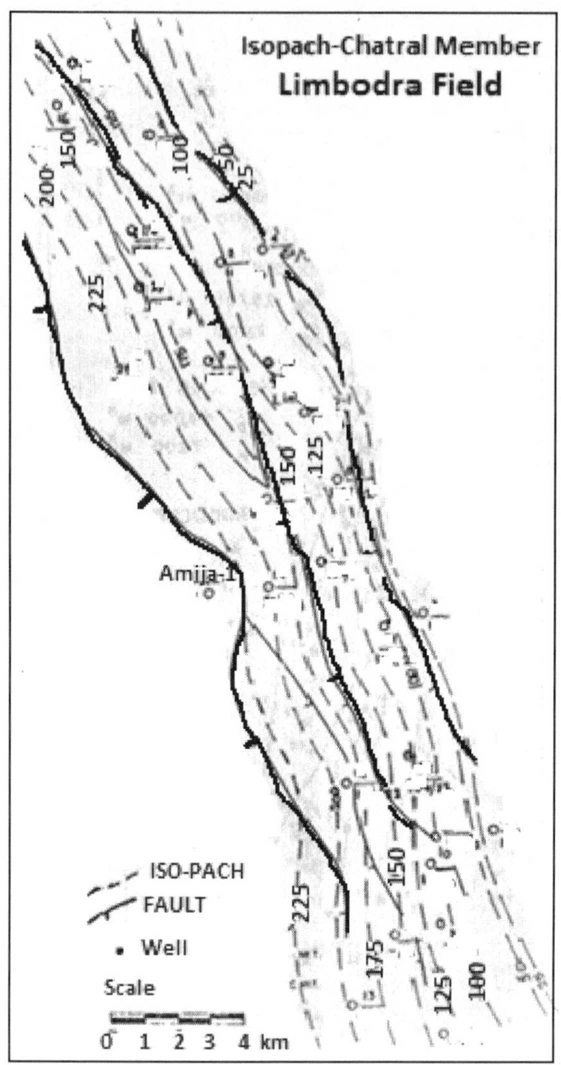

Figure 10.ca.15. Isopach Map of Chhatral Member Fields (After Panwar, P., *et al.*, 1999).

South Cambay area

Ankleshwar

The field is located in the Narmada Block towards south of the Narmada fault. The first major oil-find came in 1958 with the discovery of Ankleshwar field located about 80 km south of Vadodara. The field came on stream 1961 and in 35 years more than 55 per cent of recoverable reserves have been produced.This anticline is about 20 km long and 4 km wide. The long axis of the reservoir is 17 km and the width is

3 km. The North-Western flank has gentle dip of less than 8° whereas South-Eastern flank is steeper with maximum of 18° dip. It is a doubly plunging, asymmetrical anticline, bounded to the south by a reverse fault (Jain, *et al.*, 1982). The hydrocarbon accumulations are generally dip-closed to the northwest and fault-closed against the bounding fault to the southeast (Ahuja, *et al.*, 1990). Oil is available at depths varying from 1,000 to 1,200 meters. About 700 wells have been drilled. The Field contains a STOIIP of 134 MMt and recoverable reserves of 71 MMt with recovery factor ~ 53.3 per cent. (Pathak, *et al.*, 1995).

The field contains a total of 12 productive sandstone reservoirs (S0 to S11), which are grouped into the Lower Sands (S0 to S1), Middle Sands (S1 to S5) and Upper Sands (S6 to S11). The oil is light (40-43° API) with low viscosity (0.31-0.5 cp), Porosities range from 18 per cent to 23 per cent, while permeabilities are typically in the range 23-395 mD although fluid properties vary somewhat from reservoir to reservoir.

Oil is produced from the reservoirs under moderately strong, edge-water aquifer drive, supplemented by water injection into the reservoirs, which was initiated early in field life. Artificial lift is provided in wells by gas lift and sucker rod pumps. Highly variable flood fronts are evident in each reservoir, while increasing water-cuts are countered by water shut-offs and selective re-completions. Injection wells were initially completed below the original OWC, in a line-drive pattern. As the OWC has moved upwards, up-dip watered-out producers are converted to new injection wells.

Water injection has proved to be an effective pressure maintenance method, with reservoir pressure stabilizing near saturation pressure. As water-cuts in producing wells have risen, an increasing number of wells are flowing with the aid of gas lift and sucker rod pumps. The field since few years producing with an average water cut of 88 per cent, which was found to be uneconomical to operate. In this perspective, a CO_2 EOR has been implemented in to productive sand layers of Ankleshwar formation. A significant amount of un-swept pockets of oil is left in this reservoir, providing an ample scope for tertiary recovery operation.

Cambay

Though the Cambay field is recognized as a gas field, heavy crudes are encountered in a number of horizons both in the Eocene and Oligocene sections. Cambay crudes are heavier as evidenced by higher specific gravity values, higher contents of asphaltene-resin and paraffin wax and lower contents of light fraction. Four productive horizons are recognized in Eocene.

Palaeogene deposits in Cambay starts from transitional variegated rocks overlying the traps. They consist of tuff and shales sharply varying in thickness. The Eocene section contains the alternation of dark grey fissile shales and siltstones. Oligocene section consists of dark grey shales with bands of silts and sandstones. This section is 130 m in thickness and it contains three producing horizons. The upper part of Oligocene and the lower part of Miocene consisting of light grey quartizitic ill-sorted sandstone are also productive.

Gandhar

This very big field (800 square kilometres) is located in the rising north-western flank of the Broach depression in the Jambusar- Broach block. In the north, the field extends beyond the Dadhar River and in the south upto the Narmada fault. In the west, the field extends to the Gulf of Cambay. The field came on stream in 1987.

The hydrocarbons in the field are contained within the Middle-Upper Eocene Ankleshwar Formation, which reaches a maximum thickness of 450 m. The field contains 13 separate reservoirs, separated by shale intervals. Twelve of the reservoirs occur in the Hazad Member (termed sand bodies as GS-1 to GS-12) and one other in the overlying Ardol Member. The individual reservoirs are of the order of 3-10 m thick. It is contained within a combination trap, formed by east wards structural dip on the NW margin of the Broach depression, combined with stratigraphic wedge-out of deltaic sands towards the northwest. Porosities range from 17-24 per cent and average 19.6 per cent, while permeabilities are commonly 300 mD, ranging from 20-1800 mD. Porosity is impaired by the presence of detrital and diagenetic clay. The API gravity of the oil varies from 34° to 59° with an average pressure of 3912 psi. Recovery is facilitated by the low fluid viscosity, which is generally in the range 0.15-0.35 cp. The reservoir temperature s are 130 to 140°C. The dew points obtained from experiments were used to interpolate the dew point at 132°C and it was found to be 277.9 kg/cm^2.

Initially the oil was produced by pressure depletion drive, with partial support from aquifer edge-water and gas-cap expansion drive. With drilling of more wells, the gas production started increasing. The Gas Condensate Ratio has increased from a level of about 1700 v/v to about 5000 v/v. The average reservoir pressure has declined from the initial value of 312 kg/cm^2 to about 170 kg/cm^2 slowly. The condensate API gravity is also showing a rising trend from the initial value of 51.8°API.

The decline in reservoir pressure led to a drop in the oil rate from most reservoirs within 2-3 years of coming on stream. Water injection commenced in 1991. Oil production from the field reached a peak of over 40,000 BOPD in 1995. Development wells are drilled in a five-spot pattern.

Dahej and Pakhajan

The sand GS-3A is extensively developed in South Gandhar, Dahej and Pakhajan areas and is the main producer. It is a saturated reservoir producing oil, condensate and gas from South Gandhar and condensate and gas from Dahej field.

Patan

The field is in South Cambay Basin. Out of the 20 wells drilled in Patan area, oil has been established in S. Patan-5 and 7 from fine grained sandstone/siltstone developed within Older Cambay Shale. Influx of oil was observed during testing in S. Patan-8. Dharnoj-2 and E.Patan-1 and hydrocarbon indication in Wara-1, S.Patan-9 and W.Patan-2.

Padra

The field is located on the eastern margin of South Cambay Basin in Broach-Jambusar Block has commercial accumulation of hydrocarbon in Deccan basalt at a shallow depth, in addition to younger pays. The coring of 110m from trap top was carried out in Padra-42 to identify the actual hydrocarbon contributing zones, the study established existence of five different flows and highly fractured and altered rocks as reservoir rock. The best porosity development is observed in the interval 725-738m (10-17 per cent).

Dabka

The field was a platform area sloping towards southwest, and northwest and is marked by NE-SW fault pattern. Most of the fault delineated seems to effect upto Ardol and few even upto Dadhar level. The main reservoir rocks in this area belong to the Hazad Member of the Ankleshwar formation are known as upper and lower Dabka pays. These have been correlated as equivalent to GS-6 and GS-9 sand units of Gandhar field.

The Dabka pay sands are characterized by fine grained sandstone, moderately to well sorted, and characterised predominantly by fine to very fine grained argillaceous sandstone, with thin interbeds of oolitic claystone. The thickness varies from 1.5 to 3.5 m in upper pay while 3.0-6.5 m in the lower pays.

Kosamba

It is the third largest structure in the onland part of the Narmada block. It is a broadly WSW plunging structure with minor culminations in the crestal part again bounded on its southern limb by a major fault. Rapid variation in throw along strike of the fault is a characteristic feature as is observed in the seismic sections across the structure.

Olpad

This structure has a dominant NNE-SSW trend and comprises of a composite set of faults. En echelon structures are more in prominence here each forming minor culminations against the fault, especially at the older levels.

In addition, flanking the Olpad fault, another fault had with respect to the Olpad fault. It also develops a closure at its uplifted block named as the "H" structure. This doubly plunging structure is part of a series of culminations on the upthrown block. At the older levels, the culminations in the northern part are developed on the down thrown block of the fault but progressively along strike towards south, the closure tends to shift to the upthrown block towards east.

Hazira

It is the southern-most structure of the Narmada block, which extends over a distance of 14 km only a very small part of which lies within the onland area with the major part of the structure extending into off-shore. The structure is of ENE trend. A fault bounding the southern limb and abrupt variation along strike with thickness difference across the fault is characteristic of this structure.

Lakshmi and Gauri Fields

The block CB/OS-2 is located in offshore Cambay Basin and of pre-NELP regime. Cairn India is the operator with 40 per cent participating interest in partnership with ONGC and Tata Petrodyne Limited. Lakshmi and Gauri oil and gas fields are at the center of the block. Gas production commenced from the Lakshmi and Gauri field which commenced crude oil production.

Vadatal

It covering an area of 132.8 sq. km and have 380 line km.,of 2D Seismic, 3D seismic and with 11 drilled wells of which 5 wells were dry. It is located in south of Ahmedabad area of the Tarapur Block. The year of discovery was in 2014 by ONGC with oil discovery in Chhatral Pay in Cambay shale of Lower Eocene. The testing of these 5 wells indicated only oil or poor influx and the wells were abandoned.

D45

Area of the structure is 66.82 sq. km. and have 166.51 line km., of 2D Seismic and 3D seismic data and with proved reserves of 0.008 OEG MMt. of the 6 wells were drilled in the area 4 wells were dry. It is located towards west of Vadodara. The Year of discovery was in 2009 by RIL with oil discovery in Kalol sand of Lower Eocene.

Table 10.ca.1. Oil and Gas Fields of Cambay Basin

Field	*Location*	*Company*	*Year*	*Wells*	*Field Note*
Large		ONGC			
Ankleswar	South-S		1960	690	*
Balol	North-N		1970	120	*
Bechraji	N			175	
Gandhar	S		03/84	750	*
Jhalora	N		1976		*
Kadi-North	N		1967		*
Kalol	N		1961	380	*
Lanwa	N		1972	150	*
Nawagam	Middle-M		1963		*
Sanand	M		1962	175	*
Santhal	N		1971	160	*
Sobhasan	N		1968		
Jotana	N		1977		*
Kadi- South	N		1968	110	*
Sobhasan- South	N		1976		
Wadu	N		07/83		*
Wasna	M		1969	140	*

Field	Location	Company	Year	Wells	Field Note
Ahmedabad	M	ONGC	1967	60	*
Dholka	M		1966		
Motera	N		1988		
Nada	S		1988		
Viraj	N				
Medium					
Cambay	S		1958	20	*
Dahej	S		09/80	22	*
Kathana	S		1965		
Kudara			1985		
Limbodra	N		03/85		
Motwan-Southwest	S		1974		
Nandasan	N		1975	114	
Sobhasan-West	N		1976		
Small					
Allora	N		1967		
Andara			1989		
Asjol			1972		
Bakrol	M		1966		
Bechraji-West	N		1989		
Bhandut	S		1976		
Dholasan					
Dabka	S		1970	50	*
Gamij	M		11/82	20	*
Indodra	N		1971		
Jotana-East			1990		
Kanwara	N		1971		
Kathana-North	S		1975		
Kosamba	S		1963		*
Langhnaj	N		1981		
Linch	N		1972	82	
Lohar	N		02/81		
Mahi (High)	S		1980		
Malpur-South	N		1987		
Mansa	N		1988		
Mewad	N		1981		
Modhera	N		1982		
Motawan-West	S		1978		
Matar	S		1978		
Mohuda			1987		

Field	Location	Company	Year	Wells	Field Note
Nandej	M	ONGC	1990		
Ognaj	N		06/82		
Padra	S		1977		*
Pakhajan	S				
Paliyad	N				
Palej	S		1990		
Patan-South	N		1988	20	SF
Sanganpur	N		1990		DSF-I
Siswa	S		1974		
Sisodra			04/81		
Viraj- South	N		1985	50	
Wavel	N		1962		
Akhej	N	ONGC	1979		
Akholjuni	S			32	
Amja	N				
Ambliasan	N				
Ambliala	M				
Ambheti	M				
Asamli	M				
Anand	M				
Akhol	M				
Baad	N				
Baroda	S				
Bharkooda	M				
Atali	S				
Asjol	M				
Balol-North	N		1972		
Baola	N		1971		
Boots-West	N				
Borisana	N				
Bhalada-West	M				
Bhatlai	S				
Chandrora					
Chansama	N				
Chanchwel	M				
Changara	N				
Ckaklasi	M				
Dasawara					
Denwa	M				

Field	Location	Company	Year	Wells	Field Note
Detroj	N	ONGC			
Dholasan -North	N				
Dholasan-South	N		1984		
Dharnoj					
Dhangarwa	M				
Dhamdhuka	S				
Dhadhar	S				
Dungri	S				
Dhumas	S				
Dhantri	S				
Elas	S				
Gajera	S		1978		
Gamanpura	N				
Galodra					
Gumsan					
Galodra	N				
Govindpura	N				
Halisa	N				
Hazira	S		1969	10	*
Jambusar	S				
Jakasna	N				
Jetlapur	N				
Jhagadia	M				
Kadi	N			467	
Kalol-North	N				
Kamboj	N				
Kalamsar	M				
Kareli	S				
Kauka	N				
Kaira	N				
Karjisan	N		10/85		
Kavita	M				
Kawas	S				
Kasiyabet					
Khambhat-Lunej	N		1958	65	*
Kharach					
Kherwada	N				
Kim	S				
Kheda	M				
Kolat	N				

Field	Location	Company	Year	Wells	Field Note
Kundra	S	ONGC	04/81		
Kural	S				
Kunwara					
Langanaj	N				
Lakshmanpura	N				
Mahisagar	SG				
Malpur	N				
Matar	N		1978	14	
Mehsana City	N				
Mitli	N				
Mahuda	M				
Miyagam	S				
Mewad	N				
Mewad-South	N		02/84		
Mewad-West	N				
Nadiad	M				
Olpad	S		1963	27	*
Pansar	N				
Pakhajan	S		12/85		*
Patan	S			20	
Patan-West	N				NELP
Patan-SouthWest	N		1989		
Panjroli	S				
Panoli	S				
Piswara	N				
Rakhanpur	N				
Ratanpur	S				
Ras	S				
Rupal	N				
Sabarmati-motera	N		1988		
Sarkhej-North	N				
Sarkhej-South	N				
Sunikalla	S				
Sobhasan- North	N				
Sobhasan-East	N				
Rupavati	N				
Rayma	S				
Uber					
Unwa	N		1988		

Field	Location	Company	Year	Wells	Field Note
Uneli	S				
Umarwada	S	ONGC			
Vadatal	N				DSF-I
Vera	N				
Vaso	M				
Vithalpur					
Wadasma					
Wadi-Paligad			1983		
Wamaj	N				
Wamaj-North	N				
Wamaj-South	N				
Warosan	N				
Warosan-South	N				
Wasna-Lava	M				
Wasna -Keliya	N				
Walod	N				
Wataman	N				
Watadra	S				
Walner	S				
Balkhodra					
Chakonde					
Sajali				2	
Chakleo					
Nolkan					
Sanadkund					
Dantrai					
Talodra			Sq. km		
Elad	Gas		9.98		DSF-! Small fields
South Patan	Oil+Gas		9.88		
Khambel	Oil+Gas		9.78		
Kamboi	Oil+Gas		2.35		
West Bechraji	Oil+Gas		9.15		
Private Participation					
Lakshmi		Cairns			CB-OS-2
Gauri					
Tarapur		GSPC			CB-ONN-2000/1
PNE					
Changara-1					

Field	Location	Company	Year	Wells	Field Note
Kheda		ONGC			
ENS-1					
P-1,					
ESU-1		Essar		Essar	CB-ON/3
PK					
SE-10,1,4,3					CB-ONN-2000/2
Miroli					CB-ONN-2000/3
SPD-1		HOEC			
Parmoda					
Bheema					
Pelej					
Pasuamnia		BRPL			
Jyothi		Mercator			
ENP-1		Essar			
EEU-1					

Cauvery Basin

Introduction

The Cauvery Basin is located on the east coast of southern India, including Gulf of Mannar and Palk Bay forms the complete basin. The basin encompasses a total area of around 62,500sq.km up to a bathymetry of 200m which extends in to deep waters in the Bay of Bengal. It is about 100 to 150 km wide in E-W direction and 400 km long in N-S direction. It was divided in to five sub-Basins namely, Pondicherry-Ariyalur, Tranquebar, Nagapattinam, Tanjore and Ramnad-Palk Bay.

Exploration History

The exposed sedimentary rocks along western margin encouraged the geoscientists of ONGC to initiate exploration for hydrocarbons in the basin who carried out geological and geophysical surveys in the onland areas since 1958 (Kailasam,L.N. and Sinha, K R.M. 1963) and drilled 10 structural wells in different parts of the basin to understand subsurface continuity of outcrops.

The first deep well was drilled on Karaikal structure in 1964 which gave oil and gas shows. In Madanam-1, well non-commercial gas was tested (Ramanathan, S., 1964). In 1976, Karaikal-10, drilled to the east of Karaikal-1, struck oil. Second phase of drilling in the basin was commenced in 1984 which has led to the discovery of two important fields Narimanam and Kovilkalappal. Ramnad field was discovered in 1994 which is a major gas field and was put on production in 2002.

Though offshore geophysical surveys were commenced in 1964, detailed offshore seismic surveys could be carried out after 1975 when ONGC acquired its own survey ship 'Anweshak'.

In 1976, the Coromandal offshore basin Gulf of Mannar was awarded to Asamera Group for exploration and production. They carried out seismic profiling in Pondicherry area and drilled Mannar-1 in 1977 and relinquished the area in 1978.

In 1985 the PSC was signed with Russian team called IIEP in 1995 for northern part of the basin called Ariyalur sub basin for exploration. They carried out seismic data and drilled 18 wells during 1990-94 and due to non-commercial results relinquished the area in 1994.

Under small fields offer during 1992, PY-1 field covering 75 sq. km was offerd to Mosbacher and HOEC, but the PSC was signed in October 1995. During 1994 PSC signed for a block covering 81 square kilometres with the proven PY-3 field with Hardy as operator, ONGC, HOEC and Tata Petrodyne as partenrs.

So far about 195 structures have been probed by deep drilling of which 152 in onland and 41 in offshore (Figure 10.cu.1). This has resulted in finding 35 structures are oil and gas bearing. The development, delineation and exploratory efforts by different companies are continued. These efforts has resulted in drilling of nearly 800 wells in the Basin (Table 10.cu.1).

During NELP regime since 1997 many blocks were awarded to different operators and only RIL claimed finding oil in one of the northern blocks of the Basin (Table.10.cu.2).

Geology of the Basin

Early and Late Cretaceous sediments outcrop in the western margin of the basin. The outcrop strata have been divided into three groups, Uttatur, Trichinopoly and Ariyalur (Blanford, 1862), consisting of seven formations in stratigraphic order, *viz.* Dalmiapuram, Karai, Garudamangalam, Sillakkudi, Kallankurichchi, Ottakkovil and Kallamedu (Archana Tewari, *et al.*, 1996a,b; Sundaram, *et al.*, 2001). (Figure 10.cu.2.)

Uttatur Group

The geology and stratigraphy of the Dalmiapuram region have been documented by many researchers (Rama Rao, 1956; Gowda, 1964; Sundaram and Rao, 1979; Ramasamy and Banerji, 1991). The Dalmiapuram and Karai Formations of Uttatur Group are equivalents of the lateral facies. The Dalmiapuram Formation consists of coral algal limestone, bedded limestone and marl (Kale, A.S., *et al.*, 2000, Nagendra, R, *et al.*, 2001). The Karai Formation consists of variegated shales, veins of gypsum, phosphatic nodules (Nagendra, R, *et al.*, 2002b). Phosphatic nodules encased in a chalky shell. The Odiyam sandstone and Maruvattur clay members are the lateral facies equivalents of the upper part of Karai Formation. The Dalmiapuram Formation was developed during the Albian to Middle Turonian (Chiplonkar, 1985; Ravindran and Kalyanasundar, 1995; Raju, D.S.N. *et al.*, 1993) and unconformably (?) overlies the Terani beds. The upper contact is unconformable with the overlying Garudamangalam Formation of the Trichinopoly Group.

The Jurassic-Lower Cretaceous sediments are exposed near Uttatur and Sivaganga (Banerji, R.K., 1972). The principal alignment of these exposures is parallel to the Eastern Ghat trend. In the Sivaganga area, the boundary line of the Gondwana sediments at places coincides with the NW-SE (Dharwar) and the N-S trend as well. The western margin of these exposures in general show the NE-SW Eastern Ghat trend. The Cretaceous outcrop near Ariyalur, the largest one in the basin, exhibits

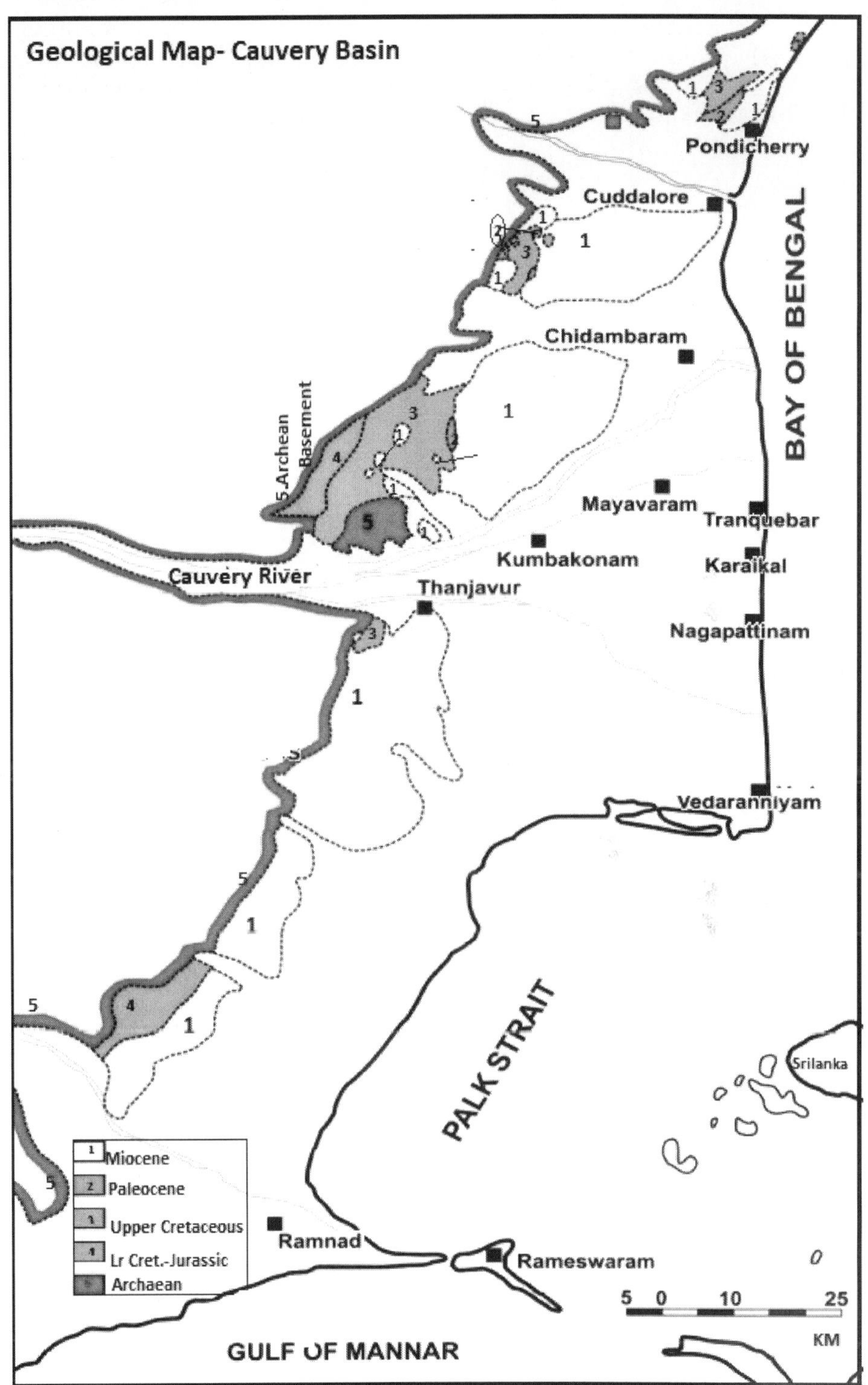

Figure 10.cu.1. Oil and Gas Fields Map-Cauvery Basin.

AGE	GROUP		OUTCROP FORMATION	LITHOLOGY
Recent-Plistocene			Alluvium	
PLIOCENE			CUDDALORE SANDSTONE	
MIOCENE				NO EXPOSURE
OLIGOCENE				
EOCENE				
PALAEOCENE	NINIYUR		NINIYUR	
MAASTRICHTIAN	Upper Cretaceous	ARIYALUR	KALLAMEDU Kallamkurichichi Ottankoil	
CAMPANIAN		ARIYALUR	KALLER CONGLOMERATE SILLAKUDI	
SANTONIAN		TRICHINO POLY	Saturbhagan sst	
CONIACIAN		TRICHINO POLY	GARUDAMANGALAM	
TURONIAN		UTTATUR (Dalmiapuram)	PARAVAY	
CENOMANIAN		UTTATUR (Dalmiapuram)	MARAVATTUR CLAY KALLAKUDI GREY SHALE KARAI SHALE	
ALBIAN		UTTATUR		NO EXPOSURE
APTIAN	GONDWANA			
BARREMIAN	GONDWANA		SIVAGANGA FORMATION: UPPER UNIT	
NEOCOMIAN	GONDWANA		SIVAGANGA FORMATION: THERANI BEDS LOWER UNIT BOULDER BED	
PRE-CAMBRIAN			ARCHAEAN CRYSTALLINE	

Figure 10.cu.2. Outcrop-Stratigraphy-Cauvery Basin.

the Eastern Ghat trend in its lowest formation only, the lower (western) boundary of each successive formation gradually swinging towards N-S until in upper part of the Ariyalur formation and in the Niniyur formation this trend is almost N-S.

An E-W trend in the outcrops is very well seen at the southern boundary of the Cretaceous sediments near Ariyalur. Formation of the basin started during Jurassic times in a NE-SW axis parallel to the Eastern Ghat trend with the deposition of the Gondwana sediments. From the two localities of Upper Gondwana exposures near Uttatur and Sivaganga it is apparent that two major NE-SW lines of subsidence were initiated one roughly between Trichinopoly and Pondicherry and another between Sivaganga and Tanjore (Rangaraju M.K, *et al.*, 1991).

The Cretaceous deposition also started more or less with the same NE-SW alignment as can be seen from the pattern of the three outcrops between Ariyalur and Pondicherry. These outcrops are separated from each other by east-west tracts of alluvium (Dutta, A.K., and T.S, Bedi, 1968).

Trichinopoly Group

The Trichinopoly Group consists of the Garudamangalam Formation, which unconformably overlies the Karai shale. This formation is divided into three members Kulakkalnattam sandstone, Anaipadi calcarenite and the Saturbhagam members. The Kulakkalnattam sandstone has extensive burrows. The Anaipadi Member contains calcarenite with shelly benthos (Sundaram, *et al.*, 2001). This member is dominated at places by fossil wood concretions at a stream section near Saturbhagam. The Anaipadi calcarenite member is overlain by fluvial sandstones, silty sandstones with reworked feldspar pebbles and cobbles. Fluvial sandstone and silty sandstones with reworked feldspar pebbles and sandstone cobbles are exposed and conformably overlie the Anaipadi calcarenite member. Large-scale trough cross-bedding (3 m thicknesses) is indicative of fluvial channel fill. The upper contact between Garudamangalam and Sillakkudi Formation are unconformable.

East-West faults, which mark the southern limit of the Trichinopoly-Pondicherry trough during Cretaceous, also marked the northern limit of the Sivaganga-Tanjore trough during the same period. The E-W trend of the northern limit of this trough is near Tanjore. The southern limit of the Tanjore trough during the same period was, however, controlled by the Dharwarian trend as can be conjectured from the gravity trend north of Devakottai and from the outcrop pattern of the Gondwana rocks near Sivaganga.

Ariyalur Group

The Ariyalur Group is represented in stratigraphic order by Sillakkudi, Kallankurichchi, Ottakkovil and Kallamedu Formations (Nagendra, R, *et al.*, 2002a). The Sillakkudi Formation is dominated by sandstones exposed along the Mettol. Sedimentological studies indicate that the sandstones are mineralogically and texturally immature and poorly sorted. The Vayalpadi sandstone exposure of the Sillakkudi Formation is a quartzarenite with calcareous cement and 15 per cent clay. The abundance of feldspar, especially acid plagioclase, indicates rapid deposition from a nearby granitic/gneissic provenance. The presence of angular grains indicates the short distance transported from a nearby provenance.

The Sillakkudi sandstone sequence is topped by a Kaller conglomerate bed exposed at a stream section, near Kaller. Orange yellow, pebbly grain stone with abundant sub-angular to angular, pink to white quartz and feldspar pebbles and cobbles characterize the conglomerate bed. The Kallankurichchi Formation unconformably overlies the Sillakkudi Formation. The presence of Kaller conglomerate between the Sillakkudi and Kallankurichchi Formation indicates a break in the deposition during the Late Campanian. The Kaller conglomerate unit is non-calcareous indicating a sub-aerial depositional regime. The Kallankurichchi Formation consists of dominantly marine carbonates.

The lower most ferruginous limestone is a fine to medium-grained massive biomicrite, rich in benthic foraminifera (Rama Rao, L., 1968). The lower arenaceous limestone directly overlies the ferruginous limestone (Raju, D.S.N, *et al.*, 2005). It is yellowish, massive, highly compact, and rich in silica. The Gryphaea limestone overlying the lower arenaceous limestone is reddish brown in colour, fine- to medium-grained and rich in carbonate. The upper arenaceous limestone lithologic unit is rich in terriginous influx and directly overlies the Gryphaea limestone. The Ottakkovil sandstone contains macrofossils and rare burrow structures. It has conformable upper and lower contacts. The cross-lamination structures in the overlying Kallamedu Formation (Rasheed, D.A., Ravindran, C.N., 1978).

Tectonics

The Cauvery basin was evolved in Late Jurassic-Early Cretaceous time as a result of rift-drift phenomenon of the Indian plate from Gondwanaland (Rangaraju, *et al.*, 1993). The Basin is also believed to have developed due to extension between India and Sri Lanka during the break-up of Gondwanaland (Burke, K. and Dewey, J.F., 1973). The extension was initiated in NW-SE direction, resulting in rifting along a pre-existing older Precambrian lineament between India and Sri Lanka. This is also seen by the presence of transverse faults subdividing the Cauvery basin into various blocks from north-east to south-west (Ramanathan, S. and Rao, V. Raghavendra, 1965).

The spreading center between India-Sri Lanka and Antarctica was aborted, followed by the rapid northward movement of India. A major unconformity separates the Upper Cretaceous sequence from Upper Palaeocene after which there was a south-eastward tilt of the basin: This tilting, which has also been recorded on the other east coast basins of India, is probably related to the uplifting of western peninsular India during the passage of India.

The earliest fractures in the area belong to the NE-SW or Eastern Ghat trend which was initiated during Jurassic. More or less simultaneous with this there originated the E-W fractures in the northern part of the basin and the NW-SE fractures (Dharwar trend) in the southern part. N-S fracture pattern started during late Cretaceous. Still younger fractures which initiated during late Tertiary and Quaternary, are those of WNW-ESWand ENE-WSW trends. Once initiated, all these different fractures have been reactivated till recent times.

The basin has undergone at least two major tectonic phases during its evolution. The early sheared rift extensional faulting initiated during Late Jurassic/ Early Cretaceous (Shyam Chand and Subrahmanyam, 2001; Scotese, 1997), and was followed by a progressive rift that seems to have continued until the end of the Turonian. The subsequent reverse movement caused by basinal uplift and contraction during the Campanian-Maastrichtian ended in the Early Tertiary (Kumar,SP, 1983 and Prakhakar, K. N. and Zutshi, P. L. 1993).

The episodes of basin evolution as:

- ☆ Extension Stage/Rift Stage: development of syn-rift sequences over the Pre-Cambrian Basement during East Gondwana rifting -Late Jurassic-End of Turonian.
- ☆ Early Thermal Subsidence stage/Early Post Rift Stage: Minor NE tilt of the basin -Late Cretaceous and Paleocene.
- ☆ Late Thermal Subsidence stage/Late Post Rift stage. Dominated by pronounced south-easterly tilt of the basin and development of Eocene and Younger sequences.

The initial rifting caused the formation of NE-SW horst-graben features. Subsequent drifting and rotation caused the development of NW-SE cross faults.

Plate reconstruction model of India and Sri Lanka, indicate that India and Srilanka were together up to Late Jurassic. The oldest magnetic anomaly south of Sri Lanka has been dated as 134Mabp. Hence it is assumed that the separation of India and Sri Lanka would have been initiated during that period (Katz, M. B. 1978). Deposition of thick 'early syn-rift' and 'late syn-rift' would have coincided with early Cretaceous period. The Cretaceous basin in Gulf of Mannar area can be divided into two sub–basins, the Mannar subbasin to the NW and Pamban depression to the SE, towards the SE part of the basin adjoining Sri Lanka, maximum thickness of sediments is seen indicating Rift initiation. Turonian is marked by widespread intrusive volcanics in the basin. The intrusive met in one well was dated as 60.8 + 3 ma while the intrusive encountered in Pearl-1 well in Sri Lankan offshore has been dated as 76 ma through k/Ar dating.

The basin is divided into several horsts and grabens, which are oriented in a northeast-southwest direction. Based on Composite, Residual Gravity and Magnetic data, the Cauvery basin has been divided into three depressions (Qureshi, M N., 1964) each separated by intervening ridges named as below (Figure 10.cu.3).

The peneplanation of horsts was completed by the end of Santonian and the present shelf-slope system appears to have existed since the Late Santonian (Scotese, 1997). Albian to Maastrichtian sediments are exposed around Ariyalur area.

- ☆ Ariyalur-Pondicherry Depression-Kumbhkonam-Madnam-Portonovo High
- ☆ Tranquebar Depression-Karaikal High
- ☆ Nagapattinam Depression-Vedarniyam High
- ☆ Thanjavur Depression-Pattukuttai-Manargudi Ridge-Mandapam Ridge
- ☆ Mannar Depression

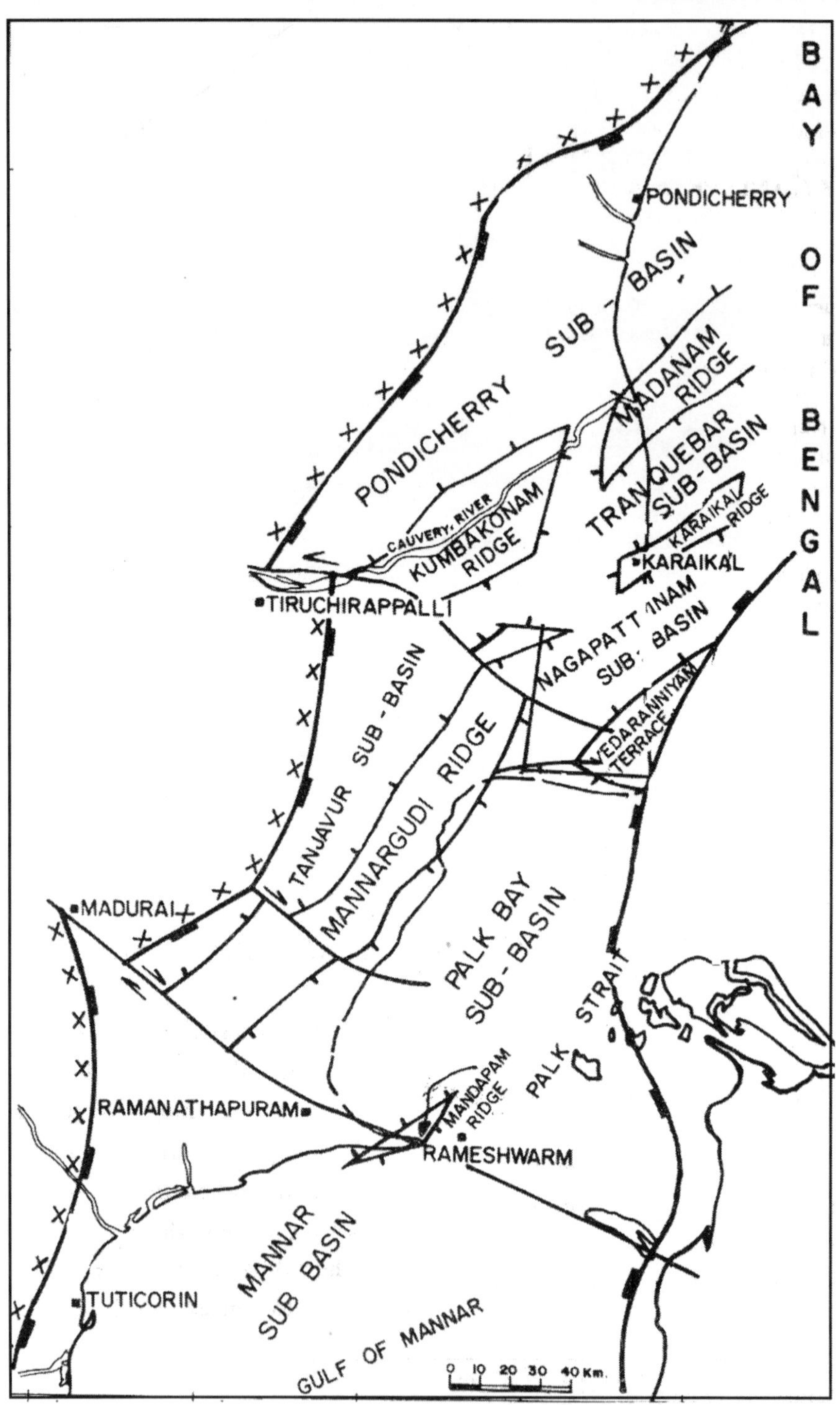

Figure 10.cu.3. Tectonic Map–Cauvery Basin (After Prabhakar, K N., *et al.,* 1993).

Ariyalur-Pondicherry depression is the northernmost sub-basin of Cauvery basin and is partitioned by Kumbakonam-Madanam ridge from the adjacent Tanjore-Tranquebar depression. This subbasin, an asymmetric rift graben trending NE-SW, is flanked by outcropping Archaean granite gneisses in the west and northwest.

The Ramnad subbasin is flanked in the West, NW and SW by the Pattukottai Mannargudi ridge, Palk bay offshore to the North and Northeast, Mandapam delft ridge to the east and Gulf of Mannar to the south (Govindan,A., 1993).

Sediments of 'late drift' phase and 'passive margin subsidence' (Paleocene to Oligocene) followed this and during the last major tectonic activity of collision of India with Asia, the basin witnessed a 'compressional' regime which is manifested as a gentle folding of Neogene sediments. The fault pattern in the Mannar basin indicates two major trends *viz.*, NNE-SSW paralleling strike and in NNW-SSE direction. The older extensional faults have been reactivated indicating a multiphase tectonic activity.

Evolution of the Basin in relation to the sediment fill can be understood to have taken place through three distinct stages:

Late Jurassic-Early Cretaceous Rift Stage

Initiation of rifting begun during the Late Jurassic/Early Cretaceous. Rift stage sediments Shivganga and Therani formations of Upper Gondwana affinity are known from exposures. These were deposited in fluvial environments. The Kallakudi Limestone, younger to the Shivganga Formation, may represent an episode of basinal deepening and paucity of clastic supply. In the subsurface, the Andimadam Formation, overlain by the Sattapadi Shale, appears to mark the peak of this transgressive episode during Cenomanian.

Late Cretaceous

Rift stage sediments, Therani formations of Upper Gondwana affinity are known from exposures. It may represent an episode of basinal deepening and paucity of clastic supply. In the subsurface, the Andimadam Formation, overlain by the Sattapadi Shale, appears to mark the peak of this transgressive episode during Cenomanian.

Post Cretaceous

Towards the end of the Cretaceous, the basin experienced a phase of upliftment and erosion and a gradual basinward tilt of the shelf. The Tertiary sequence was deposited in a general prograding environment with gradual subsidence of the shelf. This sequence can be subdivided into two groups, the Nagore and Narimanam. The Nagore Group is well developed in the south, whereas the Narimanam Group attains its full development north of Karaikal High. The Tertiary deltaic environment appears to have considerably progressed eastwards.

Sedimentation

Sedimentation is associated with the tectonic events and also known with major regional unconformities at the end of Albian, Turonian, and Maastrichtian and Middle Eocene, Oligocene/Miocene time.

Rift fills characterized by non-marine lacustrine facies of Late to Early Cretaceous period were restricted to the northeast-southwest-oriented grabens. Rift-drift transition sedimentation covered major part of the basin, including transgressive shales and arenaceous sequences. Tertiary drift phase deposition extended over the entire basin, consisting of shelf-slope prograding sequences (Ramanathan, S. 1968).

As the Indian craton drifted apart from Gondwana land, marine incursion took place into the Basin during different periods through Ariyalur-Pondicherry sub-basin in the north, Tranquebar sub-basin in the middle part and through Ramnad sub-basin in the south. Probably marine incursion was limited into Tanjore subbasin due to a small inlet between Kumbakonam and Pattukkottai ridges (Sastri,V V, and Raiverman. V. 1968).

Sedimentation in Ariyalur-Pondicherry subbasin began in Late Jurassic/Early Cretaceous period with the first definite marine influence in the Albian. Cessation of riftng phase can be marked by the end of Turonian (Sastri, V. V, *et al.,* 1973). Thickness of sediments in the basin is about five to six kilometers ranging in age from Late Jurassic to Recent in the central graben (Figure 10.cu.4).

In the wells drilled in the Mannar offshore in Indian Territory, intrusives have been encountered in Upper Cretaceous, except in the well drilled on the Mandapam Delft Ridge.

Stratigraphy

The study of the stratigraphy from outcrop geology and sub-surface information, indicate the earliest sediments deposited in this basin are those of Upper Gondwana group

These E-W faults marked the northern and southern limits of the NE-SW troughs. It is interesting to note that sediments of Uttatur and Trichinopoly stages are present at Ariyalur, but their equivalents are missing at Vridhachalam towards north. Obviously, the inferred East-West fault between these two outcrops acted as a local northern limit of the Cretaceous basin (Mishra, P.K, *et al.,* 2004). Thus, the sedimentation in the northern part of the Cauvery basin was likely initiated during lower Cretaceous or even earlier. The time map on the top of Cretaceous indicate nearly peneplanation of older tectonic setup with slope to towards East.

The Eocene outcrop at Pondicherry trends NE-SW, the axis of maximum deposition of the Tertiary sediments towards south in the area between Mannargudi and Karikal is expected to trend in N-S direction (Nair, K.M., 1980).

The Mio-Pliocene Cuddalore sediments south of Tanjore show a N-S trend for considerable length as seen from the outcrop pattern of Pleistocene sediments and recent alluvium south of Madurai and Mahabalipuram, where the trend of the coast line is N-S.

Figure 10.cu.4. Sedimentary Thickness Map–Cauvery Basin (After Rangaraju MK, *et al.*, 1993).

Integrated Litho-stratigraphic Nomenclature

The deep drilling data also added knowledge of continuity of outcrop versus subcrop data (Govindan, A., 1993). With all available geological data and other subsurface data ONGC geoscientists attempted complete litho-startigraphic nomenclature to all rock-units following the standered practices. (Venkatarangan, *et al.*, 1993). The data are presented in the Figure 10.cu.5 (Sundaram, R, *et al.*, 2001).

AGE	SUBSURFACE FORMATION		LITHOLOGY
RECENT PLEISTOCENE	TITTACHERI		
PLIOCENE			
MIOCENE	MADANAM L.ST./ TTP SST. SHIYALI CLAY STONE	TIRUTTURAIPUNDI	
OLIGOCENE	NIRAVI S.ST./ KOVILKALAPAL		
EOCENE	TIRUPPUNDI KARAIKAL KAMALAPURAM		
PALAEOCENE	KAMALAPURAM		
MAASTRICHTIAN	PORTONOVO	KOMARAKSHI	
CAMPANIAN	NANNILAM		
SANTONIAN			
CONIACIAN	KUDAVASAL		
TURONIAN	PALK BAY	BHUVANAGIRI	
CENOMANIAN		SATTAPADI	
ALBIAN		ANDIMADAM	
APTIAN			
BARREMIAN		UNDIFFERENTIATED ANDIMADAM	
NEOCOMIAN			
LATE JURASSIC			
PRE-CAMBRIAN	ARCHAEAN CRYSTALLINE		

Figure 10.cu.5. Sub surface Startigraphy–Cauvery Basin.

Andimadam Formation

In the subsurface, the formation is developed in grabens, namely, the Ramnad, Tanjore, Tranquebar and Ariyalur Pondicherry grabens. The lower boundary of the formation is marked by Archaean Basement rocks, while the upper boundary is defined by an argillaceous section. It comprises pale grey, fine to coarse grained, micaceous sandstone and micaceous silty shale.

Sattapadi Shale

This formation is widely distributed in the basin. It is absent in the southeastern part of the basin. The Andimadam Formation marks its lower boundary and an arenaceous facies of the Bhuvanagiri Formation marks its upper contact, the age assigned is Albian-Cenomanian. It comprises mainly silty shale and thin calcareous sandstone. The environment of deposition is inferred to be marine.

Bhuvanagiri Formation

The formation is developed mostly in the northern and central parts of the basin. The formation is predominantly sandstone with minor claystone and shale and of Cenomanian-Turonian age. It is inferred to have been deposited in middle shelf to upper bathyal environment.

Kudavasal Shale

The occurrence of this formation is restricted to the Palk Bay. The lithology is dominantly calcareous sandstone with a few bands of sandy claystone. The depositional environment is inferred to be shallow marine in a fan delta setting.

Late Cretaceous

The sediments in the outcrops are classified under two groups, namely, the Trichinopoly and Ariyalur groups (Govindan, A, *et al.*, 1996). The Trichinopoly and Ariyalur groups in outcrops consist of Sandstones and Limestone formations (Jyotsna Rai and Ramkumar, M., 2007) (Figure 10.cu.6).

Nannilam Formation

It is conformably overlain and underlain by the Porto-Novo and Kudavasal formations respectively. The formation consists of alternations of shale, calcareous silty shales and occasional calcareous sandstones. The formation age ranges from Santonian to Campanion.

Porto-Novo Shale

Predominantly developed in the northern part of the Ariyalur-Pondicherry Sub-basin, west of Karaikal Ridge and Palk Bay Sub-basin (Figure 10.cu.7). It is predominantly argillaceous with minor siltstone. The age of the formation is Campanion to Maastrichtian.

Komarakshi Shale

The formation has developed towards the eastern part of the basin. It unconformably overlies the Bhuvanagiri/Palk Bay Formation and underlies the

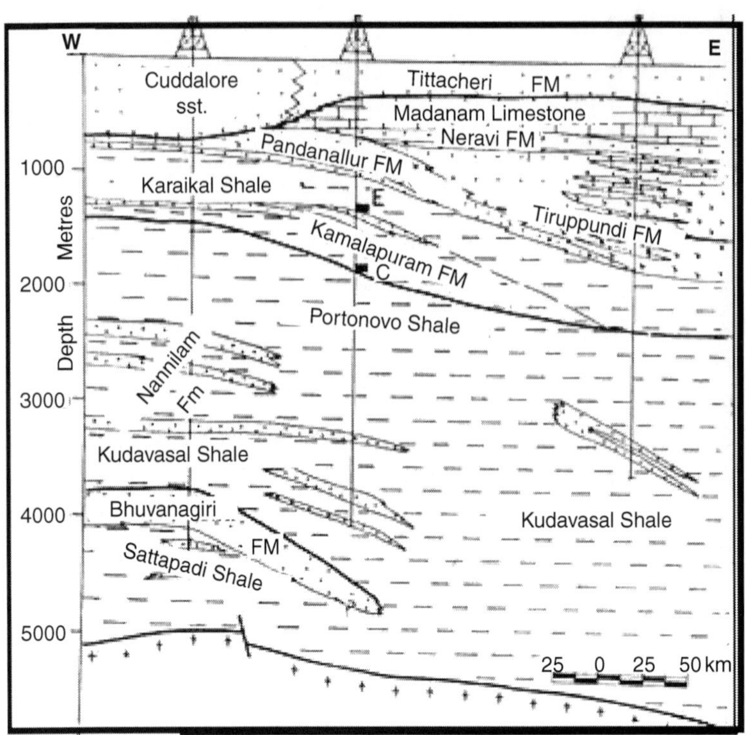

Figure 10.cu.6. Geological Section Across (West-East) Cauvery Basin.

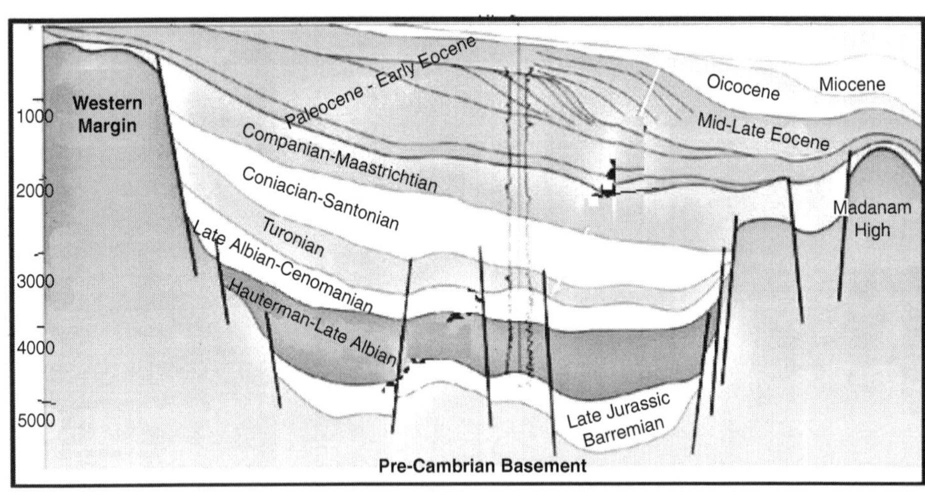

Figure 10.cu.7. Geological Section–Ariyaluru-Pondichery Sub Basin.

Karaikal/Kamalapuram formations. The formation consists mainly of calcareous silty shale. The age of the formation is Coniacian to Maastrichtian (Reddy, A.N., Yadagiri, K., 2001).

Tertiary

A complete sequence of Tertiary sediments is encountered in the sub-surface. The exposed rocks are represented by the Niniyur Formation of Paleocene age and the Cuddalore Sandstone of Mio-Pliocene age. The sub-surface section of Tertiary rocks is considerably thick and has been classified into two groups, the lower part is named as the Nagaur Group and the upper part, as the Narimanam Group.

Nagore Group

The formations of this group overlie the Ariyalur Group. The base and top of the group is marked by pronounced unconformities. The four formations recognized in this group are described below.

Kamalapuram Formation

The Porto-Novo-Komarakshi Shale unconformably underlies the formation, whereas the overlying Karaikal Shale has conformable contact. It consists of alternations of shaly sandstones and shales.

Karaikal Shale

The formation conformably overlies the Kamalapuram Formation. The formation comprises shales, which are occasionally calcareous/pyritic. The age of the formation ranges from Paleocene to Eocene (Figure 10.cu.8).

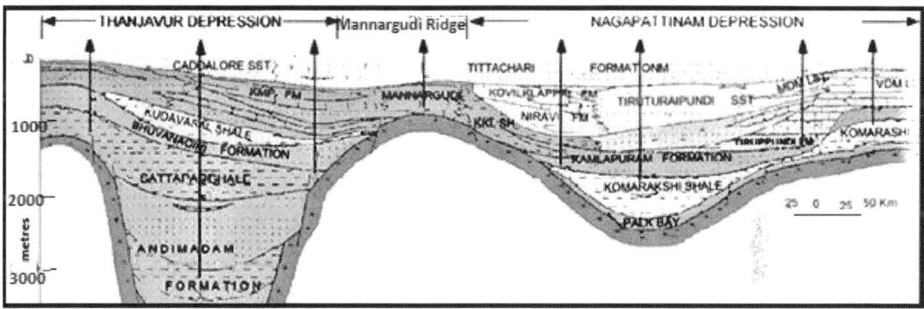

Figure 10.cu.8. Geological Section Across –Tanjore-Nagapattinam Sub Basin.

Pandanallur Formation

It has a restricted areal extension. It consists of claystone sandstone, deposited in middle shelf environment. Age of the formation is Lower Eocene.

Tiruppundi Formation

The formation is present in Pondicherry offshore, Nagapattinam Sub-basin, and south of Palk Bay Sub-basin. The formation comprises limestone, siltstone and sandstone. It is of Middle Eocene to Early Miocene age.

Narimanam Group

The youngest sedimentary sequence comprising sandstone, clay/claystone and limestone which are well recognized with distinct character is designated as a Group. This group comprises eight formations.

Niravi Formation

The formation unconformably overlies the Tiruppundi Formation/Karaikal Shale. The formation consists of grey coloured, fine to medium grained, calcareous sandstone with occasional pyrite and garnet.

Kovilkalappal Formation

It occurs in Tanjore and Nagapattinam Sub-basins and overlies the Niravi Formation, and underlies the Shiyali Claystone. It is argillaceous in nature with a dominant presence of limestone (Figure 10.cu.9).

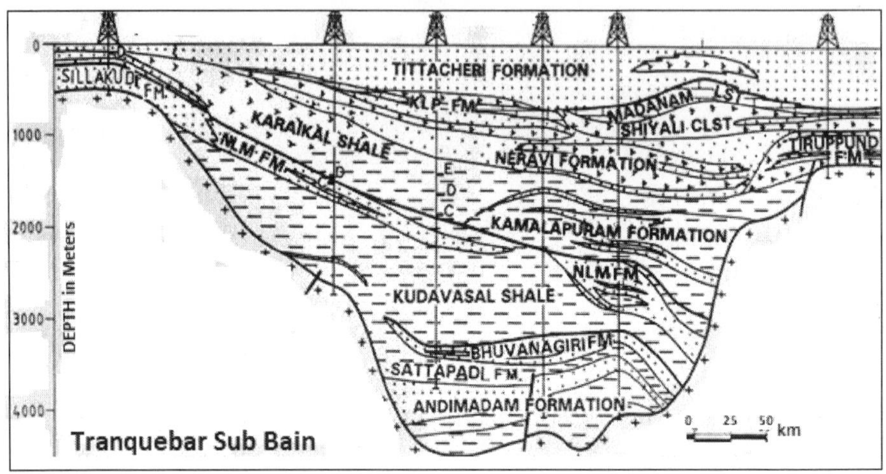

Figure 10.cu.9. Geological Section Across–Tranquobar Sub Basin.

Shiyali Claystone Formation

It is observed to occur in Madanam and Karaikal area. The age of the formation ranges from Oligocene to Lower Miocene.

Vanjiyur Sandstone Formation

The formation has limited areal extent. It is predominantly arenaceous in character and comprises dark grey, calcareous sandstone and siltstone.

Tirutaraipundi Sandstone Formation

The formation is present in the southern part of the Nagapattinam Sub-basin towards Palk Bay. It comprises mainly sandstones with minor limestone.

Madanam Limestone Formation

The formation is unconformably underlain by the Tirutaraipundi Sandstone and Vanjiyur Sandstone. It comprises mainly limestone with minor silty clays.

Vedaranniyam Limestone Formation

The formation occurs only in the southeastern part of the basin. It consists of predominantly coral limestone and minor grainstone.

Tattacheri Formation

The formation is present in a large part of the basin. It grades into the Cuddalore Sandstone Formation near the outcrops. This consists of unconsolidated gravely sandstone and earthy clays. The age of the formation is Lower Miocene to Pliocene.

Petroleum Geology

Nagapattinam sub-basin is producing commercial quantity of hydrocarbons from the sands ranging from Albian to Oligocene. The northern part of the Nagapattinam sub-basin has major hydrocarbon producing fields like Kamalapuram, Adiyakkamangalam, Tiruvarur, Kizhavalur and Narimanam, Kovilkallapal and Tulsapattinam and Tirukallar. The hydrocarbon finds have been discovered in several reservoirs *viz.* Andimadam, Bhuvanagiri, Nannilam and Kamalapuram formation. The hydrocarbon finds have been discovered in several reservoirs, Andimadam, Bhuvanagiri, Nannilam and Kamalapuram formation. Hydrocarbon discoveries have been made in the pre-Palaeogene and Palaeogene sediments, which are mainly confined to the Ariyalur- Pondichery, Nagapattinam and Ramnad sub-basins of the Cauvery Basin.

The Cauvery Basin is an established hydrocarbon province with a resource base of 700 MMT. 430 MMT for onland areas and 270 MMT in the offshore. Of which about 165 MMT (O+ OEG) in-place reserves were established with 65 MMT as recoverable.

Source Rock

The source rock facies are encountered in Andimadam Formation of Albian and Pre-Albian sediments. Within the graben areas of the basin during lower Cretaceous for some period anoxic conditions prevailed which enhanced the source rock characteristics in the sequences (Welte, D.H., 1965). The lacustrine shales of the rift phase and transgressive shales of rift-drift phase constitute the major source facies. The Cretaceous sediments (up to Santonion) have good source rock development throughout the basin. During the Early Palaeocene, fairly good source rock development has occurred in the Ariyalur-Pondichery sub-basin. The Sattapadi shale is the source rock available in the area. Thomas, N.J. and Sarma, V.N., 1991).

The effective source rock in the North part of Nagapattinam sub basin. Wells drilled in South Nagapattinam sub basin have not encountered effective source rock facies. However, potential source rocks are known in many wells in these areas. The paleo depositional studies in the area also indicated anoxic conditions prevailed in certain depositional intervals in the area (Figure 10.cu.10). The source rock development in the Kovilkalappal area is largely confined to the lows to the North of main Kovilkalappal high.

The Cretaceous sediments contain dominantly type III kerogen. Late Palaeocene and Early Eocene sediments are poor in organic carbon content, whereas sediments rich in organic matter are observed in the Nagapattinum, Tanjavur and Tranquebar sub-basins during the Late Eocene.

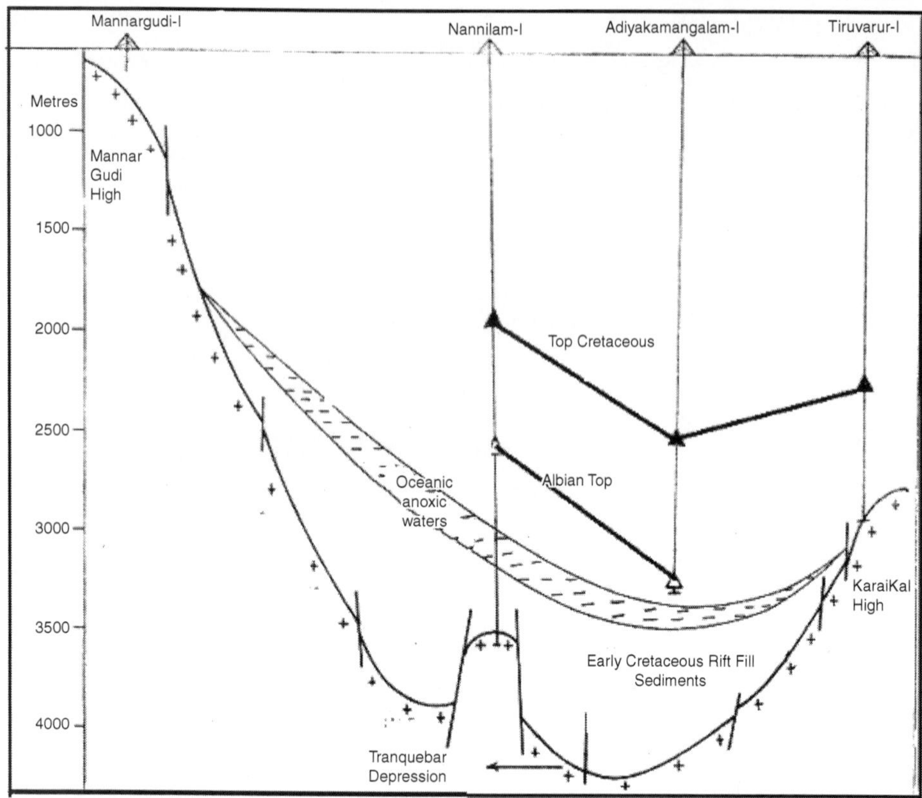

Figure 10.cu.10. Sourcefacies Window–Tranqubar Depression–Cauvery Basin.

Late Palaeocene and Early Eocene sediments are poor in organic carbon content, whereas sediments rich in organic matter are observed in the Nagapattinum, Tanjavur and Tranquebar sub-basins during the Late Eocene.

Based on source rock studies thickness of source facies in the Cauvery basin indicate 200-400 m in most of the areas (Figure 10.cu.11). Sattapadi shale of Cretaceous forms main source rock.

Maturation/Generation

The generation of hydrocarbons in the basin initially governed by the nature and process of sedimentation in the area and its thermal history for both protection and maturation of associated organic contents in the sediments (Vairavan, V., *et.al.*, 1997). Intra-cratonic nature of the basin suggests that the synrift sequences have equally high hydrocarbon potential (Vail, P.R., Mitchum, R.M., Thompson, 1977).

This subsidence might have helped in both protection and matuartion of organic matter. Further, the effective paleo-temperature gradients calculated are higher on the structural highs (around 4.7°C/100m) compared to the depression (around 2.9°C/100m).

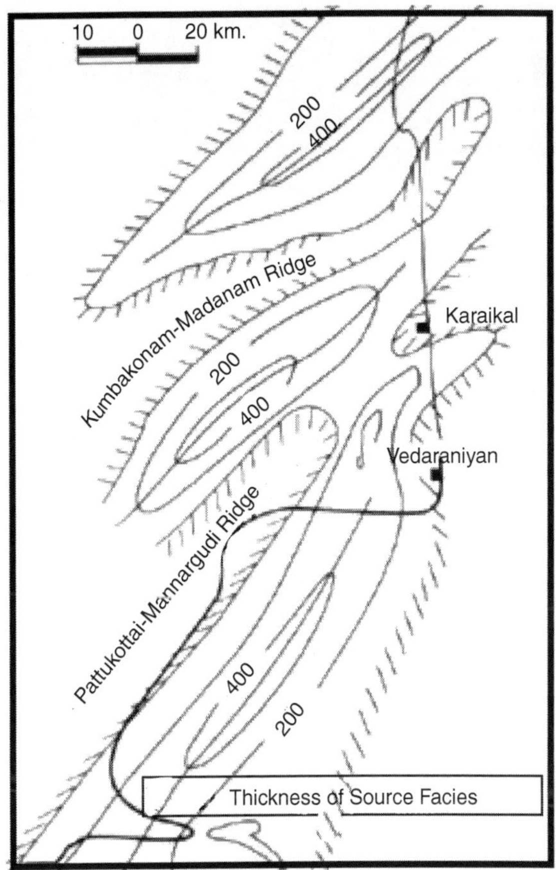

Figure 10.cu.11. Thickness of Source Facies Map Cauvery Basin.

The entire Tertiary sediments in the Cauvery Basin are not adequately matured except in some highs and ridges especially in Karaikal high, where the Eocene and Paleocene sediments are adequately matured. The most of the Cretaceous sediments are matured and they are well within the oil and gas generation zone. The maturation studies also indicate that there is a very prominent unconformity between the Upper Cretaceous and the Paleocene. The Cretaceous sediments contain dominantly type III kerogen and are mature, but the Palaeogene sediments are in the early catagenic stage of oil generation.

The subsidence history of various wells brings out the asymmetry in burial history of Bhuvanagiri Formation in the basin (Narasimha Chari, M. V, *et al.*, 1995). Initially the rate of basin subsidence was relatively faster till the time of 80 my. After which basin got uplifted during the period in between 80-75 m.y. Faster subsidence was continued upto 50 m.y. Beyond 50 m.y. rate of basin subsidence was much lesser. Burial temperatures in the Bhuvanagiri Formation, computed using thermal alteration index (TAI) shows about 90°C at 2700m., 120°C at 4000m., and about 190°C at 4800m (Figure 10.cu.12).

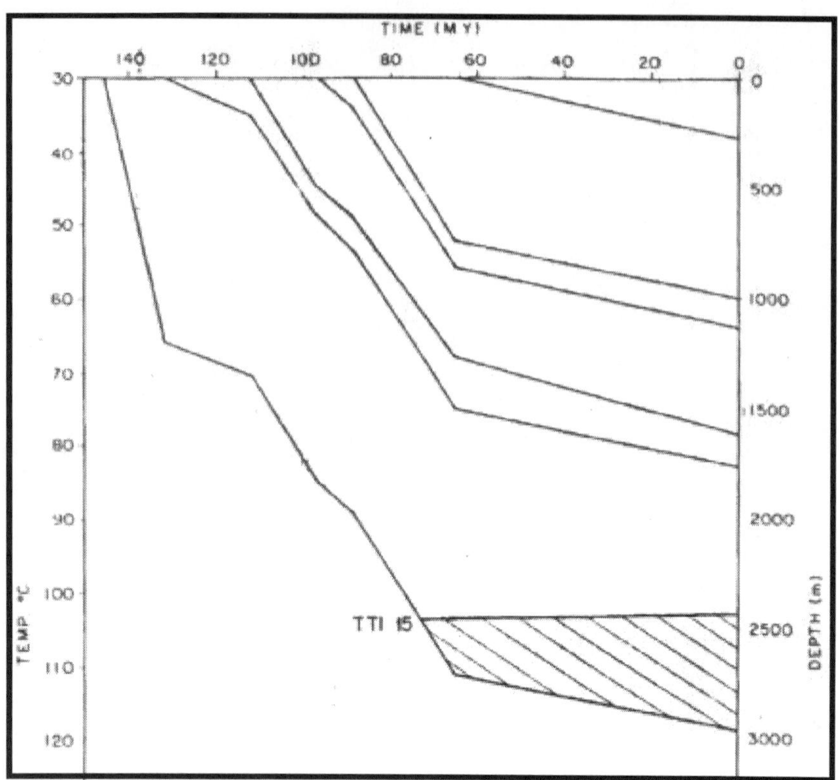

Figure 10.cu.12. Time Temperature Index of Well–TRL (After Govindan, *et al.*, 2000).

TTI calculations of wells show that the Sattapadi Shale and lower part of the Bhuvanagiri Formation entered oil window (Figure 10.cu.13). Study has indicated the possibility of synrift source facies entering the oil window stage as early as 70ma (Tissot, B. and Welt. D.H. 1978). All gases of Cauvery Basin are thermogenically generated in the catagenetic stage of maturation from a source of mixed type II and III and are dominantly oil associated gases suggesting generation of gases during peak generation time (Dow, W.G., 1977). Sediments get over-pressured and early maturation/expulsion of hydrocarbon is initiated.

Thin streaks of effective source rocks encountered within Pre Albian and Albian sequence in wells in Ramnad sub-basin have indicated proclivity to predominantly oil associated gas. Condensates of Ramnad are paraffinic in nature and might have been generated from mixed marine Type II/III organic matter deposited in suboxic to peat swamp environment. Since this sub-basin is close to Mannar Basin associated with intrusives, the prevailing high temperatures might have resulted in gasification of generated liquid phase if any.

Based on thermal maturation data on information from deep wells areas of maturation were seen in the basin. Further a depth map indicate top of maturation

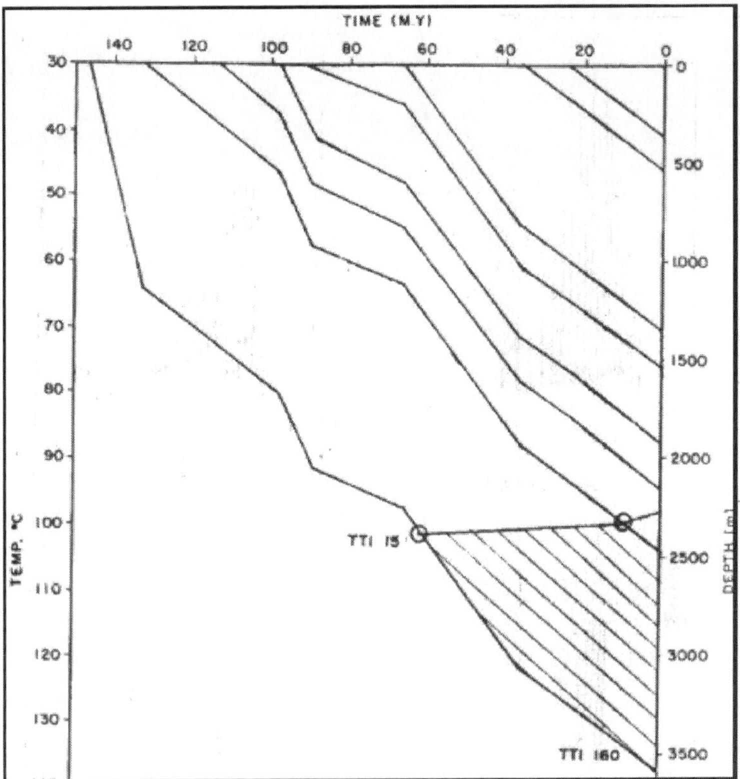

Figure 10.cu.13. TTI-well-in Ramnad Sub Basin.

sequence in the Basin which will give an idea about exploration depth window in the Basin (Figure 10.cu.14 and 15).

Migration

Considering the more thickness of rift sequences and the synrift subsidence history of the basin, the source facies at the deeper part of the basin could have reached expulsion phase in the upper Cretaceous time itself. Play analysis revealed accumulations in reservoirs juxtaposed to source facies, suggesting short distance migration.

The reactivated faults become conduit for migration and formation of traps due to inversion provide the entrapment. The reactivation of faults during Basin inversion causes leakage in the structure and releases accumulated hydrocarbons. Breaking down of liquid hydrocarbons in to gases due to basin inversion and associated shrinkage, is another significant negative effect. The closeness in occurrence of hydrocarbon locales and matured source rocks indicate limited lateral migration in the basin (Thomas, N.J. and Sharma, V.N., 1993). The base level of hydrocarbon accumulations in northern Perungulam block can be considered up to 2290 m. whereas in Periapattinam – Ramanavalasai area, it is up to 2180 m.

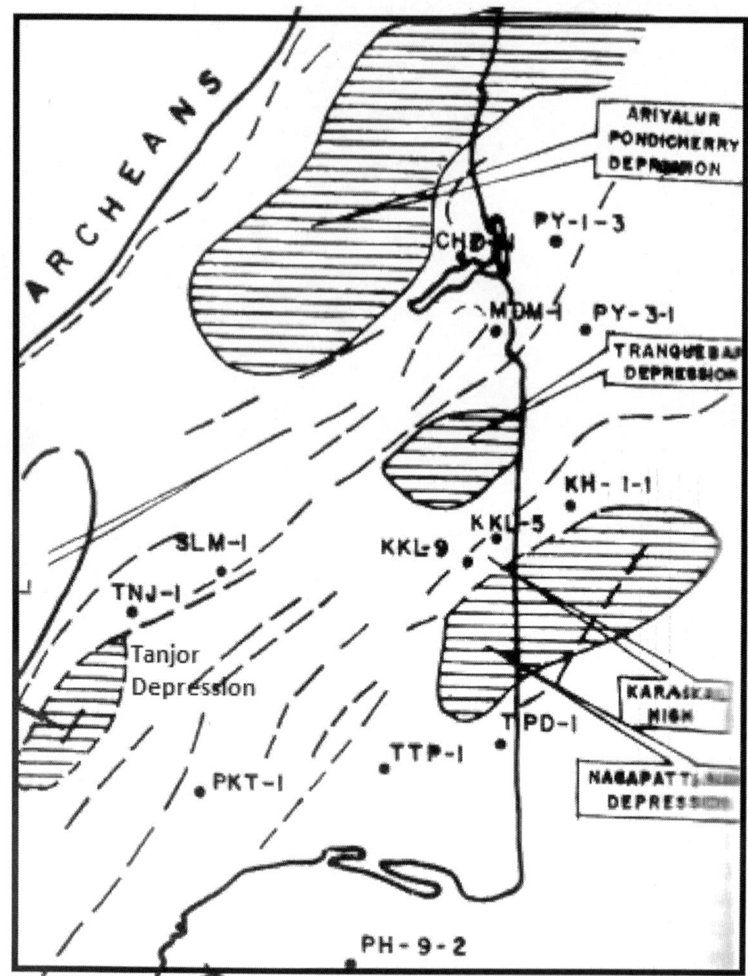

Figure 10.cu.14.Areas Maturation-Cauvery Basin

Reservoir Rocks

Tirutturaipundi sandstone splits up with two tongues from Karaikal Ridge and northwards, the lower tongue is named Vanjiyur sandstone whereas the upper one is the Neravi Sandstone. The Neravi Sandstone is overlain by Shiyali Claystone. In the subsurface,three carbonate bodies present are the Pattukottai Dolostone of Early Cretaceous, the Tiruppundi formation of Paleocene to Oligocene and the Madnam Limestone which ranges in age from Middle Eocene to Early Miocene. Pattukottai Dolostone is restricted to the subsurface and occurs as a patch a few square kilometers in area, around well Pattukottai-I.

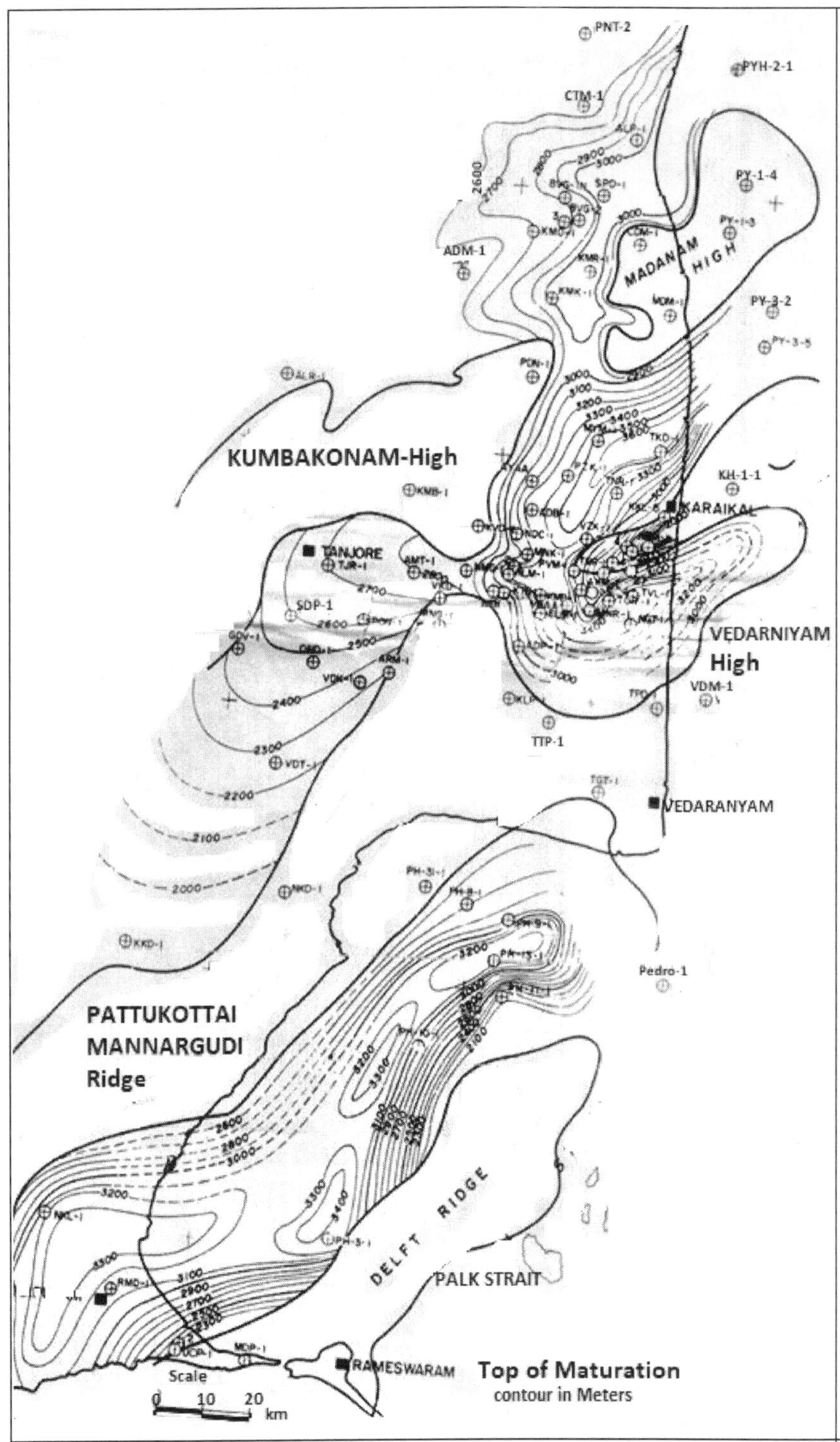

Figure 10.cu.15. Top Maturation Map–Cauvery Basin.

Entrapment

The hydrocarbon entrapment is structural with a few having strati-structural in nature. The reservoir rocks constitute sands within Andimadam, Bhuvanagiri and Nannilam Formations (Ramanathan, R.M., 1977).

The regional geological setting of the Cauvery Basin is such that compressional forces were not present. The movements are essentially vertical and tensional forces prevailed. In such tectonic set-up the type of traps are fault closures, compaction type of structures, caused by diapirism, *etc.* In fact the traps are necessarily combination type if the reservoirs are inter-fingering with clays and/or faults bring sands against clays with a good cover would be an ideal situation. The area south of Perungulam and area falling between Periapattinam and Ramanavalasai is promising and detailed anticlinal closures at pay level may be favorable for entrapment.

Petroleum System

Proven/Expected Play Types

Proximity of Cretaceous source to Cretaceous and younger reservoirs in a half graben setting appears to be the preferred hydrocarbon habitat. More petroleum accumulations are likely to be in similar settings onland and in deep-water offshore, flanked by basement ridges. Structural and combination traps in Early Cretaceous to Paleocene sequences. Stratigraphic traps such as pinch-outs/wedge-outs and lenticular sand bodies in Early to Late Cretaceous sequences.

Source: Sattapadi shale within Cretaceous– main source Kudavasal Shale within Cretaceous Basal part of Kamalapuram Fm (Paleocene).

Reservoir: Andimadam, Bhuvanagiri and Nannilam Formations within Cretaceous Kamlapuram and Niravi Formations within Paleocene Precambrian Fractured Basement.

Cap Rock: Sattapadi shale within Cretaceous Post unconformity shales like Kudavasal and Kamlapuram.

Entrapment: Structural/Stratigraphic, Combination traps.

Considering all factors one composite petroleum system is operative in the Cauvery Basin may be called Andimanam System.

Oil and Gas Fields

Kamlapuram

The field is located in the southwestern plunge of Karaikal ridge at the trijunction of Tanjore, Tranquebar Nagapatinam sub-basins.The structure was drilled in 1988 and found to be oil and gas bearing. The sequences encountered in Kamlapuram field range from Early Cretaceous to post Eocene. Kamlapuram Formation is the main hydrocarbon producing formation in the field. This formation is present in almost entire basin with two major depocentres in the Ariyalur-Pondicherry Sub-basin and the other along Karaikal ridge area. It is unconformably underlain by

Portonova Shale while the overlying Karikal Shale has the conformable contact with the formation (Figure 10.cu.16).

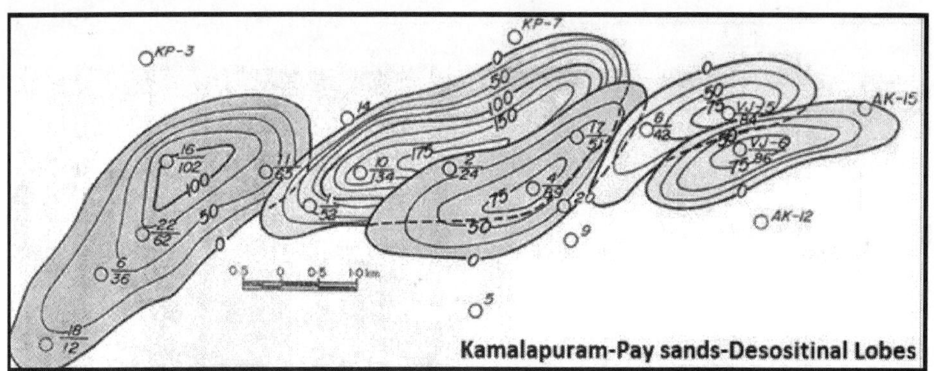

Figure 10.cu.16. Kamalapuram Pay sands-as Depositional Lobes (After Tattanchery, *et al.*, 1998).

Kamalapuram Formation has discreet sand bodies encased within thick shales and can be easily distinguished from shales above and below it. The environment of deposition of Kamalapuram Formation is outer shelf to upper bathyal. Thickness of the Formation ranges from 600 to more than 1000m. These sand bodies are thin and limited lateral extension. Sedimentological studies indicate that these sands have been deposited as discreet isolated submarine fan lobes having East-West alignment. In a well drilled #9 during 2008 interval 1923-1925 m produced oil @ 24 m^3/d. So far about 46 wells were drilled in the field.

Karaikal

The structure is basement controlled horst feature and it was the initial feature taken up for drilling by ONGC in 1964. The first well was drilled down to 1,771 m., which gave oil shows in Eocene section and gas shows in Palcoccnc. Exploration was continued over and flanks of the basement high, the well# 10 could establish oil flow from the Eocene sand. About 12 wells were drilled in the field.

Narimanam

The structure is located in the Cauvery basin, southern India for drilling in 1985 and found oil and gas. The field has multiple oil and gas pays of Oligocene and Eocene age. In the Oligocene section several different sands were identified and put on regular production. The estimated recoverable reserves are at 2.8 Mmt of oil with 5.5 Mmt of original oil in place. So far about 72 wells were drilled. Log correlation data indicated the presence of gas cap in some wells and H_2S was also observed in the gas. Whereas PVT analysis indicated under saturated crude in some wells. So on the basis of PVT data together with production behavior of the wells it was concluded that the sands comprises of two different lobes one having gas cap and another having under saturated crude.

Kovilkalappal

The structure is located in Nagapattinam sub-basin and was taken up for drilling during 1984. It was targeted to drill up to 2,500 m to basement, but since basement was met it was drilled up to 1,765m. The well was completed as gas producer. So far about 29 wells were drilled to delineate and produce from the field.

North Kovilkalappal

The structure is located towards north of field Kovilkalappal. The first well drilled has established oil flow in the interval 2120-32 @20 m^3/day. In another well Kovikalappal#3 the interval 2135–2145m (Andimadam Formation, Cretaceous) flowed oil@ 80 m^3/day through 6mm bean and Gas @ 10000 m^3/day with FTHP: 84.4 Ksc. On this structure in another well #6 tested in the zone 2143-2134m, flowed oil @ 13.4 m^3/d and water @ 5.7 m^3/d through 6 mm bean. In this well another interval 2098-2078 m., flowed oil @ 50 m^3/day and gas @ 9,500 m^3/d with water @ 4 m^3/d through 6 mm bean. Both the objects flowed oil from Early Cretaceous Bhuvanagiri Formation.

Kizavalur

The structure located in Nagapattinam sub basin and established oil baring pay sands in Nannilam formation. In a development well drilled (#7), has further established new pool in the interval 2768-2773 which flowed oil @ 69 m^3/day and gas @ 8,991 m^3/d.

Kuthalam

The structure is located in Tranquabar Sub basin. The pay sands in the structure were established in Andimnam Formation. In the development well (# 37 has identified new pay sand in the interval 2510-2518 which flowed gas @ 50,100 m^3/d with 8 m^3/3 condensate through 6 mm bean.

Bhuvanagiri

The structure is in the Tranquebar sub-basin. The area was under IIEP project and the first well was targeted to 5,000m and was drilled up to depth. Though this well was abondoned, the second well drilled to 4,795m has established gas flow with in Cretaceous sequence. So far 11 well were drilled in the field.

CY-ONN-2002/2

The Block is of area 140 sq. km is held jointly by ONGC as operator with 60 per cent PI and Bharat Petro-Resources Ltd (BPRL) as partner with 40 per cent PI. The exploration well in the block–MD 5–was drilled down to a target depth of 2,175 metres and flowed gas. This exploratory well was named ‘Madanam-3’ produced 115 cubic meters per day of oil and 11,500 cubic meters per day of gas in the interval 1430-1505 in bare-foot completion in the Basement. Subsequently, two more wells were drilled. The discovery of hydrocarbon is from fractured basement which is an unconventional reservoir of Precambrian age and from Kamalapuram formation.

The Basement high is trending major NE-SW and NW-SE normal faults with steeply dipping flanks. Kamalapuram formation deposited during Paleocene-Eocene in the passive margin set up and these sediments drape over the Madnam Horst, thus forms structural entrapment for the hydrocarbon migration.

PY-3 Field

This proven field was offered in 1994 as an exploration block named CY-OS-90/1, comprising 81sq. km. of area. It is located in water depths ranging from 40 to 380 meters. The members of the Joint Venture are Hardy Exploration and Production (HOEC,India), as operator with Hindustan Oil Exploration Company Limited and ONGC as partners.

Continued to produce oil from 4 wells at an average rate of 4397 BOPD. Annual oil production from these four wells was 0.198 MMt (1.609 MMstb). The field is being produced through a Floating Production System. The field performance review and reservoir simulation studies indicate that the PY-3 Field holds higher in-place oil than what was estimated earlier. Accordingly, based on the simulation studies, the In-place oil has been revised to 20 Mmt and the reserves to 4.6 Mmt.

The crude oil is of high quality light oil with API 50 degree and with low sulphur and low wax content. Asphaltene content values and trace metals are very low. The average GOR is 1200 cft/bbl.

PY 1 Field

It is located in Offshore Cauvery Basin and ONGC drilled the first well in 1980 and found to be oil bearing in basement (?). By 1990 in the field 11 wells were drilled by ONGC. Later it was awarded to HOEC as operator under small fields offer policy of GOI and ONGC remained as partner. The drilling of 'Surya' well could not be completed due to loss of drilling fluids while drilling. There is a reduction in the assessment of Total Proved Reserves and the same are re-estimated to be 120 bcf. In the same block anothe well drilled during 2007 has resulted in establishing oil and gas and it was named as ***Ganesha***.

Periyakudi

The first well drilled by ONGC, the interval 4350-4273 m (AndimadamFormation, Cretaceous) flowed oil @ 3.34 m^3/day, 6mm bean and gas: 5500 m^3/day with FTHP: 15.8 Ksc.

Adichapuram

ONGC drilled an exploratory well #2 to a depth of 2461m and found gas with condensate in the interval 2399 -2430 (12m perforated) @ 80,000 m^3/d gas and 20 m^3 condensate. Another higher interval 2335-2346 also flowed gas @ 59,000 m^2/d and 40 m^3/d condensate.

Neduvaasal

The structure is located in Tanjore sub Basin. The well #2 has established hydrocarbons in Andimanam Formation and in Basement. Though the intervals in the basement has flowed little oil, water along with gas, the flow reduced rapidly. However, the interval 1705-1710m has produced @ 18m^3/d oil and 5,000 m^3 gas through 6mm bean.

Pallivarmangalam

The structure is located in Tanjor sub basin. In the well (#5) oil and gas were established in the much deeper level in the interval 3278-3282 m which flowed 12 m^3/oil and 10,000 m^3 of gas through 6 mm bean.

Neyveli

The structure is located in Ariyaluru sub basin. In the well the tested in the interval 2563-2566 m has flowed gas @30,000 m^3/d with condensate @ 3.5 m^3/d.

Tulsapatnam

The structure is located in Nagapattinam sub Basin and the first well was drilled up to 2650m. This well flowed oil in the Cretaceous section @ 600 liters/ hour through 8 mm bean.

Ramanavalasai

The structure is located in the Ramnad Sub-Basin. In this well all three objects tested have flowed gas. The deeper object in the interval 2189-2194 m flowed gas @ 73,000 m^3/d along with condensate. The middle object in the interval 2061-2064 m flowed gas @80,000 m^3/day. The top object in the interval 1881-1885m flowed gas @ 48,000 m^3/day.

Kuthanallur

The structure is located in the Nagapattinam Sub-Basin. In the well the second object tested in the interval 2264-2267 m became active by nitrogen application produced oil @ 8m^3 per 14 hours through ½ " choke with little gas.

Tiruvaarur

The structure in central graben area of the basin and the first well was drilled during 1985 up to 2827 m bottomed in basement. It was abandoned without good indications. The second well which was drilled up to 2910, flowed gas in basement. Another well was activated with repeated nitrogen applications in the interval 2921-2930 m flow of oil with water @3m^3/d.

Vijayapuram

In the structure the well #2 was tested many objects, but the bottom most one in the interval 3260- 3217 (20m only) produced 7m^3/d oil with gas 21,678m^3/day. Another object in the interval 2535-2525m (4 m only) had collected about 17m^3 of oil in reversereverse circulation.

Perungulam

In the well #2 the interval tested 2103-2111 m flowed gas @87,043 m^3/d with condensate 1.2 m^3/d through 6mm bean. In another well the top interval 2049-2086 m flowed gas @ 90,000m^3/d with condensate 1.4 m^3/d through 6mm bean.

CY-ONN-2002/1

The block was awarded to Jubilant Energy as operator and GAIL (India) and Gujarat State Petroleum Corporation as consortium partners. The discovery of gas and light crude oil has been made in Andimadam and Sattapadi formation in the Cauvery Basin. A total of 110 meter of net pay was encountered within the well.

CY-ONN-2004-2

This block was awarded to ONGC in 2005 and the exploratory drilling in the block has established oil production from Basement. The well ***Panadanallur #8*** was drilled in which, the basement interval 2692 -2982 in the well when tested bare-foot has flowed oil @ 30m^3/d and gas @ 18,000 m^3/d through 6mm bean.

CY-DWN-2001/2

The block is situated 62 kilometres from the coast and in the Cauvery basin. The block, was awarded to RIL as the operator with 70 per cent interest and BP has a 30 per cent share. Well CY-III-D5-S1 was drilled in a water depth of 1,743 meters, to a total depth of 5,731 meters, with the primary objective of exploring Mesozoic reservoirs. In the very first well drilled in this block produced at the rate of 31 million cubic feet/day of natural gas and 1,200 barrels/day of oil. Another zone tested below the main zone produced 550 barrels/day of oil and 1 million cubic feet of gas. The well which had the initial reservoir pressure of 8,000 psi flowed gas at the rate of 35.2 million standard cubic feet per day with condensate. The discovery has been named D-56. Preliminary evaluation of well data and fluid samples indicated presence of gas condensate in the reservoir interval with a gross column of 143 meters.

Table.10.cu.1. Deep Wells Drilled–Cauvery Basin

Cauvery Basin	*Code*	*DD*	*Year*	*Well*	*Bottomed in*	*Remarks*
Adambavur	AB	3650	1991	2	Cret	
Adichapuram	AC	2300	1991	4	Cret	
Alangudi		3280	1991	1	Basement	
Alapakkam	AL	4000	1990	1	Cret	IIEP
Allinagaram		2693	1992	1	Basement	
Ammapettai	AP	3085	1992	1	Basement	
Anandatandavapuram	ATP	3030	1994	1	Cret	IIEP
Andimadam	AA	4500	1988	2	Cret	IIEP
Annavasanallur	AS	3150		3	BHV Fm	
Anniyur	AY	2085	1995	2	U.Cret.	
Ariyaluru		1939	1992	1	Basement	
Arumulai	AM	1735	1989	1	Basement	
Chatram	CT	4399	1989	1	Cret	IIEP

Cauvery Basin	Code	DD	Year	Well	Bottomed in	Remarks
Chidambaram	CD	3041	1970	2	Cret-Bst-3175	
Chokkanavur	CK	1869	1995	1		
Chomaitangi		2501		1	BHV Fm	
Elamangalam	EM	3000	1992	1	Cret	
Enadi	EN	2250	1990	1	Basement	
Erambur		2300		1	Nannilam	
Hariharankudal	HH	3200	1996	1	Cret	
Jayankondan	JK	2851	1995	1	Basement	
Kannandangudi	KG	1904	1992	2	Cret	
Kariyapattinam	KM2	2383		1	Basement	
Karuppampulum	KRP	2251		1	Basement	
Kattimedu	KT	2935	1989	1	Basement	
Kattumannarkoil	KM1	3724	1971	3	Cret	
Kavalakudi		2901	1991	1	Cret	
Kilkkadu		2598		1	Basement	
Kirangudi	KU	3305		1		
Kollidam	Kd1	3192	1990	1		IIEP
Koluvur		3555		3	Andimanam	
Komarakshi		5151	1987	1	Cret	
Kottarakudi		2935	1989	1	Basement	
Krishnapuram	KNP	2377		1	Basement	
Kuchipalayam	KC	2250	1996	1	Basement	
Kudavasal	KV	2674	1991	1	Basement	
Kumarankudi		4000	1990	1	Cret	IIEP
Madanam	MD	1685		2	Basement	
Madapuram	MA	2450		1	Nanilam	
Manakkal		4003		1		
Manakkudi		2901		1	Andimanam	
Manalur	ML	3300		1		
Mananar	19	3766	1977	1	Cret	
Mandapam		1679	1975	1	Basement	
Mangudi	MN	2522	1989	1	Basement	
Marungur	MG	2630	1987	2	Basement	
Mayavaram	MV	3500	1973	1	Cret	
Meluru		3005	1995	1	Cret	
Mudithalam	MM	3005	1994	1	Cret	
Mugayur	MY1	2503	1991	2	Cret	IIEP
Mullaivasal	MU	1903	1995	1	Cret	
Nagangudi	NI	2498	1997	1	Cret	
Nagapattinam	NP	4001	1967	1	Cret	

Cauvery Basin	Code	DD	Year	Well	Bottomed in	Remarks
Nagore	NG	2300	1986	1	Paleocene	
Nallikottai	NT	705	1992	1	Basement	
Narasimhapuram	NS	1768	1993	1	Basement	
Nayanarkoil		3633	1992	1	Basement	
Neduncherry	NC	3803	1990	1	Cret	
Needamangalam	NM	3205	1989	2	Basement	
Nidur	NU	2925		1	Basement	
Niduvirapattu		3040		1	Andimanam	
NRM-vengidangal		2710	1990	2	Basement	
Orattanadu	ON	3300	1986	2	Cret	
Palangudi	PG	2700	1991	1	Cret	
Palkolam		2252	1994	1	Basement	
Palluthurangarai		3002		1	Nannilam	
Panruti	PI	3050	1991	1	Cret	IIEP
Pappannacheri	OC	2702	1987	1	Cret	
Paravakottai	PK	1151	1992	4	Basement	
Peruvalandan	PV	1416	1987	3	Basement	
Ponnapur	PP1	4105	1991	1	Cret	
Pozakkudi	PZ	3302	1993	1	Cret	
Puduvur	PR	2500	1991	1	Cret	
Radhapuram				1		
Raghunadhapuram	RG	2036	1987	1	Basement	
Rajamangalam		2201	1991	1	Cret	
Rajasingamangam		3350	1996	1	PreAlbian	
Ramnad		3550	1992	1	Cret	
Rayanallur	RY	3085	1991	1	Basement	
Rettaikudi	RT	3155	1997	1	Cret	
Sakkarakottai		2505		1	BHV Fm	
SaliaMangalam	SL	1050	1971	1	Basement	
Sattanur	SR	3450		1	Andimadam	
Sattapadi	ST	4701	1987	1	Cret	IIEP
Semangalam	SM	3200	1992	1	Cret	
Siruvattur	SV	2417		1	Basement	
Sundarampatti	SP	2201	1992	1	Cret	
Tagattur	TG	2630	1990	1	Basement	
Tattankoil	TK	3088	1990	1	Cret	
Tenangudi	TN	3303	1992	1	Cret	
Thathanthottam	TO	2271	1995	1	Basement	
Thevur	TH	3503	1985	1	Cret	
Thirukkadiyur	TY	4505	1989	2	Cret	

Cauvery Basin	*Code*	*DD*	*Year*	*Well*	*Bottomed in*	*Remarks*
Thirumarugal	TL	2901	1992	1	Cret	
Thirumullaivasal	TI	2265	1994	1	Basement	
Thirunallar	TR	3753	1992	1	Cret	
Thiruthonipuram	TM	1900	1991	1	Basement	IIEP
Thiruthuraipundi	TT	2370	1967	3	Basement	
Thiruttonipuram	TU	1900		1	Basement	
Thiruvasal	TS	3500	1993	2	Cret	
Tiruvadaimraudur	TD	2364	1996	2	Basement	
Uchipuli		2985	1991	2	Basement	
Umamaheswarapuram	UM	2500		1	Komarakshi	
Uthirangudi	UG	3005		2	Basement	
Vadakkukottai	VK	2501	1994	1	Lr.Cret	
Vadavuru	VR	3500	1993	1	Cret	
Vanagiri		3750		1	Komarakshi	
vandampalai		3451	1996	1	Cret	
Vanjiyur	VN	3500	1994	1	Cret	
Vazhkudi	VZ	3985	1991	1	Cret	
VDM-1	10	1337	1988	1	Basement	-
Vettikadu	VT	2967	1989	3	Basement	
OFFSHORE	W.D				64wells	
CDW-1		2127		1	Basement	
CY-DW-F		3552		1	BHV fm	
CY-DW-N-1		4053		1	Basalt	
CY-DW-N-2		3910		1	Basalt	
GM-1	23	2689	1993	1	Basement	
GM-4	46	3604	1993	1	Cret	
GM-5		3500		1	Andimadam	
GM-6		3500		1	Andimadam	
GMS-3		2650		1	BHV fm	
GMS-9		2266		1	Karaikal fm	
KH-1	25	1316	1978	1	Basement	
KH-2	58	2408	1981	1	Basement	
KH-3	50	2750	1987	1	Basement	
KH-4	21	4000	1993	1	Lr.Cret	
PBS-11				1	palkBay Fm	
PBS-6		2800		1	PalkBay Fm	
PBS-7		2809		1	PalkBay Fm	
PH-10	12	4359	1987	1	Basement	
PH-11	12	2510	1987	1	Basement	
PH-13	12	3400	1987	1	Basement	

Cauvery Basin	Code	DD	Year	Well	Bottomed in	Remarks
PH-3	11	3761	1979	1	Basement	
PH-31	09	2351	1987	1	Basement	
PH-8	09	3070	1982	1	Basement	
PND-1	35	4491	1990	1	Cret	
PON-1	23	3862	1991	1	Basement	
PON-2	27	4476	1991	1	Basement	
PY-2	35	2596	1988	1	Cret	
PY-4	93	3350	1991	1	Basement	
PYH-1	52	4964	1991	1	Cret	
PYH-2	33	3423	1978	1	Basement	
PY-S		2566		1	Basement	
TR-A	41	4500	1993	1	Cret	
TR-B	39	4501	1993	1	Cret	
TR-C	60	4501	1990	1	Basement	
TR-D	52	3800	1992	1	Cret	

W.D = Water Depth

Table.10.cu.2. Shallow Wells Drilled-Cauvery Basin

Name	Code	DD	yr	Wells	Bottomed in	Type
Gandharvakottai		1000		1		St.well
Karaikudi		743		1	Basement	St.well
Kumbakonam		558		1		St.well
Mannargudi		777		1	Basement	St.well
Nakkudi		362		1		St.well
Pandikanmay		1000		1		St.well
Pattukollai		988		2		St.well
Tanjavur		1000		1		St.well
Thirupundi	TP	1141	1968	2	Mid.Eocene	St.well

Table. 10.cu.3. Oil and Gas Fields–Cauvery Basin

Name	code	OG	Year	Wells	Sub-Basin	Payzone	Formation
Adiyakkamangalam	AK		1989	19	Ngp-Tanjor		
Attikadai+			1990	1			
Bhuvanagiri	BV	O	1988	11	Tranquebar		
Kali	KI	O		9	Tranquebar		
Kamalapuram	KP		1988	45	Ngp-Tanjor		
Kanjangudi		G		13			
Karaikal	KL		1964	12	Ngp-Tanjor		
Kizuvalluru+			1995	8		2768-2773	Ariyaluru

Name	code	OG	Year	Wells	Sub-Basin	Payzone	Formation
Kovilkalappal	KK		1984	29	Ngp-Tanjor		
Kuthalam	KA	G		39	Tranquebar	2510-2518	Andimadam
Kuttanallur	KN		1993	12	Ngp-Tanjor	2164-2167	
Mattur	MT	O	1992	12			
Mayladuthurai	MY	G		1	Tranquebar		
Nannilam+			1987	18			
Narimanam	NR		1985	72			
Neduvasal	NV		1993	2	Ngp-Tanjor	1705-1710	
Neyveli	NY	G		2	Tranquebar	2663-2566	
NanniLlam	NL	O					
Pallivarmangalam	PM		1992	8	Ngp-Tanjor	3272-3278	
PBS-1		G	2001	8	Ramnad		
Periyapatnam	PP	G	1998	15	Ramnad		
Perungulam		G	1994	8	Ramnad	2103-2110	
PH-9			1981	8	Ngp-Tanjor		
Pundi		O	1990	3	Ngp-Tanjor		
Py-1			1980	12			
PY-3			1988	14			
Ramanavasalai		G	1997	5	Ramnad	1881-2194	
TirukKlLar	TL		1987	3	Ngp-Tanjor		
Tiruvaarur			1985	28		2921-2930	
Tulsapatnam	TA		1996	3	Ngp-Tanjor		
Vadatheru	VD		1991	4	Ngp-Tanjor		
Vijayapuram	VJ		1993	30		2525-3260	Nannilam
Tirunagar							
Adichapuram				4	Ngp-Tanjor	2335-2430	
Kanjangudi		G	2002	14	Ramnad		
Tirungari			2014			gas	
Pandanallur	PN		1971	5			

Krishna Godavari Basin

Introduction

Krishna-Godavari basin is located along the East Coast of Indian peninsula comprises of sedimentary facies from Early Permian to Recent. The basin is covered by two major river systems, namely Godavari and Krishna. The areal spread of the basin is 28,000 sq. km. in onland and 1,45,000 sq. km. in offshore.

Geographically, the basin lies between Kakinada in the northeast and Ongole in the southwest to an extent of 500 km in length along the coast. The basin extends towards southeast into the deep waters of Bay of Bengal. The basin represents a depositional setting of a well-defined delta, shelf, shelf-edge to deep water.

Achaean crystalline basement and Upper Cretaceous sedimentary outcrops demarcate the northwest basin margin (Reddy, P.R, *et al.*, 2002). A significant part of the onshore basinal area is covered by Quaternary alluvium (Figure 10.kg.1).

Exploration History

The basin attained importance since the formation of Oil and Natural Gas Commission (ONGC) in 1956 for searching oil and gas, with geological mapping at the basin margin areas. Apart from geological mapping, geo-morphological studies and detailed paleontological and palynological studies were also carried out to establish clear startigraphic boundaries. The detailed geological mapping was carried out covering 4,220 sq. km. At basin margin area about 70 shallow bore holes were drilled during 1970-75 to know subsurface continuity of outcrops in subcrop.

Gravity-Magnetic surveys in on-land part have been carried out by ONGC since 1964 over an area of 19,200 sq. km. In offshore area, M/s. Prakla Seismos and GSI acquired the gravity-magnetic data for ONGC. Initially single fold seismic surveys were started in 1965 and with technology development, 48 fold 2D seismic data were also acquired in the basin. ONGC has also carried out 3D seismic data in promising areas in onland area.

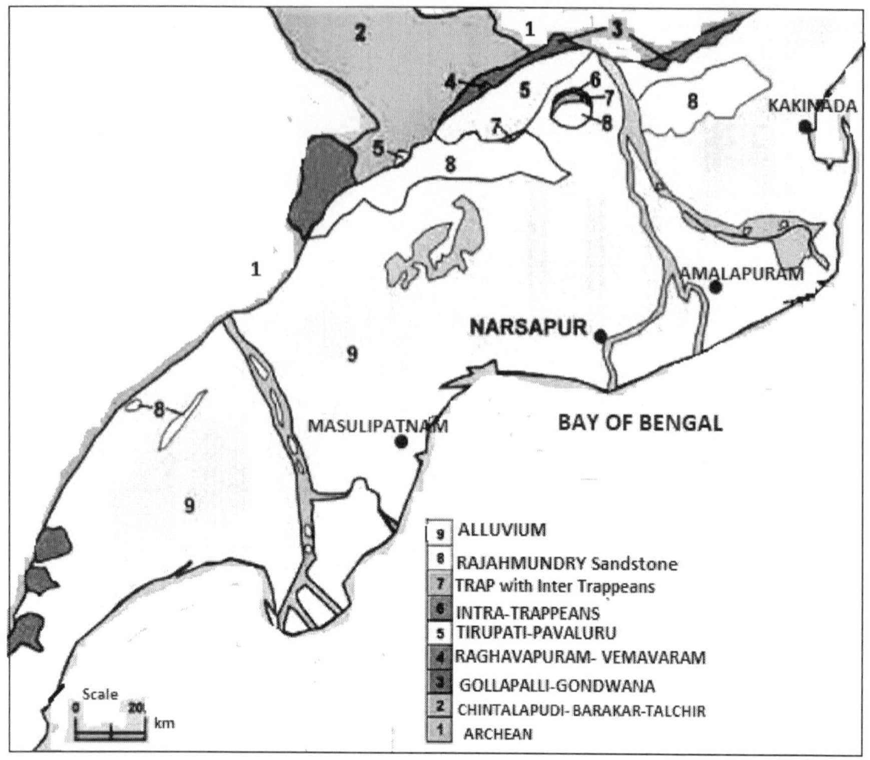

Figure 10.kg.1. Geological Map-K G Basin (Modified after Kumar,SP, 1983).

The integrated subsurface data interpretation resulted in identification of few structural anomalies in Narasapur-Razole area in 1976. Accordingly ONGC started the exploratory drilling in the basin at Narasapur. This very first well drilled has established gas accumulation in the structure and found associated with abnormal formation pressures (Shenai, K.R. and Rao, M. S., 1982).

Offshore conventional seismic surveys were carried out in the continental shelf in 1964-by Soviet expedition. The Prakla Seismos has carried out 3,380 km of 48 fold CDP survey with simultaneous gravity and magnetic observations. The offshore seismic data acquired also established good amplitude structures with in Neogene section. As a follow up action, nearly 6,700 km of extensive regional and detailed surveys in the shelf and over the continental slope were carried out between 1979 and 1981 by mvAnweshak (ONGC), and by contract mv RC Dunlop (GSI) and Western Islander (Western Geophysical). Additionally, during 1972-74, 2,028 km. of refraction data were also acquired to study the basement configuration and also shallow reflectors.

The first offshore well was drilled at G-1 structure during 1980 at water depths of about 250m. This very first well in offshore has also established oil accumulation with in Miocene section.

Interpretation of the data gave an idea of the Tertiary and Cretaceous sedimentary thickness in the offshore and the presence of Tertiary deltas, Paleogene and Neogene, of the Godavari and Krishna rivers.

The subsurface data obtained both through drilling and seismic have helped to understand geological models and the occurrence of abnormal formation pressures though initially related to under compaction below basalt, but later analysis made to understand it is sedimentation criteria which crosses geological stratigraphic boundaries. So based on depositional pattern in an area its formation pressures need to be understood.

Advances in seismic data acquisition also helped in demarcating clear litho-markers/stratigraphic boundaries, which in turn aided to bring clarity in depositional models in relation geological time which in turn helped in devising Litho-startigraphic Nomenclature to various depositional units within a given stratigraphic unit.

As we know the hydrocarbon shows during drilling is an important factor to help electro-log interpretation to identify potential zones for testing, which only can establish hydrocarbon bearing reservoirs. Sometimes, the much overbalance drill-mud during drilling may hamper to get such hydrocarbon shows. Hence, the operational team in consultation with drilling teams slowly and optimally reduced the much over-balanced mud weights in different stratigraphic levels which helped in observing the oil and gas shows in all stratigraphic levels in the basin. Now Krishna Godavari basin is the unique basin having established oil and gas reservoirs in all formations ranging from Pliocene to Permian (Rao.,G N 1991).

Further, in order to make exploration more economical and faster, drilling through deeper very high formation pressures was restricted and based on geographic location in the basin coupled with Formation pressure gradients.

During these 40 years of exploration and production activity so far about 910 deep wells were drilled (yr 2018) in different parts of the basin. More than 275 prospects have been probed, hydrocarbon accumulations have been proven in 139 of these prospects (39 oil and 100 gas) (Figure 10.kg.2).

As of now, the deepest on-land well is Bhavadevarapalli in WG Sub-Basin and in offshore GS-11-1 in water depth of 45m was up to 4,611m. In ultra-deep area of offshore UD-1 was drilled to 6,035 in water depth of 2,843m. The exploratory efforts made in the basin are shown in Tables 10.kg.1.

As of now, about 133 MMT of O+OEG reserves were established in onland area of the basin and 341 MMT in offshore part., out of 1,130 MMT of calculated resources.

Pre-NELP

During late 1980s oil exploration was carried out off Visakhapatnam coast by Chevron Company under license from GOI, after carrying out initial seismic data interpretation it was opined that, the sediments were at shallow levels all through, hence may not be prospective for generation of hydrocarbons.

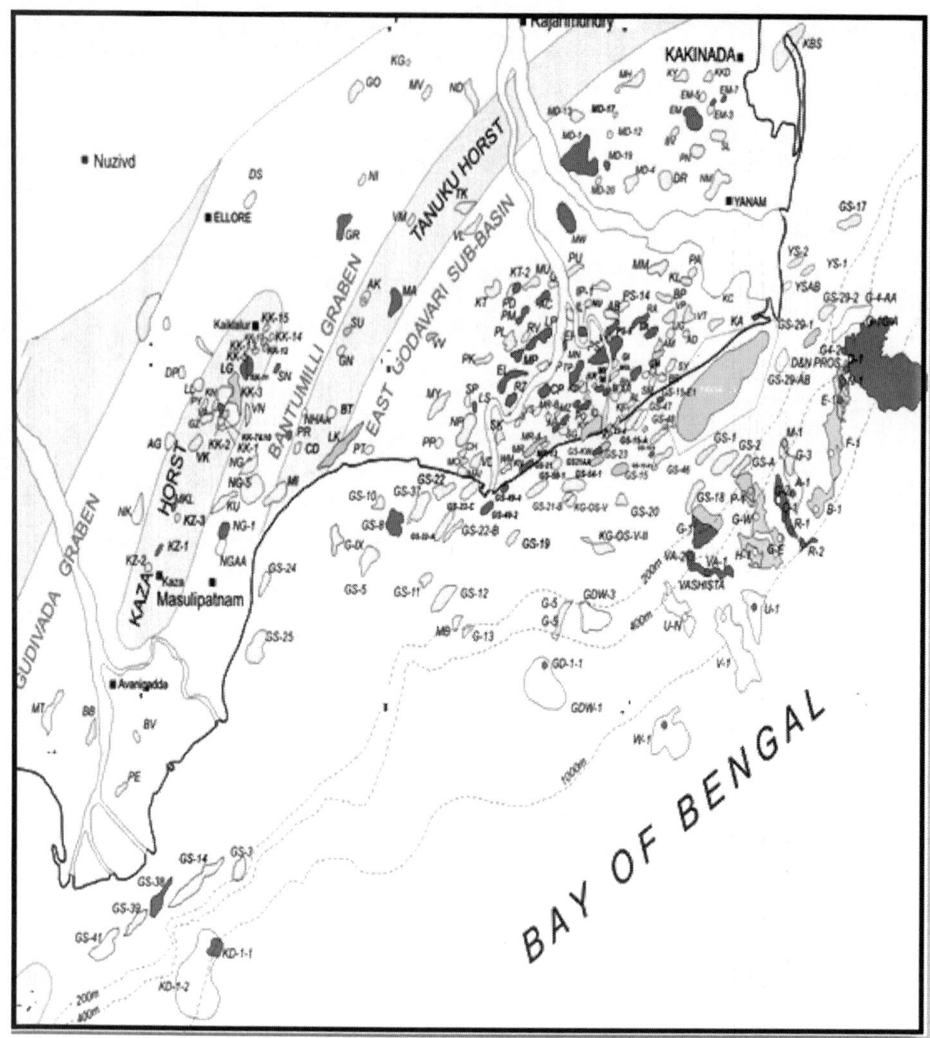

Figure 10.kg.2. Oil and Gas Fields of–K G Basin.

During 1992, GOI formed 4 blocks for exploration in the Basin, one in Pranhita Godavari Graben was awarded to HOEC consortium, one in Krishna Mouth area to Tullow India and RIL and the third one located actually off Nellore in shallow water area is in Palar basin was awarded to Hardy, HOEC and Nagarjuna Ferlizers Consortium. Last one in shallow a waters off Machlipatnam was awarded to Cairn. After initial exploration studies the blocks were relinquished and only in KG-OS-6 off Machilipatnam drilling of one well done without success which was also relinquished.

NELP

Under New Exploration Licensing Policy (NELP) rounds of bidding by the Government, the K G Basin attracted the bidders. About nineteen discoveries have been made by Pvt./JV companies so far in NELP blocks, Dhirubhai discoveries(15) by RIL in blocks KG-DWN-98/3 and KG-OSN-2001/2, three discoveries by Cairn Energy Pty. Ltd. (CEIL) in block KG-DWN-98/2 within Mio-Pliocene which was later acquired by ONGC added 3 more in the block KG-DWN-98/2 and one discovery by GSPC in block KG-OSN-2001/3 within Lower Cretaceous.

Tectonics

Geological History

The East Coast of India is defined by the rifted passive margin that evolved in response to the continental rifting and seafloor spreading processes between India, Antarctica and Australia that occurred during the early Cretaceous period (Powell, *et al.*, 1988).

The peninsular India was a part of Gondwanaland upto Early Mesozoic period (Chatterjee, S and Ilotton, N, 1986). The initial breakup of this super continent was about 200 million years ago and the Indian plate separated from Antarctica and Australia about 174 million years ago. The divergent triple junction coincides with Machilipatnam Bay of the basin (Kaila, K.L, *et. al.* 1990). At the time of break-up it was positioned at 50°south latitude (Qureshy, M. N, *et al.*, 1968). Though the rift systems of Pranhita-Godavari and Carnovar of NW Australia placed close even now, the popular continental fit indicates that the Enderbyland of Antarctica was positioned at the Machilipatnam Bay, the tri-junction of the continental separation (Krishna, K.S., *et al.*, 2009).

The Indian plate started moving and newly generated oceanic crust since Cretaceous. Early Triassic to Jurassic section is represented by red clays. Once oceanic crust was formed on either side of the Indian plate, 90° East Ridge and Carlsberg ridge in the west have become trajectories for the movement of the Indian plate (Rao, G.N. and D.S.A.N.Raju. 1992).

Because of the intrinsic distinctions between different sub-crustal blocks (Radhakrishna, 1989; Lal, *et al.*, 2009), the East Coast of India is tectonically and geomorphologically segmented into several peri-cratonic rift basins (Murty, K.V.S., Ramakrishna, M., 1980).

The K G Basin has been classified as a major intra-cratonic rift within the Gondwanaland until the Early Jurassic period (Rao, 2001), and it later transformed into a peri-cratonic rift basin (Biswas, *et al.*, 1993). The horst-graben structural configuration and other major basement trends in the onshore and the shallow offshore region of this basin are well-known because of geophysical studies and exploratory well drilling data (Sastri, *et al.*, 1973; Kumar, 1983; Prabhakar and Zutshi, 1993; Mohinuddin, *et al.*, 1993; Rao, 2001; Gupta, 2006; Bastia, 2006).

Knowledge of the structural architecture from the onshore to the deep offshore areas of the margin is useful for defining the geometry and structural parameters of the basin and for the delineation of the Continent-Ocean Boundary (COB). These elements contain a record of the early continental rift-drift history and are important for evaluating the deepwater hydrocarbon potential (Sastri, V.V., Venkatachala, B.S., Narayanan, V., 1981).

The Krishna-Godavari basin occupies a strategic position as it coincides with the triple junction of the diverging India, Australia and Antarctican plates (Nemcok, M., *et al.*, 2012). Though the southeastern continuity of PG graben could not be established with certainty, the recent marine geophysical surveys carried out by NIO have established with fairness the continuity of PG graben up to OCB (Subrahmanyam, A.S, *et al.*, 2010).

The intense magnetic anomaly located off Machilipatnam is related to a hotspot at the time of initial rift of the Indian plate (Rajaram, Mita, Anand, S.P., Erram, V.C., 2000). The Chintalapudi cross-trend appear to be older than other structural features of KG basin. Another linear trend at Yanam called Pithpuram cross-trend suggest a linear basement ridge with an average depth of 3 km and it is fault controlled. PG graben is now known to continue up to OCB between these two cross trends with axis located off Amalapuram at right angles to coastline (Mishra, D.C., 2002 and Venkateswarlu, P.D, *et al.*, 1992).

The eastern passive continental margin of India has subjected to tectonic readjustment of continental blocks with the denser oceanic crust which has resulted in to subsidence of crustal blocks (Rao, G.N., 1993a). During this phase, the Eastern Ghat trend (NE-SW) was reactivated and resulted into NE-SW aligned horsts and grabens superimposed over the pre-existing trend. Due to this passive margin subsidence southeasterly slope was developed which facilitated a NW-SE drainage with sedimentation (Murthy, K.S.R, *et al.*, 1995). More than 3000 of Cretaceous sediments are known in the peri-cratonic Krishna-Godavari basin. The Cretaceous fluvio-marine sediments are occurring as outcrops at NW basin margin areas of the basin.

The Residual gravity anomaly map shows the offshore extensions of the onshore trends, a series of linear highs and lows, parallel to the Eastern Ghat trend (Bhaskara Rao, V, *et al.*, 1971), as a result of block faulted movement of the basement (Figure 10.kg.3).

A pronounced gravity low south of Eluru (onland) and another SE of Narasapur (offshore) are more or less in general alignment with the Gondwana trend, indicating the possibility of a NW-SE downward flexure in the basement superimposed by the more pronounced Eastern Ghats trend (Radhakrishna, M., Chand, S., Subrahmanyam, C., 2000).

The magnetic data is not as diagnostic as the gravity, but corroborates the results obtained from gravity in the areas of relatively shallow basement, total magnetic 'negative anomalies' depicting basement highs in the Nizampatam bay and off Kakinada (Protacio, J.A., *et al.*, 2011).

The Bouguer gravity map of the area also clearly brought out the PG graben and its structural limits, however in the coastal onland area the gravity lows and highs are aligned in NW-SE trend. It can be seen that few small lows are aligned in NW-SE trend right up to shallow offshore area.

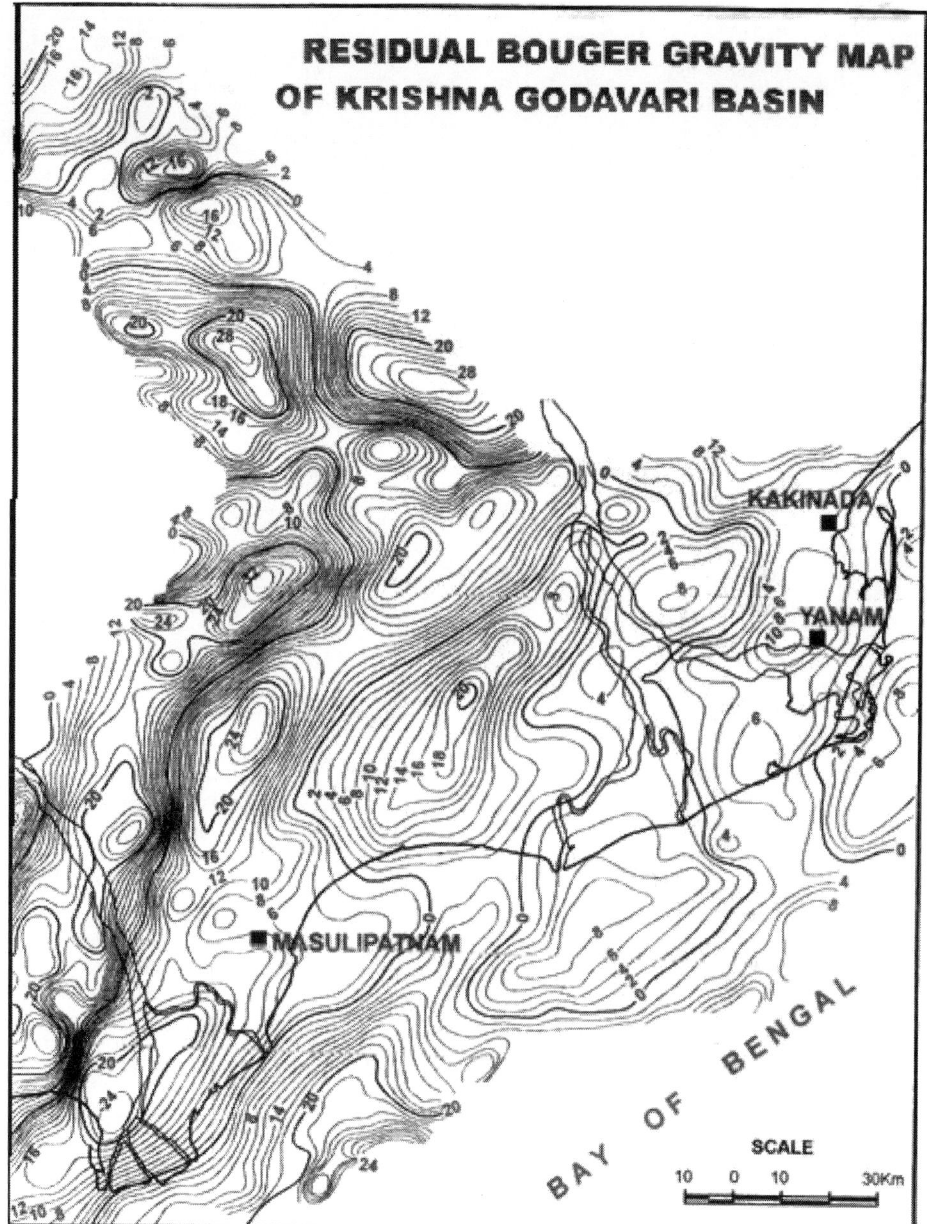

Figure 10.kg.3. Residual Gravity Map-K G Basin.

Tectonic Elements

The basin evolved as a composite of rifted grabens, beginning in Late Jurassic, and formed a part of the development of the east coast divergent margin (Subrahmanyam, C, *et al.*, 1999). The horsts and grabens were separated by vertical or steeply dipping faults (Pateria, M.L., Rangaraju, M.K., Raiverman, V., 1992). The regional basement horsts divide the basin into several sub-basins such as, Krishna, West Godavari, and East Godavari (Figure 10.kg.4).

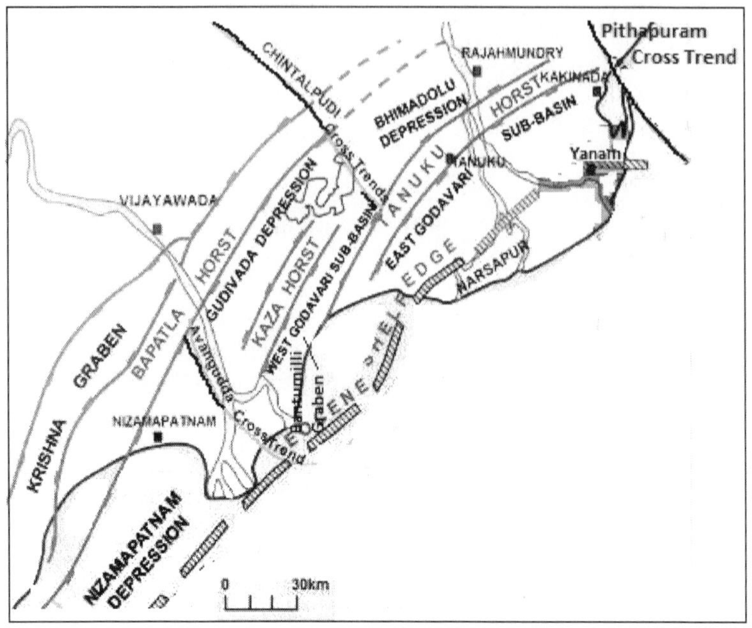

Figure 10.kg.4. Tectonic Map-K G Basin.

These sub-basins contain thick Cretaceous and older sediments above the Archaean basement, with several intervening unconformities.

Tectonically the basin can be divided into Krishna, West Godavari and East Godavari subbasins separated by Bapatla and Tanuku horst. The West Godavari subbasin is further separated by Kaza-Kaikalur horst into Gudivada and Bantumilli grabens.

The subsurface geology depicted in NW-SE geological section confirms the continuation of the PG graben underneath NE-SW trending KG basin bounded by two cross trends called Chintalapudi and Yanam. The KG Basin show the tectono-sedimentary events from Permian to Recent (Raiveman, V., Rao, M.R. and Pal, D. 1985).

The rift phase terminated by the end of Turonian in most parts of the basin, and subsequently, the post-rift sedimentary sequences prograded to the east with development of a continental shelf-slope system.

In the Tertiary, the deepwater area became structurally deformed by numerous sets of growth faults and related features. In most areas, a major de-collement surface is present near or at the top of the Eocene and marks a major tectonic event.

A series of growth fault systems progressively developed, with increased sediment influx during the Oligocene through Miocene time. The onset of the

Pliocene is marked by major sea-level fall and a prominent erosional surface. These low-stand conditions prevailed into the Pleistocene (Rammohan Rao,T., 1987).

Rift Stage: The basin got initiated through rift/syn-rift tectonics between Permo-Triassic to Early Cretaceous and is essentially characterized by lagoonal to fluvial to occasionally brackish water sediments. The northeastern part of the present onland basin was part of an intra cratonic rift set up till Jurassic that constituted the southeastern extension of NW-SE trending continental rift valley slopping northward (Verma, R. K., 1991).

The seismo-stratigraphic analysis towards northwest of Tanuku horst indicate fault patterns with southeasterly dip, suggesting thickening of Lower Gondwana sediments. However, the axis of the Palaeozoic Kommugudem graben lies towards northwest of Tanuku horst. It can be inferred that during Late Mesozoic (Late Jurassic to Early Cretaceous) there was a mild upwarping of the Tanuku horst due to reactivation of basement faults resulting into erosion of older sediments over the horst.

Syn Rift Stage: The early stage synrift sediments were deposited during early subsidence by tectonic fault systems. Basin subsidence continued along basement bound fault system accommodating synrift sediments of late Jurassic to early Cretaceous (Ramana, M.V, *et al.*, 2001).

Drift Stage: The Indian plate during its movement might have tilted slightly towards southeast resulting in the formation of major sedimentary basins along east coast of the craton (Radhakrishna, M., *et al.*, 2012). Rift to drift transition is marked by a southerly/southeasterly tilt of the basin leading to widespread marine transgression during Cretaceous and deposition of marine shale sequence followed by onset of overall regressive phase during Late Cretaceous, represented by a deltaic sequence comprising Tirupati Sandstone with dominant arenaceous facies. By Early Cretaceous period, the horst and graben morphology was peneplained by sedimentation as evidenced by a major marine transgression up to basin margin (Raghavapuram Shale).

Late Drift Stage: Initial soft collision between the Indian and Eurasian Plates and initiation of Matsyapuri-Palakollu fault appears to have greatly influenced the Paleogene and younger tectonic regiment and the consequent sedimentation pattern.

Late Cretaceous to Early Tertiary period witnessed the volcanic activity (Deccan traps) in the basin and thickness of the trap was controlled by Pre-trappean topography. These basalts occur at a depth of 100m near basin margin, 700-1200m upto Matsyapuri-Palakollu fault trend and 1500-3000m within the step-fault zone, whereas southeast of this trend, basalts occur around 3300m. This indicates a major southeasterly dipping step faulting after the volcanic activity. The top of basalt forms the basin floor for overlying Palaeocene sedimentation. During Tertiary period, the sedimentation was controlled by northwest-southeast drainage pattern with progradation of delta progressively towards southeast with a NE-SW aligned depocentre.

Geology of the Area

Towards north and west, the basin is limited by Archaean Basement consisting of gneisses, Khondalites Charnokites. Outcrops of Tertiary, Mesozoic and Permian, are seen in the Rajahmundry-Eluru. Jurassic to Cretaceous rocks are exposed in three main patches located around Eluru; Budavada-Vemavaram and Kandukur area near the western basin margin (Ramam, P.K., Murthy, V.N., 1997). Briefly, they consist of Chintalpudi formation (mainly sandstones) of lower Triassic to upper Permian as lower Gondwana, Gollapalli formation (sandstones with conglomerate bands) of middle Jurassic; Raghavapuram formation (shales with thinsandstones) and Tirupati formation (sandstones with minor shales) of lower Cretaceous of upper Gondwanas; Infra-trappeans (Limestones, marls and shales) of Paleocene; Trap of lower Paleocene and Rajahmundry formation (sandstone with shales and conglomerate) of Mio-Pliocene (Kumar, S.P., 1983). Although Eocene sediments are not exposed, these are seen in the Bhimadolu and Denduluru shallow wells Bhimadolu formation comprising silty clays and sandstones.

This thick pile of sediments have resulted in to a isostatic rise at the basin margin area, due to which lower Gondwanic sediments exposed as outcrops. Southeastern limit of Chintalapudi Sandstone formation at Paloncha neck demarcate the areas of isostatic rise towards northwestern parts of the graben and passive margin subsidence towards southeastern parts. (Figure 10.kg.5).

These upper Gondwanas are overlain by Infra-trappeans. This contact is neither exposed nor penetrated by any shallow well. About 75 metres was encountered in Dudukuru shallow well, where three trap (Deccan basalts) flows have been recognized. These basalt flows are separated by inter-trappean beds. Rajahmundry sandstone of Mio-Pliocene occur as outcrop near Rajahmundry area and also encountered in most of the shallow wells and seem to have deposited in continental to lagoonal environments. The Holocene sediments of fluvial and deltaic origin cover the Rajahmundry formation.

Sedimentation

Sediments in West Godavari and Krishna sub-basins are attached to the basement whereas towards SE of Matsyapuri-Palakollu fault of East Godavari sub-basin are basement detached, due to which syn-sedimentary subsidence was more pronounced.

During the initial rifting-drifting phase fluvio-lacustrine sediments deposited all over the basin. However, the transportation was over short distances, and therefore clastics of that time are devoid of sorting and are mainly argillaceous arkosic sandstones. The first marine incursion appears to have occurred during the Albian.

The deep drilling data indicated presence of thick Lower Gondwana sediments comprising Permo-Triassic arenaceous (Chintalapudi sandstone) and Permo-Carboniferous sand-coal-shale sequances (Barakars) towards northeastern part of the basin. It suggests extension of Gondwana graben underneath NE-SW trending Krishna-Godavari coastal basin along prominent NW-SE cross trends. In some wells the Lower Cretaceous - Upper Jurassic sediments directly overlie the basement. It

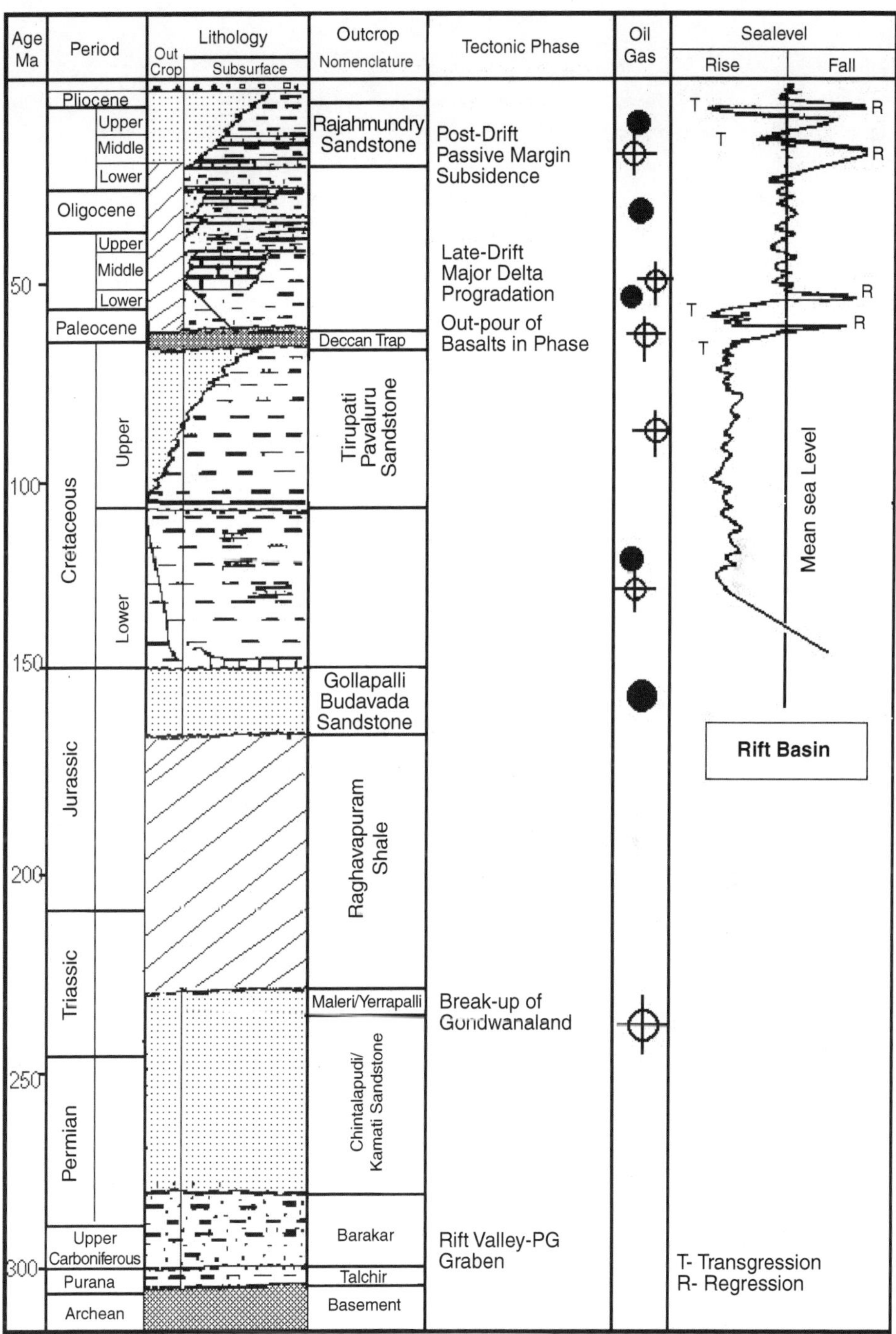

Figure 10.kg.5. Out Crop to Subsurface Correlation K G Basin.

can be inferred that up to Early Mesozoic period, the sedimentation was controlled by Palaeozoic horst and graben morphology (Rangaraju, M.K., Aggarwal, A., Prabhakar, K.N., 1993).

Mesozoic deposition during Permo-Triassic-Jurassic time was primarily non-marine and associated with horst and graben features. The rift phase continued till Lower Cretaceous time and the end of this phase was marked by a transition from non-marine to marine environment during Mid-Upper Cretaceous (Rao, G. N., K. Satyanarayana, and K. S. Mani, 1993).

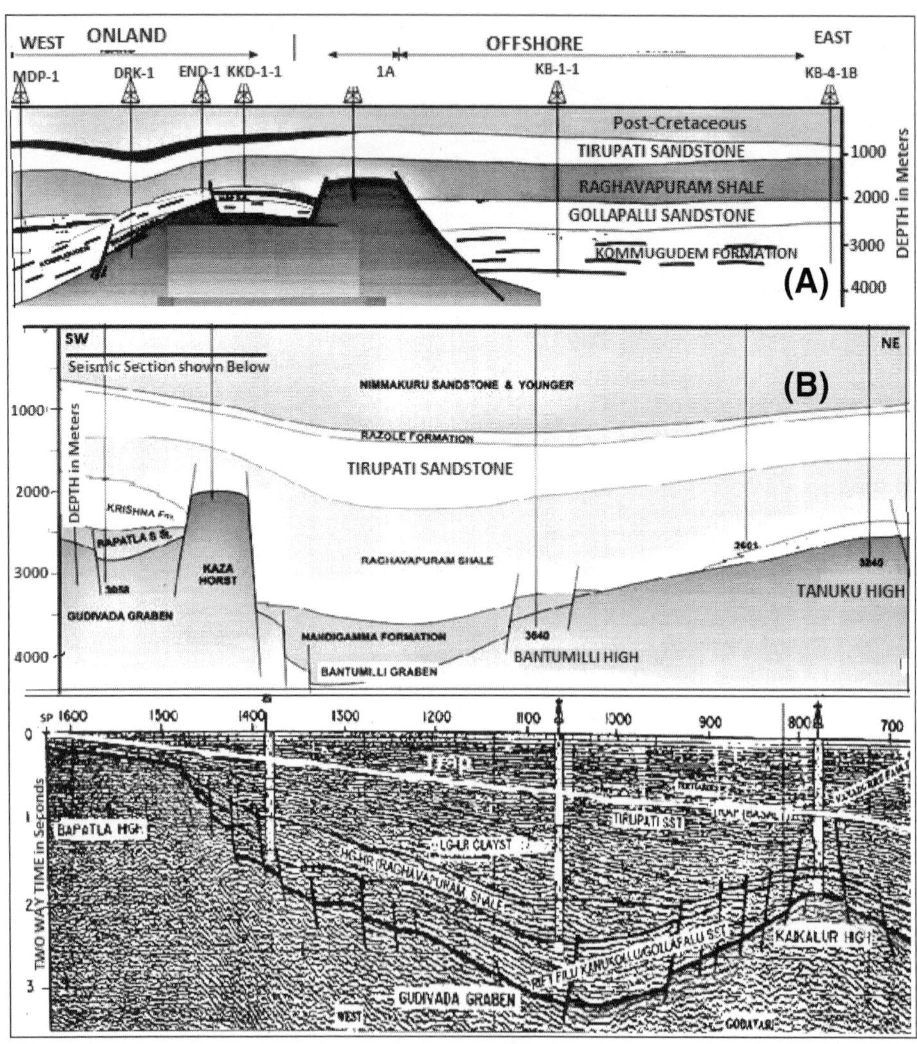

Figure 10.kg.6. Geological Section-
(A) East Godavari Sub Basin-(B) West Godavari Sub Basin.

Based on litho columns penetrated in deep wells, litho-contrast lines of different geological ages have been marked as first step indicating, towards NW basin margin side major arenaceous section and towards SE of the line major argillaceous units of corresponding age can be seen. (Figure 10.kg.6 A and B).

Increased gradients for the river systems and increased sediment load coupled with significant sea level falls during Neogene had triggered sediment induced tectonics in the shelf and slope parts of the basin.

The post-rift deposition was primarily shallow marine to marine characterized by prograding clastic sequences. The growth faults and associated rollover conditions featured the deposition. Post-Miocene tectonism resulted in the formation of smaller but sharp, as well as flat and large, anticlinal structures within the Tertiary sequences in the deep waters. Paleocene sequences onlap onto the paleo-highs of Cretaceous and Miocene sequence onto Oligocene at the shelf break. Major slope failures act as the primary source of turbidite accumulations. Most channels are filled with coarse sediments. Since Proto-Godavari is flowing through the rift graben with sediments, the Neogene sediments in the basin are of second order sediments with more of finer clastics and shales.

The rate of sedimentation calculated from the present thickness of the sequences versus time period during Triassic to Early Jurassic, is very low suggesting longer periods of non-deposition in between (Lakshminarayana, G.L., 2002). From the well data, the prominent red coloured clay-stone termed as 'Red Bed' clearly demarcates the Lower and Upper Gondwana sediments.

By Early Cretaceous, all these horst and grabens were smoothened by sedimentation. During the drift phase, the easterly tilt of Peninsula has resulted in a major marine transgression upto basin margin represented by Raghavapuram Shale during Cretaceous. The overlying regressive phase of Upper Cretaceous is represented by Tirupati Sandstone. In subsurface, this arenaceous sequence is known to be continued upto Draksharama-Bantumilli trend and further SE, it progressively changes into a thick (>2000 m) argillaceous sequence.

During Late Cretaceous to Early Paleocene, the basin experienced volcanic activity. The thickness of the basalts ranges from 10 m to 400 in with inter-trappean claystone and limestones. The post-trappean down to basin step-faulting has created a steep gradient towards SE of Matsyapuri-Palakollu fault zone.

This coastal basin may be similar to the Niger type delta (Talukdar, 1981). The coastal basin is broken up by basement ridges trending NE-SW, on which is superimposed the SE directed paleo-slope with the major part of the depositional prism was build-up.Three broad zones in offshore area, zone of homoclinal dip from the coast to 100 m bathymetric contour, zone of anticlinal reversals in the area between 100 and 200 m bathymetric contours and zone of Roll-over anticlines in deeper waters beyond 200 m bathymetric contour.

The shelf areas received deposits of clastics and carbonate sediments while the slope registered deeper-water fan sediments. This setting, during which progradation was dominant, persisted throughout the Tertiary (Ranga Raju, MK and Subba Rao, CV 1986). The Paleocene and Eocene, in part, are considered to have been deposited during sea-level low-stands, thus forming fan complexes. From Oligocene onward, sea level began to rise, and more accommodation was available. The basin is expected to have a maximum sedimentary thickness of more than 6 kilometers. It is likely composed of primarily clastic facies of shale, fine sandstone, siltstone and clay with thin bands of carbonates.

This has facilitated rapid sedimentation during Paleogene in this part of the basin. The depocentre during Paleocene lies at Tatipaka-Pasarlapudi area, whereas during Eocene the depocentre is located in Mori-Kesanapalli trend nearer to present coastline. Due to the steep gradient of the trap floor, during Paleocene to Early Eocene period, the arenaceous sediments were suddenly dumped into deeper parts of the basin. The rapid sedimentation during Paleocene and Early Eocene has reduced this gradient and facilitated smooth delta progradation thereafter. Due to this greater subsidence of the basin, Paleogene do not outcrop anywhere. The Upper Eocene is an unconformable surface, due to positive movements in the basin. This could be the reason for very minimal sedimentation rate during Oligocene (40-150 m).

The Neogene sediments are outcropping at basin margin and termed as Rajahmundry Sandstone. In subsurface, about 2000 m of Mio-Pliocene sediments are demarcated with depocentre area lying in present day shallow water area of offshore.

Depositional Model–Krishna-Godavari Basin

Towards the end of Cretaceous, due to southeasterly tilt of the Basin, developed well defined basin margin -shelf-slope system. During that period, considerable part of the area was in the slope regime resulting in deposition of thick basinal shales. However, during Paleocene, with further fall in sea level, shelf edge steadily prograded towards southeast exposing a substantial part of the shelf area. Southeasterly flowing streams constituted a network of delta system which prograded basin ward. Moderately thick limestones, sandstones and minor shales are the lithotypes in the shelf regime during this period, which grade into thick monotonous shales. Deposition of thick sand bodies in the slope regime due to turbidity currents is envisaged in the area between wells Bhimamanapalli and Amalapuram (Rao G N., *et al.*, 1996).

During Eocene, position of shelf edge had shifted marginally towards southeast. The sandstone, limestone and claystone gradually grade into thick, monotonous shales in the deeper part. Development of deepwater sandbodies are observed in the Bhimanapalli well, progradation of shelf to the south-east with further fall in sea level. Dumping of clastics by river distributaries had resulted in active channelling and submarine canyon formation. The canyon is generally associated with a fan system on the downslope. One such feature can be seen in the area beyond Amalapuram to adjoining shallow offshore area. The sedimentation pattern over the area is governed by arcuate, listric, normal faults, which have originated due to shale bulging in the basinal part during Late Paleocene–Eocene and onwards. The general trend of these faults is NE-SW with hade towards south to southeast. Time structure maps indicate a number of small roll over structures associated with these growth faults. Progradation of shelf to the south-east with further fall in sea level (Serra,O., 1985).

During a low-stand period, the accommodating space in shelf environment decreases, facilitating the basin ward migration of the fluvial delta. This process loads the shelf edge, forming shelf-edge delta (Rangaraju, M. K. 1987). The collapse of a shelf edge due to slumping or direct fluvial discharge to deep basin, carries the sediments into the deep basin, forming the lowstand deposits in the form of basin-floor fan, slope fan (channel-levee complex), crevasse splay, frontal splays, *etc.*

The base of Miocene can be demarcated with the help of faunal data, but it becomes difficult to demarcate the top of Miocene sequence with the help of faunal data (Vijayalakshmi, 1988). However, on examining electrologs, thick clay bed is observed overlying the lignite/coal bearing Miocene sequence. The base of Miocene is marked also by an unconformable surface coinciding with the top of Oligocene in onland part of the basin. The thickness of Oligocene is very minimum in the basin (40-150 m). Further, a major erosional cut/unconformity surface, as seen from many seismic lines, is another contact in demarcating the top of the sequence.

The thickness of Miocene section thus arrived at by electrolog correlation of wells is shown in isopach map. It depicts that, from basin margin, the thickness of Miocene section gradually increases from 100 m to 1600 m. The depocentre is located off Vainateyam River.

When sand isolith contours of the sequence are superimposed over isopach contours brings out three major sand prone depositional areas, one towards west of Vasistha river at Narsapur area, the other at the southeast of Amalapuram area and the third one in the deep water area of offshore. The sand concentration in the deep water area may possibly due to reworking of sand bodies by turbidity currents.

Recognition of Paleo Deltaic Regimes

The iso-pach maps of Upper Cretaceous, Paleocene and Eocene sequences indicate single depositional lobe with migration to north east located at present Godavari river mouth (Figures 10.kg.7, to 10.kg.11). The iso-pach map of Miocene and above section clearly brought out two lobes one against Godavari and other at present Krishna Mouth. It can be fairly concluded that, during late Eocene or Oligocene due to some tectonic uplift in mid land area the proto-Godavari might have been bifurcated in to Krishna and Godavari river systems. Further, the prograding sediment into sea at Krishna mouth area suggests lesser syn-sedimentary subsidence of the area.

The isopach maps in relation to geological age also suggest that Godavari Delta had more sedimentary influx and the rate of subsidence also helped in protecting the organic content taking it below oxidizing zone which in turn helped in generating more quantum of hydrocarbons, further, facilitated by higher geothermal gradient in the area (Fisher, W. L. and L. F. Brown, 1983).

In subsurface, identification of paleo-deltas can also be a primary target for exploration of oil and gas. Using the exploration data such paleo-delta positions with respect to geological age in the subsurface of Krishna Godavari basin is made

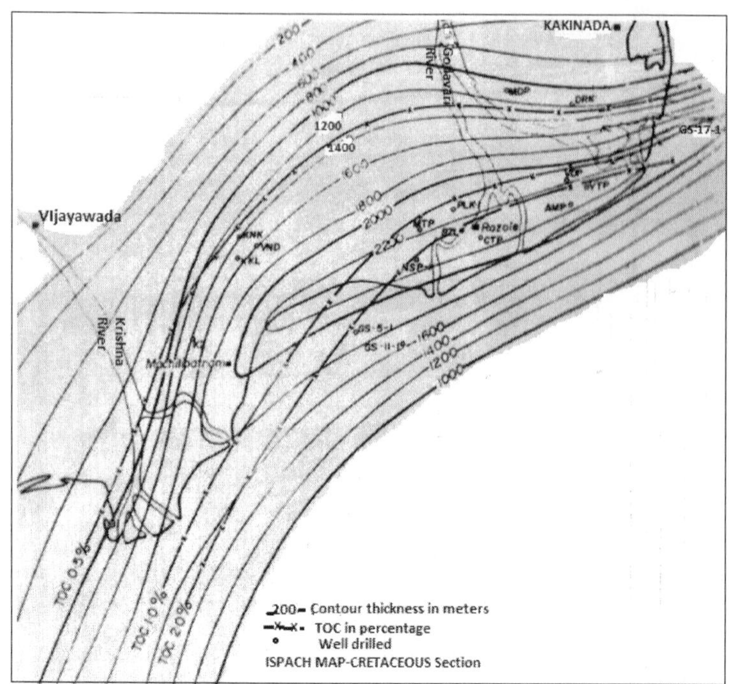

Figure 10.kg.7. Isopach Map-Cretaceous.

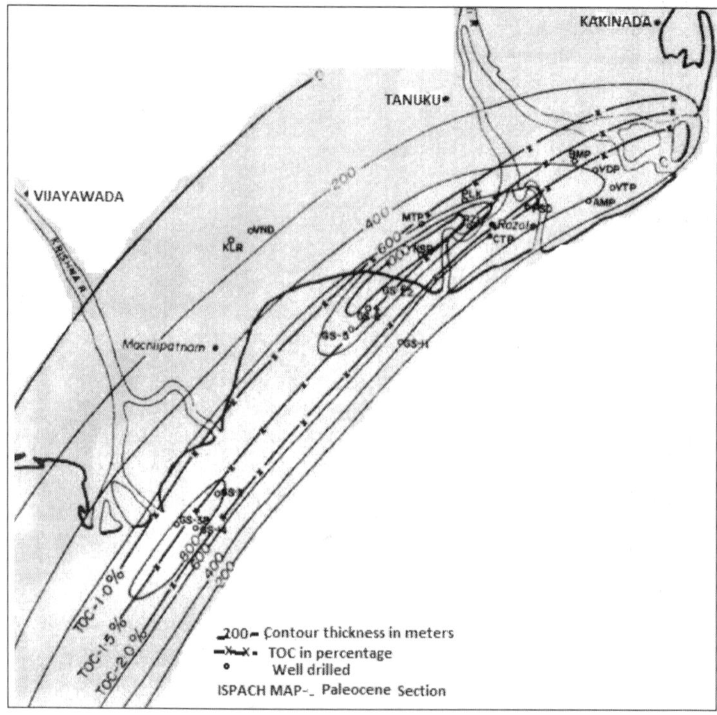

Figure 10.kg.8. Isopach Map-Paleocene.

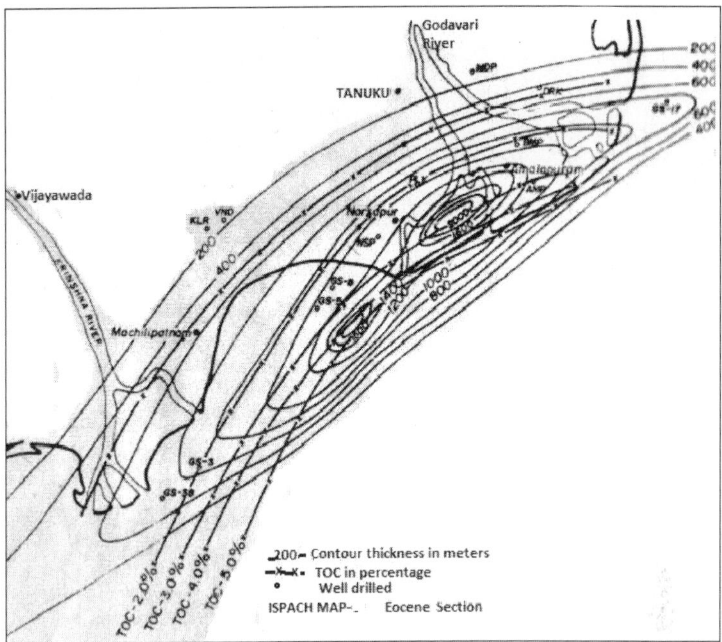

Figure 10.kg.9. Isopach-Map Eocene.

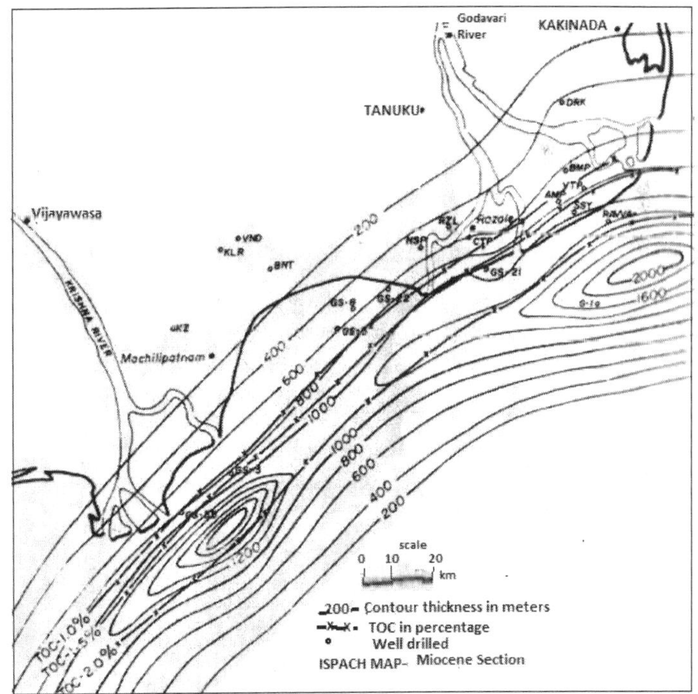

Figure 10.kg.10. Isopach Map-Miocene.

by marking paleoshore lines withrespect to geological age as mentioned above (Figure 10.kg.11).

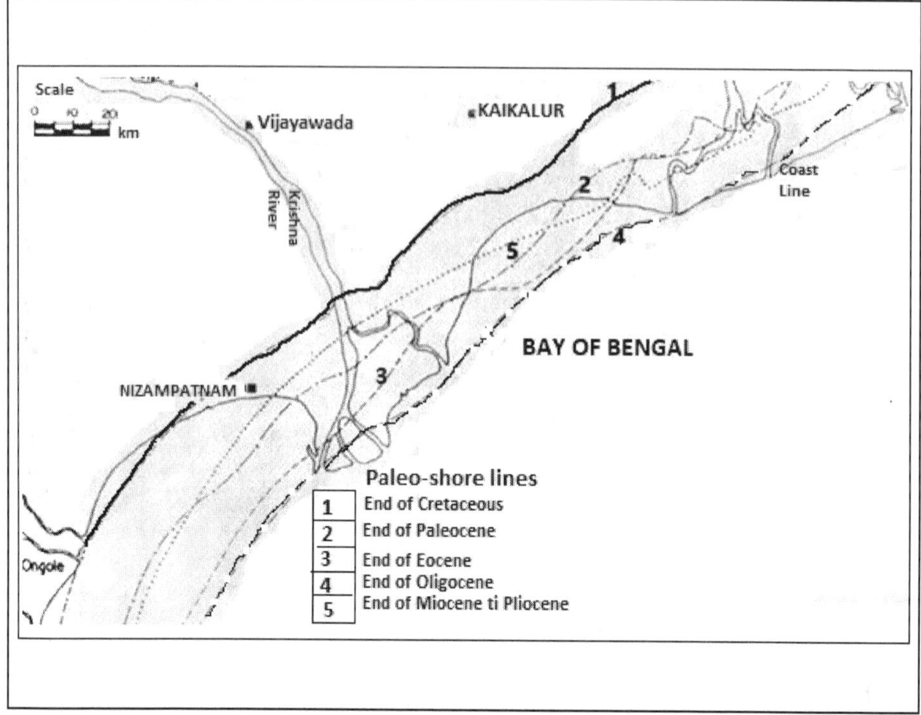

Figure 10.kg.11. Paleo-Shorelines-K G Basin.

Stratigraphy

The stratigraphic column of the basin include older Talcher Formation of Permo-Carboniferous overlain by Barakars which in turn underlie the Chintalapudi sandstone of Permian (Rao,G N., 1994). A red coloured sandstone sequence separates these fluvial pre-Cretaceous sequence and deltaic/marine Cretaceous and younger sections (Govindan, A., 1984). The Late Cretaceous-Early Paleocene basalts lithologically distinguishes the Tertiaries and Mesozoics in the basin. A thick carbonate section of middle Eocene further differentiates the stratigraphic column. The litho facies variations of a given geological time, that is, proximal arenaceous and homotaxial basinal argillaceous sequences can clearly be identifiable in the basin (Rao G N., 1990 and Venkataragan, R, *et al.* 1993). All these litho facies variations are studied in space and time and a comprehensive rock stratigraphic nomenclature was formulated as shown in Figure-10.kg.12.

Lithostratigraphy-Nomenclature

Kommugudem Formation (Early Permian) is the oldest sediment deposited over the Archean basement although, at places, argillite (Draksharama Formation) is present underneath. Kommugudem Formation is dominantly a shale sequence with interbedded coal and sand. The coal beds are generally 1–6 m in thickness. The environment of deposition is fluvial-to-lagoonal.

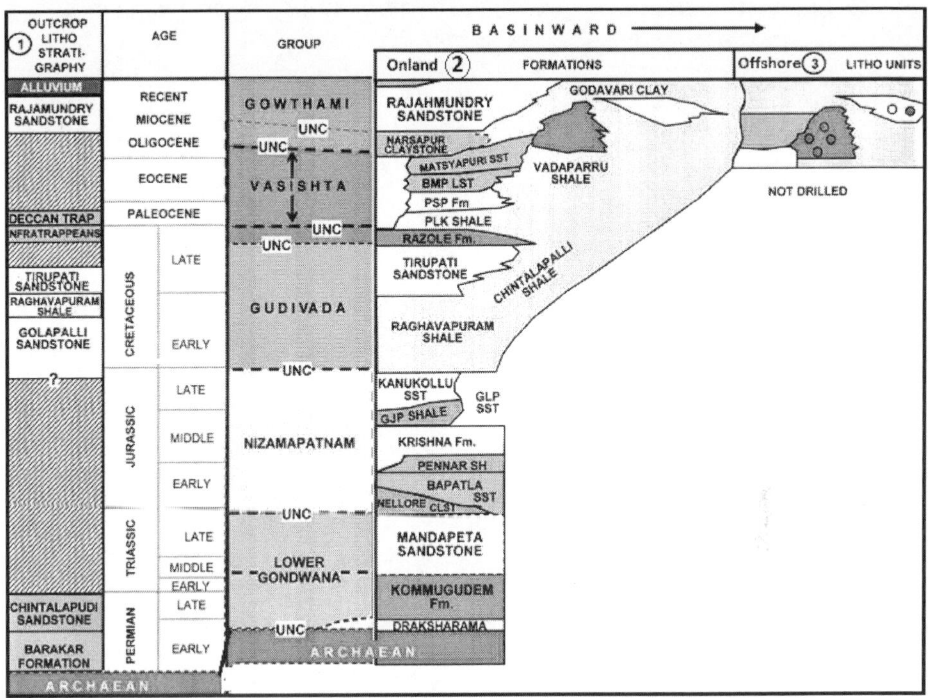

Figure 10.kg.12. Litho-Stratigraphic Nomenclature-K G Basin (After Venkataragan, R, *et al.*, 1993).

Mandapeta sandstone (Late Permian) is a thick nonmarine feldspathick and micaceous sandstone deposited in a fluvial environment. The presence of relatively thick intervening shales suggests cyclic flood plain conditions.

Bapatla sandstone (Late Jurassic) is a non-marine sandstone, clay, and shale section. It rests directly on the Archean basement. The basal section is claystone with thin sandstone layers interbedded. The overlying section is dominantly arenaceous with thin intercalations of shale and claystone.

Gajulapadu shale and overlying Kanukollu sandstone (Aptian/Albian) rest unconformably over the Bapatla sandstone in the Gudivada graben. The Gajulapadu shale, deposited in a lacustrine environment, is highly carbonaceous and rich in organic matter. A few interbedded sandstone layers are also present. The overlying Kanukollu sandstone is predominantly sandy and fairly consistent in distribution. Deposition took place in a marginal marine environment (Manmohan, M, *et al.*, 2003).

Nandigama Formation (Aptian/Albian) is primarily marine shale with thin sandstone beds. The basal part is coarser clastics with thin sandstone interbedded. Gollapalli sandstone (Aptian/Albian) is red claystone, overlain by sandstone, which rests unconformably over the Mandapeta sandstone as fill sediment. The depositional environment was shallow marine.

Raghavapuram shale (Cenomanian to Early Maastrichtian) can be subdivided into lower and upper units. The high-resistivity lower unit is rich in organic matter. The upper unit has thin interbedded layers of lenticular sand and shale. Sedimentation took place under shallow marine conditions. Tirupati sandstone (Early to Late Maastrichtian) unconformably overlies the Raghavapuram shale. It was deposited during the retreating Cretaceous sea and is predominantly sandstone with minor claystone. It is progressively shaly basinward and is called the Chintalapalli shale.

Razole Formation (Early Paleocene) consists of widespread volcanic flows over the Tirupati sandstone. There are at least 2–3 submarine trap flows. The traps at times develop fractures, and the top surface is altered and weathered into variegated clay (Kamaraju A V V S., *et al.*, 1998).

Palakollu shale (Middle to Late Paleocene) was deposited in an outer neretic to bathyal environment south of the Matsyapuri-Palakollu fault. The thickness of this litho-unit increases basinward.

Pasarlapudi Formation (Early Eocene) overlies the Palakollu shale and consists of alternating sand and shale layers with some limestone. Toward the south and east, the Pasarlapudi sands grade into shale and are called the Vadaparru shale. The depositional environment was outer neretic to bathyal. Bhimanapalli limestone (Middle Eocene) is an algal dolostone with abundant dolomite crystal and red algal fragments. Interbedded sandstones within thick carbonates are common. The depositional environment is outer neretic to bathyal.

Matsyapuri sandstone (Oligocene-Miocene) is a thick sandstone sequence with interbedded claystone, between the Matsyapuri-Palakollu fault and the Miocene growth fault close to the coast. Sedimentation occurred in middle shelf conditions.

Ravva Formation (Miocene) consists of thick and coarse clastics deposited south of the Miocene growth fault in an inner-to-middle shelf environment. Sediments were subjected to "push ups" and rotation because of the tectonics of the underlying Vadaparru shale. The sands are separated by interbedded clay (K. Satyanarayana., *et al.*, 1996).

Narsapur claystone (Pliocene) is a monotonous claystone section with a minor amount of intervening sand and silt. This unit thickens basinward where it is called the Godavari clay. The depositional environment was shallow inner-to-middle shelf. Vadaparru shale (Miocene-Late Eocene) is a thick marine clay section deposited under outer shelf conditions. It also represents the basinward clay facies of the Pasarlapudi Formation, Bhimanapalli limestone, and Matsyapuri sandstone.

Godavari Formation (Plio-Pleistocene) sediments are mainly clay with minor silt layers deposited south of the Miocene growth fault. However, the depocenter is farther south of the Pliocene growth fault where thick coarser clastics, brought in by a deepwater river-fan system, accumulated. The depositional environment was inner-to-middle shelf.Both vertical and lateral variations of these litho-stratigraphic units in different sub-basins are shown (Figures 10.kg.13 and 14).

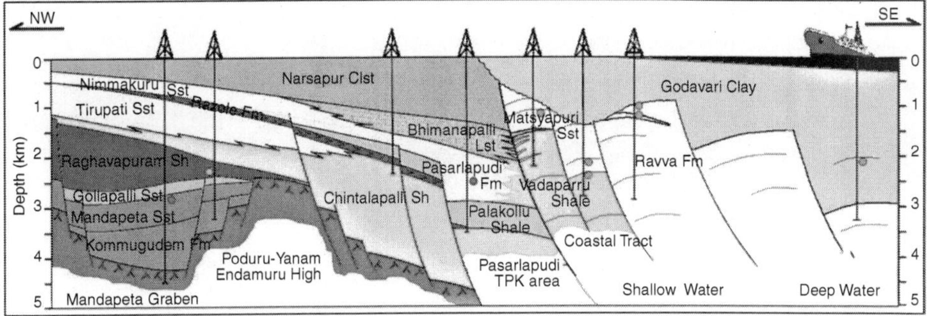

Figure 10.kg.13. Geological Section Across EG Sub-Basin with Litho-Stratigraphic Nomenclature.

LEGEND

I - Pliocene Top
H - Miocene Top
G - Mid-Miocene Top
F - Eocene Top
E - Early Eocene Top
D - Palaeocene Top
C - Cretaceous Top
B - Turionian Top
A - Albian Top

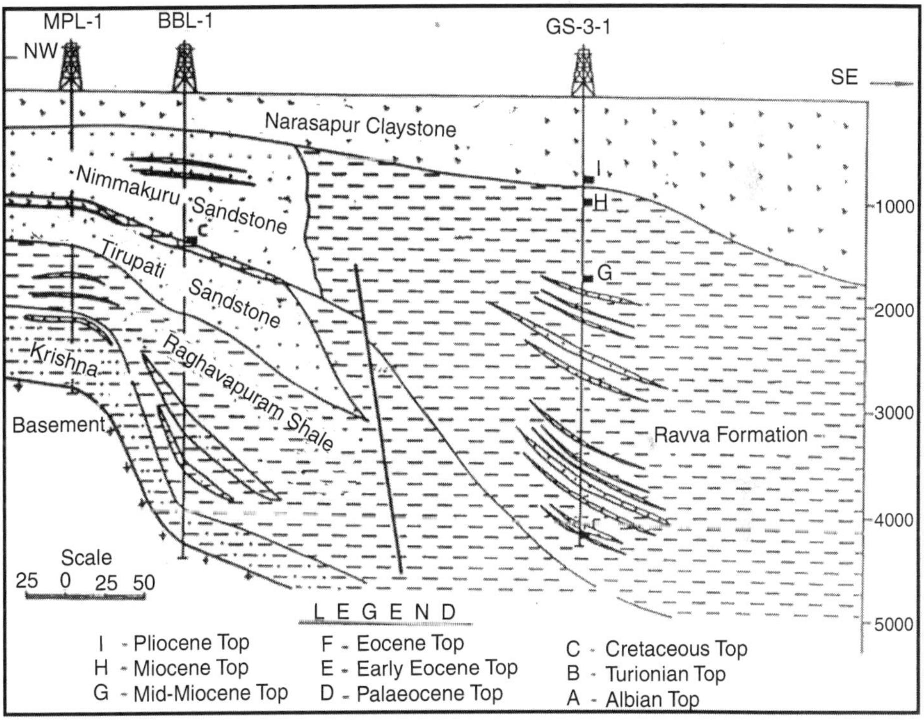

Figure 10.kg.14. Geological Section Across WG Sub-Basin with Litho-Stratigraphic Nomenclature.

Sequence Stratigraphy

Based on the major tectonic events and sedimentation relationships, three distinct sequences have been identified as first order sequences as: Intra-cratonic Permo-Triassic Gondwana sequence, Late Jurassic - Early Cretaceous Rift sequence and Late Cretaceous to Recent Passive margin sequence.

The sub-aerial unconformities identified within these sequences facilitated further division into second order sequences *viz.* two within Gondwana graben set up, two within Rift sequence and four within Passive margin sequence

The sediments of the **first order** sequence were deposited in the NW-SE trending Pranhita-Godavari graben, an intracratonic graben, during Lower Gondwana period (250 mybp) before the breakup of the Gondwana land. The lower boundary of the sequence, over the basement and the upper boundary with the Rift sequence.

The **second order** sequence of Permian age comprises of Draksharama Argillite and Kommugudem formation which are the equivalents of Talchirs and Barakars in the out crop respectively. The lower boundary of the sequence is the unconformity with the basement and the upper boundary is an unconformity between Permian and Triassic. This sequence, which is dominantly coal-shale, is an established source mainly for the gaseous hydrocarbons. The second order Triassic sequence is an established reservoir rock that hosts the main gas accumulations in Mandapeta area. Late Jurassic to Early Cretaceous second order sequence overlies the basement in the western part of the basin in the Kavitam area and eastern part of the basin. It comprises of red beds and biostratigraphically the upper most part of Mandapeta sandstone (Upper Pay). The magnitude of hiatus between early rift and late rift sequence is of the order of 3-4 Ma.

The Aptian-Albian second sequence, which unconformably overlies the early rift fill sequence marks the marine transgression during Early Cretaceous into the Mandapeta, Gudivada and Bantumilli grabens, as evidenced by the on laps of the reflectors of this sequence onto the early rift fills. The upper boundary of the sequence with Late Cretaceous. This sequence could possibly extend to the south of Matsyapuri- Palakollu fault in the offshore, in between the Avanigadda and Yanam cross trends. Then Late Cretaceous sequence unconformably overlies the Early Cretaceous Rift sequence and in turn overlain by the Early Paleocene volcanics in the onland and the upper Paleocene shales in south. The upper boundary is clearly identifiable where the volcanics are present but in the southern part of the basin, where the basalt is absent, Lower Paleocene directly rests over K-T boundary, where seismic characteristics only help to mark the same.

The Late Paleocene to Oligocene time-span comprises of marine shales, shallow marine sands and inter-bedded limestone and sandstone. The Late Paleocene to Early Eocene section constitutes the slope section and is highly condensed. The Middle Eocene limestone is developed on shelf and defines the demarcation of shelf edge.

The lower boundary of Miocene second sequence is underlying Oligocene sequence and the upper boundary is the boundary with over lying Pliocene. Regionally, at many places the upper Oligocene hiatus has been observed. The section is characterized by sandstone and claystone. The lower boundary of this upper most Pliocene to Recent sequence overlain by the Pliocene transgressive shale. On the shelfal part, sediments are deposited with rapid seaward migration of shore line. Early Pliocene witnessed huge deformation in the form of plastic flow and the shelf-slope architecture.

This **third order** sequence of Cenemonian-Campanian has its lower boundary with Rift sequence and upper boundary with overlying sequence. It comprises dominantly of shale with minor sandstone and shale alternation towards top. The Campanian to Maastrichian sequence comprises dominantly of shale with minor sandstone and shale alternation towards top. The sequence of Paleocene to Early Eocene time-span is limited below and above and comprises of sandstone and shale alternations. The third order sequence of Middle Eocene to Oligocene time-span is limited below and above and comprises of predominantly limestone and sandstone/shale.

Abnormal Formation Pressures

In the first deep well drilled at Narasapur-1, at 4035 m while drilling in Upper Cretaceous section with 1.4 specific gravity mud, heavy kick was observed which was very severe like blow-out (Figure 10.kg.15). In Narasapur, Razole, Tatipaka, Amalapuram and Pasarlapudi prospects within Upper Cretaceous argillaceous unit, the formation pressures are found to be maximum. In Tatipaka, the pressures

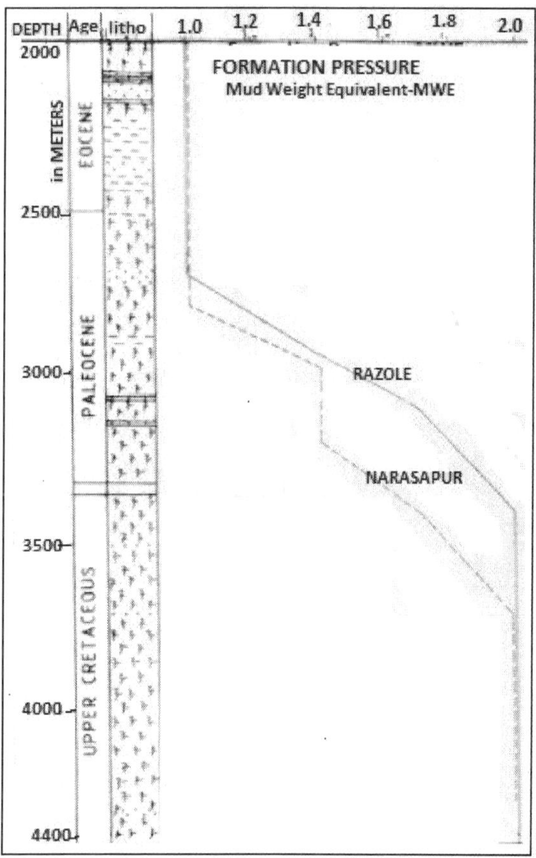

Figure 10.kg.15. TTI-Plot of Razole-Narasapur Wells Well–K G Basin.

are 2.2 times more than the hydrostatic (Ramakrishna, MJ 1987). This indicates, the transition pressure regime starts at 2700 m, there is a sharp rise of pressures upto 3300 m, thereafter, and the pressure becomes stable. In offshore area, below 2000 m, the Miocene sequence is found to be having abnormal formation pressures of more than 1.5 MWE (Mud Weight Equivalent).

In the basin, abnormal formation pressures are recorded in Upper Cretaceous, Paleocene, Eocene and Miocene sequences in East Godavari Sub-basin. The maximum thickness of these sediments are more than 2000 m, 1000 m, 1500m and 1500 m, respectively. By considering the time duration of 25, 10, 15 and 12 million years respectively, the maximum rate of sedimentation for these sequences works out to be 0.8, 0.7, 1.0, 1.25 cms/100 years (Rao, G. N., and K. S. Mani, 1993).

The quantum of formation pressures in the sequence is maximum in the depocenter areas of the respective period (Fertl, W Walter 1976). The Upper Cretaceous and Paleocene sediments showed maximum pressures towards SE of Matsyapuri-Palakollu fault zone, in the Lower Eocene sequence towards SE of Amalapuram-Mori trend and in Lower Miocene sequence below 2000 m in Ravva area and further southeast.

The study of time of generation of abnormal pressure within a given formation (Chapman, 1983), has revealed that, the Upper Cretaceous sediments in Narasapur-Tatipaka area have lost normal compaction during Eocene whereas towards further SE in machilipatnam Bay, the sequence.

In case the syn-sedimentary subsidence is faster than normal compaction time, the bottom section may become under-compacted and become abnormally pressured with in such formations. This aspect is also seen in sedimentation of Godavari basin. The upper Cretaceous section lying below basalts is found to be abnormally pressured, leading to drilling complications, during initial stages of exploration in the basin. With addition of more data the reasons for generation of abnormal pressures were analyzed and concluded that, it is sedimentation related. In a prograding deltaic sedimentation coupled with higher rate of sedimentation the level of under-compaction is found to be shallowing up and even Mid- Miocene sequence in shallow water area is associated with high formation pressures (Figure 10.kg. 16).

By integrating the available data on formation pressures by RFT and testing, iso-formation pressure contours are prepared at 500 m of depth interval for the sequence from 2000 m to 4500 m (Figures 10.kg.17 to 22). The data are being used to optimize the mud policy for drilling within an abnormally pressured regime. The balanced drilling helped in both safe completion as well as without heavy caving, which also helps in turn recording of quality electrologs and proper cementation of casing.

Petroleum Geology

The hydrocarbon resources sequence wise are 330 in pre-Cretaceous, 230 in Cretaceous, 300 in Paleogene and 270 in Neogene reservoirs. Geographically, 575 MMT in onland area and 555MMT in shallow offshore area (Awadesh Rai,

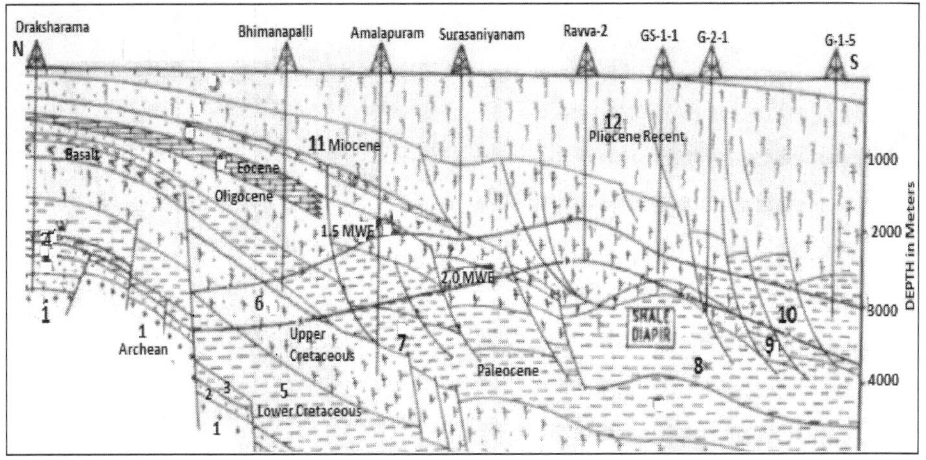

Figure 10.kg.16. Geological Section Across East Godavari Sub-Basin-KG Basin.

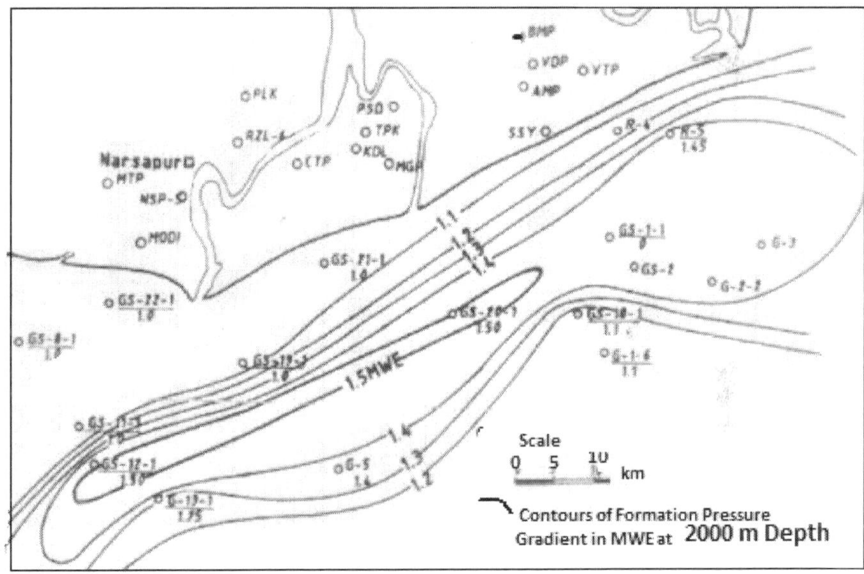

Figure 10.kg.17. Formation Pressure at 2000m Depth–K G Basin.

P., Chandrasekharan and Mishra, V.N., 1998). So far nearly 473 MMT of in place reserves were established out of 1,130 MMT of resources estimated (Rao, G.N. 1986). Further,1,020 MMT resources are visualized in ultra-deep area extending up to 85 deg east ridge area. Hence the basin bear good prospects for hydrocarbon exploration.

Deltas are favorable locale to have all the requisites for hydrocarbon generation and accumulation. Further, for maturation of organic matter to generate oil and gas sufficient pressure and temperature conditions are also needed. Geologically deltas overlying paleo-rift basins provide such conditions and found to be best sites for

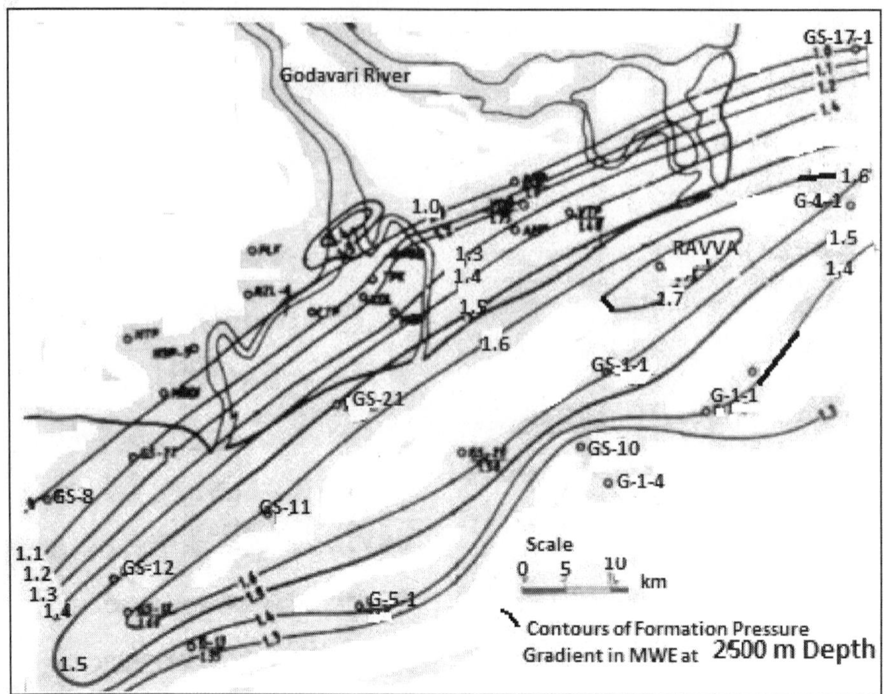

Figure 10.kg.18. Formation Pressure at 2500m Depth–K G Basin.

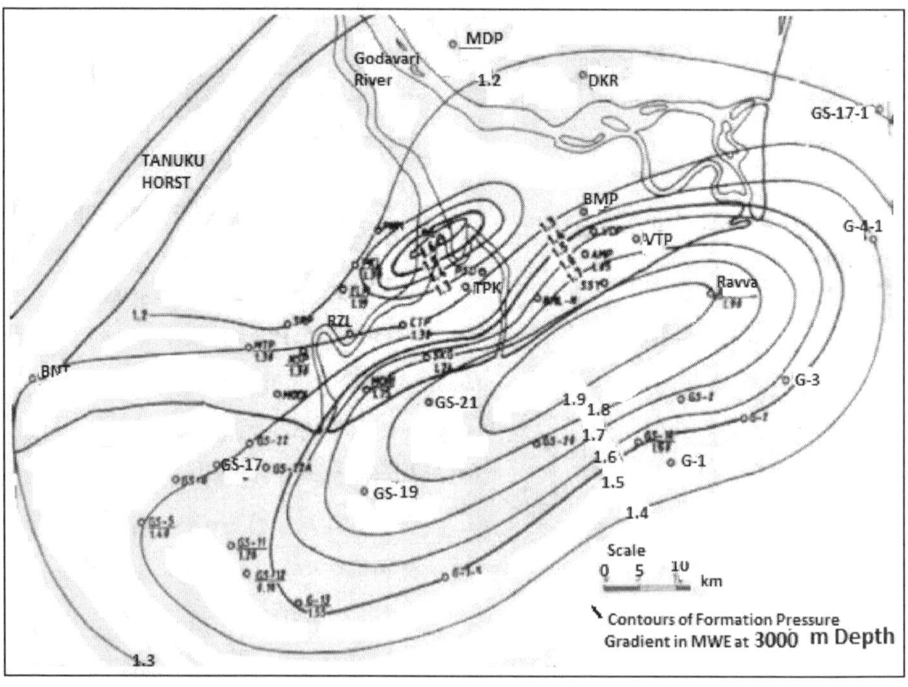

Figure 10.kg.19. Formation Pressure at 3000m Depth–K G Basin.

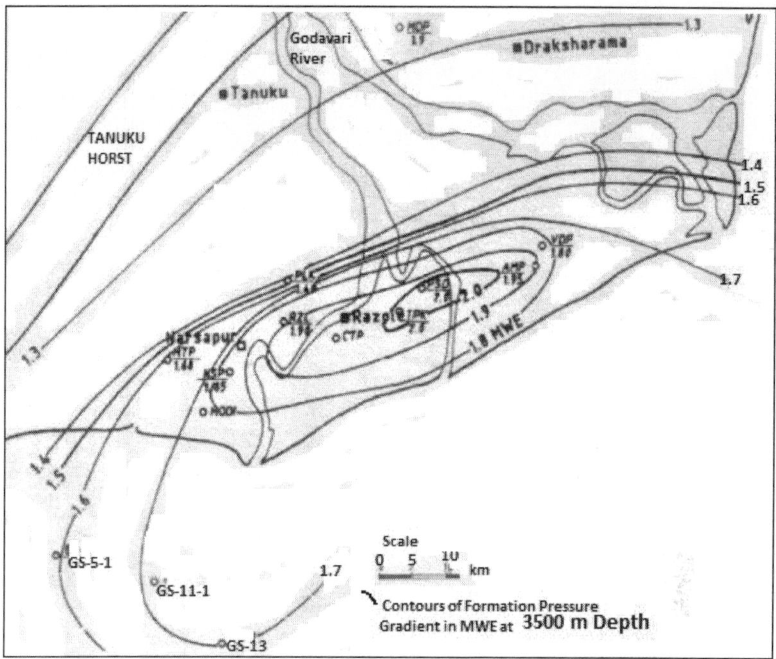

Figure 10.kg.20. Formation Pressure at 3500m Depth–K G Basin.

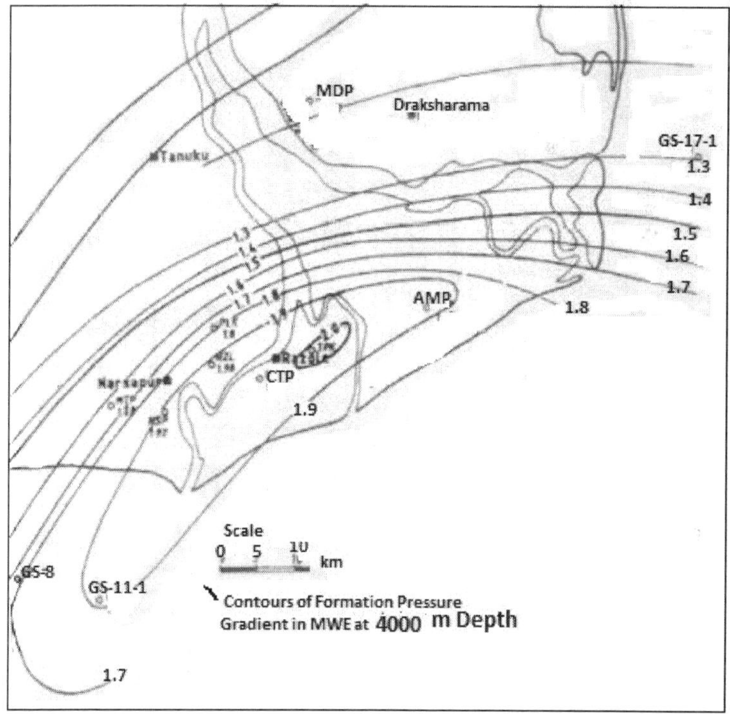

Figure 10.kg.21. Formation Pressure at 4000m Depth–K G Basin.

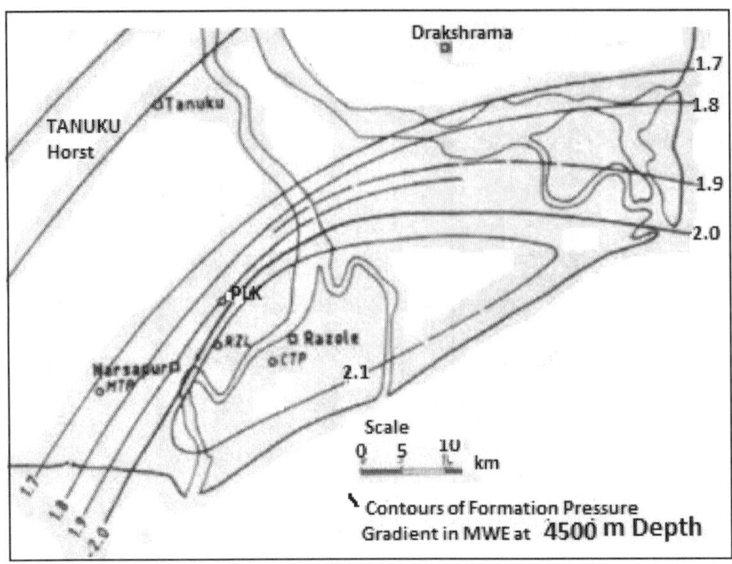

Figure 10.kg.22. Formation Pressure at 4500m Depth–K G Basin.

hydrocarbon exploration. Krishna-Godavari basin is ideally located for oil and gas search, like Niger delta in West Africa.

Delta front and pro delta areas are depositional sites of good quantity of organic matter both terrestrial and marine. If syn-sedimentary subsidence is there, such organic matter may reach reducing conditions early (below photic zone) there by protected from oxidation (Ray, D., *et al.*, 1985). If such delta front areas are in case located over transitional crust and/or paleo rift basin, more heat flow can make early maturation of source beds (Sar,D, *et al.*, 2009).

The major petroleum occurrences include Mandapeta gas field from a Permian sandstone, Kaikaluru, Lingala oil fields from a lower Cretaceous clastics, Tatipaka, Pasarlapudi gas field and Mori oil field from lower Eocene sandstones, Ravva oil field in Miocene clastics off Amlapuram coast. However, the pre-Cretaceous and Cretaceous hydrocarbon fields are more of stratigraphically controlled and hence bear relatively less potential, whereas fields in Tertiary sequences are of strati-structural in nature and hence may have higher potential.

The gas occurrence in Penumadam in a relatively thick sandstone layer with hydrostatic pressure is a significant lead to explore for Upper Cretaceous deltaic facies within the area falling between Matsyapuri-Palakollu fault trend and Draksharama-Kavitam trend.

But, the faster rate of subsidence is also unfavourable by generation of differential compaction of argillaceous matter, there by generationg abnormal formation pressures within such sequences. The sand-shale ratio in an abnormally pressured sequence, in K G Basin is about 0.1. Though the generated hydrocarbons are more in quantity, some of these thin reservoirs produced water. Possibly these arenaceous beds might have attained abnormal formation pressures by the time migration of hydrocarbons started in the sequence. The abnormally pressured reservoirs acted as pressured chambers making the entry of hydrocarbons more difficult.

Generally the reservoirs associated with abnormal pressures are found to be relatively thin, limited in areal extent and only gas bearing. The deepest flowing reservoir was at 4,444m level at Chintalapalli#6. These reservoirs when put on production are found to be fast depleting. Drilling also will be costlier due to abnormal pressures. Hence, economic viability to develop such zones found to be limited (Srinivas, M.S. and Iyer, B.R.1990).

Source Facies

The source rock studies in the Cretaceous sequence indicate poor organic contents in arenaceous unit with less than 1 per cent TOC but the argillaceous units indicate type III organic matter with sapropelic content at some places with 1.1-2.3 per cent TOC. The higher content of TOC is reported in Kanukollu (2560-2610m,to 6.6 per cent). The organic matter matured with average HI 70-169 mg Hc/gm TOC and vitrinite relfectance (Vro) values ranging 0.70-1.0. The sequence is capable of generating mainly gas with subordinate oil.

The Palaeocene sequence depicts average quantity of organic matter is about 1.2 per cent but in Bhimanapalli increases to 2.67 per cent. ISO-TOC lines indicate that towards northwestern part of the depocentre organic content is poor but it is rich towards basinal side with type II organic matter of marine algal input.

The average organic content is 2.5 per cent but due to the presence of Eocene limestone, it increases upto 5.5 per cent at some places (GS 11-1). The organic matter is type III with oil prone kerogen at places. The Oligocene sequence is very thin (40-150m) in the basin. The organic content of the sequence is less than 1 per cent rating them as very poor source rocks.

The organic content in Miocene is about 0.35 per cent which increases upto 2.18 per cent in deep basinal part. The organic matter is type III and at places even type II.

The maximum thickness of Pliocene sediment is about 2500m in deep waters. The organic carbon content is very poor with 0.5 per cent which increases upto 2 per cent in deeper parts of the basin. Nature of organic matter is type III D and even type IV consisting of dead cutinite, inertinite and fuisnite matter. The sequence is not matured anywhere to generate hydrocarbons. Maximum vitrinite reflectance value is 0.46. The hydrocarbon accumulation in these reservoirs may be due to migratory hydrocarbons.

Maturation

Geothermal gradient map with the well data in the basin showed higher than normal trend at shallow offshore area off Godavari mouth (Ravva area) which can possibly due to overlying of these sediments above the transition/oceanic crust. Another higher trend is noticed in Kaza Kaikaluru area in West Godavari sub Basin.

Gamma-ray spectroscopic logs of these wells indicated higher thorium content and hence the anomaly may be due to radioactive mineralization.

The Paleocene sequence is matured towards southeast of Palakollu-Matsyapuri fault trend with Ro ranging from 0.55 to 0.82. The Eocene sequence is generally matured below 2200m indicating only the Lower Eocene argillaceous unit to be mature than the Middle Eocene carbonate sequence. The Vro values range from 0.34 to 0.85. The limestone is likely to occur at deeper depths towards southwest of GS-22 area and hence may be favourable for maturation (Samanta, U, *et al.*, 1994). At deeper parts of the basin the organic content increases to 1.5 per cent and is of type III. Vitrinite reflectance (VRo) data suggests immature sequence except at deep basinal part where it indicates early phase of maturation (Vro 0.36-0.40). In onland part of the basin the sequence is immature, whereas in offshore part in Ravva area and further southeast, the sequence is mature below 2000m depth.

The Time Temperature Index (TTI) plots of Razole well is shown (Figure 10.kg.23.) and also few more wells indicate maturation of organic content (15-60) in most of the wells and matured during Early Miocene time.

Expulsion and Migration: Regional basin modeling and subsidence modeling suggest generation of hydrocarbon from the Early Miocene to present day. Regarding migration timing, there are two possible models: Migration and filling of the Middle Miocene reservoirs occurred during Late Miocene as this provides a seal.

The continued subsidence coupled with progradation of deltas may add sufficient over burden which will help maturation as well as expulsion of hydrocarbons from source beds that means facilitates primary migration of hydrocarbons. Land ward continuity of these prograding sequences might have helped the connectivity in sandstone beds which will form good carrier beds for secondary migration.

Trapping must have occurred against older faults or stratigraphic traps that were in place prior to Early Pliocene faulting. The main migration and filling episode occurred after Early Pliocene faulting (and possibly until recent times) filling the crest of the rollover anticlines and in any reservoir beds falling within the path of migration.

Reservoirs: The sandstone reservoirs within the Oligocene to Pleistocene sequence were deposited in a wave dominated deltaic environment where relative sea-level fluctuations through time resulted in fluvial to sub-marine slope sedimentation. The best hydrocarbon reservoirs are located in the fluvial channels, beach ridges, barriers and upper shore face environments.

Caprocks: Two types of caprocks are present *viz.* Regional highstand sequences and Intra-formational shale. The highstand thick shale sequences of Lower late Miocene and Pliocene-Pleistocene packages form the most effective top and lateral seal for Middle Miocene reservoirs and Late Miocene gas caps. The former is heavily faulted. Intra-formational shales are 15-100 m thick and are present throughout Tertiary sequence; they are proven seals in Early Miocene and Late Oligocene.

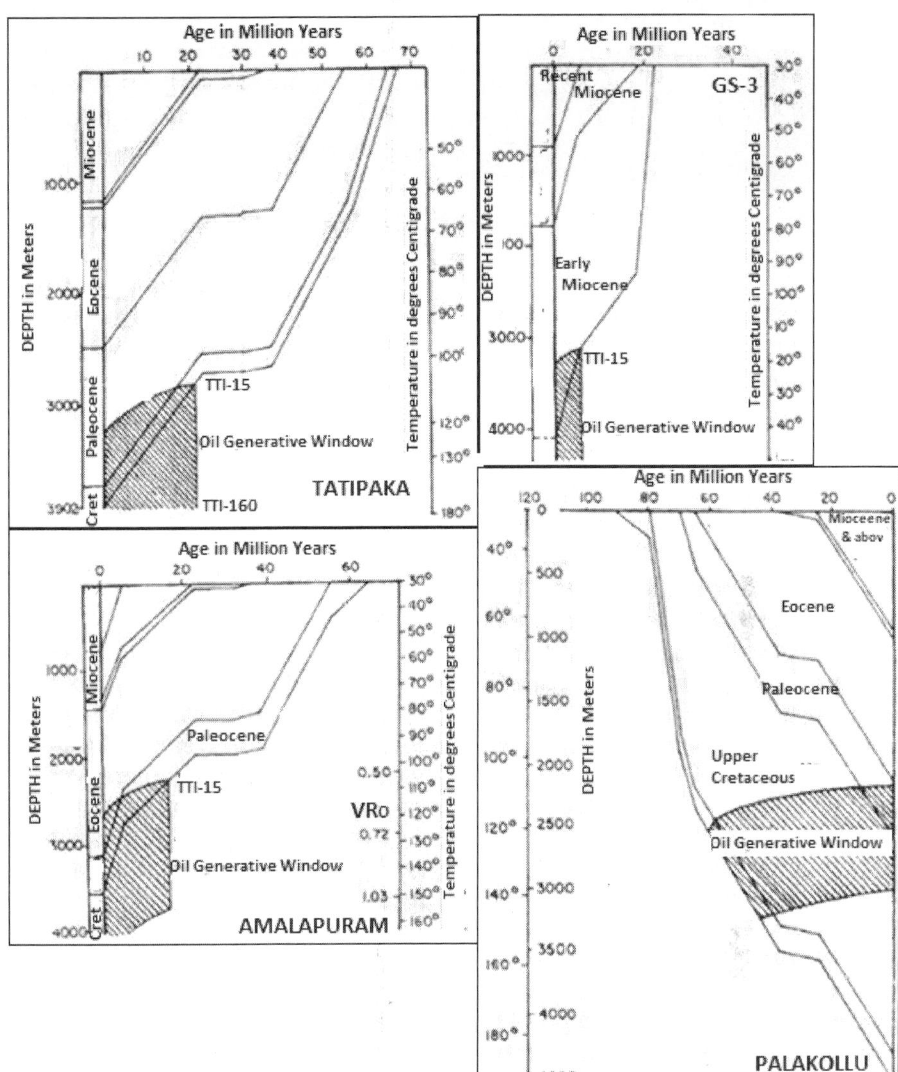

Figure 10.kg.23. Subsidence History Plot of Wells in K G Basin.

The Mio-Pliocene sequence in offshore part is promising. The commercial hydrocarbon accumulation in Ravva field is known. The prospects GS-38, G-1 and G-2 are also hydrocarbon bearing in Mio-Pliocene strata. As many as fourteen commercial finds have come from this sequence.

Entrapment Styles

Syn-sedimentary growth faults in delta front areas may form good lateral seals helping to form major hydrocarbon pools. The proportionate mix of both argillaceous and arenaceous sediments will improve the both local and regional cap rock facies which will help in forming strati-structural accumulations. Low amplitude structural closures, fault closures wedge outs, unconformity controlled structures, permeability barriers in different reservoirs. Big amplitude structural closures are not common in the basin. The identified and drilled prospects are of small areal closures.

Petroleum Systems

A Petroleum System can be summarized as an independent stratigraphic compartment within which the three primary requisites for hydrocarbon accumulation that is, source, reservoir and cap rocks can occur (Venkatarengan, R., and Ray, D., 1993).

In Krishna Godavari Basin ***four*** Petroleum Systems are identified and they are stratigraphically classified as Pre-Cretaceous, Cretaceous, Paleogene and Neogene systems.

Kommugudem-Mandapeta-Red Bed Petroleum System

It is the oldest System of Pre-Cretaceous age. In this, Barakar Coal Shale Sequence is a matured source facies sequence Type of organic matter is mainly Type-III. Possibly due to which the hydrocarbon generation is mainly gas prone. Reservoir rocks are of Mandapeta formation with good porosity and permeability. The entrapment style in the system is of strati-structural nature. The overlying Red bed sequence is acting as regional cap facies for this oldest operating system in the Basin. Mandapeta gas Field and its adjacent fields are producing gas from the system.

The reservoir facies of Permo-Triassic occur within the well identified source facies at the bottom and overlying Cretaceous argillaceous facies, which act as source as well as cap. So the Late Paleozoic sediments, imparts the older sequence a fair degree of importance.

Raghavapuram-Gollapalli-Razole System

In the system the lower Cretaceous Raghavapuram Shales and Upper Cretaceous Chintalapalli Clay stone are matured source facies and matured during Eocene (TTI data). The available overburden pressure is facilitated for primary migration of hydrocarbons from the source rocks. Since organic matter is of land derived it seems to be gas prone. Deccan Basalts of Early Paleocene act as Regional cap facies. Due to under compaction of sediments in SE part of East Godavari sub-basin, the reservoir sequences are associated with abnormal formation pressures. Entrapment style is of mainly startigraphic nature.

The Cretaceous and Early Tertiary accumulations of hydrocarbons are present in several fields, *e.g.*, Kaikalur, Bantumilli, Lingala, Narsapur, Razole Chintalapalli, *etc.* both in East as well as West Godavari Sub-Basins. The Cretaceous sequence in offshore wells, like well KB-1B, has also indicated presence of hydrocarbons during drilling. In view of this, the Cretaceous sequence holds good potential for accumulation of hydrocarbons where many commercial accumulations have been discovered.

Palakollu-Pasarlapudi-Bhimanapalli System

In this system Paleocene Vadaparru shale is a matured source facies. Sometimes carbonate section of Bhimanapalli Limestone with organic facies may act as source facies and generation of oil is a possibility with in this system. The top of Eocene carbonate section is also acting as regional cap facies in SE part of East Godavari sub basin. Paleocene sandstone facies sometimes with in the carbonate sequence are reservoir beds in the system. Entrapment conditions include strati-structural in nature, Sands are thin and multi-layered, clastic wedges, fault closures, roll-over structures, *etc.* some of the oil and gas fields of the system include Bhimanapalli, Elamanchili.

A number of gas fields are producing from Paleocene reservoirs, particularly in East Godavari Sub-Basin. The Tatipaka, Pasarlapudi, Kadali and Manepalli are the fields located onland, while, GS-8 is occurring in the offshore part of the basin.

Ravva-Godavari-Petroleum System: The analysis of oil from Middle Miocene sequence indicate shales with organic content matured during Late Miocene. Reservoirs with in the sequence are clastics with varied thickness related to depositional styles. Source rocks have kerogen of Type III with some marine input. The source rock maturity varies between 0.7 per cent -0.8 per cent. The overlying Godavari Caly of Plio-Pliestocene may act as regional seal. Many offshore fields off Godavari delta belong to this system, like Ravva, G-1, G-2, G-4.

Details of Oil and Gas Fields

East Godavari Sub-Basin

Narasapur

This was the first exploratory well drilled in basin in 1978. It is a small amplitude structural closure found above basalt. While drilling below basalt at 4,035m with mud of 1.40 MWF, blowout situation was met with and gushing of gas was observed. The productive zone is around 4200m and two or three thin sandstone reservoirs. Five wells were drilled on the prospect and wells were put on production, which were ceased after some time. EOR technique Microbial injection was also implemented sometime around the year 1993, with unsuccessful results.

Razole

It is an independent structure, reached the target depth of 4500 m passing through 3370 m of Tertiary and Quarternary sediments, 260 m of Late Cretaceous basalt and 871 m of Cretaceous strata were drilled through. Indications of gas under high pressure within Cretaceous was observed. Four wells were drilled. Production could not sustain long due to abnormal reservoir pressures.

Bhimanapalli

This prospect was identified on Mid-Eocene Limestone level. While drilling severe mud loss was observed throughout limestone section. Good oil shows were also observed. Two wells were drilled. Initially some reserves were calculated but which cannot be produced due to bad hole conditions.

Chintalapalli

On this structure 6 wells were drilled. The gas reservoir is found to be of very high pressured. The flow rate of gas from one zone around 4,444m is 6MMcm^3 per day, but after short time ceased to flow.

Tatipaka

It is a small structural closure in Paleocene level, found to be gas bearing in relatively thin Paleocene sandstone reservoirs. 5 wells were drilled and are gas bearing. Wells were connected to near by GGS. After some period few wells were ceased. Producing sands are at 3,100m level.

Pasarlapudi

It is relatively bigger structural closure and also individual sand layers are found to be thicker. The structure is extended on both sides of Vasistha River and elongated in nearly east-west direction. So far about 27 wells were drilled and many are flowing. Reservoir level is around 2,500m and of Paleocene. During 1995, due to some operational technical complexities blow out was occurred in PSD#9 well, resulting into loss to property and equipment nearby.

Koravaka

This promising area located south of Tatipaka-Pasarlapudi area was established commercial oil and gas strike with in Vadaparru formation of Eocene and the continued exploratory efforts were also successful with more deeper objective in the interval 3720-24 m has produced oil @15 m^3 and gas @39,408 m^3 through 6mm bean.

Kommarada

This discovery in Eocene sequence of Vadaparru Formation in the area South of Mori-Komarada Fault has opened up the entire fault block and also given lead to explore within the sequence. In Komarada well # 3 in the interval 3,043-3,046 metres flowed oil @ 31.9 m^3/day and gas @ 1,23,512 m^3/d through 6 mm bean.

Ravulapalem

This discovery also from Eocene sequence of Vadaparru Formation in the area South of Mori-Kommarada Fault has opened up the entire fault block and also given a lead to explore the within the sequences. The objective-I in the interval 1979.5-1975.5 m., flowed gas @ 13,350 m^3/day through 6 mm bean and the object-II in the interval 1698-1695 and 1688-1685 flowed gas @ 83,200 m^3/d and condensate @ 7.3 m^3/d with through 7.0 mm bean.

West Penugonda

The discovery of hydrocarbon accumulation in Raghavapuram Formation has opened new area to the south of Tanuku Horst for exploration but thin reservoir beds with in are proved potential producers of oil and gas. In well, the selected zone in interval 3,090-3,054 m. flowed oil @ 3.4 m^3/d and gas @ 71,275 m^3/d at FTHP: 2100 PSI through 6 mm bean.

Lakhshminarasimhapuram

The structure was located towards south east of Suryaraopeta, as separate fault block with small amplitude closures in Raghavapuram shale formation. Two objects in the interval 2,811-15 and 2,841- 45 flowed feeble gas and oil @ 16m^3/d and 8m^3/day respectively.

West Kesanapalli

This structure is located very close to coast line and success was met in the Lower Miocene section as of Ravva system. Since major part of the structure is in very shallow waters of Bay of Bengal, directional wells are being drilled to exploit the hydrocarbons. In a well (#47) the interval 2,338.5-2335 metres flowed gas @ 30,000 m^3/d and traces of condensate through 5.0 mm bean from Matsyapuri formation of Oligo-Miocene age. In Kesanapalli west-27, the object in the interval 1667-68 flowed oil @ 42 m^3/d and gas @ 3,385 m^3/d through 5 mm bean.

Gaddenapalli

This structure is located in Amalapuram block towards east of Gopavaram and Neogene play like that of Ravva field was targeted for exploration with in onland area. The structure show small amplitude closure against two fault blocks. The Object –II in the well in the interval 2,045-2,059 (TVD) has flowed oil @ 74.29 m^3/d with feeble gas through 6 mm been. In another well (GL#3) drilled though oil flowed the flowing pressure was very low and needed artificial lift.

Turuputallu

The structures is basically NE-SW oriented fault closure with small amplitude closures in between. It is located towards south west of Razole and west of Vasista River. Found gas in the Pasarlapudi pay zone equivalent zone in the interval 2,284-86 and flowed gas @61,894 m^3/day.

Vygreswaram

The structure was mapped towards NNW of Amalapuram and North of Pasarlapudi area. The field is steeply dipping fault closure. The find was in Raghavapuram formation in the interval 3,335-3,388 in which 5 thin sands of 2 to 3 metre thickness collectively produced gas @ 2,21,867 m^3/d with traces of condensate along with water through 7 mm bean.

Vygrewswaram SW

This is another fault closure towards South of Matsyapri-Palakollu Fault zone located to south of Vyghreswram gas field and produced gas with condensate in the pay zone section in the interval 3,630-3,669 pack of thin sands with 2 to 4 meter thickness each. This composite zone produced gas @74,363 m^3/d with condensate with traces of water @ 3.22 m^3/d.

Mukkamala

The exploratory efforts towards south of Matsyapuri-Palakollu Fault zone for deeper Cretaceous section gave successful result with finding of gas. Two objects tested in the interval 3,612-3,640 (4 sands) and 3,086-3,099 (3 sands) flowed gas @ 38,545 and 43,080 respectively through 6 mm bean.

Gopavaram

This field is close to coast and located towards NW of Ravva. The object –II in the (well #12) interval 2,065-68 m flowed oil @ 50m^3/d and gas @2,400 m^3/d.

Mandapeta

The field was discovered in the year 1988 and is one of the big structural closure located towards NW basin margin area. Reservoir is thick and found to be gas bearing at 2,700 m level. This reservoir is geologically located below upper Cretaceous clay sequence and devoid of Fauna. Based on floral evidences, lower Cretaceous to upper Jurassic age was assigned and hence this strike of gas is known as "Gondwana" occurrence, envisaging the subsurface extension of Pranhita-Godavari graben under the K G Basin area. Following the discovery well, the area has been covered by 3D surveys (Banik,95 and Singh, 2001) and about 23 wells were drilled. Commercial gas accumulation has been established in Mandapeta main block where accumulation is within Mandapeta sandstone. The structural closures around the field are also found to be gas bearing, *viz.* Mandapeta South and Mandapeta West areas also. In addition, Gollapalli Sandstone also proved commercial gas accumulation in the southern part of the field, which have been drilled in separate fault blocks.

The structural closure located towards South of Mandapeta was found gas bearing in Gollapalli Sandstone. In well #1, the interval 2373-85 flowed gas @ 32,472 m^3/d.

KG-ONN-2004/1

It was awarded in 2006 to OIL with 90 per cent stake as operator and 10 per cent partner is Geo-global. It is of 549 sq. km located off Yanam area. Well Polavaram #1 was drilled up to 2,526m, which was bottomed in Chintalapalli shale formation. In this well in Pasarlapudi formation gas shows were encountered in the interval 1,924-2,068m.

West Godavari Sub-Basin

Kaikaluru

This prospect located over Kaza-Kaikaluru horst which separates Gudivada graben to NW and Bantumilli graben to SE. It is oil field at Cretaceous level and depth range of 1,900 m.

Kaza

It is one of the initial prospect drilled in 1985 which located on Kaza-Kaikaluru horst in the WG sub basin. Two wells were drilled to depth of 2100. Gas was flowed @22,700 m^3/day via 12/64" bean in the interval 2,038-2,045m in Cretaceous sand below Basalt, where as in the second well only feeble flow of gas was observed. The oil in-place is 0.02 MMt., with gas 0.01 BCM.

Lingala

It is small structural closure located very near to Kaikaluru and flow of oil is from Cretaceous sandstone reservoirs.

Vanadurru-South

This structure is located south of Vanadurru field with in the Bantumilli graben on its western flank. The interval 3,628-34 flowed oil @ 25 m^3/d and gas @ 51,772 m^3/d through 6 mm bean.

Bantumilli

The structure falls in the Batumilli graben in the deeper part of the same and ONGC has found the gas field during 1,990. In 2014 ***Bantumalli South*** was discovered in exploratory well BTS-1. This field is of 18 square kilometers. Due to various technical reasons, drilling beyond 4,200 metres could not be done. The BTS-1 well was capped. In this the Nandigama arenaceous unit in the interval 4,221-32 flowed gas @98,010 m^3/d through 4 mm bean. Another exploratory well called BTS-2 was drilled in the field. It was estimated that the production could be one lakh cubic metres per day from the well.

Gokarnapuram

This field is located between Kavitam and Bantumilli-South fileds and also towards South-West of Penugonda which established hydrocarbons with in Raghavapuram formation. The first well was drilled to 3,294m in 1998. This well flowed gas @ 55,439 m^3/d through 16/64″ bean, with condensate 9.61m^3/d in the interval 1681-1684m. Where as the second well drilled to 2,106 showed only water influx. The reservoir is below Basalt trap. The reserves of Gokarnapuram Field are O+OEG=0.13 mmt, gas 0.03 BCM, condensate-0.10 mmt. With an area- 0.786 sq km, reservoir thickness is 5m with porosity-21 per cent.

Suryaraopeta

It is located near Machilipatnam area in KG Basin onland covering 49.18 sq. km. The Suryaraopeta field was found in year 2000 and sofar 6 wells were drilled. The deepest well is #6 which was drilled up to 4012 TVD m. Only well #1 in the interval 2374-2372m, Flowed oil @ 90 m^3/day and gas 2451m^3/day. The well Suryaraopeta#4 in the interval-3075-78, 3107-10, 3,113-21, 3,141-48m produced gas in commercial quantity @ 33,179 m^3/day at an FTHP of 1049 psi, SBHP – 6499 psi. Most of other wells flowed water with little gas. The reservoir is Raghavapuram shale of Lower Cretaceous.

Penudurru

This prospect with area of 49.25 sq km. is located in the Bantumilli Graben and near Chinapandrika area. The discovery was in year 2001. Gas in-place reserves are of 0.12 BCM. Two wells were drilled where in the well #2 is the deeer which was drilled to 4252m (4198m TVD). Many objects teasted in the interval 3,225-4,013 m (drilled depth) flowed feeble gas with condensate.

Offshore-Shallow Water

GS-15

The prospect is located in the adjacent to Ravva area. Discovery well GS-15-E-1(AA) encountered a total of around 40m of oil and gas bearing sands within Ravva Formation. Three objects were tested and found to be gas producer @1,35,579 m^3/d through 10mm bean.

GS-21

In the well drilled in 2010 by ONGC in the structure (# 3) the Ravva formation of Mio-Pliocene has produced gas @ 69,987 m^3/d in the interval 1,336-43m.

GS-29

This discovery is located in shallow waters and success was met with in Ravva Formation, oil find in well #6 in the interval 2,390–2,414.5m @ 617 m^3/day flowed through 32/64" choke and gas @1,23400 m^3/day. In another interval 2,237–2,245 m Oil @570 m^3/day, and gas@ 1,18,925 m^3/day.

GS-48

This structure located very close to the coast off Yanam found to be oil and gas bearing in Ravva formation in a second well drilled with in three thin sands in the interval 1,009-1034 metres flowed through DST testing @ 0.24 MM m^3/d open flow potential.

GS-59

This structure is located towards west of Kesanapalli west structure in shallow offshore area and proved promising in Vadaparru formation which produced oil with water in the interval 2,840-2,852m @ 58 m^3/d water is of 25 per cent and gas @ 13,080 m^3/d through 6 mm bean.

GS-70

The oil find in the first well in the prospect is in the interval 3,703-3,697 m, produced @ 69 m^3/d, gas @ 78,000 m^3/d and water @ 1.5 m^3/d through 6 mm bean.

Ravva (GS-16)

The exploration activity in Ravva area started with the acquisition of 2D seismic data during 1978 and 1981, which led to drilling of first well GS-16-1 by ONGC in 1983. The well penetrated Middle Miocene reservoirs, but the discovery could not be tested due to drilling complications.

In 1982-83, 2D Teleseis seismic data by Western Geophysical was acquired to cover shallow water areas of KG offshore. The interpretation of the data resulted in drilling of GS-16–2 in 1987 which has established commercial oil flow from the field and subsequently wells -3, -4 and -5, also proved presence of hydrocarbons in Early and Middle Miocene sequences. Then in 1989 "Ravva" name was adopted to this field (Bhavana, P.R., 1991).

In 1990, first 3D seismic survey covering area was acquired. By drilling of wells R-22 and R-23 wherein Late Miocene sandstones below Pliocene unconformity were found to be gas bearing. ONGC drilled 30 wells out of which 21 were found to be oil/gas bearing. Two platforms RA and RB were installed and field was put on production from 4 wells from Middle Miocene reservoirs in March 1993. The Ravva main producing reservoirs comprise porosity in the ranges of 22 per cent – 35 per cent.

Consequent to an offer in 1992 from Government of India, Production Sharing Contract (PSC) was signed for 25 years with JV comprising of ONGC (40 per cent PI), Cairn Energy India Pvt Ltd. (22.5 per cent), Videocon Petroleum Ltd (25 per cent) and Ravva Oil (Singapore) Pvt Ltd. (12.5 per cent). The area of the block is 331.26 sq km.

Ravva JV drilled another 29 wells including 2 sidetracks during 1996 to 2002 (total 59 wells). Seven water injector wells were also for pressure maintenance. Five new infill wells including a horizontal well were drilled to increase the oil production. Production wells were drilled using open-hole methods. It also adds incremental reserves and augments the water injection capacity. The work-over activities were taken up to enhance oil production and water injection. The oil production was initially 35,000 BOPD which was further raised to 50,000. With more positive results the inplace oil reserves have been enhanced to 470.3 MMB (63.2 MMT). The field has cumulatively produced about 38 MMT of oil and 394.5 BCF gas and now producing just more than 10k BOPD of oil.

Deen Dayal (GSPC)

The block is located very close to Kakinada coast with an area of 1,850 sq km was awarded to GSPC, promoted by the state government of Gujarat as the main operator, with an 80 per cent participating interest, under NELP-III, along with Jubilant Enpro Ltd and Geo Global Resources.

The area is close to earlier drilled well by ONGC called GS-17-1 and also to KBS-1-1. Both of these well were dry and in later crystalline basement was at depth about 1,700 meters. The Upper Cretaceous argillaceous section with minor sand bodies proved gas potential in the South western onland part of the Basin.

After acquiring 3D seismic data and drilled wells. The re-interpretation of data located extension of many Cretaceous fan complexes in the area. The well KG-16 was tested gas in DST@2.21 MMscfd in the interval 4,483-4,590m. Tested gas in another DST @2.2 MMscfd in the interval 4,302-4,435m. In well KG #22 in the upper Cretaceous interval 4,652-72 flowed gas @ 27 MM scfd. In well KG# 31 the same Upper Cretaceous interval 3828-38 established gas in this field named ***Deen Dayal***. The company initially found potential of 20.3 million cubic metres of gas a day. The KG-21 well is estimated to produce more than 100 million cubic feet of gas a day.

KG-OSN-2004/1

This block was awarded to consortium comprising ONGC (70) GSPC (10) HPCL (10) GAIL (10) with ONGC as operator in 2005. The well CS#1 tested in the interval from 1829.5-1832m and found to be gas bearing play for further exploration. It is called Alankari a Mio-Pliocene sand pack from and the in-place gas reserves are 3.20 BCM.

Another Lower Pliocene structure called Chandrika was tested in south channel sand in two intervals of 1,978.5- 1,975.5, 1,974-1,972m, and 1,946.5 -1,952m and found to be gas bearing which added accretion of In-place of 7.65 BCM.

Further, the recent Sarangi-1 well in the interval 1,737-1,727 flowed gas @ 40,01,03 m^3/d through 40/64" choke from Godavari Clay formation of Plio-Pleistocene age.

KG-OSN-98/3 (RIL)

The hydrocarbon strike Dhirubhai – 38 in water depth of 151 m drilled up to 2820 m in which the gross hydrocarbon column of 105 m in Miocene section. Another structure called Dhirubhai – 37 was drilled to 3500 m and found gross hydrocarbon column of 32 m., in Miocene and 4 m in Pliocene.

Offshore-Deep water

G-1

The structure In the offshore, the Neogene roll-over drilled in 268 m of water depth during 1980 produced oil and gas from Pliocene sand. The well encountered high pressures around 3,200 m, which were controlled but further drilling could not be continued due to high currents. Two assessment wells on the same structure and an independent well on an adjoining structure had to be terminated due to adverse sea-bed and current conditions, which led to the sinking of two well-heads and loss of the BOPs. The structure is oil bearing.

Four hydrocarbon bearing sands have been encountered, of which sand-8 has a maximum hydrocarbon volume (Gupta, S.K., *et al.*, 1998), which is having gas cap with 14 m oil column. The simulation studies have indicated life under depletion drive. Exploitation of upper sand through depletion drive one well only. (Suri. N.K., 2003).

G-4

The structure was earlier drilled by ONGC during late eighties and found to be mainly argillaceous section with thin sand beds. After the success in near by area by GSPC in Cretaceous section the repeated effort by the company in a well drilled #6 met with little oil and gas flows with the Cretaceous section in the interval range of 4,113 – 4,667m., and also by DST in upper horizon at 3834m.

Dhirubai (KG-DWN-98/3–D-6 Block)

This block was awarded to Reliance Group in the year 2000. Initially RIL holds 90 per cent and Niko Canada with 10 per cent PI was operator. Later RIL assigned 30 per cent to BP. About 11 gas discoveries were made and felt the hydrocarbon channels were spread over the entire area. The production from the fields with output of about 30 mmcmd. Which could be increased to 61.5 mmscmd. Now drop in pressure and increased water output.

Natural gas production from the Dhirubhai-1 and 3 gas fields and the MA oilfield in block was 38.43 mmcmd in 2012. RIL has drilled 22 wells. Out of the 18 wells put on production five wells were kept closed due to high water cut/sanding issues. Now the total in-place gas reserves of 9.460 trillion cubic feet.

Another area called Dhirubhai – 36 at water depth of 565 m, in which well was drilled to 3595 m met with gross hydrocarbon column of 21 m (Mesozoic) which produced 570 barrels of oil per day. The MA oilfield, where one of the five oil wells had been shut due to water loading, produced about 12,623 barrels of crude oil with 1,672 barrels of condensate per day.

KG-DWN-2003/1 (RIL)

In the block the discovery was named as Dhirubhai – 39 in water depth of 716 m and well was drilled to 1937 m which met with gross hydrocarbon column of 84 m in Pleistocene section and the well produced 38.1 MSCF of gas per day. Another one called Dhirubhai – 35 in water depth at well of 185 m drilled down to 4081 m found gross hydrocarbon column of 150 m in Cretaceous(?) which produced 31 MSCF of gas and 1200 barrels of condensate per day. Another deeper one produced 1 MSCF of gas and 550 barrels of oil per day.

KG-DWN-98/2 (D-5 Block– ONGC)

This block was awarded to Cairn in 2000. Earlier found oil field by ONGC, G-2 falls within this block. They carried out exploratory work and drilled wells and met with success in finding oil and gas in few structures and named those fields as Annapurna, Parvati and Kanakadurga, *etc.* In 2005, ONGC has acquired 90 per cent PI with operatorship in this block from Cairn who retained 10 per cent PI.

In 2016, the new discovery in well F-1 in the interval 1,814.5-1,812.5 m flowed oil @ 732 bpd and gas @ 13,155 m^3/d with FTHP: 548 psi through 24/64" choke from Godavari Clay formation of Plio-Pleistocene age. Further in well, M-4 (KG 98 2NA-M-AD), the interval 2908-2900 m flowed oil @ 3157 bpd and gas @ 3,19,483 m^3/d through 40/64" choke in the same formation.

Ultra-Deep Area

ONGC drilled well ***UD-1*** in water depth of 2,841metres in 2006. After drilling to a depth of 6,035 metres with the original planned target depth of the well was 6,200 meters or within 100 metres of the Basement. Preliminary evaluation of the logs and MDT (Modular Dynamic Tester) indicates some interesting zones.

Table 10.kg.1. Dry Structures Drilled

Krishna Godavari Basin	*Code*	*DD*	*Year*	*Wells*	*Reasons-Dry/Remarks*
Ambajipeta	AB			1	
Adavipeta	AD				

Krishna Godavari Basin	Code	DD	Year	Wells	Reasons-Dry/Remarks
Angaluru	AG	2500	1991	1	Early Cret.
Amalapuram#1,3	AM	4003	1983	2	
Arukattapalem	AR	2660	1994	1	Eocene
Bobbarlanka	BB	4262	1991	1	Basement
Bandarupeta	BR				
Bantumilli	BT	3655	1987	4	
Bhavadevarapalli	BV	5001	1992	4	Basement
Chinapandrika	CD	3806	1992	1	Basement
Chamakupalem	CH				
Challapalli	CL				
Dondapadu	DP			1	
Draksharama	DR	3145	1986	1	
Gajulapadu	GD	3811	1990	1	
Godi---tvd	GI	2901	1996	1	Paleocene
Gandhinagaram	GN				
Gopalapuram	GO	3001	1993	1	Early Permian
Gangadipalem	GP	3100	1993	1	Paleocene
Gannavaram	GV	2681	1993	1	Cret.
Guraza	GZ	2955	1991	1	Basement
Idarada	ID	3000	1994	1	Paleocene
Isukapudi---tvd	IP	2608	1996	1	Cret.
Jonnalanka	JL	3504	1990	1	
Kandikuppa	KA	2000	1997	1	Miocene
Kommugudem	KG	4508	1986	1	
Kaikaluru	KK	1972	1984	16	
Kakinada	KKD	2600	1993	1	Basement
Kottalanka	KL	3000	1993	1	Paleocene
KesanakuruPalem	KM				
Kanukollu	KN	3103	1987	1	
Kommarada	KR	2905	1992	1	Eocene
Kakinada-spit	KS				
Kavitam	KT	4520	1991	1	Permo-Carb.
Kuduru	KU				
Kumarapriyam	KY	3250	1993	1	Permian
Lakshmipuram	LK	4267	1996	1	Basement
Lellapudi	LL	3600	1991	1	Early Cret.
Lamellapadu	LM			1	
Mailavani lanka	MAI				
Mogalalmuri	MGL				
Mahendravada	MH	2867	1992	1	Basement

Krishna Godavari Basin	***Code***	***DD***	***Year***	***Wells***	***Reasons-Dry/Remarks***
Manginapudi	MI			1	
Makanapalem	MK			1	
Modi	MO	3638	1984	1	
Mantripalem	MT	2850	1991	1	Basement
Mulaparru	MU	2650	1996	1	Early Cret.
Madhavaram	MV	3905	1992	1	Permo-Triassic
Matsyapuri	MY	4505	1986	1	
Mattaparru	MZ	3200	1996	1	Cret.
Nidadavolu	ND	2525	1992	1	Basement
Nehrunagaram	NH				
Nimmakuru	NK	3071	1989	2	
Nellore	NL			1	
Nammavanipalem	NM	2732	1996	1	Basement
Nandampudi	NU				
Pallipalem	PA				
Poduru	PD	2502	1991	1	Cret.
Peddapalem	PE	4100	1993	1	Cret
Penkillapadu	PK	3100	1993	1	Cret.
Palakollu	PL	4501	1986	2	
Penumalla	PN	2462	1992	1	Basement
Perupalem	PP	3000	1994	1	Paleocene
Penduru	PR				
Padatadaka	PT				
Pulletikurru	PU	2300	1996	1	Up.Cret.
Polavaram	PV				
Ramachandrapuram	RP	4000	1992	2	Permo-Carb.
Ravipadu	RV	3100	1996	1	Up.Cret.
Siddavaram	SD				
Sankaraguptam	SG	3389	1991	1	Paleocene
Sakhinetipalli	SK	3300	1994	1	Paleocene
Salapaka	SL	2135	1995	1	Basement
Samanthakuru	SM				
Sanarudravaram	SN				
Saripalli#2	SP	3701	1990	2	Cret.
Sitarampuram	SR	2800	2004	1	Paleocene
Satyavolu	SV	2000	1996	1	Cret
Surasaniyanam	SY	2104	1987	1	
Tanuku	TK	3132	1991	1	Basement
Uppalaguptam	UG	2000	1993	1	Eocene
Vadali	VA	2822	1987		

Krishna Godavari Basin	*Code*	*DD*	*Year*	*Wells*	*Reasons-Dry/Remarks*
Vemuladevi	VD	2701	1994	1	Paleocene
Vinnakota	VK	3087	1991	1	Basement
Velpuru	VL	4010	1991	1	Mesozoics
Vanadurru	VN	3138	1986	1	
Vadaparru	VP	3500	1986	1	
Viswswarayapuram	VS				
Vetlapalem	VT	3300	1987	1	
Viravasaram	VV			1	
Vainateyam	VY	3093	1990	1	
West of Mori--tvd	WM	3000	1996	1	Paleocene
Gudlavalleru					
Uppidi			2008		
Varigonda					
Ramanapadu					
Turupuvipaarru					
Koravaka					
Kosuru					
KBS-1--tvd	on	2106	1994	1	Basement
KGON4-NANL#1 and 2					
OFFSHORE	**WD**				
DWN-A					
G-13	280	4101	1982	1	Oligocene
G-5	324	3007	1984	1	Miocene
GD-7			2009		
GS-1	26	2599	1982	1	Miocene
GS-10					
GS-11	21	4625	1984	1	Cret.
GS-12	39	3150	1983	2	Early Eocene
GS-14	30	3517	1983	1	Paleocene
GS-17	99	4020	1984	1	Cret.
GS-18	105	3000	1984	1	Miocene
GS-19	12	2442	1986	1	EarlyMiocene
GS-2	80	3210	1984	1	Miocene
GS-20	22	2500	1985	1	Pliocene
GS-21	7	3050	1986	2	Early Miocene
GS-21B	14	2100	1992	1	Miocene
GS-22	8	2512	1986	1	Early Miocene
GS-22A	12	3401	1991	1	Paleocene
GS-22B	10	3015	1991	1	Paleocene
GS-25	7	3100	1993	1	

Krishna Godavari Basin	Code	DD	Year	Wells	Reasons-Dry/Remarks
GS-27					
GS-28					
GS-29	122	3000	1992	2	Miocene
GS-3	113	4092	1982	2	Early Eocene
GS-37	11	3515	1989	1	Paleocene
GS-38	12	3500	1990	1	Paleocene
GS-39	11	2700	1991	1	Eocene
GS-40					
GS-41	45	3200	1991	1	Eocene
Gs-42					
GS-43	12	2300	1993	1	Miocene
GS-46	15	2200	1992	1	Miocene
GS-5	19	3919	1983	1	Cret
GS-54					
GS-59					
GS-8	14	3002	1986	1	Paleocene
GS-A		3204	1994	1	Miocene
KB-1	61	3679	1992	1	Mesozoics
KG-DW-S					
KRI-1				1	
KS-3	15	2622	1985	1	Basement
KS-4				1	
MB-1#2	71	2850	1990	2	
NB-1					
NB-2					
YS-1					
YS-2					
YS-5			2009		
YS-6					
JV-Foreign					
KG-1-B#1*Chevron				1	
P-II-C*Chevron				1	
Amoco				1	
KG-OS-V-I				1	
KG-OS-V-II	82	1690	1991	1	Early Pliocene
@GS-26*IPC					
					GSPC
Deendayal				15	
D6-KG-DWN-97/3					**RIL**
Dhirubai				22	

Krishna Godavari Basin	*Code*	*DD*	*Year*	*Wells*	*Reasons-Dry/Remarks*
MA					
D5-KG-DWN-97/2					**Cairns-ONGC**
Cairns-Videocon-ONGC					
KG-6-1	116				
South Pasarlapudi			2014-15		
YS-1			"		
GD-11			"		

Table 10.kg.2. Oil and Gas Fields-Krishna-Godavari Basin

Fields	*O/G*	*Year*	*Area sq km*	*Rec*	*Wells*	*Remarks*
Adavipalem		1995	95.00			
Bandamurulanka-N				0.10		
Bantumilli			155.67	0,27		
Chintalapalli	G	1988	18.56	0.16	6	
East-Rangapuram						
Elamanchili	G	1990	6.00			
Endamuru		1991	16.63	0.77		
Endamuru -Extn		2000				
Enugupalli		1996	7.0			
Geddanapalli					3	
Gokarnapuram		1998				
Gopavaram		2000			12	
Kadali	G					
Kaikaluru	O	1984	29.0	0.32	8	
Kavitam		2003	156.35			
Kaza	O			0.02	3	
Kesanapalli	G	1991	3.7	0.03		
Kesanapalli-West		1996			28	
Kesavadasupalem		1998	26.50			
Lakshaneswaram		2000				
Lankapalem		1995				
Magatapalli	G	1997				
Malleswaram			241.18			
Mahadevapatnam		1995	138.89	0.09	1	
Mandapeta	G	1988	40.0		27	
Mandapeta-South			6.0		19	
Mandapeta-West	G		20.0			

Fields	**O/G**	**Year**	**Area sq km**	**Rec**	**Wells**	**Remarks**
Manepalli		1990		0.52	2	
Medapadu			16.6	0.04	1	
Mori	G	1990	8.06			
Mulikipalli				0.05		
Nandigama		1992	55.0			
Narasapur	G	1982		0.14	6	
Pasaralapudi	G	1987	13.10		43	
Penumadam	G	2003	12.80	0.07	1	
Ponnamanda		1993				
Rangapuram		1996				
Razole	G	1983	18.85		4	
Seripalem						
Sirikattapalli						
Sitarampuram		2004				
Suryaraopeta		2000	56.00			
Turuputallu		2006	39.58			
Uppidi						
Vadali		1987	4.0		5	
Vyagreswaram						
Vainateyam			78.0			
G-1	O	1980			6	
G-2	O	1981			4	Now in Block D-5
G-4		2004			6	
GS-23			40.00			
GS-49		2003	130.18			
GS-69		2007				
Vasista		2005	119.0			
UD-1						
GS-KW					5	
DWN-A		2006				
GS-38	G					
GS-29			172.62			
GS-59						
GS-48	G					
YS-6						
Annapurna	O					to ONGC
Parvati	O					
Kanakadurga	O					

Fields	O/G	Year	Area sq km	Rec	Wells	Remarks
KG-DW-S						
GS-8	G				2	
KGON4-NANL#1 and 2		2013				
YS-9	Gas	2015				
GD-11	Gas	2015				
			ONG	**132.66**	**wells**	**482**
RIL						
Dhirubai-1	G				18	
Dhirubai-3	G					
Mukesh Ambani-1	O				06	
GS-16/Ravva	O	1987			59	Vedanta
Small Fields						
Bhimanapalli	O	1984	15.1		2	
Koravaka		2013	9.9			
Mummidivaram			9.73	0.04		
Palakollu		1987	9.18	0.14	2	
YS-5		2009	26.5			
GD-7		2009	40			
GS-59	O/G					
GS-70	Gas		4.3			
GD-1	Gas		40			
KD-1	Gas		55			
Sanarudravaram	Gas		9.35			
Achanta		1993	14.10	0.02	1	
Tatipaka	G	1986	62.00		6	
Sirikatapalli		2000	163.0			
GS-15			40.00		15	
South Pasarlapudi		2015				
Nagayalanka	Oil	2012				
Lingala	O	1989	17.60	0.59	18	

Mumbai Offshore Basin

Introduction

Mumbai Offshore Basin is the most important petroliferous basin in the west coast sedimentary province of India. The Bombay High field is called gaint oil field of India. The basin is located on the western continental shelf of India between Saurashtra in NNW and Kerela-Konkan in the south. The basin is bounded by the Deccan Trap outcrops to its north and east, Kori-Comorin ridge to its west and Vengurla arch in the south. It covers an area of over 120,000 sq km upto 200 m isobath.

Bombay High was discovered by a Russian and Indian oil exploration team operating from the seismic exploration vessel Academic Arkhangelsky during mapping of the Gulf of Khambat (earlier Cambay) in 1964-67. This basin is the most explored and also the most prolific amongst offshore basins of India. This basin holds more than 60 per cent of the hydrocarbon reserves discovered in the country.

Bombay (later it was renamed as Mumbai) Offshore Basin has been divided into five blocks on the basis of tectono-stratigraphic considerations. To the north Surat Depression, north-northeast is the Tapti Daman block, a major depocentre which is contiguous with the onland Cambay Basin. In the central part from west to east are the Shelf Margin, Bombay High-Deep Continental Shelf (DCS) and Heera-Panna-Bassein blocks. Ratnagiri block lies in the southern part of the basin.

The mature source rocks are present in the lower Eocene-Paleocene Panna Formation. Further, marginally mature potential source rocks within Oligocene in Tapti-Daman area and within Neogene in DCS and deeper part of the basin also exist. Hydrocarbons have been discovered in multiple reservoirs in this basin, ranging from fractured basement to middle Miocene.

Exploration History

Exploration in the area started with offshore seismic data acquired during the 1960's. The seismic data interpretation led to the delineation of numerous large and small structures.

Drilling of the first offshore well in India was started in early 1970 at Aliabet and gathered momentum after the discovery of large commercial oil pools of L-II and L-III reservoirs in Bombay High during 1974-75. The Bombay High field produces primarily from Miocene limestones and some sandstone reservoirs on a fault bounded block (Harbison, R. N., and B. G. Bassinger, 1970).

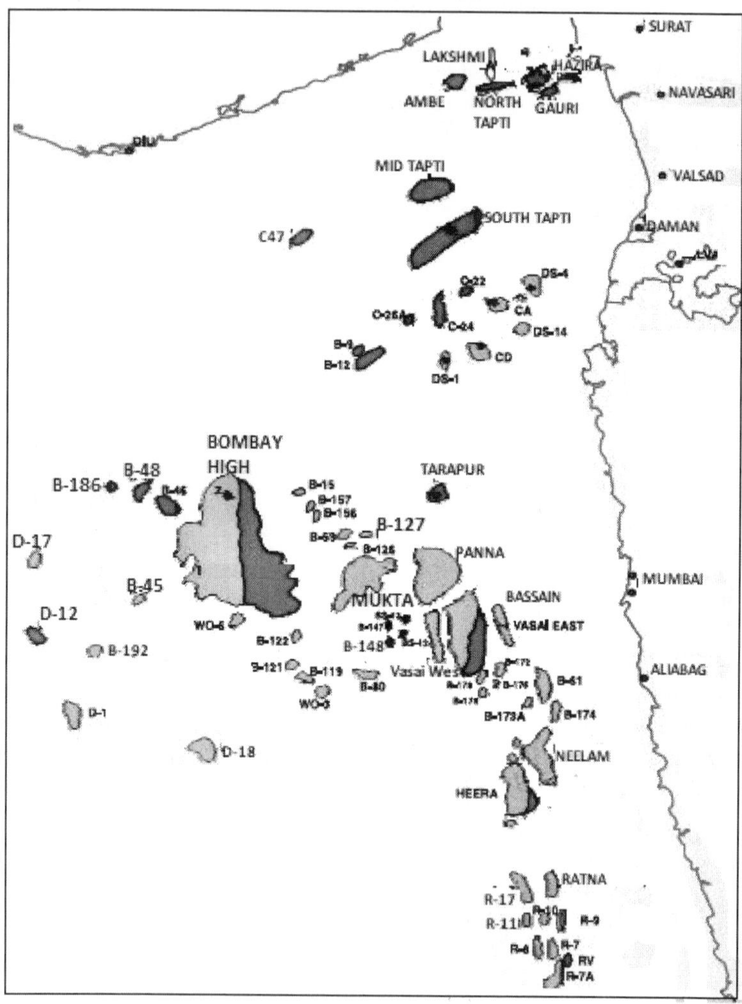

Figure 10.mu.1. Oil and Gas Fields Map-Mumbai Offshore. (After Gupta, G D, *et al.*, 1993).

Subsequent intensification in exploration and development activities in this basin have resulted in several significant discoveries including oil and gas fields like Heera, Panna, Bassein, Neelam, Mukta, Ratna, Soth tapti, Mid Tapti, *etc.* In addition number of marginal fields like B-55, B-173A, B-119/121, D-1 and D-18 have been put on production in the last decade.

In Mumbai Offshore basin 3,600 wells have been drilled on 135 structures, of which 80 are proved oil and gas fields (Bombay High, Bassein, B-38, B-37B, R-9 and R-12), four are oil-bearing non-commercial (D-1, B-37A, B-22, H1-5 and Aliabet), three are gas fields (South Bassein, South Tapti and B-55), one containing some sour gas (R-15) and three containing non-commercial gas (B-12, B-51 and B758) (Table 10.mu.1). Since 1998 NELP policy was implemented in India to invite private participation (Figure 10.mu.1).

The exploration has generated vast amount of data on structure, stratigraphy, facies distribution and petroleum geology. A sizeable amount of literature is already available on the above topics. Mohan, *et al.*, 1983; Rao and Talukdar, 1980; Ramaswamy and Rao, 1978; Basu, *et al.*, 1980, 1982; Sahay, 1985; Venkatachala, 1984; Biswas, 1982; Roychoudhary and Deshpande, 1982; Zutshi, *et al.*, 1986 are to name a few.

The basin has been covered by gravity-magnetic and close grid seismic surveys and with 3-D seismic surveys have been introduced to acquire detailed reservoir information about areas of rapid facies changes. The data enabled mapping of numerous structural prospects. (Figure 10.mu.2 and 3).

Mid-Tapti and South Tapti gas fields have also grown with more step outs drilled indicating stratigraphic play in the formation of trap. Similarly, the southern culmination of D-1 structure has come out to be a promising extension of the oil field (Mathur, R.B. and K.M. Nair, 1993).

Even the exploratory drilling had encouraging results in Daman Offshore resulting in several oil and gas strikes. The accumulations here are in combination traps with both structural and stratigraphic elements playing important roles.

Tectonics

This basin is a pericratonic rift basin situated on western continental margin of India, during Late Cretaceous, as a result of fragmentation initiated by transform faulting (Biswas, 1988). Large scale volcanic eruptions followed this episode which covered the basin excepting paleohighs.

Towards NNE it continues into the onland Cambay basin. It is bounded in the northwest by Saurashtra peninsula, north by Diu Arch. Its southern limit is marked by east west trending Vengurla Arch to the South of Ratnagiri and to the east by Indian craton (Biswas S K, Bhasin A L and Ram J 1993). On the basis of its structural configuration and its nature, as well as the type of sediment fill, the basin is divided into six tectonic blocks: Tapti-Daman, Diu, Heera-Panna-Bassein, Bombay high-DCS, Ratnagiri, and Shelf Margin blocks (Figure 10.mu.4).

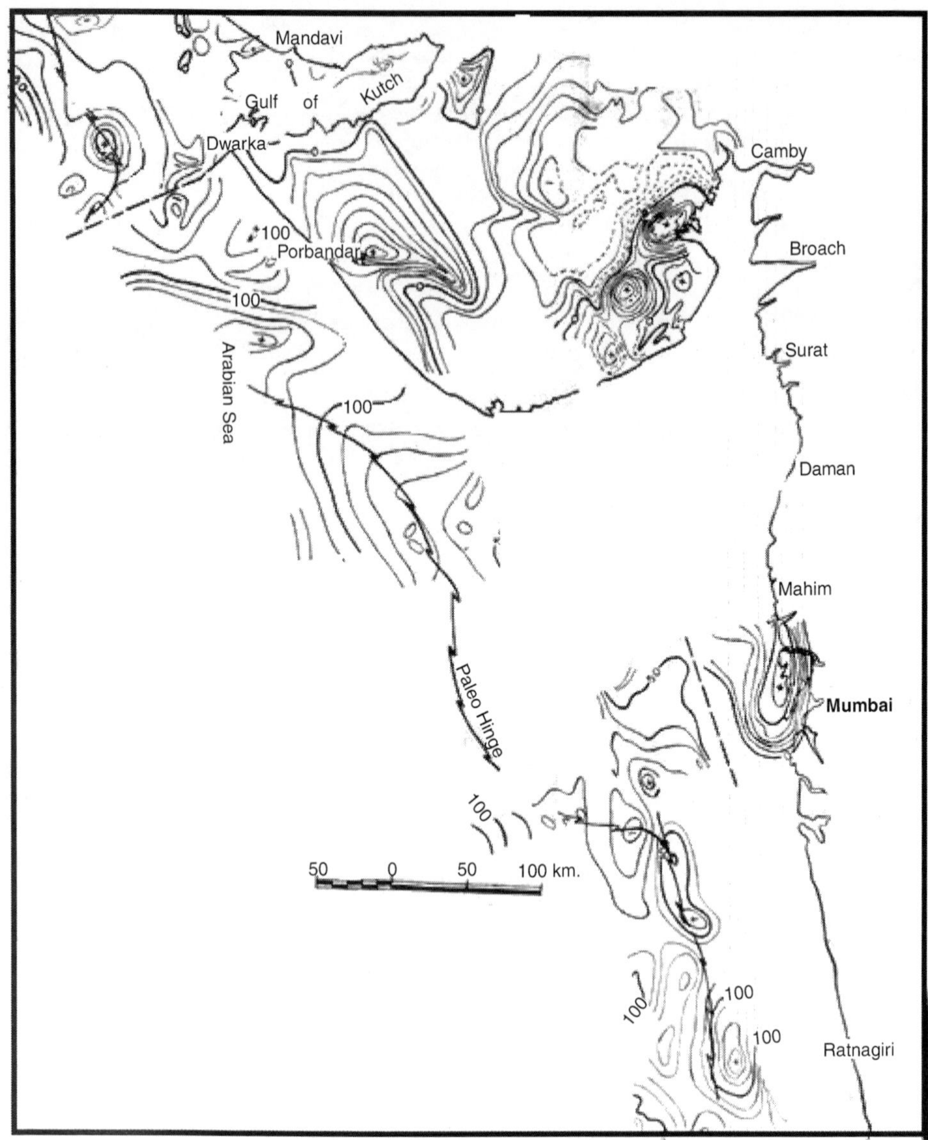

Figure 10.mu.2. Bouger Gravity Anomaly Map Mumbai Offshore Area (Afrer Murty, KVS, *et al.*, 1984).

The western passive margin of the Indian plate evolved during the separation of Madagascar-Seycheles in Late Cretaceous between 90 Ma and 65 Ma. The rifting started in Late Cretaceous with syn-tectonic Deccan volcanic activity that continued to Early Paleocene. Early Paleocene to Early Eocene hiatus was the period of the rift-drift transition marked by a widespread unconformity in all the basins. The post-rift thermal cooling resulted in sagging and the basins evolved into a marginal sag basin initiating marine transgression.

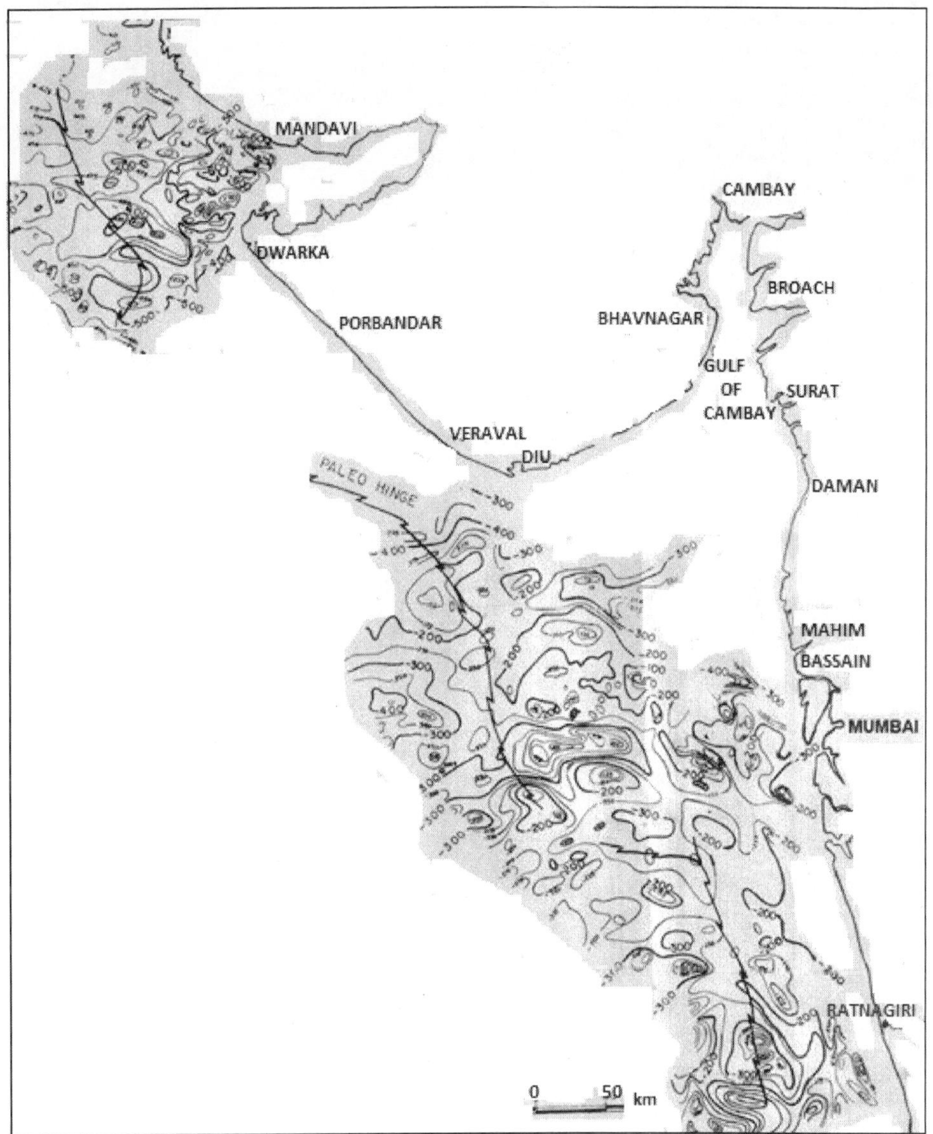

Figure 10.mu.3. Magnetic Anomaly Map-West Offshore (Afrer Murty, KVS, *et al.*, 1984).

The uppermost sequence of passive margin is recorded between Eocene top unconformity and Mio/Pliocene unconformity, covering early Oligocene to Miocene. The sequence is divided into two units early Oligocene and late Oligocene-Miocene succession separated by an unconformity. The lower unit covering the early Oligocene is present over the entire Mumbai offshore basin except Mumbai High platform. Lithologically, the sequence shows a facies variation across the basin. In many areas, it is characterized by carbonates and shale deposited in shallow

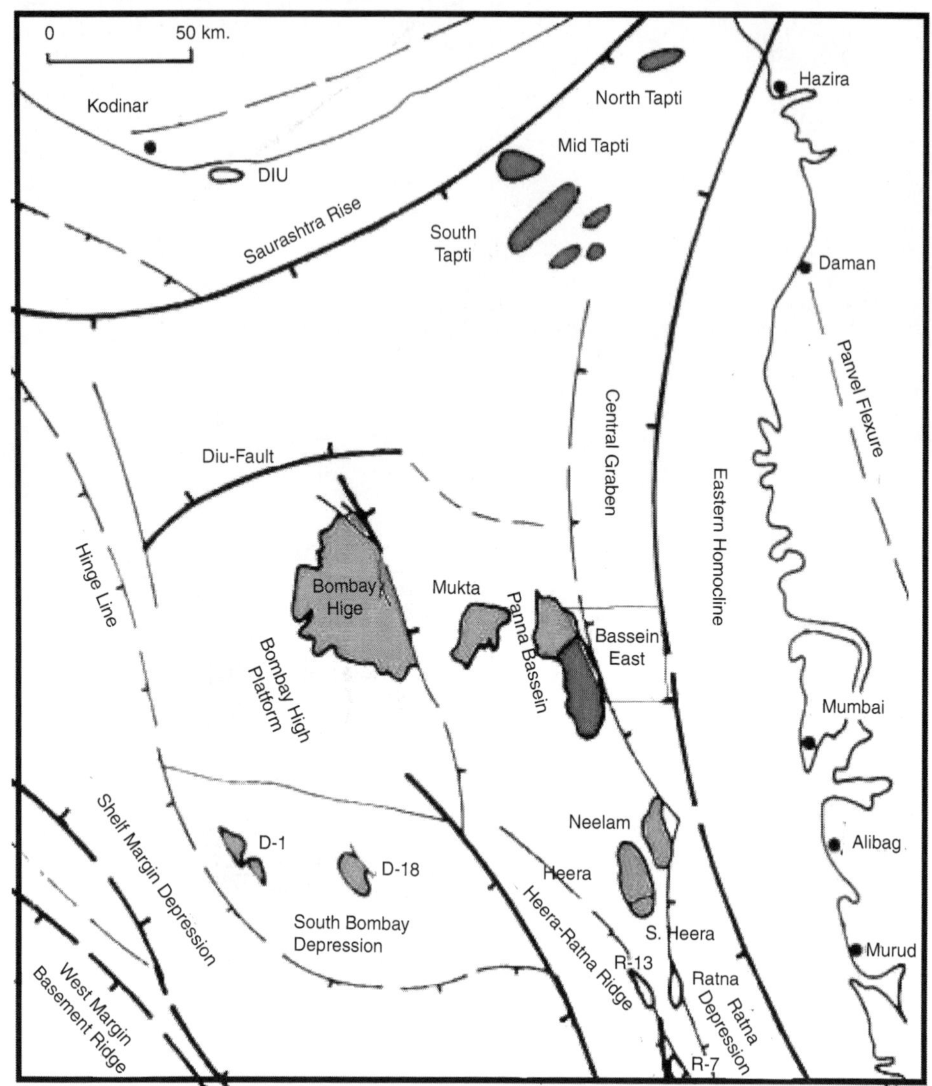

Figure 10.mu.4. Tectonic Framework-Mumbai Offshore (After Jitendriya Mishra, *et al.*, 2011).

inner-shelf setup, while in the northern areas of Tapti-Daman intertidal to shallow marine clastics characterize the sequence. The overlying, late Oligocene-Miocene epochs represent deposition of shallow-shelf carbonates and shale over most parts of the basin, except in northern parts, and where fluvial to shallow-marine clastics characterize the late Oligocene. This succession is encountered over all areas of Mumbai offshore and overlain by the finer clastics of late Miocene to Recent age (Biswas S K., 1987).

Coast parallel Ridge-Depressions, Kori-Comorin Ridge/Depression and Laxmi-Laccadive Ridge/Depression, crossed by first order transverse basement arches are the major features. The transverse arches, from north to south are, Saurashtra, Bombay, Vengurla and Tellicheri arches, which divide the shelf into five offshore sub-basins Kutch, Surat, Ratnagiri, Bombay offshore basin and Konkan-Kerala basins.

Kori-Comorin marks the edge of the present continental shelf, which is rifted developing a "shelfal horst-graben" complex. Each sub-basin extends offshore across the shelf depressions and ridges. Thus a shallow shelfal horst-graben complex, a shelf margin depression and ridge, a deep slope-parallel depression and an outer ridge (Laxmi) separating the basin and abyssal plain are the characteristic structural domains of each sub-basins.

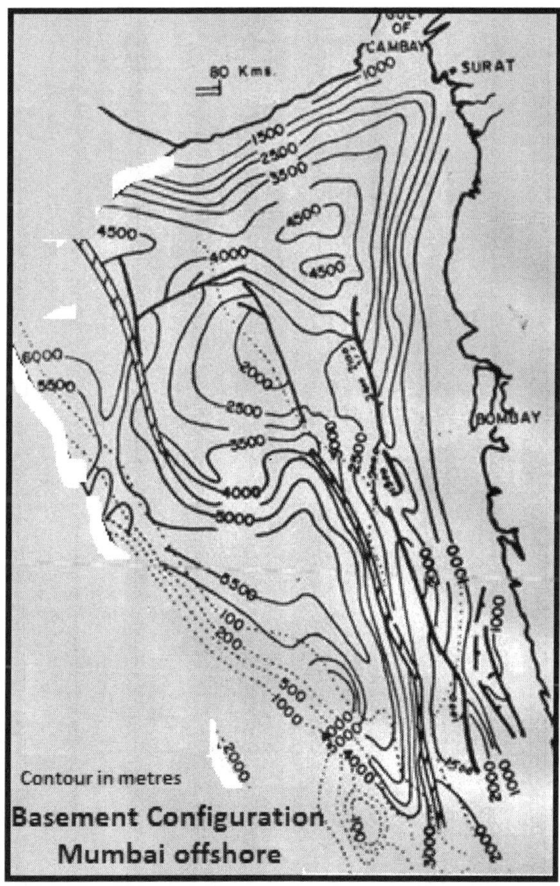

Figure 10.mu.5. Basement Configuration in Bombay Offshore Area (After Mishra, J and CRC Sekaran, 1997).

Kerala-Konkan and Ratnagiri basins are confined to offshore. The West Coast Fault terminates these basins along the coastline. Onshore extension of Surat basin is known as Narmada basin and an E-W-trending Kutch rift basin extends from onshore to offshore shelf.

The age of the basin ranges from late Cretaceous to Holocene with thick sedimentary fill ranging from 1100-5000 m. Though possibility of occurrence of Mesozoic synrift sequences in the deep-water basin have been indicated by the seismic data (Figure 10.mu.5).

The main Mumbai high block is surrounded by three depressions like, Surat depression covering Daman, Purna and Navsari lows and its southward extension through Mahim depression in the east. Saurashtra low in the northwest and southern paleo-sink and Murad depression in the southwest.

The Shelf Margin block bounded on its west by Kori-Comorin ridge and east by Paleogene hinge and its northern part includes Saurashtra offshore. Surat depression and its southward extension through Mahim depression to Vijaydurg depression in Ratnagiri block are the prime depocenter of the clastic sediments of early Eocene to Paleocene age. Murud depression and Saurashtra low had relatively more open marine environment.

Sedimentation

Based on the tectonic evolution of the basin, two sequences have been identified in the Cenozoic succession over the archaen basement or basaltic flows. The lower one of the Paleocene age represents the rift phase. The upper sedimentary succession is of late Paleocene to Recent.

The Paleocene rift sequence represents dominantly fluvial intertidal to shallow inner-shelf clastics, with subordinate carbonates, trap-wash and trap derivatives in the lower part represent post-trappean sediments as graben fills during synrift phase. The clastics comprise siltstone, claystone, shale and sandstone with coal shale. The lower boundary is an unconformity with Deccan Trap and upper boundary is also an unconformity. The late Paleocene-Recent passive Cenozoic marine sequence in wells represents the first marine transgression in the basin over the rift-fill.

The lowermost succession of passive margin corresponds to late Paleocene-early Eocene and represents the first widespread Cenozoic marine transgression in the basin. The sequence represented by fluvial-intertidal shale, coal shale and finer clastics in the lower part and shallow inner shelf carbonates in the upper parts, is recorded all over the Mumbai offshore basin except over the Mumbai High platform area.

The succeeding Middle-late Eocene succession has a very large areal extent over the basin, during this period, a large part of the basin witnessed widespread carbonate sedimentation in a shallow inner-shelf setup. The lower limit of the sequence at late Paleocene-early Eocene/Middle Eocene boundary and the upper boundary is an unconformity between Eocene and Oligocene (Figure 10.mu.6). In the northern areas of Tapti-Daman, finer clastics dominate, while in other parts of the basin, limestone and shale with minor clay are the dominant facies (Pangtey K.S, *et al.*, 1993).

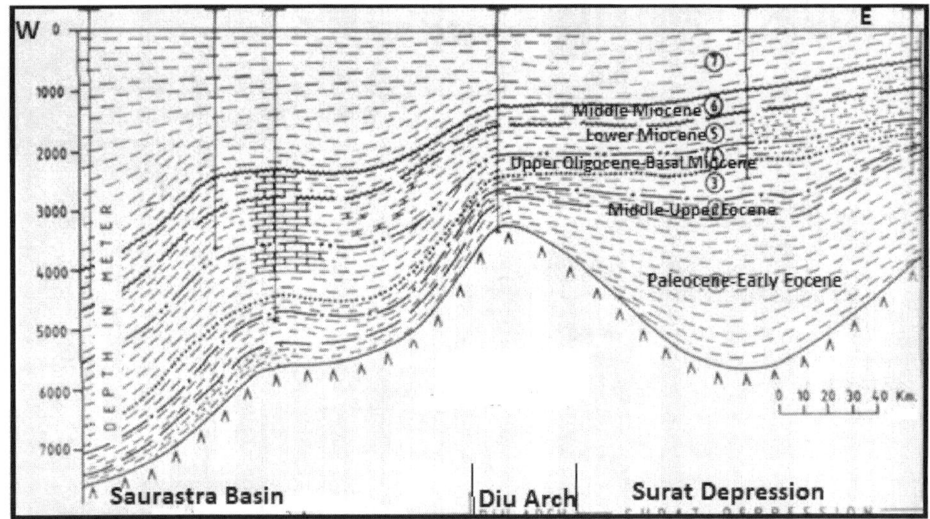

Figure 10.mu.6. Geological Section in the Northern Part of Mumbai Offshore.

As the rifting continued, the sediments deposited as alluvial fans, filled the depressions during Paleocene. This was followed by first marine incursion towards the close of Paleocene and begining of Early Eocene. The Bombay High, Heera and parts of south Ratnagiri, however, remained as paleo-high and were unaffected by this transgression. Surat depression with the southward extension through Mahim graben and the Vijaydurg graben were the main clastics depocentres. A minor hiatus is indicated towards the end of Paleocene (Figure 10.mu.7 A).

The Middle Eocene transgression once again inundated the basin almost completely except the Bombay High and southeast Ratnagiri shelf. During this period, most of the Bombay Ratnagiri paleo-shelf experienced carbonate sedimentation deposited on a broad carbonate platform. Tapti-Daman and Diu blocks remained as clastic depocentres (Figure 10.mu.7 B). Deep Continental Shelf and Mahim graben, almost uninterrupted sedimentation during Middle and Late Eocene was observed (Guha, *et al.*, 1987), whereas in other parts of the basin a major unconformity is observed at the end of Middle Eocene.

The Early Oligocene transgression covered most of the basinal area including a large part of Bombay High. Besides Tapti-Daman area which was continuously subsiding as a pronounced low, Murud Low appeared in the shelf Margin depression. High subsidence rates in Tapti-Daman block accommodated pro-delta clays and at some places shelf carbonates are developed. A major unconformity is observed at the top of Early Oligocene.

A regressive phase set in during late Early Oligocene probably in response to the westerly tilt of the basin. During this period Bombay High was completely submerged. Delta building started in the north-eastern part of the basin resulting in supply of coarse clastics which gradually prograded from northeast into Tapti-Daman block during Late Oligocene (Figure 10.mu.8A). Basal Miocene finer clastics

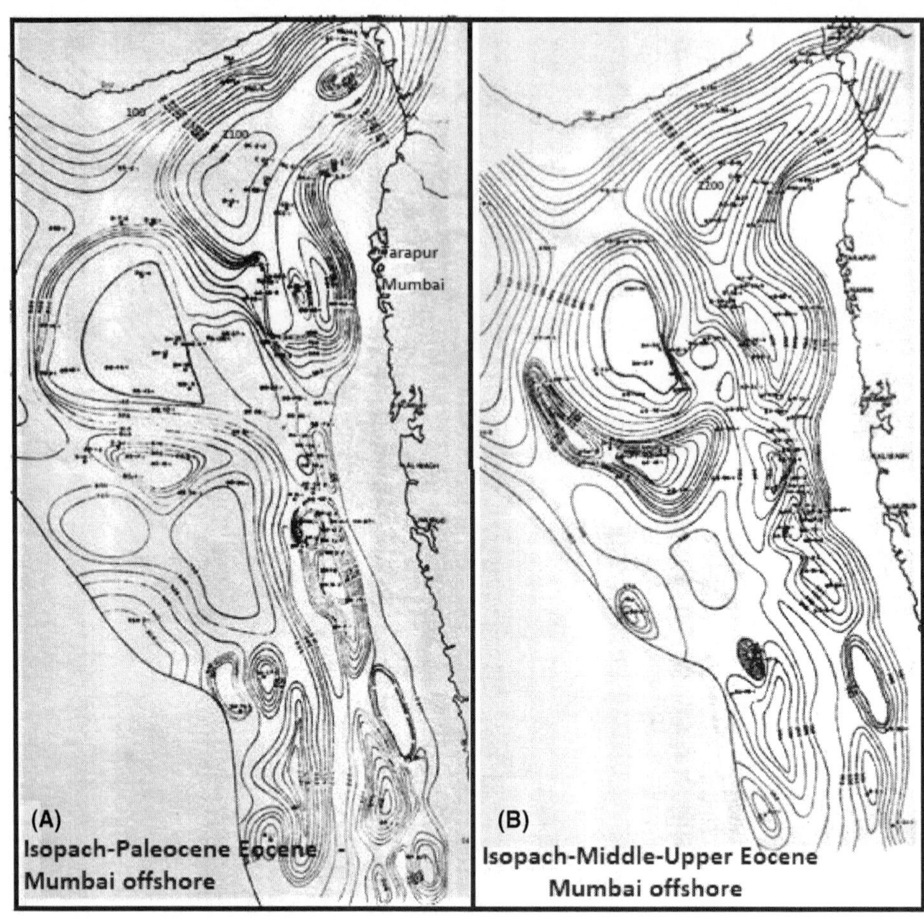

Figure 10.mu.7. (A) Isopach-Paleocene Eocene, (B) Middle-Upper Eocene-Mumbai Offshore.

were transported down to the Panna-Bassein, north Ratnagiri and Shelf Margin blocks. In response to the Early Miocene rise in sea level large areas of coastal tracts and basin fringes got inundated and the prograding Oligocene delta in the north-eastern part of Tapti-Daman receded north and north-eastward (Nair, K M, 1992). The finer clastics prograded westward and southward into the fast subsiding Shelf Margin depression (Figure 10.mu.8-B). The Lower Miocene top appears to mark another regional unconformity in the area as is suggested by biostratigraphic data and supported by seismic studies.

The Middle Miocene transgression the last phase of the widespread carbonate sedimentation in the Bombay High-DCS and Ratnagiri blocks. The outbuilding of platform which started in Early Miocene towards western part of DCS and Ratnagiri continued. The Panna-Bassein block came under sedimentation during this period. At the end of Middle Miocene, a sudden deepening of the Marvin depression is observed (Figure 10.mu.9).

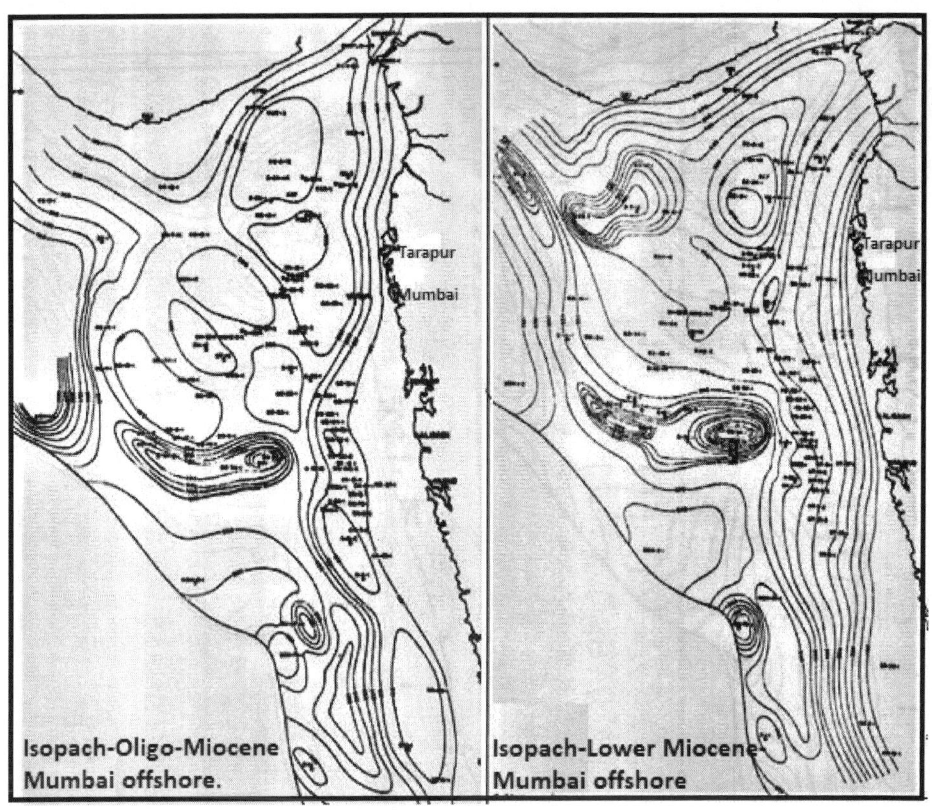

Figure 10.mu.8. (A) Isopach-Oligo-Miocene, (B) Lower Miocene-Mumbai Offshore.

The Post Middle Miocene was flooding of fresh areas along the basin fringes excessive increase in the clastic supply deepening of the basin inhibited further sedimentation, except in the form of isolated shelf carbonate build ups like the Angara Bank.

Stratigraphy

The basement consists of basalts (Deccan traps) as well as Precambrian rocks. In the Mumbai Offshore basin, number of wells penetrated Tertiary over the Archean basement without Deccan Trap or the Mesozoic. A few wells bottomed in trap, (Figure 10.cu.10). East of Mumbai, the trap is as much as 2,500 m thick (Madan Mohan., 1983).

The proposed a comprehensive stratigraphic scheme for the entire Bombay Offshore Basin and suggested a formal rock stratigraphic nomenclature (Basu, *et al.*, 1982). The regional chrono-stratigraphic correlation across the basin was carried out by identifying seven chronos-tratigraphic units traced over the entire basin (Nair, *et al.*, 1991 and Basu, *et al.*, 1982). The study of these seven sequences has

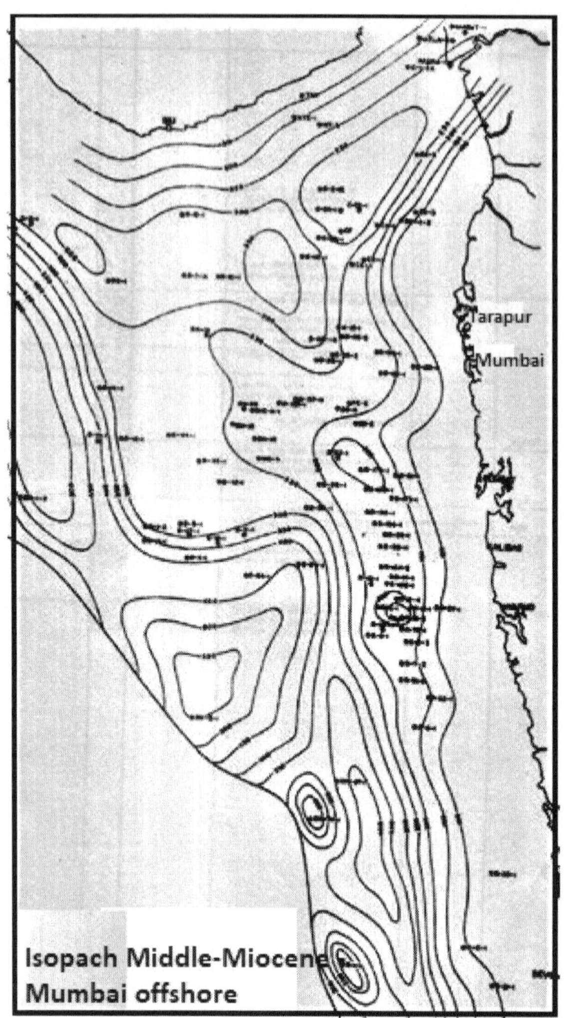

Figure 10.mu.9. Isopach Middle-Miocene-Mumbai Offshore.

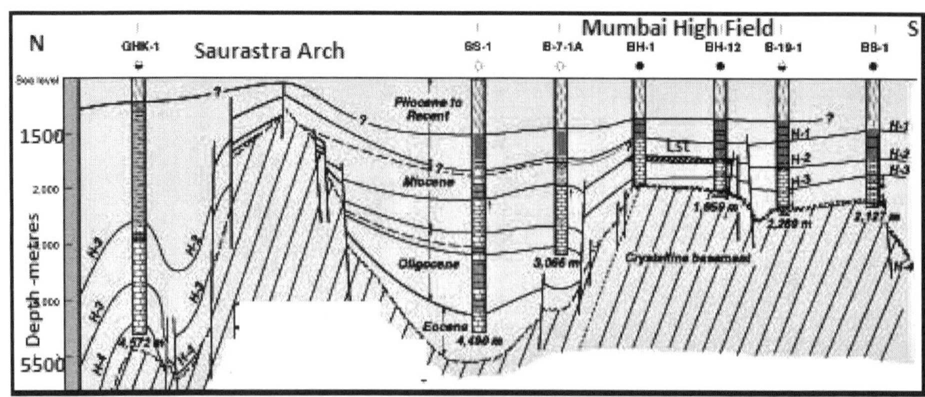

Figure 10.mu.10. North-South Well Correlation Bombay High Area.

helped in understanding the tectonics, paleogeography, paleo-geomorphology and distribution of facies with respect to geological time.

Litho-Stratigraphy-Nomenclature

Late Paleocene-Early Eocene

This phase signifies the early syn-rift stage which represented by trap-derived clastics contributed by the then existing paleo-highs essentially in continental to fluvial environment in its lower part (***Panna Formation***). It is overlain by grey to dark grey shales with subordinate sands possibly representing the first marine transgression into the basin. Presence of carbonaceous shale and coal at a few places suggest localized restricted conditions (Idem. 1982).

Main clastic depocenters like Surat Depression and the southward lows like Central Graben (Panna Bassien block) and Vijayadurg Graben (Ratnagiri Block) received these sediments in considerable thickness. The depressions in Mumbai Platform area and over some other horst blocks also received these sediments. Shelf Margin block, the facies developed are mainly claystone, argillaceous and carbonates (Figures 10.mu.11)

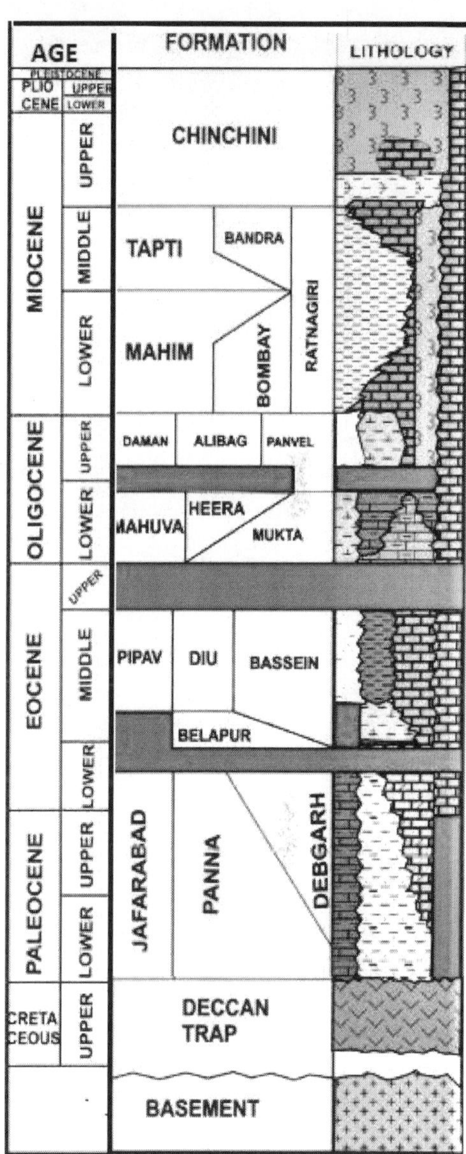

Figure 10.mu.11. Litho-Stratigraphy of Mumbai Offshore Area in different Blocks.

Carbonate facies (***Devgarh Formation***) development is observed towards the southern edge of Mumbai High in the form of muddy foraminiferal-algal banks; Deep Continental Shelf area and isolated off-shelf carbonate build-ups at a few places in Shelf Margin and Ratnagiri. The syn-rift stage got terminated with regression and development of an unconformity (Mishra, Y.K., Parida, G., and Ramani, K.K.V., 1997).

Middle –Late Eocene

After a period of peneplanation, the basin witnessed a major transgression. Extensive carbonate sedimentation occurred in the shallow shelf area of Mumbai Platform, Panna-Bassein-Heera block and Ratnagiri block (***Bassein Formation***). However, the period witnessed essentially clastic sedimentation in Surat Depression (***Belapur and Diu formations***) with occasional carbonate bands and a few thin sand streaks, argillaceous carbonates and shales in Central and Vijayadurg Grabens (Panna-Bassein-Heera block and Ratnagiri block). Shelf margin was generally starved of clastics with deposition of minor claystone and carbonates of mixed middle shelf to bathyal origin.

Bassein Formation also indicates a wide range of environments like, restricted platform, shelf lagoon with isolated shoals in Bassein area to open carbonate shelf in DCS and Ratnagiri and finally deep water carbonates in Shelf Margin area. It also formed wedge outs against the rising flanks.

Early Oligocene

During this period, Surat Depression experienced maximum subsidence accumulating thick under-compacted claystone relating to the prograding delta from northeast (***Mahuva Formation***). The Mumbai platform experienced generally shallower water depths and shale interbeds within limestone becoming more frequent. In Shelf Margin area thinner carbonates were deposited under relatively deeper conditions. End of Early Oligocene also witnessed initiation of the westerly tilt of the basin.

Late Oligocene

Close of Early Oligocene is marked by a minor period of non-deposition except in Shelf Margin area. A few brief spells of transgression followed by continuous eustatic rise in sea level up to Early Miocene marked this period.

Surat Depression witnessed lesser subsidence resulting in a regressive coastline. The sand bodies deposited in distributary channels, coastal bars, tidal deltas and other transitional environments encased in marginal marine silty and carbonaceous shale overlying over pressured pro-delta clay stone of Early Oligocene (***Daman Formation***).

There was faster subsidence in Shelf Margin and the finer clastics reaching the area were mainly deposited in the depression between Kori High and the carbonate platform. (***Alibag Formation***).

Southward by end of Early Oligocene is marked by a minor period of non-deposition except in Shelf Margin area. A few brief spells of transgression followed by continuous eustatic rise in sea level up to Early Miocene marked this period (Rao, V.K. and V.K. Mittal, 1990).

Southward, from Surat Depression, clays got dispersed over Panna-Bassein-Heera block, including the crestal areas and the northern part of Ratnagiri block as well as Mumbai Platform. While in Mumbai High-DCS area and southern part of Ratnagiri, the unit is termed as ***Panvel Formation,*** in Panna-Bassein-Heera and northern part of Ratnagiri, the unit is named as Alibag Formation.

Early Miocene

It was a period of eustatic rise in sea level punctuated by a brief spell. The finer clastics entering Surat Depression got mostly dispersed westward into Saurashtra basin and Shelf Margin area. Limited quantity of clastics got dispersed southward and entered Mumbai platform at its southeast and also up to Heera area. In response to the rising sea level, the delta being formed in Surat Depression in Late Oligocene shifted eastward.

Bassein and the area to its south experienced shoaling conditions during Eocene was the site for fine clastic deposition during Early Miocene. Mumbai High and its western part (DCS) underwent fairly thick carbonate sedimentation.

Middle Miocene

The sea level was continued to rise during this period. Clastic supply also continued into the basin. However, much of the clastic material got dispersed westward into Saurashtra and Shelf Margin areas. Considerable quantity of clastics got dispersed southward also covering the entire Panna-Bassein area and also the Mumbai High and its immediate surroundings to the west and south. This clastic unit over Mumbai High includes sheet like sand, which has also been found to be hydrocarbon bearing. Carbonate sedimentation continued in Ratnagiri and DCS areas. Toward the later part of Middle Miocene, clastic deposition almost came to a halt in Mumbai High and other areas and consequently carbonates got deposited over many areas. Uppermost part of the Middle Miocene Limestone in Heera field has been found to be hydrocarbon bearing. Close of Middle Miocene was marked by a very pronounced unconformity.

Middle Miocene- Holocene Post Middle Miocene witnessed a major transgression covering the entire basin coupled with spectacular increase in clastic supply. The earlier initiated westerly tilt of the basin also became more pronounced. All these events brought the carbonate sedimentation to a total halt. The increased clastic supply also resulted in a significant progradation of Miocene shelf at places up to 80 km (***Chinchin Formation***).

The northeastern part covering Central Graben is represented by is shaly facies with minor carbonates. The northern part to the north of Diu fault is also represented by shaly facies. In eastern side of Eastern Homocline, carbonate facies are mainly developed. In central part there is development of some clastic/shaly facies. In southeastern part of the area, thick carbonate facies were developed with patchy argillaceous facies. The clastic input direction was from the northeast, as seen in the wells of Central graben and Mahim arch area in the northeast (Mohan, M and P. Kumar, 1982). The lithofacies variations are shown in both west-east and north to south sections with geological age (Figures 10.mu.12).

Deposition in Tectonic Blocks

Saurashtra Basin and the Shelf Margin Basin appear to have undergone partial starvation due to silted conditions caused by Diu arch -Bombay High Horst. There are transgressive-regressive cycles corresponding to major eustatic changes. During the period, Ratnagiri arch acted as an effective barrier for the southward dispersal of clastics brought into North Ratnagiri block. The Middle Eocene-Lower Oligocene carbonates (Bassein Formation) occupy the entire platform area except Bombay High. The Upper Oligocene-Middle Miocene mudstones are the thickest, the carbonates constitute localised foram banks in a clay filled regime. Upper Oligocene mudstones are dispersed in maximum area in the platform and have caused shrinkage and fragmentation of the carbonate ramps as reflected by the Aliabag Formation (Biswas, S.K., and Deshpande, S.V., 1983). The Lower Miocene polycyclic carbonates of ***Bombay Formation*** are essentially restricted to platform areas.The DCS area generally has carbonates with Early Eocene to Middle Miocene shales.

Surat Depression

It forms the southward extension of the Cambay Basin and to the west it is separated from Saurashtra Basin by Diu Arch. An arm of this depression extends far south in to Panna-Bassein-Heera block (Central Graben) and further south into Ratnagiri block (Vijayadurg Graben). The Surat Depression has numerous structural features of different origins like basement-controlled anticlines (Deb S., *et al.*, 1992 and 1997). Surat have the prime depocentres of the clastics of Paleocene-Eocene age. On the platform areas, the clastics are in isolated depressions. The Surat Depression contains deltaic sand bodies encased in shales of Oligocene age (Dhar, P.C. and Pande, D.K. 1978).

Panna-Bassein-Heera Block

This block located east of Mumbai High/Platform and south of Surat Depression has three distinct N-S to NW-SE trending tectonic units which lose their identity in Miocene. The western block is a composite high block dissected by a number of small grabens. The eastern block is a gentle eastward rising homocline. It is a syn-sedimentary sink during Paleogene and Early Neogene.

Bombay offshore region is experienced thick Paleogene and Neogene sedimentation over a predominantly trappean floor/basement. The Early Miocene sequence comprises of two formations, Bombay Formation consisting of limestone with thin beds of shale is overlain by Mahim Formation which consisting of shales with thin limestone bands is with a gradational contact unconformably overlain by ***Bandra Formation***. This Middle Miocene formation consisting of limestone facies with thin shale bands is unconformably overlain by Chinchini Formation of Post-Middle Miocene age. The sedimentary section drilled in the Bombay High field can be classified into three units: Basal sands, lignite, and clays, Middle limestone and alternate shale and Upper shale and claystone.

The lithostratigraphic section appears that the entire Surat Depression consists of shale and claystone. The middle section in the Bombay Platform consists of alternate limestones and shales which grades into a limestone west of Bombay

High in the DCS area and toward the south in the Ratnagiri block (Valmiki D. and T.K.N.Gopala swamy, 1988).

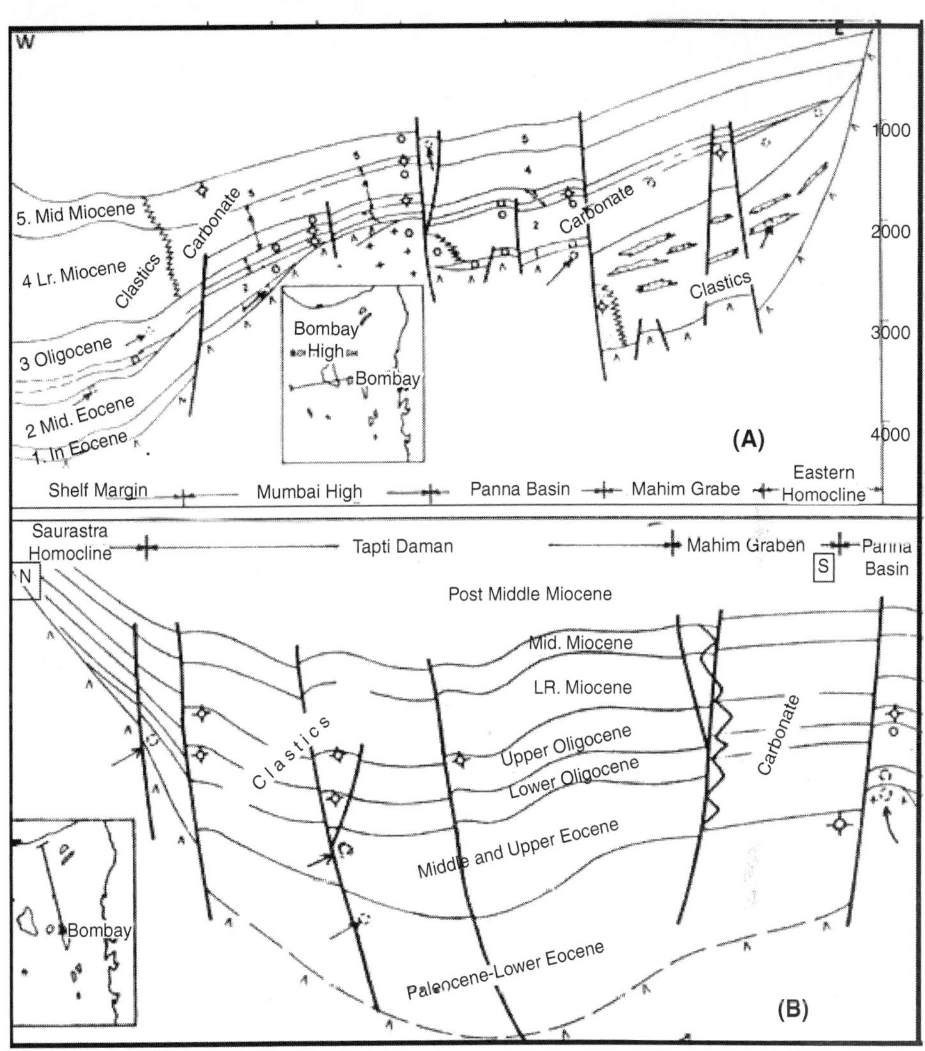

Figure 10.mu.12. Mumbai- (A) E-W Section- (B) N-S Section Showing Sediment Thickness.

Ratnagiri Block

It is the southward continuation of the Panna-Bassein-Heera block. This block is differentiated into four distinct tectonic units by three sets of NNW-SSE trending enechelon fault systems. The western block is termed 'Shrivardhan Horst' and to its east is 'Vijayadurg Graben' which is also a syn-sedimentary sink during Paleogene and Early Neogene. There is a general southward shallowing of this graben. Adjoining is the 'Ratnagiri Composite Block' with a number of 'highs' and lows' and further east, there is a gradual easterly rising homocline called 'Jaygad Homocline' (Rao V K., 1994).

Shelf Margin

Its northern boundary with Saurashtra Basin is indistinct and to its west lies Deep Sea Basin with the western boundary marked by part of a regional ridge 'Kori High'. Except for the deposition of thin carbonates during Eocene, possibly due to paucity of clastic supply into the basin during this period, the block remained a clastic depocenter throughout Oligocene and Neogene times. During post Eocene times the area experienced continuous sinking with varied intensity to accommodate the enormous clastic material that was being brought into Surat Depression by proto Narmada and Tapti river systems and getting dispersed westward into this block.

Petroleum Geology

The continued and sustained exploratory and development efforts in Mumbai Offshore Basin for about five decades have shown remarkable results with an estimated resource base of 9190 MMT. The total in place reserve are 3390 MMT (O+OEG). Of this Bombay High field alone accounts for 2122 MMT of reserve.

Some of the established hydrocarbon fields of the Basin include Bombay High, Bassein, Panna, Mukta, Neelam, Heera, South Heera, Ratna, D-1 fields, Tapti (north,mid and south), C22/24 structures.Hydrocarbon accumulations in Bombay High-Deep Continental Shelf-Heera-Panna-Bassein areas (Bhat, R.K., Pandey, A.K., and Saha, A., 1997).

These are structurally controlled and accumulations are mainly in carbonate reservoirs ranging in age from Middle Eocene to Middle Miocene. Some hydrocarbon zones were also identified in Paleocene to Oligocene plays where as in Tapti-Daman sector Daman (Late Oligocene) and Mahuva (Early Oligocene) plays are well established (Shukla, S.N., 1983). Some are found to be marginal fields in terms of economic sustainability. (Figure 10.mu.13).

Discovery of gas in the very fine grained Pliocene reservoir towards west of the Neogene Shelf margin of Mumbai Offshore block from shelf edge roll over structure opened up a new play.

Source Rocks

The clastic sediments in the lower Eocene to Paleocene sedimentary sequences (Panna Formation) are the principal source rocks across the basin. Thickness of the source rock varies from 30 m to 1000 m. In the Mahim graben, a 400-m-thick sequence in the Panna Formation contains very good/excellent oil-prone effective source-rock facies, Mukta and Heera formations in the east of Panna and Bassein fields have average TOC=2.3 to 15.4 per cent, average S2=3.5 to 50.1 mg HC/g rock, average HI=112-277 mg HC/g TOC (Das, N.C, *et al.*, 1987).

In the Tapti-Daman area, wells located in the eastern flank of Navsari low, contain about 70-metres thick oil and gas prone source-rock layers (average TOC=2.3 to 5.4 per cent; average S2=2.5 to 8.3 mg HC/g rock; average HI=91 to 154 mg

HC/g TOC), and better source rocks are more likely to occur in the Purna graben and west Daman low.

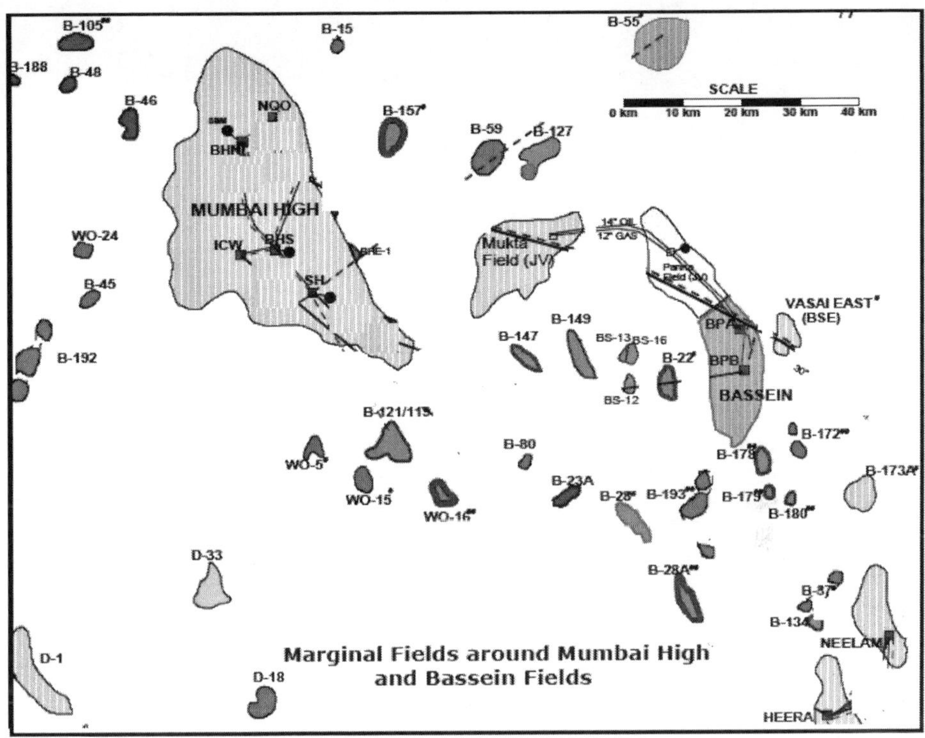

Figure 10.mu.13. Marginal Fields South of Mumbai Area.

The Surat Depression consisting of organic rich shales (Panna Formation- Paleocene to early Eocene and Belapur Formation- Middle Eocene) 3-11 per cent organic carbon and the organic matter is mixed Type II and Type III. Expected oil window is around 3000 metres. In Shelf Margin area several layers of shale/claystone have requisite TOC and have reached the oil window. The oil window from the available geochemical data appears to be between 2900m and 3850m.

Generation and Migration

Generation and expulsion adjacent to the Bombay-High area may have occurred as early as middle Miocene and continued in the Pliocene. Burial-history data indicate peak generation most likely occurred during the late Miocene and early Pliocene. Eocene to middle Miocene sedimentary sequences in the shelf-margin area west of the Bombay-High area show thermal-alteration-index (TAI) values >2.5 (Rawat, *et al.*, 1996). Time Temperature Index plots were also indicated maturity and the onset of hydrocarbon generation occurred at approximately 2.5 Ma. Based on maturity values a source maturity map was made to know relative prospectivity in the western offshore area (Figure 10.mu.14 and 15).

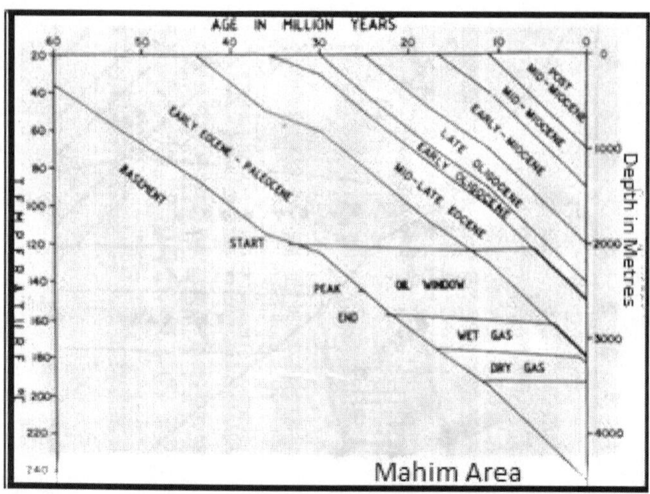

Figure 10.mu,14. Subsidence History and Maturation – Mahim Depression (After Mathur, R B and K M Nair, 1993).

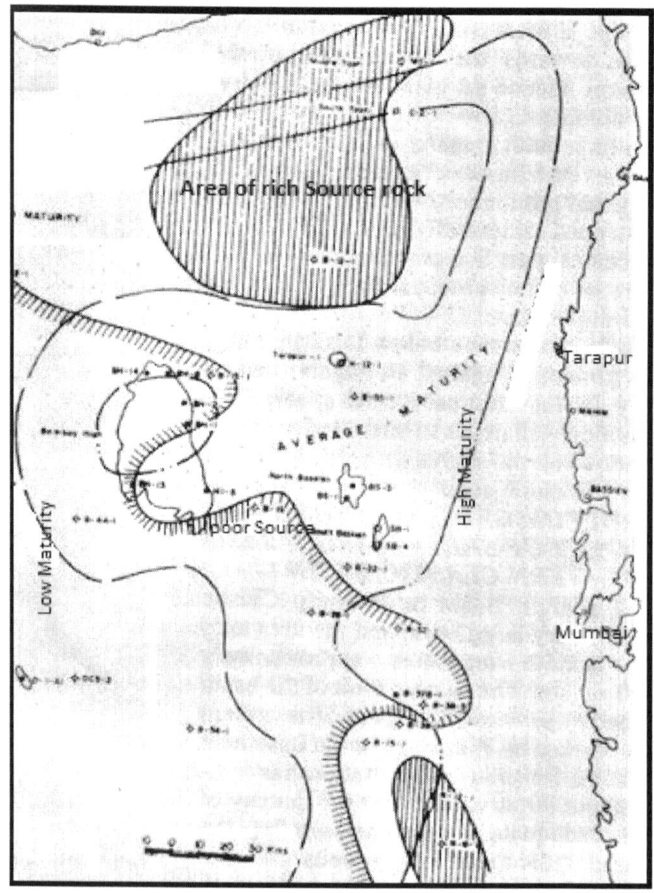

Figure 10.mu.15. Maturity Map of Source Rock-Mumbai Offshore.

Reservoir Rock

Middle Miocene carbonate sections at Ratnagiri, Mumbai high and Diu (Ratnagiri and Bandra formations). The uppermost part has been found to be hydrocarbon bearing at a few places. A sheet like sand deposited over Mumbai High is also proved to be gas bearing in commercial quantity in Mumbai High. Lower Miocene carbonates are huge reservoirs of oil and gas over Mumbai High (Bombay, Ratnagiri).The porosity is mainly intergranular, mouldic, vuggy and micro-fissures and the solution cavities interconnected by micro-fissures provided excellent permeability.

Oligo-Early Miocene sands in the central and mid-eastern part of Surat depression, *i.e.*, Tapti-Daman area, Daman formation. Carbonates adjoining Mumbai High (Panvel formation) deposited under prograding delta conditions proved to be excellent reservoirs Eocene and Early Oligocene clastics of Surat depression (Mahuva Formation). Deposition of thicker carbonate facies over the horst blocks in Panna-Basein-Heera and Ratnagiri blocks (Bassein, Mukta and Heera formations). The Paleocene coarser clastic facies developed within the upper marine shale sequence in areas of Mumbai High, Panna and Ratnagiri (Panna Formation) are proved to be excellent reservoirs in the south-west flank of Panna –Basin platform.

Cap Rock/Seals

In the Mumbai offshore areas, the most likely seals are an extensive series of thick middle to upper Miocene shales. Shale covering the coarser clastic facies in the Paleocene section, widespread transgressive shale overlying the Middle Eocene Bassein Formation, alternation of shale and tight limestone over early Oligocene Mukta Formation, widespread intervening shale layers within Early Miocene Mumbai formation over Mumbai High and in DCS area, post Middle Miocene clay/ claystone of Chinchini Formation over parts of Heera, *etc.* had provided effective seal in the hydrocarbon bearing areas.

Entrapment

The basin has been endowed with a wide variety of entrapment situations with independent four way structural closures, fault closures effective fault sealing, strati-structural features like Paleogene wedges against rising flanks of paleo-highs, mud mounds, carbonate build-ups, unconformity controlled traps, Paleogene and Neogene carbonate wedges against the rising homoclines. Anticlinal structures are also productive in the area of the Bombay High, Surat Depression. Stratigraphic traps and pinch-outs are productive in the graben.

Most of the oil and gas fields are located within the shelfal horst graben complex where the horsts provide the stratigrapbic set up for reservoir and traps like anticlines, unconformities. The hydrocarbon accumulations in many major structural features like Panna, Bassein, Heera, South-Heera lying on the western horst block suggesting a major westward hydrocarbon migration.

Petroleum Systems

The Petroleum System in Mumbai basin is Palaeocene and Early Eocene-Early

Miocene. Palaeocene and Early Eocene sequences are the source rocks and Early Miocene carbonates are the reservoir rocks containing the largest volume of in-place petroleum (Biswas S K, *et al.*, 1994).

Major Identified Play Types

Paleogene Synrift clastics (Paleocene-Lr. Eocene, Panna Fm). In the area western and south-eastern flank of Mumbai High, western flank of central graben, Heera-Panna Block Trap: structural/strati-structural (updip pinch outs). Commercial production from few wells of Heera Field. Areas SW of Bassein Field containing B-80, B-23A prospects and prospect D-33 in DCS have been identified for pre-development studies along with younger pays. In addition commercial flow has been observed in prospects like B-34, B-59, B-127, Panna East wells, *etc.*

Eocene Carbonate Platform (Bassein Formation). The commecial production from several medium sized and marginal Fields–Mukta-Panna, Bassein, Heera, Neelam, B-55, B-173A, South Heera. In addition several prospects in Heera-Panna Composite block and BH-DCS block has been tested hydrocarbon in commercial quantities.

Lower Oligocene Carbonates (Mukta and Heera formations). In the Heera -Panna composite block, MH-DCS platform. Structural traps Fields: Heera, panna, neelam, Basseim, B-55,D-18. ***Oligocene-Lower. Miocene Distal Deltaic-Coastal Play*** (Daman and Mahuva Formations) in the area of Tapti-daman Block. Trap is structural/stratistructural. Fields are South Tapti/Mid Tapti.

Upper. Oligocene carbonates (Panvel and Ratnagiri formations). In MH-DCS Block Traps are structural and fields: B-121/119.

Lower Miocene carbonate (L-III and L-IV reservoirs). In Mumbai High area structural Traps. Source rocks: Paleocene-Eocene (Panna Formation). Fields: Mumbai High, D-1 Prospects identified for development WO-24, B-45 with other pays. ***Lower-Mid. Miocene clastics*** (S1 sands) Area: Mumbai High and adjoining area. Trap: strati-structural. Fields: Mumbai High, recent discovery on BNP prospect. ***Mid. Miocene carbonate*** (L-I and L-II, Bandra Formation) Mumbai High, DCS and adjoining area where traps are structural. Fields include Mumbai High, Heera, D-1

Details of Fields

Mumbai High

Mumbai High is an offshore oil field in water depth of 80m in the Arabian Sea and 160 km west-north-west of Mumbai city. Discovered in 1974 and put on production in 1976. This giant field holds nearly 1700 MMt of oil and about 420 MMt have been produced (~25 per cent of OIIP). The good part of the main oil bearing reservoirs display premature water breakthrough, gas conning with differential pressure depletion. Major oil production comes from carbonate reservoirs in the field. Mumbai High typically built to a plateau, sustained for 4-5 years and then experienced gradual decline. With these reservoir problems the ultimate recovery from the fields is forecasted to be about 30-35 per cent (Narasimham, J.L., *et al.*, 2005).

Having reached to a peak production level of nearly 400,000 barrels of oil per day during 1989-90, the field in subsequent years experienced steep decline in the output. Increase in gas oil ratio (GOR) levels, reduction in reservoir pressures and undesirable flow of water were some of the reasons for decline in the field output. With the implementation of development drilling and treatment of reservoirs, the declining trend was arrested and upward trend in production could be re-established.

Deposited in the mid-Miocene period in a shallow marine environment, the reservoir is essentially composed of limestone layers separated by thin shale/marl or shales/marls sections. They divide L-III reservoir into a number of sub-layers (Bhandari L. L. and Jain S. K., 1984). The primarily inter-granular porosities range from less than 10 percent to greater than 30 percent, with an average between 15 and 20 percent. Permeabilities range from 0.01 milidarcies (mD) to 1,000 mD with averages in the range of 100 to 250 mD. The crude oil is one of the sweetest (highest quality). For instance, Bombay High crude has more than 60 per cent paraffinic. The field is exploited using the gas lifts method, a major artificial lift method used for oil production.

The reservoir characterization, in terms of layer-wise permeability assessment and mapping in partly weathered carbonates, has been carried out in ***Bombay High North*** (Basu, D.N., *et al.*, 1980). A successful attempt has been made to link the coefficient of permeability, determined from core samples, with the log characteristics through the help of transforms using multiple Regression Technique (Bhandari, L.L. and S.K. Jain,1984,). Having calibrated against the permeability interpreted from pressure build up data, the transforms were further used for estimation of layer-wise permeability (Azim, S.A, *et al.*, 1994). The permeability maps have brought out vertical and spatial variation delineating areas which were poorly drained and affected by early water and gas encroachment (Amitabh Pandey., 2003).

With significant advances in drilling related equipment, LWD technology and realization of higher per well productivity, drilling horizontal and multi-lateral wells (Babu *et al.*, 1989, Dikken *et al.*, 1990, Mukherjee *et al.*, 1989, Vij, SK *et al.*, 1998). Now more horizontal/multi-lateral wells were drilled and some sub-optimally flowing conventional wells were side-tracked and most of them re-completed as horizontal wells successfully (Dikken, B.J., 1990).

Besides these, concept of drilling high angle to increase the length of exposure in the pay zone well, was drilled for the first time at a high angle of 80° which resulted in the pay-zone exposure to almost four times (Drilled 344m of 6″ hole section at 80°). Fast activation using surge plug for higher draw down, has contributed to an increase of production.

WO-24

The structure is located towards west of Mumbai High field and in well WO-24#3 I which hydrocarbons were discovered. Tested nine zones in which one zone alone yielded more than 3,000 barrels per day of oil and there are deeper zones where oil and gas both were encountered. The discovery has indicated potential in-place reserves of about 29.74 million tonnes of oil and oil equivalent gas

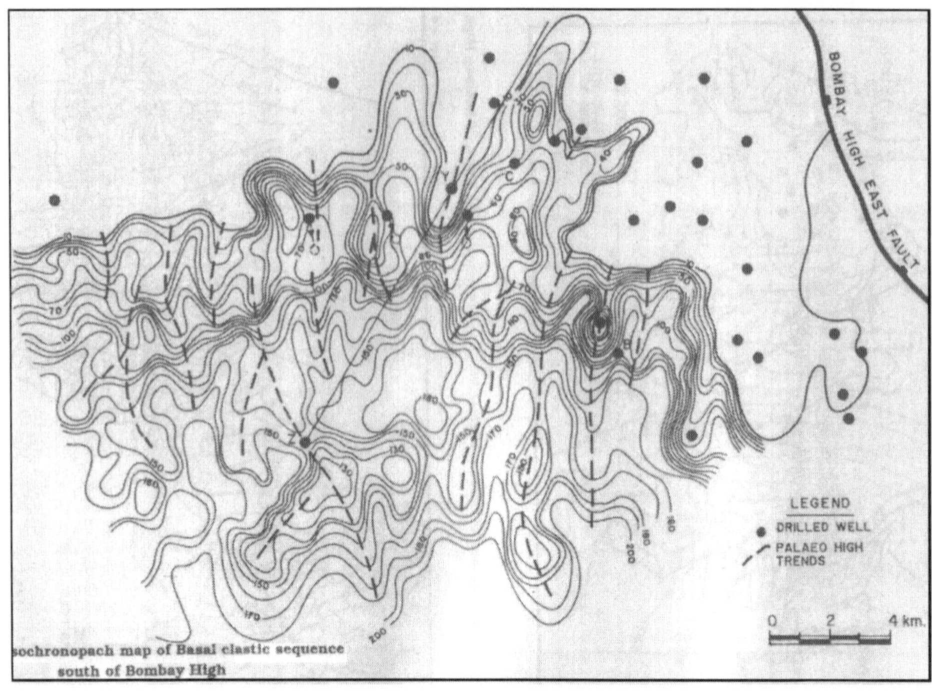

Figure 10.mu.16. Isochrono Pach of Basal Clastics Sequence-South Mumbai High (After Shyam Mohan and S Mangal, 1997).

Bassein

It is located 80 km off the Mumbai coast, the Bassein gas field, also called the Vasai field, was discovered in 1976 and put in production in 1988. Bassein field is giant gas field with 453 billion cubic meters of initial gas in place. This formation is under production since 1988 and the cumulative gas production about 248 BCM. The gas accumulation has taken place in two formations, namely Mukta and Bassein. The reservoir characteristics of Bassein limestone is better than that of Mukta. The two formations are separated by a thin tight limestone-shale alternation of thickness 10-12m (Bassain). In this gas field the Middle Eocene Limestone is a thin (10-15 m) oil column a between thick (~60m) gas cap and bottom water (Figure 10.mu.16).

Using the Geocellular model, the saturation height function along with the petro-physical data (Cherukupalli, P.K., *et al.*, 2003) has improved the volumetric estimates which were in agreement with production history of the field. The model has clearly characterized the heterogeneity of the formations and moulded the fluid distribution more accurately. The South Bassein field that produces 18 mmscmd of gas.

The **Heera and South Heera fields** have more than 172 development wells including 18 horizontal and 6 multi-lateral wells have been drilled from 15 platforms. The peak oil production rate from Heera and South field was 88,625 bopd with an average water cut of 22 percent with increase of average water cut of 50 per cent.

Few wells are under water injection. The field has produced about 41.03 MMt of oil, *i.e.*, 14.5 per cent of OIIP (Narasimham, J.L., *et al.*, 2005). The core analysis of South Heera field shows poor reservoir characteristics with porosity varying from 7 per cent to 16 per cent and permeability in the range of very less to 10 md. Different types rock types like packstone, wackstone, mudstone and shale were identified on the basis of cross plots indicates the heterogeneous nature of the reservoir.

Panna-Mukta- Tapti –PMT (JV)

The Panna-Mukta oil field and the Tapti gas field to the north were discovered by ONGC, who initially operated these fields. Following a privatization policy a Reliance-Enron consortium gained a 25-year lease on the oil field in 1994. Later enron shares the consortium.

The Panna field is located 95 kilometres towards north-west of Mumbai and has an area of 430 square kilometres. It is just north of the Bassein gas field and about 50 kilometres east of the Bombay High oilfield.

The Mukta field is about 100 kilometres northwest of Mumbai, and has an area of 777 square kilometres. Average water depth is 45 metres in the Panna field and 65 metres in the Mukta field. The estimated recoverable reserves of oil were about 38 MMt.

B-173A

It is a small field lies in Heera-Panna-Bassein Block of Mumbai offshore located 25 km north of Neelam field. The structure has a roughly NNW-SSE trend which has a rising trend in NNE.

Mukta Formation of Lower Oligocene age is gas bearing and Bassein Formation of Middle to Upper Eocene age is both oil and gas bearing. Based on log characteristics Mukta and Bassein Formation have been subdivided into two and three sub layers.

Out of seven exploratory wells two wells are oil bearing in Bassein Formation and one well shows oil indication, whereas two wells flowed gas from Mukta Formation after acidization. Subsequently, the field is being developed with five producers completed in Bassein Formation only and producing oil@ 1882 bopd with GOR and 86 per cent water cut.

NW-B-173A

This field is located 5 km northwest from B-173A field which is about 25 km north of Neelam field at an average water depth of 55 m. The well B-173A#8 produced oil at the rate of 2,246 bopd on conventional testing from Mukta pay, resulting in declaration of discovery of this field.

Neelam

The field with two structural culminations which is located in Western offshore of India was proved petroliferous, the reservoir is limestone with shaly intercalations (Dipankar Ghosh., *et al.*, 2005). The field is located approximately 45 km south west from Mumbai. It was discovered in year 1987. Its initial oil in place is 170 MMT with Ultimate recoverable reserve of 85 MMT. The cumulative production was about 26 MMT.

The production was started from this field in year 1990 with the peak production was 95,000 BOPD (Fig-10.mu.17). Since wells started ceasing and production from field declined, wells had to be put on gas lift during early period. It was also witnessed a very high GOR in wells of northern sector with high water-cut. The tight zone got preferentially fractured due to its less ductility which is believed to be a barrier, is providing sufficient transmissibility to fluid through its fracture network (Chandra Prabha Verma., *et al.*, 2006). The field was developed through years with more than 120 wells which include some water injection wells. The porosity is 26.70 per cent uniform but permeability varies very widely few milli-darcies to hundreds of milli-darcies.

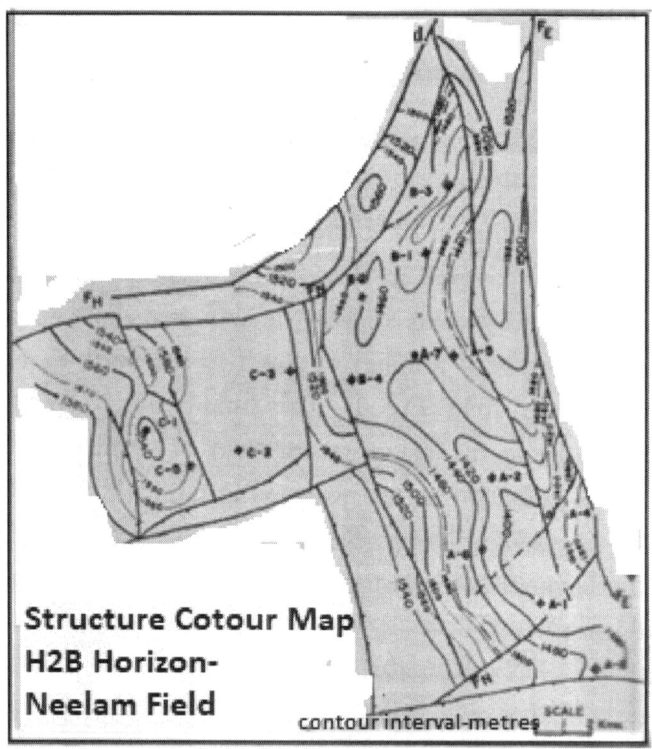

Figure 10.mu.17. Structure Contour Map -H2 B -Neelam Field (After Sen, G, *et al.*, 1993).

Ratna and R-Series Oil and Gas fields

ONGC had discovered the fields in 1979 and created facilities in Ratna R-12, at an average water depth of 45 metres and at 130 kilometres southwest of Mumbai city. The initial in-place reserves of hydrocarbons estimated fields were 58 MMT of oil. The field holds an estimated recoverable reserves of 20 (!) MMt of oil and 2

BCM of gas. ONGC had previously drilled 35 exploration and 9 development wells in the Ratna Fields.

These fields were offered to others and by then had cumulative total oil production of 12.33 million metric tonnes (MMT) and estimated cumulative total gas production of 1,285 million metric standard cubic metres (mmscm) over a project life time of 22 years. Essar Oil and Premier Oil of UK were awarded the field.

B154N-1(B-154N-A)

The exploratory well was drilled in shallow water to a depth of 3,417 m in western offshore basin. A clastic interval 3211-3217 metres in Panna Formation of Late Paleocene-Early Eocene on conventional testing flowed oil @ 72 BPD and 16147 m^3/day of gas through 1/4" choke.

D-1

The D-1 field is located in "Deep Continental Shelf' (DCS) block of western offshore, India. The structure is a prominent NW-SE trending high along the Paleogene shelf-slope break.The field is located about 80 km southwest of Mumbai High field in deep continental shelf at a water depth of about 90m (Figure 10.mu.18). The field was discovered in 1976 with the first exploratory well was drilled to a depth of 3,245m. D-1 field was later named as "NB Prasad field".

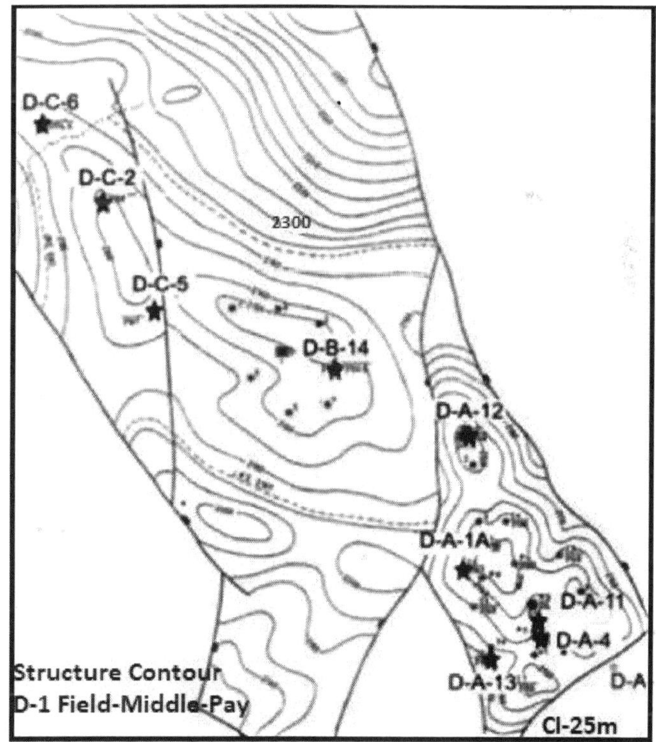

Figure 10.mu.18. Structure Contour Map – Middle Pay D-1 Field (After Maji., A K, *et al.,* 2013).

The Early-Middle Miocene carbonates belonging to Ratnagiri formation constitute the main reservoir. Major accumulations are in the carbonates of Early Miocene besides some accumulations found in Middle Miocene carbonates. The oil bearing limestone facies occur at an average depth of 2150m to 2450m. The in-place oil reserves are about 53 MMt.

In 1976, the second well DCS-2(D-C-2) was drilled in the north-western part and proved to be oil bearing in Middle Miocene and Late Oligocene. Wells were drilled in two other culminations were found dry. Later the well D-A-4 drilled south-west was proved all the pay zones in lower Miocene limestone, flowed oil in commercial quantities. Further in a well (#14) oil was discovered in Ratnagiri Formation. These finds expected to bring a total 30-35 million cubic metres (1 billion cubic feet) of gas a day. Other four deep sea satellite gas discoveries have 617 billion cubic feet of reserves and can produce gas. D-34 discovery estimated to hold recoverable reserves of 1.4 trillion cubic feet.

C-1/North Tapti

The field was discovered in 1984 and in recent well new oil and gas pool in Mahuva Formation in the interval 1102 – 1100.5 m, produced Oil @ 1105 BOPD and gas @ 1,01,203 SCMD through ½″ choke. Further, in the selective intervals tested in the zone 852 to 771 m produced gas @ 1,46,138 SCMD through ½″ choke. (Upadhyay H., *et al.*, 2000).

Table.10.mu.1. Drilled Prospects and Fields-Mumbai Offshore

Mumbai Offshore	*Area-Skm*	*Year*	*Wells*	*Remarks*
Bombay High	1500	1974	580	Gaint field
Bassein	214		35	Gas field
Neelam			120	
Heera South			175	
Panna (JV)	157		38	PMT-JV
Mukta (JV)			8	
North Tapti			-	
South Tapti (JV)	72		-	
Mid Tapti (JV)	40		-	
Ratna			45	R-series to Pvt
Bombay High East	76			
Dahanu (B-12)	69		-	
B-174			3	
B-178		1984	4	
B-179			2	
B-180			2	
B-188			1	
B-192			5	
B-37j	17		5	

Mumbai Offshore	Area-Skm	Year	Wells	Remarks
B-45			-	
B-46			3	
B-48		02/1984	1	
B-51	16		3	
B-55	60			
B-57		08/1981	5	
B-58				
B-59			6	
B-80			3	
BS-12			1	
BS-15			1	
C-1		02/1984		
C-22			2	
C-23			7	
C-24			7	
C-26			4	
C-37			1	
CA			4	
CD-1		05/1986	4	
D-1-14				
D-12	22	12/1982	2	
D-18			6	
DCS	12			
D-I			10	NB Prasad Field
KD-1		05/1984		
R-10			3	
R-12	36			
R-13		03/1983	5	
R-7	23	5/1980	5	
R-71		03/1986	1	
R-7A			3	
R-8		02/1983	1	
R-9	20		4	
SD-1			3	
SD-14			1	
SD-4			3	
WO-15			2	
WO-16			4	
WO-5			7	
NELP				

Mumbai Offshore	***Area-Skm***	***Year***	***Wells***	***Remarks***
MBOS51NAA				
MBS051NBA-A				
MBS-NBA-1				

Rajasthan Basins

The formation of rift basin was in response to global tectonic processes that resulted in separation of the Indian plate from Gondwana during Jurassic and Cretaceous times (Coulson, A. L., 1933). The growth and geometry of Mesozoic to Tertiary basins have largely been controlled by NW-SE and NE-SW trending fracture systems (Figure 10.ra-1).

It is a part of the Indus foreland which is divided into basins like, (i) Jaisalmer basin (JB) on the northwestern slope of the Jaisalmer-Mari basement arch, (ii) Bikaner-Nagaur (BN) basin on the northeast flank of the arch, (iii) Shahgarh-Miajlar (SM) basin southwest of the arch, and (iv) Barmer-Sanchor (BS) basin south of the arch (extension of Cambay basin) (Roy, A.B. and Jakhar, S.R., 2002). These Phanerozoic basins are along with alkali magmatism at Mundwara, Sarnu and Tavidar (Bhandari A., 1999).

It forms the eastern flank of Indus geosyncline and comprises the sedimentary tract to the west and northwest of Aravallis upto Indo-Pakistan border. This pericratonic basin also forms a part of the great Thar Desert. Total basinal area of Rajasthan: 1,36,000 sq. km. Basin wise distribution: Barmer-sanchor: 11,000 sq. km, Bikaner-Nagaur: 70,000 sq. km. Jaisalmer: 45,000 sq. km.

The total ***resource in Rajsthan basins*** have been prognosticated as 380 MMT (O+OEG). In Jaisalmer Basin 2.57 BCM of in place gas reserve have been established. Estimated in place reserves in Bikaner-Nagaur and Banner basins are 94MMT of oil and 116 MMT of O+OEG respectively.

Exploration History

The Baghewala Structure was identified during 1985. Further 2D seismic survey was carried out in the area during 1992 by OIL. The first well, Baghewala-1, drilled in 1991 on the structure and discovered heavy oil in rocks of lnfra-Cambrian age.

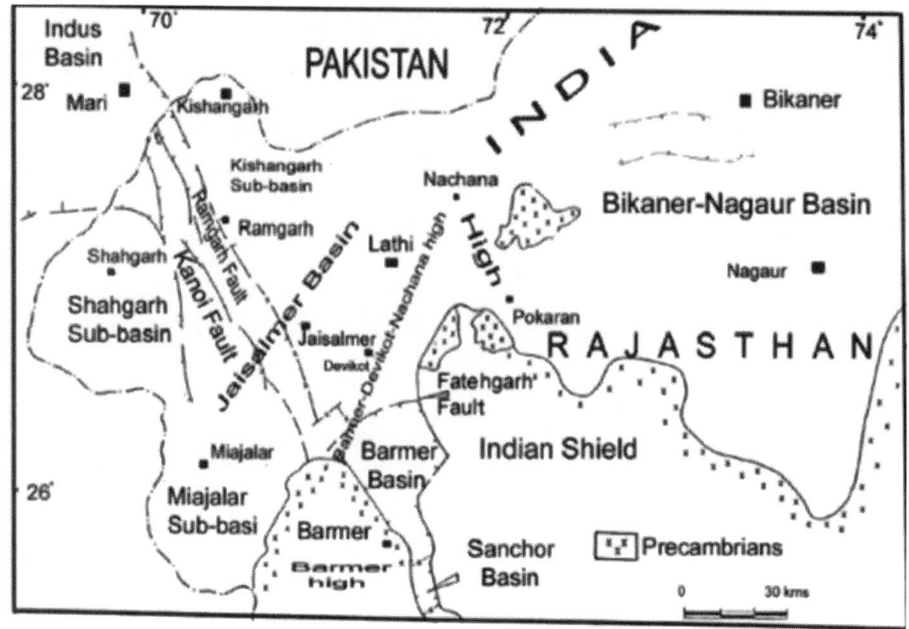

Figure 10.ra.1. Tectonic Map Rajasthan (After after Misra, *et al.*, 1993).

During pre-NELP regime, the Production Sharing Contract (PSC) for the block RJ-ON-90/1 was signed in 1995. In 1997, Cairn acquired 100 per cent of the exploration interest. It was the discovery of oil in Guda structure by Shell which indicated the hydrocarbon potential of the basin. Recently, successes in striking oil in the fields, Mangala, Aishwarya, Kameshwari, Saraswati and Raageswari has established its petroleum potential. So far oil and gas presence were established in 37 structures in the area. The assets of Cairn were taken over by Vedanta group. The geophysical data include 2D, 3D Seismic, Gravity and Magnetic. About 313 wells were drilled in the basin (Figure 10.ra.2 and 3). In the NELP regime OIL has succeeded in getting a filed and named as Punam in 2010.

Tectonics

The tectonic evolution of Rajasthan Basins took place with breaking of Indian Plate from southern continent during Jurassic, collision of Indian Plate with the Asian Plate from Eocene onwards. The Jurassic basin of Kutch and Jaisalmer in Rajasthan resulted from separation of the Indian continent from eastern Gondwana. These basins are the Jaisalmer, Barmer, and Bikaner-Nagaur basins. The uplift of Sind-Baluchistan fold belt resulting in filling up of the Indus shelf (Naqvi, 2005).

The **Jaisalmer basin** extends over 30,000 km^2 in southwest Rajasthan. It is mainly the western dipping eastern flank of Indus Basin which extends to the Mari region of Pakistan. (Datta AK, 1983). A pronounced NW-SE trending regional step-faulted Jaisalmer-Mari high zone traverses the center of the basin. The basin is controlled by wrench-fault tectonics (Misra, *et al.*, 1993) and is divided into four tectonic elements;

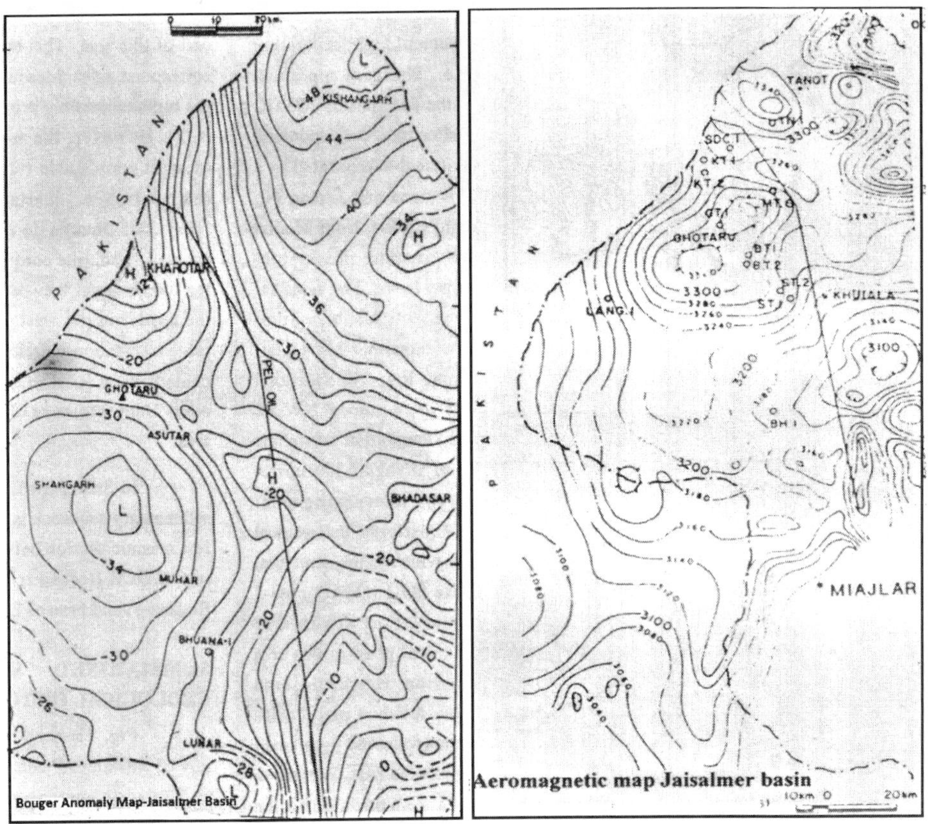

Figure 10.ra.2. Bouger Map Rajasthan (After Mukherjee, MK, *et al.*, 1995).

Figure 10.ra.3. Aero Magnetic Map Rajasthan (After Mukherjee, MK, *et al.*, 1995).

the Mari-Jaisalmer high shown by the Kanoi and Ramgarh faults, the Shahgarh sub-basin, the Miajalar sub-basin and the monoclinal Kishangarh sub-basin (Pareek, 1984). It is separated from the Bikaner-Nagaur basin by the Pokaran-Nachana high to the northwest and from the Barmer basin by the Barmer-Devikot-Nachana high in the South.

The lowermost horizon in the Jaisalmer region is the Lathi Formation, which is of Jurassic age. This horizon comprises a fluvial sequence of conglomerate and sandstone. The sandstone comprises silicified land plant fossils. After deposition of the Lathi Formation, a marine transgression occurred in the basin represented by the Jaisalmer Formation, followed by the Baisakhi, Bhadasar, Pariwar and Habur Formations. The Tertiary sequence in the basin is represented by the Sanu, Khuiala, Bandah and Shumar Formations (Das Gupta, SK., 1975).

The **Bikaner-Nagaur basin** is situated in the northern part of Rajasthan and the Tertiary sedimentation began in the basin with the Palana Formation, which was deposited during the Paleocene. The Marh Formation indicates encroachment of the sea during the Upper Paleocene to Lower Eocene. The Jogira Formation of Lower to Middle Eocene age succeeds this. After the withdrawal of the sea in Eocene, the Kolayat Formation was deposited under fluvial, arid and eolian conditions (Ghosh,R.N.,1983).

The **Barmer basin** lies in the southeast of the Jaisalmer basin, and it forms a north-south graben over 100 km length The Cambay rift basin extends northward through the Sanchor and Barmer basins. It is an Intra-cratonic basin. Prominent tectonic elements in this sub-basin are Devikot-Nachna Uplift with Fatehgarh Fault which marks the northern limit and an east-west trending Tharad Ridge marks the southern limit which also separats it from Cambay Basin.

Another cross trend along Sarnu Hill Fault divides Barmer-Sanchor Sub-basin into Barmer Depression in the north and Sanchor Depression in the south. It is a Tertiary basin and NW extension of Cambay Basin towards south and overlies the SE downwarp of Jaisalmer Basin (Figure 10.ra.4). The Pokhran High separates the Bikaner-Nagaur Sub-basin from Jaisalmer Sub-basin which is a Pericratonic Basin. Devikot-Nachna uplift separates Jaisalmer Sub-basin from Barmer-Sanchor Sub-basin

Barmer-Sanchor Basin

Exploration History

The block RJ-ON-90/1, is spread over 3,111 km^2 west of Barmer. Cairn India is the operator with 70 per cent participating interest and ONGC, has the remaining 30 per cent interest. The block consists of three contiguous development areas. The annual oil production from the block was 62 mmboe, at an average of 169,609 boepd of crude oil and gas. Gas production from the Rageswari field known to an average of 27 mmscfd with total production of 9.9 bcf. The cumulative production of the block was about 370 mmboe.

Mangala, Bhagyam and Aishwariya (MBA) are the key producing fields which have with 2.2 billion barrels of hydrocarbons in place and gross recoverable oil reserves and resources of over 1.0 Billion boe. This includes proven plus probable (2P) gross reserves of 636 mm boe with a further 308 mm boe or more of EOR resource potential.

Sedimentation and Stratigraphy

It contains predominantly terrestrial sediments ranging from Jurassic to Cretaceous in age. The presence of Cretaceous-Paleocene volcanogenic sediments are in the graben. The Paleocene to Eocene period was marked by deposition of coastal, marine, shallow-water sediments. The basal unit of the Barmer basin is the Fatehgarh Formation. This Formation comprises mixed siliciclastic, carbonate and phosphorite rocks of Cretaceous age with glassy balls, highly magnetic fine dust and micro-brecciated matrix and suggested the volcanic source (Mathur, *et al.*, 2005).

The Malani igneous suite and metamorphics constitute the basement over which the Mesozoics and Tertiaries were deposited. The Proterozoic sequence is represented by Birmania and Randha formations followed by the Lathi and Sarnu formations of Mesozoic age. The syn-rift sediments, Barmer Hill and Fatehgarh Formations are exposed at Barmer and Fatehgarh. The Fatehgarh formation constitutes the base of Tertiary sequence overlain by Barmer formation having three sub-units named as Akli Formation, Mataji ka Dungar and Kapurdi members.

AGE			BARMER BASIN	JAISALMER BASIN	BIKANER-NAGAUR BASIN
CENOZOIC	HOLOCENE		Dune sands, alluvium gravels Utarlai Formation	Shumar Formation	Sand Dunes Mar Formation
	PLEISTOCENE				
	PLIOCENE				
	MIOCENE				
	OLIGOCENE				
	EOCENE	Priabonian		Bandah Formation	Jogira Formation
		Bartonian			
		Lutenian			
		Ypresian	Kapurdi Formation Mataji ka Dungar Fm. Akli Formation	Khuiala Formation	
	PALEOCENE	Thanetian	Barmer Formation Fatehgarh Formation	Sanu Formation	Marh Formation
		Montain			
		Danian			Palana Formation
MESOZOIC	CRETACEOUS	Santonian			
		Coniacian			
		Turonian			
		Cenomanian			
		Albian			
		Aptian		Habur Formation	
		Neocomian	Sarnu Formation	Pariwar Formation	
	JURASSIC	Tithonian		Bhadasar Formation	
		Kimmeridgian		Baisakhi Formation	
		Oxfordian			
		Callovian	Jaisalmer Formation	Jaisalmer Formation	
		Bathonian			
		Lias	Lathi Formation	Lathi Formation	
	TRIASSIC				
PALEOZOIC	PERMIAN				Badhaura Formation
	CARBONIFEROUS				Bap Boulder Bed
	DEVONIAN to ORDOVICIAN				
	CAMBRIAN		Birmania Formation	Birmania Formation	Nagaur Formation Bilara Limestone
NEOPROTEROZOIC			Randha Formation	Randha Formation	Jodhpur Sandstone
			Malani rocks/Basement Complex	Malani rocks/Basement Complex	Malani rocks

Figure 10.ra.4. Startigraphy-Rajastan Basins (After Bhandari., A,1999).

The Sarnu Formation, correlatable with the sandstones of Jodhpur Group, exposed on the eastern margin of the basin, comprises thin and fining-upward sand bodies with intervening red siltstone. The Lathi Formation exposed at the northern periphery the basin, comprises medium to coarse, fining upward fluvial sands with fossils.

The Barmer Hill Formation comprises sandstone and clast supported conglomerates, exposed along the western boundary of the basin and rest unconformably on the basement. These represent rapid deposition in an alluvial fan environment with source from Malani Rhyolite.

The Fatehgarh Formation exposed at the northern boundary of the basin shows a mixed sand and mud tidal-flat environment. It comprises conglomerate at the base, overlain by sandstone. This in turn, is overlain by ferruginous phosphatic sandstone of Bariyara Member which form the base of Mataji-ka-Dunger Formation. These post-rift sediments are exposed at the northern and western margins of the basin and the fluvio-deltaic sandstone shows visible porosity with no secondary infill. The base of the sequence exposed at the south of Fatehgarh comprises mostly claystone. This is interpreted as pro-deltaic, delta-slope/delta-mouth deposition.

The Akli Formation exposed at the central part of the basin, overlying the Mataji-ka-Dunger Formation comprises bentonitic claystone, grey bituminous clay-stone, lignites and light yellow claystone.

Petroleum Geology

Source

Though it was opined that, the updip migration northwards from Cambay basin is considered to have sourced the hydrocarbons encountered in the basin, thick sediments of Tharad formation having good source rock characteristics constitute a significant source sequence within kitchen areas of the basin itself. Average TOC content varied between 1-10 wt. per cent for different aged source rocks with Type II and III Kerogen. The oil from Mangla field is waxy crude with API value varying from 25-30° and sulphur content is less than 0.2 per cent.

Reservoir

In Barmer basin oil occurs in the sandstones of Paleocene Fatehgarh and Barmer formations. The reservoir is characterised by stacked channel sand and so the vertical connectivity of the channels is responsible for the performance of the reservoir. The high-quality fluvial reservoir has good permeability and porosity, and is of shallow depth with low pressure resulting in thicker oil. The reservoir contains 150m oil column categorised as medium gravity and is of sweet grade with a low sulphur content of about 0.1 per cent.The crude has a high pour point and viscosity due to its waxy nature and its highly paraffinic nature.

Potential oil and gas bearing reservoirs range from the upper Cretaceous volcanic basalts to the Eocene-Oligocene arenaceous units. Mesozoic sediments are also possible reservoirs in the northern part of the basin.

Entrapment

Hydrocarbon entrapment is established in structural as well as strati-structural traps with possibility in updip pinchout prospect towards the basin margin. The roll-over anticlines developed on the boundary faults and fault closures against the transverse faults.

Description of Fields

Mangala

The field is located in northern part of the basin. Though the structure was identified based on 2D seismic data, the proper imaging of faults by 3D survey clarified the places of reservoirs. About 150 development wells were drilled and completed with 96 producers and 33 injectors, which include11 horizontal wells. The EOR, technique is polymer flood. The Mangala oilfield is estimated to have 1.1 billion barrels and recoverable reserves of 256-380 million barrels. An additional 100 million barrels can be produced using enhanced oil recovery techniques.

Bhagyam

It is the second largest field in the block. A total of 62 development wells have been drilled with 21 producers and four injectors operational.

Aishwariya

The field is the third largest discovery in the area. Following an assessment of higher production potential and design optimisation due to increased reserves and resources. All the 20 infill wells along with ramp up of injection and sector-wise voidage management have helped to optimise reservoir management and increase recovery.

Raageshwari Deep Gas Field

The marginal field is to supply gas to meet the energy requirements for development of fields. The average gas production from field was increased to 27 mmscfd from 16 mmscfd. Hydro-fracture programme was focused on exploiting this multilayered deep tight gas reservoir with the objective of maximising recovery from each well.

The hydro-frac job has resulted in higher well productivity. In Rageswari North#1 the interval 3535-3698m the sand is of Deccan age produced oil @50-90 bopd 12/64″ chock. Another interval 3382-3415m produced oil @30-80 bopd through 20/64″ chock. In Rageswari South #1 well the interval 2057-2067 m flowed oil @ 41 bopd through 8/64″ choke after carrying out fracture job it has increased to 350 bopd.

Saraswati

The field commenced production in 2011, is currently producing at a rate of 250 bopd. The well Saraswati #4 produced oil @ 165 bopd through 10/64″ chock from the Basement.

Bikaner-Nagaur Sub-basin

Exploration History

Exploration in the basin strated with the discovery of heavy oil in the Baghewala well from the clastics of early Cambrian age. Further oil was struck in the Nanuwala structure in this basin from the Hanseran Evaporite formation at an approximate depth of 960 m.

Several structures have been mapped and drilled in the basin without success. There is indication of higher concentration of lighter gases like CO_2, N and He, in the gas from the fields.

Sedimentation

It is a Paleozoic basin and it is floored by Malani rhyolites as basement and has a thin sedimentary cover (~ 1500m) of lower Paleozoic overlain by Mesozoic and Tertiary sediments (Agarwal, V., 1984). These sediments, overlain by a thin Tertiary/Mesozoic sediments. The Paleozoic sediments include evaporites and carbonates, and are similar to more NW Pakistan.

Facies analysis of Cretaceous Baisakhi and Bedesir formations indicates south-westerly drainage system and development of a delta complex in this part of the basin. The Cenozoic sedimentation in the basin began with the coal-bearing Palana Formation that was deposited during the Paleocene in subtropical swampy conditions on the continental part. Marine sedimentation indicates encroachment of the sea during the Upper Paleocene to Lower Eocene.

Petroleum Geology

The Mesozoic is the main prospective target in the Basin, particularly the Cretaceous Pariwar, Goni, and Path formations, which are the main producers in Pakistan. The second proven prospect is Late Proterozoic and Permian. So far exploration had been focused on the structural closures. However, possibly most of these structures are late generation structures.

The data indicate that the Cambrian sediments have generated both oil and gas. Oil is discovered in Cambrain dolomites in the well Karampur-1. Heavy oil is present in multiple zones within Cambrian sediments (Bilara Limestone and Jodhpur Sandstone) in Baghewala, Tavriwali and Kalrewara structures. Presence of light oil has been reported in an exploratory well drilled in an exploration block RJ-ON-90/5.

Source

The, carbonates of Bilara formation of lower Cambrian age constitute the source facies and the Indus foredeep in Pakistan is the main kitchen area and the generation of hydrocarbon took place in post-Miocene time. The migration is later than that, so the lighter gas fraction could reach the distal foreland margin in Rajasthan while the proximal structures in Pakistan are with gas and/or oil.

In basin oil is heavy and viscous oil (19° API gravity) is present in infra-Cambrian Bilara and Jodhpur formations. The Bilara dolostones seem to be both source and reservoir rock with evaporite as the cap. The samples of Jodhpur and Bilara formations of Karampur/Badhaura Shales, Upper Carbonate Dolomites, Bilara Shales and Dolomites indicate moderate to high TOC and HI up to 400mg HCg TOC.

Reservoirs

The main reservoirs in this basin are in the Infra-Cambrian Jodhpur, Bilara, Hanseran and from lower-Cambrian carbonate formations. Upper Carbonate Dolomites, Nagaur Sandstones, Jodhpur Sandstone and Bilara Dolomites. Potential source and reservoir rocks have been identified in the Paleozoic sediments. The Jodhpur sandstones are also the reservoir rocks.

Entrapment

The moderate to high basement-controlled anticlinal structures and fault closures. Further, early Cambrian Transpressional fault-blocks, sands developed at the toe-set end of 'clinoform sequence', Tertiary reactivated fault-blocks also better entrapment conditions in the area.

Description of Fields

Baghewala

The Baghewala Structure was identified on 2D seismic data during 1987. The first well, Baghewala-1, was drilled in 1991 and discovered heavy oil in zone at 1104-m within the Jodhpur sandstone of lnfra-Cambrian age (Figure 10.ra.5).

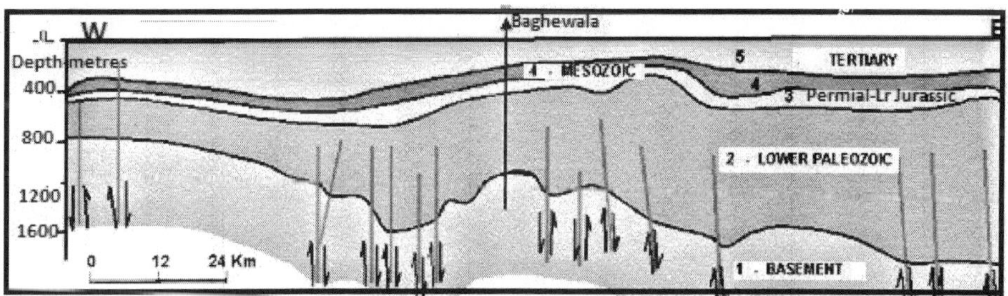

Figure 10.ra.5. Bikaner Nagaur Geological Section.

Four more wells called Baghewala-2,3, 4 and 5, were drilled in the structure and one well, Tavriwala-1 was drilled in the neighbouring Tavriwala structure. A contract was awarded to M/s PDVSA, Venezuela in November' 2002 to carry out an integrated detailed geological and reservoir engineering study. Based on the study, oil in-place estimated was 25 million tonnes (MMT) with recoverable reserve was 3.75 MMT. The entire Baghewala Structure and its adjoining areas were covered by 3D seismic survey conducted.

Punam

OIL has drilled the well in their NELP-VI block in 2012 and found oil at 1230m depth.

Jaisalmer Basin

The Jaisalmer Basin in India and part of Indus Basin in the adjacent Pakistan area. The Jaiselmer-Mari high subdivids Kishangarh in north and Shahgarh in south side. The micro-faunal studies in the Khartar Limestone and Sadewala shale indicate progressive upward swallowing of the depositional area.

The structural style of the basin was controlled by long north-west passing fault trends. Only indications of liquid hydrocarbons noticed in Ghotaru-Dandewala structures (Mitra., P 1993). The Paleo structural analysis reveled, till end of Permian, the basin have gentle slope towards north. Bhuana, Lunar and Shumarwali were initial structural areas (Figure 10.ra.6). By the end of deposition of Pariwar formation these features were nearly peneplained. By end of Goru formation time the normal faults associated with these structural features became prominent (Sinha AK, *et al.*, 1993). The Jaisalmer basin occupies an area of about 45,000 sq. km. The prognosticated hydrocarbon resources are of the order of 380 MMT.

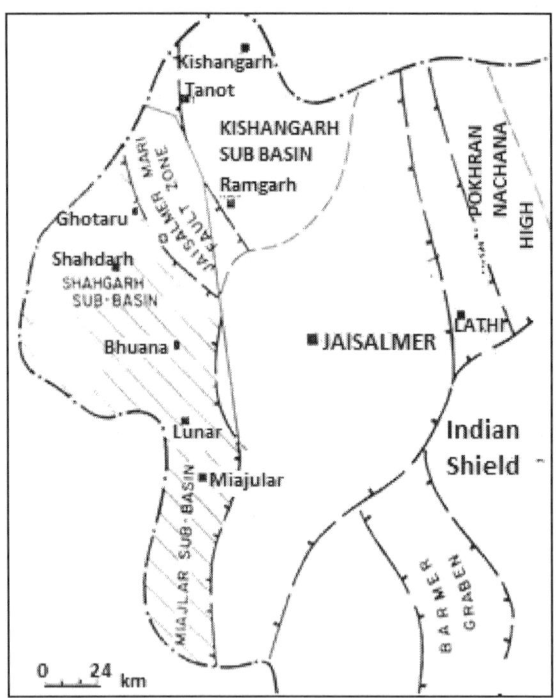

Figure 10.ra.6. Jaisalmer-Basin-Tectonics (After Mukherjee, MK, *et al.*, 1995)

Exploration History

The basin has no commercial discovery of liquid hydrocarbons except some show in Chinnawala Tibba1 and Ghotaru-2. The liquid hydrocarbon shows in these wells is from Cretaceous sequence. Gas has been found in seven structures. The gas from Manhera Tibba from Cenozoic and Chinnawala Tibba from Cretaceous is methane-rich and commercially viable. In the remaining five structures (Ghotaru, Bankia, Bhakari Tibba, Khartar and Sadewala) the gas is nitrogen-rich. Exploration by ONGC and OIL in Jaisalmer basin has resulted in discoveries of several gas fields. Oil India had also discovered natural gas in Jaisalmer sub-basin of Tanot fields.The depth of drilling varies between 1,200 metres and 2,200 meters.

Sedimentation

The basin contains Upper Jurassic to Upper Cretaceous sedimentary section with hydrocarbon finds (Copestake, *et al.*, 1997). Though litho-stratigraphy may indicate different names in both sides of the Indus Basin, the chrono events indicate exploration fairways with source, reservoir and cap facies it is known as "Lower Goru-Sembar- Pariwar" Play in the Jaiselmar Basin. The marine Tertiary outcrops with pronounced unconformity are known around Khuiala-Bandah area (Singh NP., 1984). A maximum sedimentary thickness of 7 kilo metres has been estimated in the basin. It forms the eastern flank of Indus geosyncline and comprises the sedimentary tract to the west and northwest of Aravallis upto Indo-Pakistan border (Figure 10.ra.7).

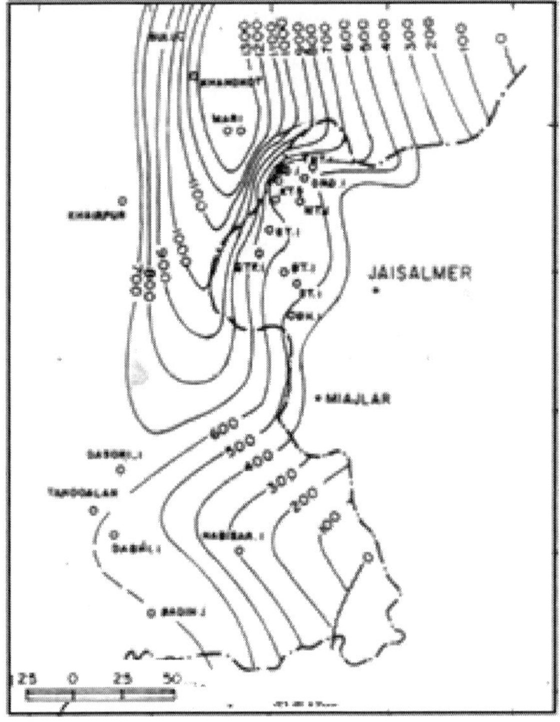

Figure 10.ra.7. Isopach Map-Samber/Pariwar Formation-Jaisalmer Basin (After Mitra, P, *et al.*, 1993)

This is the western-most part and separated from the Bikaner Nagaur basin by the Pokaran-Nachana high to the northwest and from the Barmer basin by the Barmer-Devikot Nachana high in the south. A pronounced NW-SE-trending regional step-faulted Jaisalmer-Mari high zone, marked by the Kanoi and Ramgarh faults that traverse the centre of the basin and divides it into the Shahgarh sub-basin, the Miajalar sub-basin and the Kishangarh sub-basin (Bhushan, S.K., 2000).

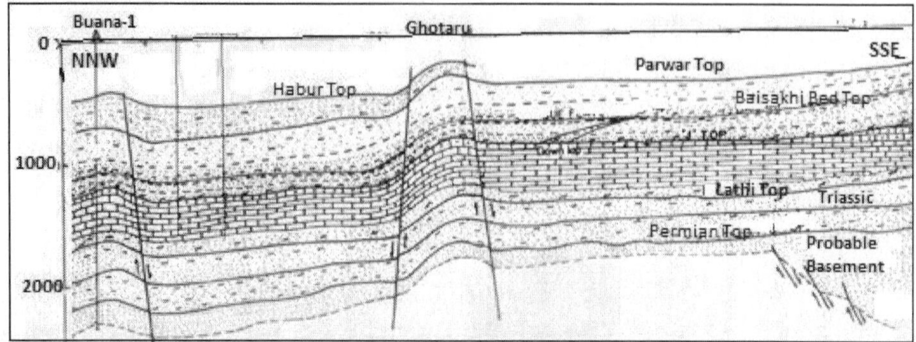

Figure 10.ra.8. Geological Section-Jaisalmer Basin.

Three distinct sequences are identified, based on the basin forming tectonic events. The Proterozoic-early Cambrian succession unconformably overlies the Precambrian basement rocks, while the unconformable upper contact of the sequence with Permian, and has a hiatus span of about 190 Ma. The Proterozoic to early Cambrian rift sequence comprising mainly the thickly bedded sandstone with shale intercalations in the lower part and dolomitic and cherty limestone with shale and sandstone interbeds in the upper part. The first cycle of sedimentation corresponding to Proterozoic-Early Palaeozoic is designated as Birmania/Randha Formation representing deeper water sediments (Figure 10.ra.8).

A prominent orogeny, probably resulting from the collision of older plates led to a major uplift and hiatus in this region. On the peneplained Birmania-Randha topography, the Karampur Formation of Permain age was deposited over the unconformable surface.

A major regression took place during the Triassic and Early Jurassic, which resulted in deposition of Shumarwali Formation and Lathi Formation. The Jaisalmer sediments have been deposited in very shallow water depths. During this period the Bikaner-Nagaur Sub-basin became positive due to differential block movement.

During the Upper Jurassic, a narrow elongated Cambay rift graben was connected to Jaisalmer Sub-basin through Sanchor-Barmer Sub-basin. The resulting clastic sequence representing Baisakhi and Bedesir formations in Jaisalmer Basin continued to be deposited in an oscillating shallow marine environment whereas in Sanchor Depression conglomerates and pebbly sandstone facies was deposited under fluvial regime designated as Serau Formation.

The Lower Cretaceous (Neocomian) Pariwar Formation represents further regression with shallow marine and brackish condition towards the lower part and complete regression with setting in of continental conditions towards the top. A major uplift of the axial belt gave rise to a prominent hiatus ranging in age from Maestrichitian to Danian and brought an end to this marine cycle. The next cycle of sedimentation began with the deposition of clastic sequence of Sanu/Akli/Balutri Formation representing brackish to shallow marine environment probably during early Paleocene and Early Eocene.

The Khuiala and Bandah formations of Jaisalmer Sub-basin, and Mandai and Kapurdi formations of Barmer Depression indicate a somewhat stable condition during Lower Eocene. However, Sanchor Depression witnessed a minor pulse of regression, resulting in deposition of Lower Tharad Formation.

In the central part of Indus Basin, clastic deposition continued under slightly oscillating shallow marine condition represented by Mari and Gaj formations of Oligocene and Lower Miocene.

Petroleum Geology

In Jaisalmer basin detailed geophysical surveys were carried out only on the Jaisalmer-Mari Arch where as the Kishangarh sub-basin and Shahgarh depression were partially surveyed. Fifteen deep wells, including 8 on Manhera Tibba structure which is a gas field on Jaisalmar-Mari Arch and one deep well Bhuwana-1 in Shagarh depression and one well Pupal-1 in Bikaner-Nagaur basin have been drilled. Reserves of gas in Manhera Tibba structure is about 770 million cubic metre. The exploration drilling has indicated presence of heavy oil in haline/carbonates. Baghewala, Nanuwala and Binjybala areas have indicated hydrocarbon occurrences. Eight small lean gas fields have been discovered over the Jaisalmer platform by exploratory efforts of ONGC and OIL. Though oil indications are common in Cretaceous strata drilled on a number of structures, commercial strike has not been materialized so far.

Source

Source Formations in Jaisalmer include Lower Goru, Pariwar, Sembar/ Bedesir - Baisakhi Shales, Karampur/Badhaura Formation Shales, Bilara Shales and Dolomites. The Early Cretaceous Baisakhi-Bedesir and Pariwar shales are considered to be the regional source rock. These shales are gas prone with Type III kerogen matters and are matured in the north-western pan of the basin.

In the basin, the source-rock evaluation suggests that in the deeper part of the basin in the west the source rock development is poor. In the northwest, good development of source rock is observed and the maturity of the source rock is not enough to generate the gases of this maturity. Two potential source rock layers that more or less extend laterally across the basin, wherever source rock development is significant have been identified at late Jurassic and early Cretaceous levels.

Baisakhi-Bhadasar formation is considered the regional source facies with total organic carbon (TOC) up to 4.99 per cent and Tmax of 415-431°C in the Manehara Tibba, Ghotaru and deeper area of Miajlar. Pariwar formation is also reported to have up to 7.92 per cent in most wells. The source-rock development in Sahagarh sub-basin is poor at late Jurassic and poor to marginal at early Cretaceous, whereas in Miajalar sub-basin fair to good source rock development has taken place at late Jurassic only, but it is at shallower depth and has not yet reached the top of oil window maturity (VRo » 0.5).

Hydro-geochemical analysis of formation waters indicate the hydro dynamic activity is fron east to northwest (Bawa, RK., 1993). The genetic correlation of known accumulation of gases in Cenozoic and Cretaceous reservoirs suggests that they have been generated from sediments at higher ***maturity*** and have been trapped at their present locale after long-distance migration.

Reservoir

The Randha and Birmania formations in Miajlar sub-basin and sandstones in Lathi, Baisakhi-Bhadasar, Pariwar and Goru formations provide adequate reservoir qualities. In addition the limestone in the Bandah and Khuiala formations also host gas-bearing reservoirs.

Traps

The other structures drilled were not found to be oil or gas bearing. The main prospects are: structural traps including fault closures in Eocene, Cretaceous and Jurassic in Jaisalmar basin, sand pinch-out and unconformity traps against Cretaceous-Eocene unconformity, delta front sands of Lathi age in Shahgarh sub-basin, probable-reefoidal bodies in Jaisalmer. Anticlinal closures, Fault related closure/traps, Unconformity related traps, and wedge outs become litho-stratigraphic traps. Jurassic to Eocene sequences in the Jaisalmer Basin contains suitable source, reservoir and cap rocks.

Hydrocarbon Fields

Chinnewala Tibba

The structural area of the field is about 73 sq km. There are two fields CT-1 and CT-2 discovered by ONGC with main pay zone (gas only) with 4 wells drilled of which one was dry. The producing horizons are in Pariwar and Lower Goru Formations of Cretaceous age. The first well drilled in 2004, flowed Gas @60,829 m^3/d-5mm bean with water (9.7gpl) in the interval 1901-1909 and also flowed Gas @53,637 m^3/d-5mm bean with water (19.7gpl) in the interval. Production testing of second well CT-2 the zone at 1900 mts (Pariwar Formation) produced gas at commercial quantities (60,000- 70,000 m^3/day) in wells CT-1 and CT-2. The lower Goru Formation is also producer in both the wells with a lower potential. Formation pressures are hydrostatic. The Inplace Reserves of 1.90 MMt O+ OEG.

Table 10.ra.1. Rajasthan Oil and Gas Fields

Field	*Basin*	*O/G*	*Block*	*Year*	*Wells*	*Operator*	*Remarks*
Manhertibba	Jaisalmar	G			8		
Khartar		G					
Bankia		G					
Bakhi Tibba		G					
Gotaru		G			2	ONGC	
Chinnewala Tibba		G		2004		ONGC	
Aishwariya	Barmar		RJ-ON-90/1-2002	2004	81	Cairn	

Field	Basin	O/G	Block	Year	Wells	Operator	Remarks
Bhagyam				2004	62		
Bhagyam South				2005			
DP-1				2015			Baramar Fm
GSV				2005			
Guda (G-R-F-1)				1999			
Guda South				2014			
Kameshwari				2003			
Kameshwari West				2014			
Mangala				2004	148		
Manhertibba		G					
N-C-West				2005			
N-I				2005			
NL-2				2015			100-120 b
NR-3-2100				2014			
Raag Deep				2005			
Raageshwari				2003			
Raageshwari-South				2013			
Raageshwari-East				2008			
Saraswati				2001			
Saraswati SW				2015			
Shakti				2004			
S-L-1				2015			
V2YChannel				2014			
Vandana/NR-2				2005			
Vijaya/NR-1				2004			
Bhagewala	Bikaner		19.5° API	1991	6	OIL	Heavy oil-
Tavriwala				1991	1	OIL	
Punam			RJ-ONN-2004/2			OIL	
SSF-2			RJ-ON/6			Focus	
SGL							
SPF-1							
SSG-1						"	
Baghewala				1991	6		
Chinnewala Tibba	BN	G					
Pupal	BN						
Sadewala		G					

Category II

Andaman and Nicobar Basin

Introduction

The Andaman Islands comprise South, Middle and North Andaman Islands. The Andaman-Nicobar basin located in the south-eastern part of the Bay of Bengal occupies an area of 8,270-km^2 onland and about 32,000-km^2 offshore beyond 200m isobaths (Curray, J.R. and Moore D.G., 1971). This basin forms a part of a major island arc system (Figure 10.an.1). The Andaman area is arcuate shaped chain of islands linking the Sumatra Archipelago to its south and the Arakan Yoma in the north.

Status of Exploration

Geological Surveys

Exploratory activities in the basin by ONGC started in 1959 with geological mapping. Since then, 5,433-km^2 of area was covered by reconnaissance, semi-detailed and detailed mapping, and 646 Line km. by traversing (Figure 10.an.2). To supplement this surface geological data with sub-surfce, 14 shallow wells have been drilled on the Neill and Havelock Islands (Chatterjee, P.K., 1967).

More than 3,000 m of out crop sediments ranging in age from Cretaceous to Recent are exposed (Tipper, G.H., 1911). Three major unconformities have been identified. The main Andaman and Nicobar islands expose mostly rocks of Cretaceous-Oligocene age and the Neogene strata are exposed mainly in the Ritchie's Archipelago (Brunnschweiler, R.O., 1970).

Surface geology indicates N-S structural trends with mainly tight folds. The structures of the main Andaman Islands are primarily of pre-Neogene age. The structures developed in the Neogene sediments are relatively gentle, and structural continuity can be traced in the adjoining islands. A series of mud volcanoes with thin film of oil occur on the main island.

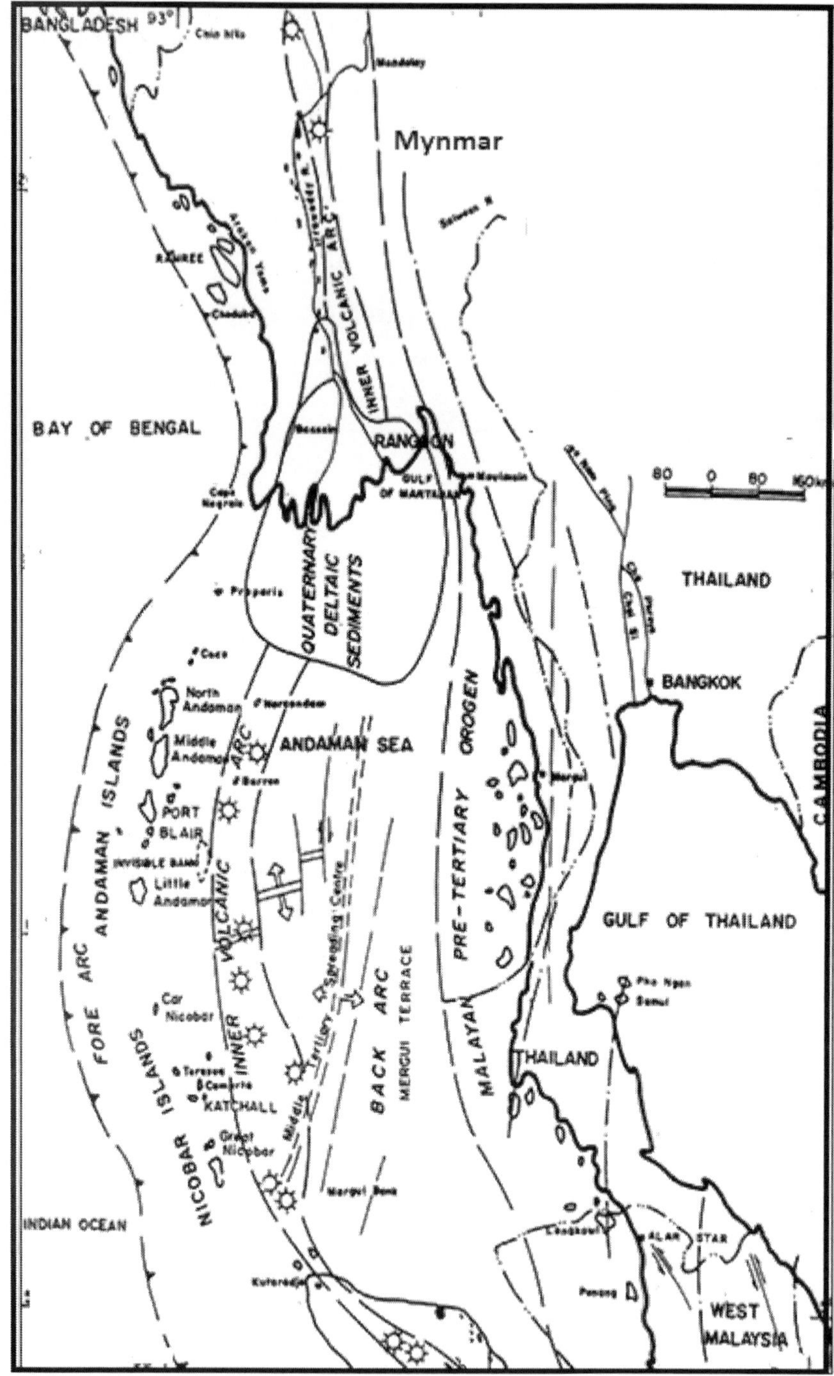

Figure 10.an.1. Andaman and Nicobar Basin-Eastern Tectonic Elements (After Biswas, S K, 2012).

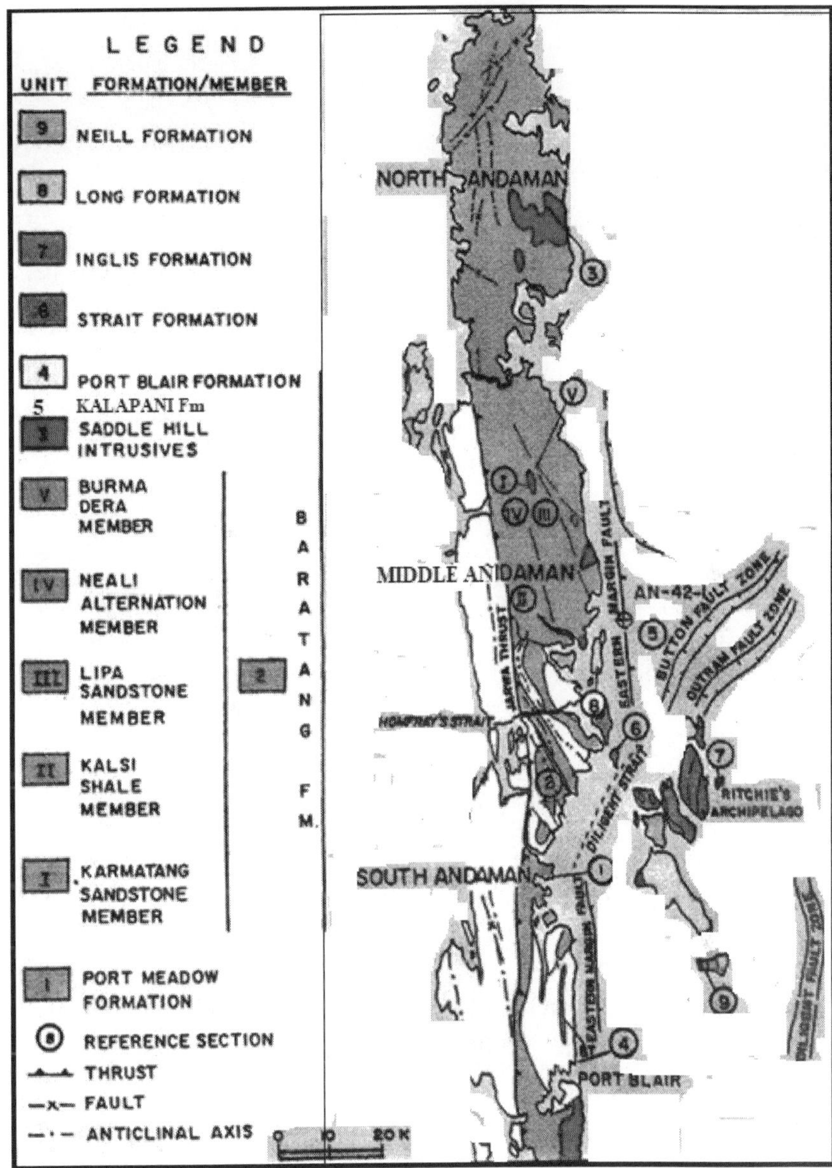

Figure 10.an.2. Geological Map-Andaman Islands.

Geophysical Surveys

The geophysical surveys in offshore areas of the Andaman-Nicobar Basin were initiated by ONGC in 1977. The existing data base include; reflection seismic surveys 66,120 L km of 2D and 3D - 13,710 sq. km. Satellite Gravity Survey of about 0.3 million sq.km and Gravity-Magnetic Surveys include 25, 000 Lkm. The seismic data, helped in detailing of the Neogene prospects identified earlier and in mapping of additional prospects within the Pre-Neogene section (Udintsev, G.B. *et al*. 1975).

In the west coast off north Andaman, one faulted anticlinal structure - 'Gamma' has been delineated by seismic survey. Off the east coast two small structures *viz.* 'Alpha' and 'Beta' on either side of the Ritchie's Archipelago islands have been mapped.

Geochemical Surveys

In 1988, 3,740 Line km. of Sniffer surveys and 20 sedimentary cores have been carried out in the Andaman offshore. Geo-chemical interpretation studies have been carried out on marine sediments. Interpretation of the results obtained from absorbed gas survey indicate the presence of wet gas in these sediments suggesting occurrence of reservoirs in the subsurface. The sediments exhibit major marine algal inputs.

The first **exploratory well** AN-I- was drilled in 1980 which resulted in discovery of gas, flowing @ 180, 000 m^3/day from Middle Miocene limestone. A total of 22 exploratory wells have been drilled, 1 of which lies in the western and the remaining are in the eastern Andaman offshore. Indications of gas from Early Miocene sediments were recorded in a number of wells. Six exploration blocks in deep waters were offered under **NELP** round have been awarded to ONGC and its consortium ENI-ONGC-GAIL. In these Blocks after initial data interpretation about ten wella were drilled.

Tectonics

The Andaman Basin extends from Myanmar to Sumatra and was formed by the oblique converging plate boundaries of the Indian Oceanic and South East Asian Tectonic plates, which was initiated in the early Cretaceous and has continued even now (Gansser, A., 1966). The effects of the easterly subduction of the Indian Ocean Plate beneath the South East Asian Plate created a classic island arc system (Weeks, L.A, *et al.*, 1967), with the formation of six discreet geotectonic units which from west to east through the Andaman area include, foredeep (Andaman Trench); inner slope/accretionary prism, island arc/outer structural high; fore-arc basin; volcanic-arc; and back-arc basin (Figure 10.an.3).

In the eastern most part lies the Malayan orogeny. The oil producing basins of Sumatra are apparently in alignment with the Andaman Sea "Back Deep", west of the "Back deep" is the Cretaceous folded belt or the "inner volcanic arc", west of the above zone, is an interesting sedimentary trough or "Inter Deep". The islands of the Ritchie's Archipalego, in alignment with the Irrawady delta containing most of the Myanmar oil/gas fields are included in this zone.

The basin formed in intra-oceanic settings as an island arc in a convergent set up where the Indian plate is subducting under the Burma-Malayan plate (Mc Kenzie, D and Sclater, J.G., 1971). The basin came into existence as a result of northward movement and anti-clockwise rotation of the island plate and it's under-thrusting beneath the Asiatic plate since Cretaceous time (Curry, J.R *et al.* 1979). The Andaman fore-arc basin is sitting on the subduction prism represented by outer sedimentary arc.

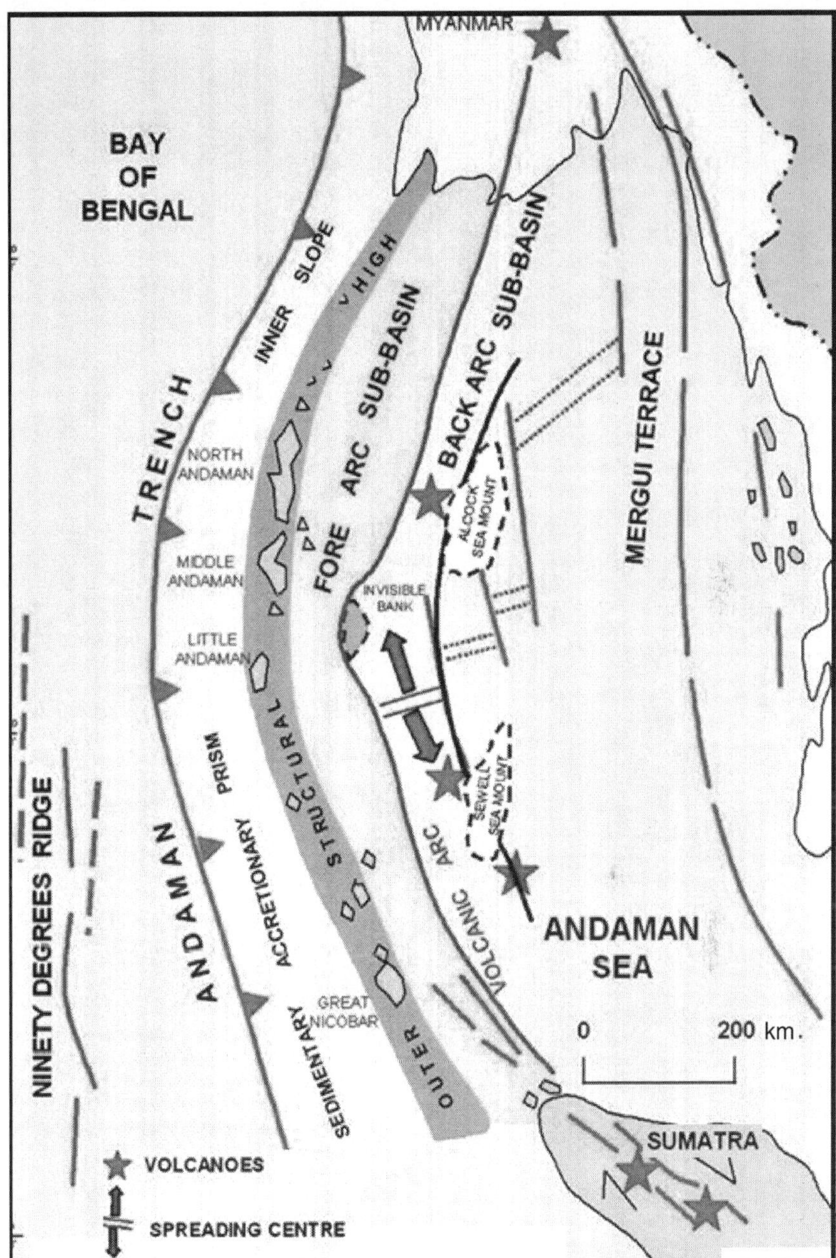

Figure 10.an.3. Tectonic Map–Andaman.

The Andaman-Nicobar basin is associated with the uplift of subduction zone related to convergent plate margins of Indian Ocean and Asiatic plates (Reading, H.R., 1982). The Andaman chain of islands form a part of geo-tectonic regime; "The Great Indonesian Island Arc system" of the East Indies and it extends up to Burmese

Arc, *i.e.,* Arakan Yoma fold belt. It is more than 5000 km. in length (Burton, C.K. and Bigell, J.D., 1969).

While subduction progressed the outer high Arc of the complex started rising steeply, thereby creating a depression or a forearc basin between volcanic arc and the outer high arc. The rapid upliftment of the outer high/arc (Karunakaran, *et al.,* 1964 a and b). At the end of Oligocene, the outer arc emerged out of the sea, culminating in the pronounced uplift of the entire arc-trench system as shown by the wide spread strong unconformity above the Oligocene Port Blair Formation (Burns, R.E., 1964). Tectonic thickening by the continued thrusting caused uplift and formation of the Andaman Nicobar Ridge as chain of Islands.

The continued subduction during Neogene caused further rise of the outer high and accumulation of sediments behind the rising high. Rapid subsidence of the fore arc basin took place along faults on both the trench and arc side (Jacob, K., 1964). The thick accumulation of sediments appear to have caused triggering of shale diapirism. Within the thick Neogene sediments there are several breaks and erosional phase which are identified in the seismic section, outcrops and wells. These are indicative of the dynamic interplay of sedimentation, and tctonics in the region (Chandra, P.K., and Guha, D.K. 1963).

A new spreading center emerged during middle Miocene in the Back Arc in an NW-SE direction. Keeping the plate tectonics concept in perspective, the various geotectonic elements in the Andaman area have been identified (Figure 10.an.4).

Outer High and Tectonic Elements Further West

In the Andaman area, main islands form the Outer high/Trench slope break. The main Andaman Islands probably represent an uplift of the subduction complex after under-riding of the Indian Ocean plate beneath the Asiatic plate. These are characterized by North-South trending thrust which divides the islands into two thrust packages recognized as the Baratang and Port Blair Formations. The islands are represented by North-South trending tight folds associated with several minor faults of same orientation (Rodolfo, K.S. 1969 a and b).

A complex series of fold packages progressively under-thrust from east to west lie to the immediate west of the Andaman Islands. To its west, lies the converging point of the Indian and Asiatic plates marked by the Andaman trench. The Andaman trench is a northward extension of the Jawa trench. The Indian plate covered by sediments of the Bengal Fan is located further west of the trench (Paul, D.D. and Lian H.M. 1975).

Tectonic Elements East of Outer High

The eastern Andaman offshore is represented by a long trough known as the Fore-arc basin. The western margin of the Fore-arc basin has been down-faulted against the Andaman Island with a significant throw. This fault is traceable over a long distance from north to south, and is known as the 'Eastern margin fault'. A prominent unconformity has been noted between the Neogene and Paleogene sediments. The Archipelago group of islands came into existence as a result of strong folding and uplift. Volcanic activity in the area is represented by Cretaceous ultrabasic outcrops and Eocene vitric tuffs in well AN-42-1. Volcanic ash beds have also been recorded from rocks of Early Miocene and Middle Miocene age in wells AN-32-1 and AN-27-1 respectively.

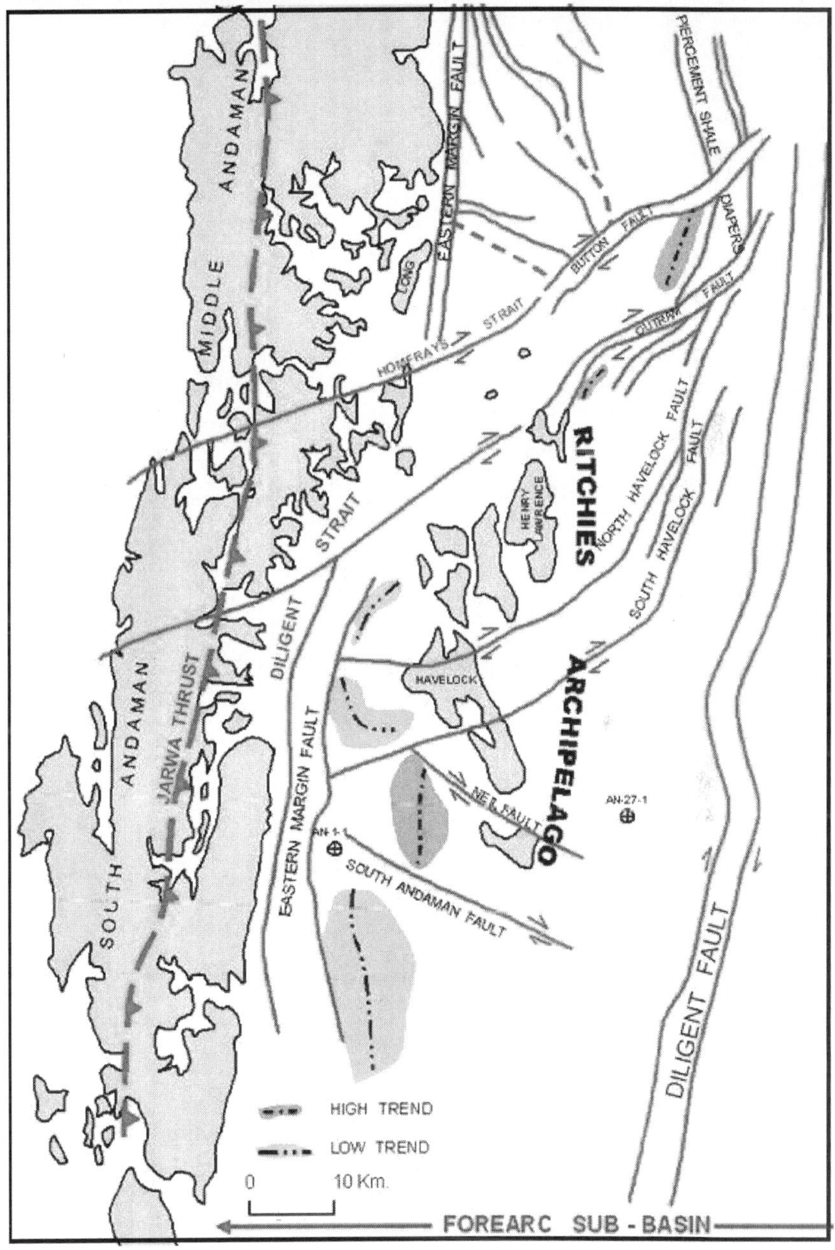

Figure 10.an.4. Major Tectonic Trends-Andaman Fore-arc Basin.

The active volcanoes of the Barren and Narcondam Islands represent a part of the volcanic arc. The trend of the arc continues southward up to south of the Andaman Islands. The Invisible Bank, a large geomorphological feature, lies to the south-east of the Andaman Islands and to the east of the volcanic arc adjoining it (Mc Kenzie, D and Sclater, J.G., 1971). This feature is capped by undisturbed Neogene sediments which thicken to the west and south. The well AN-63C-1, situated on the top of the Invisible Bank, penetrated more than 55m thick basaltic andesite.

Stratigraphy

Thick Palaeogene flysch overlain by shallow marine Neogene sediments are exposed in the islands of Andaman and Nicobar. The surface geological mapping on various islands along with 14 shallow wells contributed towards evolving the onland stratigraphy of the basin. The offshore exploratory wells contributed to the knowledge of subsurface stratigraphy (Rodolfo, K.S. 1970).

Litho-stratigraphic correlation has been evolved for the basin linking outcrop data to the subsurface from the drilled offshore well information. Port meadow formation the oldest sedimentary section in the outcrop of pre-Cretaceous age is not penetrated in any of the offshore wells. The next younger formation namely Baratang, with intrusion and Port Blair are not differentiable in subsurface in the drilled offshore wells. These formations have been grouped together and designated as Kalapani Formation. The Straits and Long formations of the surface outcrop of the Andaman Islands are also recognized in the subsurface from the well data.

The conceptualized Middle Miocene palaeogeography suggest, the Main Andaman Islands emerged during Late Oligocene. The shelf adjoining the Main Andaman Islands, particularly in Diligent Strait was very narrow and carbonate prone (Bukry, D. and Bramlette, M N., 1970). The generalized stratigraphy of Andaman Islands is shown (Figure 10.an.5).

Andaman Goup

Port Meadow Formation

The outcropping Port Meadow Formation is the oldest sedimentary formation encountered in the Andaman Basin. It consists of cherts, jaspers, slates, chalky limestone, marble, quartzite and chert. The base of the formation is not encountered, but it is inferred to unconformably overly the basaltic rocks. It has an unconformable relation with the overlying Baratang Formation.

Baratang Formation

The formation unconformably overlies the Port Meadow Formation in the outcrops and has an unconformable relation with the overlying Port Blair Formation. The formation consists dominantly of argillaceous shales, inter-bedded limestones and siltstone in Baratang Island and shales with minor sandstones and limestone in South Andaman. The formation is of Late Cretaceous to Eocene age. It is undifferentiated in North, South Andaman and Baratang Island but has been subdivided into five members in Middle Andamans (Karunakaran, *et al.*, 1968 and 1975).

Age	Formation		Thick. (m)	Lithology	Lithological Desription
Pleistocene	Neil FM.		45		Modhard Buff Coloured. Fine Tomed Grained for Lisalgal Block Carnite
MID. Miocene Pliocene	Long FM.		1450		Grey Massive Mid Stone Silty Shale Hard Limestone
Early-Mid. Miocene	Inglis FM.		280		Nanno Form Chalk, Silty Shale, Shale
Early Miocene	Strait FM.		90		Sandy Limestone SCFT Maral Tuff Slices Chalk Siltcalc Bands
Oligocene	Portblair Formation		1000		GT. BW. Buff to Light Grey Sand Stone with Grey Buff Water Beded Siltstone and Shales, Grits and Conglomerate at the Base
Eocene to Cretaceous	Baratang Formation	Burma Dera	410		Mainly Shale with Sandstone Marl Siltstone with Waterbeded Claystone with Minor Limestone Occasional Ultrabatric Intrusion
		Neali Alternation	318		Alternation of Sandstone and Shale with Minor Limestone, Conglomerate with Minor Coal and Occasional Ultra Basic Intrusion
		-ipa Sandstone	452		Sandstone, Conglomerate Grit Beds and Minor Shales with Occasional Ultra Basic Intrusion
		Kalsi Shale	260		Grey to Dark Grey Carbonaceous Shale, Claystone. Siltstone and Intercalation Sandstone
		Karmatang Sandstone	846		Sandstone Carbonaceous and Phyllitic Shale & Limestone with Ultrabasic Intrusion
Pre Cretaceous	Port Meadow FM.				Quartizites, Phyllite, Chert. Jasper Slate Cherty Limestone and Marble

Figure 10.an.5. Andaman Stratigraphic-Section.

Karmatang Sandstone Member

The greywacke type sediments forming the basal part of the Baratang Formation in Middle Andamans have been named as Karmatang Sandstone Member. It has an unconformable relation with the underlying Formation and towards the top of the member. The maximum thickness of the unit in Middle Andaman is 846 m. In general the Karmatang Sandstone consists dominantly of sandstone, with carbonaceous and phyllitic shales and minor limestone. The alternation of sandstone and shale is feature of this unit in Middle Andaman, where five informal units have been observed within the member.

Kalsi Shale Member

This unit is exposed on the western part of the Middle Andaman and consists of thick sequence of dark grey carbonaceous shales with claystone, siltstone and intercalations of sandstone. Base of the member is not exposed. It has been considered as homotaxial with the Karmatang Sandstone. Maximum thickness of the member is 260 m.

Lipa Sandstone Member

The dominantly arenaceous unit, lying unconformably over the Karmatang Sandstone, having much coarser fragments in comparison with Karmatang Sandstone Member, has been named as Lipa Sandstone Member. It has a conformable relation with the overlying Neali Alternation Member. Maximum thickness of the unit is 452 m. The member comprises of sandstone, conglomerate, grit beds and minor shales. Occasionally ultrabasic rocks are also observed.

Neali Alternation Member

The alternating sequence of sandstone and shale overlying the Lipa Sandstone and limited by Burma Dera Member on the top, has been named as Neali Alternation. In the lower part, it shows conformable relation with the Lipa Sandstone. The upper contact with the Burma Dera Member is disconformable, as evidenced by the presence of conglomerate and pebble beds at contact of the two members. Maximum thickness of the member is about 320 m. The Neali Alternation comprises of alternation of sandstone and shale with minor limestone, and a few bands of conglomerate and thin coal beds.

Burma Dera Member

The dominantly argillaceous sequence, lying disconformably over the Neali Alternation constitutes the Burma Dera Member. It is overlain by Port Blair Formation with a pronounced unconformity representing a significant time gap. Maximum thickness of the unit is about 410 m in Middle Andamans. The Burma Dera Member consists dominantly of shale with thin beds of sandstone, siltstone, argillaceous sandstone, claystone and calcareous marl.

Saddle Hill Intrusives

In middle Andaman Island, all the members of Baratang Formation are intruded by basic and ultrabasic intrusive. In other islands also, such intrusives are observed in the undifferentiated Baratangs. These intrusive have been named after the Saddle Hill Range of North Andaman and consist of dunite, peridotite, pyroxinite, gabbro along with rare volcanic rocks. The intrusives have not been observed in the Port Blair Formation and are considered to be Post Baratang - Pre-Port Blair, *i.e.*, Pre-Oligocene in age.

Port Blair Formation

The formation is well exposed in the main Andaman chain and has both lower and upper contact unconformable with Baratang and Strait formations respectively. The formation is best exposed near Port Blair town, where it attains a thickness of more than 1000m. Lithologically it comprises of alternating sequence of sandstone and shale. In the basal part, at the contact with Baratang Formation, the sequence comprises of conglomerate and grit beds. The upper contact with Strait Formation is observed in the Strait Island. The formation has been assigned an Oligocene age. Sandstone and shales of this formation are typical of geosynclinal facies and represent flysch deposition.

Kalapani Formation

This sequence comprising mainly of shales and clays with minor sandstones and streaks of limestone has been named in the present work as Kalapani Formation. The formation is homotaxial with the Baratang and Port Blair formations of the outcropping areas. The type section of this formation is in well AN-42-1. The well was terminated in the intrusives. The formation has a pronounced unconformity at the top, where it is overlain by Neogene formations. The formation is assigned a Cretaceous to Oligocene age.

Archipelago Group

Strait Formation

The formation has been encountered both in the surface and in the subsurface in Andaman Basin. It is underlain unconformably by the Port Blair Formation and is conformably overlain by Inglis Formation. The formation consists predominantly of moderately hard, grey to cream coloured sandstone, sandy limestone with hard calcareous band and white to greyish white tuff and ash layer, siliceous chalk and silt in the upper part (Jacob, K., 1964 b). The formation is about 90m thick in the Strait Island and has been assigned an Early Miocene age (Sastri, V.V. and Bedi, T.S. 1962). An upper to middle bathyal depositional environment has been inferred for the calcareous sandstones and sandy limestones. The Early Miocene submarine volcanism is a significant event in the Andaman Basin and has been widely recorded in the surface and subsurface sequences (Jacob, K. and Sastri, V.V., 1951).

Inglis Formation

The formation is encountered only in the outcrops and has not been recorded in the well sections. It is widely recognised in the Islands of fore arc basin and dominantly consists of Nanno-foram chalk with interbedded calcareous silty bands and rare feebly calcareous shales (Srinivasan and Azmi, 1976). Maximum thickness of the formation is about 280m in the Havelock Island. It has conformable lower contact with Strait Formation and is succeeded by Long Formation with a minor break. The formation has been dated as late Early to early Middly Miocene and deposited in middle to lower bathyal depths.

Long Formation

This formation is well developed in the surface and subsurface of Andaman Basin and includes upper part of Round, Guitar, Long, Sawai Bay formations and Neill silty mudstone Member. Its lower boundary with Inglis Formation is marked by a small hiatus and upper contact with Neill Formation is disconformable. The formation dominantly comprises of calcareous, grey to bluish clays, silty clays, hard to calcareous bands in the lower part. The maximum thickness of this formation is 1450m. The clays and silty clays are very rich in foraminifera and have yielded an assemblage of Middle Miocene to Pliocene age.

Neill Formation

This is the youngest formation recognized in the Andaman Basin and is exposed on the west coast of Neill Island, where it disconformably overlies the Long Formation. The formation has also been recognized in the well AN-63C-1 on the Invisible Bank. It consists mainly of fine to medium, compact to porous, buff colored foraminiferal, algal limestone and attains a thickness of about 45 m. The Neill Formation is unconformably overlain by Sub-Recent to Recent coral rags, coraline limestone and beach sands.

Data of Drilled Wells

AN-1-1 well was drilled up to 3,734m and bottomed in the sediments of late Cretaceous age (Kalapani Formation).The claystone and shale between 2200 - 2300 m and 3200-3300 m had specks of flourecence and bituminous material. During drilling severe gas cut problem was encountered below 1815m.Two objects were tested. Object-I in the interval 3372 - 3392 m is within the Paleocene and it comprises sandstone and yielded little gas and object-II in the interval 2345 - 2365 m belongs to Miocene age and consists of limestone. On acidisation, this object has produced 1,80,000 m^3/d of gas with 45 bbls of water having salinity 47 gpl.

AN-1-2 well was drilled only upto 2882m due to technical problem. A 147m thick of middle Miocene sandstone in the internal 2735-2882m was encountered. Thick argillaceous limestone has developed from 2340m to 2405m. The remaining intervals are characterized mainly by clay/claystone and siltstone. Two objects were tested in the interval between 2758-2735m and 2374 -2350 m were proved to be dry.

AN-1-3 well was drilled to delineate AN-1-1 pay, but could not beyond 3412m, due to drilling problem. The bottom of the limestone met in AN-1-1 was 100m deeper in this well, and it has not produced any hydrocarbons. Three objects were tested in this well. The first Object-I in the interval 2450-2465m produced water with salinity of 19. 9 gpl. The Object-II in the interval 2402-2418m gave no influx. The Object-III in the intervals 2324 and 2370 m had produced only water with salinity 24.57 gpl.

AN-1-4 well was also on the highest part of AN-1 structure as a step-out well. The well was drilled down to 2720 m. The thick limestone section (220m) of AN-

1-1 well has changed into finer clastics with few thin limestone beds in this well. Traces of dead oil in CC-1 (2245 - 2263 m) and asphalt in cuttings below 2100 m were noticed. Three objects were tested in this well. Object-I (2240 - 2244m and 2227-2232 m) in limestone did not become active even after acidisation. Object-II (2215-2221m) in argillaceous sandstone produced 44 bbls of water with salinity of 13.455 gpl. The Object-III (2170 - 2180 m) in calcareous sandstone had produced 25.5 barrels of water with salinity about 13 gpl.

AN-2-1 well was drilled upto 4088 m. The sediments in the interval 4088 - 3099 m probably belonged to the late Cretaceous to middle Eocene Baratang group. The Port Blair formation of late Cretaceous to Oligocene age is represented by the section from 3099 to 2450m. The sediment encountered between 2450m to surface is probably belonged to the Mio-Pliocene, Archipelago group. The upper clay/ claystone section up to 946 belong to the Pleistocene and Recent. No hydrocarbon shows were observed however the three objects were tested and dry.

AN-14-1 well was drilled only up to 2538m. The lithology of the drilled section is mainly argillaceous with limestone beds in the interval 852-899m and few thin argillaceous limestone bands in the interval 2150 - 2300m. No oil gas indications were observed.

AN-27-1 well was drilled down to a depth of 2556 m and bottomed in Early Miocene section. The drilled section is predominantly argillaceous with very thin streaks of sandstone and limestone. No hydrocarbon shows were observed in cuttings, SWCs, and conventional cores. A very small quantity of gas (about 0.05 cft) was collected during FMT at 1907 m.

AN-32-1 well, located north of well AN-27-1, was drilled down to 2510 m. The entire sedimentary section between 580m and 2510m is of Early Miocene age. Some ash beds were encountered below 1450m. Gas shows were recorded up to a maximum of 70 per cent (C1) in the interval 1,225 m - 2,285 m during drilling.

AN-42-1 well was drilled up to 3972m. Basement was encountered at 3765m in this well. The thick reservoir limestone of AN-1- is absent in the well. No hydrocarbon shows were observed.

AN-63C-1 well was drilled to a depth of 1853 m.

AN-70-1 well was drilled to depth of 3500 m. The sedimentary section consists of 963m of Upper Cretaceous 2037m Mio-Pliocene and 350m of Pleistocene to recent. The hiatus below Miocene is well pronounced. Four objects were tested in the Miocene section. Object-III on testing has produced little gas.

HLO well was drilled as an exploratory test well on the southern extension of Henry Lawrence anticline which extends into the offshore. The well was terminated at 3078m due to drilling difficulties. The well has penetrated Pliocene claystone, Miocene calcareous claystone, Oligocene claystone and micrite limestone. Three objects were tested. The object-III (1632-1626m) flowed little gas with formation water and hence abandoned.

Petroleum Geology

The first drilled well discovered gas in Miocene limestone, which flowed on test. This fault system has possibly hindered westward migration of hydrocarbons generated in the lows. Considering the source potential of the lows, structurally favorable locations between the lows but to the east of the fault are prospective locations. (Chatterjee SM, *et al.*, 2010). The prognosticated resources for this basin have been estimated to be about 180 Mmt.

Several good surface seeps of oil and gas have been mapped in the Andaman Islands. Besides, several mud volcanoes keep emitting significant traces of oil and gas along the fault system separating the Andaman Islands from the offshore basins.

Source Rocks

The lows filled with thick sediments in the tectonic set-up are likely to be good source areas. Geo-chemical sniffer survey has indicated the presence of positive anomaly. Light hydrocarbon components measured in the survey area shows low average concentration of hydrocarbons with significant higher concentration near carbonate reef structures.

The geo-chemical analysis from the exploratory wells reveal that most predominant organic matter is of Type II and Type III is also present in a few sequences. Potential source rocks, including shales and mudstones of Late Cretaceous to Oligocene age with TOC (Total Organic Content) up to 5.2 per cent have been reported onshore in the Andaman Islands and from surrounding offshore areas; the Baratang Formation can be considered a potential source for thermogenic generation. Geo-chemical study of AN-1-1 well indicated fairly rich organic content in the Paleogene section.

Sedimentary packages, especially in the Fore-arc basin, comprise strata of fluvial, shelf and deep marine regime. Accumulation of significant organic matter is considered a strong possibility. Possible reefal growths within the Neogene section enhance further scope of enrichment of organic matter. Proximity of the Fore-arc basin to the volcanic arc is likely to ensure optimum heat flow for hydrocarbon generation. The Inner slope has been the locale for Paleogene turbidite deposition with organic matter.

Maturation

Well AN-1-1 is known to have encountered mature organic sediments are of type III Kerogen, which is gas-prone in the pre-Neogene section. The section attained maturity below 2,500 m (Ro = 0.5 per cent), studies also indicate top of oil window at around 2700m and 2800 m in wells drilled in the eastern shelf. The Baratang Formation is in the early phase of maturation and lies within the principal phase of oil generation up to 5,000 m (Figure 10.an.6). Similarly, the Back Arc zone has been an area with significantly high heat flow and thus considered favorable for hydrocarbon generation.

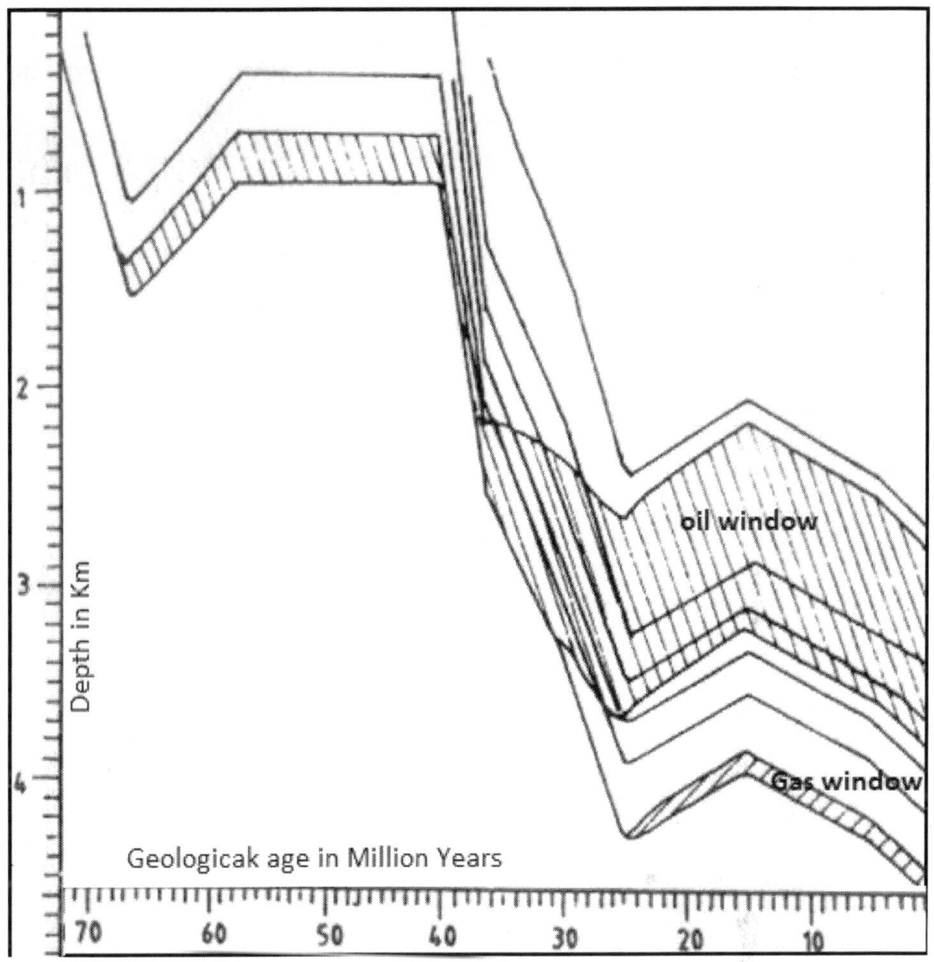

Figure 10.an.6. Thermal Maturation Plot of Well Drilled. Andaman Basin.

The Plio-Miocene Archipeago group indicate immature nature of sediments with good to very good richness of organic matter and fair generation potential. Owing to relatively low geothermal gradient, the level of maturation is lower in the western shelf compared to the maturation level in the eastern shelf.

Reservoir Development

The Paleogene section in the outcrops of the Outer high and in the subcrops of the Forearc basins and Inner slope contains turbidite deposits. Frequent occurrence of arenaceous bodies within such sequence is known to be present.

In the Neogene section, association of calcarenite and sandstone along with conglomerate and marl is common in the Strait Formation of Lower Archipelago Group. The sandstones and calcarenites are known to have excellent porosity and permeability. Presence of carbonate deposits at higher stratigraphic level are also well known. Porosity pods within Tertiary limestones constitute excellent reservoirs in offshore Andamans like in well AN-1-1.

Entrapment Conditions

A large number of anticlinal structures within the Cretaceous-Oligocene sequence with fairly good source and reservoir facies, form excellent entrapment conditions.To the east of South Andamans, onlapping sedimentary sequences over the slope of magmatic arc are likely to provide ideal entrapment conditions for accumulation of sizeable hydrocarbon deposits. Occurrence of fairly large anticlinal structures provides suitable entrapment conditions for hydrocarbon accumulation. In the Invisible Bank area, the onlapping sedimentary sequences on the flanks are ideal locales for pinch-out/wedge-out features favoring entrapment.

Habitat of Oil and Gas

The convergent margin Andaman-Nicobar basin is extremely complex in nature being associated with folding, faulting, thrusting and volcanic activity. From the presence of commercial hydrocarbons in the adjoining basins of similar geologic setting, this basin is considered to have fairly good hydrocarbon potential.

Tectono-sedimentation studies carried out suggest presence of various petroleum systems in Andaman Basin like, deeply buried thermogenic system in a ponded fore arc, thermally induced petroleum system around the volcanic arc and Biogenic/shallow gas system.

Table 10-an.1. Drilled Wells–Andaman Basin

Well	*WD-m*	*Drilled m*	*Year*	*Bottomed in*	*Remarks*
AN-1-1	95	3734	1980	Late Cretaceous	Gas well
AN-2-1	33	4088	1981	Late Cretaceous	
AN-1-2	93	2882	1981	Mid Miocene	
AN-1-3	101	3412	1981	Pre-Miocene	
AN-32-1	56	2510	1985	Early-Miocene	
AN-14-1	77	2538	1985	Oligocene	
AN-1-4	99	2720	1985	Mid-Miocene	
AN-63C-1	74	1853	1986	Early-Miocene	
AN-42-1	52	3972	1986	Late Cretaceous	
AN-27-1	175	2556	1986	Early-Miocene	
AN-70-1	90	3500	1987	Late Cretaceous	
HLO	30	3078	1987	Oligocene	
AOE-1					Drilled by Oil
AOE-2					-do-

Kutch-Saurastra Basins

Introduction

The Kutch-Saurashtra basins are located towards the west in Gujarat state, and extends to offshore to water depth up to 3,000 m. The extension of the ENE-WSW trending faults, and the Saurashtra arch into the Kutch deep-waters suggest the initiation of rifting in the area in early Jurassic. Basins in this region evolved after break up of eastern Gondwanaland. As a result of block faulting and continuous subsidence, marine phase prevailed during the Middle Jurassic (Figure 10.ks-1). The Kutch-Saurashtra deepwater area has bathymetry range of 400m - 3000m and is located to the west of the present day shelf break, off Gujarat state of Western India

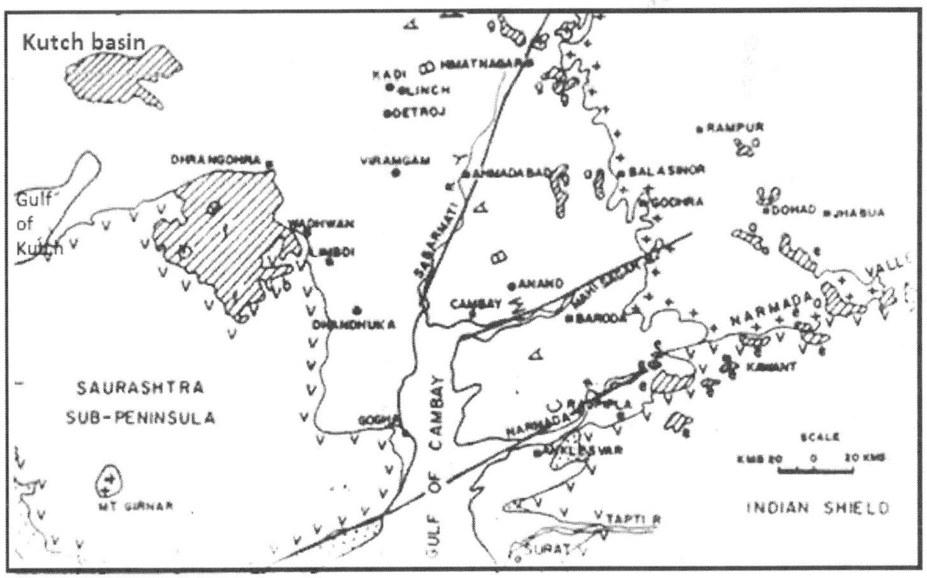

Figure 10.ks.1. Geological Map-Western-Onland.

(Biswas, S. K. and Khattri, K. N., 2002). Mesozoic, Tertiary exposures in Saurashtra peninsula and in offshore area of the basin have been covered by detailed to semi-detailed geological mapping.

Kutch Basin

Kutch Offshore Basin is located between the Mumbai Offshore Basin in the south and south east and petroliferous Sind-Indus geological province in the north. Kutch Offshore is a passive margin basin situated in the northern part of Arabian Sea as an extension of onland Kutch Basin. The basin is separated from Bombay Offshore Basin by a prominent arch at the level of Eocene and Miocene sediments. This basin covers an area of about 35,000 sq. km., onland and around 40,000 sq. km. offshore. Exploratory efforts have led to the discovery of oil and gas in the offshore part of the basin.

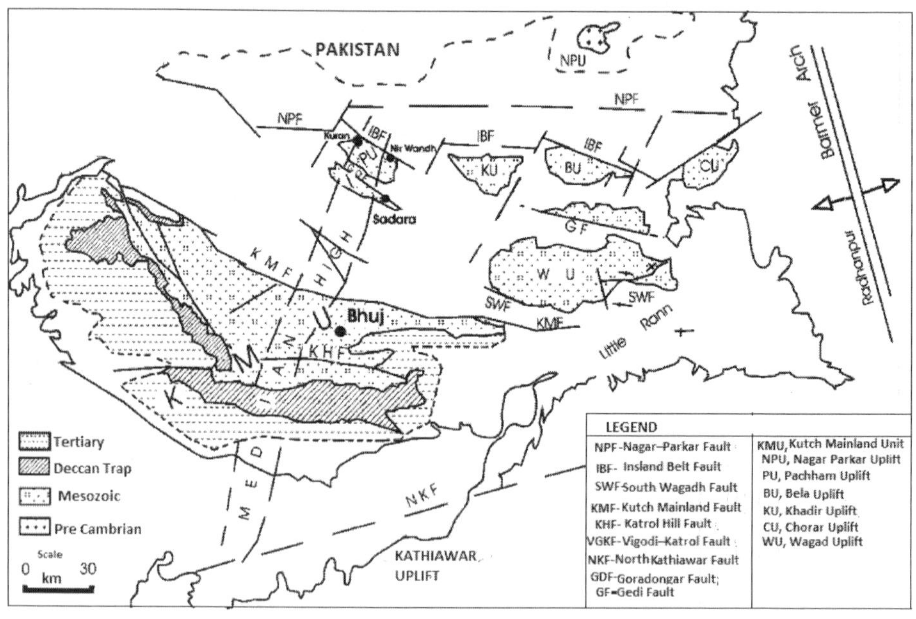

Figure 10.ks.2. Kutch-Tectonic Map (After Biswas, SK 2012).

Tectonics

It forms the north-western part of the western continental margin of India and is situated at the southern edge of the Indus shelf at right angles to the southern Indus fossil rift. It is the earliest rift basin that initiated as a result of north and northeast drifting coupled with counter clockwise rotation of the Indian plate after its detachment from the Gondwanaland during Late Triassic or Early Jurassic (Biswas,

S. K., 1999). This mid-Mesozoic E-W rift occurs between the Nagar Parkar-Tharad basement ridge on the north and Saurashtra Plateau on the south. On the east, the rift is constrained by the NNW-SSE trending Radhanpur-Barmer basement arch, which is also the western rift shoulder of the Cambay rift basin (Figure 10.ks-2).

To the south occurs another parallel rift basin, the western offshore extension of Narmada rift, with the Saurashtra horst between them. To the west, the rift is open, merging with the continental shelf. The basin slopes towards the southwest. The striking feature of the basin is the occurrence of a NNE–SSW trending high across the middle of the basin, so that the uplifts plunge bilaterally and the sub-basins have a central high. The raised ground of Banni above the Rann surface is an expression of this subsurface high. The median high controls the sedimentation along the hinge line of the basin (Biswas SK., 2005).

Sedimentation

In both Kutch-Saurastra areas is the edge of a vast apron of continental Mesozoic rocks derived from the uplift and erosion of the Aravallis. Surface outcrop mapping and stratigraphy has indicated the extensive development of Deccan Trap covers the highest Mesozoic succession in the outcrops and continues in the subsurface of the Kutch shelf (Figure 10.ks-3).

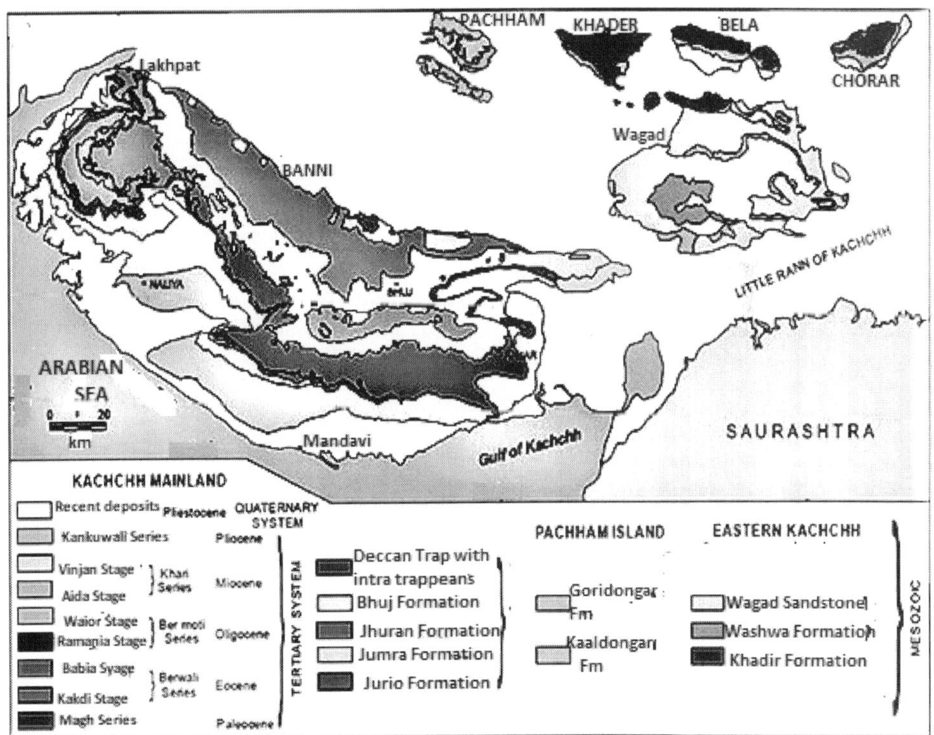

Figure 10.ks.3. Geology of Kutch Area.

The sediment fill thickens from less than 500 m in the north to over 4000 m in the south and from 200 m in the east to over 2500 m in the west. To the north, Precambrian basement rocks are exposed in Meruda Takkar and Nagar Parkar (in Pakistan) hills. To the east, the Precambrian rocks, under cover of alluvium, extend over the North Gujarat plains. To the west the basin extends across the continental shelf while to the south, the uplifted Saurashtra platform covered by Late Cretaceous sediments and Deccan Trap lava delimit the basin (Bisen, M., Sriram K., Gupte, S.S., 2010).

The infra-trappean succession comprises clastic sediments of Middle Jurassic to Late Cretaceous. In the Gulf, however, the Late Cretaceous is mainly a carbonate facies. Two distinct sequences have been identified: the early Middle Jurassic to Middle Cretaceous, and the upper late Cretaceous to Recent sedimentary succession.

Middle Jurassic, late Jurassic and later successions have been encountered in several domes in mainland Kutch. Middle Jurassic period consists of non-marine-estuarine sandy shale, followed by shallow-marine sand, shale, and carbonates marking the first major marine event and the golden oolite, shale and minor carbonates are overlying.

The succeeding sequence is bounded by the Middle/late Jurassic unconformity at the base and by the overlying Cretaceous boundary comprising carbonates, shale and minor sands in lower part and dominantly shale and finer clastics in the upper part.

A large thickness of non-marine sandstones (*Bhuj sandstone*) constitutes the uppermost unit of the Mesozoic sequence. This, unconformity bounded succession is encountered both on land and in subsurface.

In the outcrops, the early Cretaceous succession is unconformably overlain by Deccan Traps but in the subsurface, it is overlain by late Cretaceous carbonate dominated succession. The span of hiatuses in the outcrops, covering entire late Cretaceous. Deccan basalts, extend from Lakhpat in west to Anjar in the east, and occur in more or less linear outcrops.

Paleocene rocks in the on-land Kutch are exposed consistently bordering the outcrops of younger Cenozoic rocks and Deccan Trap. These outcrops near Matanomadh village consist of basalt derived sediments, variegated clay, tuffaceous shale and sandstone. Limestone interbedded with shale is observed in the upper part.

The Middle – late Eocene succession is recorded extensively from outcrops and several wells in Kutch offshore, characterized dominantly by shallow inner shelf limestone and argillaceous limestone with thin interbeds of shale. The Middle Eocene witnessed major transgression and northeastern extension of the shelf, with coast line shifting north to lie close to Kutch mainland fault. Extensive outcrops of Middle Eocene carbonates extend as homoclinal shelf over an extensive area in offshore.

The overlying Oligo-Miocene succession is represented by limestone, argillaceous limestone with thin shale in lower part and sandstone, shale, claystone in the upper part. A minor hiatus is inferred between early and late Oligocene. The early Oligocene succession with claystone and coralline limestone is well exposed in the outcrops and extends over the entire offshore. In the offshore, however, the carbonates are associated with finer clastics deposited in an inner shelf setup. The late Oligocene-Miocene succession is relatively conformable over the outcrops and in offshore. It dominantly comprises limestone with thin shale in the lower part and sandstone, shale and claystone in upper part.

A significant uplift of the Murray ridge commenced after Early Miocene, which has induced the southern movement of Indus sedimentation. The continental shelf of western India, made of prograding sediments, is generally wide (100-200 km) and the continental slope is smooth.

The Mio/Pliocene correlative conformity defines the upper limit of the succession that is overlain by Pliocene to Recent sedimentary succession.

Sub-surface

These sediments are of considerable thickness with basement occurring at more than 6000 m depth. The clastic input for the Mesozoic to Lower Miocene section was from the east while there after was brought from the north by the Indus River. A thick sequence of Jurassic-Cretaceous (about 2400m) and a complete Tertiary sequence (about 700m) is separated by about 600m of Deccan Trap of Late Cretaceous age in the south-western part of Kutch peninsula adjoining offshore basin. The basin is filled up with approx. 1550 to 2500m of Mesozoic sediments and 550m of Tertiary sediments in onland and upto 4500m in offshore region (Figure 10.ks.4).

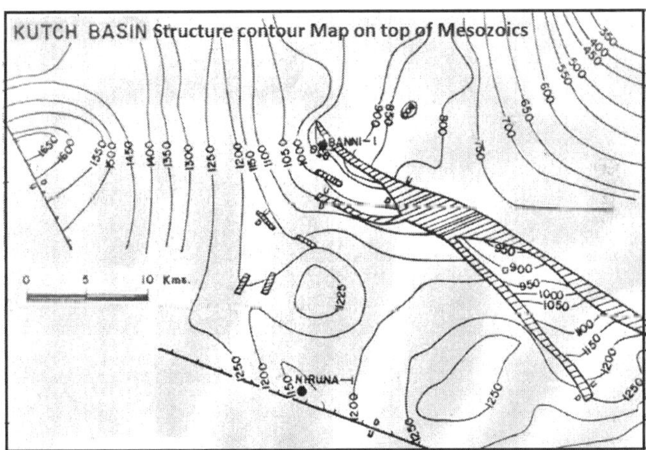

Figure 10.ks.4. Structure Map on Top of Mesozoics-Kutch Basin.

The late Cretaceous succession is encountered only in the subsurface of wells drilled in Kutch offshore, characterized dominantly by shelf carbonates. In the wells drilled on land, like the outcrops, Deccan trap directly overlies the Early Cretaceous and the entire late Cretaceous is absent. The thickness of these basalts encountered in offshore wells increases from south to southwest. In the wells of shelf and deeper areas, the carbonates of this sequence are dolomitic. In the outcrops, the succession is represented by claystone, shale, limonite and bands of sideritic concretions with limestone and shale in upper part.

One well drilled in Kutch offshore basin encountered a thick limestone sequence from Middle Miocene to early Eocene but was found to be devoid of significant hydrocarbon accumulation (Figure 10.ks.5).

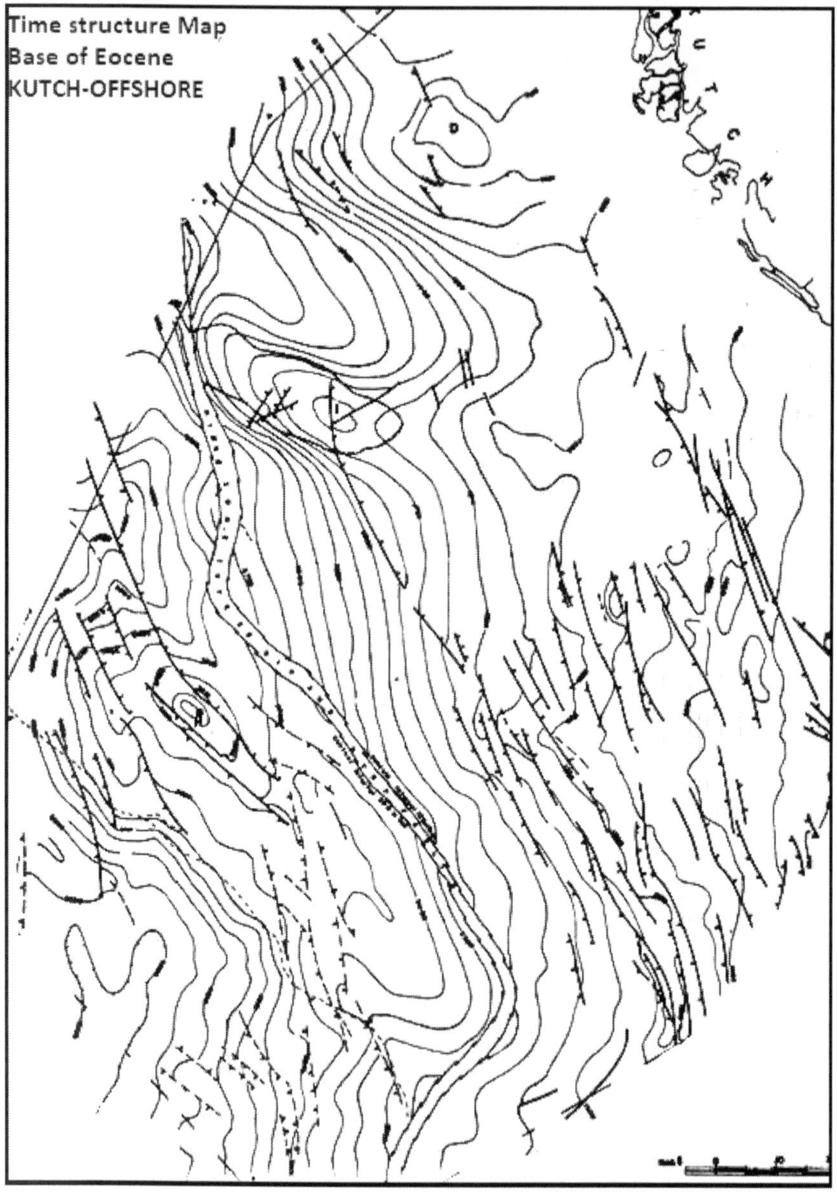

Figure 10.ks.5. Time Structure Map Base of Eocene-Kutch Offshore Basin.

Stratighraphy

The stratigraphic succession, ranging in age from Middle Jurassic to Holocene, is exposed in the highlands of the Kutch Basin. The basin extends across continental shelf to the west while to the south, the uplifted Saurashtra platform covered by Late Cretaceous sediments and Deccan Trap lava delimit the basin.

Mesozoics

Dharangdhara Formation of Lower Cretaceous and consists of feldspathic, argillaceous sandstones with subordinate shales and clays. Coarse to fine grained, cross bedded, conglomeratic, pebbly, ferruginous, fining upwards sandstones. Shales occur as lenses with carbonaceous matter. Sedimentary environment of deposition appears to be fluvio-deltaic. In the well Dhanduka-1 this formation lies directly over the Precambrian Basement.

Wadahawan Formation of Upper Cretaceous (?) which is divided into 3 members. Lower member consists of brick red ferruginous, medium-fine grained, pebbly, moderately sorted sandstone with lenticular shales. Middle member consists of sandy shale with impure but fossiliferous limestone bands. The Upper Member consists of current bedded, pebbly, gritty Quartzose Sandstone. Exposed in the north-eastern part of Saurashtra Peninsula overlying Dharangdhara formation and is likely to be present westwards underneath the basalt cover. The contact between the two formations is gradational.

Deccan Trap Volcanics consists of columnar basalts and the inter-trappean clastic beds of shales/clays, siltstones and sandstones. Its thickness varies from few hundred metres upto 2000m in the onland part.

Cenozoics

Jafrabad formation is of Upper Paleocene to Lower Eocene with alternating sequence of Limestone/Shale with few quartzitic sandstone layers in the lower part of shallow marine. Formation is present over the Diu Blocks and it pinches out against Saurashtra Arch It lies unconformably over Deccan Traps and is 165m thick in well OS-II-G-1.

Nakhtarana Formation is of Paleocene which is white-buff compact limestone with alternations of silty claystone. Lower part contains dark to greenish gray, moderately hard, calcareous, pyritic claystone/shale, foraminiferal wackestone/ packstone, weathered basalts and minor sandstone. The lower part of this formation is generally sandstone shale facies. Formation is present only the deep waters of Saurashtra offshore and is present in Kutch offshore as well. Unconformably overlying the Deccan volcanics in the deep offshore. Formation has an average thickness of 250m.

Belapur Formation is of Lower to Middle Eocene which is dominantly calcareous shale/claystone with a limestone marker bed at the bottom. Calcareous shales of Belapur formation changes to silty shales of conformably overlying Diu formation in the western part of Diu Arch. The formation is present over Diu block in Mumbai offshore basin and extends over Saurashtra shelf and pinches out against Saurashtra arch and is also absent over the Saurashtra Peninsula. The formation is 424m thick in well B-12-1 (Mumbai offshore) and reduces to 61m in the Saurashtra shelf.

Jakhau Formation is of Lower Eocene and consists of Limestone with minor claystone, dominantly wackestone with chalk, foraminiferal wackestone/packstone and sideritic siltstone of inner neritic. Formation is present in Kutch offshore and is reasonably thick in the Saurashtra deep offshore.

Diu formation is Late Eocene and consists mainly of shales with few thin limestone bands. Towards Saurashtra arch lithology changes from silty shales to sandstone facies of prodelta. The formation extends over most of the Tapti Daman block and Diu block. The formation unconformably overlies Belapur Formation and is unconformably overlain by Mahuva Formation.

Fulra Limestone is of Upper Eocene in the well OS-II-G-1, limestone is white to medium gray, moderately hard to hard, very fine to fine grained with abundant carbonaceous matter. Thickness of the formation is exposed in Kutch mainland, extends into the Kutch offshore and beyond Saurashtra arch to Saurashtra Shelf and conformably overlies the Belapur formation. Maximum thickness is about 60m in onland Kutch.

Tuna Limestone is of Early Oligocene, in the well OS-II-G-1, the formation consists of white to grey, moderately hard to hard limestone with clayey carbonaceous material. Towards the Diu arch, limestone facies of this formation changes into dominantly shaly facies littoral environment. It is present in Kutch offshore and extends over Saurashtra shelf.

Narayan Sarovar Formation is of Upper Oligocene and consists of white to medium grey, moderately hard to hard, very fine to medium grained grainstone/ packstone with abundant carbonaceous material. It is present in Kutch offshore and extends over Saurashtra shelf with unconformably overlie and underlie the formations. It is of thickness of 378m in well S-2-1.

Gaj Formation is of Early Miocene and consists of dominantly gypseous clay and limestone. This formation changes into dominantly limestone facies in offshore where it is called as Navibandar formation. Sedimentary Environment is shallow marine. It is exposed in the coastal part of Saurashtra Peninsula with a maximum thickness of 50m. It lies unconformably over the Deccan traps and is conformably overlain by Dwarka formation. Thickness in the coastal part is 40m which increases to 865m in the offshore well OS-II-G-1.

Navibandar Formation of Lower Miocene and consists of white, buff to grey, moderately hard mainly mudstone and wackestone, fossiliferous and pyritiferous of Inner shelf environment. It is subsurface or offshore equivalent of Gaj formation and is present over Saurashtra shelf. It is of 280m thick in well OS-II-G-1

Mitti Nadi/Chhasra Formation is of Lower – Middle Miocene and dominantly mudstone with grey claystone with calcareous bands, Chhasra formation is present in the Kutch mainland (south and eastern part of Kutch) and Mitti Nadi is present in deeper parts of Kutch offshore.

Dwarka Formation is of Lower to Middle Miocene and consists of silty clays and argillaceous fossiliferous limestones. In the offshore Saurashtra, it contains limestone and shale alternations of inner shelf/shallow marine. It is exposed in the northwestern coastal part of the Saurashtra Peninsula and extends into offshore. Changes into Mangrol Formation in the deeper shelf and conformably overlies

Mangrol Formation it is of Early Miocene to Recent which consists of silty clays and argillaceous fossiliferous limestones. In the offshore Saurashtra, it contains limestone and shale alternations of inner shelf shallow marine transgressive. It is present in the deeper part of the Saurashtra shelf and western part of the Diu arch. Its thickness in the well OS-II-D-1 is 3600m.

Saurashtra Basin

It is located in the northern part of western continental margin of India which trends NNW-SSE. The onland part of the basin is also known as Saurashtra Peninsula. This basin lies north of Mumbai Offshore and south of Kutch basin. ENE-WSW Satpura trend influences the structural grain of Saurashtra offshore basin. The environmental framework of the Saurashtra Basin and basins south of it got differentiated into shelf, slope and basin floor. The shelfal horst-graben complex formed the site for development of shelf depositional systems.

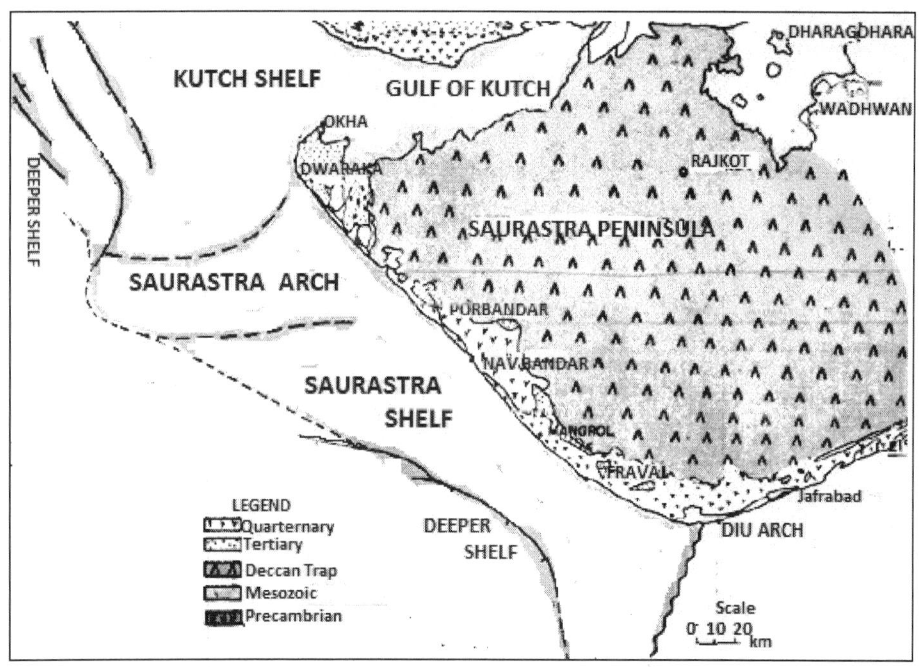

Figure 10.ks.6. Tectonic Map-Saurashtra.

The onland part of the basin borders with the Cambay Basin on its eastern flanks. The deeper offshore Saurashtra basin borders with the Indus fan of the Arabian Sea (Figure 10.ks-6). The Goa rift is situated immediate to the southwest of the Saurashtra deep waters. The Saurashtra Peninsula occurs as a horst block between the three intersecting rifts namely Kutch, Cambay and Narmada (Biswas, 1987). Mumbai offshore shelf extends south of the Narmada rift along the West Coast fault. Mesozoic, Tertiary and Quaternary sediments fill up the Kutch basin.

Saurashtra is largely covered by basalt, but the existence of a negative gravity anomaly in the centre of the peninsula suggests that the Mesozoic sediments, which are exposed in the north-eastern corner of Saurashtra near Dhrangadhra are likely to thicken towards Rajkot.

In the onland part of the basin it is distinguished by highlands which are oriented east-west along five principal faults: - Nagar Parkar fault, Island Belt fault, South Wagad fault, Kutch Mainland fault, North Kathiawar fault. To the west the basin extends into the western offshore shelf. E-W oriented horsts, grabens and half-grabens feature the structural style of the onshore basin.

Saurashtra Arch: - The ENE-WSW trending Saurashtra arch separates the area into two distinct geological entities *viz.*, the northern Kutch and southern Saurashtra deepwater areas. The length of the arch is around 250 km and width of around 60 km in the deep water area. Intrusion of volcanic material had taken place along the ENE-WSW weak zone to form a linear uplift as a volcanic ridge during Late Cretaceous to Early Paleocene.

The crestal part of the arch, in the deep water area, is marked by a striking presence of graben and horst complex all along its axis. The rifting/faulting continued till Middle Eocene. The southern limb witnessed tilting with rotatation at high angle. This has caused the southern limb of the Arch steeper and the southern binding fault is steeper and has larger throws than the northern binding fault.

Dwarka Fault: - The Dwarka fault is a major linear tectonic feature trending NNE-SSW towards the north of the Saurashtra Arch. The Dwarka fault is having a length of around 190 km. It is a normal fault with down thrown to the west and is basement controlled. The up- thrown block of the fault is characterized by presence of number of carbonate buildups all along the fault trend.

Girnar Fracture Zone (GFZ): - The Girnar fracture zone as a strike slip zone in the area north of the Laxmi ridge which is seen extending eastward in the Saurashtra deep water area. South of the Saurashtra Arch this is the most prominent linear feature trending ENE-WSW with volcanic rocks have erupted all along the fracture. This facilitated carbonate growth activity along the Girnar fracture zone over the highs till Middle Eocene times.

Somnath Hills: - A number of clustered carbonate highs are located to the south of Girnar fracture zone. These carbonate highs look like buried subsurface hills. These carbonate highs mark the junction between two major tectonic features *viz.*, ENE-WSW trending Girnar Fracture Zone and NWSE trending West Margin Basement arch.

Kutch Low is situated towards north of the Saurashtra arch and northwest of the Dwarka fault in the Kutch deep-water area. ***Chorwad Low*** is between the Saurashtra Arch and the Girnar fracture zone ENE-WSW.

Continental Slope: - The present day continental slope bordering the deepwater regime to the east, has a general NW-SE trend and is smooth. An offset in its trend is observed to the west at the Saurashtra Arch. The change in the trend of the shelf edge reflects tectonic movements associated with the formation of this arch.

Sedimentation

Sedimentation during the Mesozoic in Saurashtra Basin accumulated over 1500m of fluvio-deltaic sequences resting over the Pre-Cambrian granitic basement (Figure 10.ks.7). The Mesozoic sediments were then intruded and covered by Deccan Trap flows in Late Cretaceous.

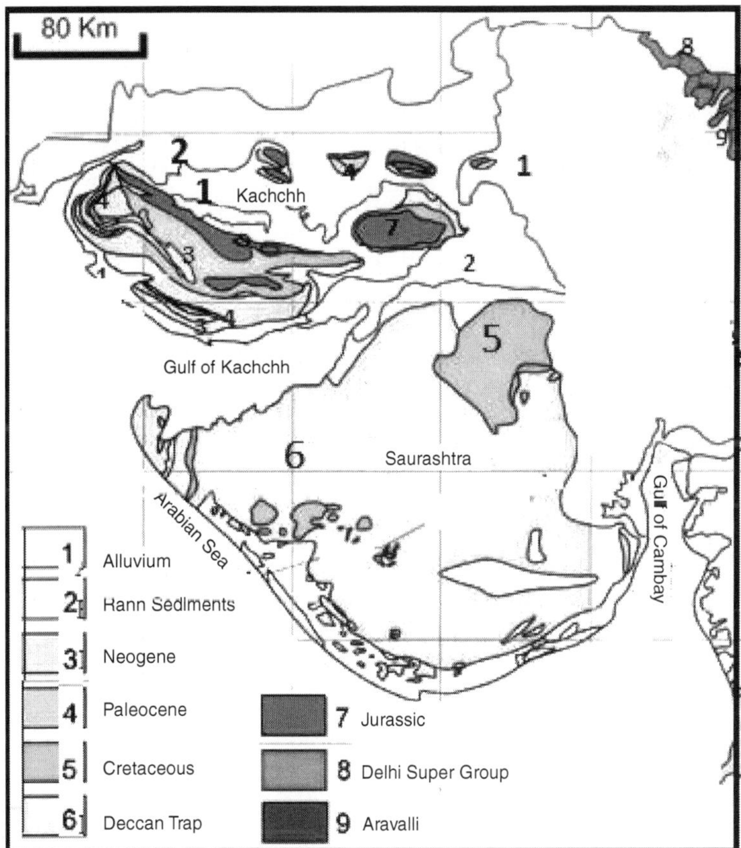

Figure 10.ks.7. Geological Map–Saurastra-Kutch Basins.

The sedimentation in the eastern part of Saurastra started in early Createceous concomitant with deltaic sedimentation in Kutch area (Biswas S K and S V

Deshpande 1983). The sediments were deposited in two major cycles - a Middle Jurrasic transgressive cycle and a Late Jurassic-Early Cretaceous regressive cycle.

The oldest Mesozoic rocks exposed in the north eastern part (Dharangdhara Formation) of this basin. However, major part of the Saurashtra Basin is covered by Deccan Traps. Rocks of Early Cretaceous are exposed in the onland part of the Saurashtra Peninsula. The Tertiary sediments are exposed only on the periphery of the Peninsula which consisting dominantly of Carbonates and fine grained clastics.

Saurashtra onland part remained a stable horst block throughout the Tertiary with deposition of Tertiary sediments on the peripheral lows. On the other hand the shelf had good development of deltas and other related clastic depositional environments, carbonate tidal flats and extensive carbonate ramps. The slope-basin regions formed the site for development of submarine fans, hemipelagic and pelagic deposits.

Stratigraphy

The Saurashtra Basin is mainly a clastic basin and development of limestone is noticed over Saurashtra arch and other paleohighs. Limestone sequences of Eocene to Miocene age which attain considerable thickness over Mumbai platform are found to be shaling out in the Saurashtra depression. The Tertiary sedimentary section is interrupted by several unconformities.

Petroleum Geology

The rift basins particularly the marginal ones are favourable habitat of petroleum. The synrift Jurassic marine sequence with sandstone-shale-carbonate association appears quite promising (Biswas, S. K., Bhasin, A. L. and Ram, J.1994). The hydrocarbon resource base of 210 MMT onshore and 550 MMT offshore have been prognosticated for this basin. Hydrocarbons have been encountered in the early and late Cretaceous succession in the offshore Kutch indicating presence of Mesozoic petroleum system. Hydrocarbon strikes are also recorded from early-Middle Eocene carbonates but commercial accumulation is still to be found. The Cretaceous delta plays (Bhuj Formation) yielded both oil and gas in commercial quantities.

ONGC drilled five wells in onshore area namely Lakhpat-1, Banni-2, Nirona-1, Sanadra-1, Suthari-1 and few wells in offshore area namely GK-22C-1, GK-29A-1, GK-28. Gas show observed in onshore well Nirona-1 and oil discovery in offshore well GK-28 has proven Mesozoic prospectivity and certainly helped to intensify exploratory efforts (Pandey USD,2010). Recent discoveries of oil and gas in Eocene and Early Miocene proved the hydrocarbon potential of the offshore part of this basin. In the area about 130 wells have been drilled with untiresome efforts to strike commercial oil and gas.

Source

The Middle Jurassic and early Cretaceous sediments have organic-rich interval with TOC values within good source rock interval. During Paleocene organic-rich shale and coal were deposited in coastal and lagoonal swamps have a good source potential. However, organic rich shales occurring onland are immature. The Lower

Eocene black shale and byline rocks are good source rocks. It is an enigma that in spite of rift and hot spot related magmatic activities the shales are immature.

Reservoirs

The Early Cretaceous post-rift deltaic sequence is highly promising exploration target and yielded gas at a few shallow offshore locations. Recent discovery of oil from Early Cretaceous sandstones confirmed the expected high potential of this deltaic prospect particularly in the mid-shelf region where delta- prodelta transition is expected.

Traps

Both structural and stratigraphic plays are present. The onshore Tertiary sequence is not promising due to limited thickness but offshore it appears highly promising, as the sequences and facies are same as in Mumbai offshore basin. Mid-Eocene and Oligocene limestones are capped by extensive Miocene shales.

Hydrocarbon Prospects

GK-28

The structure is located towards Okla and Dwaraka offshore and found promising in oil and gas occurrences. So far about 10 wells were drilled and most them flowed oil and gas at different levels shallow level in Miocene to deeper level in Basalts. The well (C-2) flowed oil in the interval 2,558.5 to 2,557 m.@672 with 30 per cent water cut through ½ " choke in the Cretaceous play. Well (L-10) in the interval 1452-1520 m flowed gas @ 1,26,226 m^3/day with ½" choke in the Deccan Basalt section. Well (D-3) the interval 1,342-39 m flowed oil @ 68m^3/day and gas @ 18,000 m^3/d with ½ " choke. The interval 1229.5-1216.5 flowed gas @ 2,34,884 me/d. The interval 733.5-732 and 728.75-717.25 flowed gas @1,45,684 m^3/d. (oil in Nakhtarna formation and gas in Jakhau Formation). Well A-1 in the Interval 682-671 and 697.7 -696.2 m flowed gas @ 1,26 926 m^3/d through 12mm bean (shallow Miocene play). In Well 9-G, the interval 1,249-1,251 m flowed gas @2,21, 950 m^3/d through ½ " choke.

GKS 101 NAA (GK-OSN-2010/1-NELP-IX)

ONGC found gas in (A-1) the interval 1,291-1,300m through ½ " choke from the Jhakau Formation of Early Eocene.

GK-42-1

This prospect located in Kutch offshore and lies towards south GK-28. The tested interval 1,369-1,372 m in Nakhtarna formation of Paleocene flowed liquid @ 36blpd with 20 per cent emulsion and gas @ 9,898 m^3/d with 3/8 " choke.

GSS041NNA (NELP Block- GS-OSN-2004/1)

Well #B-2: the gas was found in the interval 4483.59-4434.22 m which flowed @ 1,56,563 m^3/d with water @ 4,771 bpd through1/2 " choke in Mesozoic sequence in Saurastra offshore.

Table 10.ks.1. Kutch-Saurastra Offshore

Field	*O/G*	*Operator/Area*	*Wells*
GKS092NAA	Gas	ONGC	offshore
GKS091NDA	Gas		
Gulf A			
Aliabet			
GK-28			12
Gulf D			
GK-22C			
GK-29			
GK-42			1
GS-OSN-2004-1			
KD-1	O,G		1
SD-1			3
GK-29			
GK-22C			
SD-14			
GK-39	gas	22.2 sq km	

Mahanadi Basin

Introduction

Mahanadi Basin is located in the northeastern part of the eastern passive continental margin of India. Geographically the basin lies between Jagannathpur in the north-east and Chilka Lake in the south west. Archean crystalline basement and Late Jurassic to Early Cretaceous sedimentary outcrops (Athgarh Sandstone) demarcate the northwestern basin margin.

Mahanadi River rises from Madhya Pradesh and flows through Odisha state before it joins the Bay of Bengal. The coastal geomorphic province is drained by the Mahanadi River system (Mahanadi, Brahmani, Baitarani and Dhamara Rivers) with sufficient sediment load.

The onland basinal area is about 18,000 sq. km. towards south the basin extends into Bay of Bengal. The basin covers a total area of 55,000 sq. km. out of which about 14,000 sq. km. lies in the shallow offshore area. The shelf break occurs approximately along 150 m isobath. The deep-water part of the basin covers a much larger area (Radhakrishna, I., 2000).

Status of Exploration

ONGC carried out Gravity-Magnetic survey during 1965 in onland part of the Basin (Anand,S.P.2002). Aero-magnetic survey was done by NGRI both onland and offshore during 1978. ONGC acquired 2D Seismic Survey 1300 Lkm during 1970. The seismic data acquisition in Mahanadi in shallow offshore part, took place in 1978 of about 3,200 line km. (Murthy, K.S.R and TCS Rao,1990). OIL drilled first offshore well up to 2740m to basement and also drilled more wells. A study of Deep Seismic Sounding (DSS) of NGRI in 1983 indicated the presence of thick Pre-Tertiary sediments in deeper parts of the basin. Geochemical Sniffer Surveys revealed a number of geochemical sniffer anomalies in the shallow offshore area in 1988.

In North East Coastal (NEC) area, two wells *viz.* BB-A-1R and BB-B-1 were drilled as part of south-western Bengal basin and later, few wells drilled by Reliance gave encouraging results. Gas reserves of about 2.5 trillion cubic feet have been estimated but subsequently the block was relinquished. Some more seismic data were acquired by DGH and blocks were awarded under NELP programme. Further, after the award, differernt oil companies also acquired seismic data as part of their committed work-program.

The well drilled by Gazprom to a depth of 2,445 m. indicated abnormally high reservoir pressure but after drilling two more wells the block was relinquished. ONGC as an operator was also awarded five blocks under NELP. In one well with the discovery of gas from Mio- Pliocene age. The ONGC is continuing its exploration activities because of the discovery made in two blocks, one shallow and other deepwater. So far about 22 wells have been drilled in different parts of the basin.

Geology of the Area

Geological Survey of India (GSI) carried out extensive geological survey and mapping in outcrop area in onland part of the basin. Geological mapping indicates Archean basement of comprising granites of Eastern Ghat complex, and Athgarh sandstone which at places directly overlies the Achaean basement. Towards north-eastern part of the basin Baripada Beds of Mio-Pliocene occur as out crops (Chetty, T.R.K., 1995).

A major part of the onshore Mahanadi Basin is covered with alluvium with various rock exposures of upper Gondwanas, khondalites, granites, gneisses and laterites (Figure 10.ma.1). Volcanic rocks equivalent in composition to the Rajmahal Traps are present throughout the basin. Sedimentary sequences resting on the metamorphic basement range in age from Cretaceous to Holocene (Sambasiva Rao, M et. al. 1978).

Oldest exposed sediments in onshore part of Mahanadi Basin belong to Athgarh Sandstone Formation of Late Jurassic to early Cretaceous age. These sediments directly overlie Pre-Cambrian Basement rocks along the western basin margin near Cuttack and are concealed under a widespread cover of laterite and alluvium of Pleistocene to Recent age at places.

Tectonics

Mahanadi Basin is a peri-cratonic basin along the eastern continental margin of India which is separated from Krishna-Godavari basin by 85° E ridge. The main tectonic trends of Mahanadi Basin includes NE-SW trending Late Jurassic-Early Cretaceous "main rift phase" which is orthogonal to the older NW-SE trending Permo-Traissic Mahanadi-Gondwana trend (Burke, K. and J. F. Dewy,1973 and Ramana, M.V., *et al.*, 1997). The main rift phase is characterized by a predominantly volcanic sequence with inter trappeans (Krishna, K.S., 2003).

The Eastern Continental margin of India represents a rifted passive margin evolved in response to continental rifting. India separated from southern super-continent Gondwana Land during Late Jurassic-Early Cretaceous, although rift initiation was during the Permo-Triassic prior to continental splitting. Rift initiation was in the form of linked rift triple junctions. One of the failed arms of these rifts forming aulacogen north of the Mahanadi Basin. Rifting and synrift infill of the Mahanadi Graben was continuing in Jurassic and Early Cretaceous (Yoshida, M, *et al.*, 1992).

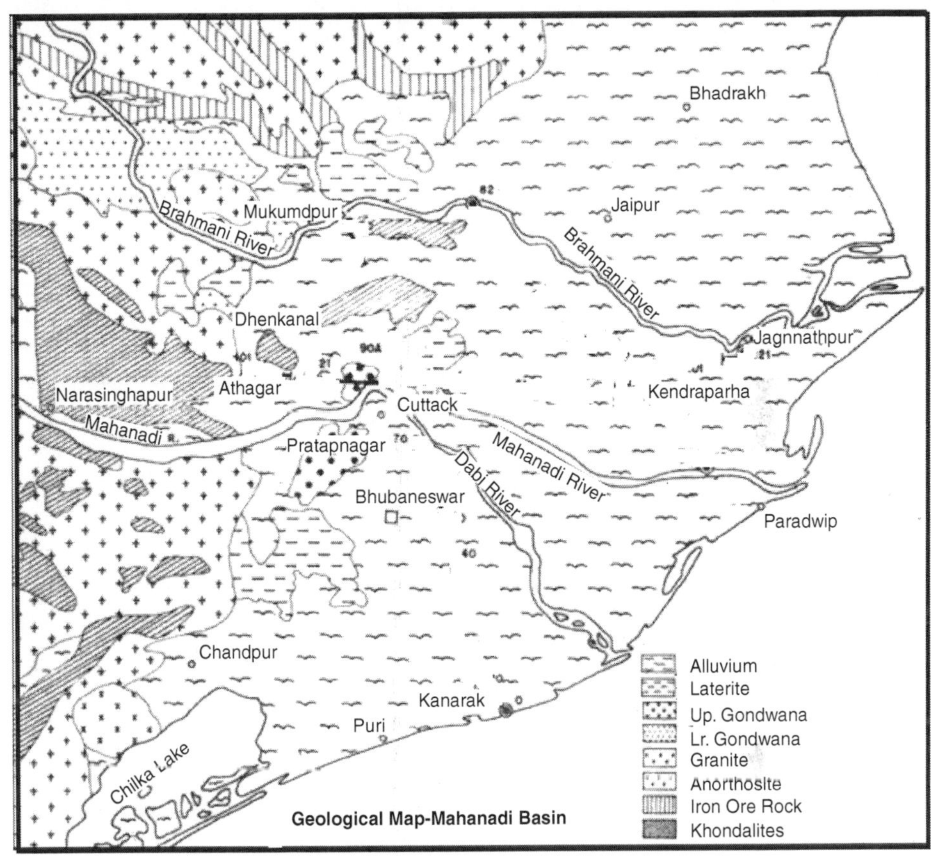

Figure 10.ma.1. Geological Map of Mahanadi Basin.

At the time of rifting during Jurassic the eastern margin of Indian plate was positioned at 50°south latitude and oriented in East-West direction. Initial subsidence of the basin was dominated by differential block adjustment in NNE-SSW direction which created localized graben or half graben where continental, fluvial and paralic sediments were deposited during Early Cretaceous. These differential vertical block movements facilitated lava flows to form Rajmahal Trap (Curray, J.R., Munasinghe, T., 1991).

The basin located on the northern part of the present eastern continental shelf is the southwestward extension of the paleo-shelf of the Bengal basin. It is separated from the latter by a NW-SE trending fault bounded basement ridge (Murthy, K.S.R, *et al.*, 1993).

Available regional geo-scientific information suggests that the basement in the Mahanadi onland and shallow offshore area is of continental type to the north of the Eocene Hinge Zone. It may be mainly of oceanic type further to the south of it. A transition zone from continental to oceanic crust of varying width may be present in different parts of the basin. Continental Basement is made up of Eastern Ghat granulites and gneisses of Pre-Cambrian age (Kaila K.L. Tewari H.C. and Mall D.M.1987).

The onset of drift phase is marked by regional basin-ward tilting and transgressive event also coincide with cessation of rift generated fault activity. From Late Cretaceous the initiation of shelf edge started with marine regressive phase at around 65 Ma where proto Mahanadi, Devi, Brahamani, *etc.* were the main channel system for transporting clastics to deltaic complex at the shelf edge.

Tectonic lineaments consist of NE-SW trending horst graben features in the basin with cross oriented transverse faults (BabuRao,V, *et al.,* 1982). Hinge zone which is considered to be a major tectonic element in this part lies as a prominent NE-SW trending slope break within offshore area. During the course of tectonism volcanic intrusion took place at early stage of sedimentation all over the basinal area providing a significant time marker in the region.

Late Cretaceous surface is an erosional unconformity indicating rate of sea-level fall faster than the rate of subsidence of the shelf. During end of Paleocene to Late Eocene there were marine transgressive and regressive phases. Indian plate was at the Equator during the period indicating warm climatic condition conducive for carbonate development. Limestone with minor alteration of some clastic as well as some transported or reworked limestone can be expected in the deeper part of the basin.

With the onset of collision of Indian plate and Eurasian plate the major sedimentation activities prevailed during Post-Oligocene period till recent when channel activity was the major contributor of sediments in the deeper part of the basin. In the basal part of Miocene there was a major transgressive phase with small pulses of regression and the upper part of Miocene and Pliocene are mark progradation and numerous channel activities (Ray, S.B. and Mohanti, M. 1989).

Onland part of the Mahanadi basin consists of two NE–SW trending half grabens. In the inner graben lower Gondwana sandstones may occur. In the outer Puri graben, thick Tertiary sequence are likely to occur. Seismic survey does not indicate any significant structural features in Puri graben.

During its evolution since the Cretaceous, the basin experienced a number of marine transgressions and regressions followed by development of several elongated depressions and intervening uplifts not only onshore, but also offshore (Fuloria, R.C., 1993). Several deep-seated faults in the basement were also reported (Dasgupta,S *et al.,* 2000) beneath the delta.

The reconnaissance seismic data indicated monoclinal basin-wards dip and more or less uniform thickening of sediments towards south. Due to rapid thinning of sediments from offshore towards the coast, large scale wedge outs are expected (Jagannathan, C.R, *et al.,* 1983).

Rift Stage

Mahanadi Basin during Late Jurassic by rifting and subsidence of the Pre-Cambrian Basement by a number of major faults with a dominant ENE-WSW trend and subordinate NNE-SSW and NNW-SSE trends (Figure 10.ma.2).

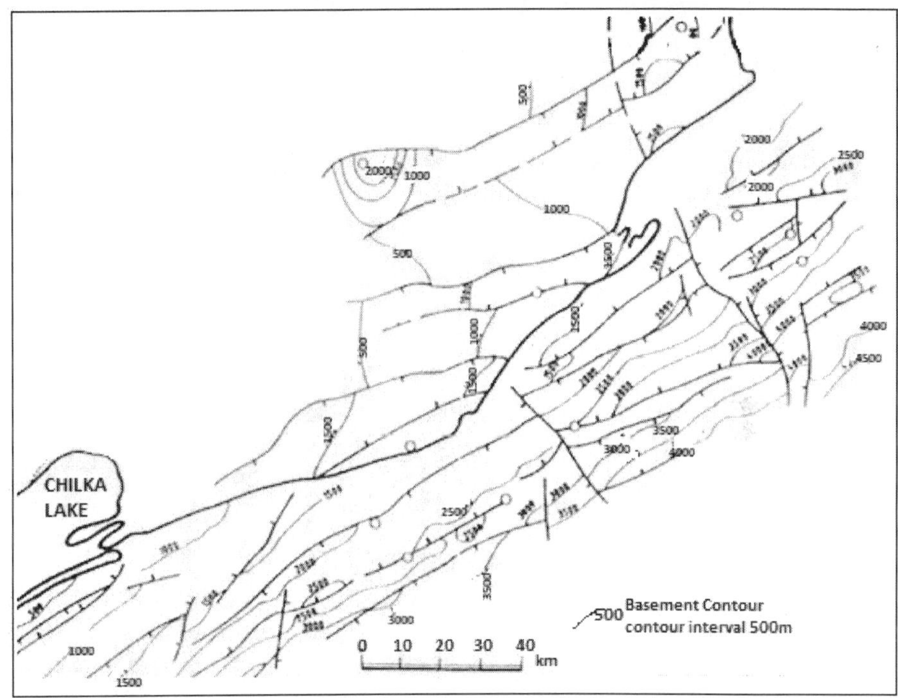

Figure 10.ma.2. Time Map on Top of Basement-Mahanadi Basin (After Fuloria, RC 1993).

These faults, which are parallel/sub-parallel to the present-day coast line divided the basin into a number of linear depressions and uplifts Bhadrak Shallow Basement Block, Cuttack- Chandbali Depression, Bhubaneswar-Kendrapara Uplift, Paradip Depression, Nimapara-Balikuda Uplift, Puri Depression and Konark Uplift from north to south in the onland part. The Konark Uplift extends into shallow offshore part of the basin.

Further south-eastward, the shallow offshore part up to the Eocene Hinge Zone has also been differentiated into a number of ENE-WSW trending uplifted blocks and depressions. Further to northeast, a NNW-SSE trending fault is present probably, due to which the Eocene Hinge Zone takes a sudden NE-SW swing and further northeastward, it assumes almost N-S trend. North of this part of Hinge Zone is designated as NEC area (Figure 10.ma.3).

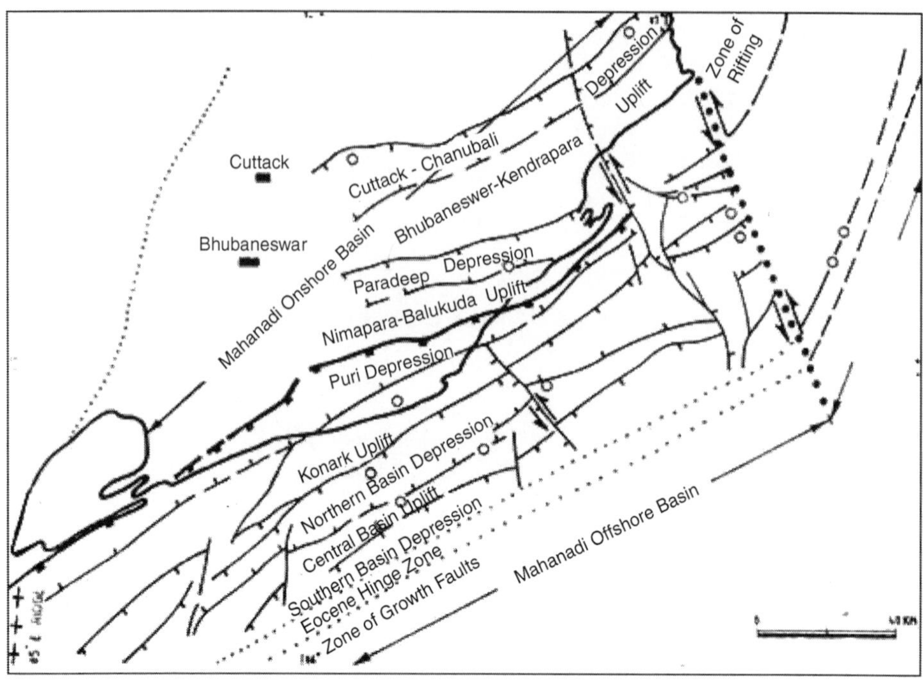

Figure 10.ma.3. Tectonic Map –Mahanadi Basin (After Fuloria, RC 1993).

The Gravity Anomaly Map of offshore part indicates that the lineament from which ocean spreading initiated was coast parallel implying that rifting and spreading was controlled at least in part by the NE-SW trending Achaean Eastern Ghat trend (Figure 10.ma.4 and 5).

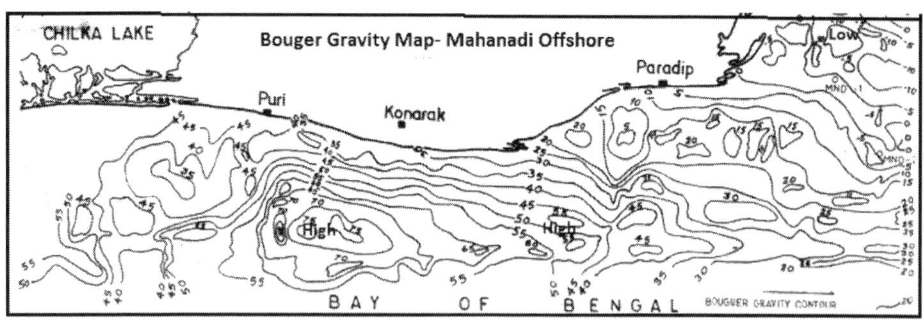

Figure 10.ma.4. Bouger Gravity Map of Offshore-Mahanadi Basin (After Jagannadhan C R, *et al.*, 1983).

NNW–SSE trending faults divide the basin into transverse blocks. Significant lateral movement has taken place along these strike slip faults during the course of basin development. Both the basin strike and the alignment of the Eocene hinge takes turn to NE along one of these faults suggesting a post-Eocene movement.

Figure 10.ma.5. Gravity Map Onland and Offshore-Mahanadi Basin.

Early Thermal Subsidence

The initiation of early thermal subsidence during Late Albian is marked by a southerly/southeasterly tilt of the continental margin leading to marine transgression. The structural configuration continued to be of horst-graben type.

Unconformably overlying the volcano-clastic rocks of Early Cretaceous is an appreciably thick sequence of sandstones and shales with minor limestone at places deposited under shallow inner shelf conditions in the present day shallow offshore part of the basin. The thickness of this stratigraphic sequence varies widely upto 500m. However, the equivalent sediments in the onland part were deposited under fresh water conditions due to restricted marine transgression.

Late Thermal Subsidence

Eastward flowing rivers along the East Coast of India deposited Paleocene sediments that filled Cretaceous topography (Goutam Kumar Nayak, *et al.*, 2006). Crustal shortening initiated in Late Eocene - Oligocene continued into Early Miocene during the northward drift of the Indian plate and the consequent upliftment of Himalayas. During this time sedimentation of the great Bengal Fan in the offshore part was initiated and built up with time. These sediments might have overlapped the earlier sediments in the NEC area of the Mahanadi basin (Figure 10.ma.6).

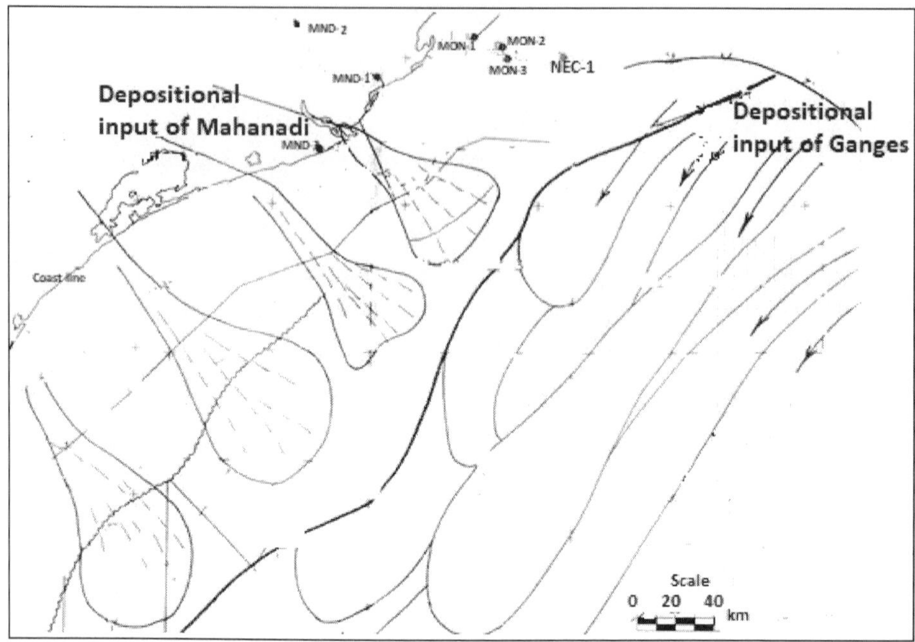

Figure 10.ma.6. Sediment Input Neogene-Mahanadi Basin.

Sedimentation

The Lower Gondwana sedimentation during Permian represent Talchers followed by Karharbari, Barakar and Raniganj coal measures. These are overlain by Kamthi Formation with a marked unconformable contact. The Lower member of Kamthi sandstone is included in Lower Gondwana whereas the Upper member of Kamthi and younger sequences up to the basaltic flow are included in Upper Gondwana. Surface geological studies reveal that, Athgarh Sandstone directly overlies the basement. It is composed of white to gray, hard, cross-bedded, poorly sorted, medium to coarse grained, ferruginous and feldspathic sandstone with intercalations of lenticular grayish white to pinkish clays or carbonaceous shale.

Predominantly coarse grained sandstone with subordinate shales and thin coal streaks of Late Jurassic to Early Cretaceous age have been encountered in the onland wells. These sediments have been inferred to be deposited under fluvial conditions with occasional marine incursions (Casshyap S.M. and Srivastava, V.K, 1988). Several flows of basic volcanic rocks of different thickness are met within this sequence. The thickness of volcanics varies from around 25 m in MND -3 to 850m in MND -7 indicating the variation in volcanic activity over the basin (Bharali, B., *et al.*, 1991). Based on palynological evidence, these volcanics are correlated with Rajmahal Trap of Neocomian-Aptian age (133 to 144 Ma). The inter-trappeans are mainly argillaceous and their thickness vary from 150 m in MND-1 to almost nil in MND-7.

The thickness of pre-basalt section shows two differently trending lows with one median high. The sand percentage map depicts the same (Figure 10.ma.7). The Lower Gondwana pre-basalt sedimentary section was deposited in a WNW-ESE trending rift valley, these two depositional directions are in conformity with the geological set-up. Adjacent horst blocks might have acted as provenances and filled the rift valley (Casshyap, S.M. and Tewari, RC. 1984).

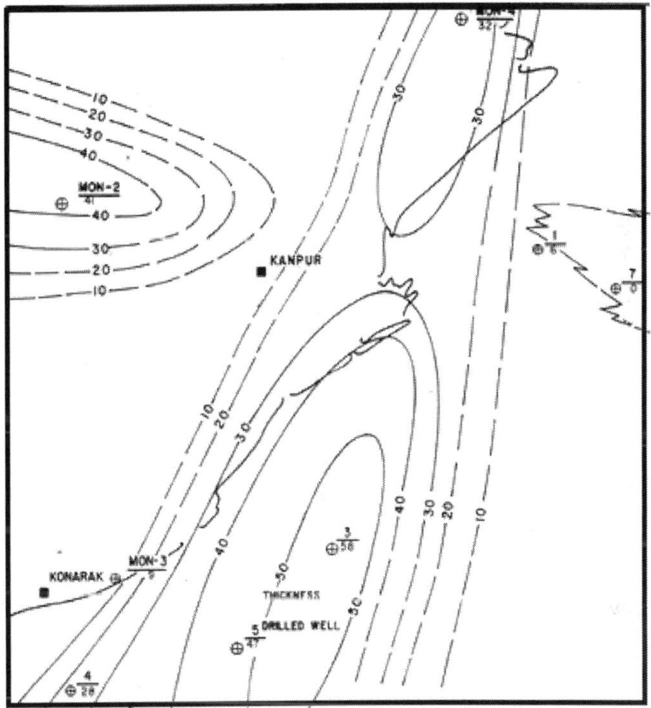

Figure 10.ma.7.Litho-Facis (Sand Percentage) Map-pre-Basalt-section Mahanadi Basin.

The occurrence of Paleocene sediments penetrated in onland wells suggest that, the marine transgression during Paleocene was restricted to the basinward periphery of the present day onland part of the basin. While the sequence is represented by shales, sandstones and thin limestones deposited in the shallow marine environment, in the offshore part it is represented by a dominant argillaceous limestone, shales, siltstone and sandstone deposited under deltaic to shallow marine conditions (Badhe, S, *et al.*, 2003). There was a paucity of clastic supply into the area during Early Paleocene resulting in the deposition of limestone in the shallow offshore area. However, during Late Paleocene the deposition of sands and silts with minor limestones may be under deltaic conditions in the southwestern part of the offshore basin (Fuloria,R.C., 1999). However, northeastern part of the offshore part of the basin (MND-2 and 7) continued to be deprived of clastic supply leading to uninterrupted deposition of limestone.

The sequence between Cretaceous/basalt to Middle Eocene limestone shows a NE-SW trending depocenter with two definite depositional trends, one with NW-SE trend towards northern part and another with E-W trend located towards southern part of the basin with a median high. It is observed that, the Paleocene-Cretaceous sequence is thicker in south central part than in the northeastern part suggesting basinal dip was towards south, while Eocene section is thicker in northeastern part indicating change in regional slope of the basin after Paleocene. This could be possible due to reactivation of a NNW-SSE trending fault.

Eocene Hinge can be clearly identified through seismic data (Subrahmanyam A. S, *et al.* 1994). Clear evidence of deep erosion into the upper part of the Eocene sequence is reflected in the form of uneven topographic surface and erosional channels and total absence of Oligocene section in both onland and shallow offshore part of the basin indicates a major Late Eocene-Oligocene marine regression from the basinal area (Figure 10.ma.8 and 9). In the NEC area, the well NEC-1 located 40 km away from Mahanadi offshore well MND-2, encountered 567m of Oligocene. However, in deep offshore, Oligocene sediments seem to wedge out against Eocene shelf edge.

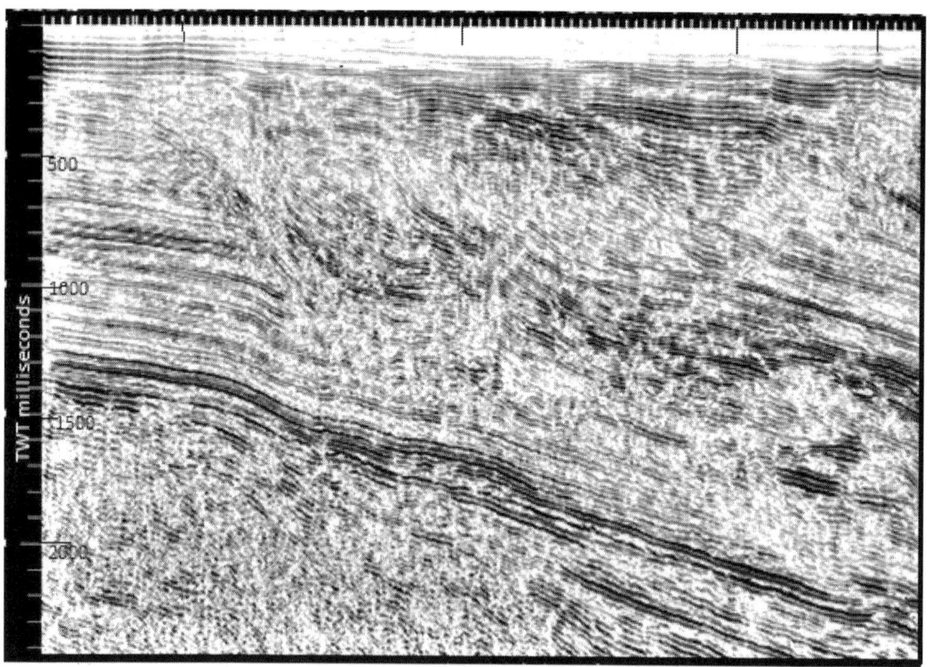

Figure 10.ma.8. Seismic Section-Eroded Shelf-Mahanadi Basin.

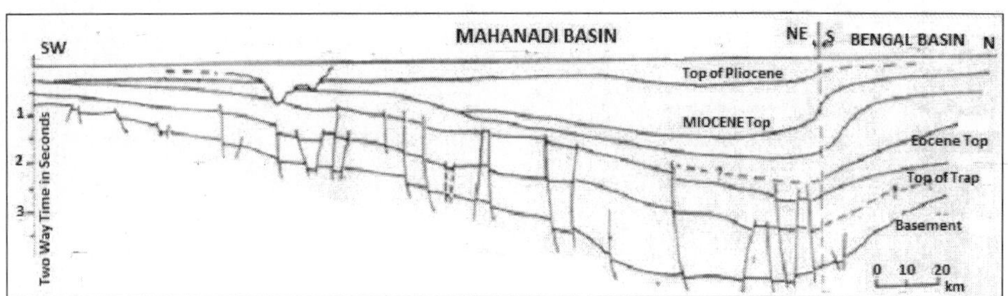

Figure 10.ma.9.Seismo-Geological Section from Mahanadi to Bengal Basin.

During sea level lowstands, sediments were deposited to the shelf edge initiating the formation of shelf edge deltas, levee-channel and basin floor fan complexes.

Occurrence of medium to fine grained sandstone and shales deposited under inner shelf conditions in a lone onland well and total absence of this sequence in other wells suggest that major portion of the onland part possibly remained uplifted area (Fuloria, R.C., *et al.*, 1992).

The general trend of delta building indicates dominant clastic supply axis is in NW-SE direction with some influence from the northeast to southwest in the area. The Middle to Late Miocene delta progradation was much more active in the eastern part suggesting the dominant clastic supply. A number of erosional cuts and fills are observed beyond the shelf edge within Miocene sequence indicating a period of major unconformities which may be the depositional surface for the turbidity mounds deposited in the deep water area.

Early Miocene and part of Middle Miocene is a period of major transgression. From Middle Miocene onwards the shelf edge shows continuous seaward progradation and the present day shelf edge is located close to Eocene shelf slope break (Rao, G.N., *et al.*, 2002 and 2003). The Mahanadi-NEC area is so greatly influenced by Bengal deltaic sedimentation system with its northeastern boundary with Bengal Basin.

Seismic expression of both Upper Miocene and Pliocene sequences exhibits deltaic facies in the lower part while the upper part of the sequences exhibit deep canyon cut and fill features. The canyon cuts are steep sided, narrow to wide and at times 1-2 km deep (Posamentier, H.W. and R.D. Erskine). These canyons have created locally deep-water conditions over the shelf (Subrahmanyam, V, *et al.* 2006).

Channels within these canyons acted as conduits for transporting coarser clastic sediments into deep-water areas (Catuneanu, Octavian, 2002,). As the canyons got filled up, they gradually came under shallow water environment. It is observed that those canyons fills, in general, are of fine clastics with some sands representing the channel facies within them. The cyclic nature of occurrence of shelfal canyon complexes clearly suggest periodic spurt in the fluvial energy coupled with sea level fluctuations (Emmel, F.J. and J.R. Curray, 1985).

Stratigraphy

Chrono-Stratigraphy

The sedimentary succession in the basin ranges from Permian to Recent, and three first order sequences *viz.* Permo-Triassic Gondwana graben-fill sequence, Early Cretaceous Rift fill sequence and Late Cretaceous to Recent Passive margin sequence are known. This Rift sequence is bounded by sequence boundaries Gondwana top at the base and top of Early Cretaceous Athgarh Formation at the top.

Two second order sequences namely Pre-trappean sedimentary sequence and Post-trappean Rift sequence (Athgarh Formation) have been identified within it. These two sequences are separated by Rajmahal basalt. The six second order sequences are identified within the first order Passive margin sequence. These are Late Cretaceous, Paleocene, Eocene, Oligocene, Miocene and Pliocene sequences. Three third order sequences, Early Eocene, Middle Eocene and Late Eocene have been are known within the second order sequence.

The third order sequence is bounded by Oligocene top unconformity at the base and top of Early Miocene at the top.

With the drilling deep wells in the basinal part, the stratigraphy of the region had been established with minor differences between offshore and onland lithologies (Figure 10.ma.10). In general offshore lithology is dominated with carbonates and argillites as compared to onland part where a dominant arenaceous section was encountered.

Petroleum Geology

Significant hydrocarbon shows have been encountered in the wells drilled in the Mahanadi Offshore basin. Gas recovered during Formation Interval Testing (FIT) of Miocene sandstones and Eocene limestones in the well MND-2 contained significant proportions of higher hydrocarbons. Specks of brownish yellow fluorescence was observed in MND1 well in the Eocene interval 1839-1913m and dull yellow fluorescence, faint cut/Bright yellow fluorescence in Eocene section in the interval 2635-2980 m. and on Gas with water in FIT was collected in Early Miocene and Eocene in the well MND2. Patchy light yellow fluorescence, faint cut observed in MHD3 with in Paleocene section. In MND6 Patchy yellow fluorescence with bluish white cut and gas in RFT was seen. In well MND7 bright yellow fluorescence with cut was noticed.

The total resources of the block are estimated to be 248 to 367 million tons of equivalent oil, including 319 billion cubic meters of natural gas, 37.6 million tons of condensate and 18.8 million tons of oil. The proved resource base of 45 MMT onshore and 100 MMT offshore have been prognosticated for the basin.

In the exploratory wells drilled in the Block NEC-OSN-97/2 by RIL proved to be gas bearing within Upper Miocene to Pliocene sequence and the initial gas reserve estimates were about 2.5 trillion cubic feet. ONGC also estimated the initial in-place reserve for the two discoveries in deep water area as 0.362 to 1.03 TCF with an additional 1.89 tcf.

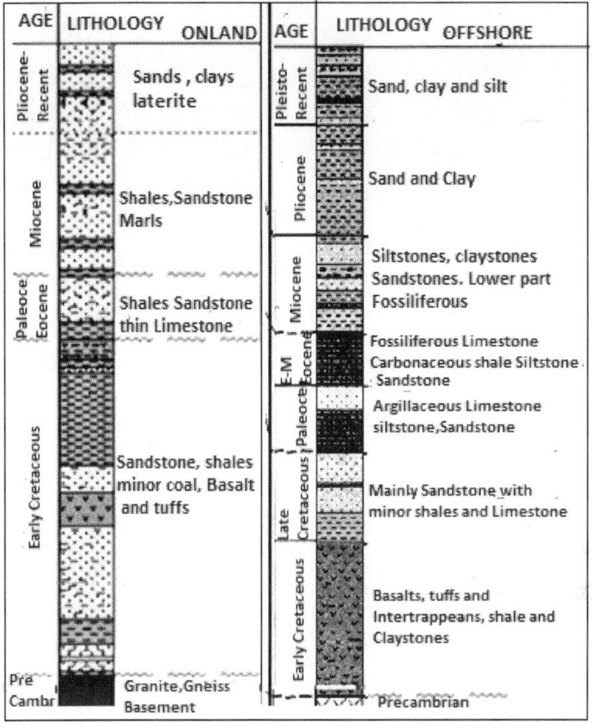

Figure 10.ma.10. Stratigraphy-Mahanadi Basin (After Fuloria, RC 1993).

The sedimentary facies indicate more of araenaceous than argillceous. It could be possible if proto Mahanadi flowed through the Gondwana grabens. If so, the second order sediment input will be of good percentage of argillaceous components, like that in the adjacent Krishna- Godavari Basin. This inturn resulted in lesser source facies in the basin. Major part of the Basin seems to be resting on Continental crust below with less heat flow.

As it is observed most of the deep water area of the basin is covered with Mio-Pliocene Bengal Fan. The huge thickness with sediments with less time might have facilitated to generate thermogenic gas (?).

Source Facies

The geochemical Sniffer Surveys have revealed the presence of a number of anomalies. Basin modeling studies carried out indicated presence of mature source rocks in some of the onshore depressions, suggesting similar possibilities in deeper parts of various depressions in the basin (Fuloria,R.C., *et al.*, 1990).

The onland well MON-2 drilled in the Cuttack Graben has a high content of TOC (upto 9.5 per cent) in the Early Cretaceous sequence and the well MON-1 drilled in Paradip Depression has adequate TOC of more than 2 per cent in the Miocene sequence. A total thickness of 960 m of source rocks have been encountered in the well MON-2 with an average potential. Most of the organic matter is of type II and III.

Source rock studies in drill cuttings and sidewall cores from the wells drilled in shallow offshore indicate presence of organic matter of TOC- 1.5-3.0 per cent, suggests that better source sediments can be present in the deeper parts of the basin.

Maturation

In the onland part of the basin, the top of maturation in the Cuttack Graben was around 2300 m depth. The organic matter within Paleogene sequence of well MND-2 has reached adequate maturity (Vro of 0.55 at about 2800 m and 0.65 at about 3300 m). Geochemical studies show that, in MND2 oil window begins at Eocene top at 2696m.with VR_0 of O.65 and maturity increases to 0.76 per cent at 2799m.

In the well MND-7, the organic matter is type-III and type-II. In well MND7 VR_0 reaches to 0.6 per cent at Paleocene level at depth 2711m.The organic matter in the Paleogene and older sequences in MND-5 and MND-6 shows marginal maturity.

Expulsion and Migration

The Rock-Eval studies show that migratory hydrocarbons are present in the Miocene and older sediments of Mahanadi offshore wells. Faults within the underlying Cretaceous and Paleocene sequences can act as effective migration pathways. The eastward basinal tilt in post Late Cretaceous could have facilitated hydrocarbon migration towards northwest from the basinal depocenter.

Reservoir Rocks

Sandstone reservoir facies have been encountered in the wells drilled in Cuttack Depression in the onshore part of the basin. Average porosity in these sands is in the range of 15 to 25 per cent. The inter-trappean sequences with associated fractured and weathered volcanic flows are also potential reservoir rocks.

The subsurface data in the shelf part of the basin suggest availability of good reservoir rocks within Early Miocene and older sediments. Porous and permeable sandstones and carbonates within Late Cretaceous, Paleogene and Early Miocene are the potential reservoir levels.

Reservoir sands within channel/Levee complexes of Pliocene and Pleistocene and Late Oligocene Wedge-outs against the Eocene Hinge in deep offshore part of the basin are also the potential reservoir targets (Sudipto Datta, *et al.*, 2012).

Cap Rocks

Presence of claystones and shales at different stratigraphic levels reported from the drilled wells are likely to provide cap rock in the basin and in particular a relatively thick sequence of Miocene comprised of claystones provides a good regional cap rock for hydrocarbon accumulation.

Entrapment/Trap Types

The potential traps may be sealed vertically by Paleocene/Eocene shales. Complex development of this sequence has brought out diverse potential situations like, local faulted anticlines and positive erosional features over the Cretaceous surface. Carbonate buildups can be envisaged along/over the Eocene hinge.

The 85° East Ridge to the southwest acts as a barrier onto which Paleocene, Eocene, Oligocene and Miocene sediments onlap and can have entrapment potential. These units may consist of thin sands, carbonates and shales terminating against the Ridge as updip wedges facilitating entrapment of hydrocarbons. Oligocene wedge-outs against the Eocene hinge can also have entrapment potential.

Submarine canyon and turbidite fan Systems are more characteristic of Pliocene-Pleistocene sequences with slope failures caused by significant increase of sediment input into the basin. (Niyogi, D., 1975).

Results of Exploration

Block MN-OSN-2000/2

The conventional production testing has established a production rate of 41,000 m^3 of gas per day through a 12/64" choke with a Flowing Tubing Head Pressure (FTHP) of 1,945 psi. Three Mio-Pliocene sequences like, 1,690-95, 1,645-1,655 and 1,411-1,419 m. the Modular Dynamic Test (MDT) have found that gas flowed with a pressure of 2,324 psi at a depth of 1,694 m.

Table 10.ma.1. Drilled Wells–Mahanadi Basin

Well	*Operator*	*Drilled m*	*Year*	*Bottomed in*	*Remarks*
MON#1	OIL	1393	1987	Basement	
MON#2		2993		Cretaceous	
MON#3		1914		Cretaceous	
MON#4		2408		Basement	
MND-1	OIL	2740	1980	Basement	
MND-2	OIL	3650	1980	Basement	
MND-3	OIL	3284	1983	Basement	
MND#4		2240	1983	Cretaceous	
MND#5		4026	1984	Basement	
MND#6		2248	1984	Eocene	
MND-7	OIL	3888	1884	Basement	
NEC-1		5280	1985	Basalt	
NEC-2		3620	1985	complication	
NE		4642	1987	Miocene	
M-1	ONGC				Block- MN-OSN-2002/2
M-2					
M-3					
M-4			2006	Dry	
M-5				Gas	

Well	Operator	Drilled m	Year	Bottomed in	Remarks
M-6				Gas	
M-7				Gas	
M-8				Dry	
M-9				Gas	
MDW-13					
MDW-5				MDT- gas	**MN-DWN-98/3**, Packer-3753m
MDW-4A					
MDW-2A				Gas- 1644-59	flowed@2,12,555m^3/d 12/64 "
MDW-10				Flow gas	DST interval- 1351-1964-2877 m
MMA-1	**RIL**		2003	Gas	MN-DWN-98/2 and **MN-OSN-97/2**
	Gazprom				NEC-OSN-98/7
		2445	2006		
		?			High pressures TD 4775m
		3200	2007		

Category III

Bengal Basin

Introduction

Bengal Basin, having an area of 89,000 square kilometers and sedimentary fill of 10-15 km, is the north-eastern most of the east coast basins of India. Indian Shield and Shillong massif form the western and northern limits of Bengal Basin. Eastwards the Basin extends into Bangladesh and is bounded by Arakan Yoma uplift. Southwards Basin plunges into Bay of Bengal.

Status of Exploration

The preliminary studies of data by the Stanvac from 1949-53 led to signing a joint-venture project with ONGC called Indo-Stanvac Petroleum Project (ISPP) in 1953. The contract covered an area of 25,900 square kilometers in NW part of Bengal Basin. It has initiated exploration of hydrocarbons in Bengal Basin by geological and photo-geological surveys. This was followed by aero-magnetic survey of shelf and geosynclinal trough of Bengal basin. This helped in preliminary demarcation of the basin configuration and gave an idea about sedimentary fill. Detailed gravity survey with 12000 stations was followed by 500 line kilometer of refraction survey and 8000 line kms of reflection seismic survey.

Ten identified prospects were drilled by ISPP during 1959-1962. Sedimentary sequences of Pliocene to Cretaceous age were encountered in these wells, but no commercial hydrocarbons could be established. So the efforts were stopped by the company (Biswas,B, 1961).

Natomas-Carlsberg Group of USA carried out 5600 line kms of seismic survey in offshore area of West Bengal and subsequently 2 wells, BB-A-IR and BB-B-I, were drilled during 1975-76. Though no commercial discovery could be made from these wells but only the indication of gas and the presence of high pressure in BB-A-IR kept continuation of the exploration interest.

ONGC has taken up the exploration activity in the western part and to offshore part (Kumar, MNV 1963). Many parts of the basin were extensively covered by 24 and 48 fold surveys from 1975 to 1988. 3D seismic data of 210 SKM was acquired in 1989-91 for confirming the presence and mapping the extension of reefs and other features/prospects which were inferred to occur in Amtala, Golf Green and Ichapur areas. Offshore areas were also covered by 24 fold seismic surveys conducted in 1975.

This was followed by additional CDP survey in offshore areas and in the Sunderban estuaries during 1979-81. 3D seismic survey in offshore was in 60 sq.km.

So far about 47 deep wells in the western part of the basin have been drilled (Table.10.bb.1). The commercial hydrocarbon strike could not be established in the basin barring a few gas shows and oil flow in Ichapur well. The deepest well in onland was drilled upto 5826m and to 5841m in offshore.

ONGC took two exploration blocks one in onland and other in very shallow waters of offshore facing Sunderbans under new NELP policy of government in 1998. During that period, ONGC attempted to drill one well in offshore block during 2005, but due to tilting of one leg of Jack up rig it could not be completed.

Geological Setting

Most of the Bengal Basin area is covered by alluvium and there are only a few exposures of rocks towards western fringe of the basin. Gondwana rocks are exposed in the western part as outliers on the Archean basement. Other exposures include basaltic rocks, known as the Rajmahal Trap, lying over the Gondwana sediments (Figure 10.bb.1).

Tectonics

The Bengal Basin is bounded on the west by the Peninsular shield of India, on the north by Shillong Plateau, on the eastern side by the folded belt of Tripura and towards south it opens into deep waters of Bay of Bengal. The Garo-Rajmahal gap between Peninsular shield and Shillong Plateau was created due to strike slip movement of Shillong Plateau by Dauki Fault.

Indian plate got separated from Gondwanaland during lower Cretaceous moved northwards and during upper Eocene had first collision with Eurasian Plate in the north and with Burmese plate in the northeast. Since then Bengal basin, attained the status of convergent margin basin.

The basin had its origin during the collision of India with Eurasia and Burma, building the extensive Himalayan and Indo-Burmese Ranges and thereby loading the lithosphere to form flanking sedimentary basins. Basin development began in the Early Cretaceous when the Indian plate drifted away from Antarctica. The Cretaceous to Holocene Bengal Basin forms a "remnant ocean basin" at the juncture of the Indian plate and the Burma plate (Hammer, S. and Anzoleaga, R, 1975).

It lies on the north-eastern part of the Indian subcontinent, between the Indian Shield to the west and north, and the Indo-Burmese Ranges to the east, and occupies most of Bangladesh, parts of West Bengal state of India and the Northern part of Bay of Bengal. Tectonically from Mesozoic to upper Eocene time, it was a divergent margin basin and was initiated as a response to the breakup of the Gondwanaland along the rifted margin of the Indian plate (Biswas, S.K. *et al.*, 1998). It depicts characteristics related to marginal sag basin.

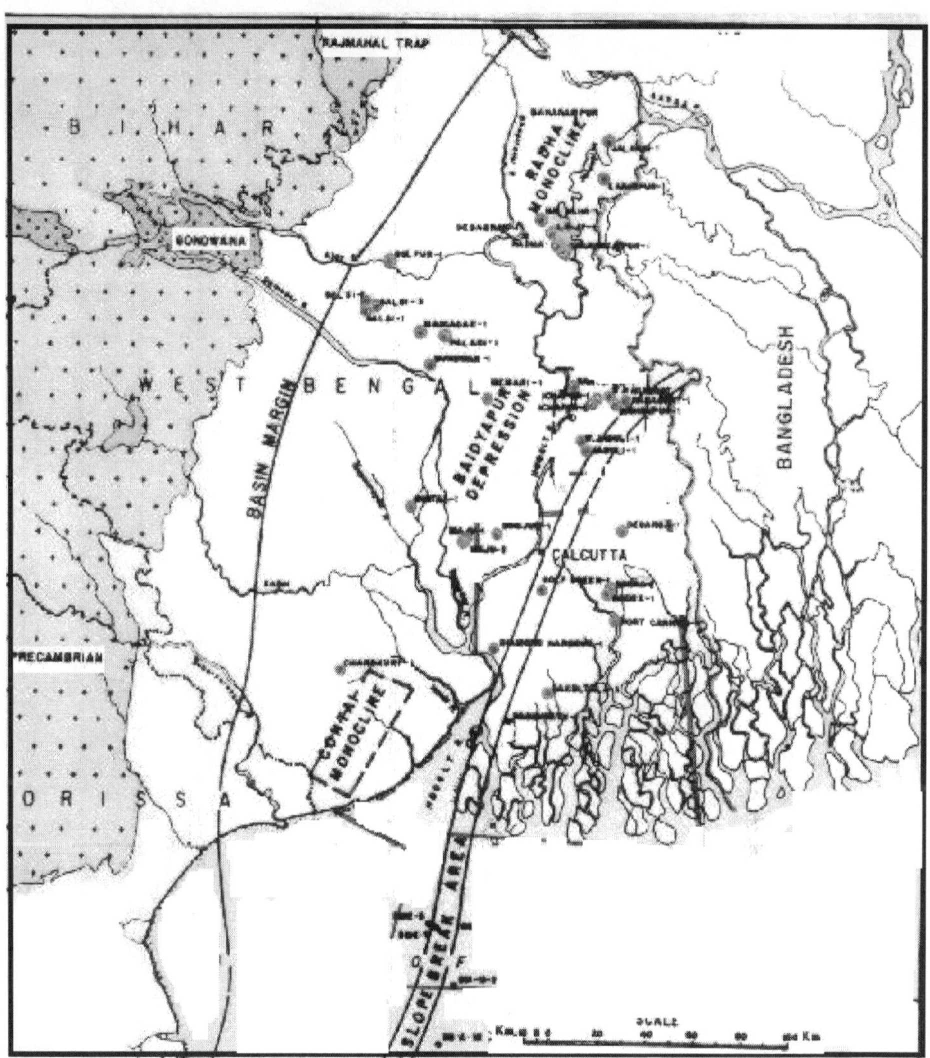

Figure 10.bb.1. Geological Map-Bengal Basin (Modified after Mukherjea, A and S Hazra, 1997).

The basin constitutes the shelf part of the Assam-Arakan geosyncline till Miocene and later developed into large delta of the Ganges-Brahmaputra river system. Both onland and offshore part of the basin has been divided into the western fault scrap zone, the shelf area, the hinge zone and the deep basin (Figure 10.bb.2). In offshore part of Bengal basin large-scale roll-over structures are developed, east of the hinge zone (Alam, *et al.*, 2003).

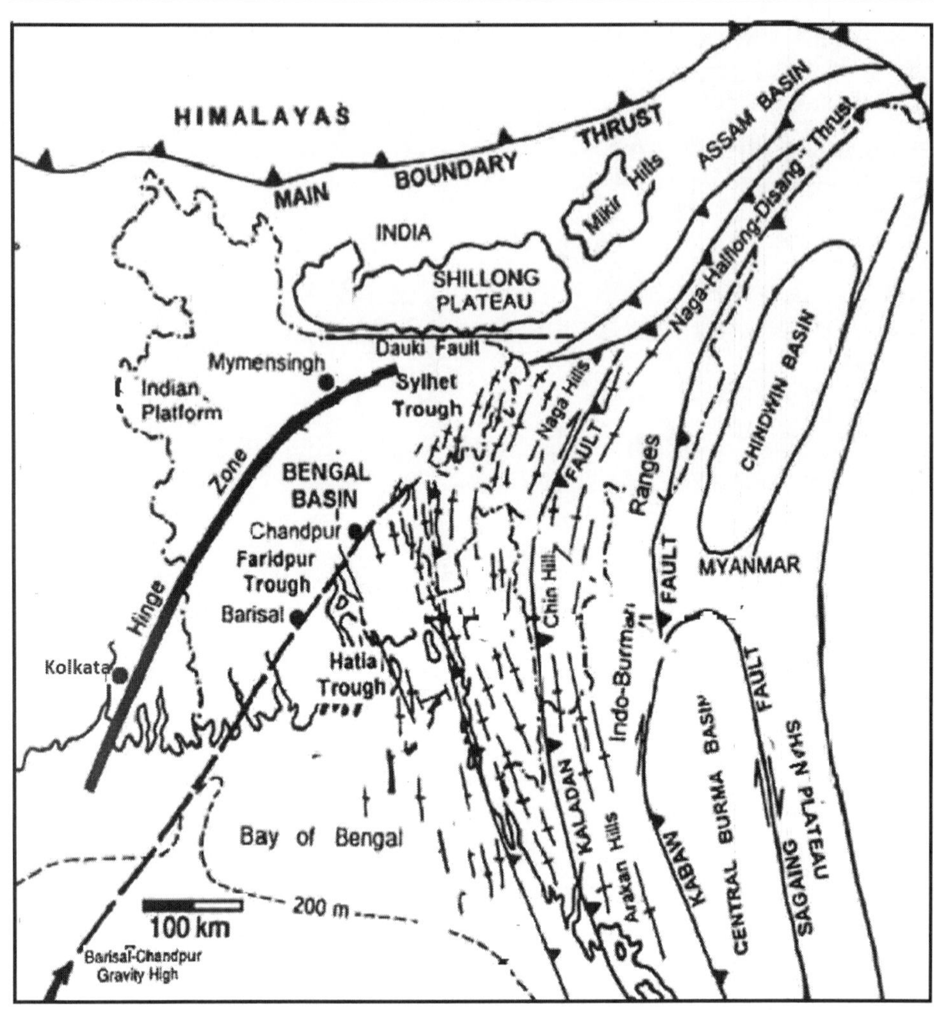

Figure 10.bb.2. Tectonic Map–Bengal Basin (After Biswas, S K and Abinash Agrawal, 1990).

However, as seen by gravity and magnetic data the basin is oriented in a dominant north-south trend instead of northwest-southeast trend as in other east coast basins (Figure 10.bb.3 and 4). The positive gravity anomaly towards west demarcates the basin margin area. A north-south trending low is marked nearer to the basin margin (Prasad, BN, Murty, K.V.S., 1981). After a brief volcanic activity resulting in outpouring of basaltic lava, the basin evolved in a passive margin setup during post-Cretaceous period (Choudhury, SK and A.N. Dutta,1973).

The basin can generally be divided into two parts; the western and northwestern Pre-Cambrian platform slope with thin sediment fill, which is a peri-cratonic rift, and the foredeep of geosynclinal area with thickness of sediments more than 15 km to southeast (Supriya Sengupta,1960).

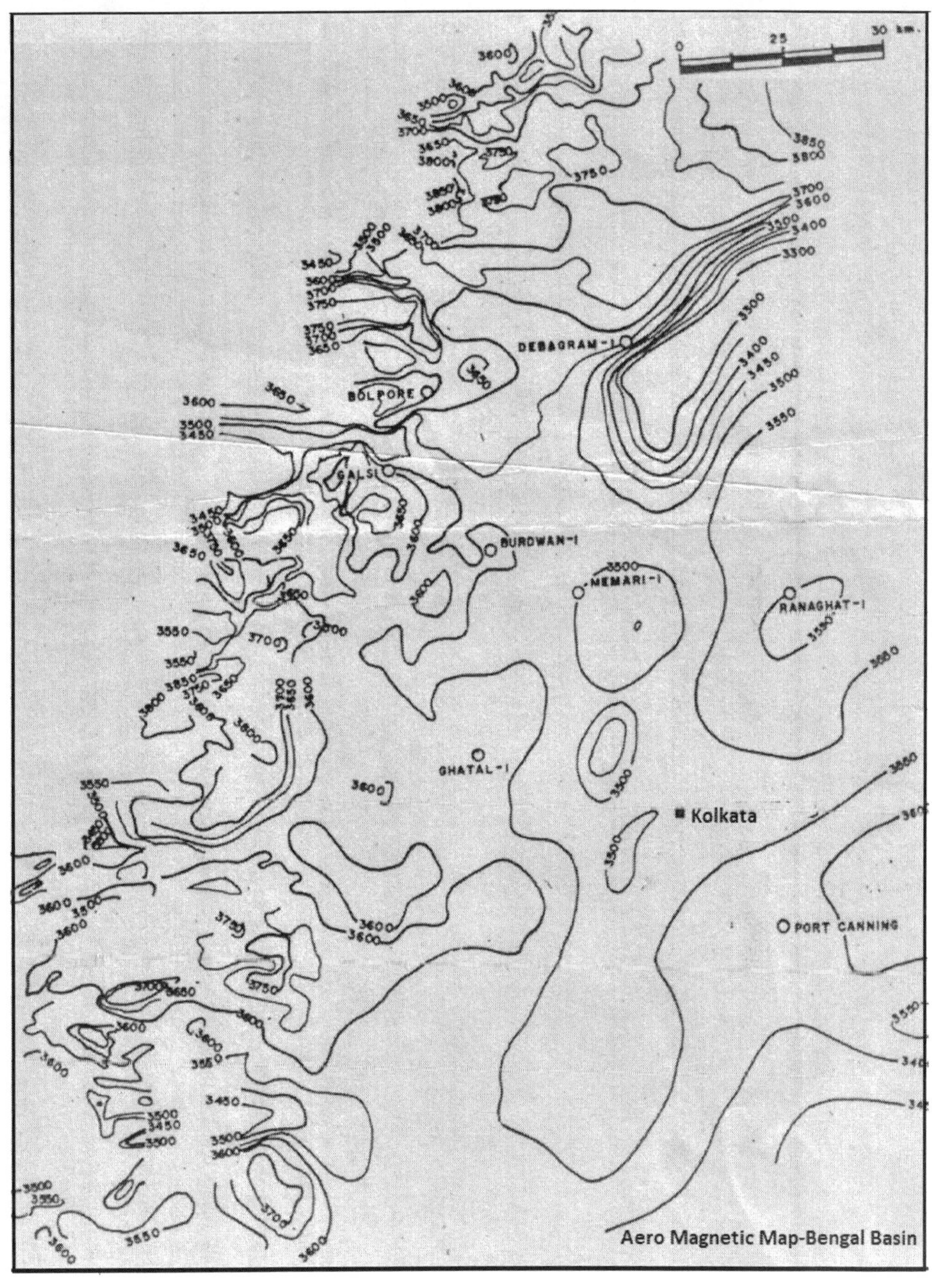

Figure 10.bb.3. Aero Magnetic Map–Bengal Basin (After Venkataramnan, D, 1984).

In the northwestern part, the N–S running Malda–Kishengarh Fault detected in the satellite imagery may be considered as the likely northwestern margin of the Bengal Basin (Figure 10.bb.5). The Bengal Basin in its western and southwestern

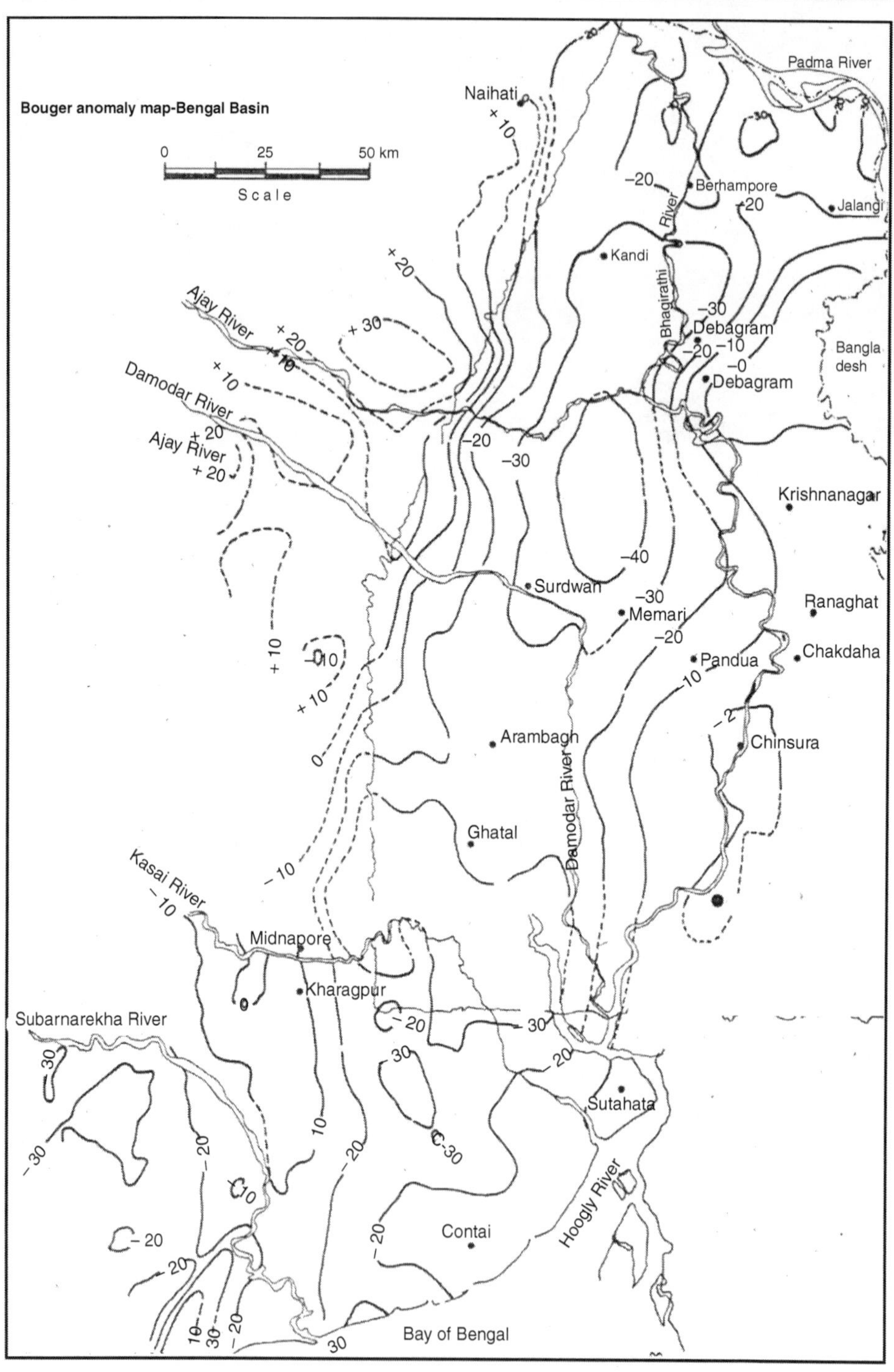

Figure 10.bb.4. Bouger Gravity Map–Bengal Basin (After Venkataramnan, D, 1984).

parts consists of an easterly inclined shelf, separated from the Precambrian Shield (Roy Burman, A and K.V.S. Murty 1982). The basin shows two distinct phases of development, Gondwana phase and post–Gondwana phase (Roy Burman,A, 1992).

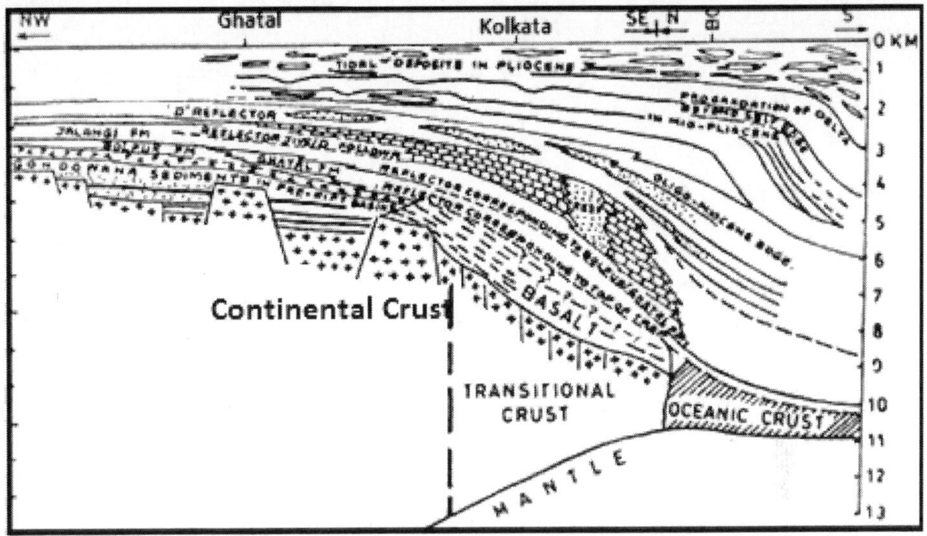

Figure 10.bb.5. Crustal Section–Bengal Basin (After Roy Burman, A 1992).

The Bengal Basin is constrained on all sides by different fault-controlled tectonic features, except in the south where the rivers drain into the Bay of Bengal. In the northeastern part, the Bengal Basin abuts against the Shillong Plateau. The E-W trending subvertical Dauki Fault marks the southern boundary of the Shillong Plateau.

Structure

The subsurface structural configuration of the basin is well recognizable with the help of Bouguer anomaly map, gravity and seismic configuration, the basin was mainly divided into shelf area, NNE-SSW trending shelf edge, slope and deep basin. The hinge zone forms a curvilinear feature extending up to Shillong massif from offshore Mahanadi basin to Calcutta to Mymansingh with a NE-SW trend. Within the shield area the Damodar Graben came into existence in the Permian along E-W trending faults extends underneath the Bengal basin at the Basin margin area up to Hinge-zone. The main tectonic and structural zones identified in the basin are, *Basin Margin Fault Zone, Shelf Zone, Hinge Zone and Deep Basin*.

Basin Margin Fault Zone

This NNE-SSW trending fault zone demarcates the western crystalline/ metamorphic complex of Precambrian age from the shelf sediments. The fault zone is, apparently, the result of down warping of the shelf region during Early Cretaceous, probably, concomitant with the eruption of Rajmahal basaltic lavas.

West of this feature are the exposures of Gondwana sediments resting on the Precambrian basement (Venkataraman. D.1984)

The geometry of the Bengal Basin is asymmetric having thinner, gently sloping sediment cover in the western and northern parts, which thicken in the basin-ward direction. The geo-scientific data indicate considerable heterogeneity in different parts of the basin in terms of stratigraphy, sedimentation, tectonics and basin evolution (Figure 10.bb.6).

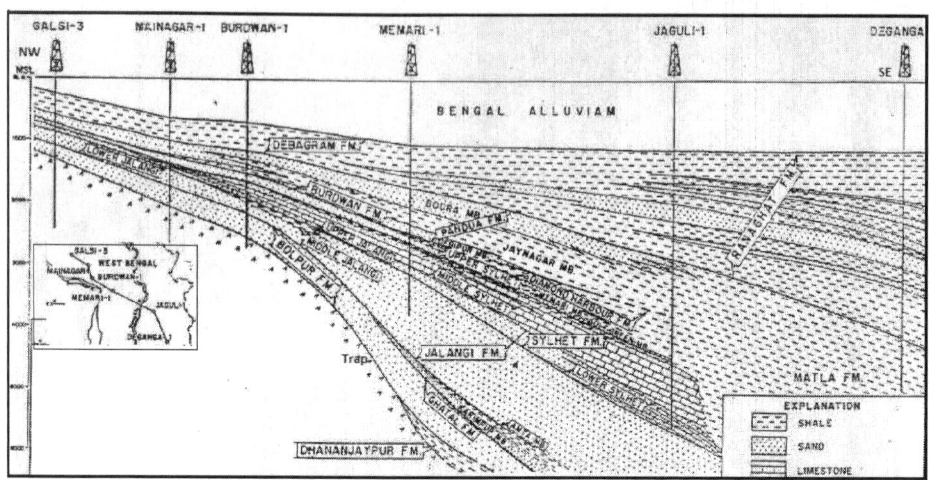

Figure 10.bb.6. West to East–Wells (Galsi–Deganga)-Geological Section (After Hajra P.N., *et al.*, 1997).

Shelf Zone

The Bengal foreland shelf is more than 100 km wide in north and narrows a little towards south. The Tertiary sedimentary prism thickens towards east and merges with the deep shelf beyond the hinge. Most of the exploration activities more or less confined within this zone. West of Burdwan the break in the basement slope is conspicuous. The basement slopes gently towards east and is marked by numerous step faults with small displacements (Figure 10.bb.7). The post-Paleocene period witnessed a gradual deepening of the basin.

Hinge Zone

Bengal 'Hinge Zone' is a narrow elongated zone separates the thick Post-Eocene sediments in the east from the shelf zone of the west. This feature runs in NNE-SSW direction. This created necessary environment for the development of shelf carbonates during Eocene. Development of a linear zone of thick limestone along the shelf edge, interpreted as a reefal build-up along the shelf break. Eocene limestone further east of hinge zone possibly grades to shales and other deep basin sediments. This zone has given rise to a belt of possible flexures and faulting during Oligocene and Miocene. Truncations and wedge-outs against the hinge zone are characteristic. Golf Green and Ichapur wells are observed to fall in this trend (Mukhrjee,B. and

A.K.Dey1997). The hinge zone is associated with gravity high and magnetic low. However, seismic data indicates only a change in basement slope along this zone. All the sequences show gradual change in depositional environment from fluvial through deltaic to marine conditions from west to east.

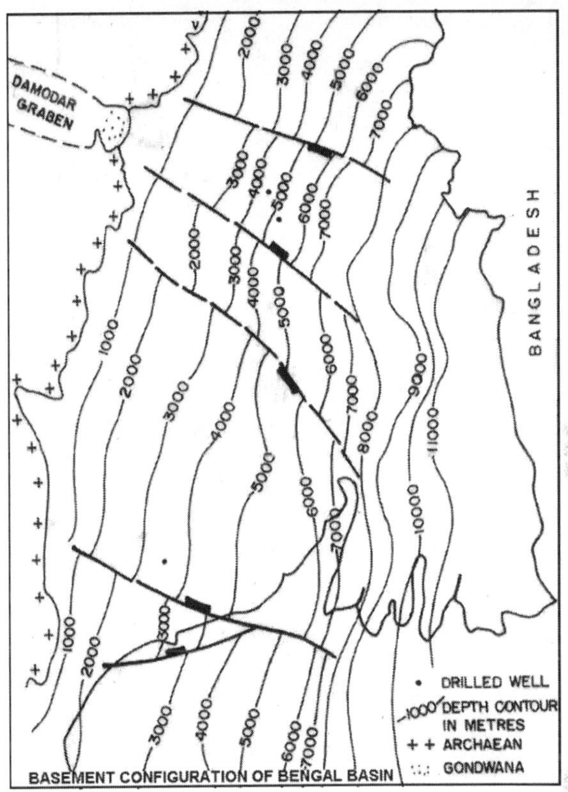

Figure 10.bb.7. Basement Configuration–Bengal Basin.

Deep Basin

It extends towards east and southeast in to Bangladesh territory leaving small portion within Indian boundary. Constrained between Eocene shelf-break in the west and the easternmost margin of the Chittagang Tripura Fold Belt (CTFB), the Southeastern Sub-basin shows a huge sediment thickness ranging between 10 and 18 km. A Continent–Ocean Crust Basalt (COB) line has been recognized between the shelf-break in the west and the Barisal–Chandpur Gravity High (BCGH) in the east. The basinal part located towards southeast of the Eocene hinge zone, hosts a thick prism of 10-15 km of sediments which are largely made-up of post–middle Miocene deltaic deposits (Maroof Khan M.A and Musherraff Husain 1980).

Sedimentation

During the Gondwana phase, non-marine sediments got deposited within a graben oriented in N-S direction. This initial phase of sedimentation was followed by a period of peneplanation and initiation of a general tilt of the shelf towards east. After the outpouring of Rajmahal lavas, an uninterrupted phase of marine sedimentation in a subsiding basin started in Late Cretaceous. Marine inundation covered practically the whole of the basin up to its western margin. A thick sedimentary prism was deposited during Tertiary.

During Permo-Carboniferous and Early Mesozoic times, the rift-related grabens were filled with land-derived coarser clastics, carbonaceous shale and coal. The arenaceous facies comprising ill-sorted sand, derived from granitic provenance in the west filled the graben. After the break-up of the Gondwana land, the outpouring of Rajmahal basaltic lava during Early Cretaceous.

The first major phase of sedimentation started with the Cretaceous marine transgression and continued up into the Oligocene period. Since the Cretaceous, sedimentation in the greater Bengal Basin was controlled by the movement of the Indian plate with the Burmese and Tibetan plates, and by the uplift and erosion of the Himalayas and Indo-Burmese mountain ranges (Benerjee, Ranjit K., 1983).

The first marine transgression over the shelf took place during Late Cretaceous with the deposition of the Dhananjaypur Shale followed by the Bolpur Formation. Deposition of thick coastal, shelf and marine sediments during Late Cretaceous and Tertiary periods took place in several cycles of transgression and regression The Bolpur formations of Upper Cretaceous were the first to be deposited by the proto-rivers flowing from western highland. During Late Paleocene and Early Eocene, a thick shale-sand alternation, interspersed with carbonates were deposited in a transgressive sea.

Development of carbonate facies of Eocene marks a relatively quiet and warm period during which subsidence was slow or partially arrested. Subsequent to the deposition of the Kalighat Limestone, a major phase of clastic sediment deposition started. This is attributed to the initiation of collision of the Indian Plate with the Eurasian Plate towards the end of Eocene. This event also marks the beginning of a marked shift in the direction of drainage system from easterly to southerly, consequent upon the emergence of Himalayas (Figure 10.bb.8).

The Oligo-Miocene sedimentation condition was followed by a strong deltaic environment. Early Miocene-Pliocene epoch is characterized by the deposition of a thick succession sediments brought by the Brahmaputra-Ganga river system as delta fan deposits over the lower shelf-slope system (Biswas S.K and Abhinas Agrawal,1990). The Miocene collision of the Indian plate, there was a change from flysch to molasse sedimentation.

The ***depositional environments*** of the western part of the Basin can be summed up as: The Gondwana sediments are directly overlying the basement and deposited in continental environment. The basaltic flows (Rajmahal Trap) overlying the Gondwana sediments are partly subaqeous. The sediments overlying Trap were deposited mainly in continental environment. The first evidence of transgression in the basin is during Cretaceous period and upper most Cretaceous sediments are deposited in regressive phase. During Paleocene period first transgression and later a regressive phase is identified. The transgression during this time is not widespread as during Cretaceous time. Eocene period is having widespread transgression covering the area as shallow as Galsi. The end of Eocene is marked by the onset of regressive phase. Early Oligocene is represented by a regressive phase.

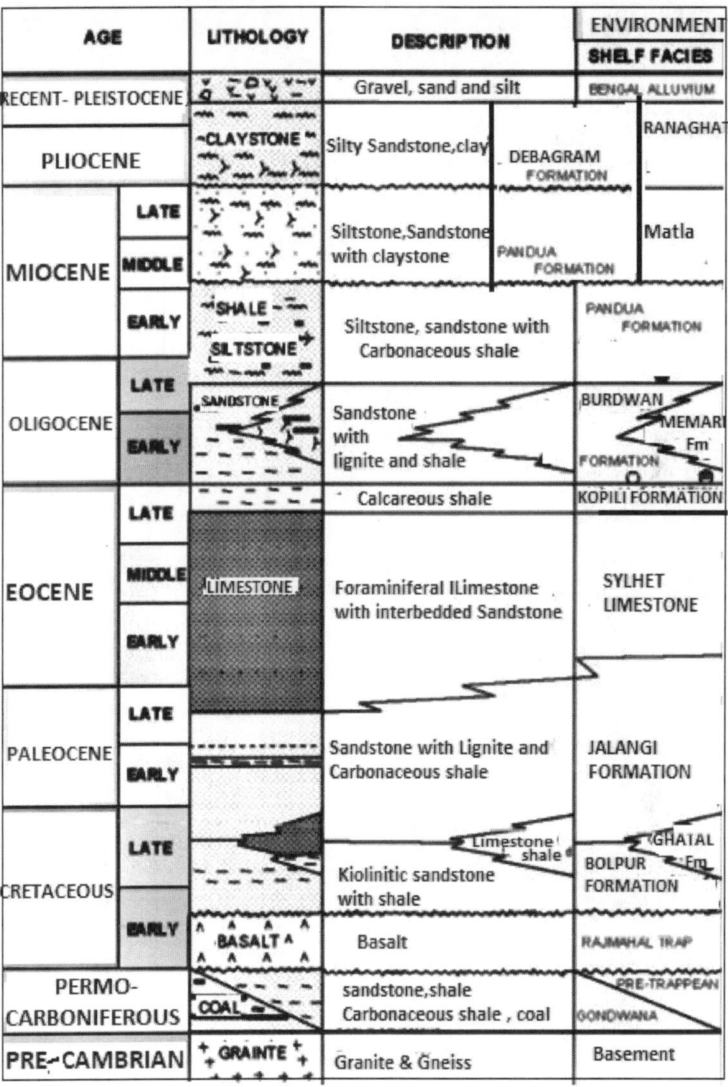

Figure 10.bb.8. Litho-Facies Variations in different Tectonic Blocks of Bengal Basin (After Das Sharma, S, *et al.*, 1993).

For a short interval during Middle Miocene, there is evidence of marked regression. Otherwise, a widespread transgression appears to have taken place covering the whole of the basin even beyond Galsi. The upper Miocene is again identified by a regressive phase. Pliocene is having two phases of transgression with a regression in between. Quaternary sediments appear to have deposited in oscillatory environment but in general in a regressive phase (Saxena, I.P. C.K. Singh and Sukhadarsan, Kumar 1984).

Stratigraphy

The data obtained through deep wells drilled different parts of the basin revealed a thick sediment fill of continental to marine clastics and carbonates ranging in age from Early Cretaceous to Miocene (Raman, K.S., Kumar, Sukhdarshan, Neogi, B.B., 1986). The depth of the Archaean basement in Galsi well is at 1233m towards northwestern part of shelf area of the basin.

Four wells (Chandkuri, Galsi-3, Mainagarh and Palashi) in shelf area and two in Bogra shelf of Bangladesh (Kutchma and Singra) have penetrated pre-basalt Lower Gondwana. Thick alluvium of Ganga and Brahmputra rivers cover the sub-surface stratigraphy in the whole area of the basin.

Based on initial drilling data the ISPP has formulated the detailed stratigraphic classification (refer Figure 10.aa.8). Some changes have been made using available information from time to time which include the identification of pre-trappean (Mesozoic) sediments below Rajmahal Trap in Galsi-3 and a thick section of Diamond Harbour Formation, the equivalent of Burdwan formation of Oligocene age renaming of Sylhet Limestone as Kalighat Formation, identification of Dhananjaypur Formation as a post-basalt clastic wedge and renaming of Kopili as Hoogly Formation. The present litho-stratigraphic column is shown in Figure 10.bb.9.

Cretaceous

The Bolpur Formation consists of lithic sandstone and red clays and rest over an erosional unconformity. The greenish and grayish white quartzose rocks represent deposition in fluvial to estuarine conditions and in the prominent grabenal areas. The marine sedimentation was commenced in Late Cretaceous is continued up to Early Paleogene.

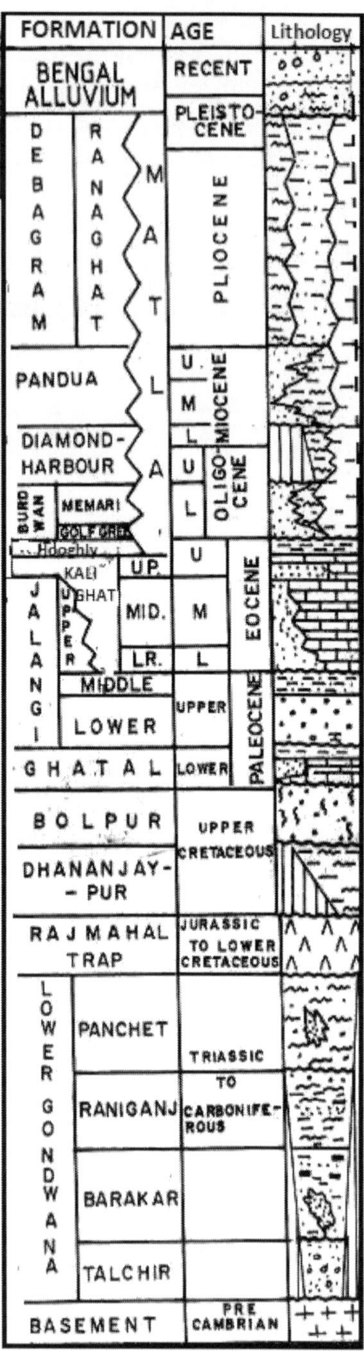

Figure 10.bb.9. Litho-Stratigraphic Nomenclature–Bengal Basin (After Mahesh Chandra, *et al.*, 1993).

The Ghatal Formation is represented by calcareous shale, shaly limestone, siltstone and fine sandy intercalations. The isopach map of these formations indicates thickening towards east and depocenter is located east of Damodar graben. The data also suggest the passive margin subsidence of newly formed continental margin. These varied lithofacies are indicative of oscillatory movements of basin margin area during deposition in probably brackish, lagoonal and littoral environment. The sequence stratigraphic studies carried out in Bogra shelf area indicate low stand deposition. The sequence also indicates updip pinch-out and the top is marked by a carbonate bed which is possibly of 68 Ma.

Paleocene

The Jalangi Formation is composed of sandstone with carbonaceous shale and is represented by fluvial in lower part and brackish to shallow marine beds towards upper part, which represent gradually shallowing upward carbonate platform of the high stand system tract and part of the transgressive phase. The boundary between overlying Sylhet (**Kalighat**) Limestone and Jalangi Formation seems to be diachronous.

Eocene

The Eocene represents three carbonate bands separated by two coarse clastic inter-beds of varying thickness. During Middle Eocene the depositional conditions appear to be stable. The fauna consisting of corals and algae suggest warm open marine conditions of deposition and the presence of burrows indicate tidal environment at northern part of the basin margin (Debagram) area. The lesser terrigenous matter and evidences of bio-turbation can be related to lower rate of sedimentation. These conditions are favorable for reefal growth. The thicker Paleocene and Eocene succession in southeast suggests major depocenter in the area. It is observed that the effect of Damodar graben during sedimentation is minimized and NE-SW trending shelf edge has taken its shape. The variations in thickness of Kopili sedimentation is due to filling of geomorphic lows over the Sylhet top which seems to be an erosional surface at places. Though the shales represent a transgressive phase, the thin sequence indicates the delta was starved of sedimentation (Gayatri Ghose, *et al.*, 1999).

Oligocene

The sedimentation rate has increased since Oligocene followed by major regressive phase which helped in major delta progradation. The rate of deposition was low compared to the rate of subsidence hence the area experienced transgression. There were also intervals of regression caused either due to shallowing up of the basin or increase of sediment supply due to uplift in the provenance area. The latter seems to be more probable because major upheaval of Himalayas had taken place during that time. This has resulted in a basinward progradation of depositional sequences.

Miocene

The Pandua Formation exhibits a wide variation in depositional environment (estuarine to littoral). The thickness of the formation increases rapidly towards southeast, indicating preferential swing of the basin because of more sediment input from north to south due to rise of Himalaya and/or syn-sedimentary growth faulting. The delta build up is also had swung from WNW-ESE to north-south. Since Late Miocene onwards, the sediment supply from north was more than the rate of subsidence.

Pliocene

The Debagram Formation was deposited in marginal marine conditions. This was a period of continuous regression (Saxena, I.P. C.K. Singh and Sukhadarsan Kumar 1982). The isopach map indicates enormous increase in sediment input from north due to which the spread of delta are was increased towards west. Due to frequent changes in sea level during Pleistocene there were intervals in delta-progradation (Mahesh Chandra, *et al.*, 1993).

During Holocene, enormous sediments brought by the Ganges and Brahmaputra rivers had given rise to the tide dominated delta (Gayatri Ghose., *et al.*, 2004).

Abnormal Pressures

The data from the drilled wells on the stable platform suggest hydrostatic to sub-hydrostatic pressures down to basement whereas near the slope break, the lower part of Miocene shows abnormal pressure slightly more than hydrostatic. The pressures in the deep basinal part are of more than hydrostatic in the lower part of Pliocene and become abnormally high near the top of Miocene. The high rate of sedimentation represented by dominant clays and interbedded sandstones (high shale/sand ratio) under rapid bottom subsidence are the likely reasons of high pressures in Bengal Basin (Magara, K. 1978).

The well **Port Canning-1** situated in the same area was drilled under Indo-Stanvac Petroleum Project, but could not reach the target depth due to high pressure troubles and was finally abandoned. Bakultala-1 in the nearby area was abandoned before reaching the high pressure zones (Fertle, H.W, 1976).

The well **Bodra-2** is situated in the deep basinal part of Bengal Basin, east of Eocene slope break. It is the second well drilled on the Bodra structure. The earlier well (#1) could not be successfully completed to target due to the pressure of very high pressure resulting in various complications and was abandoned at a depth of 4179.5 m (Majumdar. S K., and Saibal Basu 1985).

In Bodra well #2 which was targeted to drill up to 5500 m has experienced drilling complications like kick, mud-loss, stuck-up at depth 4324m. After several attempts to recover the well, it was side-tracked by placing cement plug over the fish left inside the well with top at 3777m. However, again while drilling at 3800 mud loss happened while drilling with 1.96 mud weight and same was reduced to 1.94 and drilled down to 4075 (Rao, G.N 1999). Again mud weight was increased to 2.02 because formation pressure was calculated to be 2.01 MWE below. As self-flow was observed mud weight was increased in stages upto 2.02MWE, with this total mud loss was observed. Well was further drilled down to 4951m with all precautions. At this depth decided to lower 5 ½″ casing but during that operation total mud loss was observed. Well some times blew out when reduced the mud weight. Ultimately well was capped and abandoned.

Abnormal pressure phenomenon in Miocene sequence of Bengal Basin indicate that the associated shale sections have not yet reached the stage of equilibrium in compaction. During Upper Miocene and Pliocene due to higher rate of Sedimentation by major Ganges-Brahmaputra combine in the basin suffered normal compaction due paucity of time coupled with syn-sedimentary subsidence in the deeper part of the Basin.

Further, this has also impacted in lesser fracture pressure in litho-sequences. As seen by well behavior of Bobda#2 well, there is very less difference between formation pressure and its fracture pressure (Tiwari, S, *et al.*, 1986).

Petroleum Geology

The prolonged exploration in Bengal Basin did not result in a commercial strike of hydrocarbons. For commercial accumulation of hydrocarbon the essential criteria include, good source facies, maturation and expulsion, reservoir and seal facies, migration path and finally trapping mechanism.

Theses essentials were examined to certain extent it was observed that, Oligocene and Eocene carbonaceous marine shales are considered to be the potential source rocks of the Bengal basin (Dow, W.G.1977). Early to Middle Miocene sandstones and siltstones are the potential reservoir rocks in Bengal basin. Porosity ranges generally from 10 to 20 percent. Reservoir sands range from thick channel-fill and littoral or marine bar deposits to sandstones thinly interlaminated with shale and siltstone. Structural and combination traps of Miocene age occur along stratigraphic boundaries, in sandstone-filled channel deposits and in sandstone beds sealed laterally by shale-filled channels and these comprise major traps in the eastern part of the basin (Hantschel, T., and A.I. Kauerauf, 2009).

Mio-Pliocene shale sequence acts as an effective cap rocks for the gas reservoirs of the Bengal basin. Intra-formational shales and claystones are common seal for the gas fields of Bangladesh.

It is well known that for generation of hydrocarbons the maturity of organic matter is the most important factor. Maturity of organic matter is a function of geologic time, temperature, and depth of burial. The source rock maturation for various stratigraphic units and probable oil window zone across the shelf. Maturation values for various stratigraphic sequences such as Paleocene, Eocene, and Miocene show that the hinge zone and deeper basinal areas form favorable areas for maturation of organic matter (Chawla,B.K., *et al.*, 1985). The temperature curves of the four wells (Debagram, Radha, Abhay and Dhananjaypur) indicate vary low temperature gradient of the order of 1.8° C to 2° C per 100m. Recently one well Asoknagar#1 toward shallow northern part of the basin flowed good amount of gas. It has to be examined in detal the nature of gas and if there is any local kichen in the area.

The *prognosticated hydrocarbon resources* in Bengal basin are placed at 190 MMt of oil +gas equivalents of which 30 MMt are in onland prt and 160 MMt in offshore.

Source Facies

The Lower Gondwana sediments are contain upto 4 per cent organic carbon with Type-II and Type-III kerogen in wells Galsi-1, Mainagar-1, Palasi-1. The Vitrinite reflectance values of the source material within the Gondwana sequences range between 0.47 per cent and 3.29 per cent (Mukherjee, J and A.S. Chopra (1990).

Cretaceous sediments show maturation in Bolpur, Ghatal and Dhananjaypur Formations. Paleocene sediments of VRo maturation value of 0.65 per cent and shows a good source development with type II and III kerogen in the Jalangi Formation. Source beds within the Ghatal Formation, having good to moderate percentage of TOC are in the early stage of maturation.

The Early Eocene carbonate facies are observed to change basin ward into highly organic-rich marine shale in wells (Venkatachala, B.S. and J.Mukherjee., 1982)

The geo-chemical studies carried out in the basin indicated the presence of reasonably good quality type II organic matter in Pre-Cretaceous Lower Gondwana sequence in Palashi, in Cretaceous sequence of offshore, in Paleocene sequence at Chandkuri, in Eocene section at Domjur and offshore area in a few samples of Oligocene at Ichapur, and in Miocene section in off-shore In other successions the organic matter is of essentially of type III nature (Hunt, J. M., 1979). The concentration of organic matter is generally about 0.4 per cent in most of the sections. However, good concentration of more than 1 per cent is found in Paleocene section of Palashi area, in Eocene at Maju-Domjur-Debagram-Radha area, in Oligocene interval at Jaguli area, in Miocene section of Diamond Harbour-Raghunath and offshore area (Hajra P.N., *et al.*, 1997). The lipids from the kerogens had mixed terrestrial-algal source input.

Maturation

The geothermal gradient is low in Naldah area, 1.8°C per 100 meters with an average of 2.4° and maximum of 3.5° at Chandkuri area. The source facies in the Lower Gondwana section in Palashi-Chandkuri area, Paleocene–Eocene section in Maju-Domjur area, and the Miocene sequence in Bodra, Raghunath and offshore area have indicated the early phase of maturation (Kapoor P.N. 1997).

The Time Temperature Index (TTI) studies in the basin here revealed that, the Lower Gondwana sediments have matured to generate hydrocarbons during Late Tertiary period, the Miocene sequence in deep basin area during Pliocene (Thomas and Thomas1985).

The TTI value of 15 was reached only in the upper Createaceous section in the deeper part of the Radha-1 and Abhay-1 wells through seismic data. The onset of oil generation from the upper Cretaceous section started vary late (2.5 to 5 million years before present) during Pliocene time. This suggests that the source rocks are not adequately matured to give sufficient amount of hydrocarbon to form oil/gas pools.

The deeper sections in Palashi well indicated the initiation of gas generating window around 4800m and below. The expulsion efficiency during primary migration of hydrocarbons from source facies is expected to be good because of sufficient overburden. Significant maturation (Ro=0.65) has commenced in the entire Bengal Basin around 4000 m. Jaguli-1 has revealed a new mature source rock sequence in the Oligocene 3915-4457 m having wet gas with subordinate oil prone type III-A kerogens.

The Miocene succession in deep basinal part, in Bodra-Portcanning area, is abnormally pressured below 3500m and attained high formation pressures during Pliocene. The formation pressures up to double the hydrostatic pressure were recorded in the area. Hence, the expulsion of hydrocarbons in that part of the basin may be poor.

Oligocene to Miocene sediments are, by and large, immature within the Bengal Shelf. Beyond the 'Hinge zone', maturation of Oligocene sediments is expected to increase with the increase in the depth of occurrence of such sediments. This is also observed in case of the Jalangi Formation which forms an important source sequence in the Bengal Basin.

Reservoir Facies

Reservoir facies are available in abundance within the Gondwana and Paleocene sequences are of arenaceous character. Sandstone reservoirs occur in abundance within the Bolpur, Lower Jalangi and Upper Jalangi formations and higher up within the Burdwan Formation of Oligocene age. A sand within the Burdwan Formation is found oil bearing in well Ichapur–1. In the Miocene sequence, sandstone reservoirs are present in abundance within the Bodra and Pandua formations (Roy Burman, A., 1983).

Cap Rock and Entrapment

Gondwana rocks have been deposited as an alternating sequence of sand, shale and coal in Galsi, Mainagar and Chandkuri areas. Graben setting, together with composite lithology, may provide suitable conditions for hydrocarbon accumulation in structural, stratigraphic and strati-structural traps. Barakar and Raniganj shales should serve as good seals.

The argillaceous Amta Member of the Ghatal Formation can serve as seal for Bolpur sand reservoirs. Argillaceous Middle Jalangi unit also may serve as seal for the reservoirs of the Lower Jalangi unit. Hydrocarbon accumulation in this section is expected to be mostly in stratigraphic traps. Cretaceous and Paleocene: Low amplitude structural highs, broadly in a NE-SW trend through Contai–Ghatal-Jalangi–Karimpur areas.

Oligocene: Onlapping reflectors of good continuity and moderate to high amplitude are observed in the seismic sequence corresponding to the Oligocene section.

Tight limestone of the Eocene section and the Hoogly Shale may serve as seal for the porous reefoidal Eocene reservoirs. Accumulation in the Eocene section is expected to be in stratigraphic and strati-structural traps.

In the deeper part of the shelf, the Matla Formation is observed to thicken considerably.The Memari Member of this formation may act as a good seal for reservoirs within the Burdwan Formation and the Jayanagar Member can be a seal over Miocene reservoirs. The deltaic system with composite lithofacies within the Oligocene and Miocene sections has a potential to form stratigraphic hydrocarbon plays in the area near and beyond the hinge zone (Mukherjea A. and Neogi, B.B. 1993).

Growth fault related low amplitude structural prospects in the depositional low in Krishnanagar–Ranaghat area. Eocene shelf edge forms a favorable locale for the development of carbonate reef prospects.

Hydrocarbon Plays

Gondwana source rocks at the shelfal area are presently in dry gas generation maturity, if suitable traps are available hydrocarbons might have accumulated in Gondwana reservoirs itself for which trap volcanics might have provided the seal resulting in conventional/unconventional accumulations similar to coal bed methane.

Effective source rocks in late Cretaceous Bolpur Formation, Paleocene Jalangi Formation, Eocene Kalighat-Hoogly Formation and Oligocene Burdwan Formation are presently in wet to dry gas to over mature stage and Mid to Late Miocene Pandua Source rock layer is in main oil to wet gas window towards basinwards, east of hinge zone where significant hydrocarbon generation and expulsion has taken place.

Gondwana Entrapment conditions are expected to exist in basement related structures and stratigraphic wedgeouts in the central, northern and southern parts of the Gondwana graben.

Velocity analysis suggests that some anomalies correspond to sand-shale alternations within Oligocene and can be target for exploration. The wedgeout zones are equally interesting for stratigraphic accumulation.

Mio-Pliocene Prospects: Seismic facies analysis and basin modeling indicate development of two types of prospects in the Mio-Pliocene section, deltaic, prospect associated with turbidite facies along the shelf-slope area south of Sunderbans.

Petroleum System Modelling

The 2-D petroleum system modeling along a dip section of Bengal Basin used to assess hydrocarbon charges and its migration for evaluation of possible hydrocarbon prospectivity.

The source rocks charecteristics in the Gondwana sequence are very good. Fair to good source rocks are present in Creataceous and Tertiary sections with mostly type-III kerogen. In the basin, main oil window extends from Gondawana sequence at shelf area and up to bottom part of Miocene at eastern shelf edge and beyond. The source rocks are within main oil to dry gas window at present analysis. But the hydrocarbon charge from the Cretaceous and Tertiary sediments since 30 Ma is not sufficient enough for existence of commercial volumes of hydrocarbon (Pahari, S, *et al.*, 2010).

Possible Reasons for hither to Unsuccessful Exploration in West Bengal Basin

For commercial occurrence of hydrocarbon generation with avaialability of rich organic content and its preservation are important, which in turn is controlled by rate of sedimentation. For preservation of organic carbon in sediments faster subsidence below oxidizing conditions (wave Base/photic Zone) is important. In shelf part of the Basin most of the sedimentary columns in were deposited in above photic zone.

In Bengal Basin, during Mio-Pliocene time the rapid sedimentation probably diluted the organic content, which might have been responsible for non-generation of enough hydrocarbons for commercial accumulations.

Further, the temperature needed for maturation is low due to lower geothermal gradient. Though other source rock and maturation data indicate some favorable conditions for maturation, some of them may have its limitations. When palynological data show TAI value for maturation, at the same depth where paleontological data indicate forams were deposited under oxidizing conditions. As it was known, the TAI value is the coloration of spores and pollens due to effect of temperature with geological time. So proper collective synthesis is needed for effective conclusion (Moore, L.V., and T.L. Leyenberger, 1994).

Along the shelf edge nearly 1000m of thick limestone may have local porosity pods. Though, the algal nature is attributed to these limestones and reefal build-up in the basin is to be probed in detail through seismic data (Tiwari. S and S.k. Banerjee 1985). Reflection free in a bump area though drilled did show any major algal content. Those sections are found to be compact Limestone sequences (Krishnan, P.V. and G.N. Rao 1982).

In the deep basinal area within the Neogene section the reservoirs are thin and seal-cum-source facies are thick. But the below 3700 m in deep basinal area the section is abnormally pressured for optimal economic drilling of wells knowing that only occurrence of gas is the possibility. The high formation pressures prevailing in the deep basinal part might have caused the gasification of possible liquid hydrocarbons generated if any, in the deep Basin (M. Julleh jalalur Rahman and Shigeyuki Suzuki 2007).

For expulsion of hydrocarbons generated if any is based of overburden pressure after maturation of source facies. In the area of shallower part of the Basin it may not be sufficient, but towards east of shelf in the deep basin area it may be sufficient, but maturation is also simultaneous during Pliocene section.

The direction of migration play major role in filling traps in that path-way. As it seems, in the greater Bengal basin the deeper part is having criteria for hydrocarbon generation, but major commercial accumulations are seen to east of Barisal –Chandpur High, where clear high amplitude structures are available, whereas the available small traps to the west of this high are devoid of commercial accumulations (Md. Noor Alam., *et al.*, 2006). It is giving an indication that, in the basin, the migration path was towards east. No major pools could be located in the shelf part of the basin which may indicate further that, hydrocarbons might not have vertically migrated from deep Basin area in to the shelf area through steeply dipping hinge zone (Rao, G.N., 1984).

The continued exploration in the West Bengal basin about four decades, more wells were drilled in shelf part. In this area in many stratigraphic levels the penetrated section is more areanceous. The traps are mainly very low amplitude fault closures. In such case fault prospects with less throw may not be sealing to arrest hydrocarbons with in the envisaged traps.

The Paleogene section is found to be occurring as clastic wedge against steep dipping hinge zone. Possibly such wedges might not have been located in the migratory path of hydrocarbons if any, after the formation of such traps (Rao, G.N., 1985).

The prospects in Bengal Basin lie in seismically uncovered Sundarbans area, possible development of reefs in Eocene carbonates and clastic wedges in Palaeogene and Early Neogene close to kitchen areas.

Table 10.bb.1. Data of Drilled Wells in Bengal Basin

Well-Bengal Basin	*Company*	*Depth. m*	*Bottomed in-Fm*	*Completed*
Burdwan	ISPP	2726	Rajmahal@2525	09-08-57
Galsi#1		1296	Rajmahal@1261	06-10-57
Jalangi		3576	Rajmahal@3380	06-05-58
Debagram		3106	Rajmahal@3072	30-08-58
Bolpur		1524	Rajmahal@1286	24-12-58
Ranaghat		3965	Pandua@2612	18-05-59
Memari		3807	Ghatal@3794	11-09-59
Port Canning		4042	Matla@3942	15-02-60
WestRanaghat		2604	Pandua@2520	28-03-60
Ghatal		3109	Rajmahal@3008	24-05-60
Bodra#1	ONGC	4197	Pandua@3600	14-05-69
Bakultala		3700	Ranaghat@ @292	01-03-76
Galsi#2		1238	Rajmahal@1222	01-03-77
Galsi#3		2484	Basement@2455	01-01-78
Diamond Harbour		5555	Jalangi@5200	03-02-81
Radha		3506	Ghatal@3180	06-02-81
Abhay		2909	Ghatal@2730	29-01-82
Jaguli		5637	Jalangi@5303	26-07-83
Bishnupur		3006	Pandua@2500	24-07-62
Naldah		3506	Rajmahal@3375	02-06-83
Bodra-2		4950		18-12-87
Maju#1		4020	Jalangi@3820	
Dhananjaypur		5204	Rajmahal@4425	
Maju#2		4009		

Well-Bengal Basin	**Company**	**Depth. m**	**Bottomed in-Fm**	**Completed**
Deganga		5007	Matla@3125	
Domjur		5189	Jalangi@3849	
Raghunadh		5339	Matla@3424	
Palasi		5640	Basement@5623	Jul-89
West Jaguli		3010	Pandua@2680	
Mainagarh		4221	Basement @4087	
GolfGreen		5826	Slhet@5320	
Chandkudi		3862		
Ichapur#1		5613		
Karimpur		3702	Bolpur@3633	
Barhampur		2052	Rajmahal@2023	
Ichapur#2		4500		
Santipur		4000		
Rajganj		3045		
Karandighi		3143	Lr.Gondwana@1130	
Lahil		3336		
Ashok Nagar			Gas flow	2018
offshore				
BB-A-1R	carlsberg	4073		1976
BB-B-1	"	4697		1977
SME-1	ONGC	5079	Kopili@4707	
SMM-1	"	4951	Matla@3374	
SMM-2	"	4975	Matla @3224	
SME-5	"	5841	Ghatal@5555	
SME-4	"	5575		

Vindhyan Basin

Introduction

The Vindhyan Basin, named after the Vindhyan mountains in Central India occupies an area of about 1, 65, 000 km^2 of which 65,000 sq.km., underlie the Deccan volcanic flows in the southwestern part. It is the the largest Proterozoic sedimentary basin of India which is arcuate shaped with thick sediments.

The basin also extends into the Ganga valley in the north beneath the Tertiary cover of the Himalayan foredeep. It has been divided into the Chambal sub basin lies west and north-west of the Bundelkhand massif while the Son Valley sub basin is in the east.

Exploration History

Initially the GSI mapped the basin area with sample collection for detailed studies. NGRI conducted, the Deep Seismic Sounding (DSS) survey from Hirapur over Bundelkhand Massif to Mandla, south of Jabalpur. Aero-magnetic survey to the tune of 59,771 Lkm have also been carried out in this basin. ONGC commenced geological surveys in the basin in 1962 followed by the acquisition of Gravity-Magnetic data and Remote Sensing data in 1983. Seismic surveys were initiated in 1988 with CDP reflection survey covering 6,802 Lkm. Further, DGH also carried out vibroseis reflection survey amounting to 566 Lkm.

ONGC's efforts in exploring for hydrocarbons resulted in drilling of first well at Jabera#1 to a depth of 3,597 m during 1990 which encountered a gas show. This attracted intensification of exploratory efforts with drilling of two more wells, Damoh # 1 and Kharkhari # 1. Within the ONGC's area covering Damoh-Jabera-Katni, after the Jabera results, carried out both 2D and 3D seismic in Nohta area. So far about 14 deep wells were drilled and detiled seismic data interpretation is being carried out.

Geology of the Area

A thick sequence of shallow marine clastic and carbonate sediments are found in Vindhyan Basin. The basin which is the biggest intra-cratonic basin situated on the Indian platform with the total sedimentary thickness in the deepest part of the basin is of the order of 5,500m ranging in age from the Upper Proterozoic to Early Cambrain. The sediments comprise of limestones, sandstones, shales and conglomerates deposited in a shallow marine environment (Srivastava, B.N. *et al.*, 1983). The sedimentary succession, which is divisible into carbonate dominated Lower Vindhyan (Semri Group) and clastic dominated Upper Vindhyan (Kaimur, Rewa and Bhander groups) sequences, separated by a large hiatus. The strata are exposed in three major sectors: Son valley, Bundelkhand and Rajasthan (Fig10.vb.1). Substantially thick Vindhyan rocks have also been recognized under the Gangetic alluvium. The Vindhyan Basin, containing more than 5,000 m thick sequence of sandstones, shales and limestones.

The initial transgression of the sea from the north is inferred to have taken place in the eastern part of the basin over the Bijawars. The shallow sea appears to have established lagoonal conditions near the coastal part during the subsequent regressive phase.

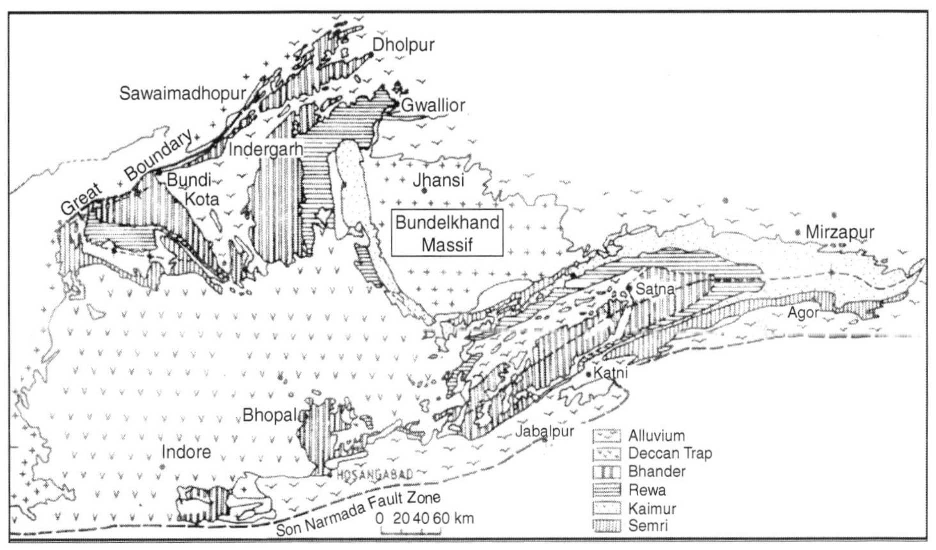

Figure 10.vb.1. Geological Map Vindhyan Basin.

Tectonics

It is a Proterozoic intra-continental basin that developed in the central part of the Indian shield along with several other basins such as Cuddapah and Chattisgarh. The Vindhyan Basin is bordered by the Aravalli–Delhi Orogenic belt along the NE–SW trending, Great Boundary Fault (GBF) in the west, the Monghyr-Saharsa Ridge in the east, the Narmada-Son lineament marked by a major ENE-WSW trending in the south-west and the Satpura-Orogenic belt in the south-east. Bundelkhand

Massif divides this basin into two parts – the Son Valley on the south-eastern side and the Chambal Valley.

The Vindhyan basin was formed as a result of large scale downwarp in the northern part of Indian platform after the Delhi orogeny. In the remote sensing data some geomorphic anomalies possibly representing sub-surface structural features are observed at Hoshngabad north of Domah and Rewa and South of Hanumana village (Dey, *et al.*, 1995).

A large number of ENE-WSW and NW-SE trending faults have been mapped. There are a large number of rift and wrench faults in the Son valley that are geomorphologically expressed as offset in the ridges, depressions, straight drainage channels, termination of trends and shift in lithology. Many of these form boundaries of structures. The fault coincides with the trend of Son-Narmada fracture (Bhoj, R. 2001).

The Vindhyan strata define a broad, regional syncline trending ENE-WSW. The axis of the syncline is slightly curved and plunges gently towards west. The average dip of the southern limb is greater than that of the northern limb rendering the axial plane to dip southerly.

Initially, the Son-Narmada Lineament was dormant, but at the onset of Vindhyan sedimentation later, the fault system along this downwarp became active with the formation of the southern limit of Vindhyan deposition (Figure 10.vb.2). Intense folding with trending roughly with north-south trend with asymmetrical trend are observed in the area (Srivastava, *et al.*, 1983).

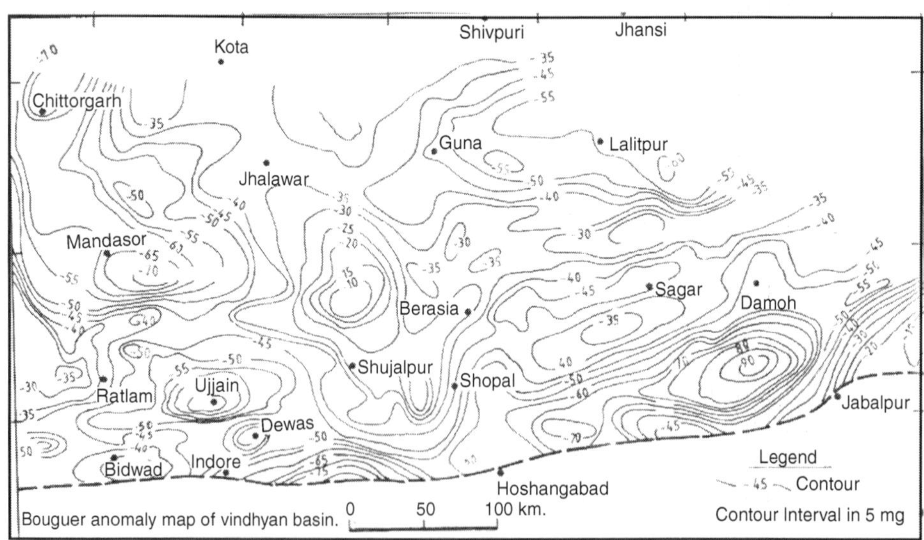

Figure 10.vb.2. Bouger Map Vindhyan Basin.

Except in the vicinity of the Great Boundary Fault towards the western margin of the Vindhyan Basin and the Son-Narmada fault towards south, the rocks of the Vindhyan sequence are largely undisturbed. The eastern and western parts of the basin around the Bundelkhand granite were apparently parts of a continuous basin which the deepest part possibly, covered by the Deccan traps towards south.

The progressive and successive angular unconformities suggest that the deformation pattern shown by the Vindhyan strata is a reflection of syn-sedimentary tectonic activity. The individual sequences of the Vindhyan succession are related to discrete episodes of tectonic activity that induced the subsidence necessary for accumulation of sediments and subsequent deformation of the older strata.

The Domah basin located towards south of Jabera, is bounded by two major faults trending NW-SE, whereas Jabera dome is associated with ENE-WSW trending faults which are in close proximity of Sonata rift. The gravity data show the lateral extent of the basin is much further towards south.

Sedimentation

The Vindhyan basin is known to contain shallow marine deposits as thick as 6000 m belonging to Meso-Neoproterozoic age. Ken and Son rivers form the major river systems that drain the Son valley, while the Chambal River along with its tributaries Kali Sindh and Parwan drain the Chambal valley towards northeast prior to its confluence with Yamuna River. Precambrian gneisses and meta-sediments border the Vindhyan basin in the east and south-southeast. A large tract of Chambal valley is concealed under the Deccan Trap basalt towards the south.

The basement rocks of the Vindhyan sediments belong to two different ages, Aravallis and Bundelkhand of Archean and Bijawar,Delhi Gwalier rocks of Algonkian age (Srivastava, V. *et al.*, 1983). The Basal Conglomerate horizon consists of quarzites and Jasper.These conglomerates are glacial tillites found in Ken valley. The thickness map of upper Vindhyan group indicates the maximum thickness of 4,500 m and the depo-center is found to be shifting towards east (?). The extensive existence of Stromotolites indicate the presence of algal life in the sediments (Valdiya, K.S., 1968). Trace fossils are also found in the Rohtas limestone near Katni.

The deposition of Kajrahat Limestone, the Son-Narmada Lineament again became active resulting in emission of volcanic material, which was deposited as the Jardepahar Porcellanite. In the subsequent regression, the shore line shifted towards northwest. The Fawn Limestone was deposited over the shelf in a tidal flat environment (Figure 10.vb.3). This was followed by shallowing of the basin as is evident from the overlying Glauconitic Sandstone. The fresh marine transgression resulted in the deposition of marine shales followed by Rohtas Limestone.

There is no evidence that the sediments have been affected by volcanic activity though there is evidence of volcanic activity during that period in the volcano-clastics (tuff and agglomerates of porcellanite member) of the Son valley and dolerite dykes of Kajrahat and nearby areas. The sediments are unmetamorphosed. There is some evidence that in the eastern part of the basin euxinic conditions prevailed. The stromatolites in the Fawn limestone and Lower Bhander limestone and compared them with the lesser Himalayan rocks. The environment of deposition of various sediments in the basin vary from beach through tidal falts to lagoon areas. The facie are found to be repeated vertically possibly due to shifting of shoreline.

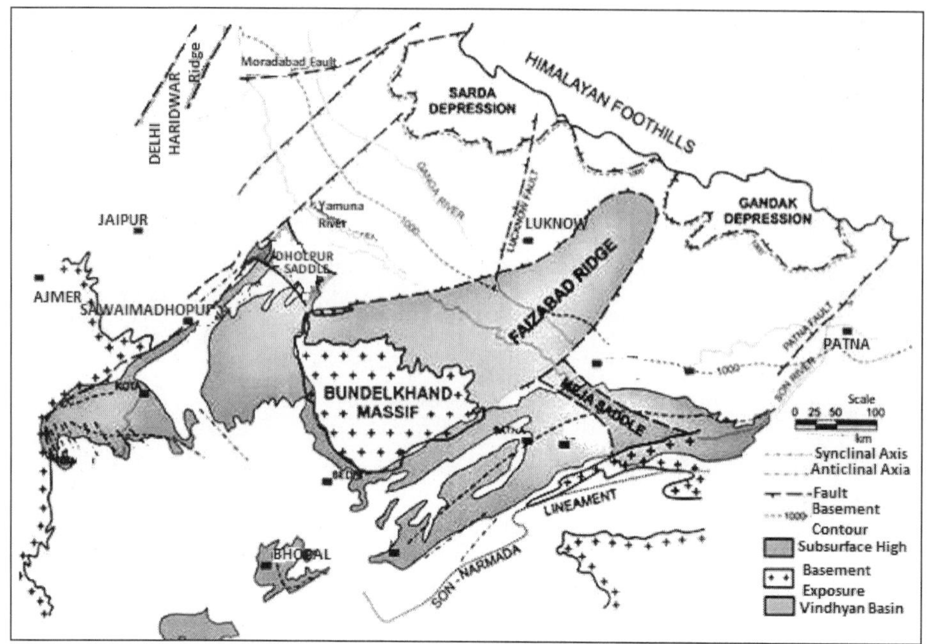

Figure 10.vb.3. Tectonic Map Vindhyan Basin.

Stratigraphy

The entire Vindhyan succession, maximum thickness estimated to be around five kilometres, and comprising mainly sandstone, shale and limestone is assigned as the Vindhyan Supergroup. The Supergroup is divisible into four groups: Semri Group, Kaimur Group, Rewa Group and Bhander Group.

Each group is again subdivided into several formations. The Semri Group in the Son valley rests unconformably on a variety of pre-Vindhyan rocks such as granites and metamorphics. In the Bundelkhand area, the group overlies the Bundelkand Granite Gneisses and Bijawar Group of metamorphics, whereas in the southern Son valley, In some localities the basement is granite. The Semri succession of the Bundelkhand area has two detached outcrops around Chitrakut and Bijawar respectively, and is only a few s of meters thick.

The different depositional systems recognized in the Vindhyan succession are: alluvial fan, fan delta, braid delta, braidplain, eolian sand sheet, tidal flat (carbonate as well as siliciclastic), shoreface (tide and storm dominated), storm dominated shelf, homoclinal carbonate ramp, distally steepened carbonate ramp and epeiric peritidal flat (siliciclastic). The overall paleocurrent directions of the depositional systems in the Son valley are northerly suggesting a source towards south. The unconformities divide the Vindhyan succession of Son valley into five sequences. Each sequence consists of several systems tracts representing different paleogeographic settings and marking paleogeographic shifts. The radiometric dating of Sibru shales of Bhander group of upper Vindhyans indicate about 750 Ma. (Rathore, S.S., *et al.*, 1999).

Angular unconformities resulted due to erosion of the uplifted crest of the anticlines on which the next sequence of strata was deposited with an angular discordance. There are sediment packages at the northern part of the Vindhyan basin developed from a northerly source and thus representing different tracts and sequences from those of the southern part (Figure 10.vb.4).

STRATIGRAPHIC UNITS Vindhyan Basin				THICKNESS m	LITHOLOGY
Mesozoics			LAMETAS/ DECCAN TRAP		Basalt
Paleozoics	Vindhyans	Upper	Bhander	450	
			Rewa	150	Sandstone
			Kaimur	850	
		Lower	Semri	2250 4000	Clay/Shale, Limestone
Pre-Cambrian	GWALIORS BIJAWARS				Phyllite/Schist
	BUNDELKH AND GRANITES				Granites

Figure 10.vb.4. Stratigraphy–Vindhyan Basin.

These packages are represented intermittently in the succession and have been interpreted as representing periods of uplift of the Bundelkhand Granite and subsequent erosion in the north. However, the source for the clastics occurring within the Semri and the lower parts of the Kaimur and Rewa Groups in the Bundelkhand sector was perhaps the Bundelkhand Granite Gneiss, Bijawar and Gwalior Group of rocks as manifested by the southerly paleocurrent.

The first order sequence comprises of the Rewa and Bhander groups of upper Vindhyan. Age of Rewa and Bhander groups is estimated beyond the glaciations (635 to 542 Ma). The Lametas of Late Cretaceous age overlie this sequence, which show limited extent and small sporadic outcrops are seen in Son valley northeast and southeast of Sagar town. Deccan Trap lava flows overlie the Lameta sediments.

Petroleum Geology

The exploratory drilling in the Son Valley has established the presence of gas in the carbonates of Lower Vindhyan. The Proterozoic basins such as the Tunguska basin of Siberia and Amaedus basin of Australia have produced hydrocarbons. Vindhyan basin being of Proterozoic age can be hydrocarbon bearing due the gas shows in the well Jabera#1. The Damoh basin, situated in the north of Jabera also indicated the presence of gas.

The presence of hydrocarbons generating conditions in this area as in the Nimbahera limestones has also been indicated by chemical analysis. There are possibilities of oil and gas shows leaking towards the edges of the Vindhyan Basin, adjacent to the Great Boundary Fault. The presence of hydrocarbon indications have been noted in the limestone in the Vindhyan Basin of the trans-aravalli side in Rajasthan.

Source Rocks

Lower Vindhyan shales, particularly, Hinota and Pulkova shales with organic carbon contents of upto 3.89 per cent and 1.86 per cent respectively are considered to be having good potential as source rocks. Organic matter in Vindhyan sediments is assumed to be Type I and II as the source at that age can only hold algal bacterial, suitable for generating liquid hydrocarbons.

The deep well has penetrated upto 681m is characterised by black shale and siltstone intercalations. TOC ranges between (0.5-10.14 per cent). Charkaria Shale (TOC 0.42-1.84 per cent) and Rampur Shale (TOC 1.14-1.78 per cent) can be source beds.

The black shales from Kaimur Sub-group (Upper Vindhyan), Hinota shale and Pulkova shale member from the Lower Vindhyan sequence are of interest as potential source rocks. On organic matter maturation scale, the organic matter is well within the preservation limit of wet to dry gas.

Reservoirs

Sandstones and limestones of Lower Vindhyan and Upper Vindhyan sequences are potential reservoir rocks. The known value of total porosity including fracture porosity is 5-4 per cent. Primary porosity is negligible, so far as regards reservoir rocks (Petres and Anand, 1982 and Pendkar and Peters, 2002), it is apparent that reservoirs in the Vindhyan basin could be fractured reservoirs.

Entrapment

In the central part of the basin, there exist a number of step faults. It is envisaged that these faults have created favourable trapping conditions for hydrocarbon accumulations. There are many structural features like the Jabera dome in the Son Valley and a number of anticlines near Chittorgarh in the Chambal valley. Updip truncations against regional unconformities may also constitute hydrocarbon plays. A regional and thick (490 m) Charkaria shale appears to be the regional seal.

Data of Hydrocarbon Field

Jabera Structure

Jabera dome is a prominent structure, which is located about 40 km NW of Jabalpur. This dome is an oval shaped feature extending in ENE-WSW direction and developed during the early stages of upper Vindhyans (Jokhan Ram, *et al.*, 1996). It occupies an area of about 320 km^2. Axially, it is about 20 km in length and 16 km in width. In Jabera #4 well the interval 1359-1279 flowed gas @ 1536 m^3/d through 6 mm bean in a barefoot completion. In Nohta#2 well the interval 1727-1702 flowed gas @ 3724 m^3/d through 6 mm bean in barefoot testing. It is from Rohtas formation of meso- Neo Proterozoic age. In Domah#4 well interval 1544-1550 m flowed gas @ 443 SCMD through 4 mm bean from Mohana Fawn Limestone of Proterozoic age

Table 10.vb.1. Wells Drilled-Vindhyan Basin

No.	*Well*	*Depth (m)*	*Year*	*Remarks*
	Jabera-1	3597	1990	
	Jabera-2	5324		
	Jabera-3			
	Jabera-4			
2	Damoh # 1	3501		
	Damoh#2			
	Damoh#3			
	Damoh#4		2014	
3	Kharkhari # 1			
4.	Nahota#1	4395		
	Nahota#2			Son Valley-1702-27 flowed gas @3,700m^3
	Nahota#3			
	Nahota#4			
	Nahota#5			

Kerala-Konkan Laccadive Basin

The Kerala- Konkan-Lakshadweep basin occupies areas off the coast of Konkan, Kerala and Lakshadweep islands in the western coast of India. The basin is separated from the Mumbai offshore basin in the north by the Panjim fracture zone around 15°N latitude. In the south, the basin merges with the east coast basins around the southern tip of Indian peninsula. Geographically, the Konkan-Kerala shelf covers an area of about 77,000 sq.km upto 200 m isobath. The shelf areas around Lakshadweep islands are not clearly defined due to very rapid increase in water depth all-round the island chain, the Lakshadweep area covers about 92,000 sq.km.

Exploration History

Geological mapping of the outcropping coastal belt, consisting of Neogene and Quaternary clastic rocks was carried out by geologists of ONGC and other scientific organizations. In the onland area of Kerala ONGC also carried out reconnaissance geological surveys during the period 1957-60, over an area of about 4,000 km^2 and also geomorphological and neotectonic surveys. During 1958-59, gravity magnetic surveys were also conducted over an area of about 1000 km^2 extending from Ernakulam in north to Quilon in south and from Kottayam in the east to Vempanad-Allepey in the west. Based on interpretation of the acquired geo-scientific data it was inferred that over a major part of coastal region, the nature of sediments and their meagre thickness are not favourable for commercial accumulation of oil and natural gas in the coastal onland areas. Therefore, ONGC has taken up the exploration work in the offshore.

Preliminary seismic surveys in the Konkan-Kerala shelf areas were initiated by Soviet ship 'Akademic Arkhangleisky' in collaboration with ONGC during 1965. With the discovery of Mumbai offshore area prompted ONGC to restart seismic surveys in the shelf areas along the Konkan and Kerala coast in 1977. In the area over 62,500 line km of seismic surveys by ONGC and over 12,000 line km of seismic surveys by international oil companies have been carried out. Subsequently, 3240 line kms of regional CDP surveys have been conducted in the deeper waters of

Lakshadweep area in 1981. The regional seismic coverage in Lakshadweep area has not indicated any structures. The DSDP well data is also available for geological understanding of the area.

The Geophysical data include Gravity and Magnetic Surveys which covers an area of 45,000 LKM. The 2D Survey covers an area of 146,445 LKM and 3D Survey, covers an area of 9,734 S km.

ONGC drilled 10 exploratory wells in the shallow and deepwater parts of the basin since 1977 and 4 drilled by Shell and BHP, of which 3 wells penetrated basalts. In some wells hydrocarbon shows were observed.

Since 1998, under NELP regime though some blocks were taken by ONGC and Reliance, but owing to unsatisfactory results they were relinquisghed (Bastia. Ravi, *et al.*, 2008).

Tectonics

The western continental margin basin of India evolved through the break-up of eastern Gondwanaland in the Late Triassic/Early Jurassic and the subsequent spreading history of the eastern Indian Ocean (Biswas, SK and Singh, NK,1988).

As the area was subjected to three phases of tectonism, incipient rifting, with the separation of Madagascar and Seychelles. Each stage may be having limited circulating conditions and with stagnating conditions during the basin fill. During the Early Cretaceous (Norton and Sclater, JG 1979) the Proto-Mozambique Ocean was a restricted basin with no open ocean circulation and hence very likely had an anoxic environment (Biswas, SK 1982 and 1987).

The western margin evolved through early rift and post rift phases of divergent margin development. A series of regional and local horsts and grabens resulted in response to rifting along the dominant basement tectonic trends (Figure 10.kk.1). The area lying south of Mumbai offshore basin is bounded by Vengurla arch in north and extends beyond Cape Comorin into the Indian ocean to the south and divided into two sub-basins the Konkan basin between Vengurla and Tellichery arch and Kerala basin between Tellichery and Trivandrum arch.

The northernmost part of the western continental margin was the first to be subjected to continental rifting and crustal subsidence in the Late Triassic. The process of rifting gradually advanced towards south and by Cretaceous time almost all the rift-related horsts and grabens came into existence (Kumar, Anil, *et al.*, 2001).

Kerala-Konkan basin is spread over both shallow and the deep water areas. The East-West extension of the basin is Kori Comorin deep offshore in the West and coastal area of Kerala, Karnataka and Goa in the East (Arts, K.P, *et al.*, 2003) (Figure 10.kk.2).

The structural elements, from east to west include: Shelfal Horst and Graben Complex, Alleppey Plateau, Lakshadweep Depression, Laccadive Ridge and Arabian Abyssal plain.

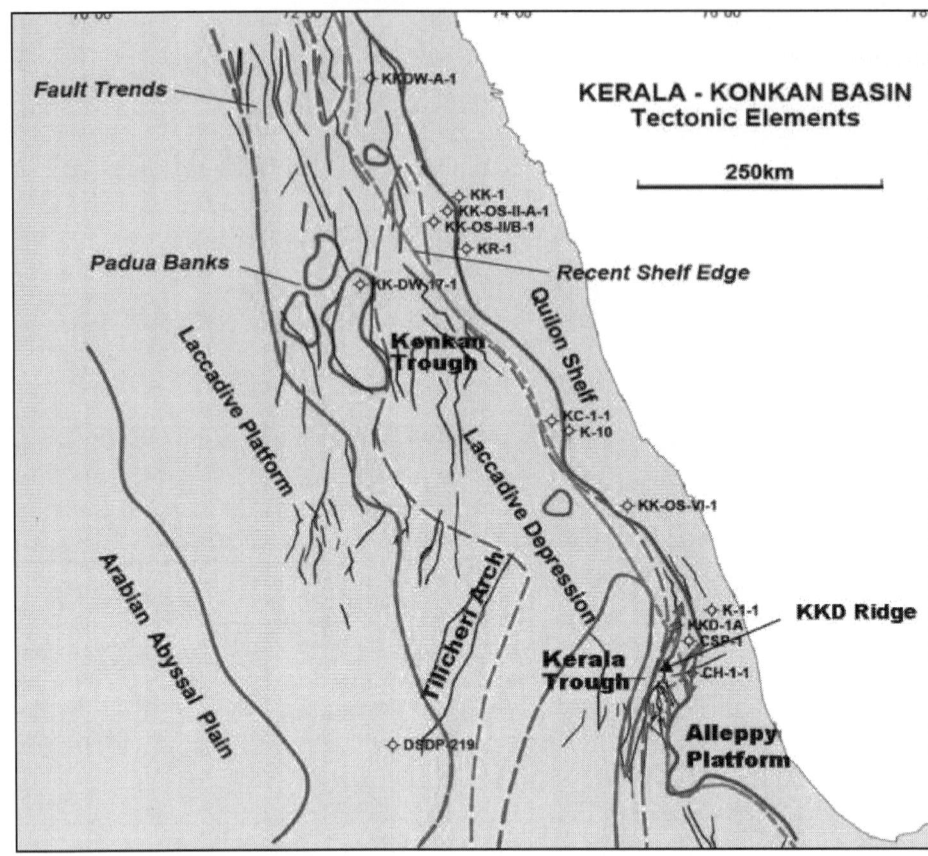

Figure 10.kk.1. Tectonic Map of Kerala–Konkan Basin (Modified after Surajit Gohain, 2012).

Shelfal Horst and Graben Complex

It constitutes the part of basin lying between the coast and 1500 m Isobaths. In the north, it merges with Mumbai shelf.

Alleppey Plateau

It is well-outlined between 200 m and 1800 m isobaths, and extends up to 2000 m down-slope. It developed in the southernmost part of the basin, south of Cochin. Morphologically it is a typical marginal plateau, the part of the continental margin, thrown down to deeper to a continental shelf terrace.

Lakshadweep Depression

It is situated between continental shelf and the Laccadive Ridge. The basement of this depression is by normal faulting and a series of basement tilted blocks. The width of the depression gradually decreases from approx. 215 km in the south to approx. 70-65 km in the north. The thickness of sediments varies from 200 to 3200 m reaching a maximum about 5500 m at places. This depression displays no sign of its spreading nature.

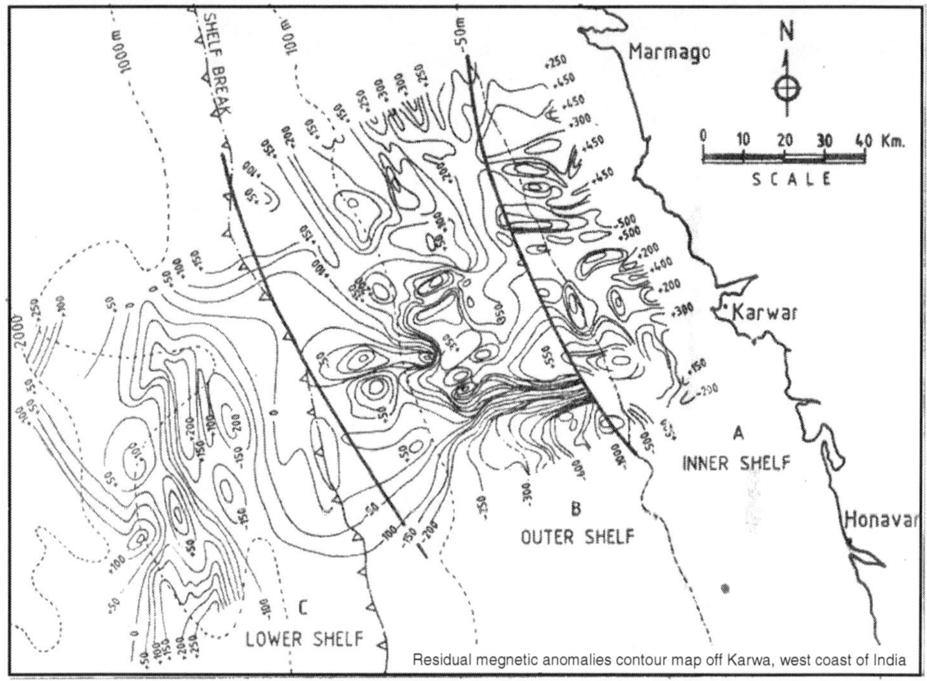

Residual megnetic anomalies contour map off Karwa, west coast of India

Figure 10.kk.2. Residual Gravity Map Karwar Coast-Kerala–Konkan Basin.

Laccadive Ridge

It is limited by the Arabian abyssal plain in the west and by the Lakshadweep depression in the east. The ridge has a complex block-like basement structure, by a system of grabens, half grabens and single normal faults. The rift extends along eastern part of the ridge in N-S directin and named as Cannanore Rift System. The origin of the Laccadive Ridge has been attributed to different processes, such as a hotspot related volcanism or transforms faulting and is considered as a continental remnant rifted away from the western continental margin of India. For most part of early Tertiary period, Alleppey Plateau and Laccadive ridge were subjected to relative differential rise during entire Paleogene, thereby creating depressions around them.

Arabian Abyssal Plain

This oceanic depression, created in Cenozoic times, through the sea-floor spreading along the Carlsberg Ridge with a set of well-pronounced magnetic anomalies indicating the beginning of the spreading basin was around 61 Ma.

Geological Set-up-Deposition

The basement arches control the architecture and north-south limits of the area. The northern most Vangurla arch is differentiated into three shallow depressions by transverse basement arches. These are Konkan depression, Cochin depression and Cape Comorin depression.

The seismic sections in Kerala Offshore area showed the sediments to be gently dipping to the west, with down to basin faults extending to the basement. A broad anticlinal nose faulted on its updip side was noted about 40 km northwest of Cochin at the level of a reflector which is the top of Paleogene below an unconformity surface.

The deposition started with continental environment, changed gradually to paralic and to pulsating marine conditions followed by basic lava flows (Deccan Trap) in the terminal stages towards the end of Cretaceous and Early Paleocene. Towards the end of the early rift phase, most of the rift related grabens and horsts, located in the deeper parts were covered up with sediments.

There were progressive transgressions, accompanied by syn-sedimentary positive radial movements of low to medium amplitude. Lakshadweep Depression was characterised by rejuvenation of faulting and subsidence of basement, where continental, deltaic, lagoonal and shallow marine deposition was taking place. The lithological sequence shows upward gradation from continental sand-clay section to sand-shale-coal section and finally terminating in thin sand-shale-coal limestone section.

The depocenters kept shifting from west to east during Paleocene to Mid-Eocene period. During late Eocene-early Oligocene period finer differentiation into second order horst and graben features became more pronounced. The radial movement could have resulted in minor epochs of alternating transgression and regression and lateral shifting of the depocenters. Thus, the late Eocene-Early Oligocene period witnessed dominance of paralic to shallow marine depositional environment in Lakshadweep Depression.

The shelfal horst-graben complex formed the site for development of shelf depositional systems, whereas the Lakshadweep Depression and Laccadive Ridge lay in slope and basinal region. The shelf depositional systems comprised deltas, clastic and carbonate tidal flats, strandplains and carbonate ramps, platforms and banks. Lows within the shelf provided the bulk of the terrigenous clastic sediments, whereas the distal and relatively positive shoal areas favoured development of thick limestone sequences. Extensive carbonate sequences thus developed over the northern part near Vengurla arch, Alleppy platform, distal parts of paleo-shelves and around Laccadive and Maldive group of islands are the regions of favourable buildups.

In the basin slope regions, west of the shelfal horst-graben complex, deposition of submarine fans and hemi-pelagic and pelagic deposits is visualised. Association of shallow and deep marine fauna in limestone bands inter-bedded with clastics of Early Eocene age in a few drilled wells in the Lakshadweep Depression suggests, that submarine canyons traversing platform limestone areas caused intermittent deposition of turbidites composed of limestone in basin slope transition zone.

In the post-Oligocene period, the basin acquired a marked westerly tilt, evidenced by numerous sigmoidal/progradational features observed in the Mio-Pliocene sections. Numerous channels, levees and turbidities observed on southwesterly slope have accommodated thick pile of post-Miocene sediments. A significant part of this pile is likely to belong to the Indus River fan in the major part of the Lakshadweep Depression (Chatterjee S.M. *et al.*, 2006).

Konkan Basin

The Konkan basin is situated offshore is separated from it by an east-west trending subsurface basement arch and covers an area of about 83 900 km^2. The basin evolved through early rift and post rift phases and contains more than 5 km of the sedimentary sequence is comprised of Mesozoic and Tertiary sediments (D'Silva, K *et al.*, 2008) (Figure 10.kk.3). The basin development was initiated during the Cretaceous with a series of north-south trending step faults with throws towards the west. There is a gradual increase in thickness of the Cretaceous and younger strata westwards, except over the basement highs.

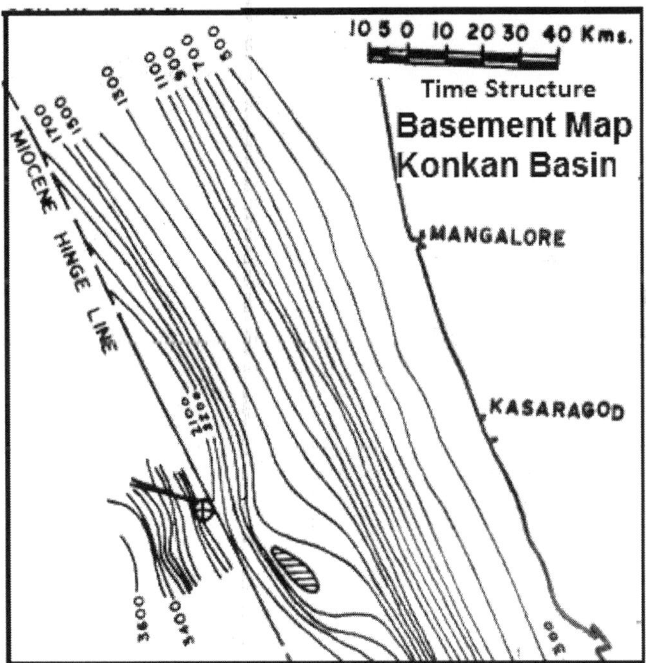

Figure 10.kk.3. Basement Map–Konkan Basin.

Kerala Basin

These sediments consist of marine fossiliferous limestone (Quilon Beds) of lower Miocene age, and continental felspathic sandstone, kaolinitic clay end lignite beds called Warkali Beds of probable Mio-Pliocene age. The entire sedimentary sequence of the area comprises mostly of gritty sandstone interrupted by basalt flows. Shales with source rock potential are significantly absent. Hopefully, the sedimentary sequence may change from arenaceous to argillaceous and calcareous towards the deeper part of the basin.

Stratigraphy

Out-crop

In the on land, along the Kerala coast between Calicut and Trivandrum, some 300-400 m of the Neogene sediments are exposed in a series of small linear depressions, possibly pre-existing grabans, within the Archaean basement. In the eastern side the outcrops are limited by the exposures of Archaean schists and gneisses. Neogene and Quaternary sand, clay, lignite and limestones overlying the basement form the outcrops. In the subsurface offshore, Palaeogene and Neogene sediments mostly quartzose sandstone and clays are found overlying the Deccan Trap.

Offshore

The Cretaceous-Recent sedimentary succession in the basin is divided into two sequences. The lower, corresponding to rifting in early Cretaceous to late Cretaceous sequence and overlying late Cretaceous to Recent is referred to as the passive margin sequence. The lowest sequence limited between older and younger basalts is encountered in wells drilled in offshore. Several flows have been observed in the upper part of this sequence in some wells which indicated that the youngest traps in this part of the basin are of late Paleocene age and evolved within a shallow marine setup. The gross lithology is sandstone interbedded with clay/claystone and siltstone with a few streaks of limestone. The succession comprises sand, shale and siltstone deposited in an inter-middle shelf setup.

The succeeding sedimentary succession overlies the Deccan trap and represents the Paleocene. Its upper boundary with early Eocene is an unconformity. The overlying unconformity-bounded succession, represents the early Eocene. The gross lithology varies from dominantly carbonates succession to clastics deposited in supra-tidal environment to shallow inner-shelf setup. Middle-Late Eocene succession is also represented by limestone with minor shale beds. Overlying Early Oligocene sequence is bounded by hiatuses. The dominant lithology of the sequence is limestone with minor sandstone beds.

With gradual sediment loading, the basement tectonics become more pronounced during Oligocene and the Tellichery arch became more prominent separating the northern and southern areas as two separate depositional regimes. Late Oligocene-Miocene is dominated by carbonates with thin beds of claystone in some parts, while in other areas well-sorted sands with interbedded shale dominate the sequence. Differentiation of sedimentary regimes in Konkan and Kerala basin continued during this period with the former receiving dominantly carbonates and the latter more clastics. Near the shelf edge, a sedimentary section from Eocene to

Pliocene and Recent has been found in the subsurface, mainly consisting of micritic dolomitic Iimestones.

Lithostratigraphy

Archaean rocks composed of granites, gneisses and charnokites are exposed along the eastern margin in the Konkan and Malbar coastal areas. The outcrops are quite close to the coast near Calicut.The rock units which are identified in Kerala-Konkan Basin are described below (Figure 10.kk.4).

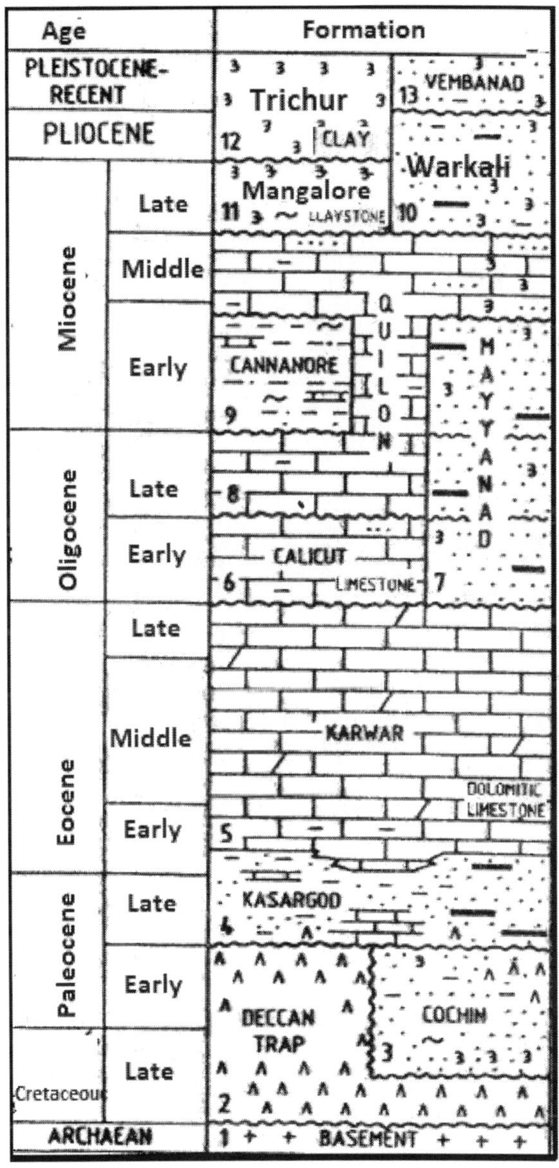

Figure 10.kk.4. Stratigraphic Nomenclature–Kerala-Konkan Basin.

Deccan Trap

The group consists of sub-aerially effusive rocks and shallow intrusives alongwith thin clastic rocks. Lithologically it consists of dark grey to dark greenish grey, hard fractured basalts of Late Cretaceous to Early Paleocene, with abundant cavities with clacite and zeolite minerals.

Cochin Formation

This formation is a clastic sequence of sandstones, siltstones, shales and claystones deposited in continental to outer neritic environments with thickness of about 919 m of Late Cretaceous to Early Paleocene.

Kasargod Formation

This formation of Early Paleocene to Early Eocene is unconformably overlying Cochin formation consists of argillaceous limestones in the lower part and sandstones and shales in the upper part, represented by fresh water to marginal marine and shallow inner neritic deposits on the deeper continental shelf.

Karwar Formation

Lithologically this formation consists of carbonate facies comprising grey, light grey and brownish grey limestone with occasional thin shale bands of Early to Middle Eocene deposited varying environments from lagoonal and restricted marginal marine towards basin margin.

Calicut Formation

It includes limestones between the Eocene and the Early Oligocene unconformities. The sediments were deposited under open marine inner neritic to shallow shelf and marginal marine environments.

Mayyanad Formation

The formation comprises coarse to medium grained sandstones with interbedded white and grey clays and lignites in the coastal area of Kerala belong to Early Oligocene to Early Miocene. The sediments were deposited under shallow fluctuating marginal marine and swampy environments.

Quilon Formation

Lithologically this formation comprises of brownish white to dirty white, buff and grey foram algal wackestones and packstones and dolomites of Late Oligocene to Middle Miocene age, which were deposited in very shallow marine to inner neritic environments (Figure 10.kk.5).

Cannanore Formation

This formation comprises of a sequence of shales and sandstones with a few thin limestones of Early Miocene. The depositional environment of this formation is a varying between middle neritic and upper bathyl.

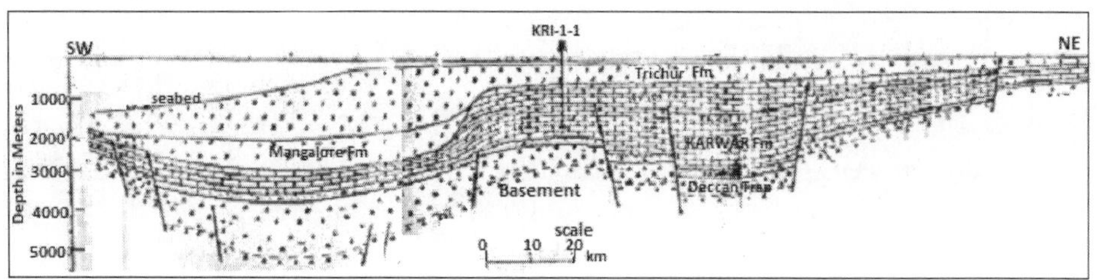

Figure 10.kk.5. Seismo-Geological Section (SW-NE) Passing through Well KR-1-1 off Cochin.

Warkali Formation

It is a sequence of grey pebbly and coarse sands variegated and mottled clays, white plastic clays and peat/lignite, which was deposited under extremely shallow marginal marine environmentand belong to Late Miocene to Pliocene.

Mangalore Formation

It comprises a sequence of claystones and shales of Late Miocene.

Trichur Formation

It comprises a sequence of light to dark grey and brownish grey, soft, sticky plastic clays and claystones with abundant shell fragments and occasional thin limestone, dolomite or siltstone bands. The formation was deposited under middle to outer neritic environment and of Late Miocene to Pliocene.

Vembanad Formation

This formation of Pleistocene to Recent, comprises sands, peat, black, sticky and sandy calcareous clays, gravel beds with lateritic gravels, lime shell deposits and alluvium in Kerala coastal region.

Petroleum Geology

Hydrocarbons have been discovered all through coastal areas of India except Kerala-Konkan basin. Moreover the establishment of the hydrocarbon kitchen area, migration path and entrapment need to be established. About 14 wells were drilled in Kerala-Konkan has shown the presence of hydrocarbons while drilling but production testing results not encouraging this fact (Surajit Gohain,2012). The three initial wells drilled by ONGC (Karwar-1,Kasargod-1, Cochin-1) without success. But wells like Kasargod-1, KK-OS-V1-1 and CSP-1 showed faint oil fluorescence. The prognosticated resources of Kerala-Konkan offshore basin are estimated to be 660 MMt.

Source Rocks

Cretaceous to Recent sediments. Post-rift mature sediments with sufficient organic carbon content are present in the basin (Naini, BR and Talani, M,1982).

The source rock is mainly Paleocene to Eocene shales.The Karwar Formation (Eocene) has rich organic matter (TOC upto 29 per cent) and Kasargod Formation

(Late Paleocene) has fair to good source potential. Paleocene to Middle Eocene shales, have fairly high TOC content mainly of type II and III organic matter.

Maturation

Paleocene to Middle Eocene shales are thermally mature. The shelfal depressions and Lakshadweep depression are considered to be kitchens for hydrocarbon generation. Exploration from wells indicate possible kitchen areas in deep water area.

The nearness of oceanic crust and the effect of sub-aqueous trap flows are favourable factors for maturation (Sweeney, J. J., and A. K. Burnham 1990). Apart from some minor gas shows and fluorescence in KG-I-1 well between 900 to 1200 m depth, no significant hydrocarbon indications have been observed in the basin.

Reservoir Rocks

Paleocene to Middle Miocene Sandstones are the possible reservoirs. Extensive carbonate platforms and shelf margin reefal bodies developed during Eocene to Middle Miocene post rift phase over almost the entire western part of the shelfal horst-graben complex which serve as the reservoirs. Carbonate platforms around Laccadive Islands are likely to provide abundant reservoir bodies in the region of Laccadive Ridge.

Traps

Both structural and stratigraphic and combination traps are expected in the basin. Several structural closures have been mapped in the shelf areas. In the Kerala shelf, where sands are the predominant lithology, may represent the fan delta, distributary channels, offshore bars on the shelf and pinch-outs beyond the hinge zone which can provide excellent statigraphic traps. Reef development around the hinge zone can also form good traps particularly in the northern part of the basin (Figure 10.kk.6).

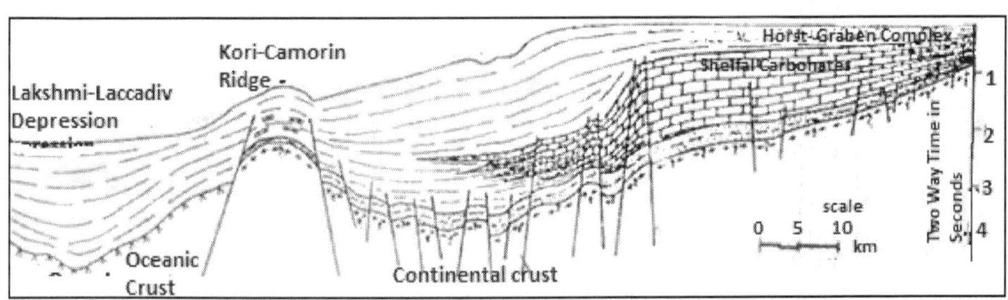

Figure 10.kk.6. Geological Section Showing some Entrapment Conditions–Kori Comorin (After Singh, N K and N K Lal, 1993).

The accumulations are mostly related to horst blocks, shale tectonics, normal anticlines, and strati-structural unconformity controlled entrapments. Basement highs in the shelf margin are the sites of carbonate buildup at several places. Pinchouts of Paleogene sequence and wedge outs against Vengurla arch are also acting as trap.

Challenges

The focus towards the Mesozoic exploration, the thick basalt masks deeper seismic events below. Long offset data has yielded better sub-surface images below the trap which has been masked earlier. Heat flow studies done in few blocks have yielded better results. Multi-beam survey, done in few RIL blocks, have given the direct indication of hydrocarbons. Sea bed logging survey has given indication of hydrocarbons. Reprocessing of data gives better resolution which helps in identifying interesting structures (Shah. J, *et al.*, 2009).

Table 10.kk.2. Wells Drilled–Kerala-Knokan

Well Drilled	***Opera***	***D D (m)***	***Remarks***
Karwar-1	ONGC	1678	Drilled upto Eocene
Kasargod-1	ONGC	2970	Minor Gas shows in Miocene
K-1-1	ONGC	1755	Encountered Basalt, drilled upto Cretaceous
CH-1-1	ONGC	4627	Terminated in Basalt/Trachyte section
CSP-1	ONGC	3833	Oil fluorescence in Eocene sands and carbonates
Quilon-1	ONGC	413	Abandoned prematurely
KK-DW-17	ONGC	2999	Drilled to Paleocene- Cretaceous
KK-1	ONGC	2762	Terminated in trap. Mesozoic not encountered
KK-DW-A-1	ONGC	4213	Only Basalt encountered from 3200m upto bottom
KKD-1	ONGC		Dry
KK-OS-II/A-1	SHELL	2975	Encountered Basaltic flows -
KK-OS-II/B-1	SHELL	2020	Gas shows in Late Miocene Claystones
KK-OS-V1-1	BHP	3203	BHP-Terminated in Basalt section
KK-B	RIL		

Himalayan Foreland Basins

Introduction

The Himalayan Foothills basin is a NW-SE aligned and located in the Himalayan region. The basin borders with Pakistan to the NW and Nepal to the SE. Northern and southern limits are demarcated by the Main Boundary Thrust (MBT) and the Himalayan Frontal Fault (HFF). The total area of the basin is about 30,300 sq. km. The sub-Himalayas extending over a distance of more than 1500 km from Jammu in the west to Arunachal Pradesh in the east separate Indo-Ganga plain in the south from the Himalayan mountain ranges in the north.

The hill ranges of WNW-ESE trending mostly constituted of Tertiary rocks rise to about 1800 to 2700 m. beyond the southern slopes of HFF, the plain area is having a cover of alluvium. Many rivers along with their tributaries form the main drainage system of the area. The basin occupies a zone of varying width from a maximum of 80 km in Bilaspur area to a very narrow width of about 20 km in the Kumaun area.

The Himalayan foothill belt, is characterized by each major lithologic unit is bounded by thrusts which follow the regional strike of the beds. Moreover, the rock groups maintain, on either sides of each thrust zone with uniformity for long distances.

In the Simla Hills area the major thrusts in a southwest to northeast direction are successively-the first one demarcating the boundary between plains and the Siwaliks, the second between the Siwaliks and the Subathus-Dharmsalas, the third between the Subathus-Dharmsalas and Krols and so on. The third one is defined as the "Main Boundary Thrust".

Exploration Status

Geological surveys of the Foothills were conducted by Geological Survey of India (GSI). (Figure 10.hf.1).

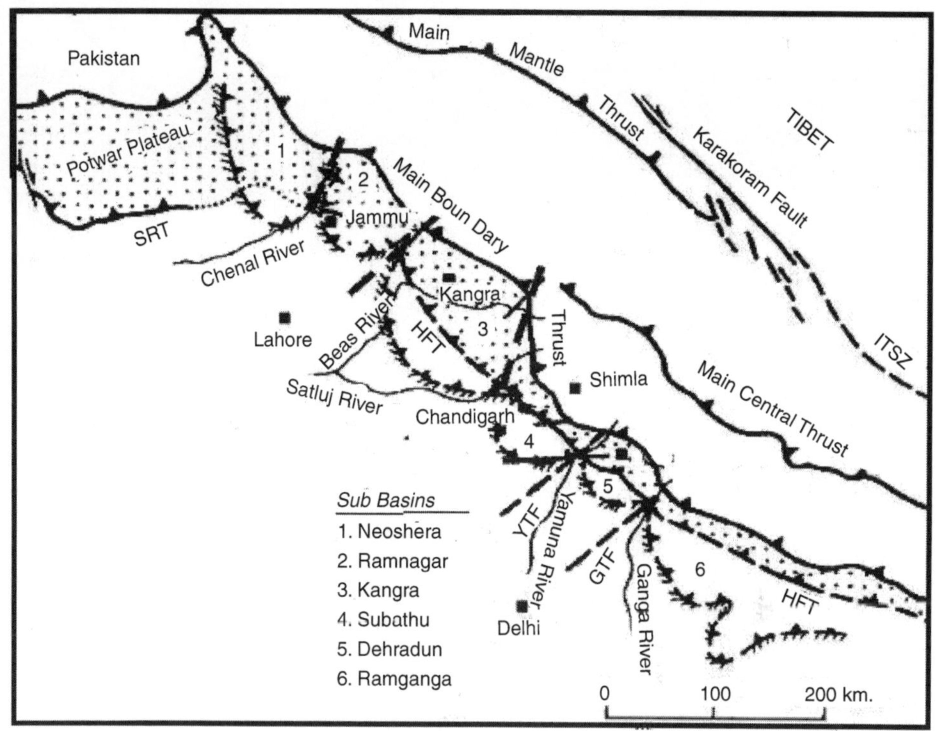

Figure 10.hf.1. Himalayan Foothills Sub-Basins.

The exploration work by detailed surveys and investigations was started by ONGC since 1956. Including detailed and semi detailed geological mapping, special studies on sedimentary features like current bedding, reconnoitory traversing and investigations of oil and gas shows were carried out. Geochemical samplings were undertaken in Subathu (Eocene) and Lower Dharmsala (Oligocene) sediments for laboratory analysis. Through this mapping, detailed stratigraphic framework and tectonics of the foothill belt has been established that helped in knowing the subsurface geology.

Major part of the basin is covered by Gravity and Magnetic surveys. Initially, seismic surveys conducted in the Punjab plains were of refraction and conventional type which were followed by more sophisticated CDP surveys during seventies. The foothill area has been partially covered by CDP surveys.

Initially few structural wells were drilled to know continuity of outcrop terrain and its variations and thickness. Exploratory drilling in the basin was initiated with the drilling of well Jawalamukhi-1 (JMI-1) during 1957-58. Two minor gas pools were discovered in the Lower Siwalik sub-Group section above Jwalamukhi thrust. Based on the map, three wells were drilled. One well, Ramshahr-1, Chari-1 lies on the northern side and Mohand-1 towards the south-eastern side of the area. Again ONGC took one block under NELP-II (HF-ONN-2001/1) in 2003 and one well Kasuli#1 was drilled to 3500 meters and the block was relinquished (Table.10.hf.1).

Tectonics

Tectonically, the Himalayan foothills forms a part of the Himalayan foredeep superimposed on the northern slopes of the Indian platform. The Himalayan mountain system is the product of collision between the Indian and Asian Plates. The initial consequence of this collision resulted in the withdrawal and closure of the Tethys Sea. Extensive marine sedimentation represented by the Subathu Group in the Lesser Himalayan province and in the marine sequences of Ladakh terminated in the early Middle Eocene (Bhatia, 1986). After the closure of the Tethys, deformation spread southwards across earlier shelf. The well data has helped to know palaeo-current from the west and the north suggests that orogenic processes on the continental margin had reversed the palaeo-slopeduring Middle Eocene (Figure 10.hf.2).

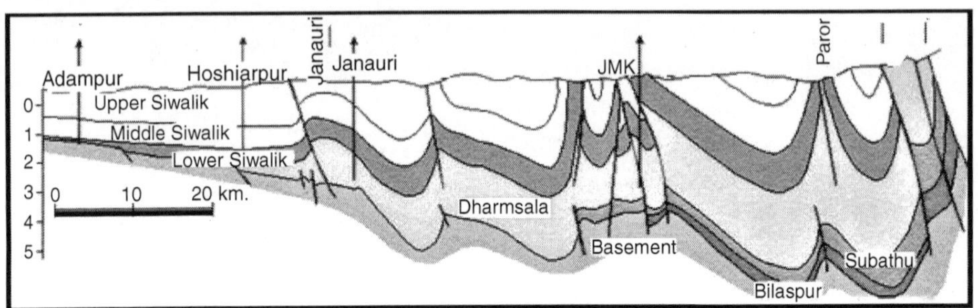

Figure 10.hf.2. Himalayan Foothills Basin-Section-WSW-ENE.

This changeover of the mode of collision from interactive continental/oceanic island arc systems to a purely continent-to-continent type has resulted in intra-continental shortening. The basin consists of several tectonic blocks demarcated by deep seated faults. During Late Tertiary, the major Himalayan orogeny brought significant structural deformation and reactivated some of the deep seated faults giving rise to the thrusted blocks. This enables, the division of the area into three tectonic belts.

The "**Inner Tectonic Belt**" is wide and structurally complex. In Himachal and eastern Jammu foothills, a number of emergent imbricate thrusts have developed, exposing various sequences of Tertiary sediments (Siwalik, Dharamshala/Murree and Subathu). These thrusts are associated with broad synclinal structures in between. Structural highs corresponding to the related tectonic setup are Sarkaghat, Paror, Rihlu and Najot, Jwalamukhi and others.

The "**Median Tectonic Belt**" encompasses the area between the northwest hading Jwalamukhi and Kishanpur-Mandili thrusts and the south hading Barsar and Deragopipur thrusts. The major anticlines in this belt are Seri, Suruin-Mastgarh, Balh, Chilli, Cheri and Changartalai.

The **Outer Tectonic Belt** encompasses the structures of the outer foothills. The northern limit of this belt is marked by the Barsar thrust, and the southern limit by the Himalayan frontal thrust. The major structures in this belt include the Deragopipur, Bharwain, Dhionsar, Ramgirdhar and Janauri anticlines.

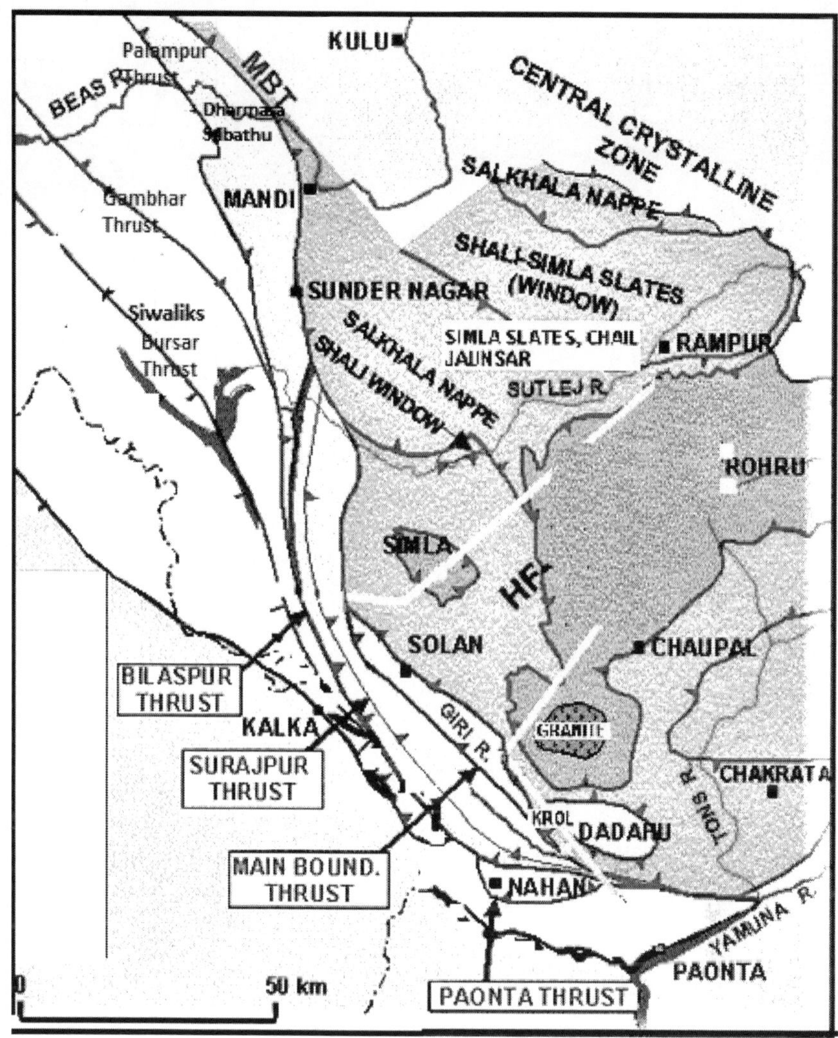

Figure 10.hf.3. Himalaya-Map.

Adjoining the foothills area, due south are the Punjab plains. Here the tectonic deformation in Tertiary sediments is minimal which show a clear basin ward thickening due north.

Both Gravity and Magnetic maps clearly show that NW-SE Himalayan trend is the dominant lineament in the basement. However, there is a NE-SW (Aravali trend) lineament in the area lying between Kalka and Dehradun.

Sedimentation

The Himalayan Foreland is an area in which sequences of the order of 10 km thickness have accumulated. The Early Cambrian to Precambrian limestone forms the base for the Tertiary sediments. Sedimentation was initiated in the Paleocene time under paralic conditions at the beginning of transgressive phase in which carbonaceous shale and coal were deposited followed by shallow marine conditions with the deposition of green shales and foraminiferal limestones of Upper Paleocene to Eocene age.

The basin started with a shallow marine environment which changed to estuarine and deltaic with time. By mid-Miocene, continental sedimentation marked by fluvial environment.

The sea regressed thereafter at the end of Eocene, during which red shales, laminated mudstones, grey, blue siltstones and purple calcareous sandstones of Dharmsala were deposited under transitional brackish water environment. The closing phase of marine incursion commenced due to uplifting in Proto-Himalayan source area and as a consequence, a coastal fluviatile environment was established (Figure 10.hf.4).

Deposition of Lower Siwalik started in Middle Miocene associated with Himalayan orogeny. The northern margin of the basin also got uplifted at this stage and this axis of deposition shifted southwards, sub-parallel to Himalayan trends.

During Upper Miocene times (middle Siwalik sub-Group), a thick succession of pebbly sandstones and claystones were deposited mainly by braided and meandering streams with a concomitant shift of the depositional axis southwards. This sequence is very thick at places with sediments supply from main provenance from the rising Himalayas.

During Plio-Pleistocene orogeny, the northern part, presently the Foothills area of the basin was further uplifted and as a result of which conglomerates of Upper Siwalik sub-group were deposited as alluvial fans at the foot of mountains. Tectonic movements of the Himalayas later produced the present topography and it was further witnessed by Quaternary glacial and interglacial processes with the deposition of high level river terrace conglomerates.

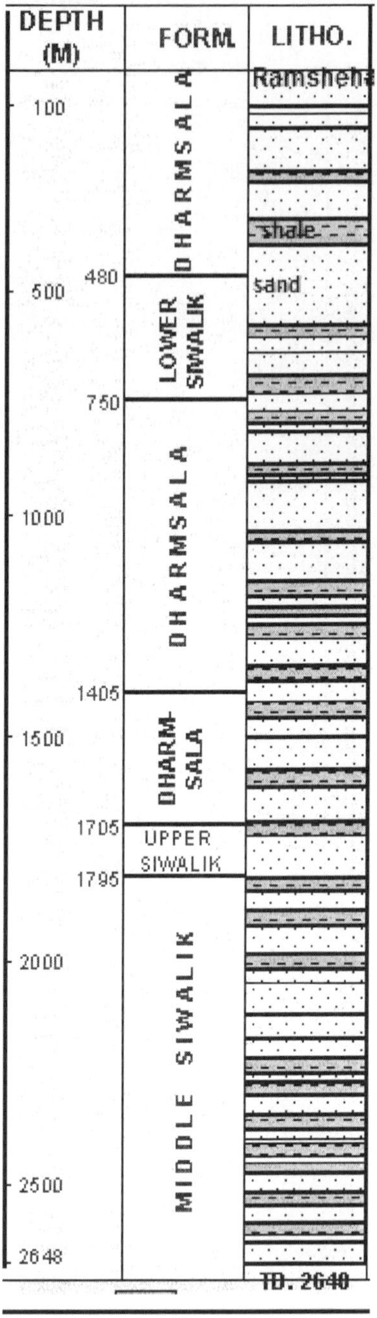

Figure 10.hf.4. Himalayan Foothills Basin-Stratigraphy.

The Murree/Dharamsala Group extends along the southwestern flank of the Pir Panjal through Punch and Naoshera into Jammu Hills. Below Dalhousie in Himachal Pradesh, it is reduced to narrow strip. The main belt expands in width in Himachal Pradesh below Subathu where they are referred to as Dagshai and Kasauli formations. Further to the east, the development of these sequences is negligible.

Stratigraphy

Stratigraphically the area underlain by Early Cambrian to Pre-cambrian rocks. The Pre-Tertiary sequence in the Himalayan foothills is mostly composed of dolomitic limestones, quartzites and shales. Stratigraphy of the Simla Krol belt, north of MBT, comprises of metamorphic and sedimentary rock types of Precambrian to early Paleozoic age. In sub-surface, granitic basement has been encountered in well, Saharanpur-1, in southern part of the basin where the sedimentary thickness is very less.

Pre-Tertiary

Jutogh Group consists of black, carbonaceous, garnetiferous phyllites, slates, quartzites and dolomites intercalated with biotite schists and hornblende gneisse. The exposures are known from Lesser Himalayas.

Simla Group comprises of bluish grey slate, micaceous shale, sandstone, quartzites intercalated with microcrystalline and oolitic limestone. These rocks are exposed around Simla hills.

Shali Group is dominantly a carbonate sequence made up of dolomite, shale, stromatolitic limestone and occasional magnesite and quartzite. Rocks of Deoban Group and Bilaspur/Bandla Limestone are considered equivalent to Shali Group. These rocks occur in tectonic windows under nappe of metamorphic rocks. At places thin Nummulitic outliers are found on these rocks. These are homotaxial to Simla slates.

Jaunsar Group of rocks is exposed north of Main Boundary Thrust (MBT). It is composed of low grade metamorphics such as slate, phyllites and quartzites. In the field, it is divisible into *Mandhali, Chandpur and Nagthat formations* which are named after localities in Garhwal.

Mandhali Formation consists of quartzites, shales, crystalline limestone and marble interbedded with slates, phyllites, gritty quartzites and bounder beds. Chandpur Formation; consists of mainly quartzites, phyllites and tuffaceous slates. Nagthat Formation comprises sandstones, grits, quartzites, conglomerates, purple and green slates and phyllites.

The top of the group represents a strong unconformity followed by a succession of rocks classified as ***Mussoorie Group*** which can be subdivided into *Blaini, Infra-Krol, Krol and Tal Formations.*

Blaini Formation consists of boulder beds, limestones and shale with characteristics of a glacial till. The boulder bed consists of dark grey to greenish grey clay matrix with pebbles of dark slate, greenish quartzite, grey sandstone, green siltstone and occasionally microcrystalline limestone. Limestones are characteristically pink, dolomitic and siliceous and grade into pink and purple calcareous shale and slates. These rocks along with Infra-Krol and Krol are exposed north of MBT. *Infra-Krol Formation* comprises dark shale, black carbonaceous shales and quartzites. *Krol Formation* consists of limestone, often stromatolitic, shale and sandstone. The limestone is grey to greyish white, dolomitic, cherty and at places microcrystalline. The shales are red, orange and dark gray in color. The sandstones are dirty white, poorly bedded with orange stained quartz grains. *Tal Formation* composed of calcareous greywacke, carbonaceous shale, micaceous shale, arkosic quartzite and grey limestone. In the subsurface, it has been identified in well Mohand-1, where it consists of shale, dolomite and silty shale.

The Krol series are of Permo-Triassic to Cretaceous, based on glacial nature of Blaini boulder bed to represent deposits of early Carboniferous glacial period. From the discovery of conodonts in these rocks in Mussoorie area, may assign a Cambrian- Ordovician age to these formations.

Following a long hiatus, the sedimentary sequence belonging to Tertiary age was deposited south of the central Himalayan axis. The Tertiary sequence is thickest in the inner foothills and becomes gradually thinner towards southwest.

Tertiary

Subathu Group consists of olive green to dark green shales interbedded with thin limestone bands and coal seams is assigned Upper Palaeocene to Eocene age. Rich assemblage of Nummulite has been reported. Subathu sequence represents coastal swamps and shallow to open marine environment deposits.

Dharamshala Group sediments unconformably overlie the Subathu group. These sediments consist of claystone, siltstones, calcareous shales and sandstones. The Dharamshala group is divisible into Lower and Upper Dharamshala. The Lower Dharamshala sediments were deposited under transitional brackish water environment and upper Dharamshala mainly represent fluvial system. The equivalent of Dharamshala group is known as *Murree Group* in Jammu and Kashmir.

Siwalik Group comprises of sandstone, siltstone and clay/claystone alternations in the lower part, whereas the upper part is represented by conglomerates, pebbly bands and sandstone lenses. These are broadly divided into Lower, Middle and Upper Siwaliks, ranging from Middle Miocene to Lower Pleistocene. The sediments were deposited as channel bars, piedmont bars, flood-plain and alluvial fan deposits.

The Alluvium consists of loose sand, silt, clay, pebble, gravel, boulder and kankar. It unconformably overlies different formations in different areas.

Jammu and Kashmir Basin

The North Eastern limit of the Kashmir valley is defined by the Himalayan range, a strong antclinorium, with south estern and southern by Pir Panjal Range (Figure 10.hf.5). Along the Pir Panjal thrust the Eocene and Paleocene sediments are thrusted over by the Kashmir neppe (Datta,N K,1983). On the eastern slope of Pir Panjal lie thick Karewas which are of lacustrine sediments. With oldest Cambrian

Dogra slates Near Wttargram there is eastrely plunging asymmetrical anticlinal fold to Plio-Pleistocene Shopian Formation.

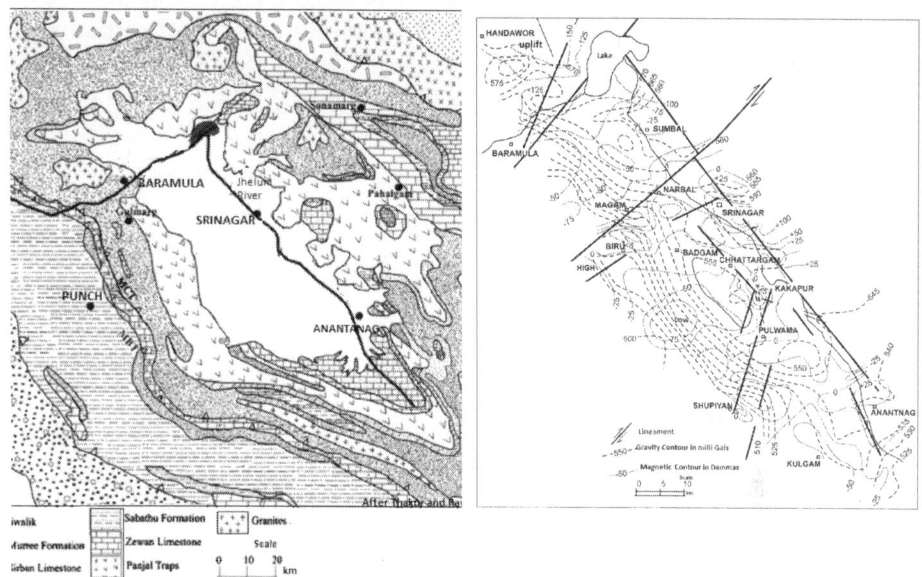

Figure 10.hf.5. Geolological Map Kashmir.

Figure 10.hf.6. Gravity and Magnetic Map-Kashmir Valley.

The gravity data indicate the deepest part is around Pulwana (Figure 10.hf.6). The structural configuration indicates inliers of older rocks are around Hundwana with also occurrence of peat and lignite indication marshy conditions.

The integration of lineaments with geological and geophysical DSS data suggests deep seated faults. Differential movements along such trends influenced the basin (Ganju J L and B M Khar, 1984).

Hydrocarbon Prospectivity

The samples along Kalkot and Triyat indicate over maturation of sediments with possibility of gaseous hydrocarbons (Mukherjee, M K, 1999). In Kahsmir Valley with more than 6,600 m of marine sediments with Paleozoic to Pleistocene with even drak grey shales and limestones merit attention for hydrocarbon exploration. But the vigorous exploration so far did not indicate any oil show only gas seeps are noticed in Karewas (Figure 10.hf.7). Though the Chattargram and Narabal wells drilled in the areas did not indicate the occurrences of oil and gas.

Suketi Basin

The Sundernagar area of Himachal Pradesh is also referred to as 'Suketi Basin' after the Suketi Khad, a north flowing tributary of the river Beas. The Lesser Himalayan part of Sundarnagar-Mandi area is a thrust-accreted Proterozoic carbonate-argillaceous sequence, resting tectonically over the Tertiary mollasse.

The tectonic contact is known as Main Boundary Thrust, which is manifested by a steep ramp at surface and gradually attains lower angle at depth. Hanging wall sequence is the part of Proterozoic passive margin deposits of Northern Indian craton, acted as basement for the foredeep sediments during Tertiary period. A series of in-sequence thrusts developed during Himalayan orogeny represents tectonics for Lesser Himalaya.

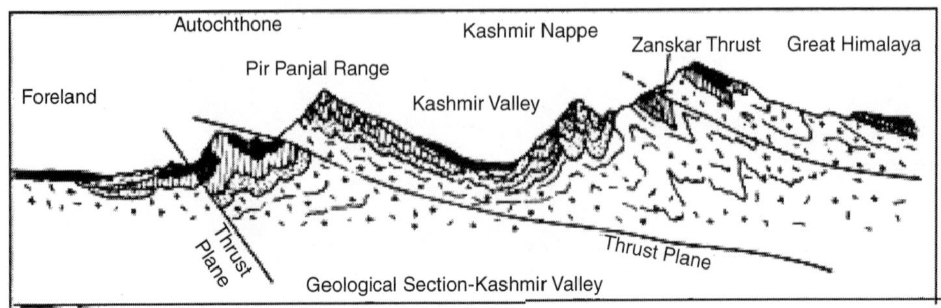

Figure 10.hf.7. Kashmir Section.

Tectonically, the area constitutes the Lesser Himalaya and is generally referred to as the 'Shali Structural Belt' or the "Lesser Himalayan Terrace". Sundernagar depression is a small and young (Pleistocene?) fault bound inter-montane basin. Subsidence was primarily controlled by faults. The valley fill was initiated by lacustrine sediments and ended up with boulder beds of proximal alluvial fan.

The stratigraphy of the area can be explained by imbricate thrusts and associated folds, periodically modified by thrusting. Within the Chail metamorphics, forelimb shear thrusts are commonly developed with compression. The shear faults are commonly associated with secondary conjugate faults and most of the thrusts are connected down the section in the anticlinal core.

Hydrocarbon Prospectivity

In Sundernagar area, there are two hydrocarbon shows, Chowmukha oil show south west of Sundernagar and Himriganga gas seep to its north. However, the section encountered in the Sundernagar well does not contain any source facies. Outcrop data suggests Subathu, the potential marine source facies, to be present subthrust of MBT. Shali Carbonates are generally low in organic content.

In the Palaeogene sequences, in the subthrust of MBT envisaged below 3,400 m, potential source facies of Subathu are expected. These are inferred to be rich in organic matter with kerogen of Type-II which is predominantly sapropelic to humic, adequately matured due to substantial overburden depth of more than 4 km.

Thermal modelling carried out for the deepest well drilled in Jawalamukhi area has indicated depths of 4000 to 5000m for oil generation temperatures (120°C) and depths of 6000 to 7000m for gas generation temperatures (150°C). It is likely to have resulted in the generation of hydrocarbons with more gaseous with subordinate oil.

The petroleum system inferred that significant petroleum generation and expulsion would have taken place prior to thrusting. In view of uplifting of the potential source facies along the thrusts in the subsurface of this area, hydrocarbon accumulations can be expected at structurally higher levels

Karewa Basin

It lies in Tethyan zone in the inter-montane valley of Kashmir located between Pir panjal Range in the NW and Zanskar range in the SW. The basin rests over the crystallines of the Salkhala Group. The envisaged petroleum systems in Karewa basin speculates that the black shale horizons in Khunamuh and Tindira formations, and Dubjian and Rembiara members (Hirpur Formation) could be the potential source. The Pir Panjal Volcanics and Pindabol Formation could be suitable reservoir rocks. For the post panjal Volcanics, the Khunamuh-Tindiara formations, and the sandstones of the Karewa Group could be the possible reservoirs. The Synclinorium and the small intra anticlines, *viz.* Shamsabari syncline, Riddi anticline, Madamati anticline and the Lidder valley anticline could act as the entrapments (Figure 10.hf.8).

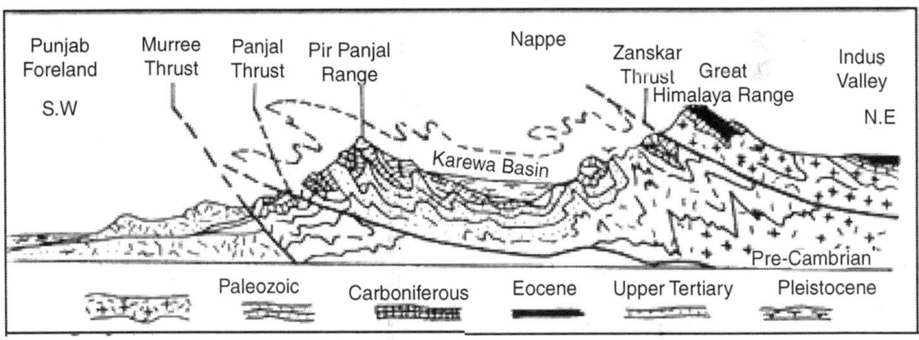

Figure 10.hf.8. Geological Map-Karewa Basin.

Borehole gas shows had already been noticed in wells drilled for hydrocarbons in Karewa basin. Gas contains 50-90 per cent methane indicating a marshy source. Also, hydrocarbon gas seepages have been reported from the Karewa, no oil show has been reported.

Shali-Basin

The term 'Shali Basin' has been used to refer to the palaeo-depositional system of the 'Shali Belt' and 'Shali Structural Belt'.The studies in the sector between the Rivers Ravi and Beas in Himachal Pradesh show the litho column as below:

Holocene	Alluvium, Terraces, Soil
Up. Pleistocene to Mid. Miocene	Siwalik Group
Lr. Miocene to Up. Eocene	Sirmur Group
Palaeocene	Kakara Formation

~~~~~~~unconformity~~~~~
~~~~~~~

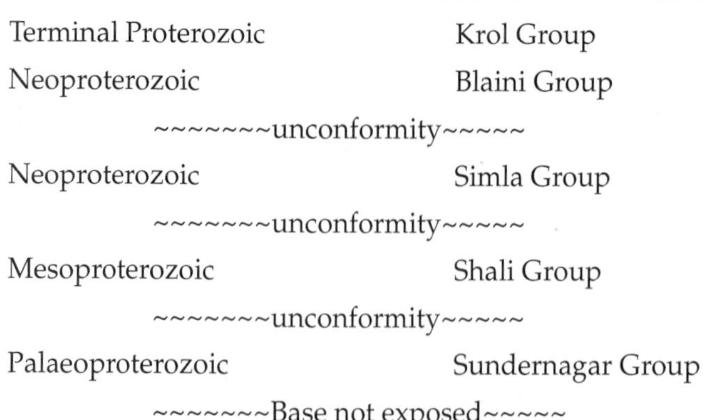

Terminal Proterozoic	Krol Group
Neoproterozoic	Blaini Group
~~unconformity~~~~~	
Neoproterozoic	Simla Group
~~unconformity~~~~~	
Mesoproterozoic	Shali Group
~~unconformity~~~~~	
Palaeoproterozoic	Sundernagar Group
~~Base not exposed~~~~~	

Pir Panjal

The south-western Pir Panjal represents part of the eastern limb of the Jhelum syntaxis. The Murree and Panjal thrusts are two major bounding faults in this area. At places small faults with minor throws are known.

The Murree thrust is a regional fault trading towards northeast at angles of 30°–40 °. Different horizons of the Gamir Formation are juxtaposed, across this fault, against the Murree. This thrust canbe described as a regional reverse fault without much horizontal displacement.

The Panjal thrust, limiting the parautochthon in the north, truncates successively older units/horizons towards east. The Rajpur Formation is terminated by this thrust on the right bank of the Suran River towards south-west of Thannamandi the Zewan beds are completely cut off. The hade of the thrust, always towards north east, varies between 50°and 40°.

Within the parautochthon there are no regional folds or faults. The folds here can be as local warps that do not extend for more than 1-1.5 km and are seen at different stratigraphic levels, agglomeratic Slate and the variegated shales of the Rajpur Formation. The wider outcrop of the Baila Formation seen south of Thannamandi is due to these warps.

The Panjal Traps, which are exposed on the flanks of the Srinagar valley, are considered to be the source of magnetic anomalies observed possibly due to remanent magnetization, having magnetite which got very low measured magnetic susceptibility values. This may be due to the subaqueous flow.

Between the Murree and Panjal thrusts, successively younger units outcrop towards north. The Jokan Limestone is exposed, south of the Murree thrust, as a narrow faulted inlier within the Lower Murree. The details of stratigraphic sequence in this area.

Gamir Formation outcropping immediately north of the Murree thrust. This unit comprises alternations of shales, limestones and quartzite. The shales of this unit are maroon, grey and green, slaty or splintery rarely graphitic. The limestones, alternating with shales, are cryptocrystalline, laminated, flaggy, marly, grey, pale-purple and cream coloured.

Towards top of this unit the quartzites change to fine grained, rusty weathering, siltys, sandstones interbedded with light grey, silty, shales. It is older to Agglomeratic Slate which are of Upper Carboniferous age.

Baila Formation Good sections of the unit are seen near Baila, Gamir and along Suran River. The formation can be divided into three members-the lower and upper shale with a massive limestore in between.

The lower shale is greenish grey to ash grey in colour, thin bedded and not infrequently traversed by thin quartz veins. The shales in the lower part are graphitic, at times gypseous.

Massive, light to dark grey, nodular limestone constituting the middle member. The beds are lenticular, especially at the contact with the shale above and below. The shales above the limestone are similar to the shales below it, expect that they are greenish grey, laminated and well jointed. The conformable contact of this with Agglomeratic Slate above suggests that this formation is older than Upper Carboniferous.

Panjal Trap: The traps are fine grained, dark grey to black, green ard purple in colour porphyritic by numerous epidote veins. The wide variation in the thickness of the Trap suggests that the individual flows are lenticular.

Zewan Beds: Varying thickness of variegated ash beds occur at the base of this unit. Above the ash beds a sequence of fossiliferous sandy limestone overlain by grey fossilferous shale is noticed. The age of this unit ranges between Middle and Upper Permian and the upper most argillaceous limestone horizon seen locally as near Mandi, may be Triassic in age.

Rajpur Formation is divisible into members-the lower limestone and the upper variegated shales. The limestone is massive, nodular, light grey to sooty black in colour and foraminiferal. The calcareous shale partings within the limestone are highly foramiriferal.

Joken Limestone is exposed as a narrow faulted inlier-thin strip in the axial region-within the Lower Murree in Sangla area. The unit consists thick band of grey, bituminous, nodular foraminiferal (Upper Eocene) limestone with variegated shales both underlying and overlying it. This unit is yourger than, and not homotaxial with, the Rajpur Formation.

Spiti-Zanskar Basin

The Tethyan Himalaya made up of the Neoproterozoic-Eocene succession resting as detached synclinoria. The main synclinoria often referred to as basins, from west to east are: Kashmir, Chamba-Bhadarwah-Tandi, Spiti-Zanskar and Kinnaur-Kumaon.

The Spiti Valley, located north of the Pir Panjal Range, exposes an excellent section of Neoproterozoic- Cretaceous rocks in the Tethyan Himalaya of Himachal Pradesh. The Spiti Tethyan rocks towards NW through the Lahaul Valley extend into the Zanskar area where Paleocene-Eocene sequences are preserved. The north western Zanksar Block highly deformed with steep dipping faults towards north where as southern Spiti basin is less deformed and exposed sedimentary sequences (Biswas AK., 1999). The Jurassic Spiti shale with Vitrinite reflectance (VRo) suggests good gas generation potential.

Occurrence of High Pressure in Himalayan Foothills

High pressure was encountered in all the structures in the drilled wells in the area except Janauri and Ramshahar. Janauri structure is situated at a greater distance towards plains and structurally less uplifted. In Ramshahar the pressure might have released through possible conduits. Interestingly the major pressure breaks are recorded in the lower part of Lower Siwalik to Upper Dharmsala/Muree in all the wells.

The litho-stratigraphy of *Balh-1* as Lower Siwalik 0-2130m, Upper Dharamsala 2130-3306m, Lower Dharamsala 3306-4470m (+). The abnormal pressures encountered at 2960m in this well and controlled by a maximum mud weight of 2.16.

Jwalamukhi

The first exploratory well in Jwalamukhi was spudded in 1957. The small gas find in two shallow zones (Siwaliks) led to drilling of a number of wells in the area to delineate its lateral extent and hydrocarbon prospects. In the Jwalamukhi-1 well; free gas (100 per cent Methane) was produced at an approximate rate of 43,000 m^3/ day from the interval 885-887m. So far, 6 deep and 5 structural wells have been drilled in this area. The well Jwalamukhi#B is the deepest well in the basin drilled to 6720m. Lower Dharamsala 4013-6,720 with Jhor thrust at 4425.The occurrence of abnormal formation pressures which necessitated to increase in mud weight from 1.54 to 1.71g/cc.The gas seepages observed since the bottom zone could not be cemented due to abnormal pressures (Mudiar and Shukla, 1991).

The gas shows in the wells, namely, JMI#1, Nurpur#1 (both Dharamsala Formation) are methane rich with small nitrogen concentration and are dry gases of thermogenic origin. Carbon isotopic composition of methane in these wells suggests a deep over mature source for these gases. Coals/carbonaceous shales within the underlying deeply buried Subathu are believed to be the probable source rocks for this gas as Dharamsala Formation has no source potential.

Petroleum Geology

Two surface shows of oil and several shows of gas, both in surface and in wells, have been observed in this belt. The other known oil show is near Chomukha village on the eastern side of Alsed Khad in Sundernagar area. This show is observed in a sandstone bed at the base of Dharmasala sequence. Droplets of light crude oil in cavities and joints of thin limestone bands/lenses have also been noticed in the Alsed Khad. This limestone is of Upper Paleocene to Eocene age (Subathu Formation).

Numerous gas shows have been observed in the median and inner tectonic zones in Himachal Pradesh, Jammu and Kashmir, the most famous being the gas shows in Jawalamukhi temple. The gas is mostly methane (87.8 per cent) with small amounts of CO_2, N2 and O_2. The other main gas shows in this belt include those in Dehar Khad River near Kotla village, Suruinsar and Mansar lake areas in J&K state and from the flank of Sarkaghat anticline in Himachal Pradesh. Some gas seeps including the one at Himriganga are characterized as those of thermal origin, whereas some others are of bacterial origin.

The gas seepage is near the Himriganga temple which is located near Mandi town in Himachal Pradesh. The gas seepage is occurring in a spring with gas escaping in the form of bubbles intermittently of variable intensity.

The prognosticated resources of the basin are 140 MMT (OIL +OEG).

Source Rocks

The pre-Tertiary sequences in Himalayan Foothills are mainly lean in organic carbon and are over mature. The outcrop samples from Vaishnodevi limestone in Jammu and Kashmir and Bilaspur limestone in Himachal Pradesh contain 0.01 to 0.33 per cent of organic carbon.

The Subathu rocks are the only marine sediments in the stratigraphic sequence present in the basin. The overlying Dharmsala/Murree and Siwalik sequences represent brackish and fluviatile sediments deposited in an oxidizing environment and have organic carbon content (0.02 to 1.02 per cent).

Dominantly argillaceous lower Dharmsala Formation in Jwalamukhi well has indicated a maximum value of VRo around 0.49 per cent, Tmax in the range of 425-441°C and TAI values between 2.5-2.75, which suggest that the Lower part of Dharamsala Group has reached an early stage of maturation (Tandon, *et al.*, 1998).

Reservoirs

The pre-Tertiary section of the Himalayan Foothills is primarily composed of mixed clastic carbonate sequence. The sandstone and carbonate reservoir facies generally have low primary porosity. However, fractures are often observed in this section. The Lower Tertiary reservoir facies in out crops also have low primary porosity and fair permeability. The Upper Tertiary (Upper Dharmasala and Lower Siwalik) reservoir facies have better development of primary porosity (8 to 15 per cent) and permeable.

Entrapment

Traps of two main types, structural and combination are likely to be present in this area. Further, the chances of structural entrapment in the sub-thrust block are considerably high, where the deformation was continuing with the accompanying possible thermal maturation of source rocks and generation of hydrocarbons. Alternations of arenaceous and argillaceous lithologies are observed in the entire Tertiary sequence from Lower Dharmsala onwards which could provide good reservoir and cap rocks.

Table 10.hf.1. Well Drilled in Himalayan Foot hills

	Well	*Depth*	*Year*	*Remarks*
1	Jawalamukhi-1	3067	1957	Flowed gas in the 879-923m @4,30,000 m^3/d
2	Jawalamukhi-2	5047		
3	Jawalamukhi-3	994		
4	Jawalamukhi-6	1274		
5	Balh-1	4474		Below 2100m high pressures -2.14 MWE known
6	Janauri-1	4531		
7	Janauri-2	5027		
8	Ramshahr-1	2648		
9	Mohand-1	5264		
10	Cheri-1	3238		
11	Jawalamukhi-B (JMI-B)	6720	1988	High formation pressures up to 1.71 MWE
12	Kasauli-1	3500	2011	Under NELP-VII
13	Adampur	2538		

Punjab Basin

The Punjab basin, located on the northern slope of the Indian Platform occupy an area of about 72,000 sq. km., with in the states of Jammu-Kashmir, Himachal Pradesh and Punjab. The basin is marked to the north by the metamorphites of the Lesser Himalaya, to the northwest and west by the Jhelum transversal uplift, to the southwest by the Sargoda basement massif and to the southeast by the Delhi-Kalka ridge (Figure 10.pb.1).

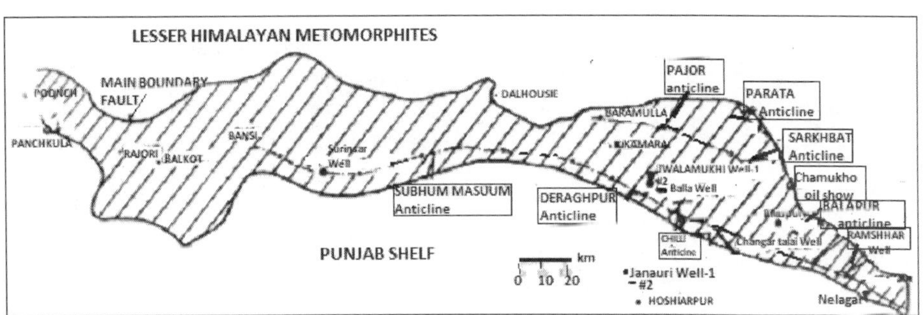

Figure 10.pb.1. Punjab Basin.

Exploration History

Systematic search for hydrocarbons in the Punjab basin was started by ONGC in 1956 with the geological mapping in the exposed foothill belt in Jammu-Kashmir, Punjab and Himachal Pradesh. The mapping work helped in knowing the stratigraphic framework and structure of the foothill belt.

Aero-magnetic surveys carried out in 1957 gave broad indications of basinal configuration which formed the basis for initiation of seismic and gravity-magnetic surveys. The initial seismic survey conducted in the Punjab plains was of refraction followed by reflection survey in the seventies. So far about 5710 line kms of refraction and conventional reflection surveys of 1084 LK m of 2D and 549 SLK of 3D survey have been carried out in the basin. So far, 11 deep wells and 7 structural wells have been drilled in this basin without any success. Hosiarpur well was drilled up to 3439 m (Table.10.pb.1).

Geological Set-up

In Kohat-Potwar basin in Pakistan about 1000m thick Palaeo-Mesozoic rocks have been encountered which were essentially deposited in marine conditions (Figure 10.pb.2).

The possible correlation of the Sirban Limestone in Jammu and Kashmir and Bilaspur Limestone in Simla area with the Vindhyan rocks of the peninsular area suggests the earliest subsidence and consequent deposition in the basin during the Riphaen time. Probably, this Riphaen basin extended to the west up to Salt Range (now in Pakistan).

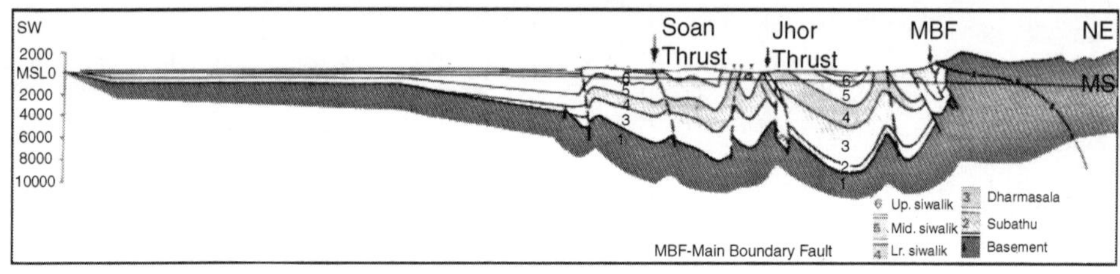

Figure 10.pb.2. Geolological Section-Punjab Basin.

During early Tertiary time, the subsidence and consequent accumulation of Palaeogene sediments, Subathu and Dharamasala covered a large area of the Punjab basin. In the major part of this basin, the Subathu sediments appear to have been deposited mainly under marine to shallow marine conditions At the close of Upper Eocene, the north eastern part of the Subathu basin was uplifted creating a basin to the south of the Main Boundary Fault where brackish and fresh water sediments belonging to Muree/Dharamsala group were deposited.

During the end of Palaeogene time, a foredeep in front of the rising Himalaya was formed in which molasse type of sediments, Siwaliks accumulated. This phase terminated during the end of Siwalik times due to folding and consequent uplift.

Tectonics

The peninsular Indian shield (Gansser, 1964) consists of metamorphic and crystalline rocks of Proterozoic/Archean age. The Himalayan range can be subdivided into two regions relative to the location of Main Central Thrust (MCT). North of MCT is the Tethys Himalaya which includes the Tibetan Plateau. South of MCT exists the sedimentary zone, which includes the complex nappes and thrust sheets of the Lesser Himalaya and both autochthonous and allochthonous units of the sub Himalayas. The Indus-Tsangpo suture, which is characterized by an ophiolitic complex, marks the first collisional contact between India and a continental or island arc terrain.

In Punjab Plains, the Indian shield slopes gently northwards, interrupted by Precambrian exposures in the Delhi-Sargodha High, a basement ridge that has an NW-SE trend, *i.e.*, Himalayan trend (Farah, *et al.*, 1977; Yeats and Lawrence, 1984). Subsurface data show that the section beneath the Jhelum Plains, Pakistan consists of Himalayan erosional products, deposited on Cambrian strata while in Potwar Plateau.

Based on gravity data it is inferred that, the foreland area of Punjab Plains is a consequence of the flexure of the elastic Indian plate right from the Jalandhar area towards northeast that underthrust beneath the Outer Himalayas. The basement fault of big throw might have played a great role in forming of thrust propagated folding of rocks in the Punjab Plains in Pathankot-Jammu area under the cover of Alluvium. The refraction section indicates the rolling nature of the subsurface strata in Punjab Plains northeast of major basement fault (north of Gurdaspur). The subsurface strata are folded and axes of the folds run parallel to the Himalayan strike.

Tectonically, the basin forms a part of the Himalayan foredeep superimposed on the slope of the Indian Platform and is divided into, folded zone with thrusts and an alluvial plains area. Frontal folded zone is the flank of Punjab foredeep situated between the Main Boundary Fault and the alluvial plains. The folded zone extends about 450 km from NW to SE with the outcropping part stretching from 20 to 100 km in width where Siwalik, Dharamsala and Subathu sediments are exposed. The basin essentially consists of several tectonic blocks demarcated by down-to-basin faults, the reactivation of which during the Late Tertiary Himalayan orogeny has given rise to several thrusted blocks. The thrusted blocks in turn are complicated by several linear anticlinal and synclinal features which are also faulted and thrusted. In general, the intensity of tectonism decreases from Lesser Himalaya to the alluvial plains.

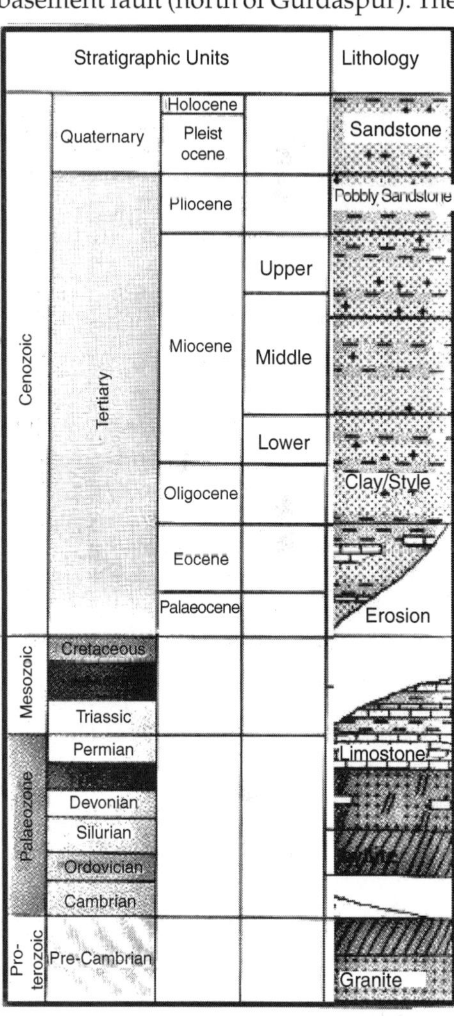

Figure 10.pb.3. Stratigraphy-Punjab Basin.

Stratigraphy

The stratigraphic sequence comprises Pre-Cambrian metamorphic/

igneous basement rocks overlain by Palaeozoic, Lower Mesozoic and Tertiary sediments ranging in thickness from 500 m in the south-west to more than 5000 m in the north-east of the basin. The generalized stratigraphic column of the basin is given (Figure 10.pb.3).

Petroleum Geology

The nearest hydrocarbon bearing basin to the Punjab basin is the Potwar basin of Pakistan in which several oil fields were discovered. The Palaeocene-Eocene Subathu Formations which contain probable source rocks, and Dharamsala and Lower Siwalik sediments can under favourable conditions provide suitable traps for accumulation of hydrocarbon.

Oil and Gas Shows

In the Punjab basin, oil shows in the surface has been observed in the Subathu limestone near Satra (SE of Poonch) and near Sundernagar (Himachal Pradesh). Several surface shows of gas are also found in the Lower Siwalik sediments exposed in the frontal folded belt.

Source Rocks

The Eocene Subathu rocks are the only marine sediments in the stratigraphic sequence present in the Punjab basin. Geochemical analysis of Subathu rocks indicate organic carbon content ranging from 0.02-13.0 per cent. The organic matter was found to be sapropelic and humic type with high degree of metamorphism. Geochemical study of organic matter in Dharamsala sediments shows very less (0.01-0.11 per cent) organic carbon, which is not considered a favourable for generation of hydrocarbons.

Reservoir Rocks

Two types of reservoir rocks can be identified in this basin *viz.* Limestone and Sandstone. Bioclastic and fractured limestones have been developed with large thicknesses and appreciable porosity, in Subathu, Dharamsala/Murree and Siwalik formations.

Cap Rocks

Subathu, Dharamsala and Siwalik formations contain numerous shale/clay band which can act as cap rocks.

Type of Traps and Prospects

Two main types of traps are likely to be present in this basin *viz.* structural and structural cum strati-graphic. The structural traps include a large number of exposed anticlinal structures running roughly NW-SE. The successive pinch-out of the Tertiary sediments against the rising basement ridges in the area may also form suitable pinch out traps. No commercial and significant oil/gas occurance has been discovered in the basin.

Table 10.pb.1.Wells Drilled-Punjab Basin

Well	*Drilled (m)*	*Remarks*
Bhimbergali		
Surinsar-1	3665	Tectonical-abnormal pressures observed
Surinsar-2	5448	
Narpur	4935	
Sarkaghat		
Parata		
Paror		
Changartalai	4968	
Bharwin		
Hoshiarpur	3439	

Ganga Basin

Introduction

The vast alluvial sediments of Ganga River and its tributaries occur between rugged upland of Peninsular India and the rising mountain province of Himalayas. This Indo-Gangetic plain represents more than 8 km thick sediments and still deepening foreland basin, sediments range in age from Meso-proterozoic to Recent. The Basin extends from Delhi-Kalka ridge in the west, to the Monghyr-Saharsa ridge in east, and covers an area about 320,000 sq km (Figure 10.ga.1).

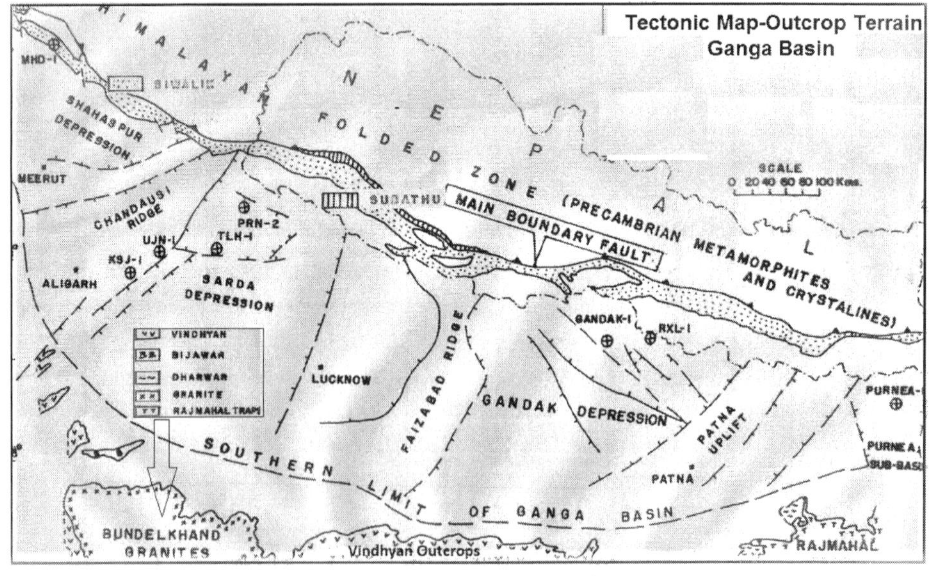

Figure 10.ga.1. Tectonic Map–Ganga Basin.

Exploration Work

The seismic work and the drilling conducted by the ONGC in this area suggests that it consist of Siwalik sediments underlain with a pronounced unconformity by folded and much older marine and non-marine strata referable to the Vindhyans.

Few deep wells and shallow wells have been drilled in this area but without any encouraging results. In the *Tilhar deep well* a quartz arenite sequence below the limestones of 100 m thick also present at 2100 m. Another *Ujhani deep well* drilled to 2119m met with metamorphics at 2062 m below the Siwalik/PreSiwalik unconformity at 1020m comprise limestones, arenite, grey shales. A thin conglomeratic zone is present immediately below the Siwaliks at 1011-1071 metres in *Ujhani structure well no. 1*, which was drilled close to Ujhani deep well no. 1. This conglomeratic zone has not, however, been encountered in the deep wells drilled near Ujhani and Tilhar (Table.10.ga.1).

No commercial hydrocarbons have been discovered in Ganga Basin so far, even though hydrocarbon shows have been encountered in few wells. During NELP regime five blocks were explored by ONGC, but the areas were relinquished due to low prospectivity.

Tectonics

The Ganga basin is a poly-history basin, with four main phases of geological evolution. Basement tectonics played an important role during the sedimentation and deformation of the basin. The basin evolved through multiphase geological history initiated with rifting during Proterozoic, followed by deposition of shallow marine sediments of early Devonian to passive margin during Late Eocene-Oligocene and finally to foreland basin since Middle Miocene. The Bundelkhand massif (granite) constitutes the basement for sedimentation in Ganga basin.

As a consequence of the continental collision, a peripheral molasses basin named Ganges Basin was evolved, which resulted in the deposition of Neogene clastics onto older Paleozioc/Preterozoic and Precambrian basement.

Sedimentation

The Bundelkhand massif (granite) constitutes the basement for sedimentation in Ganga Basin superimposed by Cenozoic sedimentation. The marked feature of the basin is the presence of several depressions in the northern part. The Mesoproterozoic sequence represents metamorphic rocks characterized by phyllitic shale and orthoquartzite. The Lower Paleozoic sequence unconformably overlies this and represents deposition in typical platform limestone-orthoquartzite-shale facies. End of this sequence is marked by a phase of compression during Hercynian orogeny (?).

The overlying Tertiary cover, however, extends far beyond the subsurface limit of the Vindhyans in the Ganga Basin, both to the west and to the east. In the west the Tertiary cover directly overlies the Precambrian basement in the Punjab shelf, and in the east, it directly overlies Gondwana sequence in Purnea Basin.

This was followed by a period of extensive erosion and deposition of Paleogene sediments in passive-margin continental setup. This is followed by thick blanket of overfilled siliciclastics sediments of Miocene and younger age followed by alluvium (Figure 10.ga.2).

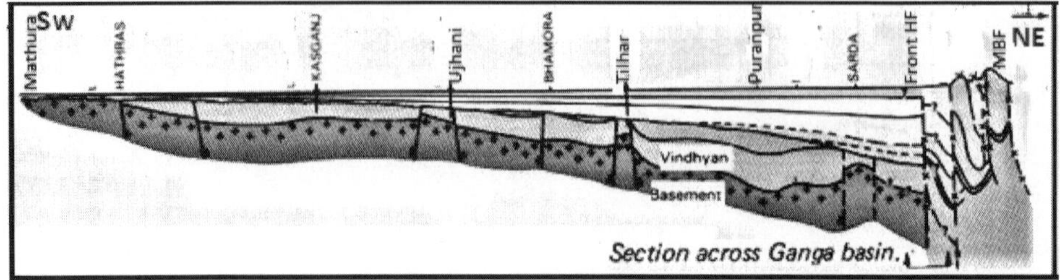

Figure 10.ga.2. Geological Section-Ganga Basin.

This sequence forms a gentle homocline dipping towards the foothills without much structural deformation. The southern limit of this deposition shifted progressively southward with time so that along the southern margin of Ganga Basin, where the youngest part of this sequence and even alluvium, rest directly over metamorphic basement.

The Palaeozoic rocks occur immediately below the unconformity which separates Siwalik and pre-Siwalik sediments and also points to a marine environment of deposition under neritic conditions (Figure 10.ga.3).

The Sub-Himalayan belt between Ganga and Sarda rivers, shows three structural zones separated by "fold-faults". The outer zone and to a lesser degree the middle zone show, in the area between Saneh river and Sona Nadi, a remarkable swing in the strike which he describes as "S-shaped" (Sastry, V.V. *et al.*, 1971). Later referred to the cross-fault as the "Laldarwaza fault" and called the reverse fault between the outer and middle zones as the Boksar thrust (Ganguly and Jagtap,1966).

Stratigraphy

The Indo-Gangatic plain represents sediments of more than 8 km thick and still deepening foreland basin, filled with sediments ranging in age of older than Cambrian (!). The lower boundary is unconformable contact between metamorphic rocks of Neo-proterozoic age and Precambrian. The upper boundery is also unconformable (angular unconformity to disconformity. This sequence represents the metamorphosed sedimentary rocks o the Ganga Supergroup and characterized by the phyllitic shale and ortho-quartzite.

A thick sedimentary package overlying Lower Paleozoic in the eastern part and Proterozoic sediments in the western part of the Sarda depression is referred to the Passive margin sequence. The sequence is identified primarily as non-marine and is characterized as lacustrine to fluvial sedimentary succession.

The Paleogene (Late Eocene-Oligocene) is further divided into two second order sequences. The Late Eocene - Oligocene sequence is present over a very limited part of the Sarda depression.

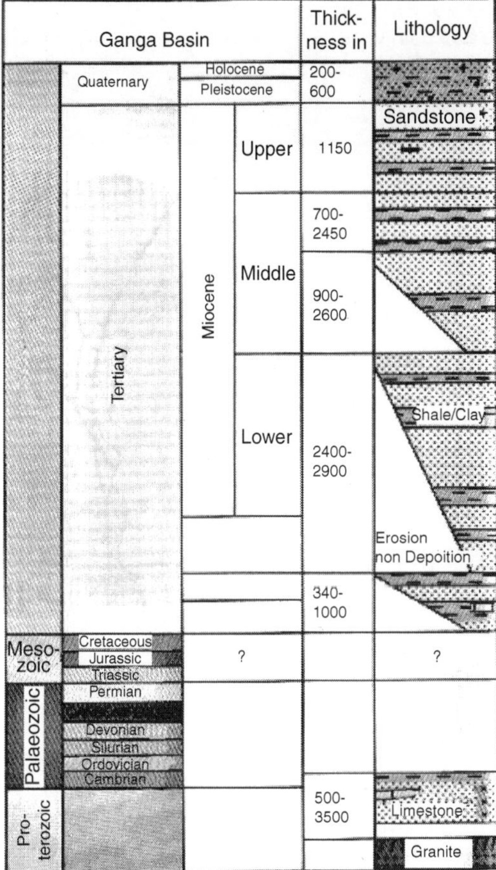

Figure 10.ga.3. Stratigraphy–Ganga Basin.

Neogene (Middle Miocene-Pleistocene) sequences comprise thick sedimentary section overlying Lower Paleozoic sediments (in some part of the basin, overlying Late Eocene-Oligocene sequence) and underlying the alluvium. The sequences are -Middle Miocene. The sequence has been encountered in Sarda, Gandak and Madhubani depressions.

Petroleum Geology

The organic content (TOC) in the source rocks are very poor except in some section in well Puranpur #2 up to 0,44 per cent was noticed. Though reservoir facies are good there are no regional cap rocks available. The Lower Ordovician sequence has good TOC, with thin organic-rich beds with good to fair hydrocarbon generation potential.

The geochemical studies carried out on samples to know the depositional environment indicated Pre-unconformity sediments in the basin except in Gandak and Purnea indicated the biogenic activity was in marine environment where as in the Gandak depression it was fresh water environment. In the post-unconformity section it was dominantly fresh water environment (Uniyal., AK, *et al.*, 1994).

Potential structural and stratigraphic traps are well developed, and good quality, highly fractured reservoirs have also been identified. The extent and burial history of potential source rocks remain unclear.

So far no good hydrocarbon indications are seen, however the resources estimated considering the vast area of the basin as 369 MMT, it seems to be speculative.

Table 10.ga.1. Wells Drilled-Ganga Basin

No.	*Well*	*Drilled (m)*	*Year*	*Remarks*
1	Tilhar-1	2225		Small gas in Vindhyan sequence
2	Ujhani-1	2127		Low TOC-0.2 per cent. 3structural wells were also drilled to shallow depth
3	Kisaganj-1			
4,5	Puranpur-1, 2	4234		# 2 TOC varies to 0.44 per cent
6,7	Sahranpur-1,2	2362		
8	Gandak-1	5314	1984	
9	Raxul-1	4901		Low TOC-0.2 per cent porosity upto 23 per cent
10	Ganuli-1	5422		
11	Madhubani-1	5957		
12	Matera-1	3927		
13	Kadmaha-1	5372		
14	Shajahanpur-1	3352		

Category IV

Pranhita–Godavari Basin

Introduction

Pranhita Godavri Basin occurs at the junction of Dharwar and Bastar cratons and developed after amalgamation of the craton in palaeo-proterozoic. Three sub-basins are identified in the basin, namely Albaka, Pakhal, and Chandarpur (Agarwal, R. P. and Bansal, S. B.1986). The western sub-basin is called the Pakhal belt or Mallampali belt and the eastern sub-basin as the Albaka belt. The Pakhal belt resting uncomformably on the Archaean Dharwar craton and the Albaka belt lying with a tectonic contact on the Bastar craton (Chakraborty, T., 1994).

The Pakhal belt extends from Khammam in the south to Adilabad in the north and is the better studied (Sreenivasa Rao, T., 1987). The Albaka belt extends from Bhadrachalam in the south to Chandarpur in the north (King, W., 1881 and Srinivasa Rao, K, *et al.*, 1979).

Pranhita Godavari graben occupies a NW-SE trending linear tract extending from north of Chandrapur in Maharastra to south of Aswaraopeta. The Pranhita-Godavari graben measuring an area of 25,000 sq km is nearly 600 km long and 40 km wide zone extending from Chicholi in the Northwest to Zangareddigudem in the Southeast. To the further NW, the extension of the graben below the Deccan traps is unknown, but to the SE it is truncated along the East coast by ENE-NSW trending faults (Figure 10.pg.1).

Stratigraphy

The aggregate thickness of the sediments of the P-G basin is estimated at about 6,000 m. The P-G Basin sediments deposited in a time span of 200 Ma from late Carboniferous/early Permian to Cretaceous (Ramakrishnan, M., 1987). Generalised litho-stratigraphic succession of the Gondwana sediments includes Talchir, Barakar, Barren measures and Kamthis are called lower Gondwana group and Maleri, Kota, Gangapur and Chikial formations called upper Gondwana group (Saha, D., Ghosh, G., 1998).

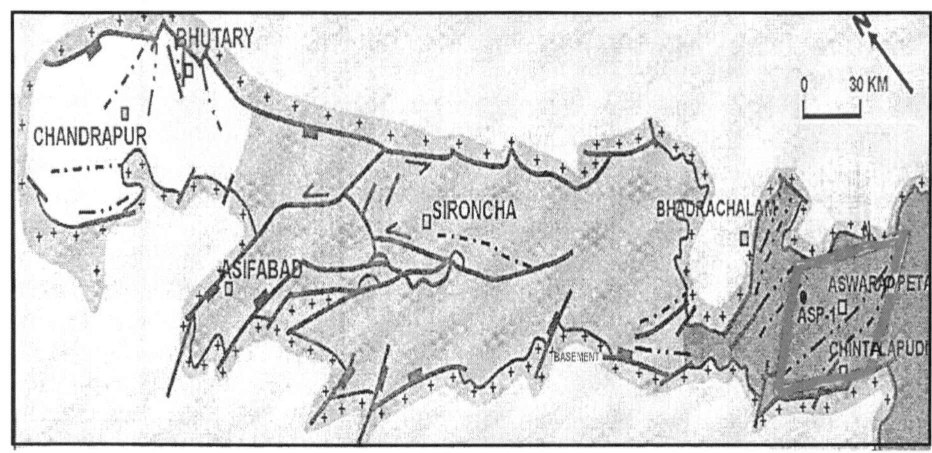

Figure 10.pg.1. Location Map–Pranhita-Godavari Graben.

A 60-800 m thick, middle to late Permian Barren Measures Formation overlies the Barakar Formation with a normal contact. Light yellow feldspathic sandstone, siltstone and grey shale are prominent lithounits (Chaudhuri, A.K., 1985).

The Kamthi Formation with thickness range from 50-400 m., is widespread distribution and covers over two-thirds of the basinal area. This overlie the older Talchir, Barakar and Barren Measures Formations. It contains conglomerate, conglomeratic sandstone and shale of early Triassic age.

Gondwanas are fluviatile to lacustrine in nature, occur in the Pranhita - Godavari Valley and continue underneath the Trap. Rocks of Precambrian age form the floor on which Gondwana sediments were deposited. Gondwana sediments of Lower Permian age consist of diamictite, conglomerates, shales, turbidites, rhythmites and coal seams. Middle Permian sediments are essentially coal free and consist of coarse sandstones and red clays (Raju, P. S. R., 1986). Gondwanas of upper Permian age are again coal bearing.

The Upper Gondwana Group begins with 250 m thick Maleri Formation comprising reddish brown to greenish grey clay, siltstone, argillaceous sandstone and lime-pellet rock along with middle to late Triassic fauna. The succeeding Jurassic Kota Formation includes large-scale cross-bedded sandstone bearing limestone in the middle and upper sandstone and clay. White sandstone, buff siltstone and light grey clay characterise the Gangapur Formation. Litho-assemblage of the Chikiala Formation includes unfossiliferous conglomerate sandstone and ferruginous sandstone (Basumallick, S., 1967). The Maleri Formation with red coloration demarcate the upper and lower Gondwanas. The Kota, Gangapur and Chikiala beds represent Upper Gondwanas (Figure 10.pg.2).

The equivalents of upper Gondwanas at the basin margin area are termed as Gollapallis, Raghapuram and Tirupati sandstone which show their sub-surface continuation in the coastal Krishna Godavari basin.

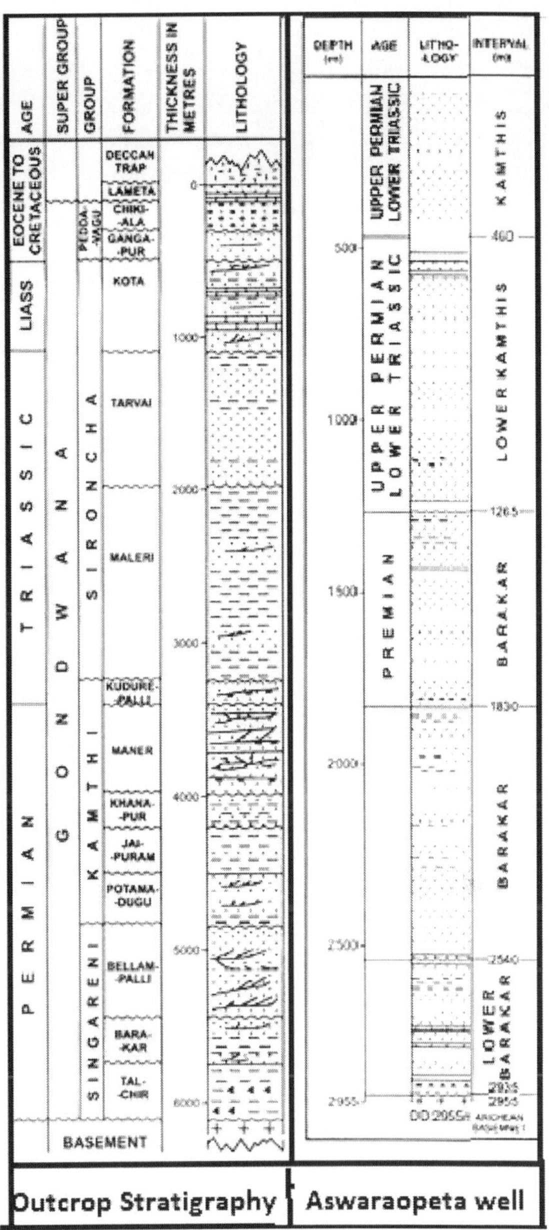

Figure 10.pg.2. Stratigraphy–Well section–Aswaerao Peta–Pranhita-Godavari Basin (After Raiverman V, *et al.*, 1983)

In the wells drilled on either side of Tanuku horst, thick lower Gondwana sediments are encountered below the Red Bed sequence. They are lower Talchirs, middle coal bearing Kammugudem Formation and upper areneaceous Mandapeta Sandstone in the sub-crop sections in the KG basin.

Tectonics

The Pranhita-Godavari graben is a linear intra-cratonic rift located in the south-central part of Peninsular India (Chaudhuri A.K. 2003 and Swamy.P, and K.David 2016). The Indian Peninsula was separated from the Gondwanaland during 141 Ma (Triassic?). At the time of break-up it was positioned at 50° south latitude. The late Carboniferous rifting within the erstwhile Gondwanaland along the weak zones, that is Dharwarian trend (NW-SE) has resulted in horst and graben morphology (Fox, C.S., 1931). These grabens were locale for lower Gondwana depositon. The prominent among such grabens include Pranhita Godavari graben (Ahmad, F., 1958).

Though the rift systems of Pranhita-Godavari and Carnovar of NW Australia placed close, the popular continental fit indicate that the Enderbyland of Antarctica was positioned at the Machilipatnam Bay, the tri-junction of the continental separation (Mishra, D.C, *et al.*, 1999 and Chakrabarthi, S. K. and Chakrabarthi, R. K., 1994).

The Indian plate started moving with newly generated oceanic crust since Cretaceous. Early Triassic to Jurassic section is represented by red clays. Once oceanic crust was formed on either side of the Indian plate 90° East Ridge and Carlsberg Ridge in the west have become trajectories for the movement of the Indian plate.

The graben in all possibility extend below the Deccan syncline towards NW which diagonally abut against E-W trending Narmada-Son rift system (Mishra, D. C, et. al,1971).

The intense magnetic anomaly located off Machilipatnam is related to a hotspot at the time of initial rift of the Indian plate. A three-armed radial rift system in which the Godavari Valley remained as aborted rift or aulacogen in Early Cretaceous, while drifting about the remaining two arms led to the development of the east coast of India (Ramakrishna, T. S. and Chayanulu, A. Y. S. R., 1988).

The eastern passive continental margin of India has subjected to tectonic readjustment of continental blocks with the denser oceanic crust which has resulted in to subsidence of crustal blocks. During this phase, the Eastern Ghat trends (NE-SW) were reactivated and resulted into NE-SW aligned horsts and grabens superimposed over the pre-existing trends (Srinivasa Rao, K, *et al.*, 1979). Due to this passive margin subsidence with south-easterly slope facilitated a NW-SE drainage of proto-Godavari River with sedimentation towards south-east. (Biswas, S. K., 2003).

The Pranhita-Godavari Gondwana basin consists of a series of NNW-SSE grabens and half-grabens (Fig10.pg.3). Three prominent fault patterns are identified *viz.* (i) NNW-SSE trending syn-depositional faults that controlled sedimentation and graben/half-graben development, (ii) NW-SE faults that imparted en echelon fabric to southwestern margin, and (iii) the NE-SW transverse faults, oriented parallel to the grain of the Eastern Ghat Mobile Belt transecting the Godavari valley basin into different sub-blocks (Murthy, K.V.S., Ramakrishna, M., 1980). It is considered that the progressive subsidence of these blocks in a south-easterly direction paved the way for the deposition of coastal Gondwana track (Kaila, K. L, *et al.*, 1990).

The Bouguer gravity map of the area also clearly brought out the PG graben and its structural limits, however, in the coastal onland area the gravity lows and highs are aligned in NW-SE trend (Venkata Raju D.Ch., R.S. Rajesh, D.C. Mishra 2002). It can be seen that, few small lows are aligned in NW-SE trend right upto shallow offshore area.

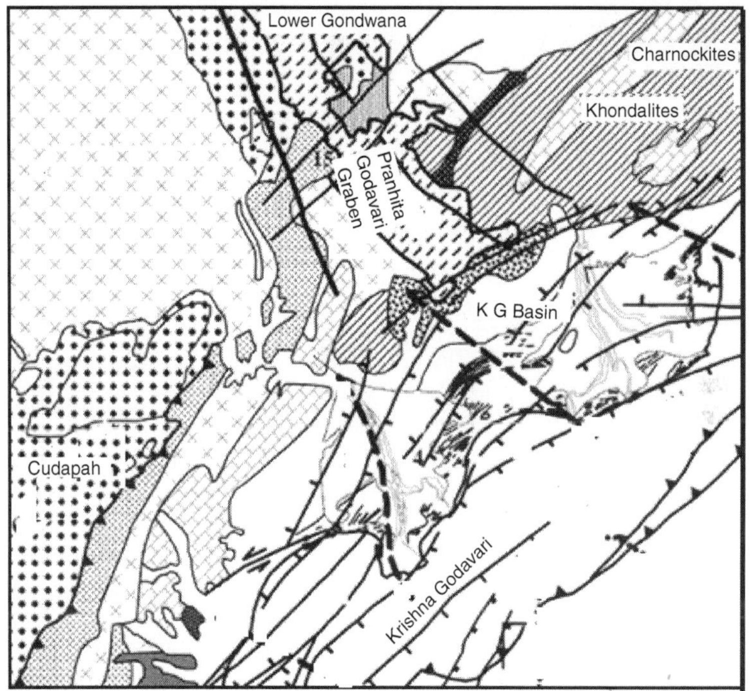

Figure 10.pg.3. Tectonic Map Showing–Pranhita Godavari and K G Basin.

The coal exploration towards northwestern part and hydrocarbon exploration towards southeastern part has enhanced the acquisition of subsurface data like gravity, magnetic, seismic and deep drilling upto 4500 m (Murthy, B. S. R. and Venkateswara Rao, M.1996).

The basement relief maps for the Godavari, Chintalapudi and coastal sub-basins showing several basement uplifts and depressions (Mishra,D.C. *et al.,* 1989 and Sarma B. S. P and M. V. R. Krishna Rao., 2005, Sarma B. S. P and M. V. R. Krishna Rao., 2005). The maximum thickness of sediments of 5–6 km in the Godavari sub-basin and 3.0 km in the Chintalapudi sub-basin that shallows towards the west (Figure 10.pg.4). A Bouguer anomaly 'low' over the Godavari and Chintalapudi sub-basins is due to low density Gondwana sediments (Qureshy,M.N. *et al.,* 1968).

The sub-basin, located northwest of the Mailaram basement "high", with in the mega Graben depicts the characteristics of a half graben. The Chinnur "high" in the centre of the basin probably represents a remanence of the areal doming characterizing the rift valleys.

The Chintalapudi basin is bounded by the Mailaram "high" and the coastal fault towards the south. This part of the basin has faulted margins on both the sides as indicated by sharp gradients in the Bouguer anomaly with 3.0 km of sediments in the central part.

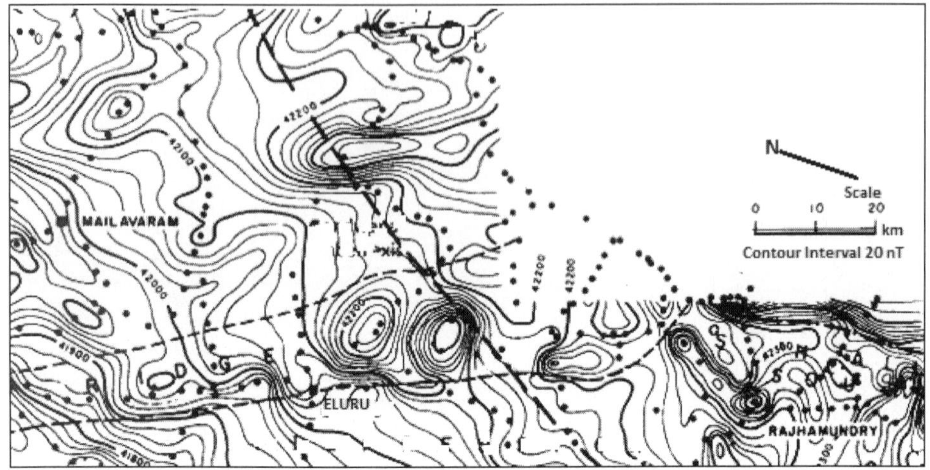

Figure 10.pg.4. Magnetic Intensity Map Pranhita Godavari Basin.

The Pranhita Godavari Gondwana Graben while extending underneath coastal KG basin it also dip towards south-east and hence its western limit occur at shallow depths whereas towards south-eastern onland part seen in to deeper depths.

The Chintalapudi cross-trend appear to be older than other structural features of KG basin. Another linear trend at Yanam called Pithpuram cross-trend suggest a linear basement ridge with an average depth of 3 km and it is fault controlled. PG graben is now known to continue up to OCB between these two cross trends with axis located off Amalapuram at right angles to coastline (Murthy, K.S.R, *et al.*, 1993 and Tewari, H.C *et al.*1996). The PG basin is divided into three sub-basins, namely the Godavari, the Chintalapudi and the coastal sub-basins separated by transverse ridges.

Sedimentation

The Paleozoic sedimentary exposures towards NW of Chandrapur and SE of Aswaraopeta demarcate the known limits of the graben. Similarly the Deccan basalt exposures at Rajahmundry demarcate the observable PG graben towards south-east. Its presence below the coastal Tertiary sedimentary basin is known due to the presence of Permian sediments at Mandapeta as evidenced by deep drilling in Krishna-Godavari basin.

This NW-SE trending Pranhita-Godavari Valley is unique as it preserves about 3000 m thick sediments deposited in a time span of 200 Ma from late Carboniferous/ early Permian to Cretaceous (Mukhopadhyay, J., Chaudhuri, A.K., 2002). Generalised lithostratigraphic succession of the Gondwana sediments includes Talchir, Barakar, Barren Measures,Kamthi (Lower Gondwana Group), Maleri, Kota, Gangapur and Chikiala formations (Upper Gondwana Group). The erosion of lower Gondwanic sediments over the horst blocks and sometimes its absence over the basement ridges and at places thickness varies in the lows.

The sedimentation in the rift valley started in a cool climate represented by coarser clastic beds (Bose, P.K., Sarkar, S., 1991). The climate gradually warmed up and paralic environment of deposition might have prevailed and represented by coal beds (Barakars). The overlying sequence is a thick sandstone which indicate fluvial conditions of deposition. Presence of many unconformities with wedging out of strata indicates that, the depositional areas were oscillatory water depths (Chaudhuri, A.K., 1977). The depositional direction in the graben was towards NW (Figure 10.pg.5).

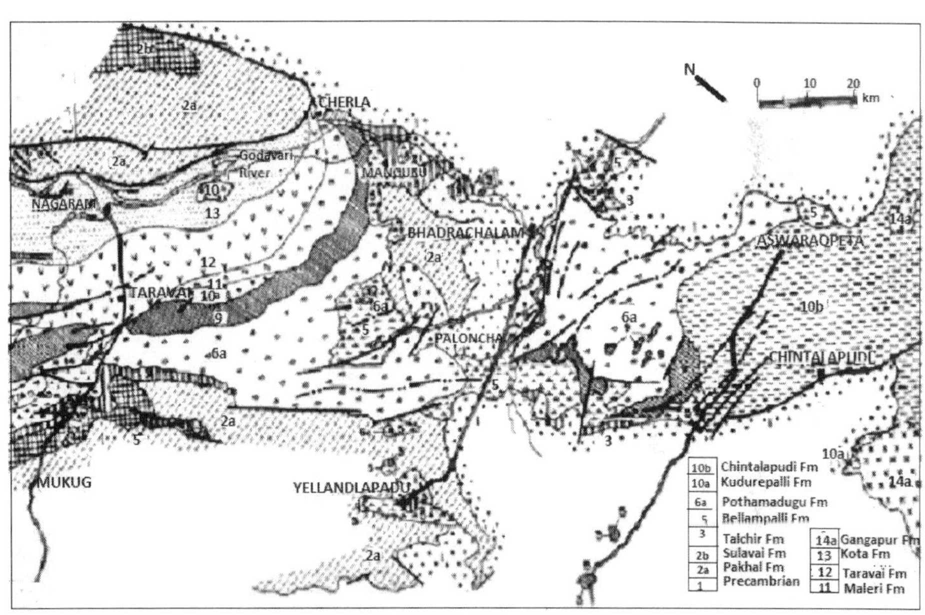

Figure 10.pg.5. Geological Map-Pranhita Godavari Graben (After Raiverman V, *et al.*, 1983).

The Cretaceous fluvio-marine sediments are occurring as outcrops at NW basin margin areas of KG basin. These are termed as Upper Gondwanas and represented by lower coarser clastics (Gollapallis), middle argillaceous (Raghavpuram) and Upper arenaceous sequences (Tirupati). This thick pile of sediments have resulted in to a isostatic rise at the basin margin area, due to which lower Gondwanic sediments exposed as outcrops (Mohinuddin, S.K., Satyanarayana, K., and Rao, G.N., 1993). Southeastern limit of Chintalapudi Sandstone formation at Paloncha neck demarcate the areas of isostatic rise towards northwestern parts of the graben and passive margin subsidence towards southeastern parts.

Petroleum Prospects

The paleo rifts and open marine peri-cratonic basins are known to have high thermal history. World over such areas where older rift valleys underlie coastal Tertiary deltas are found to be promising for hydrocarbon exploration (Kumar, S.P., 1983). In case of KG basin, most of the oil and gas bearing prospects are located in between two cross trends and within the Cretaceous and Tertiary sediments. These areas facilitate syn-sedimentary subsidence because of these areas are basement detached due to thick sedimentation. The situation also help in getting reducing environment for protection of organic matter and sufficient overburden pressure for good expulsion of hydrocarbons from the source. Hence, East Godavari sub-basin is ideally located is found to be more prolific than the other areas. Further, these deep rooted bounding faults of the graben act as entrapment condition for lateral migration of hydrocarbons.

Gas has been struck in the western part of the Krishna-Godavari basin at Mandapeta well in Chintalpudi Sandstone of Upper Gondwana, which is an encouraging factor for exploration in Gondwana sediments. These gases contain high percentage of Methane (94 per cent +) and are considered to be of thermogenic origin. Lower Gondwana sediments (Barakar Formation In particular), containing abundant organic matter, are thought to constitute good source rocks. Vitrinite reflectance studies indicate that, the shale sequences are within the zone of hydrocarbon generation (Pranhita Godavari: TOC 3.58 to 6.41 per cent, VRo 0.47 to 0.67).

However, the NW part of the graben where in NE-SW trending KG basin not in existence, the hydrocarbon occurrence may be poor, Aswaraopeta well was drilled in such area no hydrocarbon shows could be observed. The possible reasons could be, due to lack of syn-sedimentary subsidence of the organic matter and hence most of the time remain in oxidised conditions and so more prone for destruction rather than retention and maturation. Further, in such parts, the component of argillaceous facies will be less than the arenaceous facies. So source and regional cap facies may be lesser to generate and retain hydrocarbons if any.

Though block was awarded in the basin in 1993 to partners Hindustan Oil Exploration Company and Mafatlal Industries Limited (GN-ON-90/3) for hydrocarbon exploration while working some social unrest problems were experienced and it was discontinued. The block, PG-ONN-2001/1 was awarded to ONGC under the NELP in 2003.

Palar–Pennar Basins

Introduction

It is located on the east coast of India between Krishna-Godavari basin in the north and Cauvery basin in the south and covers an area of about 16,200-km^2 including the offshore part upto 200m isobath.

Palar basin is an intra-cratonic rift basin with its rift axes oriented in an N to NNE and areal extent of 36,000 km^2 located in Tamil Nadu and adjoining Andhra Pradesh with its northern part extending into offshore. The basin is separated from adjacent Pennar Basin in the North by a basement high called Nayudupeta High whereas in the south it is separated from the Cauvery Basin by, the Chingelput High (Figure 10.pp.1). The basin is sub-divided into the Pennar in the north-west and the Palar in the south-east separated by the horst.

Exploration History

About 5,000-km^2 of geological surveys were carried out by ONGC in the Palar basin. The area to the north of Palar River was covered during early sixtees and the Lower Gondwana outcrops, south of Conjeevaram were mapped during late seventees.

Geochemical Sniffer Survey was conducted during 1988-'89 over 240 km^2 area onland part of Palar graben and further studies indicated that the gas chromatograms of the adsorbed gas show methane in all the samples. Higher hydrocarbons (ethane to butane) are present in some of the samples.

The gravity-magnetic surveys were initiated in the onland part of the basin in sixteees, while offshore surveys were started in 1978. CDP seismic surveys have been taken up in the onland part of the basin since 1987. An area of about 8,000-km^2 was covered by regional gravity and magnetic surveys. The data were acquired in three phases, one by ONGC in the onland area and two companies (Prakla seismos and Chevron) in the offshore during 1988. The reflection seismic surveys of about 36,000 Line km. acquired in offshore part of the basin.

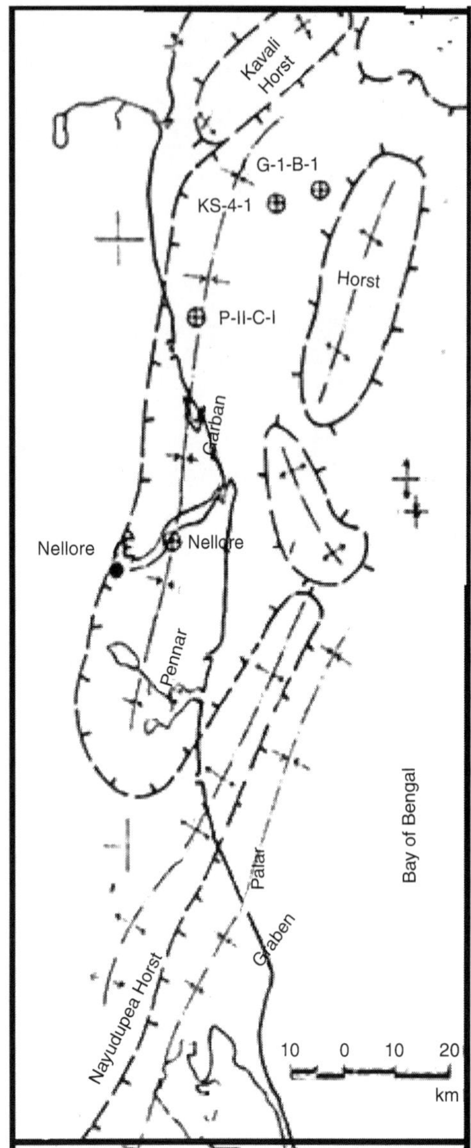

Figure 10.pp.1. Tectonic Map–Pennar Basins (After Rangaraju, MK., *et al.*, 1993).

Exploratory Drilling

Five exploratory wells, two onland and three offshore, have so far been drilled in this basin. The first well, KS-4-1, in the offshore was drilled by ONGC in 1984 down to a depth of 2,986 m against a target depth of 3,000 m. The well was bottomed within Pre-Albian sediments. KG-I-B-1 in the offshore was drilled in 1989 by M/s.

Chevron International Limited down to a depth of 2,747 m. The well was terminated within Permian sediments. KS-4-1 and KG-I-B-1 are located in the northern part of the Palar basin.

The third well, P-II-C-1, in the offshore was drilled in 1989 by M/s. Chevron International Limited down to a depth of 3,747 m ending within Jurassic section in the northern part of the basin (Figure 10.pp.2). Nellore-1 was drilled in 1992 by ONGC down to a depth of 3,000 m ending within Archean basement and Siddavaram-1 (SD-1) drilled in 1997 down to a depth of 2,791 m also bottomed in basement.

Cairn Energy has initiated work on its much awaited 3D seismic programme in the offshore Palar Block PR-OSN-2004/1, awarded to them in 2005 which is located off the east coast of India and covers approx. 9,400 sq km onshore in the Palar basin.

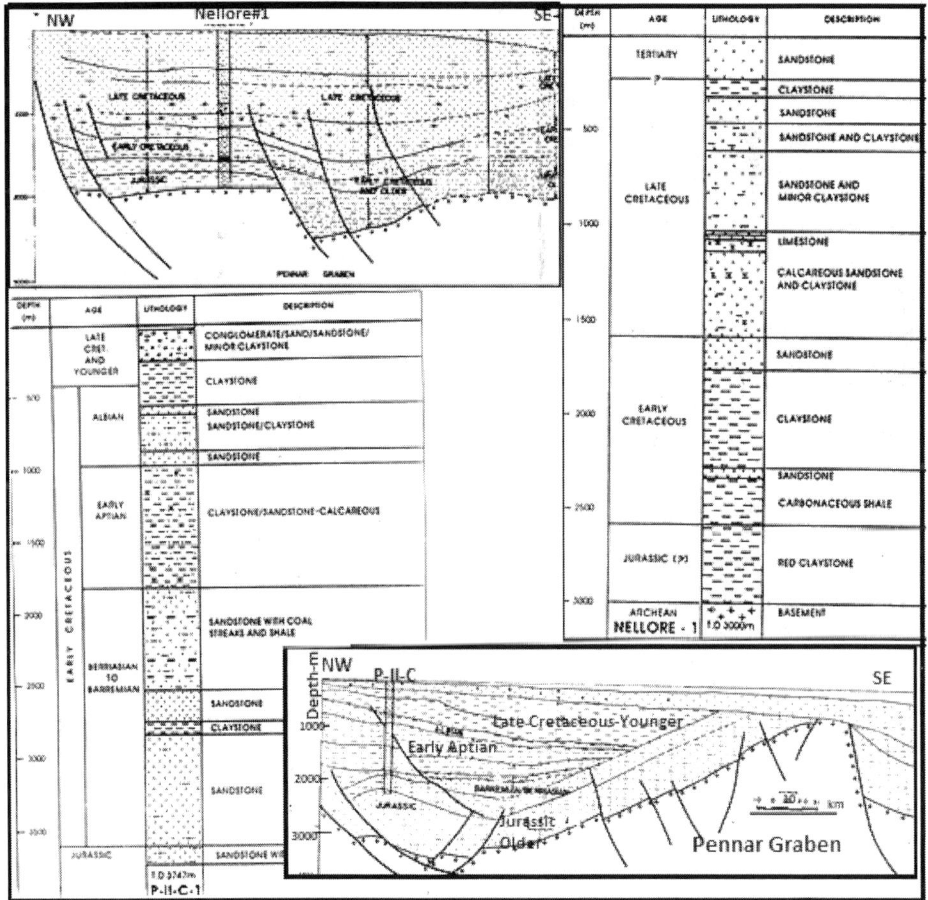

Figure 10.pp.2. Geological and Well Sections–P-II-C and Nellore-1-Pennar Basin.

Tectonics

The Palar basin is a rift basin formed during the Permian time due to fragmentation of the Archean crust. It is probably dissected into a northern and southern parts by a broadly E-W oriented subsurface basement ridge. The E-W

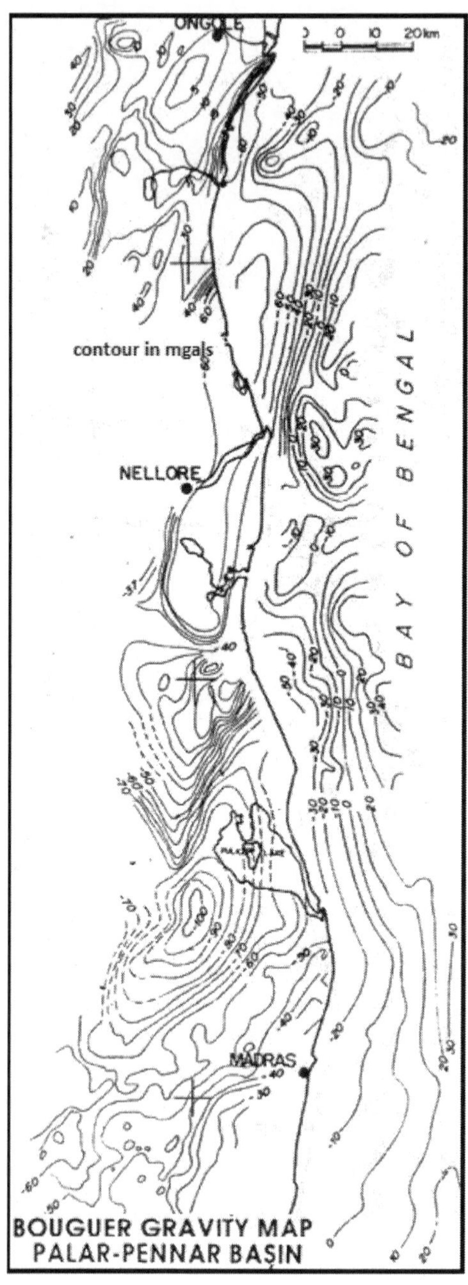

Figure 10.pp.3. Bouger Gravity Map–Palar-Pennar Basin.

ridge also acts as a tectonic divide with the area south of it showing an appreciable N-S tectonic trend that is very subtle in the north (Harinarayana T, *et al.*, 2006). This N-S trend is associated with the Lower Gondwana rifting that probably formed the earliest part of Palar Basin whereas the area north of it is dominated by Late Jurassic-Early Cretaceous India-Antarctica rift related NE-SW trend (Mazumder S, K, K.S. Pangtey and D.S.Mitra, 2013). Though the entire southern part of eastern coast is constituted of a number of NE-SW oriented basins the N-S trending basement ridge might have formed as a result of continental rifting (Vairavan V., 1993).

Gravity data indicate that the structural elements extend into the offshore. Maximum sedimentary thickness in the onland part is anticipated to be 4,000 m comprising mainly Gondwana (Figure 10.pp.3) (Kumar N, Singh A.P, Singh B., 2009 and Murthy K.S.R., *et al.*, 1995).

The basin came into existence during the Early Permian time with initial deposition in narrow linear troughs roughly aligned in the NNE-SSW direction. Two such narrow linear patches of outcropping sections are observed to the south-west of Chennai city (Figure 10.pp.4 and 5). The Palar Basin shows evidences of

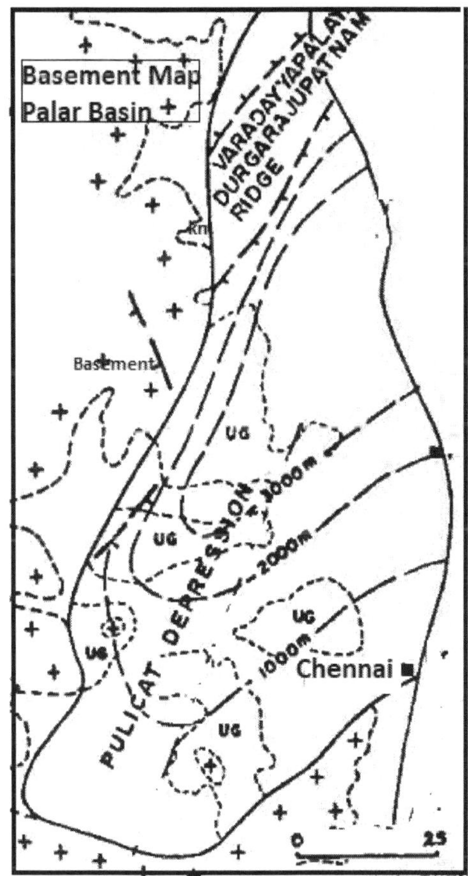

Figure 10.pp.4. Basement Contour Map–Palar Basin.

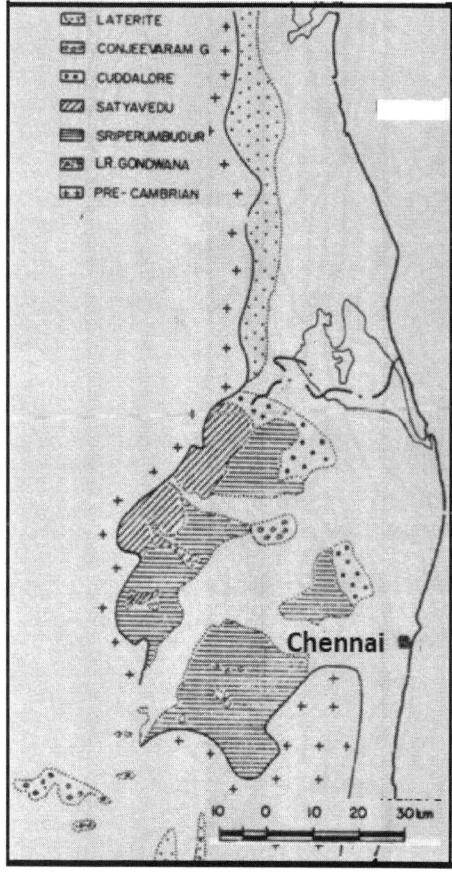

Figure 10.pp.5. Geological Map–Palar Basin.

neotectonics (Ramasamy,S.M., *et al.*, 2011, Narasimhan, T.N., 1990) implying the area is still tectonically active though it overlies a stable South Indian shield. Fault plane solutions indicate thrust faulting along with strike slip component along an E-W striking nodal plane which defines the present day or the post-rift state of stress in the region (Murty, G.P.S., et al., 2002).

Sedimentation

A major unconformity separates the Permian sequence and the overlying Upper Gondwana. The Gondwana sediments are observed to terminate against Nayudupeta horst in the east and basin margin in the west. The overlying Early Cretaceous (Aptian/Albian) section marks first marine transgression in the area followed by a major unconformity and deposition of thick marine shale. Only a thin veneer of Tertiaries were deposited over the area. The basin consists mainly of Gondwana sediments, ranging in age from Permian to Early Cretaceous. Thickness of sediments is expected to be of the order of 4,000 m in the deepest part of the basin around Satyavedu town.

Stratigraphy

Exposures are spread over an area of about 2,000-km^2. Lower Gondwana sediments of Early Permian age and Upper Gondwana sediments of Jurassic to Early Cretaceous age (Sriperumbudur and Satyavedu Formations) are exposed in the southern part of the basin. The surface and subsurface data indicate that, the Upper Gondwana section forms the bulk of the sediment-fill in the basin. The stratigraphic column is constituted as: Pundi Shale (Permo-Triassic), Sriperumpudur Formation (Upper Jurassic–basal Cretaceous), Satyavadu Formation (Aptian), Tirupati Sandstone Formation (Upper Cretaceous), Paleogene undifferentiated unit and the Cuddalore Sandstone of Neogene.

The deposition of Ongur Formation in a fluvio-glacial environment with a marine influence directly over the Precambrian crystalline rocks. The formation of horsts and grabens led to deposition of synsedimentary Sriperambudur Formation in a restricted basinal setup in lacustrine environment.

The tectonic deepening of the latest stage of rifting and led to the deposition of a transgressive Satyavedu Formation. Palar rifting probably terminated in Early Cretaceous after which the area suffered wide spread positive movements in Late Cretaceous (Figure 10.pp.6).

Hydrocarbon Potential

Hydrocarbon indications were noted within the Mesozoic section in KG-I-B-1, P-II-C-1 and KS-4-1.The presence of hydrocarbons C1-C4 indicates the probable existence of upward migration of hydrocarbons from deeper sources. The generation of hydrocarbons could be from catagenetic process. The commercial hydrocarbons have yet to be established in this basin. Prognosticated resources of oil and oil equivalent of gas are estimated to be about 40 MMT.

Statigraphy			Lithology	Thick-m	Description
PLEIST REC	Conjevaram			100	GRAVELS
Mio Plio	Caddalore			150	SANDSTONE
Cret	Upper-Gondwana	Satyavedu		200	SANDSTONE
Jurassic- Early Cretaceous	Upper-Gondwana	Sriperumbudur		1100	SHALES AND FELDSPATHIC SANDSTONE, COAL STREAKS/LENSES
Early Permean	Lower Gondwana			300	GREEN SHALES AND BOULDER BEDS
Basement					METAMORPHIC

Figure 10.pp.6. Straigraphy–Pennar Basin.

Source Rocks

Coal and carbonaceous shales within the Gondwana sediments are known to form good source rocks. Similar coal-shale sequence within the Sriperumbudur Formation of Jurassic to Early Cretaceous age in Palar basin is expected to form an important source rock.

The samples from KS-4-1 well indicate that Pre-Late Cretaceous section is organically rich with organic carbon percentage varying between 1.28 and 5.61.The Jurassic shale section in the interval 2,039-2,368 m in the well KG-I-B-1, contains average to high (upto 3.3 per cent TOC) quantities of organic matter and yields poor to good quantities of S2 hydrocarbons upto 10.5 mg hydrocarbons/gm of rock. The sediments in the intervals 2,194-2,231 m and 2,286 - 2,340 m which are characterised by higher TOC content. Source rock studies have also been carried out for the samples of P-II-C-1. The samples in the intervals 1,536-1,865 m, 2,423- 2,576 m and 2,747-2,900 m within the Early Cretaceous section contain mostly Type I, III and IV kerogen. The organic matter is of high quality and TOC as high as 25 per cent has been recorded within the coaly layers.

The source rock evaluation of samples from the onland well Nellore-1 in the Pennar graben indicated that the interval 2,300 - 2,600 m within Early Cretaceous section contains organic matter of Type III and has potential for generation of gaseous hydrocarbons with sub-ordinate oil. In the well Siddavaram-1, hydrocarbon shows have been encountered in the interval 1250 - 1310 m. Pennar Shale in this well host good oil prone source rock. The sequences are equivalent to the Sriperumbudur Formation in age.

Regarding maturation is concerned the geochemical studies of samples of KS-4-1 well, Hydrogen Index values range from 157 to 333 and Vitrinite reflectance from 0.38 to 0.59 per cent. S2 yield and Hydrogen Index are capable of generating mixed high-wax oil and gas. Source Potential Index in these intervals (SPI) is about 0.6. in P-II-C well whereas in Nellore-1 well these yield good quantities of S2 hydrocarbons upto 68 mg hydrocarbon/gm of rock. These intervals contain degraded Type I kerogen derived from algae and higher plant organic matter that was deposited in a lacustrine, sub-oxic environment. This type of Kerogen, having moderate Hydrogen Indices (generally about 200 mg hydrocarbon/gm TOC) is capable of generating mixed high-wax oil and gas.

Reservoir Development

Sriperumbudur Formation contains fairly good amount of sandstone facies along with coal-shale. In the well P-II- C-1, sediments of Early Cretaceous section constitute major reservoir facies. Porosities computed from logs in this well range from 11 to 15 per cent in the Early Cretaceous sandstones (2,815- 3,150 m) and 11 to 12 per cent in the Jurassic-Early Cretaceous sandstones (3,618 - 3,747 m). In the well Nellore-1, Early as well as Late Cretaceous sections indicated good development of sands. Log computed porosities in Early Cretaceous section range from 12 to 18 per cent. In some of the intervals the porosity development is as high as 23 per cent. In the Late Cretaceous sands/sandstones, log-derived porosities are in the range of 9 to 16 per cent.

Entrapment Conditions

Entrapment is likely to be within the anticline and fault closures related to basement tectonics during and after the Gondwanic period with Aptian shale, Jurassic and Early Permian shale acting as cap rock.

Purnea Basin

The Purnea Basin is located in eastern part of Bihar and north Bengal, and bordered in south by hills of Rajmahal volcanics and to the north by the Siwaliks along sub-Himalayan foothills. Further, it is bounded by Bhawanipur fault in west and Kishanganj fault zone in east. The southern limit of the basin, covering nearly 18,000 sq. km. remains undefined in view of its possible continuity with Bengal, basin (Figure 10.pu.1).

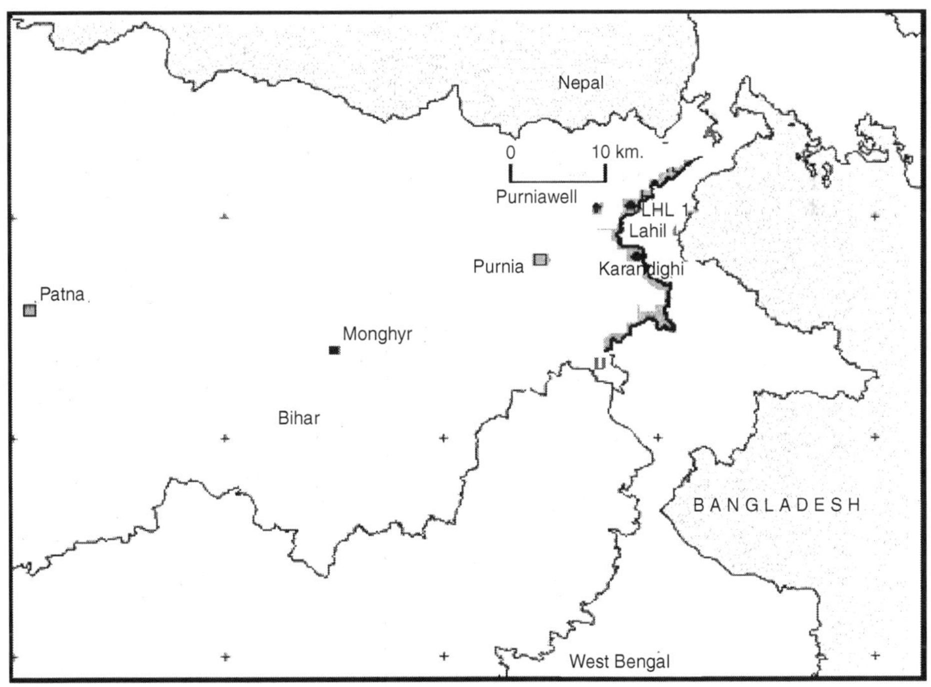

Figure 10.pu.1. Location Map–Purnea Basin.

Exploration History

Hydrocarbon exploration activities in Purnea Basin were initiated since 1956. The Gravity-Magnetic survey was carried out by ONGC during 1960-63 followed by preliminary CDP seismic work. One deep well was drilled *Purnea#1* to a depth of 5,200m in 1982 which was bottomed in metamorphic basement. Though during NELP regime three blocks were awarded to ONGC in 2004 and 2005 only Gravity Magnetic data were acquired and interpreting the same.

Tectonics

The Purnea basin is a polycyclic rift basin located in the eastern part of Indo-Gangetic plain in Bihar and West Bengal continuing further eastwards. Tectonically it could be classified as an intracratonic rift basin with an approximate thickness of about 4000m of sediments ranging in age from Permo-Carboniferous to Recent (Figure 10.pu.2).

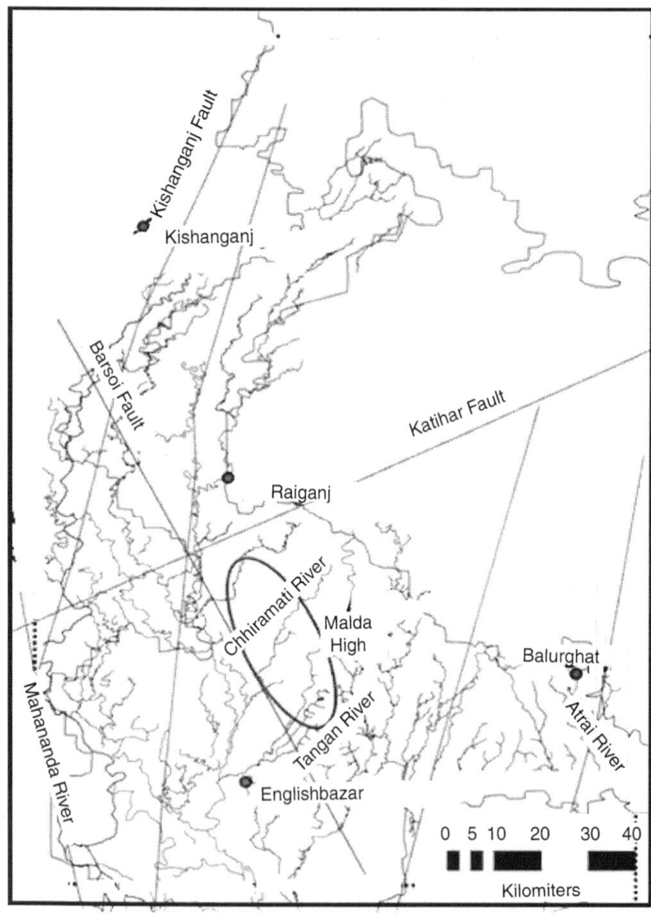

Figure 10.pu.2. Tectonic Map Purnea Basin (After Majumdar, S, *et al.*, 2011).

The basin is bounded and separated from the Ganga Basin in the west by Monghyr Saharsa ridge whereas in the east it is bounded by the Kishanganj Fault. The south and northern extents of the basin area bounded by the Malda High and Main Boundary fault respectively. The geological history of the basin initiated with early Permian Gondwana fragmentation. The absence of basaltic magmatism suggests that the depression had not yet progressed to the stage of rifting.

Precambrian gneisses and metamorphics constitute the basement for deposition of graben-fill rift sequence, referred as the Gondwana sequence. Late Jurassic-early Cretaceous breakup of Gondwana land, resulted in cessation of Gondwana sedimentation, which were later folded, uplifted and eroded.

Subsequently, as the Indian plate began its northern movement across Tethyan ocean, this basin was on the northern edge of Indian plate. During collision, the basin was downthrown along E-W fault and developed for deposition of early Paleogene. With the onset of Himalayan orogeny, the north of the basin experienced development of flexural foreland depression due to tectonic overloading of sediments.

Bouguer gravity data show abrupt slopes on northern and western sides plunging north-westward into a gravity low near Katihar-Purnea region suggestive of fault control with the southward side of the slope being more gradual.

The seismic data interpretations with overlay of faults on geomorphic data shows some match in the NE-SW trending faults. However, faults affecting the Siwaliks show a more ENE-WSW trend that is not detected from the surface morphotectonic data. The regional faults and associated geomorphic highs on the gravity high are probably effects of the upliftment due to movement along surrounding major faults during basin formation. Faulting along these deep seated NW-SE as well as NE-SW trends might lead to the development of fault related structural closures, manifestations of which are found to occur on the surface as geomorphic highs in some of the cases.

Sedimentation-Stratigraphy

The sedimentary package in the basin, comprising Gondwana and Cenozoic sediments, directly overlies the crystalline basement. It was in this setup, that the upper sedimentary sequence with provenance in the rising Himalayas to the north, has been deposited (Figure 10.pu.3).

The oldest sediments known are Talchirs of Permo-carboniferous and the youngest one are Siwaliks of Mio-Pleistocene. Recent Alluvium covers the basinal area (Figure 10.pu.4).

Petroleum Geology

Geochemical data from the four drilled wells in Purnea Basin indicate that Lower Gondwana sequence has generated hydrocarbons whereas a recent gas discovery in Salbanhat, Bangladesh close to the Indo-Bangladesh border also suggests that the equivalent sediments would be prospective.

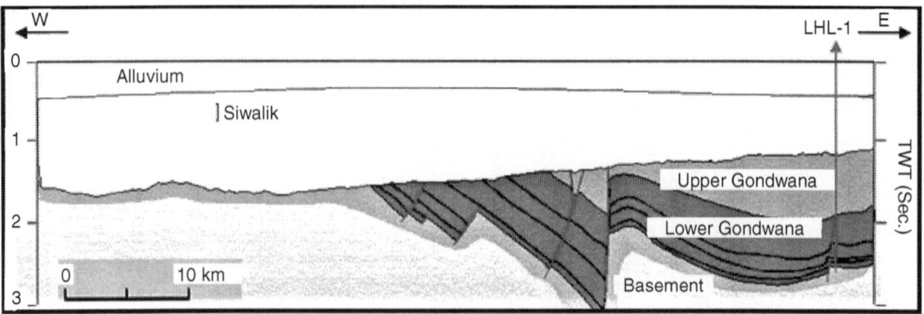

Figure 10.pu.3. Seismo-Geological Section–Purnia Basin.

AGE				GROUP	SUB-GROUP / FM.		GENERALISED LITHOLOGY	THICK. (M)
CENOZOIC	QUATERNARY			ALLUVIUM				500
CENOZOIC	TERTIARY	NEOGENE	LATE MIOCENE TO PLEISTOCENE	SIWALIK				1000
MESOZOIC	JURASSIC			GONDWANA	UPPER	PANCHET		400
MESOZOIC	TRIASSIC			GONDWANA	LOWER	RANIGANJ		600
PALEOZOIC	LATE PERMIAN			GONDWANA	LOWER	RANIGANJ		
PALEOZOIC	EARLY TO MIDDLE PERMIAN			GONDWANA	LOWER	BARAKAR		800
PALEOZOIC	EARLY PERMIAN			GONDWANA	LOWER	KARHAR-BARI		400
PALEOZOIC	PERMO-CARBO.			GONDWANA	LOWER	TALCHIR		400
PRE-CAMBRIAN	ARCHAEAN							

Figure 10.pu.4. Stratigraphy–Purnea Basin

The arenaceous facies in Upper and Lower Gondwana act as suitable reservoir rocks whereas presence of clay/shale beds in Gondwanas and Siwaliks may act as suitable cap rocks.

As such fault closures of basement related faults, pinch outs and wedge outs in Lower Gondwana sequence can act as interesting elements for exploration. These geomorphic highs reflective of basement related fault closures and falling in vicinity of depocentre areas might be seen as prospective locales for hydrocarbon accumulation in the Purnea Basin.

Satpura Basin

The Satpura basin with nearly 200 km long and 60 km wide is rhombus-shaped, relatively long in the ENE-WSW direction and its longer sides are marked by the ENE-WSW trending Son-Narmada south fault and Tapti north fault in the north and south respectively. It covers an area about 12,000 km^2 of exposed Gondwana sediments in the Central part of the Indian sub-continent. The northern limit of the basin is marked by westward flowing Narmada River marking the prominent geological feature-the Son Narmada geo-fracture while the southern limit is marked by the plateaus of Betul and Chindwara. It is the westernmost pull-apart Gondwana basin exposed in the peninsular region (Figure. 10.sa.1). It is unique among all the Indian Gondwana basins by having the longest range of stratigraphic record spanning from Permian to Cretaceous.

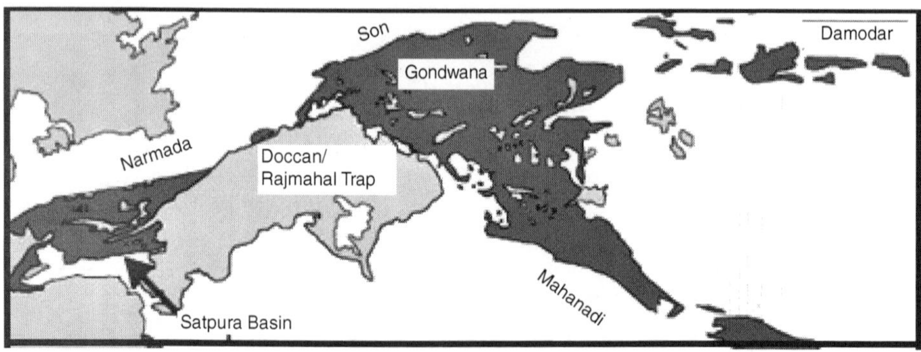

Figure 10.sa.1. Satpura-Location Map.

Exploratory Work

ONGC initiated its exploratory surveys in this basin in the year 1963. However, systematic geological mapping was taken over only from 1983 onwards. Geophysical investigations were started in 1984 with the start of GM surveys and later with seismic surveys in 1986, Geological survey (semi-detailed) 9,396 km^2, GM Survey 8,451 stations, seismic survey 73.4 SLK (Reflection) 43.38 SLK (Refraction). NGRI also carried out some regional gravity studies in the basin. No exploratory drilling has so far been done in the basin.

Tectonics

The evolutionary history of the basin indicates that it formed through, the pre-rift, rift and an incomplete post-rift cycles. Stress regimes related to episodic wrenching along the Narmada-Son lineament have a strong effect on the basin. Tectonic phases associated with the evolution of the basin are related to the evolution of the Indian subcontinent, especially post-Triassic and post-Deccan Trap phases. These phases have been related to important tectonic events like break-up of the subcontinent from Gondwanaland and collision of the Indian plate with the Asian (Biswas,S.K., 1987).

The Gondwana basins of Peninsular India are intracratonic in nature surrounded by Precambrian terrains. They are disposed along ENE-WSW trending Narmada-Son-Damodar valley, NNW-SSE trending Pranhita Godavari valley, NW-SE trending Mahanadi valley and E-W trending Satpura basin.

The Gondwana basin-fills of peninsular India are unmetamorphosed and the original basin architecture is more or less preserved. The strata crop out along three distinct linear belts are defined by the present day river valleys of Narmada-Son-Damodar, Pranhita-Godavari and Mahanadi (Choudhury, A.1977).

This basin is presently preserved as an interior fracture with block tilting causing a slight asymmetry towards the north. Basin subsidence to a depth of about 4,000 m and net emergence and erosion of about 1,900 m of sediments is envisaged. The basin shows imprint of tectonic trends related to the Narmada and Godavari rifts which in turn may perphaps be inherited from Satpura and Dharwar tectonic gains. Some of the faults and fractures along these tectonic trends display strong evidence of strike-slip movement (Figure 10.sa.2).

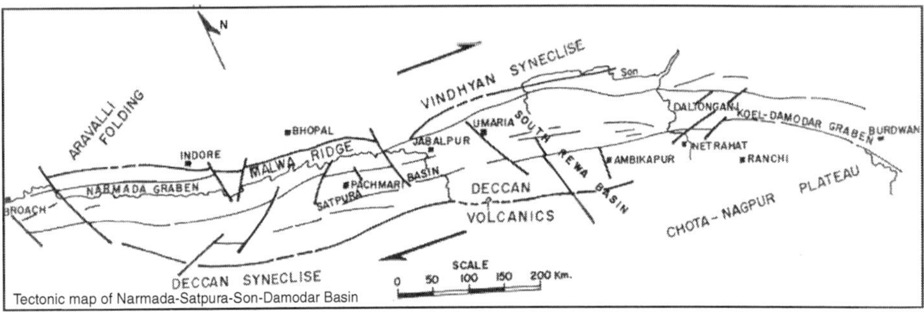

Figure 10.sa.2. Tectonic Map–Satpura Basin.

The basin-fills in some cases are asymmetric with an overall increase in thickness towards one of the boundary faults indicating that the basin bounding faults were active during sedimentation inducing subsidence and creating space for stratigraphic preservation of sediments. The basin-fill strata are also affected by intra-basinal gravity faults reflecting syn-sedimentary downward displacements that generated accommodation space for sediment accumulation throughout the history of basin evolution. Also, the subsidence rate varied across the basin resulting in an asymmetric basin fill with the thickness increasing towards the north. The Tamiya anticline located on the Malani median high can be a positive feature in the basin.

The Bouguer gravity anomaly contours, indicate significant thickness of subsurface Gondwana strata in all the basins implying syn-sedimentary tectonic subsidence. Bouger gravity anomaly map indicates two sublatitudinal trending lows of the order of 145 milligals in the west-central part of the basin while its peripheral portions show rising trends of basement. The northern margin of the basin also shows a rising trend along surface and inferred subsurface faults. (Figure 10.sa.3).

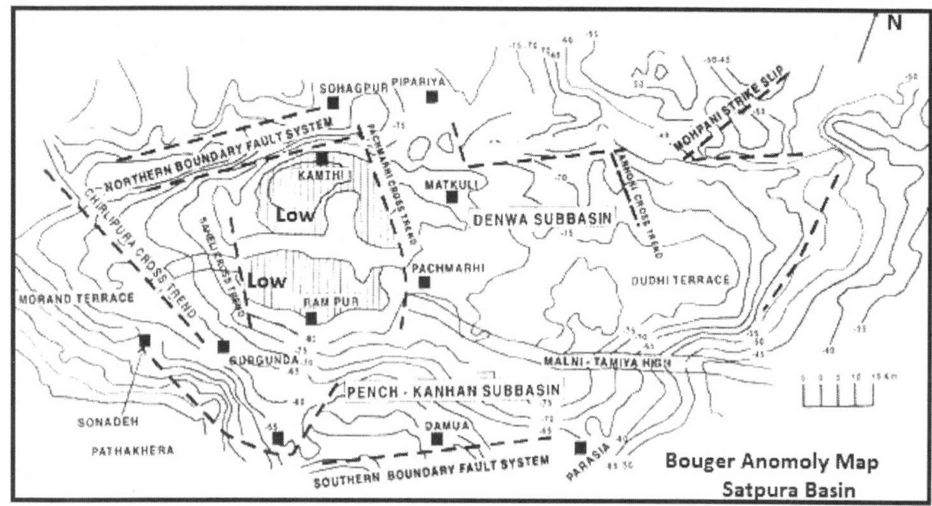

Figure 10.sa.3. Bouger Gravity Map–Satpura Basin.

The northern boundaries of many of the Son-Narmada-Gondwana basin occurring along the valley are marked by the Son-Narmada lineament. This lineament is considered to have formed in the Precambrian, re-activated many times subsequently and is still a zone of seismicity. The ENE– WSW bulk extension of the crust gave rise to the Pranhita-Godavari and Mahanadi basins, and the bulk extensional regime was terminated in the north by strike-slip zones along the ENE–WSW trending Son-Narmada lineament.

Sedimentation

The outcrop pattern shows that faults have played a dominant role in the evolution of the basin, its extent and depth. Most of the fractures are aligned parallel to the ENE-WSW Narmada structural trend. These fractures are intersected by NW-SE trending transverse fractures. The Satpura basin is filled with around 5 km thick siliciclastic succession. (Figure 10.sa.4).

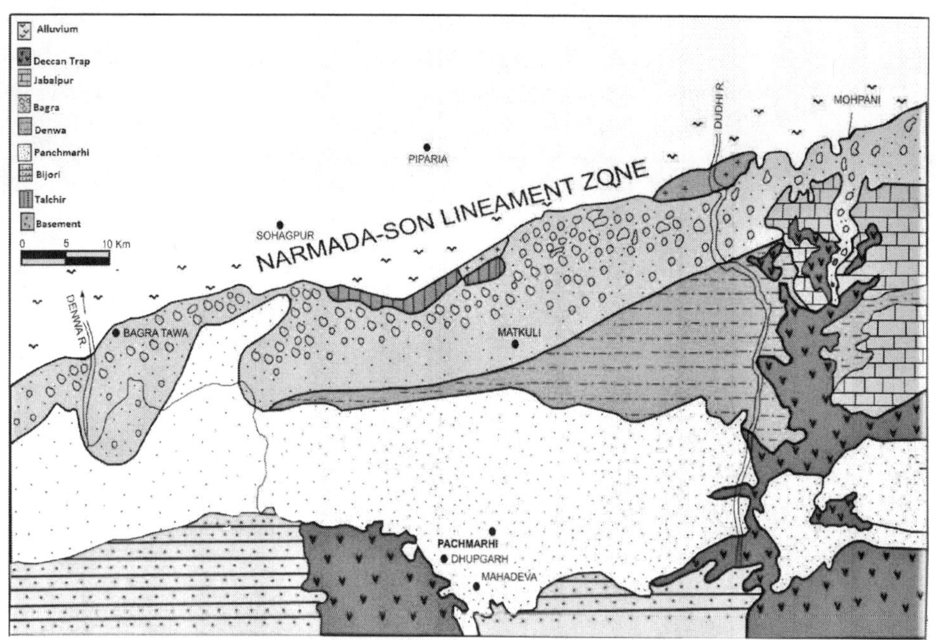

Figure 10.sa.4. Geological Map–Satpura Basin (Modified after Anis, *et al.*, 2012).

The regional strike of the basin-fill strata is NE–SW, and the regional dip of 5 degrees is northerly. There are numerous faults aflecting the basin-fill strata. These faults are intra-basinal and are of syn-sedimentary and is reflected in the thickening of hanging wall strata.

The succession about 5000 metres thick comprises of Talchir, Baraker, Motur, Bijori, Pachmarhi, Denwa, Bagra and Jabalpur formations. Gondwana succession unconformably overlies the Precambrian Basement. The lowermost Talchir unit of Satpura succession comprises of glacio-marine and glacio-fluvial deposits. The Talchir and Barakar Formations are affected by faults more than the other formations. The intra-basinal faults are normal in nature. The cross-sectional profile reveals asymmetric half-graben configuration with fault-controlled subsidence which is also reflected in the profile of the Barakar Formation. The Pachmarhi Formation defines a monocline with its anticlinal bend underlain by a subsurface gravity fault.

In the western part, the sandy Motur strata of braided fluvial origin sharply overlie the coal-bearing deltaic succession of the Barakar Formation, whose the upper boundary also be an unconformity. These unconformities divide the Satpura succession into four distinct packages: (a) Talchir–Barakar, (b) Motur–Bijori, (c) Pachmarhi–Denwa and (d) Bagra–Jabalpur.

It may be noted that transverse sediment transport prevailed till the accumulation of Motur strata, and axial sediment transport began to take place along with transverse transport since accumulation of the Bijori strata and continued thereafter. The Bagra Formation of alluvial origin indicating received sediments from all sides. The Bagra sediments are restricted along the northern margin of the half-graben and are observed to unconformably overlie Denwa, Pachmarhi, Bijori and Talchir Formations at different places. The infra-Bagra formations possibly suflered exposure and denudation before initiation of Bagra sedimentation. It appears that Bagra-Jabalpur strata represent such a tectonic stage of the Satpura basin when the southern half of the basin was uplifted and the northern half was accumulating sediments.

Stratigraphy

Gondwana sediments ranging in age from Permo-Carboniferous to Early Cretaceous were deposited intermittently in glacio-marine, marine, fluvial, fluviopaludal, lacustrine and alluvial fan environments. The Jurassic and Middle to Late Cretaceous witnessed periods of mild uplift and non-deposition. Provenance of the Gondwana sediments upto Late Triassic was the metamorphics in the south and southeast.

Each lithounit probably pinched out over the older one and was subsequently overlapped by the younger formations. This was due to a progressive shift of the depocentre towards the north. The Satpura basin has remained exposed since post Deccan Trap period (Figure 10.sa.5).

Age	Group	Formation	Thick-ness in	Lithology
Recent		Alluvium		
Paleocene to Cretaceous		Occean Trap		
		Lameta	50	
	Upper Gondwana	Jabalpur	250	
Upper Triassic		Bagra	520	
		Dewva	350	
M-Li Trassic		Panchmari	900	
Upper Permean	Lower Gondwana	Bugri	1350	
		Motur	1450	
		Barakar	800	
Lower Permean		Talchir	400	
Precambrian		Dharwar Bijawar		

Figure 10.sa.5. Stratigraphy–Satpura Basin.

Petroleum Geology

Except few gas seepages no definite presence of hydrocarbons has been indicated in the basin in the absence of any exploratory drilling so far. Analysing the gas seepage data, organic carbon ranges, kerogen composition, and TAI values, it is brought out that the indices as above are quite favourable for generation of hydrocarbons in the sediments.

A fair amount of source rocks and maturation data has been generated for Barakar and Bijori formation of the basin. The coals associated with Carbonaceous shales of Barakars contain 48- 69 per cent TOC. The source rock studies carried out also indicated Type –II and III organic matter with generation potential about 100 Hc/mg (James Peters and P K Bhatnagar,1995).

Clastic sequences occur in Barakar, Motur, Bijori and Pachmarhi formations, as well as Pachmarhi sandstone show good development of porosity, may form good reservoir rocks.

Cap rocks in the form of regional seals likely to exist in the basin formed of thick clays and shales in the sequence may be effective in Motur Formation, Bijori Formation and middle to upper parts of Denwa Formation.

South Rewa Basin

South Rewa Basin occupies the northern part of Son-Mahanadi graben in Central Indian shield, covering about 28,500 km^2 of exposed Gondwana sediments. It is geologically delimited by Narmada-Son geofracture zone in the north and towards south it merges into NW-SE trending Mahanadi basin near Manendragarh. Towards east, it is separated by the Pre-Cambrian/Archaean gneisses from the Damodar-Koel valley. Towards west, it is covered by Deccan basalts (Figure 10.sr.1).

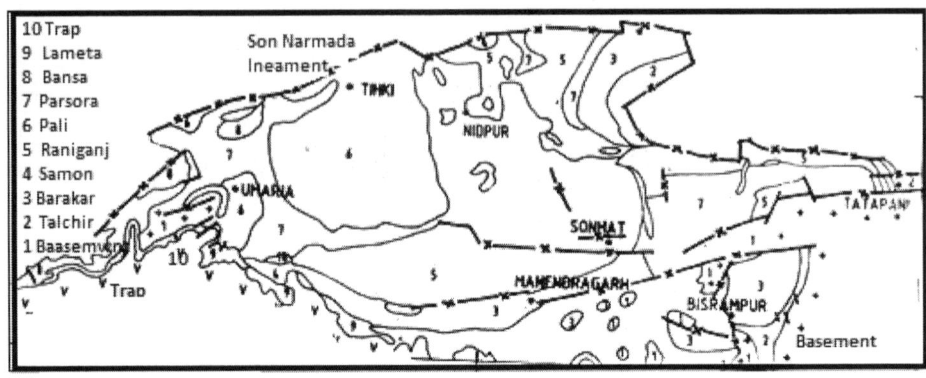

Figure 10.sr.1. Geological Map–South Rewa Basin.

Exploratory Work

The GM surveys have been carried out by GSI and NGRI. ONGC initiated exploratory surveys in this basin in 1963. However, systematic geological mapping of semi-detailed was done in 10,910 sq. km. Gravity and Magnetic surveys and later regional seismic surveys were also carried out. Refraction Seismic Survey in 1,137 line kilometres and reflection seismic data were also acquired. No exploratory drilling has so far been done in the basin.

Tectonics

The basin is having paleo-slope towards north and NW, Son-Narmada lineament marking the northern limit of the basin with a narrow zone of en-echelon faults. ENE-WSW trending faults controlled the development of various subbasins. NE-SW trending faults parallel to Eastern Ghat trend are of lesser frequency. NW-SE trending faults parallel to Mahanadi trend are of minor occurrence.

The sediments have mostly homoclinal dips towards NW with some anticlinal and synclinal features like Tetka anticline and Kharibari syncline (Figure 10.sr.2). Bouguer anomaly map show the Son-Narmada lineament, prominent gravity low striking E-W, lies to northwest of village Dehwa called, Khadda low having an extension of 20 sq km. and another low trending WNW-ESE has been observed around Kuwarpur known as Pather low, broad basement high trending E-W is seen around Tetka, known as Tetka ridge (Figure 10.sr.3).

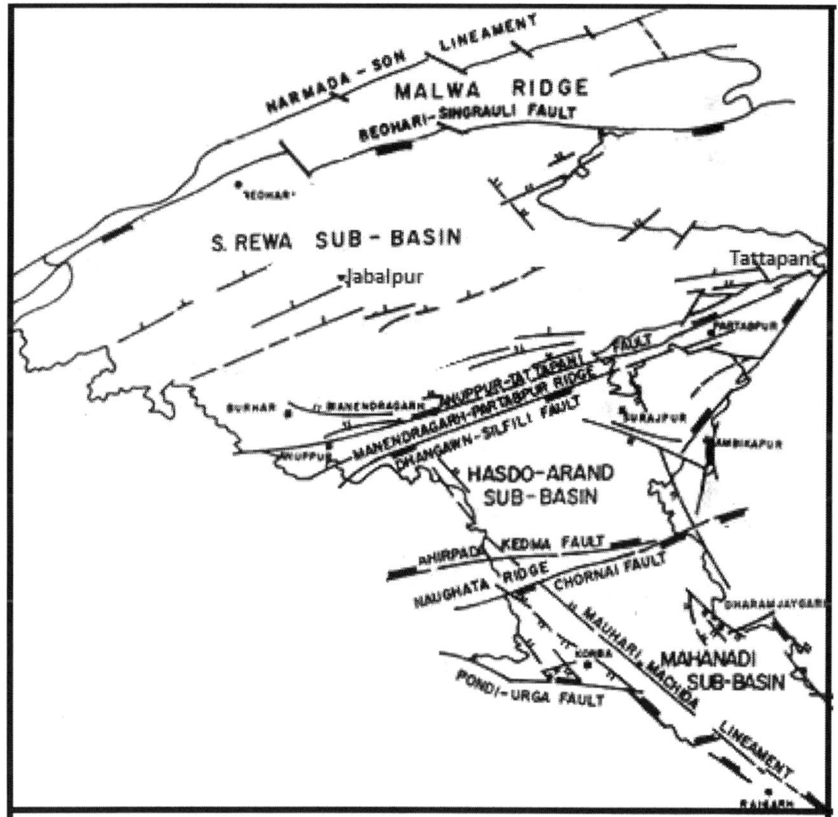

Figure 10.sr.2. Tectonic Map South Rewa-Mahandi Basin.

Another prominent feature is east-west gravity high between these two prominent lows (Das, S.K. and Nazar Abbas, 1995). The remote sensing data also suggest that the South Rewa Basin is bounded by two major basin margin faults with ENE-WSW direction that is Beohari-Singrrauli on the northern margin and Annupur-Tattapani fault along the southern margin (Suchetra Dotiwala, *et al.*, 1995).

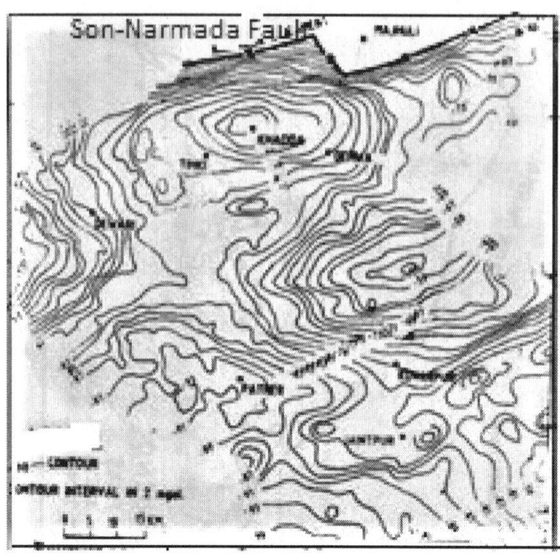

Figure 10.sr.3. Bouguer Anomaly Map–South Rewa Basin.

The identified sub Basins are Burhar, Jhager-Khand, Singrauli, Mohar-Jinahgudah, and Kusiara. The differential rates of deposition and paleo slopes might have contributed to the development of these sub basins (Asthana, M. and Valluri, R.C.1984).

The total intensity magnetic anomalies are influenced by the nearest available dykes/sills which are either exposed on the surface or are in the subsurface. There are large numbers of circular to semi-circular features, probably indicating presence of intrusive. The contour pattern in the southern part also suggests the continuation of South Rewa Basin into Mahanadi Basin.

Sedimentation and Stratigraphy

The pre-Gondwana rocks comprised of calcareous-crystalline slates, phyllites and schists act as basement over which Gondwana rocks were deposited (Figure 10.sr.4). The sedimentation was initiated with some restricted marine transgression near Manendragarh and Saran area. Lower Gondwana (Upper Permian to Upper Carboniferous?) rocks with another restricted marine transgression. Southern part of the basin has been extensively covered by Deccan Trap flows during Upper Cretaceous/Palaeocene period.

Borehole data of well indicates presence of three intrusives of thickness 228 m, 193 m and 35 m respectively (Qidwai, H., *et al.*, 1997). The maximum Gondwana sediments thickness in the basin is of the order of 4.7 km. in Khadda low; 4.4 km in Pathar low and 2 to 2.8 km. on Tetka basement ridge (Figure 10.sr.5). The depth to the basement is debatable due to lack of sufficient subsurface control and errors involved in large lateral variation in density of the rocks, presence of numerous intrusives and varied topography. Network of sills on or near the Tetka ridge is predicted and sills are likely to be close to the surface.

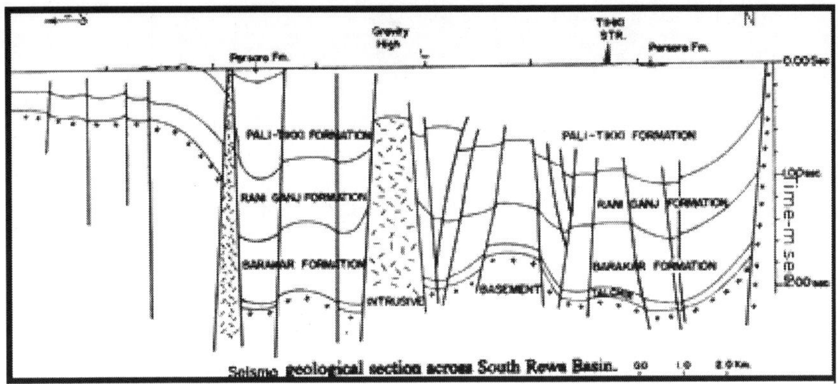

Figure 10.sr.4. Seismo-Geological Section–South Rewa Basin.

AGE	GROUP	FORMATION	THICKNESS (m)	LITHOLOGY
PALEOCENE		DECCAN TRAP		
UP. CRETACEOUS	UPPER GONDWANA	LAMETA	30-50	
LR. CRETACEOUS		BANSA CHANDIA	125-190	
UP. TRIASSIC TO LR. JURASSIC		PARSORA	390-650	
LR. TRIASSIC		PALI- TIKKI	400-650	
UP. PERMIAN	LOWER GONDWANA	RANIGANJ	800-1250	
MID. PERMIAN		BARREN MEASURES	[illegible]	
LR. PERMIAN		BARAKAR	175-900	
		[illegible]	[illegible]	
UP. CARBONIFEROUS TO LR. PERMIAN	MANENDRAGARH BED	[illegible]	[illegible]	
		TALCHIR	15-250	
ARCHEAN		BASEMENT		

Figure 10.sr.5.Stratigraphic Column–South Rewa Basin

Petroleum Geology

South Rewa Basin is classified as Category IV basin, on account of its uncertain petroleum prospects. Except for few gas seepages, no definite presence of hydrocarbon has been indicated in the basin.

Source Rock and Maturation

Barakar sediments have good source potential. Barren Measures and Raniganj shales also contain adequate amount of organic carbon and are considered good source rocks. Analysis of gas seepages data, rock-eval studies, TOC values indicate that, Lower Gondwana rocks hold good prospects for generating hydrocarbons (Shukla. S.N. *et al.*, 1993).

Proterozoic Basins

The Proterozoic hydrocarbon plays of the world such as Riphaean Basin of Siberian Platform of Russia, Amadeus Basin of Australia, Sichuan and Bohai Bay basins of China and their hydrocarbon habitat may be perspective for exploration in the Indian Proterozoic basins. The plate tectonics during Proterozoic time and its bearing in evolution of Intra-cratonic basins, Paleo-ecological and Paleo-environmental studies, may help to understand the possibilities of Precambrian petroleum prospects in India. In this context, the geotectonic origin, sedimentary history and radiometric dating of the Proterozoic basins such as Chattisgarh-Bastar, Cuddapah, Kaladgi, Badami, Bhima, Pakhal have been extensively studied (Pichamuthu, 1967, Krishan, 1968, Naqvi, S.M. and Rogers, 1987).

Tectonic Outline

The Indian Precambrian geology constitutes mainly the Peninsular shield in the south and Himalayan foothill belt in the north. Broadly the cratonic shield area has been divided as Dharwarian, Aravalli and Singhbhum Protocontinents. Proterozoic basins are mainly confined to these protocontinents and the cratonic basins are broadly termed as cratonic sag basins. (Biswas,S.K., *et al.*, 1993). These basins are evolved mainly due to crustal down-warping associated with thermal action. The cratonic basins are in association with the Proterozoic mobile belt (Radhakrishna, B.P. and Naqvi,S.M., 1986) and more so with thrust contact with the Proterozoic mobile belts (Harinarain, 1987).

The Vindhyan basin is surrounded by Delhi-Satpura orogenic belt and that of the eastern margin of Cuddapah by Eastern Ghat Belt. North-west margin of Chattisgarh basin and south-west margin of Pakhal basin. The Pakhal Super-group of Pranhita-Godavari rift along with un-metamorphosed sediments of other cratonic sag basins (Radhakrishna, B.P., 1984).

Proterozoic rock assemblages of Indian sedimentary basins are broadly characterised by mature sandstone, limestone, shale with minor quartzites and conglomerates. Cuddapah and Kurnool groups of Cuddapah, Kaladgi, Badami suites of Kaladgi-Bhima, Indravati group of Bastar, Lower Vindhyan group of Vindhyan, Pakhal Super group of Pranhita-Godavari rift basin and suites of lesser Himalayan autochthonous zones are stratigraphically correlated to Proterozoic age.

Vindhyan sediments are mainly deposited under influence of tidal or shoal environment (Singh, 1973). Presence of organic matter are observed from Vempalle formation in Cuddapah basin and the associated stromatolites indicate intertidal mud flat environment (Logan, *et al.*, 1964). Organic remains are also reported from Badami and Kaladgi group of Kaladgi Basin (Venkatachala, B.S. and Rawat, M.S. 1973, Akhatar, K., 1977). The sedimentary structure and litho-association of the Chandarpur and Raipur Groups of sediments of Chattisgarh basin also infer influence of shallow marine to intertidal environment (Muni, 1987). Carbonaceous shale of Pakhal basin is characteristics of deposition under lagoonal condition (Rao, T., 1987).

Narmada Basin

Subsequent to the initial stages of Cambay rifting, activation of the ENE-WSW striking Narmada transform fault opened up the western Narmada rift basin. This rift basin extends offshore, south of Saurashtra. The Narmada rifting was restricted in the western part of mega lineament. The coastal basin formed in the Narmada valley region by this rifting was transgressed during Late Cretaceous deposited Bagh and Lameta formations over Early Cretaceous Nimar Group.

Deccan Syncline

The Deccan Syneclise is located in the western and south-central part of India. The main rock exposures in the basin consist of a number of basaltic lava flows which have been poured out during Cretaceous-Paleocene period, covering all pre-existing rocks. It is bounded to the north by the Narmada-Son rift and to the south and east by the Precambrian sediment and Archaean metamorphic exposures. The western limit may be defined as the longitudinal fissures from which the basaltic lava erupted and is marked by a highly increased thickness of Deccan volcanics, even though the Trap is known to extend upto the shelf margin of Mumbai offshore basin. Recent surveys, carried out by National Geophysical Research Institute (NGRI) have revealed the presence of a few sedimentary basins buried under the lava flows. These sedimentary basins and those on its periphery are believed to hold good hydrocarbon prospects. Gravity data and DSS profiles reveal several rift basins buried below the Deccan Trap. Two rifts, namely, the Koyna rift, the northwest continuation of the Kaladgi basin and the Kurduvadi rift, the northward continuation of the Bhima basin filled up by low density material, were identified under the Trap separated by the Sangola uplift.

Bhima–Kaladgi Basin

Bhima-Kaladagi Basins are the northern extension of Cuddapah Intra-cratonic Foreland, evolved due to pulses of thermally driven crustal sagging, punctuated by extensional stretching. The NE trending irregularly sinuous Bhima-Kaladgi basins consists dominantly of limestone. It overlies the granitic basement of Eastern Dharwar craton with a profound unconformity and has faulted contacts at many places.

Bhima basin is overlain by Deccan Trap in the north. The sediments have an aggregate thickness of about 325 metres. The basin is well known for its limestone and uranium occurrence near Gogi.

Kaladgi basin is an E-W trending irregular basin underlain by the basement granitoids of the Dharwar craton in the south and east and overlain by the Deccan Trap in the north. The basin is made of an older Kaladgi sequence and younger Badami sequence. Kaladgi basin is not marginally deformed, as it is not spatially associated either with mobile belt or with terrane boundaries, instead the deformation is concentrated in the centre of the basin. Kaladgi basin hosts vast resources of limestone and dolomite, as well as building and ornamental stones, besides minor iron ore.

Being part of Cuddapah palaeo-seas, the basin has experienced three cycles of sedimentation equivalent to Kurnool mega sequences indicating gradual deepening of marine basin followed by gradual regression. Nearly horizontal to low dipping over large areas, the basin has experienced some tectonic disturbances around faults and Deccan Trap junctions. Alternating clastic and chemogenic sedimentary rocks are exposed on both sides of the Bhima River.

Considering vast areal extent with less sedimentary thickness, these basins although appears lucrative from hydrocarbon exploration point of view, but still the mega sequences south of Deccan Synclinse and concealed by traps may be impediment for complete subsurface understanding for hydrocarbon exploration.

Cuddapah Basin

The Cuddapah is a crescent shaped, easterly concave and N-S trending basin in the east central part of Dharwar craton. The basin extends for a length of about 450 km along the arcuate eastern margin with a mean width of 150 km. It is characterized by quartzite-carbonate-shale cycles having thickness vary between 6 and 12 km.

The early sediments of the basin are interspersed with basic volcanics and sills. The basin is well known for its mineral potential in the form of limestones and dolomites, bedded and vein barites, chrysotile asbestos and steatite, besides occurrences of base metals, diamond, phosphorite, uranium and abundant building and ornamental stones. The eastern half of the basin is metamorphosed and the western half of the basin is undeformed and consists of four sub-basins, the Papaghni, Kurnool, Srisailam and Palnad.

The Papaghni and Kurnool sub-basins are geographically interlinked, but were sites of deposition at different times (Figure 10.cd.1). The Papaghni sub-basin containing lower Cuddapah sediments and the Kurnool sub-basin containing the younger Kurnool sediments overlie the Cuddapah sequence with a major unconformity. Srisailam sub-basin contains upper Cuddapah sediments and Palnad sub-basin exposes only the younger Kurnool sequence. The eastern half of the

Cuddapah basin is occupied by the Nalamallai fold belt (NFB) consisting of upper Cuddapah sediments which is demarcated from the undeformed western sub-basins by the prominent fault lineament called the Rudravaram Line.

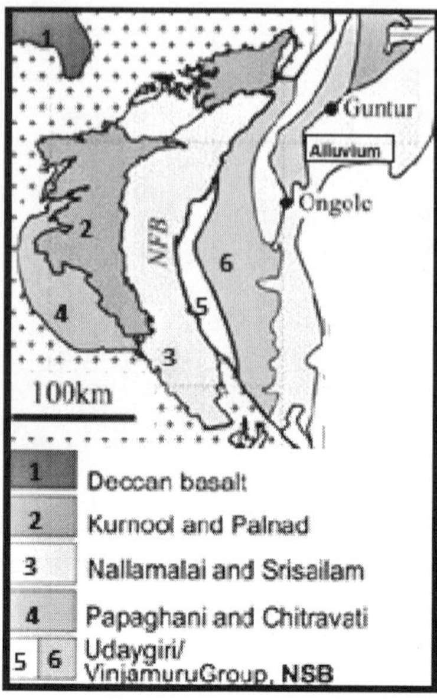

Figure 10.cd.1. Geological Map–Cuddapah Basin.

Bastar Basin

The Bastar craton in the central India is one of the oldest cratons in the Indian shield and is bounded by two Meso-proterozoic belts of Eastern Ghat mobile belt in the east, Satpura mobile belt in the northwest, the Palaeoproterozoic mobile belt in the northeast, and the Godavari graben in the southwest.

Chattisgarh Basin

Chhattisgarh basin is the third largest basin after the Vindhayan and Cuddapah, and extends for about 300 km in the E-W and 200 km in the N-S directions, with an irregular configuration. This basin is situated on the northern edge of Bastar craton. It is bounded by the Kotri-Dongargarh orogen in the west, Satpura mobile belt in the north, Gondwana graben of Mahanadi in the northeast and the Eastern Ghats mobile belt in the southeast. In the south, the basin overlies the basement gneiss enclosing the Sonakhan greenstone belt. Vast limestone and dolomite resources of the basin are known. The granitic terrain surrounding the basin with number of kimberlite pipes. The basin contains about 2500 m thick sediments of orthoquartzite-carbonate-pelite suite, deposited in multiple sedimentary cycles, intercalated with minor felsic volcanics and pyroclastics and by unconformities. Each cycle starts with arenaceous facies and ends up with shale-limestone facies. The basin is divided into two sub-basins, the Hirri in the west and Baradwar in the east separated by the Sonakhan high.

References

Abdul R. Ismal, James M. Peden and Azmi (1994)-"The Effect of Solid Concentration And Formation Characteristics On Formation Damage And Permeability Recovery", SPE- 28762.

Acharyya, S.K., and Ray, K.K. (1982). Hydrocarbon possibilities of concealed Mesozoic-Paleogene below Himalayan Nappes-Re-appraisal, Am. Assoc. Pet. Geol. Bull., v.66, pp. 57-70.

Adams, N.J.: (1985)- Drilling Engineering (A Complete Well Planning Approach)., Tulsa, Oklahoma.: PennWell Books- Penn Well Publishing Company.

Agarwal, G.C. (1984). Structural Elements along Narmada River in Gujarat Proc, Symp. Quat Episodes. Deptt Geol, M.S. University, Baroda.

Agarwal, R. P. and Bansal, S. B.(1986). The structural evolution of Pranhita–Godavari Graben, India. Bull.ONGC., 23, pp 93–108.

Agarwal, R.G., Al-Hussainy, R. and Ramey, H.J., Jr. (1965). The Importance of Water Influx in Gas Reservoirs. J. Pet. Tech., November: 1336–1342. Trans. AIME.

Agarwal, V. (1984). Geochemistry of the volcanic rocks around Tavidar, District Jalore, Rajasthan. Unpublished thesis, University of Rajasthan, Jaipur, 110 p.

Aggarwal R. P. D. N. Prasad, U. Samanta and C. M. Berry (1994). "Hydrocarbon potential of Siwalik Basin", Jour. Him. Geol., v 15, 301-320.

Aharon, P., Schidlowski, M. and Singh, I.B. (1987). Chrono-stratigraphic markers in the end-Precambrian carbon isotope record of the Lesser Himalaya, Nature, v. 327, pp. 699-702.

Ahmad, F. (1958). Palaeo-geography of central India. Geological Survey of India Records 87, 513-548.

Ahmed, F (1964). The Permian Basin of Peninsular India. In 4th International Geological Congress, 22nd Session, 123-138.

Ahmed, F (1965). An aspect of Tectonism in the Indian sub-continent. In Proceedings Upper Mantle Symposium, Delhi, 48-59.

Ahmed, F (1966). Post Gondwana faulting in Peninsular India and its bearing on the tectonics of the Subcontinent. Annals of the Geol. Dept., A.M.U.

Ahmed, F (1970). Marine transgression in the Gondwana of Peninsular India - A Reappraisal. In 2nd Gondwana Symposium, S. Africa, 179.

Ahmed, F and Ahmed, ZS (1977). Tectonic frame-work of the Gondwana basin of Peninsular India. In 4th International Gondwana Symposium, Calcutta India, 2, 720-733.

Ahmed, U., Cray, S.F. and Coates, G.R. (1989) "Permeability estimation: the various sources and their interrelationship", SPE-19604 in SPE annual technical conference proceedings on formation evaluation: SPE, 649-662.

Ahuja,A. D., S.V. Rao, B.K. Patel, A.K. Singh S.S. Saini and B.J. Thattacharry (1991). Structural styles in south Cambay Basin-Genesis and impact on Hydrocarbon Accumulation.,ONGC Bull vol 25 no.1.

Ajay Kumar Dwivedi (2016) –Petroleum Exploration in India A perspective and Endeavours -Proc Indian Natn Sci Acad 82 No. 3 July Spl Issue pp. 881-903.

Ajay Kumar, Monu Bhatnagar, A.K. Barthwal, H.S. Mishra, P.K. Bhatnagar, Krishan Pal and N.P. Joshi. (2003)-Novel Approach to Minimize borehole erosion through interplay of hydraulic impact force and Drilling Fluid inhibition -Proceedings of PETROTECH-2003 January 9-12, New Delhi.

Ajay Suri and Mukul M. Sharma (2001)-Strategies For Sized Particles In Drilling And Completion Fluids" SPE-68964.

Akbar, M. (1993). Middle East well evaluation review-14, pp25-43.

Akhtar, K. (1977). Depositional environment, Dispersal pattern and paleo-geography of the elastic sequence in the Bhima Basin, Ind-Mineralogist, 18, 65-72.

Alam, M, Alam MM, Curray JR, Chowdhury ALR, Gani MR. (2003). An overview of the sedimentary Geology of the Bengal Basin in relation to the regional tectonic framework and basin-fill history.– Sedimentary Geology. 155: 179-208.

Al-Chalabi, M. (1979), Velocity determination from seismic reflection data: In Developments in Geophysical Exploration methods, Vol. 1, Fitch, ed., pp 1–68, Amsterdam: Elsevier.

Alden, J., et al., (1997); Characterisation of Petrophysical Flow Units in Carbonate Reservoirs, Bulletin AAPG,Vol.81, No.5, P.734-759.

Alger, R P., Raymer. L.L., Hoyle, W.R., and Tixier,P.(1963): Formation Density Log Applications in Liquid-Filled Holes," J. Pet.Tech.

Alger, R.P.: - Modern Logging Programs and Interpretation Methods. Schlumberger Well Surveying Corporation, O'Huston, Texas.

Allaud, L.A. and Ringot, J.: (1969)-The High Resolution Dipmeter Tool," The Log analyst 10,No. 3.

Allen, J. R. L. (1965), A review of the origin and characteristics of recent alluvial sediments: Sedimentology (spec. issue), v. 5, no. 2, p. 191.

Allen, P.A. and Allen, J.R. (1990.) Basin analysis – principles and applications. Blackwell, Oxford.

Al-Sadi, H. N. (1973), Dependence of the P-wave amplitude spectrum on focal depth. Pure and Applied Geophysics, vol. 104, p 439–452.

Al-Sadi, H. N. (1980), Seismic Exploration, Technique and Processing. Birkhäuser Verlag, Basel, 215p.

Al-Sadi, H. N., and Baban, N. (2014), Introduction to Gravity Exploration Method. 203p, Sulaymaniyah University, Kurdistan Region, Iraq.

Amalendu Roy Barman (1992). Geological History and Hydrocarbon Exploration in Bengal Basin- Indian Journal of Geology Vol 64 N0. 3 pp-235-258.

Amani, M. (2012). The Rheological Properties of Oil-Based Mud Under High Pressure and High Temperature Conditions. Advances in Petroleum Exploration and Development, 3(2), 21-30.

Amani, M., and Al-Jubouri, M. (2012). The Effect of High Pressures and High Temperatures on the Properties of Water Based Drilling Fluids. Energy Science and Technology, 4(1), 27-33.

Amitabh Pandey. (2003) -Dynamic up-scaling of Laboratory Relative Permeability curves using pseudo functions for Simulation of a Giant offshore Carbonate Reservoir– Proceedings of PETROTECH-2003.

Amyx, J., Bass, D., and Whitney, R. (1960) "Petroleum Reservoir Engineering". New York: McGraw-Hill Book Company.

Anand, S. P., Erra V. C. and Rajaram, M. (2002). Delineation of crustal structure of Mahanadi basin from ground Magnetic Survey, Jour. Geol. Soc. India, V. 60, pp. 283-291.

Anand, S.P., Vinit, C. Erram., Mita Rajaram. (2002). Delineation of Crustal structure of Mahanadi Basin from ground magnetic survey. J. Geol. Soc. India. 60, 283-291.

Anderson, R.A., Ingram,J.S., and Zanier, A.M.(1972): Fracture Pressure Gradient Determination from Well Logs, paper SPE 4135 presented at the 1972 SPE Annual Meeting.

Andreson, B.C., et al. (1999), "Complex Inhibitor Drilling Mud For Drilling Deep Wells In Complicated Conditions", Petroleum Engineer International, pp 51-57.

Anil Kumar, Prem Prakash, K.S. Pal, Aditya Kumar and Chaman Lal. (2005)-, Reservoir Engineering aspects of PVT analysis., PetroTech2005.

Anuja Mishra, C.S. Mishra, A.K. Katiyar and A.K. Mittal (2003). Petroleum Geochemistry of Borholla-Champang, Khoraghat and Rudrasagar Area of Upper Assam Basin-Proceedings of PETROTECH-2003.

Anis, Abdullah Khan, Shahnawaz, Sarwar Rais (2012). Sedimentary Facies and Depositional Environment of Early Triassic Pachmarhi Formation, Satpura

Gondwana Basin, Central India, Madhya Pradesh-International Journal of Scientific and Research Publications, Volume 2, Issue 12.

Archana Tewari, Hart, M.B., Watkinson, M.P. (1996a). A revised lithostratigraphic classification of the Cretaceous rocks of the Trichinopoly District, Cauvery Basin, and Southeast India. Contributions XV Indian Colloquium of Micropaleontology and Stratigraphy Dehradun, 789-800 pp.

Archana Tewari, Hart, M.B.,Watkinson, M.P. (1996b). Foraminiferal recovery after the mid- Cretaceous oceanic anoxic events (areas) in the Cauvery Basin, southeast India. In: Hart, M.B. (Ed.), Biotic Recovery from Mass Extinction Events. Geological Society, London, Special Publication, vol. 102, pp. 237-244.

Archie, G. E. (1942),- The electrical resistivity log as an aid in determining some reservoir characteristics: SPE-AIME Transactions, v. 146, p. 64-62.

Arora, H.K. and N.J. Thomas.,(2003) Geochemistry of Recent Sediments of the West Coast of India: Offshore Region of Dwaraka--Proceedings of PETROTECH-2003.

Arthur H. Hale, Hon Chung Lau, Louis A. Bernadi (1996)–Selection and Qualification of Drill-In Fluids For Horizontal Wells In Unconsolidated Sands' SPE37077.

Arts, K.P., Radhakrishna, M., Murthy, B.V.S., Arora, S.K. and Nambiar, C.G. (2003). Gravity anomalies, structure and tectonics of the Kerala Basin, Southwest Continental margin of India. Mem. Geol. Soc. India, no.54, pp.203-216.

Arun Garg and Sanjoy Goswami.(2003), Development of Lakwa Barail sand-2 in the Lakwa-Lakhmani oil field of Upper Assam, India., PetroTech 2003.

Aschoff, J. L. and Giles, K. A. (2005), Salt Diapir-influenced, shallow-marine sediment dispersal patterns: Insights from outcrop analogs: American Association of Petroleum Geologists Bulletin, v. 89, p. 447-469.

Ashok Kumar (2006)., Formation Evaluation of Unconventional Basaltic Deccan Trap Basement Reservoir of Gamij Field, Cambay Basin, India., 6th International Conference and Exposition on Petroleum Geophysics "Kolkata.

Ashok Kumar, B.K. Tiwari, Pushpa Sharma, K.N. Jha and S.K. Sanyal (2005) -Rock type and Flow unit identification– a case study on reservoir characterization.,PetroTech.

Ashok Kumar, J.P. Narayan, S. Mahanti, S. Lakhera and R.P. Verma (2005) Formation Evaluation of Low Resistivity Kalol pay sands of Gamij field, Cambay Basin, india – A case study., Proceedings -article number 431.

Ashok Kumar, N. Pendkar, T. Chattopadhyay, J.P. Narayan, S. Lakhera and R.P. Verma (2003). Inhomogeneous Deltaic Reservoirs -Examples of Hazad sand units GS-3A and GS-12A from Gandhar and Dahej fields, Cambay Basin, India., Petrotech Proceedings.

Ashok Kumar, T. Chattopadhyay, N. Pendkar, K. Rai, V. K. Sud (2002). "Salinity estimation and analysis, through well logs, for GS-1, 2 3A, 3B and 11 Sand Units of Hazad Member in South Gandhar and Dahej Fields, Cambay Basin – A case study". 4th Conference and Exposition on Petroleum Geophysics of the Society of Petroleum Geophysicists, India, 13-19.

Ashok Kumar, Yogesh Chandra and Shri Pati Das,(2010) -The logs vis-à-vis production characteristics of TS2 paysand of Lakwa Field, India- SPG-Hyderabad.

Asit Kumar Samadder (2010)., Advanced Perforation Techniques for Production Maximization., 8th SPG.

Asit Kumar, N. Bhattacharyya and Shyam Mohan (2006).,Cross-equalization for Time-lapse Study in Balol Field, India., 6th International Conference and Exposition on Petroleum Geophysics Kolkata".

Asquith, G. B. (1982), Basic well log analysis for geologists: American Association of Petroleum Geologists, Tulsa, OK, 216 p.

Asquith, G. B. (1985), Log evaluation of shaly sandstones: a practical guide: American Association of Petroleum Geologists Continuing Education Course Note Series No. 31, Tulsa, OK, 59 p.

Asthana, M. and Valluri, R.C. (1984). Geology and Hydrocarbon Potential of South Rewa Basin, Petroleum Asia Journal-SPE- 1 Vol 3, PP-200-206.

Atchuta Rao, D., Ch. Rama Rao, and D. S. Bhaskara Rao (1999). Extension of the Intracratonic Mahanadi rift towards the east coast of India–A reference from aeromagnetic and gravity, J. Geol. Soc. India, 53, 71–77.

Atwood, D.K.: (1964) "Restoration of Permeability to Water Damaged Cores" JPT, 1405.

Aubouin, J. (1965) Geosynclines_Elesvier Publication.

Aud, B. W. (1974)- Abnormal pressure zones can be predicted by seismic data. World Oil, Aug-74, pp. 37– 39.

Auden, JB (1933). Vindhyan sedimentation in the Son Valley. Geol. Survey: Ind. Memoir, 62, Pt. 2.

Augustine Manju and R.S. Chauhan (2005). Introduction of production logging in high pressure well of Assam asset – Case study., Proceedings –article-373 abstract

Avasthi D N and S. Venkataraman (1979). Exploration of subtle traps in North Gujarat Vol 1. 16, No. 2.

Avasthi D.N. and Varma, S.K. (1973) Analysis of the statistical structure of seismic reflection for delineation of stratigraphic traps for oil; Indo-Soviet Synip. Recent Trends in Expl. Min. Oil and Ground Water, Ind. Nat. Sci. Acad., New Delhi.

Avasthi D.N., and D. Das.(1979), Gravity Map series of India,(five maps on the scale 1: 5,000,000,with an explanatory brochure)-published by the National Geophysical Research Institute, Hyderabad 500007, first edition, 1978; ONGC Bull vol. 16, no. 1 pp.107.

Avasthi, D. N., R. Nath and A.N. Datta (1969).,Operational and Interpretational Methods of Geophysical surveys in sub-Himalayan foothills and their limitations., ONGC.bull vol.6 no. 1.

Awadesh Rai, P., Chandrasekharan and Mishra, V.N. (1998). Indian sedimentary basins and their hydrocarbon potential, Proc. Nat. Symp: Recent researches in sedimentary basins, 91-103.

Azad, J. Bhattacharya. S. and Dutta B.D (1971). Hydrocarbon accumulations in the Nahorkatiya oil field. Assam. Proc. of 8th W.P.C.

Azim, S.A, S.P. Rao Hebbare and P.B.Pati (1994).–Assessment of Permeability in a Multi-layered Heterogenous Reservoir of Bombay High Oil Field, India.ONGC Bulletin Vol. 31, No. 1.

Azmi, S A., and Mangla V.K. (1997)-"Introduction Of PHAP To Drill Caving Prone Barail Coal-Shale Sequence Of Geleki Field Of Upper Assam". Petrotech-97.

Babu D.K. and Odeh Aziz S.(1989), "Flow Capabilities of Horizontal Wells", Journal of Petroleum Tecnology.

Babu P.V.L.P. (1984). A structural synthesis of the Lower Narmada River Basin and its bearing on the Mesozoic Oil prospects, Petroleum Asia Journal, petroliferous Basins of India, v.2.pp.97-101.

Babu Rao V, Atchuta Rao, D. Sankar Narayan, P.V. and Ratnam C. (1982). Aeromagnetic Survey over parts of Mahanadi basin and Adjoining offshore region. Geophy. Res. Bull.40, pp 219-226.

Badhe, S.,D.V. Ramana, T.K. Mathuria and R.K. Phukan (2003). Hydrocarbon prospect evaluation of northeastern part of Mahanadi offshore basin- Proceedings of Petro-Tech-2003 January 9-12, New Delhi

Bahuguna, R.M.,S.S.Pabla, Mohan Lal and Hans Raj (2006). Impact on Estimation of Water Saturation Values Using Laboratory Determined 'a', 'm' and 'n' Parameters – A Case Study.,6th International Conference and Exposition on Petroleum Geophysics.

Bahuguna, V.K., R.K. Singhal, Rajendra Kumar, D.S. Negi, Aditya Kumar and Chaman Lal (2003)-Improving Productivity of Nandej oilfield by stimulating with new acid system –Fluoboric acid., Proceedings of PETROTECH.

Bais, P. S. (1994). Drilling Operations Manual, Institute of Drilling Technology, 1st. edition, ONGC,. Dehradun.

Baker, D.M., Lillie, R.J., Yeats, R.S., Johnson, G.D., Yousuf, M.,and Zamin, A.S.H. (1987). Development of the Himalayan Frontal Thrust zone: Salt Range, Pakistan, Geology, v.16, pp. 3-7.

Bakshi S S and G. Lahiri (1994). Study of Hydrocarbon Accumulations in Cambay Shales of Wadu Area ONGC Bull Volume 31 Number 2.

Bakshi S.K, (1965). Stratigraphy of Barail series. Shilling Plateau, Assam. India, Bulletin AAPG, Vol 49 No. 12.

Bakshi., S. S. (1992). Study of Electro-facies Over G-1 and G-2 Structure in Godavari Offshore., ONGC Bull., Volume 29,No.1.

Bandaru A Rao (2010).,Clay-typing, Depositional Environment and Framing a Petrophysical Model for Formation Evaluation in Vashista Field, Krishna - Godavari Basin-a Case Study, 8th SPG.

Banerjee, A., S. Pahari, M. Jha, A.K. Sinha, A.K. Jain, N. Kumar, N.J. Thomas, K.N. Misra, and K. Chandra, (2002). The Effective Source Rocks in the Cambay Basin, India: AAPG Bulletin, v. 86, p. 433-456.

Banerjee, P and K. P. Singh.(2006), Shallow Refraction Survey Avoiding Explosive Permanently and Consistently 6th International Conference and Exposition on Petroleum Geophysics.

Banerjee, Ranjit K. (1981). Cretaceous - Eocene sedimentation, Tectonics and Litho facies in the Bangal Basin, India.

Banerjee.,A S Pahari, M jha, A K Sinha, AK Jain N Kumar, NJ Thomas KN Misra and K Chandra (2002). The effective Source rocks in the Cambay Basin, India, AAPG Bulletin Vol 86, no.3, pp-433-456.

Banerjee.A. et al. (2000). The Effective Source rocks of Cambay Basin, India, AAPG International Conference and Exhibition.

Banerji, R.K. (1972). Stratigraphy and micropaleontology of the Cauvery basin, Part. I. Exposed area. Journal of the Palentological Society of India 15, 32-41.

Baranov, V., 1957:)-A new method for the interpretation of aeromagnetic maps: Pseudogravimetric anomalies, Geoph., Vol. 22, No. 2.

Barnes, A. (1993), Instantaneous spectral bandwidth abd dominant frequency with applications to seismic reflection data, Geophysics, vol. 58, P.119–428.

Barnes, A. (2001). Seismic attributes in your facies: CSEG Recorder, 26, Sept., pp. 41–47.

Barua, I., S.S.Deb, R.Borgohain, B.N.Talukdar.(2006). Modelling Fluvial Sandstone Architecture in a Mature Oilfield of Upper Assam Basin using Old Geophysical Logs.,6th International Conference and Exposition on Petroleum Geophysics "Kolkata.

Basin, N. C. (1995). Amplitude-Versus-Offset analysis over a part of Dabka-Gajera-Jambusar oil and gas fields of Cambay basin, Gujarat, India, ONGC Bulletin, 32, 1-36, 1.

Bassi. K.D. and R.K. Gang (2003). Identification and Evaluation of suitable Defoaming Agents for crude oil of Viraj Field to enhance efficiency of surface equipment thereby improving oil productivity.,Proceedings PETROTECH.

Bastia R. (2003). Depositional Model and Reservoir Architecture of Tertiary Deep Water Sedimentation, Krishna-Godavari Offshore Basin, India, Journal Geological Society of India, Vol. 64, July 2004, p. 11 – 20.

Bastia, R. (1999). Indian deepwater basins holds the key for the new millennium- 36 annual convention and meeting on earth system sciences in the new Millennium. Indian Geophysical Union.

Bastia, R. (2004). Depositional model and reservoir Architecture of Tertiary Deep Water sedimentation, Krishna, Godavari offshore basin, India. Jour. Geol. Soc. India, V. 64(1), pp. 11-20.

Bastia, R., Chakraborty, S. K., Guha, J. and Mishra, B. S. (2006). Depositional architecture of Mio-pliocene sequences in offshore Bengal Basin, Jour. Geol. Soc. India, V. 67(5), pp. 575-583.

Bastia, R., Singh, P. and Nayak, P (2006). Linking shelf delta to deep waters; Krishna-Godavari basin, Jour. Geol. Soc. India, V. 67(5), pp. 619-628.

Bastia. R., B. Das, M. Acharya, P. Gabriels, M. Sams, P. Mesdag. (2005). Reservoir Characterization and Modeling of thin beds in a Deep water Gas field, offshore India., PetroTech.

Bastia. Ravi, Duggirala. Nathaniel. Moses.,Tenepalli. Srinivas. Akella. Muralikrishna., Verma Ravi,D'silva Kenneth (2008). "Mesozoic of South Western margin of India: A frontier exploration". 19th WPC, Spain, Forum 01.

Basu AR, et al. (1993). Early and late alkali igneous pulses and a high- 3He plume origin for the Deccan flood basalts. Science 261.

Basu D.N., Banerjee, A and D.M. Tamhane, (1982). Facies distribution and petroleum geology of Bombay Offsore Basin, India: Journal Petroleum Geology, v.5, no.1, p. 51-75.

Basu, D.N. Banerjee, A. and D.M. Tamhane, (1980), Source area and migration trends of oil and Gas in Bombay Offshore Basin, India: AAPG Bulletin, v.64, no.2,p.209-220.

Basu, D.N., Banerjee, A. and Tamhane. D.M. (1980): Source area and migration trends of oil and gas in Bombay Offshore Basin, India, Bull. AAPG, vol. 17 pp. 109-220.

Basumallick, S. (1967). Problems of the Purana stratigraphy of the Godavari valley with special reference to the type area in Warangal district, Andhra Pradesh, India. Quaternary Journal of Geological Mining and Metallurgical Society India 39,115-127.

Bateman, R. M. (1985). Open-hole log analysis and formation evaluation: IHRDC Publishers, Boston, MA, 647 p.

Bates, C. C. (1953). Rational theory of delta formation: Am. Assoc. Petroleum Geologists Bull., v. 37, no. 9, p. 2119–2162.

Bates, R. L. and Jackson, J. A. (1983): Dictionary of Geological terms. –3rd edition, Anchor Books.

Battacharya, G.C., et al. (1994). Evidence of seafloor spreading in the Laxmi Basin, Northeastern Arabian Sea. Earth and Planetary Science letters 125 pp 211-220.

Bawa R K (1993). Hydrogeochemical Studies in Jaisemar Basin in relation to Hydrocarbon Entrapment Conditions -Second Seminar on Peroliferous Basins of India –Vol-2 Indian Petroleum Publishers, Dehradun., pp-335-341.

Beard, D. and P.Weil,(1973), Influences of texture on porosity of unconsolidated sands: AAPG Bulletin, v 57, p.349-369.

Behera, C., K.P. Ramachandran and S. Bateja (2003). Pressure Transient study facilitates Exploitation strategy for Marginal fields in Geologically complex basin: a case study., Proceedings of PETROTECH.

Belov, V. P. and Ghildiyal, R. D. (1964). Effect of Caving of the Hole upon Rate of Drilling, O. N. G. C. Bulletin, vol. 1, No. 1.

Bennion, D.B. and Thomas, F.B. Bennion, D.W. and Bietz R.F (1995.). Underbalance drilling and formation Damage- Is it a total solution?, The Journal of Canadian Petroleum Technology.

Berg, R.R. (1970). " Method for determining permeability from reservoir rock properties", in Gulf Coast Association of Geological Societies Transactions, 20, 303-317.

Bhandari A. (1999). Phanerozoic stratigraphy of western India: A review. In: P. Kataria (ed.) Geology of Rajasthan: Status and perspective. MLS University, Udaipur. p. 126-174.

Bhandari L. L. and Jain S. K. (1984). Reservoir Geology and its role in the development of the L-III reservoir, its petroleum geology Amer Assoc Petro. Geo. V. 99 pp 433-465.

Bhandari L.L. and L.R Chowdhary. (1975). Stratigraphic analysis of Kalol and Kadi Formation, Cambay Basin, India; AAPG Bull. Vol. 59 No. 5.

Bhandari L.L., Fuloria R.C. and Sastri V.V. (1973). Stratigraphy of Assam Valley, India, Bull A.A.P.G., v. 59 (s), pp. 642-656.

Bhandari, L.L. (1968). Occurrence of Syeoite in subcrop in the North Cambay Basin; Mineral Wealth; Vol.4,No. 2.

Bhandari, L.L. and Kumar, S., 1970; Palaeo-current analysis of Dhrangadhra Formation, Gujarat, India, Bull. ONGC, India, 7(2), 101-108.

Bhandari, L.L. and Mathur R.B., 1966. Strand line sand–Eocene prospects in northern Cambay basin; Second Indian Petroleum Contcrence, Baroda.

Bhandari, L.L. and S.K. Jain (1984). Reservoir geology and its role in the development of L-III reservoir Bombay high field, India: Journal of Petroleum Geology, V.7,p.27-46.

Bhandari, L.L., and L.R. Chowdhary, (1975). Stratigraphic Analysis of Kadi and Kalol Formations: AAPG Bulletin, v. 59, p. 856- 871.

Bhanja, A K, Rajesh Chandra, Asim Samanta, (2010). Reverse modeling based on testing results using log overlay technique provides insight into non Archie Reservoirs of Ankleshwar field in Cambay Basin, India.

Bhanja,R., A. Saha, and T. Friedel (2010). Mumbai High Field Heterogeneity Addressed Through Quantitative Seismic Reservoir Characterization. SPG 8th Hyderabad.

Bharali, B., N. Barua, U. C. Das, B. N.Talukdar (2006). Environment of Deposition and Sand Development Pattern From Image Log- A Case Study from North Chandmari Area of Upper Assam Basin.,SPE 6th International Conference and Exposition on Petroleum Geophysics.

Bharali, B., Rath, S., Sarma, R. (1991). A brief review of Mahanadi delta and the deltaic sediments in Mahanadi Basin. In: Vaidyanadhan, R. (Ed.), Quaternary deltas of India. *In J. Geol. Soc. India.* Bangalore, Memoir 22, 32-49.

Bharali, B., S.K.Srivastava, V. Ravichandran (1992). Seismo-stratigraphic analysis of Cretaceous –Tertiary Sequence of the Mahanadi offshore basin. Geol.Surv. Ind.Spl.Pub.

Bhargava, O.N. U.K. Bassi and R.K. Sharma (1991). "The crystalline thrust sheets, age of metamorphism, evolution and mineralization of the Himachal Himalaya", Indian Minerals, 45(1 and 2), 1-18.

Bharwan Sahay and Walter H Fertl (1988). Origin and Evaluation of Formation Pressures-Allied Publishing Private Limited, pp-292.

Bhaskara Rao, V., Satyanarayana Murthy, B.V. and Venkateswarlu, P.D. (1971). Gravity anomalies and tectonics on a part of the lower Gondwana basin, Symp. Gond. Systems, AMU, 148-156.

Bhat, M.I., S. Claesson, A.K. Dube and K. Pande (1998). "Sm-Nd age of the Garhwal -Bhowali volcanics, Western Himalayas: Vestiges of the Late Archaean Rampur flood basalt province of the northern Indian Craton", Precam. Res., v.87, 217-231.

Bhat, R.K., Pandey, A.K., and Saha, A. (1997). Exploration and exploitation of hydrocarbons in western offshore, India, in Proceedings second international petroleum conference and exhibition: New Delhi, PETROTECH-97, p. 269–273.

Bhatia M.R. and K.A.W Crook,(1986). "Trace element characteristics of graywackes and tectonic setting discrimination of sedimentary basins", Contrib. Mineral. Petrol., 92,181-193.

Bhattacharjee, D. Ghosh, S. K., Roychoudhuri, S. and Biswas, N. R. (2006). Quaternary sedimentary environments in Ganga prodeltaic region, Northern part of Bay of Bengal, Jour. Geol. Soc. India, V. 67(5), pp. 585-593.

Bhattacharya, S. and Chaliha, S. (1973). Mapping of Channel-Fill deposits as a tool in oil exploration in Moran oil field, Assam, Indo-Soviet Symposium.

Bhattacharya, S., S.K. Shrivastava, R. Talukdar and J.S. Saluja (2003). Effect of Reservoir Description using Geostatistics in Simulation studies.,Proceedings of PETROTECH-2003.

Bhavana, P.R., M.V. Rao and S.S. Yalamarty (2005). Sedimentary Features and Depositional Environments Interpreted from FMI images, case studies from Krishna-Godavari Basin, India.,Proceedings-article- 574.

Bhoj, R and Peters, James (2002). Geology of Aloni – Kharkhari and contiguous area, Vindhyan Basin – A re-look through integrated Remote sensing and field studies, Proceedings of the first conference of APG, Vol 1, pp.291-296.

Bhowmick P.K.: Phanerozoic Petroliferous Basins of India, Glimpses of Geoscience Research in India -pp253 268.

Bhuin. N.C.(2003). Amplitude–versus–offset (AVO)–a Seismic signature of lithology and its Petrophysical Characteristics.-Proceedings of PETROTECH.

Bhumija Agarwal, K. Baskaran, M.V.M. Rao, C.P. Daswal, Gita Singh, M.Das, D.V.R.Murti (2006). Model Based Processing in Gojalia Area, Tripura., 6th International Conference and Exposition on Petroleum Geophysics.

Bhushan, S.K. (2000), Malani Rhyolite - A Review, Gondwana Research, v. 3(1), p. 65-77.

Bhushan, Y., A.K. Das and B.L. Lohar (2003). Estimation of Gas and condensate Reserves in a gas condensate reservoir - an EOS approach.,Proceedings of PETROTECH-2003.

Bielstein, W.J. and Cannon, G.E.(1950). "Factors Affecting the Rate of Penetration of Rock Bits", API Drilling and Production Praxes, 61.

Billings, M.P. (1960). 'Structural Geology'. Modern Asia edition, Prentice Hall Inc., N.T.

Billings, M.P. (1979). Structural Geology. Prentice of India Pvt Ltd-pp 1-606.

Bird, K.J. (1994). Ellesmerian- petroleum system, North Slope, Alaska, USA, in L.B. Magoon and W.G. Dow, eds., The Petroleum System–From Source to Trap: AAPG Memoir 60, p. 339–358.

Bisen, M., Sriram K., Gupte, S.S. (2010). Structure and Tectonics of Deep-Water Kutch-Saurashtra area, Western India; SPG Conference Hyderabad.

Bishnu Kumar and Mahendra Kishore. (2006). Electrofacies Classification – A Critical Approach 6th International Conference and Exposition on Petroleum Geophysics "Kolkata.

Biswas A K (1999). Hydrocarbon Exploration Perspectives of frontier Basins of India- Proc PetroTech 99 pp-31-39.

Biswas, B. (1961). Geology of Bengal Basin with special reference to stratigraphy and Micropaleontology unpublished report of ISPP.

Biswas B. (1962). Stratigraphy of the Mahedeo,Langpar, Cherra and Tura Formations, Assam, India, Bull. Geol. Min. Metal. Soc. India, v.28,pp. 1-48.

Biswas S K (1987). Regional tectonic framework, structure and evolution of the western margin Basins of India Tectonophysics 135 307-327.

Biswas S K (1989). Hydrocarbon exploration in western offshore Basins of India Geol Surv India Spl Pub 24 184-194.

Biswas S K (1998). Overview of sedimentary Basins of India and their hydrocarbon potential In: R N Tiwari (ed) Recent Basin analysis and petroleum exploration with some examples from Indian sedimentary basins: Geol. Soc. India Jour., v. 20, p. 42-60.

Biswas S K (2008a). Petroliferous Basins of India - Fifty years' history and perspective Mem Geol Soc India 66 159-202.

Biswas S K (2008b). Geodynamics of Indian plate and evolution of Mesozoic-CenozoicBasins Mem Geol Soc India 74 247-260.

Biswas S K (2012). Status of Petroleum Exploration in India –Proc Indian natn Sci Acad 78 September 2012 475-494.

Biswas S K and Singh N K (1988). Western continental margin of India and hydrocarbon prospects of deep sea Basin Proc 7th Offshore SE-Asia Conf Singapore OTSC 88119, p 170-181.

Biswas S K, Rangaraju M K, Thomas J and Bhattacharya S K (1994). Cambay-Hazad (!) petroleum system in south Bombay High Field, India. Journal of Petroleum Geology, 7, 1, pp. 27-46.

Biswas S K., and SV Deshpande (1983). Geology and Hydrocarbon Prospects of Kutch Saurashtra and Narmada Basins - Petroleum asia Journal SPEB-I Vol.I,pp-111-126.

Biswas S.K. (1991). Exploration in Bombay Offshore Basin-An overview: Proceedings IERAP, Conference, ONGC, Dehradun, pp. 15-26.

Biswas SK. (2005). A review of structure and tectonics of Kutch basin, western India, with special reference to Earthquakes' (Current Science, 2005).

Biswas, B. (1961). Geology of Bengal Basin with special reference to stratigraphy and Micropaleontology unpublished report of ISPP.

Biswas, S. K. (1980). Structure of Kutch–Kathiawar region, Western India. Proceedings of the 3rd Indian Geological Congress, Pune,pp. 255–272.

Biswas, S. K. (1999). A review on the evolution of the rift basins in India during Gondwana with special reference to Western Indian basins and their hydrocarbon prospects. Publ. INSA, 65, 261–283.

Biswas, S. K. (2002). Structure and tectonics of the Kutch Basin, Western India, with special reference to Earthquakes. 8th Foundation Lecture, Indian Geological Congress, Roorkee.

Biswas, S. K. (2003). Regional tectonic framework of the Pranhita– Godavari basin, India. J. Asian Earth Sci., 21, 543–551.

Biswas, S. K. and Khattri, K. N. (2002). A geological study of earthquakes in Kutch, Gujarat, India. J. Geol. Soc. India,60, 131–142.

Biswas, S. K., Bhasin, A. L. and Ram, J.(1994). Classification of sedimentary basins of India in the framework of plate tectonics. Proceedings of the Second Symposium on Petroliferous basins of India, KDMIPE, Dehradun, vol. 1, pp. 1–42.

Biswas, S.K. (1965). A New Classification of theTertiary Rocks of Kutch, W. India; Bull. Geol. Min. MesOc. India, vol 35.

Biswas, S.K. (1982). Rift basins in western margin of India and their hydrocarbon prospects with special reference to Kutch Basin.Amer. Assoc. Petrol. Geologists. Bull. (AAPG), 66, 1497-1513.

Biswas, S.K. (1982). Rift basins in western margin of India and their hydrocarbon prospects with special reference to Kutch Basin: AAPG Bulletin, v.66, no.10, p.1497-1513.

Biswas, S.K. (1987). Regional tectonic framework, structure and evolution of the western marginal basins of India. Tectonophys., 135, 302-327.

Biswas, S.K. (1998). Overview of sedimentary basins of India and their hydrocarbon resource potential, Proc. Nat. Symp: Recent researches in sedimentary basins, 1-25.

Biswas, S.K. and Srivastawa, B.P. (1977). Exploration in Tripura. 5th Oil and Gas Conference, Dehra Dun.

Biswas, S.K., M.K. Rangaraju, J. Thomas, and S.K. Bhattacharya, (1994). Cambay-Hazad (!) Petroleum System in the South Cambay Basin, India, in L.B. Magoon and W.G. Dow, (eds.), Petroleum System-from Source to Trap: AAPG Memoir 60, p. 615- 624.

Biswas,A.K. (1992). Abnormal high pressure in Adamtila-F, Cachar area. *Bulletin* ONGC. Vol.29.no. 1 pp

Biswas,S K and Abinash Agrawal, (1990). Tectonic evolution of the Bengal Foreland Basin since the early Pliocene and its implications on the development of the Bengal Fan –Geological Survey ind. spl. Pub No.29 pp-5-19.

Biswas,S.K, and Abinash Agrawal (1990). Tectonic Evolution of the Bengal Foreland Basin since the Early Pliocene and its Implications on the development of the Bengal Fan.,Geo.surv. ind. Spl pu.no.29. p 5-19.

Bizanti, M.S. and Blick, E.F. (1984). "Proper Nozzle Configuration Helps Clean Around the Bit", World oil, December 100.

Blair, P.M. and Weinaug, C.F. (1969). Solution of Two Phase Flow Problems Using Implicit Difference Equations. Soc.Pet.Eng.J., December: 417-424. Trans. AIME.

Blandford, H.F. (1965). On the Cretaceous and other rocks of the South Arcot and Trichinopoly District; Mem. Geol. Surv. India; Vol. 4.

Bloch, S. (1991). Empirical prediction of porosity and permeability in sandstones: AAPG Bulletin v.75 No,7.p1145-1160.

Bose, P.K., Sarkar, S. (1991). Basinal autoclastic mass-flow regime in the ChandaLimestone Formation, Adilabad, India. Sedimentary Geology 73,299-315.

Bose, V. Singh, A.K. Tandon, B.S. Josyulu and S.K. Gupta (2005). Thin Reservoirs Delineation Through an Integrated Approach: a case study.–, PetroTech.

Bourdet, D.; Ayoub, J.A. and Prirard, Y.M.(1984). "Use of Pressure Derivative in Well Test interpretation", SPE- 12777.

Bourgoyne, A.T., Millheim, K.K., Chenevert, M.E. and Young, F.S. (1991). Applied Drilling Engineering.

Bradley, J.S. (1975). Abnormal Formation Pressure. AAPG Bulletin, Vol. 59, No. 6, June: 957–973.

Braj nandan-(1971)-A Simple Formula for Finding Average Reservoir Pressure in a bounded circle reservoir having a fracture job., Bull.ONGC.,Vol. 8, No. 1.

Brajnandan (1980). A 2-D Mathematical Model for Gas Reservoirs., Bull.ONGC., Vol. 17, No. 2, December.

Brookefield M.E. and R. Kumar (1985). "Trace fossils from upper Simla Group, Himachal Pradesh, India", Bull. Ind. Geol. Assoc., v. 18(1), 25-27.

Brown, A. R., 2011. Interpretation of three dimensional seismic data, 7th. edition, AAPG Memoir 42 (9), 646p.

Brunnschweiler R.D. (1966). On the geology of the Indo-Burma Ranges, J.Geol. Soc., Australia, v.13 (1) pp. 137-194.

Brunnschweiler, R.D. (1970). Geology of Andaman Basin and Burma, Bulletin Geological Society of America Vol. 81, pp. 1847-49.

Bryant S,. Cade C.and D. Mellor (1993). Permeability prediction from geological models: AAPG Bulletin, v.77, no 8, p,1338-1350.

Bukry, D. and Bramlette, M N. (1970). Coccolith age determinations, Leg. 3, Deep-Sea Drilling Project, Initial reports, Vol. 3, text fig. 1, tables 1-3.

Burger, H. R. (1992). Exploration Geophysics of the Shallow Subsurface,- Prentice Hall P T R.

Burke, K. and J. F. Dewy, (1973). Plume generated triple junctions: Key indicators in applying plate tectonics to old rocks, J. Geol., 8, 406–433.

Burns, R.E. (1964). Sea Bottom heat flow measurements in Andaman Sea Jour. Geoph. Res., Vol. 69, pp. 4918-4919.

Burton, C.K. and Bigell, J.D. (1969). Cretaceous-Tertiary events in S.E. Asia. Geol. Soc. America, Bull. Vol. 80. Pp. 681-687.

Calvat, T.J., Rau, R.N., Wells, L.E. (1977). "Electromagnetic Propagation- A New Dimension in Logging,"paper SPE6542 presented at the SPE California Regional Meeting.

Campbell, I. H. (2005). Large Igneous provinces and the mantle Plume Hypothesis Elements: v.1, p. 265–269.

Campbell, R.L., Jr (1968). "Stratigraphic Applications of Dipmeter Data in Mid-Continent," Bull., AAPG Vol. 52, No. 9.

Carcagno, G. (2005). The Design of Tubing and Casing Premium Connections for HPHT Wells. SPE-97584, presented at 2005 SPE High Pressure High Temperature Sour Well Design Applied Technology Workshop, The Woodlands, Texas, USA, 17/19.

Carlos Silva, David Spooner, and Corpus Christi. (1991). " High Resolution Induction Logging – A comparison with conventional Induction as used in thin sands in the Texas Gulf Coast region". SPWLA 91-WW.

Carr, B.S., Matcaffe, S.L. and N.G. Saleri (1989). Estimating in-situ permeabilities for reservoir simulation using multi-variable transforms: two case studies: SPE 1843.p.351-403.

Casshyap S.M. and Srivastava, V.K, (1988). Glacial and Pro-glacial sedimentation in Son-Mahanadi Gondwana basin, Paleo-geographic re-construction, American Geophysical union, Geophysical Monograph,v 41,pp.167-182.

Casshyap, S.M. and Tewari, RC. (1984). Fluvial models of the Lower Permian Coal measures of Son-Mahanadi and Koel-Damodhar valley basins, India. Special publications of the International Association of sedimentology V.7, P.121-147.

Casshyap, SM (1977). Patterns of sedimentation in Basins. In 4th International Gondwana Symposium Calcutta, India, 2, 525-551.

Catuneanu, Octavian, (2002). Sequence stratigraphy of Clastic systems: concepts, merits, and pitfalls: Journal of African Earth Sciences, 35,p 1-43.

Chakrabarthi, S. K. and Chakrabarthi, R. K. (1994). Nature of the Precambrian crustal assembly and evolution of the Godavari valley Gondwana basin of Peninsular India. In Gondwana Nine, Ninth Int. Gondwana Symp., Hyderabad, vol. 2, pp. 599–612.

Chakraborty A. (1972a). On the rock stratigraphy, sedimentation and tectonics of sedimentary belt in the southeast of Shillong plateau, Meghalaya, Bull. ONGC, v.(9) No. 2, pp. 133-141.

Chakraborty, A. (1972b). Stratigraphy and structure of Garo-Western Khasi Hills. ONGC, Bull. v.9.

Chakraborty, T. (1994). Stratigraphy of the late Proterozoic Sullavai Group, Pranhita-Godavari Valley, Andhra Pradesh. Indian Journal of Geology 66,124-147.

Chandan Chakraborty and Sanjoy Kumar Ghosh (2005). Pull-apart origin of the Satpura Gondwana basin,central India J. Earth Syst. Sci. 114, No. 3, June, pp. 259-273.

Chandan Chakraborty and Sanjoy Kumar Ghosh. Pull-apart origin of the Satpura Gondwana basin,central India-Geological Studies Unit, Indian Statistical Institute, 203, B.T. Road, Kolkata 700 108, India.

Chandra Kuldeep, et al. (1991). "Petroleum source rock potentials of the Cretaceous transgressive-regressive sedimentary sequences of the Cauvery basin". Journal of South East Asian Earth Sciences, Vol.5, no 1-4.

Chandra P.K. and L.R. Chowdhary (1969). Stratigraphy of Cambay Basin; ONGC Bull., Vol.6, No.2 pp 37-50.

Chandra Prabha Verma, J.K.Negi, Dev Singh, Aditya Kumar and U.S.Prasad (2006). New Technique for Modeling Movable Water Saturation Distributions: Application in LII Reservoir of Mumbai High., 6th International Conference and Exposition on Petroleum Geophysics "Kolkata 2006".

Chandra, P.K., and Guha, D.K. (1963). The Neogne rocks from Andaman, Science and Culture Vol. 29 No. 4, pp 202-203.

Chapman, RE (1983). Petroleum Geology. El-sevier Scientific Publishing Co., Amsterdam.

Chatterjee S.M., Sar, D., Boruah. S., Nabakumar. Kh and Sawai. Sanjeev (2006). "Exploration Leads from Gravity and Magnetic Data in Kerala-Konkan Basin, India". 6th SPG Conference Kolkata.

Chatterjee, C L., P.K. Tirkey, C. Prakash, S.S. Rana and M. Shanmukhappa–(2003) Identification of subsurface Deccan Basalt flows and their Hydrocarbon potential -a case History from well Padra-42, padra field, South Cambay basin, India– Proceedings of PETROTECH-2003 January 9-12, New Delhi.

Chatterjee, P.K. (1967). Geology of the Main islands of Andaman area, Proc. Symposium on Upper Mantle Project, National Geophysical Research Institute, Hyderabad, India, pp. 348-360.

Chatterjee, S and Ilotton, N (1986). The paleo-position of India. Jour. Southeast Asian Earth Sciences, 1(3), 145-189.

Chatterjee, S.M., C.S. Bahuguna, D. Saha, Kh. Nabakumar, D. Sar, and C.L. Baloni, (2010). Subduction Tectonics and its Gravity and Magnetic response in Andaman Sea, 8th SPG Kolkata 2010.

Chatterjee., H. S., M. Shukla and T. S. Rawat (2003). "Shali Basin, a Neoproterozoic back arc basin in Lesser Himalaya, Himachal Pradesh: a new paradigm for hydrocarbon exploration in Himalayan Fold Thrust Belt", In Press, 2003.

Chattopadhyay T. and D. K. Ghosh (2006). Structural Interpretation of Dipmeter Log – A Case Study from Baramura Field of Tripura.,6th International Conference and Exposition on Petroleum Geophysics

Chaudhuri A.K. (2003). Pranhita-Godavari Valley, India. Journal of Asian Earth Sciences 21 595-611 Geological Society London Special Publication 72,127-141.

Chaudhuri, A.K. (1977). Influence of eolian processes on Precambrian sandstones of the Godavari Valley, South India. Precambrian Research 4, 339-360.

Chaudhuri, A.K. (1985). Stratigraphy of the Purana Supergroup, Andhra Pradesh. Journal of Geological Society of India 26, 301-314.

Chawla. B.K., B.N. Sarma and U. Samantha. (1985). Regional Source Rock development in Bengal Basin. Bull. ONGC.vol.22. no.2 pp 145-158.

Chawla, B. R., B. N. Sharma and U. Samanta (1985). Regional source rock development in Bengal Basin, Bull. ONGC., Vol 22 No.2.

Chemerinski, Bill (1995). "Hydraulic Wellbore Erosion While Drilling" SPE Paper No.30497.

Chen, Q. and Sidney, S. (1997). Seismic attribute technology for reservoir forecasting and monitoring: The Leading Edge, 16, no. 05, 445,447–448,450,453–456.

Cherukupalli. P.K., R.B. Senapati, Nilesh Kumar and R.N.S.G. Rao (2003). Application of Log derived Saturation-height Function in Reservoir Characterization of Bassein field., Proc. PetroTech-2003.

Chetty, T.R.K. (1995). A Correlation of Proterozoic Shear Zones Between Eastern Ghats, India and Enderby Land, East Antarctica, Based on LANDSAT Imagery. In: Yoshida. M., Santosh. M. (Eds.), India and Antarctica during the Precambrian. In J. Geol. Soc. India. Bangalore, Memoir 34, 205-220.

Chilingar, G.V. (1964). " Relationship between porosity, permeability, and grain size distribution of sands and sandstones", in L.M.J.U. Van Straaten (ed.) – Deltaic and shallow marine deposits, Elsevier, Amsterdam.

Chiplonkar, G.W. (1985). Attempts at litho- and biostratigraphic subdivision of the upper Cretaceous rocks of South India a review. Quarterly Journal of the Geological, Mining and Metallurgical Society of India 57, 1-32.

Chiplonkar, G.W. and Borkar, V.D. (1971). A preliminary note on the Wadhwan Formation of Saurashtra, Curr. Sci., 40 (15); P. 433-434.

Chiplonkar, G.W. and Borkar, V.D. (1975). Stratigraphy of the area around

Cholet, H., Baron, G., Lazayers, R. and Brisac, J. (1977). "Unique Bit Design Improves Hydraulics and Performance", World Oil, October 63.

Choubey, V.D. (1997). Narmada Son lineament, India, Nature Phys. Sci, v. 232 (28) pp., 38-40.

Choudhari, R. S. (1992). Sedimentology of the middle Bengal Fan. Geol. Soc. India. Spl. Pub. No. 29, pp. 181-189.

Choudhury, A. (1977). Tectonic and paleogeomorphic evolution of Son-Mahanadi Gondwana basin belt of M.P., India, IV Internat. Gond. Symp., Calcutta, pp. 786-804.

Choudhury, S. K. and Datta, A. N, (1973). Bouguer Gravity and its Geologic Evaluation in Western part of the Bengal Basin and Adjoining area; India. Geophysics Vol. 58, p. 691-700.

Chowdhary, L.R. (1975). Reversal of Basement-Block Motions in Cambay Basin, India, and its Importance in Petroleum Exploration: AAPG Bulletin, v. 59/1, p. 85-96.

Clark, G. and Van Golf-Racht, T.D. (1984). "Pressure Derivative Approach to Transient Test Analysis: A High Permeability North Sea Reservoir Example, SPE-12959.

Clavier, C., G. Coates, and J. Dumanoir (1977). Theoretical and experimental bases for the Dual-Water Model for interpretation of shaly sands: Society of Petroleum Engineers Journal, v. 24, no. 2, p. 153-168, SPE-6859.

Coates, G.R. and J.L. Dumanior (1973). A new approach to log derived permeability: trans. SPWLA.

Coats, G.R. and Dumanoir, J.L. (1974). A New Approach to Improved Log Derived Permeability," The Log Analyst 15, No. 1.

Coats, K.H., Dempsey, J.R. and Henderson, J.H.(1971). The Use of Vertical Equilibrium in Two Dimensional Simulation of Three Dimensional Reservoir Performance. Soc. Pet.Eng.J., March: 63-71. Trans. AIME.

Cole, F.W. (1961). Reservoir Engineering Manual. Gulf Publishing Company, Houston, Texas: 200-213.

Condie, Kent. C. (1983). Plate tectonics and crustal evolution. Pergamon Press Inc. USA.

Copestake, P, M Slatford and S Vanstone (1997). Sequence stratigraphy to

Coulson, A. L. (1933). The Geology of the Sirohi State, Rajasthan: Memoir Geological Survey of India, v. 63, 166 p.

Cox, J.W. (1970). "The High-Resolution Dipmeter Reveals Dip-Related Borehole and Formation Characteristics,"T rans., SPWLA Annual Logging Symposium.

Crawford, A.R. (1978). Narmada - Son lineament of India traced into Madagascar Jour Geol Soc. India v. 19 (4) pp. 144-153.

Crookshank, H. (1936). Geology of the northern slopes of the Satpura between the Morand and Sher rivers. Mem. Geol. Sur. India, 66, pp. 173-381.

Cross, T.A. and Lessenger, M.A. (1988). Seismic stratigraphy. Ann. Rev. Earth Planet. Sci., 16: 319-354.

Cunningham,R.A. and Goins, W.C. (1957). "How Mud Parameters Affect Drilling Rate", Petroleum Engineer, May B-119.

Curray, J.R. and Moore D.G. (1971). Growth of the Bengal deep sea fan and denundation in the Himalayas. Bull. Geological Soc. Of America Vol. 82 pp 563-572.

Curray, J.R., Munasinghe, T. (1991). Origin of the Rajmahal Traps and the 85oE Ridge: Preliminary reconstructions of the trace of the Crozet hotspot, Geology. 19, 1237-1240.

Curry, J. R. and Moore, D. G. (1971). Growth of Bengal deep – sea fan and denudation in the Himalayas. Bull. Geol. Soc. Am. V. 82, pp. 563-572.

Curry, J.R. Moore, D.G., Lawyer, L Emme, F.J., Rain, It W., Henry. M. and Kiecl fer. R. (1979). Tectonics of the Andaman Sea Burma, in Joel S. Watkins, et al. (ed), Geolo and geophysical investigations of continental n gins: AAPG Memoir 29 p. 189-198.

D'Silva, K., D.M. Nathaniel, Bastia R., Verma R., Tenepalli, S., Sreedhar S.V. (2008). Southwestern Continental Margin of India: A Poly-Rift Basin; Society of Petroleum Geophysicists, Hyderabad, India.

Dake, L.P. (1977) Fundamentals of Reservoir Engineering.

Dana, J.D. (1973). On some results of the earths contraction from cooling including a discussion of theorigin of mountains and the nature of the earth's interior. Am. Jour. Sci. Vol. 5, Vol. 6.

Daniel A. Busch. Published by Stratigraphic Traps in Sandstones – Exploration Techniques- AAPG Memoir 21.

Das Gupta A.B. (1969). Some problems of petroleum exploration and developments in India; AMN Ghosh Memorial Lecture, ONGC Bulletin v. 6 no. 2.

Das Gupta A.B. (1976). Geology of Assam-Arakan Region Quaternary Jour. Geol. Min.Met. Soc. India.

Das Gupta A.B. (1986). Some problems related to the Geology of Bengal-Assam Basin: Quarterly Journal of the Geological, Mining and Metallurgical Society of India, v. 58 no., 3.

Das Gupta A.B. (1996). A Historical Perspective of Oil and Gas Exploration in India., ONGC Bullettin., Vol 33 No.2

Das Gupta A.B. (1997). Geology of the Bengal Basin: Quarterly Journal of the Geological Mining and Metallurgical Society of India. v. 58, no. 3.

Das Gupta, A.B. (1976). Growth of technology in the field of petroleum exploration in India: 19th Sir Thomas Holland Memorial Lecture 14.2.1976. Mining Geological and Metallurgical Institute of India.

Das Gupta, A.B. (1977). Geology of Assam Arakan region, Oil Commentary, V.14 (17) pp 4-35.

Das Gupta, A.B. (1977). Geology of Assam –Arakan region: Quarterly Journal of the Geological, Mining and Metallurgical Society of India, v. 49,nos. 1 and 2.

Das Gupta, A.B. (1994). Exploration strategy for petroleum: Journal of the Geological Society of India, v. 44. p. 100-110.

Das Gupta, S. K. (1975). Revision of the Mesozoic–Tertiary stratigraphy of the Jaisalmer Basin, Rajasthan: Ind. J. Earth Sci., v. 2, p.77–94.

Das S K and Nazar Abbas (1995). Structural Style and Hydrocarbon Prospects in the South Rewa Basin- Proc PetroTech 95 pp207-212.

Das Sharma, S, M K chakrabarty and SS Bisht (1993). Shallow Marine Area Bengal Basin –The site of an Ancient Fan - Proc Second Seminar Petroliferous Basins Indian Petroleum Publications Vol.2 pp-61-82.

Das, A. K. K. Dhandapani, P. B. Pandey (2006).,What Should I Correlate? Peak Or Trough!., 6th International Conference and Exposition on Petroleum Geophysics.

Das, A.K. R.P. Gupta and G.C. Sati. (2008). Seismic Stratigraphy and Hydrocarbon Prospects of Gondwana Sediments in Kavitam-Bhimavaram Area, Southeast of Tanuku Ridge, Krishna-Godavari Basin., ONGC Bull., Vol.45. no.1.

Das, A.K., Baruah, R.M., Bisht, S.S., Agrawal, B. (1999). An integrated analysis of late Proterozoic lower Vindhyan Sediments for Hydrocarbon exploration in western part of Son Valley, Central India, Jour. Geol. Soc. India, v.53, pp.239-253.

Das, Bhavan and Patel, N.P. (1984). Nature of Narmada - Son lineaments, India. Jour. Geol. Soc. India, v. 25 (5), pp. 267-276.

Das, K., and Singh, M. and Guha, B, Jan. (2001), "Optimised Drilling Fluid Formulation Through Computer Aided Hydraulic Modelling And Field Trial Thereof To Combat Hole Erosion Problem For Drilling Of Gollapalli and Mandapeta Sandstones In K.G. Project".

Das, K., Rastogi, R., Kundu, D., Singh, M. and Guha, B., Jan. (2001), "A New Generation Thermally Stable Drilling Fluid For HTHP Wells", Petrotech-, New Delhi, India.

Das, M.,Maharaj Singh, M.Muruganandan and Dr.D.V.R. Murti (2006). Multiple Attenuation - A Case Study., 6th SPG,Kolkata.

Das, N.C., Saxena, P.K., Singh, S.P., and Kumar, Ruby K. (1987). Influence of terrestrial source material on the composition of crude oils of Bombay offshore basin, India, in Kumar and others, eds.,Petroleum geochemistry and exploration in the Afro-Asian region: Rotterdam, Balkema, p. 393–399.

Das, S.R., K.S. Pangtey and M.D. Nasir (1995). Use of Foam in drilling sub-hydrostatic formation and its activation., ONGC Bullettin., vol 32 no.1.

Dasgupta U. and Rath S (1987). Seismo-Stratigraphic evaluation of Miocene depositional system in Mahanadi Basin (offshore),AEG Bulletin, 81-898.

Dasgupta, S., Pande, P., Ganguly, D., Iqbal Z., Sanyal, K., Venkatraman, N.V., Sural, B., Harendranath, L., Mazumdar, K., Sanyal, S., Roy, K., Das, L.K., Misra, P.S., Gupta, H. (2000). Seismo-tectonic Atlas of India and its Environs. -Geological Survey of India.

Dasgupta,S.,S. Volterrani, R. Bachrach, A. Carrillat, S. Sharma, N. Dutta, M. Pratap (2006). Seismic Reflection Survey at Himalayan Foot Hill Area, 6th International Conference and Exposition on Petroleum Geophysics "Kolkata".

Datta A K. (1983). Geology Evolution and Hydrocarbon Prospects of Rajasthan Basin- Petroleum Asia Journal SPEB-I Vol.I,pp-93-100.

Datta N K (1983). Geology Evolution and Hydrocarbon Prospects of Kashmir Valley- Petroleum Asia Journal SPEB-I Vol.I, pp-171-177.

Datta, A. K., Baneni. R.K., Bedi. T.S., Soodan, K.S., and Talwalker, P. M, (1965). Note on the Foraminiferal Biostratigraphy of the Subathu sediments in the Simla and Nahan-Dadahu areas. Bull. O.N G.C., vol 2, No. 1, pp. 21-26, pl. 111.

Datta, A.K. (1964). Faraminifera from the outcrop and subcrop samples of the Cambay Basin withspecial reference to the Cambay structure, India; Bull. ONGC., Vol. 1.

Datta, A.K. (1965). Biostratigraphic significance and Vertical Zonal distribution of foraminfera in theCambay structure; Bull. Oil and Natural Gas Comm., Vol. 2, No 2.

Datta, N.R., Mitra, N.D. and Bandopadhyay, S.K. (1983). Recent trends in the study of Gondwana basins of Peninsular India, Petrol, Asia Jour., v. 6(4), pp. 159-169.

Datta., A. K. and R. K. Banerji.,.S. Bedi,K.S. Soodan and P.M. Talwalker (1965). Palaeocene species of Dictyoconoides, lockhartia and Dictyokathina from the Subathu subgroup in Simla and Nahan-Dadahu areas., ONGC Bulletin Vol. 2, No. 1.

Dave Ridyard, Cara Kiger, Paul Brettwood and Jim Musser (2006). Data Acquisition Technology for Full Wave Imaging 6th International Conference and Exposition on Petroleum Geophysics.

Davison, I., Alsop, I., Evans, N., Safaricz, M. (2000). Overburden deformation patterns and mechanism of salt diapir penetration. Marine and Petroleum Geology, 17, 601-618.

De Sitter, I.U. (1956). Structural Geology, Mc Graw Hill Book Com., Inc., N.Y.

Deb G C., V. Pratap and S.C. Gawali (2003). Role of faults in the skewed development of Jotana field, North Cambay Basin, India suggestions for an offbeat approach Proceedings of PETROTECH-2003 January 9-12, 2003, New Delhi.

Deb S., Ramani KKV, and Mishra J. (1997). Oligocene sediments of Surat Depression: Depositional System and Petroleum Prospects.ONGC Bull.Vol.34,No.2.

Deb, G.C., V. Pratap and S.C. Gawali (2003). Role of faults in the skewed Development of Jotana field, North Cambay Basin, India -suggestions for an offbeat approach.–,Proceedings of PETROTECH-2003.

Debabrata Panda and S. Mahapatra (2003). Reservoir Management of Kharsang oilfield – problems and challenges.–,Proceedings of PETROTECH-2003.

Deepak Singh (2010). AVO Techniques: Advantages, Limitations and Future Prospects., 8th SPG Hyderabad.

Deepankar Borgohain, Krishna Kumar, U G Marathe, Pradipta Mishra, Deepak Kumar- (2015). Sub-basalt exploration in the Kutch-Saurashtra basin using EM Geo India 2015.

Deepti Tiwari, A.K. Basant, N.S. Tanwar and S. Roychaudhury (2003). Dynamic Reservoir Management of a complex reservoir – Jotana– Proceedings of PETROTECH-2003 January 9-12, 2003, New Delhi.

Deepti Tiwari, A.K. Basant, N.S. Tanwar and S. Roychaudhury (2003). Dynamic Reservoir Management of a Complex Reservoir – Jotana.,Proceedings of PETROTECH-2003.

Deshpande, S.V., et al. (1993). Lithostratigraphy of Indian petroliferous basins, Assam-Arakan basin, v. I, II document X: KDMIPE, ONGC.

Desikachar S. V. (1971). Review of the tectonic and geological history of Eastern India in terms of "Plate tectonic theory", Bull-Geol. Soc. Ind.,VoL 15 No.2. pp. 137-149.

Desikachar. S.V. (1984). A note on the exploration in the Vindhyan basin of India vis-à-vis new concepts in exploration of carbonates. ONGC Bulletin Vol.21 No.1.

Desikacharr., S. V. (1984). Exploration Plays in North Eastern Indian Hydrocarbon bearing Province. Petroliferous Basins of India., Petroleum Asia Journal April 1984 SPEB-1 Publication. Pp-51-59.

Despande, S.V., et al. (1979). Minerology, Petrology, Sedimentology studies of well DCS-2. (unpub. IPE report) Ednos Paul, 1983 Carborate Deposit-ional environments. (AAPG, Mem.33, PP.268-295).

Dev Dutt and C.L. Bhan –(1993). An Appraisal to Shale Reservoir Prospects-A New Discovery in Sanand-Jhalora Area, North Cambay Basin Vol. 30, No. 1, June.

Dewan, J. T. (1983), Essentials of modern open-hole log interpretation: PennWell Publishing, Tulsa, OK, 361 p.

Dey B K, A Maitahni and R P Agrwal (1995). Exploration Thrust in Vindhyan Basin,India Using Remote sensing studies- Proc PetroTech 95 pp-293-300.

Dey, A.K., R. Senapati, A.J. Duggal, Nilesh Kumar, R.N.S.G. Rao, B.S. Bisht, A. Mukhopadhyay, K.T. Kumar and B.L. Sharma (2005). Geocellular Model: an aid to Describe Reservoir Heterogeneity –a case study from Bechraji field, Cambay Basin, India.,Petro-Tech.

Dhandapani,K., S S Sawkar, A K Das, D S Randhawa, P B Pandey (2006). Degam Wedge Sediments- A New Petroleum Play- An Outcome of Sequence Stratigraphic Analysis of Narmada- Broach Block, South Cambay Basin, Gujarat, India –6th International Conference and Exposition on Petroleum Geophysics "Kolkata 2006".

Dhar P. C. and D. K. Pande, (1978), Rock stratigraphy of Bombay offshore basin and Paleogeographic and tectonic considerations, a discussion: 7th Colloquium Micropaleontology, Stratigraphy (Madras, India).

Dhar,P C., M.K. Mukhrjee, Mukesh Jain,Sanjay Chawla and Anil Sood (1987). Hydrocarbon prospectus of the western broach depression, Cambay Basin, Gujarat. ONGC Bull. Vol. 24 No. 2.

Dhruvedra Singh, D K Bharkatiya, CL Chatrajee, P K Minz V Nagsudha and C Prakash (1999). Implications of Stratigraphic analysis and Depositional System for Prospect Evaluation of Olpad Formation in Ahmedabad Block, Cambay Basin Gujarat- Proc 3rd Petro Tech, New Delhi pp-117-122.

Dickinsion, W.R, and Yarborongh, H. (1990). Plate tectonics and hydrocarbon accumulation AAPG continuing Education. Note series 1.

Dickinson,W. R. (1974).Plate tectonics and sedimentation. In: Tectonics and Sedimentation W.R. Dickinson,Ed.), Soc. Econ. Paleontologists and Mineralogists, Special Publication, V. 22, pp. 1-27.

Dikken, B.J. (1990). "Pressure drop in horizontal wells and its effect on production performance", JPT.

Dipankar Ghosh, Sahadev Hazra and Ram Avtar (2003). A Diagenetic Model for the Tight layer within Bassein formation in Neelam Field and its bearing on Reservoir Characterization and performance., Proceedings of PETROTECH-2003.

Dixy, F. (1970). The Geomorphology of Madhya Pradesh, India. Prof. West Comm. Vol., Univ. of Sagar, Sagar, pp. 195-224.

Dobrin, M. (1977), Seismic Exploration for Stratigraphic Traps, AAPG Memoir, P329–352.

Dobrin, M. B., and Savit, C. H. (1988). Introduction to Geophysical Prospecting. McGraw Hill Book Company, New York, 867p.

Doll, H.G (1949). Introduction to induction logging and application to logging of wells drilled with oil base mud,JPT,Vol.1, pp.148-162.

Dow, W.G. (1977). Kerogen studies and geological Interpretation. J. Geochem. Explor. 7 No. 2, pp. 79-99.

Doyen, P. (1988). Permeability, conductivity, and pore geometry of sandstone: Journal of Geophysical Research, v.93, p. 7729-7740.

Dumore, J.M. (1974). Drainage Capillary Pressure Functions and their Computation from One Another. Soc.Pet.Eng.J., October: 440.

Duncan Clarke (2007). Empires of Oil, corporate oil in Barbarian worlds, Profile Books, pp 1-334.

Dutta, A.K., and T.S, Bedi (1968). Faunal Aspect and the evolution of the Cauvery Basin. Mem.Of geol. Soc Ind No. 2 168-177.

Earlougher, R.C. Jr. (1977). "Advances in Well Test Analysis", Monograph Series, SPE.

Eckel, J.R. and Nolley, J.P. (1949). "An Analysis of Hydraulic Factors Affecting the Rate of Penetration", API Drilling Production Practices, 9.

Ecket, J. T. and Bielstein, W. J. (1951). "Nozzle Design and its Effect on Drilling Rate of Penetration", API Drilling and Production Practice, 28.

Edward J Lynch-(1976) Formation Evaluation –English book depot-pp422.

Embry, A.F. (2009). Practical Sequence Stratigraphy. Canadian Society of Petroleum Geologists, Online at www.cspg.org, 79 p.

Emmel, F.J. and J.R. Curray, (1985). Bengal fan Indian Ocean. Frontiers sedimentary Geology Submarine fans nd related turbidite stems – edited by Bourma, Normark and Barnes.

Endresen, E., and Aarrestad, T.V. (1995). "HT/HP Programme Seminar," Statoil, Jan. 17-18.

Enright, D.P., et al., March, (1991). "New Fluid System Substitutes For Oil Based Muds", World Oil pp 92-98.

Eremenko., N. A., A. K. Datta (1968). Tectonic Framework of the outer Himalayas., ONGC Bulletin Vol. IV, No. 2.

Eric Davidson and Susan Stewart (1997). "Open Hole Completion: Drilling Fluid Design" SPE- 39284, 1997.

Evans P.(1964) – The tectonic framework of Assam, Jour. Geol. Soc.India, v.5, pp. 80-96.

Evans, P. and W. Crompton, (1946). Geological factors in gravity interpretation

Ewing, M., Jardetzky, W., and Press, F. (1957). Elastic Waves in Layered Media,. McGraw-Hill, 380p.,New York.

Fassihi, M. (1986). "New Correlations for Calculation of Vertical Coverage and Areal Sweep Efficiency," SPERE, Nov., pp. 604–606.

Faust, L. Y. (1951). Seismic Velocity as a Function of Depth and Geologic Time, Geophysics, vol. 16,pp. 192–206.

Fertl, W Walter (1976). Abnormal formation pressures. Elsevier Scientific Publishing Co., Amsterdam.

Fisher, W. L. and L. F. Brown, (1983). Basin analysis and seismic stratigraphy: English Book Depot, Dehradun, Student edition.

Fitzgerald, T.A. (1980). Giant field discoveries 1968-78: An overview. In M.T. Halbouty (ed), Giant Oil and Gas Fields of the Decade 1968-76. Amer. Assoc. Pet. Geol. Mem., v. 30, Tulsa, Oklahoma.

Folk, R. L. (1962). Of skewnesses and sands (disc.): Jour. Sed. Petrology, v. 32, p. 145–146.

Fox, C.S. (1931). The Gondwana system and related formations. Memoirs of Geological Survey of India, 58.

Friedman, G. M. (1961). Distinction between dune, beach, and river sands from their textural characteristics: Jour. Sed.Petrology, v. 31, p. 514–529.

Fuloria R.C., R.N. Pandey, B.R. Bharali, J.K.Mishra, (1990). Stratigraphy,structure Petroleum system in Mahanadi basin (offshore).AEG V- III (2), pp 81-88

Fuloria, R. C., Pandey, R. N., Bharali, B. R. and Mishra, J. K. (1992). Stratigraphy, structure and tectonics of Mahanadi offshore basin. Geol. Surv. India Spl. Pub. No. 29, pp. 255-265.

Fuloria, R.C. (1993). Geology and hydrocarbon prospects of Mahanadi Basin, India. In: Biswas, S.K., et al. (Eds.), Proceedings of Second Seminar on Petroliferous Basins of India. In Indian Petroleum Publishers, Dehra Dun, India, Vol. 1, 355-369.

Fuloria, R.C., Pandey, R.N., Bharali, B.R., Mishra, J.K. (1992). Stratigraphy, Structure and Tectonics of Mahanadi offshore Basin. In: Recent Geoscientific studies in the Bay of Bengal and the Andaman sea. In J. Geol. Soc. India., Spl. pub. no. 29, 255-265.

Fuloria, RC (1993). Geology and Hydrocarbon Prospects of Mahanadi Basin, India- Proc Second Seminar Petroliferous Basins Indian Petroleum Publications Vol.2 pp-355-369.

Fuloria,R.C. (1999). Geology and hydrocarbon prospects of Mahanadi Basin India. Synthesis of laboratory investigations. Wells 1–8. Mahanadi Offshore. –Proceedings of the second seminar on Petroliferous Basins of India, December. KDMIPE.

Gadalla, M. R. and Fisher, R. (2009). Exploration Geophysics, Springer-Verlag, Berlin, 262p.

Galperin, E. I. (1973). Vertical seismic profiling, SEG,Tulsa, Oklahoma, USA.

Ganguly, S (1983). Geology and Hydrocarbon Prospects of Tripura –Cachar Mizoram Region, - Petroliferous Basins of India.,Petroleum Asia Journal April 1984 SPEB-1 Publication, pp105-109.

Ganguly, S. and Jagtap, P.K. (1966). A note on the structure of the Garhwal Himalayan Foothills, Uttar Pradesh; Science and Culture, Vol. 32.

Ganguly,S. (1997). Petroleum geology and Exploration History of the Bengal basin in India and Bangladesh. Ind. Jour. Geol.Vol.69 no.1 p 1-25.

Ganju J.L. Khar B.M. and Chaturvedi J.G. (1986). Geology and Hydrocarbon prospects of Naga hills based on integrated remotely sensed data, Petrol. Asia Jour., V.8, pp. 142-151.

Ganju, JL., and B M Khar (1984). Tectonics and Hydrocarbon Prospects of Kashmir Vally-Possible Exploratory Tragets- Petroleum Asia Journal SPEB-I Vol. II,pp-207-217.

Gansser, A. (1964). Geology of the Himalayas, London Interscience Pub., pp. 8-20, 235-256.

Gansser, A. (1966). The Indian Ocean and the Himalayas-a geological interpretation. Eclog, Geol. Helv. Vol. 59 pp 831-848.

Gardner, G. H., Gardner, L. W., and Gregory, A. R. (1974). Formation velocity and density-the diagnostic basics for stratigraphic traps, Geophysics, 39, 770–780.

Garg. H.S- (1982). Liners and Liner Cementation Techniques, ONGC Bullettin Vol. 19 no.1.

Gatlin, C. (1960). "Petroleum Engineering. Drilling and Well Completions", Prentice-Hall Inc., 129.

Gayatri Ghose, G.N.Rao and A.K.Balyan (2004). Hydrocarbon Exploration in Bengal Basin- A Seminar on Modern deltas of India, Andhra university.

Gayatri Ghose, P.K.Srivastava and D.Banerjee. (1999). Prospect Evaluation of Eocene Carbonate Sequence from 3D Seismic Studies in Golfgreen Area of Bengal Basin., Bull.ONGC. Vol 36.No.1 P 153-159.

Gazdag, J. (1978). Wave equation migration by phase shift, Geophysics, vol. 43, p.1342–1351.

Geol 99 433-465.

Gerald M Friedman and John E Sanders (1978). Principles of Sedimentology – John Wiely and Sons-pp-1-792.

Ghosh B.N.et. al; (2003). Reservoir Re-classification and Geological Re-Modeling of N.Kadi Field, North Cambay basin, India; 4th Conference and Exposition on Petroleum System, Mumbai.

Ghosh D.K. and J.P. Narayan (2006). Gas Identification in Shaly Sand Reservoirs Using Neutron Porosity vs. Sonic Travel Time Log Overlay Technique – Case Studies from Tripura Basin.,6th International Conference and Exposition on Petroleum Geophysics.

Ghosh, B.N., Soma D. Sarkar and J.P. Lohia (2006). Reservoir Characterization and Management of an Aquifer Driven Reservoir -A Case History.,6th International Conference and Exposition on Petroleum Geophysics "Kolkata".

Ghosh, P.K. and Bandopadhyay (1967). Paleogeography of India during the Lower Gondwana Times. IUGS, Symposium on Gondwana stratratigraphy, Buenos Aireas, pp. 523-536.

Ghosh, R.N. (1983). Tertiary clay deposits of Kolayat and adjacent areas in Bikaner district: Rajasthan. Indian Minerals, v. 37(4), p. 56-69.

Gill, T.S. Joseph Koithara and D.L.P. Raju (1973). Permeabality Estimation for Rudrasagar field from well logs., ONGC Bull.,Vol. 10, No. 2.

Goswami, B.G., R.S. Bisht, K.L. Pangtey, H.C. Pande, J.H. Ganjoo and S. Varshney. (2005). Geochemical Evaluation of the Paleogene and Neogene Sequences and Oil-Source Correlation in the Yanam Area, K.g. Basin, India--PetroTech 2005.

Goswami, P and Goswami, P.C, (2008). Tectonic Controls on Sedimentation in Upper Assam Foreland Assam-Arakan Basin, –Proceedings of 2008 Geo India Convention and Exhibition, New Delhi, India.

Goutam Kumar Nayak, Ch. Rama Rao, and H. V. Rambabu (2006). Aeromagnetic evidence for the arcuate shape of Mahanadi Delta, India National Geophysical Research Institute, Hyderabad-500007, India Earth Planets Space, 58, 1093–1098.

Goutam Kumar Nayak, Ch. Rama Rao, and H. V. Rambabu (2006). Aeromagnetic evidence for the arcuate shape of Mahanadi Delta, India, Earth Planets Space, 58, 1093–1098.

Goutam Kumar Nayak., Rama Rao, Ch. (2002). Structural Configuration of Mahanadi offshore Basin, India: An Aeromagnetic Study. Marine Geophysical Researches, 23, 471-479.

Govindan, A. (1984). Stratigraphy and sedimentation of East-Godavari sub-basin, Petroleum Asia Journal, 7(1), 132-146.

Govindan, A. (1993). Cretaceous anoxic events, sea level changes and microfauna in Cauvery Basin, India. In: Proceedings of Second Seminar on Petroliferous Basins of India, vol. 1. Indian Petroleum Publishers, 161-176 pp.

Govindan, A., Ravindran, C.N. and Rangaraju, M.K. (1996). Cretaceous stratigraphy and planktonic foraminiferal zonation of Cauvery basin, India, Mem. Geol. Soc. India, v.37, pp: 155-187.

Govindan.,A, S.Anantanarayanan and K.G.Vijayalakshmi (2000). Cretaceous Petroleum Systems in Cauvery Basin, India- Memoirs Geol Coc Ind No.46 pp- 365-382.

Gowda, S.S. (1964). The foraminifera of the South Indian Cretaceous-Eocene', Exlogae. Geologica Helvetica 57 (No. 1), 299-313.

Grant, F. S., and West, G. F. (1965), Interpretation Theory in Applied Geophysics. McGraw-Hill, 584 p.,New York.

Green, C. H. (1938), Velocity determination by means of reflection profiles. Geophysics, vol. 3, p: 295–305.

Griffin J.M., Hayatdanoudi A. (1986)."Design of Chemically Balanced Polymer Drilling Fluid Leads To Reduction In Clay Destabilization" SPE drill Engg.

Gudemane. S.M. (1989). Application of 'd' exponent in Predicting Overpressures and Demarcating pressure regimes in Krishna- Godavari Basin (a.p.), India- ONGC Bullettin, Vol. 26 No.1.

Gunter, G.W., et al. (1997). Early determination of Reservoir Flow units using an integrated Petrophysical Method, SPE-38679.

Gupta, D.K. et al. (1985). Regional Geological Model around structure S-2 and Hydrocarbon Prospects. Indent, Ind Jour Petro Geol Vol.1 pp 181-202.

Gupta, G D., DC Tewari and A Jaiswal (1993). Carbonate Facies Reservoir Petrography Porosity nad Production Trendes In Bassain Limestone of Ratnagiri Field Bombay Offshore Basin- Proc Second Seminar Petroliferous Basins -Indian Petroleum Publications Vol. 2 pp-421-433.

Gupta, N.C.,Yogesh Chandra, Debashish De and R.P. Verma.(2005). Effect of Surface Conductance on the low resistivity Ttipam sands of Assam – a Core study,Proceedings-article- 459.

Gupta, S.K., Sharma, S.K. and Ananthakrishna, K.(1998). "Stratigraphic and Subtle Trap Exploration in Krishna-Godavari Basin", Bulletin of the Oil and Natural Gas Corporation Ltd., V-35,No-2, PP 49-53.

Gupta. A.,and B.C. Jaiprakash. (1993). Analysis of Water Cut Behaviour in a Heavy Oil Field – A Case Study from Santhal Field, Cambay Basin., Bull.ONGC.,Vol. 30, No. 2.

Gupta.,S.K.(2006). Basin architecture and petroleum system of Krishna Godavari Basin, East Coast of India, THE LEADING EDGE pp-830-838.

Gupte SS, Kothari V, Bisen Madhu, Waraich RS (2006). "Structure and Evolution of Saurashtra Arc in Kutch-Saurashtra DeepWater Area, Western India" SPG-2006.

Hagedoorn, J. G. (1954). A process of seismic reflection interpretation. Geophysical Prospecting, vol. 2,p. 85–127.

Hagoort, J. (1974). Displacement Stability of Water Drives in Water Wet Connate Water Bearing Reservoirs. Soc. Pet. Eng.Journal., February: 63-74. Trans. AI ME.

Hajra,P.N., T.K Barua, S.Guha, M.Rudra, M.Purkait and M.Bhattacharya (1997). A Geochemical Assessment of Bengal Basin., Ind Jour. No.2 p 122-145.

Halbouty M.T. (1972). Rationale for deliberate pursuit, of stratigraphic, unconformity and Paleogeomorphic traps; Bull. AAPG, Vol. 56/3.

Halbouty, M.T. (2003) Giant oil and gas fields of the 1990s: an introduction; in Giant oil and gas fields of the decade 1990-99, AAPG memoir 78.

Haldorsen, H.H. (1986). Simulator parameter assignment and problem of scale in reservoir engineering, in L.W. Lake and H.B. Carrol Jr., eds., Reservoir Characterization: Academic Press, orlando, p, 293 340.

Hall, K.R. and Yarborough, L. (1974). How to Solve Equation of State for Z–Factors. The Oil and Gas Journal, Feb. 18th: 86–88.

Hamid N. Alsadi- Seismic Hydrocarbon Exploration 2D and 3D Techniques -Springer.

Hammer, S. and Anzoleaga, R, (1975). Exploring for stratigraphic Traps with Gravity Gradients: Geo.physics, Vol. 40, p. 256-268.

Handique G K and B Bharali (1984). Hydrocarbon accumulations in Mio-Pliocene Formations in Kharsang Oil field, Petroliferous Basins of India.,Petroleum Asia Journal April 1984 SPEB-1 Publication.

Handique G. K. and Bharali B. (1981): Temperature Distribution and its relation to hydrocarbon accumulation in Upper Assam Valley, India AAPG Vol. 65 No.9, pp. 1933-41.

Handique, G.K. and Bharali,B.(1984): Hydrocarbon Accumulations in Mio/Pliocene Formations in Kharsang Oilfield, India; Petroleum Asia Journal, April, pp 45-50.

Handique, G.K. and Mallick, R.K. (1989). Neogene Sediments of Upper Assam Intermontane Basin and their Hydrocarbon Potential; Proceedings of International Symposium on Intermontane Basins, Chiang Mai, Thailand.

Hantschel, T., and A.I. Kauerauf (2009). Fundamentals of Basin and Petroleum Systems Modeling: Springer-Verlag, Berlin, Germany, 476 p.

Haq, B., Hardenbol, J., and Vail, P, (1988). Mesozoic and Cenozoic chronostratigraphy and cycles of sea-level change, SEPM special publication no.42.

Haq, B.U. (1991). Sequence stratigraphy, sea-level change and significance for deep sea. Internat. Ass. Sediment. Spec. Publs., 12: 3-39.

Harbison, R. N., and B. G. Bassinger, (1970). Seismic reflection and magnetic study off Bombay, India: Geophysics, v. 35, no. 4, p. 603-612. 1973.

Hardage, B. A. (1983). Vertical Seismic Profiling. Handbook of Geophysical Exploration. Geophysical Press, London.

Hardas, M.G. (1980). Textural parameters and depositional processes of the sub-surface Serau Formation, Cambay Basin, Jr. Palaeont. Soc. India, 25: 106-109.

Harding T.P. (1973). Newport 0 Inglewood Trend, California - An Example of wrenching style of deformaltion, Amer. Assoc. Pet. Geol. Bull, v. 57 (11), pp. 97-116.

Harding, T.P. (1974). Petroleum traps associated with wrench faults. Bull. Amer. Assoc. Pet. Geol., v. 58(7), pp 1290-1304.

Harding, T.P. (1984). Graben hydrocarbon occurrences and structural. styles Amer. Assoc. pet. Geol. Bull, v. 68(3), pp. 333-362.

Harding, T.P. (1985). Seismic Characteristics and identification of negative flower structures, positive flower structures, and positive structural inversion. Amer. Assoc. Pet. Geol. Bull. v. 69 (4), PP. 582-600.

Harding, T.P. and J.D. Lowell (1979). Structural styles, their Plate Tectonic habitats and hydrocarbon traps in Petroleum Provinces: AAPG Bulletin, v. 63/7, p. 1016-1058.

Hargraves, R.B. and Bhalla, M.S. (1983). Precambrian Paleomagnetism in India through 1982: A Review, GSI Mem-4 (Ed., S.M. Naqvi and John J.W. Rogers) pp.491-524,GS1 Publication, Bangalore.

Hari Narain (1987). Geophysical constraints on the Evolution of Purana Basins of India with special reference to Cuddapah, Godavari and Vindhyan Basins. GSI, Mem-6 (Ed. B.P. Radhakrish-nan) pp. 5-32, GSI Publication, Bangalore.

Hari Narain. (1975). Bouguer Gravity anomaly and Regional Geology Map of India (NGRI/GPH-5). Published by Director NGRI, Hyderabad. IAGA Division I, Working Group 1, 1986. International Geomagnetic Reference Field Revision 1985, EOS Trans. AGU, 67. 523-524.

Harilal, S. K. Biswal, A. Sood, and V. Rangachari (2006). Identification of Reservoir Facies within Carbonate and Mixed Carbonate-Siliciclastic Sequence: Application of Seismic Stratigraphy, Seismic Attributes and 3D Visualization, 6th International Conference and Exposition on Petroleum Geophysics.

Harinarayana T., Nag Anjaneyulu K. and Patro B.P.K. (2006). Detection of a collision zone in south Indian shield region from magnetotelluric studies, Gondwana Research, Vol. 10, pp 48-56.

Harrold, D., Ringle, D., Taylor, M. (2004). Reliability by Design – HP/HT System Development. Presented at the Offshore Technology Conference, Houston, Texas, 3–6 May. OTC-16393-MS.

Harvir Singh, A.K. Sinha and Narendra Kumar (2005).Thermal Maturation Modeling in Basinal Depocenters of Ahmedabad and Broach Blocks of Cambay Basin, India-PetroTech 2005.

Hasan N., Varshney S., and Pangtey K. (2004). Source rock evaluation and geochemistry of Punjab, U.P., Bihar and Purnea sectors of Ganga Basin (Punjab plains-PART-II, well Janauri#A), KDMIPE, ONGC, Dehradun, internal report.

Haseldine, G.M. (1974). "A method of averaging capillary pressure curves", The Transactions of the Society of Professional Well Log Analysts, 15th Annual Logging Symposium, June 2-5.

Haskell, K.H. and Hanson, R.J. (1981). An algorithm for linear least squares with equality and non-negativity constraints, Math. Programming, VOL. 21, P. 98-118.

Helio Santos and J. Queiroz, Petrobras. " How Effective is Underbalanced Drilling at Preventing Formation Damage? (SPE/IADC 58739).

Heron, A.M. (1949). Synopsis of Purana formation of Hyderabad. Journal of Hyderabad Geological Survey 5 (2), 1-129.

Hood. A., Gutjahr, C.C. M. and Heacock, R.L. (1975). Organic metamorphism and the generation of petroleum. Am.Assoc. Pet. Geol. Bull. 59: 986-966.

Huchital, G.S., Hutin, R, Thoraval Tixier, M.P.: (1949) "Evaluation of Permeability From Electric-Log Resistivity Gradients," Oil and Gas Journal.

Hukku, B.N.(1972). "A note on the probable origin and structure of the Suketi Basin, Mandi district, Himachal Pradesh", Geol. Surv. India, Misc. Pub. no. 15, 163-166, 1.

Hunt, J. M. (1996). Petroleum Geochemistry and Geology, 2ndEdition. Freeman, New York.

Hurst, W.: (1953)"Establishment of the Skin Effect and its Impediment to Fluid Flow into a wellbore", Pet.Eng. (Oct-) 25, B-6.

Hyne, J.N., Laiding, L.W. and Cooper, W.A. (1979). Prodelta sedimentation on a lacustrine delta by clay mineral flocculation. J. Sed. Petro. 49, p. 1209-1216.

Iain Buchan and Andrew Long (2006). Exploring the Benefits of Streamer Overlap Shooting for 3D and 4D Seismic Surveys., 6th International Conference and Exposition on Petroleum Geophysics "Kolkata 2006".

Ibkyo. Kumar, R., and S. N. Bhattacharya (1977). Geochemistry of Bombay High crude: Int. Cong. on Organic Geochem., Moscow.

Idem (1977). Evolution of the Indian subcontinent in terms of the new global tectonic and its implications for exploration of Hydrocarbons, IV International Gondwana Symposium, published.

Idem (1983a). "Oil and Natural Gas resources of North East India" - Special Miscellaneous publication No. 30 Pt. IV of the Geological Survey of India.

Idem (1983b). Oil and Gas deposits in the Deltaic sequence of sediments in India. Geol. Soc. of India.

Idem. (1982). Facies distribution and the Petroleum Geology of the Bombay Offshore Basin, India; Journal or Petroleum Geology. Vol. 5, No. 1, pp. 57-75.

Inderjit Singh, M.S. Gulati and M.K Bhatta (1980). Some Relative Permeability studies on Assam cores., Bull.ONGC.Vol. 17, No. 2, December.

Ingersoll, R.V. (1988). Tectonics of sedimentary basins, Geol. Soc. Amer. Bull, v. 100, pp. 1704-1719.

JACKS, H.H., SMITH. O.J.E., and MATTAX, C.C.(1973). "The modeling of a Three-Dimensional Reservoir with a Two Dimensional Reservoir Simulator –The Use of Dynamic Pseudu Functions", SPEJ.

Jacob, K. (1964 b). Sedimentary environment of the formation of Andaman flysch, Andaman Islands, India, XXII International Geological Congress Vol. 11, pp 79-100.

Jacob, K. (1964a). A new probable into the tectonic history of the Andaman and Nicobar Islands. XXII International Geological Congress Abstracts P. 45.

Jacob, K. and Sastri, V.V. (1951). Tertiary Foraminifera from Sawai Bay, Car Nicobar Island. Science and culture, Vol. 17, No. 4, pp 181-182.

Jagannathan, C.R., Ratnam, C., Baishya, N.C. and Dasgupta, U.(1983). Geology of the offshore Mahanadi Basin: Petroleum Asia Journal, Vol. IV (4): pp. 101- 104.

Jain, M K.,A.K.Sarkar, B.K.Barve, U.K.Chatterjee, S.Panigrahi (2006). Advantage of Digital Sensors Over Analog Sensors in Enhancing Seismic Resolution – A Case Study from Cambay Basin., 6th International Conference and Exposition on Petroleum Geophysics "Kolkata".

Jain, S.K., S.C. Sachdeva and Y.B. Sinha-(1982). Reservoir Simulation studles for Ankleshwar sand stone., Bull.ONGC.Vol 19 no.1.

James Peters and P K Bhatnagar (1995). Geology and Hydrocarbon Potential and Prospects of Satpura Basin- Proc PetroTech 95 pp -213-222.

James Peters.(1998). Structural Framework and Tectonic Evolution of Satpura Basin: Madhya Pradesh., ONGC Bulletin Vol.21 No.1.

Jena A. K., R. Matta, A. J. Duggal, J. S. Bhosale and Ram Avtar (2006). Fault Seal Analysis - Its Bearing on Development Plan of Clastic Reservoirs, Kalol Field, Cambay Basin, India SPG 6th International Conference and Exposition on Petroleum Geophysics "Kolkata 2006".

Jenson, J.L. and L.W. Lake, (1985), optimization of regression based porosity permeability predictions: Trans, Canadian Well logging Soc, Formation Evaluation Symposium.

Jitendriya Mishra, Gajraj singh, Rama paul, B.k. Rath, and R. k. Phukan, (2011). Facies Analysis of Early Miocene Bombay Formation in Panna-Bassein-Heera area, Mumbai Offshore Basin-The 2nd South Asian Geoscience Conference and Exhibition, GEO India.

John F.Lindsay., Douglas W Holliday and Andrew G Hulbert (1991). sequence Stratigraphy and Evolutionof the Ganges Braahmaputra Delta Complex. AAPB Bull.v 75.no.7. p-1233-1254.

Johnson, R. and Evans, C.J. (1983). "Results ofRecent Electromagn and c Propagation Time Logging in the North Sea," Trans., SPWLA European Formation Evaluation Symposium.

Johnston, D. H. (1986). VSP Detection of Fracture-induced Velocity Anisotropy, Expanded abstracts, 56th Ann. Int.SEG Meeting, Houston, 464-466.

Joi, D. and Sharma, M.M.-(1993). Dynamic Filtration Of Invert-Emulsion Muds, "SPE Drilling And Completion", pp. 165-169.

Jokhan Ram and Venkatraman (1984). Tectonic framework and hydrocarbon prospects of Mizoram, Petrol. Asia Jour, v. 7(1), pp. 60-65.

Jokhan Ram, Shukla, S.N., Pramanik, A.G., Varma, B.K., Gyanesh Chandra, Murthy, M.S.N. (1996). Recent investigations in the Vindhyan Basin: Implications for the basin tectonics. In: Recent advances in Vindhyan Geology, A. Bhattacharya (Ed.), Mem. Geol. Soc. India, v.36, pp.267-286.

Jones,P ark.I. (1945). "Production Engineering and Reservoir Mechanics (Oil, Condensate and N atural Gas)," Oil and Gas J.

Jyotsna Rai, Ramkumar, M. (2007). Calcareous nanofossils from Ottakovil Formation (Late Maastrichtian), Cauvery Basin, South India. (Abs.) XXI Indian Colloquium on Micropaleontology and Stratigraphy, pp. 144e145.

Kaila K.L. Tewari H.C, and Mall D.M.(1987). Crustal Structure and delineation of Gondwana basin in the Mahanadi delta area, India from Deep Seismic Soundings. Jour. Geol Soc. India Vol.29, pp. 293-308.

Kaila, K. L., Murthy, P. R. K., Rao, V. K. and Venkateswarulu, N. (1990). Deep seismic sounding in the Godavari graben and Godavari (coastal) basin, India. Tectonophysics, Vol, 173, pp-307–317.

Kaila, K.L. (1986). Tectonic framework of Narmada -Son lineament - A continental rift system in central India from deep seismic soundings. Reflection Seismilogy. A Global Perspective Geodynamics Series, v. 13, pp. 130-150.

Kaila, K.L. (1989). Mapping the thickness of Deccan Trap flows in India from DSS studies and inferences about a hidden Mesozoic Basin in the Narmada-Tapti region, India.

Kailasam,L.N. and Simha, K R.M. (1963). A reflection seismic traverse across the coastal sedimentary belt of South Arcot District, Madras state. Bull. Nat. Geoph. Research Inst., 1, PP. 151-166.

Kak, H.L.,R. Bastia, Anand Sahu, R.K. Sharma and S. Malik (1994). Exploration for Subtle Trap with Paleogeographic Maps: Hazad Member, Gandhar – Jambuser Area ONGC Bulletin Volume 31 Number 2 (December, Special Issue).

Kale, A.S., Lotfalikani, A., Phansalkar, V.G. (2000). Calcareous nanofossils from the Uttatur group of Trichinopoly Cretaceous, South India. In: Govindan, A. (Ed.), Cretaceous stratigraphy -An Update. Memoirs of the Geological Society of India 46, 213e227.

Kamal K. Sharma K T. Magmatism and basin tectonism in western Rajasthan, India, results from extensional tectonics and not from Reunion plume activity- Government Postgraduate College, Sirohi (Rajasthan)-India.

Kamaraju A V V S, Man Mohan M, Rao M R R, Powal D S and Yalamarty S S (1998). Mapping of basalt as an effective guide for the exploration of cretaceous hydrocarbon prospects in the north of Matyapuri fault in Krishna–Godavari Basin; Bulletin of the ONGC. V.35 195–204.

Kamleshwar Rai, Lalaji Yadav, B.S. Haldiya (2006). Validation of Shaly Sand Model using Electrical Core Measurements in Low Resistive Reservoirs of Upper Assam, India.,6th International Conference and Exposition on Petroleum Geophysics "Kolkata.

Kandpal, M.C., V.N. Sharma, Y. Subrahmanyam, B. Das.(2006). Pre-stack Merging of Different 3D Seismic Data Sets in Gandhar, Jambusar and North Sarvan Fields of South Cambay Basin: A Case Study 6th International Conference and Exposition on Petroleum Geophysics.

Kanhaiya Kumar Jha, B.K. Saikia, Pawan Kumar and S. Chandrasekaran (2005). Pressure Transient studies a Dependable Analytical Tool for updating geological/reservoir model for effective Reservoir Management- a case history., PetroTech proceedings.

Kapoor P.N. (1997). Organic matter maturation, Interpretation by Fluorescence Microscopy and maturation Profiles. Proc. Pet. Tech 97 Vol.II pp. 141-150.

Karig, D.E., Suparka, S., Moore. G.F. Hehanussa, P. E. (1979). Structure and Ceno evolution of the Sunda arc in the Central Sun' region, in Joel S. Watkins, et al. (ed.), Geolo; and Geophyiscal investigations of continental margins: AAPG Mem 29 p. 223-238.

Karmalkar, N. R. and Sarma, P. K. (2003). Plume affected upper mantle beneath Kutch: Evidence from spinel lherzolite xenoliths and host alkali basalt. Deep Continental Studies in India, 13, 2–3.

Karunakaran, C., Pawde, M.B., Raina, B.K., Ray, K.K. and Saha S.S. (1964). Geology of South Andaman Island XXII. International Geological Congress, Vol. 11, pp 79-100.

Karunakaran, C., Ray, K.K., and Saha, S.S (1964b). Sedimentary environment of the formation of Andaman flysch, Andaman Islands.Ibid. Vol. 15 pp 226-232.

Karunakaran, C., Ray, K.K., and Saha, S.S. (1964 a). A new probable into the Tectonic History of Andaman and Nicobar Islands. Rep. 22 Session Inter. Geol. Congress Vol. 4, pp 507-515.

Karunakaran, C., Ray, K.K., Saha S.S. (1968 a). Tertiary sedimentation in Andaman Nicobar Islands, Jour. Geol. Soc. India, Vol. 9, pp 32-39.

Karunakaran, C., Ray, K.K., Sen, C.R., Saha, S.S. and Sankar, S.K. (1975). Geology of Great Nicobar Island, Jour. Geol. Soc. India, Vol. 16, No. 2, pp 135-142.

Katz, D. L. and Firoozabadi, A.(1978). "Predicting Phase Behavior of Condensate/ Crude-oil Systems Using Methane Interaction Coefficients," JPT, Nov., pp. 1649–1655.

Katz, M. B. (1978). Sri Lanka in Gondwanaland and the evolution of the Indian Ocean Geol. Mag. 115,237-244

Kaufman, H. (1953), Velocity functions in seismic prospecting. Geophysics, vol. 18, pp. 289–297.

Kaul S.K. and Mangla V.K. (1999). "Formulation of Reservoir Friendly Non-Dispersed Polymer Drilling Fluid Systems For Upper Assam", Petrotech.

Kaul., H. N.(1991), KD Malviya The Evolution of India's Oil policy., Allied Publishers Limited., pp 1-296.

Kaul.S K, Krishan Pal and N.P. Joshi (2003). Borehole instability analysis and remedial measures in Barail-coal-shale sections of wells in Khoraghat area –Proceedings of PETROTECH, New Delhi

Kay, M. (1947). Geosynclinal nomenclature and the Craton. Am. Ass. Pee. Geol. Bull V. 31.

Kayal, J. R., Reena De, Sagina Ram, Sriram, B. V. and Gaonkar, S. G.(2002). Aftershocks of 26th January Bhuj Earthquake in Western India and its seismotectonic implication. J. Geol. Soc. India,59, 395–418.

Kent, W.N., Hickman, R.G., and Dasgupta, U. (2002). Application of a ramp/flat fault model to interpretation of the Naga thrust and possible implications for petroleum exploration along the Naga thrust front, AAPG Bulletin, v.86, No.12, p.2023-2045.

Khanna S.N. (1976). Use of low solids-low weight mud improves Grilling Performance., ONGC Bullettin., Volume. 13 Nos.1 and 2.

Khar, B.M. and Ganju, J.L. (1984). Tectonics of Tripura fold belt probable mechanics of folding and faulting. Pet. Asia Jour., v. 7.

Khelil, C. (1983). "A Correlation of Optimum Free Gas Saturation," SPE Paper 01983.

King, W. (1881). Geology of the Pranhita-Godavari Valley. Memoir Geological Survey of India 18,151-311.

Kinji Magara (1978)/ Compaction and Fluid Migration. Elsvier scientific Publishing-pp1-319.

Klootwijk, C. T., Gee, J. S., Peirce, J.W., Smith, G.M., and Mcfadden, P.L. (1992). An early India- Asia contact: Palaeo-magnetic constraints from Ninety east ridge, ODP leg 121.Geology, v.20 p. 395-398.

Knebel GM and Rodriguez-Eraso G(1956). "Habitat of Some Oil," Bulletin of the AAPG 40, no. 4 547–561.

Koithara,J, K. H. Hashmy, H.S. Rawat and Y. M. Mehra (1967). Effect of Shaliness on Formation Factor and Saturation exponent for the Kalol productive horizons (Eocene).,ONGC Bulletin.,Vol. 4, No. 1.

Koithara. J., and S.K. Sharma.(1987). Note on the Resistivity and Radioactivity Properties of the shale separating the middle and upper pays in Ankleshwar Field., ONGC Bull.,Vol. 14 No. 1 and 2.

Kolekar A.S., Prasad M.S., Gupta S.K. (1977). Sobhasan Subsidiary Sands-Can they be elevated to the status of subtle traps from serendipity ?. Paper presented at Oil and Gas Conference.

Kolla, N., D. G. Moore and J. R. Curray (1976). Recent Bottom current activities in the deep western Bay of Bengal, Marine Geology 21 255-270.

Kolla, V. and Coumes F. (1987). Morphology, internal structure, Seismic stratigraphy, and sedimentation of Indus Fan. Amer. Asso. Pet. Geol. Bull, V. 71 (6), p. 650-667.

Kono, M. (1974). Gravity anomalies in east Nepal and their implications to the crustal structures of Himalayas, Geophys. J.R.Astron. Soc., v. 39, pp. 283-300.

Kontorovich A.Eh., et al. (1992). The earth's upper Proterozoic deposits - A new prospective level for petroleum exploration. In procc. of the 13th World Petroleum Congress pp. 67-69.

Krishna Sarma, A.,A.K.Gogoi, R.Acharjya (2006). Static Correction for Vibroseis Data over South Kharatar, Rajasthan, India-A Case History 6th International Conference and Exposition on Petroleum Geophysics "Kolkata".

Krishna, C.V.G., B.M.Sinha and K.K.Nath (2006). Improvement of Signal/Noise Ratio and Fault Imaging with Full 3D Prestack Kirchoff Time Migration: A Case Study from Upper Assam, 6th International Conference and Exposition on Petroleum Geophysics "Kolkata".

Krishna, K.S. (2003). Structure and evolution of the Afanasy Nikitin Seamount, buried hills and 85oE Ridge in the northern Indian Ocean. Earth and Planet. Sci. Lett. 209, 379-394.

Krishna, K.S., Michael, Laju., Bhattacharyya, R., Majumdar, T.J. (2009). Geoid and gravity anomaly data of conjugate regions of Bay of Bengal and Enderby Basin e new constraints on breakup and early spreading history between India and Antarctica. Journal of Geophysical Research 114 (B03102), 21.

Krishna, M.S. (1968). Geology of India and Burma. Higginbothams (P) Ltd. Madras.

Krishnan, MS and Swaminathan, J (1959). The great Vindhyan Basin of northern India. Jour. Geol. Soc. Ind., 18, 13.

Krishnan, P.V. and Rao, R.P. (1970). Tectonic framework of the Tertiary belt of the Sub-Himalayas of Jammu; Pub. Centre of Advanced Study in Geology. Punjab, Univ. Chandigarh No. 7.

Krishnan, P.V. and Singh, A. (1969). On the Jokan bituminous limestone of Punch; Bull. Oil. Nat. Gas Comm., Vol. 6, No. 1.

Krishnan,P.V. and G.N.Rao (1982). Why wells in Bengal Basin Dry,– ONGC unpublished report.

Kroner, A. (1981). Precambrian Plate tectonics. In 'Precambrian Pate Tectonics' (Ed. by A.Kroner) Amstredam: Elsevier.

Krumbein, W and G. Monk, (1942). Permeability as a function of the size parameters of unconsolidated sand: AMIE Petroleum Trans, V. 151, p. 153-163.

Krumbein, W C. and L L Sloss (1963). Stratigraphy and Sedimentaion-WH freeman publication., pp1-660.

Kumar N, Singh A.P, Singh B. (2009). Structural fabric of the Southern Indian Shield as defined by gravity trends, Journal of Asian Earth Sciences, Vol 34, 577-585.

Kumar, A., Pendkar, N., Sangeeta (2002). Delineation and Evaluation of Basaltic Deccan Basement Reservoir of Padra Field, Cambay Basin, India – A Field Study, SPWLA 43rd Symposium.

Kumar, Anil, Pande, K., Venkatesan, T.R. and Bhaskar Rao, Y.J. (2001). The Karnataka Late Cretaceous dykes as products of the marion hotspot at the Madagascar-India breakup event: evidence from 40Ar-39Ar geochronology and geochemistry. Geophys. Res. Lett., v.28, pp.2715-2718.

Kumar, S.P. (1983). Geology and hydrocarbon prospects of Krishna - Godavari and Cauvery basins, Petroleum Asia jour., 8, 1, 57 –65.

Kumar.R.,and M.E. Brookefield (1987). "Sedimentary environments of the Simla group (Upper Precambrian), Lesser Himalaya, and their palaeotectonic significance", Sed. Geol., v.52, 27-43.

Kundu, J, MR Wani, and R K Thakur (1993). Structural Styles in South Cambay Rift and its control on Post rift Deltaic Sedimentaion- Proc Second Seminar Petroliferous Basins Indian Petroleum Publications Vol.2 pp-81-96.

Kundu, J. and Wani, M.R. (1992). Structural style and Tectono-stratigraphic framework of Cambay rift basin, Western India: Indian Journal of Petroleum Geology, Vol. I, No.2, pp 181-202.

Kunte S.V. and Rao S.V. (1989). Tectonic model for hydrocarbon exploration in Naga foot hills, Bull, ONGC, v.26(1) pp. 1-16.

Kunte. S.V.(1985). Tripura-Cachar Fold Belt genesis of structural style. (Bulletin ONGC Vol. 22.no.2 pp.

Kutty, P.S.N., A.K. Tyagi, J.V.S.S.N Murty; Jill Lewis,(2010) Troika-International; Paul Maton, Energistics. Reducing pitfalls in Seismic Data Management 8th SPG, Hyderabad.

Kvale, E., Fjagesund, T., Self, J., and Bjorvik, T.L. (1994). "HPHT Well Testing Project: Design Verification Report: Downhole Test Tools and Permanent Packers," spring.

Lafond, E.G., and Prasada Rao, R. (1956). On the erosion of the beach at Uppada. The Port Engineer, 5, pp.4-9.

Lake L.W., et al. (1991). Reservoir characterization -Academic press.

Lakhera, J.P., R.K. Shukla and R. Vijayarangan.(2005), -D-1 capillary pressure and Trapping Mechanism, Reasons for Higher sor/lower we, GS-1 and 2, Gandhar Field, India., PetroTech.

Lakshminarayana, G.L. (2002). Evolution in basin-fill style during the Mesozoic Gondwana continental break-up in the Godavari triple junction, SE India. Gondwana Research 5, 227-244.

Lal, N.K., Siawal, A., Anil, K.Kaul, (2009). Evolution of East Coast of India e a plate tectonic reconstruction. Journal of the Geological Society of India 73, 249-260.

Lalaji Yadav, Narender Pal and R.K. Bhardwaj (2003). Evolution of a Generalized Porosity – Permeability Transform for Upper Assam oil Fields. Proceedings of PETROTECH

Larner, K. and Hatton, L. (1976), Wave equation migration. Two approaches. 8th Annual Offshore Technology Conference, Houston, Texas.

Lavergne, M. (1989), Seismic Methods, 172 p., Editions Technip, Paris.

Laxmidhar Behera, and Prakash Khare (2010), Improvement of stacking image by anisotropic velocity analysis using P-wave seismic data, SPG, Hyderabad.

Lee, J., Shadravan, A., and Young, S. (2012). Rheological Properties of Invert Emulsion Drilling Fluid UnderExtreme HPHT Conditions. Paper SPE-151413-MS presented at IADC/SPE Drilling Conference andExhibition, San Diego, California, USA.

Lee, S.T. (1989). "Capillary-gravity equilibria for hydrocarbon fluids in porous media, 64th Annual SPE Tech Conference", SPE 19650.

Levin, F. K. (1971). Apparent velocity from dipping interface reflections, Geophysics, vol. 36, p.510–516.

Liang C. Shen, University of Houston,(1988). " Effect of Skin-effect correction and three point Deconvolution on induction logs" SPWLA, June-1988,Vol –1.

Lillie, R.J., Johnson, G.D., Yousuf, M., and Zamin, A.S.H and Yeats, R.S. (1987). Structural development within the Himalayan foreland fold and thrust belt of Pakistan, Can. Soc. Pet. Geol., Mem. 12.

Lindsay John F, Duglas W. Aollioby and Andrew G. Hubert (1991). Sequence Stratigraphy and Evolution of the Ganges and Brahmputra Delta complex Bull AAPG 75, 7 1223-1254.

Lubiniski, A. (1953). Factors affecting the angle inclination and dog-legging in rotary bore-holes; O.G.J.,Vol. 3, No. 14.

Lucias, F.J. and G.E. Fogg (1990). Geologic/stochasic mapping heterogeneity in a carbonate reservoir: Journal of Petroleum Technology, V. 42. P. 1298-1303.

Lunne, T. Berre, S. Strandvik, K.H. Andersen and T.I. Tjelta (2001). "Deepwater sample disturbance due to stress relief". Proceedings: OTRC Conference. International Conference on Geotechnical, Geological and Geophysical Properties of deepwater Sediments. College Station, Texas 2001, pp. 64-85.

Lunne. T.(2001). "In situ testing in offshore geotechnical investigation". Proceedings: International Conference on In Situ Measurement of Soil Properties and Case Histories, Bali, pp. 61-82.

Lydekker, R. (1883). The Geology of the Kashmir and Chamba territories and the British territory of Khagan, Mem. Geol. Surv. India, Vol. 22.

Lynch, E.J. (1964). Formation Evaluation. -Harper and Row, New York.

Lyon-Caen, H., and Molnar, P. (1983). Constraints of the structure of Himalaya from gravity anomalies and a flexural model of the lithosphere, J. Geophys. Res., 88, 8171-8191.

M. Julleh Jalalur Rahman and Shigeyuki Suzuki (2007). Geochemistry of sandstones from the Miocene Surma Group, Bengal Basin, Bangladesh: Implications for Provenance, tectonic setting and weathering. Geochemical Journal, Vol. 41, pp. 415 to 428.

Ma, S, Jiang, M.X., Morrow, N. R. (1991)" Correlation of capillary pressure relationships and calculation of permeability", 66th Annual SPE Tech Conference, SPE 22685.

Macdonald D.I.M. and P.W.G. Tanner (1983). "Sediment dispersal patterns in part of a deformed back-arc basin on South Georgia, South Atlantic". Jour. Sed. Petrol., v.53(1), 83-104.

MacDonald, R.C. and Coats, K.H. (1970). Methods for Numerical Simulation of Water and Gas Coning. Soc.Pet.Eng.J., December: 425-436. Trans. AIME.

Mackenzie, A.S. (1984). Application of biological markers in petroleum geochemistry, In Advances in Petroleum Geochemistry, V. 1, (J. Brooks and D.H.Welte, eds) Academic Press, London, p. 115-214.

Madan Mohan (1983). Geo-history Analysis of Bombay High Region –ONGC Bull Vol.20 no.2

Madan Mohan (1995). Cambay Basin –A promise for oil and Gas potential Jour. of the Paleontological society of India Vol 40, pp-41-47.

Magara, K. (1978). Compaction and fluid Migration,– Elsevier Publishing Company.

Magoon LB and Dow WG(1994) "The Petroleum System," in Magoon LB and Dow WG (eds): The Petroleum System–From Source to Trap, AAPG Memoir 60. Tulsa: AAPG: 3–24.

Magoon, L.B. (1995), The play that complements the petroleum system–a new exploration equation: Oil and Gas Journal, vol. 93, no. 40, p. 85–87.

Mahapatra, G.P., Hari Prasad, M and Anil Kumar, A (2004 and 2005). Morphology of the continental margin of India off the area between Puri and Point Calimere in the East Coast of India. Marine wing, News letter, Geol.Surv.India, V.XVII (2) and XIX (1), pp.63-64.

Mahendra Pratap. (2010). Is Mumbai High a Naturally Fractured Carbonate Reservoir? SPG.

Mahesh Chandra, et al., (1993). Lithostratigraphy of West Bengal Basin. Unpublished ONGC Document.

Maity, H.S. Dr.S.S.Prasad and S.S.Bakshi (2006). 'NewerUnderstanding the Anomalous Behaviour of Dual Induction Log Against Highly Conductive Thin Bed – A Case Study 6th International Conference and Exposition on Petroleum Geophysics.

Maji, A K, D Z Badwaik and S Chandrasekaran (2013). Transformation of a Marginal Field into a Major one - the Challenges and Success: A case study from Western Offshore Basin', India SPG 10th Biennial International Conference and Exposition Kochin.

Majumdar, S.,H. D. Dave, J. K. Samal, D. S. Mitra. (2011). Delineation of Basement related Fault Closures in Eastern part of Purnea basin based on Morphotectonic analysis- 2nd South Asain Geoscience Conference and Exhibition, GEO India.

Majumdar,S, H. D. Dave, J. K. Samal, D. S. Mitra (2011). Delineation of Basement related Fault Closures in Eastern part of Purnea basin based on Morphotectonic analysis., 2nd South Asain Geoscience Conference and Exhibition, GEO India 2011.

Majumdar. S.K., and Saibal Basu (1985). Compaction trend pressure analysis and their significance with respect to Miocene sequence of West Bengal– ONGC Bull. ONG

Malay Rudra, P.N. Hajra and A.B.S. Mahapatra.(2005). Surface Prospecting in Bengal Basin - A Review-Petro Tech. 2005 C Vol 19 No.2.

Maker,P S, K L Mehrotra, Dayawant, Shree Prakash, P K Gupta and K N Mishra (1984). Geochemical studies of Fine grained sediments of offshore Andaman wells for Evaluation of Hydrocarbon Potential _Petroleum Asia Journal pp-175-185.

Maldonado, B., Arrazola, A., and Morton, B. (2006). Ultra-deep HP/HT Completions: Classification, DesignMethodologies, and Technical Challenges. Paper OTC-17927-MS presented at Offshore Technology Conference,Houston, Texas.

Malhotra, N.C., Berry, C.M., Nautiyal, D.D., and Rawat, R.S. (1995). Palynostratigraphic and source rock evaluation studies on Oligocene-Miocene subsurface sediments of Surat Depression, in Proceedings of the first international petroleum conference and exhibition: New Delhi, PETROTECH, v. 2, p. 43–52.

Malik, J. N., Sohoni, P. S., Merh, S. S. and Karanth, R. V (2001). Active tectonic control on alluvial fan architecture along Kachchh mainland hill range, Western India. Z. Geomorphol. N.F., 2001, 45, 81–100.

Malod J. A., Droz L., Kemal Mustafa B., Patrait P., June (1997). Early spreading and continental to oceanic basement transition beneath the Indus deep-sea fan: northeastern Arabian Sea. Marine Geology 141(1997) pp 221-235.

Manas K Chakrabarty, (1992). On the Surmas of Tripura and Cachar- Bull. ONGC Vol.29 no.1 pp-39-60.

Mandal, P., et al. (2004). Characterization of the causative fault system for 2001 Bhuj Earthquake of Mw 7.7. Tectonophysics, 2004, 378, 105–121.

Mandal,K L., K. Dasgupta, Dr. R. Dasgupta and A.K.Khanna (2010). Multicomponent Seismic (3C) Data Acquisition– A case study from Oil India Ltd. operational area in Upper Assam Basin., SPG 8th Hyderabad.

Manke, K.R., Harkins, G.O. (1991). "Drill Stem Testing Techniques," OILGAS European Magazine, March/April.

Manmohan, M., Rao, M. R. R., Kumar Raju, A.V.V.S. and Yalamarthy, S.S. (2003). Origin and occurrence of Lower Cretaceous High gamma – High Resistivity Shale - A key to stratigraphic sequence for hydrocarbon exploration in Krishna – Godavari basin, A.P., Jour. Geol. Soc. of India, 62, 271 - 289.

Manoharan,P., Pulak Banerjee (2006). Integrating Field Processing Unit (FPU) with Seismic Data Acquisition System (SN388) for Facilitating Real Time QC of Field Data.,6th SPG, Kolkata.

Manoj K. Mukhejee, (1997). ''Petroleum prospects of Createceous Sediments of the Cambay Basin, Gujarat,India'' Journal of Petroleum Geology, Volume 5, Issue 3.

Maria F. Quijada and Robert R. Stewart (2008). Well log and synthetic seismogram analysis of an oilfield in Assam, India: A 3-C seismic development feasibility study- -CREWES Research Report – Volume 20.

Marland P. Billings (1979). Structural Geology= Prentice and Hall of India-pp-1-606.

Maroof Khan M.A. and Musherraff Husain (1980). A look at the Geology of Bangladesh gas fields, O and GJ Aug.4 pp 92-95.

Marussi, A (1964). Geophysics of the Karakorum, Italian expedition to karakorum (k20 and Hindu Kush, Sci. Rep. 2, 1, pp. 242.

Mathews, C. S., Brons, F and Hazebroek, P. (1954). " A Method for Determination of Average Pressure in a Bounded Reservoir", Trans., AIME 201, 182.-91.

Mathur L.P. and Evans P. (1964). Oil in India,– 2nd Internatl. Geol. Cong., New Delhi, India, 85 p.

Mathur, A D., Sanjay and Soma Chatterjee.(2006),Two Dimensional Display of Well Logs of High Deviation Wells., 6th International Conference and Exposition on Petroleum Geophysics.

Mathur, L. P., K. L. N. RAO and A. N. Chaube (1968). Tectonic Framework of Cambay Basin, India., Bulletin of the Oil and Natural Gas Commission., Vol. V, No. 1.

Mathur, L.P., Rao., K.L.N. and Chaube, A.N, (1968). Tectonic framework of Cambay Basin; Bull. Oil and Natural Gas Comm., Vol. 5, No. 1.

Mathur, P., et al. (1987). Geochemical studies of the Precambrian sediments of Vindhyan Basin, India. In Petroleum Geochemistry and Exploration in the Afro-Asian Region, (Ed. Ruby Kumar et al. Balkema, Rotterdam.

Mathur, R.B. and K.M. Nair (1993). Exploration of Bombay Offshore Basin, in S.K. Biswas, et al., eds., Proceedings Second Seminar Petroliferous Basins of India: Indian Petroleum Publishers, Dehradun, v. 2., p 365-396.

Mathur, S.C., Gour, S.D., Loyal, R.S., Tripathi, A., Tripathi, R.P. and A. Gupta (2005). Occurrence of magnetic spherules in the Maastrichtian bone bed of the Fatehgarh Formation, Barmer Basin, India: Current Science, V. 89, p, 1259-1268.

Mathur,L P, KLN Rao and AN Chube (1980). Tectonic Framework of Cambay Basin-India, Bull. ONGC Vol 17 No.3., pp203-222.

Mathur,R B and K M Nair,(1993). Exploration of Bombay offshore Nasin-Proc Second Seminar Petroliferous Basins Indian Petroleum Publications Vol.2 pp-365-398.

Mattey, S.K., M.S. Gauma and K.A. Shah (2003). An Integrated exploitation strategy of South Kadi field, Cambay Basin, India – A case study– Proceedings of Petrotech-2003 January 9-12, 2003, New Delhi.

Matthews, C.S., and Russell, D.G. (1967). "Pressure Buildup and Flow Tests in Wells," Society of Petroleum Engineers of the American Institute of Mechanical Engineers, Vol. 1.

Mayer, C. and Sibbit, A. (1980). GLOBAL, a new approach to computer processed log interpretation. SPE of AIME, ann. Fall Mtg., Dallas, SPE 9341.

Mayne, W. H. (1967). Practical considerations in the use of common reflection point technique, Geophysics, vol.32, p.225–229.

Mazumder,S K.,K.S. Pangtey, D. S. Mitra (2013). Delineation of a Possible Subsurface Ridge in Onshore Palar Basin based on Morphotectonic Studies and its Implications 10th SPB Conference.

Mc Kenzie, D and Sclater, J.G. (1971). The evolution of the Indian Ocean since late Cretaceous. Geophys. Jour. Vol. 25, pp 437-528.

McCain, W.D. Jr. (1973). The Properties of Petroleum Fluids. Petroleum Publishing Co., Tulsa.

McClay, K. R., Dooley, T. and Lewis. G. (1998). Analog modelling of progradational delta systems. Geology. V. 26, pp. 771-774.

McDougall, J.W., Hussain, A. and Yeats, R.S. (1993). The Main Boundary Thrust and propagation of deformation into the foreland fold-and-thrust belt in northern Pakistan near the Indus river, In. Treloars, P.J. and Searle, M.P. (Eds). Himalayan Tectonics, Geol. Soc. London, Sp. Pub. no. 74, pp. 581-588.

McIver, R.D. (1967). Composition of Kerogen due to its role in origin of Petroleum, Proc. 8th World Petroleum Congress, Mexico, Vol.II, P. 25-26. Merh, 1995: Geology of Gujarat.

McQuillin, R., Bacon, M., Barclay, W. (1984). An Introduction to Seismic Interpretation. Graham and Trotman, 287p., London.

Md. Noor Alam, Dr. Kazuo Nakayama, Dr. Matsuoka Toshifumi, Dr. Tamio Yohroh, (2006). Petroleum System of Bengal Basin in Bangladesh. 6th International Conference and Exposition on Petroleum Geophysics "Kolkata".

Meade, R. H. (1996) In: Sea-level rise and coastal subsidence, J. D. Milliman and B. U. Haq (eds.), Kluwer Acad. Publishers, London, pp. 63-85.

Meade, R.H. (1972). Transport and deposition of sediment in estuaries. Geo. Soc. Ame. Mem., 133, pp.91-120.

Meddlicott, H B (1873). The Shahpur Coal Field with a note of coal exploration in Narmada region. Geol. Surv. Ind. Records, 18, pt. 3, 65-86.

Meddlicott, HB (1859). Vindhyan rocks and their association in Bundelkhand. Geol. Sun: Ind. Memoir, 2 (1), 1-95.

Mehra,G., T. P. Singh and S.K. Laskar (1989). Aeromagneti Survey in Tripura-Mizoram-Manipur-Cachar area for new lead in Hydrocarbon Exploration., ONGC bull VOL.26 NO.2(DEC.1989).

Mehrotra, R.B., and C.L. Bhan (1979). Subtle traps of south Kadi field –Bull., ONGC., Vol. 16, No. 1.

Melrose, J.C., and F.C. Brandner-(1974). Role of capillary forces in determining microscopic displacement efficiency for oil recovery by water flooding, Journal of Canadian petroleum tech.

Meyerhoff, (1980). Geology and petroleum fields in Proterozoic and Lower Cambrain strata, Lena-Tunguska province, Eastern Siberia, USSR-AAPG, Memoir 30, Giant oil and gas fields of the Decades, 1968-1978.

Miller, C.C., Dyes, A.B. and Hutchinson, C.A. Jr. (1950). "Estimation of permeability and Reservoir pressure from Bottom Hole Pressure Build up Characteristics", Trans., AIME 189, 91-104.

Miller,C., U. Klotzli, W. Frank, M. Thoni and B. Grasemann (2000). "Proterozoic crustal evolution in the NW Himalaya (India) as recorded by circa 1.80 Ga mafic and 1.84 Ga granitic magmatism", Precam. Res., v.103, 191-206.

Millhem, K. (1979) Behaviour of multiple-stabiliser bottom-hole assemblies; O.G.J., Vol. 77, No. 1.

Mishra S.N., Mitra S., Dasgupta A. (1993). Integrated evaluation of Purnea Basin East of Ganga Valley.

Mishra, D. C., Atchutarao, D., Agarwal, P. K. and Sarma, B. S. P. (1971). Results of airborne magnetometer profile from Hyderabad to Puri (India). Bull. Natl. Geophys. Res. Inst. (India),9, 97–101.

Mishra, D. N., Anand Prakash, A.K.Nauriyal, K. Parsuraman and R.B.Singh.(2006). Optimizing Spread Geometry for Effective 3D Survey in South Nambar, A and AA Basin, Assam., 6th International Conference and Exposition on Petroleum Geophysics "Kolkata".

Mishra, D.C. (2002). Crustal structure of India and its environs based on geophysical studies, Indian Minerals, 56, 27 - 96.

Mishra, D.C., Chandra Sekhar, D.V., Venkata Raju, D.Ch., Vijay kumar,V. (1999). Crustal structure based on gravity-magnetic modelling constrained from seismic studies under Lambert rift, Antarctica and Godavari and Mahanadi rifts, India and their interrelation ship. Earth and Planetary Science Letters 172, 287–300.

Mishra, D.C., Gupta, S.B., Venkatarayudu, M. (1989). Godavari rift and its extension towards the east coast of India. Earth and Planetary Science Letters 94, 344–352.

Mishra, P.K., Rajnikanth, A., Jauhri, A.K., Kishore, S., Singh, S.K. (2004). Albian limestone building algae of Cauvery Basin, South India. Current Science 87 (No. 11), 1516-1518.

Mishra, R.C. (1969). The Vindhyan system, Procc. Address, 56th Ind.Sc. Cong. Bombay, 1-32.

Mishra, Y.K., Parida, G., and Ramani, K.K.V. (1997). Stratigraphic and depositional model: An Important tool for stratigraphic trap exploration of basal clastic unit in Bombay Offshore Basin, inProceedings, second international petroleum conference and exhibition: New Delhi, PETROTECH, v. 1, p. 463–474.

Mishra,D.C., et al. (2004). Gravity Studies over Saurashtra Peninsula, India- Some New Insight,5th Conference and Exposition on Petroleum Geophysics (SPG-2004).

Mishra., J and CRC sekaran(1997). Sand Distribution ans entrapment situation in C23 and adjpining area Surat Depression- Proceedings of Second Int. Nat,Pet. conference PetroTech 97 pp-499-510.

Misra K.N., Kansara A.B., Kayal P.B. and Philip P.C. (1981). Redox and Paleo-salinity environment of deposition of Subathu outcrops (H.P.). KDMIPE, ONGC, Dehradun, internal report.

Misra, P.C., Singh, N.P., Sharma, D.C., Upadhyay, H., Kakroo, A.K. and Saini, M.L. (1993). Lithostratigraphy of west Rajasthan basins Dehradun, unpublished report.

Misra, RC (1969). The Vindhyan System– Presidential address. In Proceedings Indian Science Congress, 56th Session.

Misra, RC and Bhatnagar, GS (1950). On carbonaceous discs and "Algal Dust" from the Vindhyans.– Curr. Sci., 19, 88.

Mitchell A.H.G. and H.G. Reading (1986). "Sedimentation and Tectonics", In H. G. Reading (ed.) Sedimentary environments and facies, Blackwell Scientific Publications, Oxford,pp 471-519.

Mitchum, R. M. and Vail, P. R. (1977). Seismic Stratigraphy Interpretation Procedure, AAPG Memoir 26,p. 135–144.

Mitchum, R.M. (1977). Seismic stratigraphy and global changes of sea level part 11: glossary of terms used in seismic stratigraphy. In: Payton, C.E. (Ed.), Seismic Stratigraphy Applications to Hydrocarbon Exploration. AAPG Memoir 26, 205-212.

Mitchum, R.M., Jr. (1985). Seismic stratigraphic expression of submarine fans. In: Berg, et al. (eds.) Seismic Stratigraphy II: An Integrated Approach to Hydrocarbon Exploration. Am. Ass. Petrol Geol. Mem. 39: 117-135

Mithilesh Kumar, V. K. Kulshreshtha, A.N. Borthakur, Shafique Ahmad,(2010). Geochemical Exploration-Near Surface Expression of Hydrocarbon: A Case Study from Frontier Block of Assam-Arakan Basin-8th SPG Hyderabad

Mitra S. (2002). "Fold accommodation faults", AAPG Bull., v. 86, no.4. 671-693,

Mitra, A K., S.A. Oganov and R.D. Badgayan. Causes and Nature of Deviation of vertical wells and its Prevention., ONGC Bulletin.

Mitra, A.P., et al. (1983). Exploration in western offshore basins. (Petroleum Asia Journal, PP.15 29).

Mitra, P, M K Mukherjee, BK Mathur, SK Bhandari. SM Qureshi and GC Bahukhandi (1993). Exploration and Hydrocarbon Prospect in Jaisalmer Basin, Rajasthan-Proc Second Seminar Petroliferous Basins Indian Petroleum Publications Vol.2 pp-235-284.

Mitra, P, MK Mukjrjee, B K Mathur S K Bhandari SM Qureshi and GC Bahukhandi (1993). Exploration and Hydrocarbon Prospects in Jaiselmar Basin, Rajasthan – Proc. Second Seminar on Peroliferous Basins of India –Vol-2 Indian Petroleum Publishers Dehradun.,pp235-284.

Mitra, P. and P.L Zutshi (1983). Exploration in western offshore basin: Petroleum Asia Journal, v.6, no.4, p.15-24.

Mitra, R.N. (1971). Tectonics of frontal folded belt, ONGC Bull., v. 8(2).

Mitra., P.P. (1988). Migration Tools for Enhancement of Seismic Data-Bull. ONGC., Vol. 25 No. 1.

Mittal. S.C. (1978). Development Drilling of Bombay High with particular reference to Directional Drilling., ONGC Bulletin., Volume 15., Nos. 1 and 2.

Mittal., P. K., (1969). Effect of Shot Depth and Lithology at the shot hole on the frequency content of Seismic Waves.,ONGC Bull Vol. 6, No. 2.

Mittal., P.R. (1982). A New Technique of Basement Investigations through susceptibility studies with the aid of Seismo Magnetic Data and Computer Modelling., ONGC Bull Vol 19 no.1 June.

Mohan, M and P. Kumar, (1982). Paleogene and Early Neogene sedimentation rates in an Arabian Sea sedimentary basin: Geoscience Journal, v.3, no.2, p.151-164.

Mohan, M. (1982). Geohistory analysis of Bombay High region: ONGC Bulletin, v.20, no.2, p. 23-46.

Mohan, M. and Pandey, J. (1973). Early Paleno-stratigraphy of Upper Assam. Bull. Ind. Geol. oc. Vol. 6 No. 1.

Mohan, M. Kumar, P., Narayanana. V., Govindan, A., Soodan. K.S. and Singh, P. (1982). Biofacies, Palaeoecology, Geological history of Bombay Offshore region; Bull. ONGC, Vol. 19, No. 1 pp. 13-27.

Mohan, Madan, Narayanan, V. and Kumar, P (1978). Palegene and Early Neogene Biostraigraphy of Bombay Offshore Region –Madras University– pp. 180-194.

Mohan., M.C., C.A. (1815). Some further notes on the Geology of Chamba Rec. Geol. Surv. India, Vol 18.

Mohapater, P. T E Sivarama Krishnan S Rajagopalan and J R sethi (1996). Producing Indian Basins and their Reserve Potential –Bull ONGC., Vol 33, no.1 pp 45-76.

Mohapatra, G. P., Rao B. R. and Biswas, N. R. (1992). Morphology and surface sediments of the eastern continental shelf off Peninsular India. Geol. Surv. Ind. Spl. Pub. No. 29, pp. 229-243.

Mohapatra, G. P., Vaz, G. G. and Rao, P. B. S. (2002). Morphology and surface sediments off southern part of eastern continental margin of India Proc. Four. Decads of Marine Science in India – A retrospect, Spect. Publ. Geol. Survery. India., V. 74, pp. 24-31.

Mohinuddin, S.K., Satyanarayana, K., Rao, G.N. (1993). Cretaceous sedimentation in the subsurface of Krishna Godavari basin. Journal of the Geological Society of India 41 (6), 533-539.

Mokkelbost, K.H. S. Strandvik (1999). "Development of NGI's deepwater gas probe, DGP".Proc. Conference on Offshore and Nearshore Geotechnical Engineering, Geoshore, Panvel, India,. pp. 107-112.

Momper, J.A. (1980). Generation of abnormal pressures through organic matter transformation (abs), AAPG, Bull., v.64, pp. 753.

Moor, P.L. 1974. "Drillings Practices Manual"? Penn Well Books, 258.

Moore, L.V., and T.L. Leyenberger, (1994). The hydrocarbon potential of the West Bengal Basin of eastern India and western Bangladesh, Proceedings of the First South Asia Geological Congress, Islamabad, Pakistan, February 23-27, 1992.

Morley, C. K., Nelson, R. A., Patton, T. L. and Munn, S. G. (1990). Transfer zone in East African rift system and their relevance in hydrocarbon. Bull. Am. Assoc. Petrol. Geol., 74, 465–480.

Morris, R.L. and Biggs, W.P. (1967). "UsingLog-DerivedValues of Water Saturation and Porosity," Trans., SPWLA Amual Logging Symposium.

Morrow, N.R. (1976). Capillary Pressure Correlations for Uniformly Wetted Porous Media. J.Can. Pet. Tech., October-December: 49-69.

Mukherjea,A and S Hazra (1997). Changing Paradigm of Petroleum Exploration in Bengal Basin- Indian Journal of Geology Vol 69, No.1 pp-41-64.

Mukherjee A and S. Hazra.(1997). Changing Paradigm of Petroleum Exploration in Bengal Basin.,Ind.Jour.Geol. Vol 69. No.1 p 41-64.

Mukherjee A. and Neogi, B.B. (1993). Status of Exploration in Bengal Basin. Proc. SPEB-II Vol.I pp. 93-119.

Mukherjee B K (1999). Geology and Source Rock evaluation of Subathu sediments in Kalakot and Triyathi Areas Jammu and Kashmir- Proc PetroTech 99 pp-73-78.

Mukherjee MK (1981). Evolution of Ankleshwar anticline, Cambay basin, India. Geologic Notes, AAPG, pp 336–345.

Mukherjee, H and Economides, M.J. (1989) "A parametric comparison of Horizontal and vertical well performance", paper SPE 18303.

Mukherjee, J and A.S. Chopra (1990). Study of organic matter types in relation to the Depositional Environments in well of Bengal basin. Bull.ONG. Vol. No. p81-99.

Mukherjee, Manoj K. (1981). Evolution of Anklesvar anticline, Cambay Basin, India, Amer. Asso. Pet. Geol. Bull. p. 336-343.

Mukherjee, MK, SK Bhandari and D purkayastha (1995). Hydrocarbon Prospects and Evidence of Presence of Proterozoic Basin in Lunar-Maijlar Area Rajashtan- Proc Petro Tech, New Delhi, 1995 pp-133145-187.

Mukherjee,M K., S K Bhandari and D. Purkayastha (1995). Hydrocarbon Prospects and Evidence of Presence of Proterozoic Basin Lunar-Miajlar area Rajasthan –Proc Petrotech 95 vol 1 pp-133-145.

Mukhopadhyay, A.,B.K. Tiwari, O.P. Sinha and R.K. Srivastav (2003)., An Integrated Approach for permeability modeling in carbonate reservoir.,Proceedings of PETROTECH.

Mukhopadhyay, J., Chaudhuri, A.K. (2002). Proterozoic Penganga group, Pranhita-Godavari Valley, south India: depositional setting and paleogeography of a deep-water cratonic basin succession. Journal of Asian Earth Sciences (2003) in press.

Mukhopadhyay, M., and S. Dasgupta, (1988). Deep Structure and tectonics of the Burmese arc: Constraints from earthquake and gravity data: Tectonophysics, v. 149, p.299-322.

Mukhrjee,B. and A.K.Dey (1997). Sequence startgraphy of Upper Paleogene Section from well log Analysis of Ichapur Area, Krishnanagar-Raanghat Depression. Bengal Basin. Ind. Jour. Geol Vol.69.no. 2 p 146-160.

Mundy Brink (2006). High-Resolution 4D Acquisition., 6th International Conference and Exposition on Petroleum Geophysics "Kolkata.

Murthy K.S.R, Venkateswaralu K., and Rao T.C.S (1995). Basement structure beneath the inner shelf off Mahabalipuram to Palar River, east coast of India, Indian Journal of Marine sciences, Vol. 24, pp 223-225.

Murthy, B. S. R. and Venkateswara Rao, M.(1996). The thickness of sediments and their impact on the structural analysis of gravity anomalies of Ramagundam subbasin, Godavari valley, Andhra Pradesh, India. In Gondwana Nine, Ninth Int. Gondwana Symp., Hyderabad, vol. 2, pp. 697–711.

Murthy, K. S. R. (1989). Seismic stratigraphy of Ongole – Paradip continental shelf, East Coast of India, Indian Jour. Earth. Sci, V. 16, pp. 47-58.

Murthy, K.S.R and TCS Rao. (1990). Acoustic wipeouts over the Continental Margins off Krishna, Godavari and Mahanadi River Basins, East Coast of India Jour. Geol.Soc.Ind.Vol. 35, June 1990, pp. 559 10 568.

Murthy, K.S.R., Rao, T.C.S., Subrahmanyam, A.S., Malleswara Rao, M.M.,Lakshminarayana, S. (1993). Structural lineaments from the magnetic anomaly maps of the Eastern Continental Margin of India and NW Bengal Fan. Mar. Geol. 114, 171-183.

Murthy, K.S.R., Subrahmanyam, A.S., Lakshminarayana, S., Chandrasekhar, D.V., Rao, T.C.S. (1995). Some geo-dynamic aspects of the Krishnae–Godavari basin, east coast of India. Continental Shelf Research 15 (7), 779-785.

Murthy, K.S.R., Subrahmanyam, A.S., Laxminarayana, S., Chandrasekhar,D.V., Rao, T.C.S. (1994). Some geodynamic aspects of the Krishna–Godavari basin, East Coast of India. Continental Shelf Research 15, 779–788.

Murthy, K.V.S., Ramakrishna, M. (1980). Structure and tectonics of Godavari–Krishna coastal sedimentary basins. Bulletin of ONGC 17 (1), 147–158.

Murthy, M.V.N., S.C. Talukdar, A.L. Bhattacharya, and C. Chakrabarti, (1969). The Dauki fault of Assam: ONGC Bulletin, v. 6/2, p.57-64.

Murthy. V.R., and A.K. Gupta (1982). Evaluation of Formation Porosities in Bombay High Field using cased hole Compensated Neutron Log., Bull ONGC.,Vol. 19, No. 2.

Murti, K.S. (1987). Stratigraphy and sedimentation in Chattisgarh Basin. In GSI Mem-6 (Ed. by B.P. Radhakrishna) pp-239-260. GSI publication, Bangalore.

Murty G.P.S, Subrahmaniyam A.S., Murthy K.S.R and Sarma K.V.L.N.S, 2002, evidence of fault reactivation off Pondicherry coast from marine geophysical data, Current Science, Vol. 83, No-12, pp 1446-1449.

Murty K.N. (1970). Evaluation of hydrocarbon prospects of new structures. Upper Assam, OGC. ONGC.

Murty KVS, B N Prasad and G S Rawat, (1984). Structure of Continental Margin off Bombay- Preceedings of Seminar on Geological Interpretation of Geophysical data ONGC Seminar-pp-41-48.

Murty, K.N,Tongaonkar, C.S and Kaul, S.K. (1979). Lithostratigraphy and hydrocarbon exploration, ONGC Bull vol 16 No. 2.

Murty, K.N. (1968). Assessment of Rudrasagar, Dikhomukh area, Oil and Gas Conference, ONGC.

Murty, K.V.S., Ramakrishna, M., 1980. Structure and tectonics of Godavari-Krishna coastal sedimentary basins. Bulletin of Oil and Natural Gas Corporation 17,147-158.

Murty,K.N. (1983). Geology and Hydrocarbon prospects of Assam Shelf-Recent advances and present status, Petrol. Asia Jour. V. 6(4),pp. 1-14.

Murty., K S R., A S Subrahmanyam and V Subrahmanyam (2012). Tectonics of the Eastern Continental Margin of India, TERI publication.

Murty.K.N. Tongaonkar C.S. Srivastava. D.K. and Sharma, S.D. (1979). Paleo-structural approach to oil exploration. Geleki field. Assam. Bull., ONGC. Vol 16 No. 1.

Nagendra, R., Bhavani, R., Dinakaran, V., Nallapa Reddy, A., Jaiprakash, B.C. (2001). Outcrop sequence stratigraphy of Kallankurichchi Formation of Velliperinjiyam mine, and its correlation with Tancem mine, Ariyalur Group, Tamil Nadu.– Indian Journal of Petroleum Geology 10, 23-36.

Nagendra, R., Nagendran, G., Narasimha, K., Jaiprakash, B.C., Nallapa Reddy, A. (2002b). Sequence Stratigraphy of Dalmiapuram Formation, Kallakkudi Quarry e II, South India.– Journal of the Geological Society of India 59, 249-258.

Nagendra, R., Raja, R., Nallapa Reddy, A., Jaiprakash, B.C., Bhavani, R. (2002a). Outcrop sequence stratigraphy of the Maastrichtian Kallankurichchi formation, Ariyalur Group, Tamil Nadu. Journal of the Geological Society of India 59, 243-248.

Naidu, B.D., and Panda, B.K. (1997). Regional source rock mapping in upper Assam Shelf, in Proceedings of the Second International Petroleum Conference and Exhibition.

Naik G C., and Sucheta Dotiwala (1996). Genetic Stratigraphy and the Lost Axial Drainage of North Cambay Basin Volume33 Number 2 December.

Naik G.C., et al. (1991). Paleo-latitude, continental drifting and hydrocarbon prospects in India. ONGC Bull, 28, 2, 17-36.

Naini, B.R. and Talwani, M. (1983). Structural framework and the evolutionary history of the continental margin of western India. In: Studies in Continental Margin Geology (J. S. Watkins and C. L. Drake, Eds.), AAPG Memoir, V. 34, pp. 167 – 191.

Naini, BR and Talani, M (1982). Structural framework and the evolutionary history of the continental margin of western India. AAPG Memoir 34, 167-191.

Nair K.M. Singh N.K. and Ram J. (1992). Stratigraphy and sedimentation of Bombay Offshore Basin, Jour. Geol.Soc.Ind. vol.40 Nov.1992 pp.415-442.

Nair, K.M. (1980). Lithology and stratigraphy of Carbonates of Tertiary Era, Cauvery Basin; Bull. ONGC,Vol. 17, No. 1, pp. 159-182

Najman, Y., Clift, P., Johnson, M.R.W. and Robertson, A. H. F. (1993). Early stages of foreland basin evolution in the Lesser Himalaya, N. India, In. Treloar, P.J. and Searle, M.P.(ed.) Himalayan Tectonics, Geol. Soc. Sp. Pub, no. 74, pp. 541-558.

Nakagawa, E., Santos, H. and Cunha, J,C. (1999). "Application of Aerated Fluid Drilling in Deep Water" World Oil, June Web site of weather ford company.

Nandy, J. and R.Chatterjee (2010). An Approach to Water Saturation Estimation using NMR Data in Water-Wet Rocks: A Case Study ISMU, Dhanbad, India.

Nandy, J. R.Guru, A.Ghosh (2010). An Integrated Approach for Saturation Validation: A Case Study, SPG.

Naqvi, S.M, and Rogers. (1987). Precambrian geology of India John, J.W. –Oxford University Press. Inc.

Naqvi, S.M. (2005). Geology and evolution of the Indian plate. (From Hadean to Holocene-4 Ga to 4 Ka) Capital Publishing Company, New Delhi, pp 1-448.

Narasimha Chari, M. V. J. N. Sahu, B. Banerjee, P. L. Zutshi and Kuldeep Chandra (1995). Evolution of the Cauvery basin, India from subsidence modelling Basin Research Group, KDMIPE, Oil and Natural Gas Corporation Ltd, Dehradun 248195, Marine and Petroleum Geology. Vol. 12, No.6. pp. 667-675.

Narasimham, J.L., Kharak Singh, S. Ramanan and Rajesh Kumar (2005). Exciting Brown Field Development opportunities - Mumbai High, Heera and Neelam Fields, western offshore, India., PetroTech 2005.

Narasimhan T.N. (1990). Paleo-channels of the Palar river West of Madras City: Possible Implications for vertical movement, Journal of Geological Society of India, Vol 36, pp 471-474.

Narayan., J. P. and Dr Lalaji Yadav (2006). Application of Adaptive Processing Technique for the Inversion of Open Hole Logs Recorded in Oil Fields of Indian Basins.,6th International Conference and Exposition on Petroleum Geophysics "Kolkata.

Narayana Rao., G. (1994). Sedimentation and Subsurface Startigraphy of Krishna Godavari Basin, East Coast of India., Ph.D., Thesis submitted to Andhra University.

Narender Pal and M.M. Ahmed (2006). Value Addition Using Cement Evaluation by Ultra Sonic Imaging Tool In Upper Assam Oil Fields., 6th International Conference and Exposition on Petroleum Geophysics "Kolkata".

Narendra. K. Verma and Manoj V. Thachaparambil, Ajay Gupta (2006). Generation of High-Resolution Synthetic Seismogram from Image Logs for Advanced Reservoir Applications, SPG.

Naresh Kumar, B. V. Jacob (2005). Formation evaluation behind casing – some examples from Bombay Offshore-Proceedings-article-280.

Nautiyal, R.,A. Deb, A.M. Tripath.(2010), Seismic guided Well Placement in Karstified Carbonate Reservoir: A case study., SPG 8th Hyderabad.

NecMetin Mungan: (1965)"Permeability Reduction Through Changes in pH and Salinity"SPE 1283.

Nelson, P. H. (1996). "Permeability-porosity relationships in sedimentary rocks", The Log Analyst, 35,3,38-62.

Nemcok, M., Sinha, S.T., Stuart, C.J., Welker, C., Choudhuri, M., Sharma, S.P.,Misra, A.A., Sinha, N., Venkatraman, S. (2012). East Indian margin evolution and crustal architecture: integration of deep reflection seismic interpretation and gravity modeling. In: Geological Society of London, Special Publications. doi: 10.1144/SP369.6.

Nestvold, E. O. (1987). The use of 3D seismic in exploration, appraisal and field development, Proc.12th World Petroleum Congress, Houston, 1987, 2, 125–132, Wiley, Chichester.

Newberry, B.M., Nelson, R.F., and Aluned, U. (1985). "Prediction of Vertical Hydraulic Fracture Migration Using Compressional and Shear Wave Slowness," paper SPE/DOE 13895.

Nihalani, M.C., H. Dubey and T.G. Kulkarni (2003). Super Clean Gel Plug for Arresting loss in highly sub-hydrostatic wells – a case study– Proceedings of PETROTECH-2003, New Delhi.

Nimisha Vedanti, Ravi P. Srivastava, BjØrn Arild Mythen, Martin Landro and V.P. Dimri. (2006). Feasibility Study of Time-Lapse Seismic: A Case Study.,6th International Conference and Exposition on Petroleum Geophysics "Kolkata".

Niyogi, D. (1975). Quaternary Geology of the coastal plain of West Bengal and Odisha, Indian J. Earth Science, 2(1), 51–61,. No 29, pp 247-254.

Nolen, J.S. and Berry, D.W. (1972). Tests of Stability and Time-Step Sensitivity of Semi-lmplicit Reservoir Simulation Techniques. Soc. Pet. Eng.Journal., June: 253-266. Trans. AIME.

Nolley, J.P., Cannon, G.E., and Ragland, D. (1948). "The Relation of Nozzle Fluid Velocity to ROP with Drag Type Rotary Bits", API Drilling and Production Practices, 22.

Norton and Sclater, JG (1979). A model for the evolution of the Indian Ocean and breakup of Gondwanaland. Jour. Geophys. Rev., 84, 6803-6830.

O'Dell, P.M. (1974). Numerical Reservoir Simulation: Review and State of the Art. Paper presented at 76th National AlChE Meeting, Tulsa, Oklahoma. March.

Oil Field – A Case Study from Santhal Field, Cambay Basin Vol. 30, No. 2 (December).

Ojha, Artatran.,K. Ramakrishna, G. Sarvesam.(2010). Improvement of signal to noise ratio by Group Array Stack of single sensor data, SPG 8th Hyderabad.

Oort, E. van, et al., June, (1997). "Improving HPHT Stability of Water Based Drilling Fluids", JPT.

Padhy P.K. (1994). Proterozoic Hydrocarbon Prospect–An Indian Scenario. In (Proce. second seminar on Petroliferous Basins of India Vo.III, Ed. S.K. Biswas, et al., Indian Petroleum Publisher, Dehradun India.

Padhy P.K. and Naik C.C. (1991). A feasibility study of oil shale exploration in India and its environmental impact. Abstract vol. 'International Conference on Mineral Development and Environment, ISM, Dhanbad.

Pahari, S.,Harvir Singh, I.V.S.V. Prasad, B. L. Sharma, A. K. Sinha and R. R. Singh (2010). Petroleum system modeling along a dip section of Bengal Basin, India, 8th SPG Kolkata

Pahari, S., A. Banerjee, Meenu Varshney, Ravi Misra and K.N. Misra (2003). The Potential Source Rocks of Kutch Offshore-Proceedings of PETROTECH-2003.

Pahari, S., Tapas Mitra and Jayanita Baral (2005). Hydrocarbon Generation and Expulsion in Southern Part of Nagapattinam Area, Cauvery Basin, India--Petro Tech 2005

Pal, P.C. and Bhimasankaran, V.L.S. (1976). Tectonics of the Narmada-Son Brahmaputra Lineament. Semi. Tect. Met. of South and East Asia. Misc. pub. Geol. Surv. India, v.34(1), pp. 133-140.

Pallav Baruah, Kewal Krishan, Dianuddin Ahmed (2011). An overview of TS-5A reservoir sand of South West Geleki field, Upper Assam, India – an attempt to present an integrated geological modal, 2nd South Asain Geoscience Conference and Exhibition, GEO India 2011.

Panda, B.K., B.S. Mohapatra, U.K. Ghosh and N.S. Tanwar (2005). Building Reservoir Model through Geo-statistical Reservoir Characterization Tool - a case study on Viraj Field, India., Petro-Tech.

Pandey A. N. and P.K.Minz (1993). Depositional Environment Reservoir Characteristics And Sedimentary Dynamics of Dabka Pay Sands in Dabka field Cambay Basin Vol. 30, No. 1, (June, Vol. 30, No. 1.

Pandey P.B., and Strack., K.M. (2007). Exploration with controlled-source electromagnetics under Basalt cover in India: The Leading Edge, Vol. 26, No. 3.

Pandey USD (2010). Mesozoic exploration in Western Onshore, India ONGC Bulletin issue Petroleum exploration in Mesozoic basins in Western Onshore India, ONGC SPG-2010.

Pandey, A. N. and H.Upadhyay (1995). Sedimentary Facies, Depositional Environments and Reservoir Characteristics of the Gandhar Pay Sands in Subsurface Section in Southwestern Part of Gandhar Field, Cambay Basin, India,Volume 32 Number2 December.

Pandey, A.N. and Gupta, G.D. (1983). B sand Prospect in Bombay High and Rats: Shelf areas. Bombay Offshore Basin, India published KDMIPE. ONGC.

Pandey, J., Singh, N.P., Krishna, B.R., Sharma, D.D., Parakh, A.K. and Nath, S.S. (1993). Lithostratigraphy of Cambay basin: Document III, KDMIPE publication, ONGC, Dehradun.

Pangtey K K S,D N Prasad A Maitahni and R P Agrwal (1995). Paleo-magnetis studies of the Lower Tertiary Sedimebts of Himachal Pradesh and its relevance to Hydrocarbon Exploration - Proc PetroTech 95 pp-309-318.

Panwar,P,NN Badola, MS Gauma, OP Dutta, JG Chaturvedi and K A Shah (1999). Integrated Geological Modeling of Limbodra Field- A case study-- Proc 3rd Petro Tech, New Delhi pp-97-110.

Pardeep Kumar, D.K. Ghosh, B.S. Haldia, U.S. Roy and V.K. Sud (2003). Identification of source of high water cut from well log data integrated with core studies and production data - a case history of Demulgaon Field, Upper Assam, proc of Petro Tech, 2003.

Pareek, H.S. (1984). Pre-Quaternary Geology and mineral resources of North-western Rajasthan: Memoir Geological Survey of India. v. 115. 1-95p.

Parr, J. O., Jr., and Mayne, W. H. (1955). A New Method of Pattern Shooting, Geophysics, vol. 20, p. 539–565.

Partyka, G, et al. (1999), Interpretational, Applications of Spectral Decomposition in Reservoir Characterization. Leading Edge, March-1999, P.353.

Pascoe, E.H. (1964). Manual of Geology of India and Burma. G.O.I. Publication Civil Lines, Delhi.

Passega, R. (1964). Grain size representation by CM patterns as a geological tool. J. Sed. Petro., 34, pp.830-847.

Patel S. K. (1999). "Sublayering of Sand GS-3A in Gandhar Field, Cambay Basin", Unpublished Report by Ankleshwar Asset, ONGC, 1999.

Patel, S. S. K. (1999) "Sublayering of Sand GS-3A in Gandhar Field, Cambay Basin", Unpublished Report by Ankleshwar Asset, ONGC.

Pateria, M.L., Rangaraju, M.K., Raiverman, V. (1992). A note on the structure and stratigraphy of Bay of Bengal sediments. In: Recent Geoscientific Studies in the Bay of Bengal and the Andaman Sea. Spec. Publ. Ser. Geol. Surv. India, vol. 29, pp. 21-23.

Paul F. Worthington (2003)...... Maximizing the Effectiveness of Integrated Reservoir Studies. –Proceedings of PETROTECH-2003.

Paul, D.D. and Lian H.M. (1975). Offshore Tertiary basins of the SE Asia, Bay of Bengal to South China Sea, IX World Petroleum Congress Vol. 3 pp 107-121.

Pendkar, N and Peters James, (2002). Characterization of Jardepahar porcellanite, Unconventional fractured reservoir in Vindhayan basin, Proceedings of the first conference of APG, Vol 2, pp.105-112.

Pendkar, N., Kumar, A. (1999). Delineation of Reservoir Section in Deccan Trap Basement, Example from Padra Field, Cambay Basin, ONGC Bulletin, V.36, No.1 (June 1999), p. 83-88.

Pennebaker, E. S., 1968. Seismic data indicate depth magnitude of abnormal pressures. World Oil, June 68,pp. 73–78.

Peter, G. Weeks, L.A. and Burns, R.E. (1966). A reconnaissance geophysical survey in the Andaman Sea and across the Andaman Nicobar Island Arc. Jour. Geoph. Res. Vol. 71, No. 2, pp 495-505.

Peters, K.E. (1997). Modern geochemical tools for efficient exploration and development, O.G.C.I. Training report, Oct. 20-24, Mussoorie, India.

Pettijohn, (1984). Sedimentay rocks, 3rd Ed., Harper and Bros., Delhi, India, p.628.

Pichamuthu, C.S. (1967). The Precambrian of India. In the geologic system: The Precambrian (Ed., K. Pankam) pp.1-96.

Piers GE, Perkins J and Escott D. (1987). "A New Flow m e t e r for Production Logging and Well Testing," paper SPE 16819, presented at the 62nd SPE Annual Technical Conference and Exhibition, Dallas, Texas, USA.

Pinakadhar Mohapatra.(2006). Sequence Stratigraphic Approach for Identification of Hydrocarbon Plays in Mahanadi Offshore Basin., 6th International Conference and Exposition on Petroleum Geophysics.

Poelchau, H.S., Baker, D. R., Hantschel, th., Horsfield, B., wygrala, B. (1997): in: welte. D. H., Horsfield, B., Baker, D. R. (eds.). Petroleum and Basin Evolution. –Springer Verlag, Germany.

Pomila Garga, S.K. Sharma and D.K. gupta (1982). A New Model for the Interpretation of logs of TS-1 and TS-2 sands of Lakwa and Lakhmani –ONGC Bulletin, Vol. 19, No. 2.

Posamentier, H.W. and James, D.P. (1993). An overview of sequence-stratigraphic concepts: uses and abuses. In: Posamentier, et al. (eds) Sequence stratigraphy and facies associations. Spec. Publs. Int. Assoc. Sediment., 18: 3-18.

Posamentier., H.W. and R.D. Erskine. Seismic expression and recognition criteria of ancient submarine fans. AAPG Memoir.31.

Poupon, A. (1958). Interpretation of Cement Bond Logs. Schlumberger Overseas. S.A. Schlumberger Well Surveying Corporation, Schlumberger Document No. 8.

Poupon, A., Lay, M.E., and Tixier, M.P.: (1963). "A Contribution to Electrical Log Interpretation in Shaly Sands," J. Pet. Tech.

Powell, C.M., and Conaghan, P.J. (1973). Plate tectonic and the Hiamalayas, Earth and Planetary Sci. letters, v. 20, pp. 1-12.

Powell, C.M., Roots, S.R., Veevers, J.J. (1988). Pre-breakup continental extension in East Gondwanaland and the early opening of the eastern Indian Ocean. Tectonophysics 155, 261-283.

Prabhakar, K N, A K Awasthi, SK Roy, Anand Prakash Renu Gupta and I Jaya kumar- (1993). Sequence Stratigraphy and Systen Tracts Analysis of Nagapattinam sub Basin, Cauvery- Proc Petroliferous basins. Vol-1.

Prabhakar, K. N. and Zutshi, P. L. (1993). Evolution of southern part of Indian East Basins. Jour. Geol. Soc. India. V. 41(3), pp. 215-230.

Prabhakar, K.N., Pratap, V., and Kumar, H.C.V. (1999). Re-entry drilling, an effective improvement in cost of exploration and production - a case study of Mehsana project area, -Proceedings PETROTECH-99, Third International Petroleum Conference, New Delhi, v. 5, p. 37-45.

Prabhakar, KN, AK Awasthi, Sk roy, Anand Prakash, Renu Gupta, and I Jaya kumar (1993). Sequence Startigraphy and System Tracts of Nagapattinam Sub Basin Cauvery Basin- Proc Second Seminar Petroliferous Basins Indian Petroleum Publications Vol.2 pp-201-214.

Prakhakar, K. N. and Zutshi, P. L. (1993). Evolution of the Southern Part of Indian East Coast Basins J. Geol. Soc. India 41,215-230.

Pramanik, A.G. V. Singh and A.K. Srivastava (2003). Seismic Attributes and their role in Acquisition.,Petrotech 2003.

Prasad K.K. and Chatar Singh. (2005). For Hydrocarbon Detection, SP Logging in redox mode is worth a trial –proceedings-article-95.

Prasad, BN, Murty, K.V.S.(1981). Basement configuration of West Bengal Basin, Sym. Geol. Int.of Geophy. data IPE, pp. 151-157.

Prasad, Dr S. S., C S Sajith and S S Bakshi. (2006). Evaluation of Low Resistivity Laminated Shaly Sand Reservoirs.,6th International Conference and Exposition on Petroleum Geophysics.

Prasad, G.K., Onkar Prasad, L.M., Gupta, S.K., Ramakumar, Jain, D.P. and Ekambasram, I. (1998). "Exploration Need and Scope of Subtle and Stratigrphic Traps in West Godavari Sub-basin, Krishna-Godavari Basin", Bulletin ONGC.,V-35,No-2, PP 145-57.

Prasad, N. B., G. Ramaswamy, and K. L. N. Rao, (1978). Development continues in India's Bombay High field: World Oil, v. 186, no. 5, P. 81.

Prasad. K K (2006). Field Examples Elucidate the Value Addition Through NMR Logging in Western Region of ONGC.,6th International Conference and Exposition on Petroleum Geophysics.

Prasada Rao R., and D.C. Srivastava. (1984). Regional Seismic Facies Analysis of western offshore, india., ONGC Bull Vol. 21 No. 1 (June).

Pratap Singh, (1972). Geology and micropalaeontology of the Subathu Group of the Jammu Himalaya, Jammu and Kashmir, Wadia, Instt. Him. Geol. Abstracts 8a, (Third symposium).

Pratap Singh,(1970). Larger Foraminifera from the Subathu of Beragua-Jangalgali area, Jammu Himalaya, Jammu and Kashmir. State,Jour, Geol. Soc. India, Vol.11 No.1.

Prickeu, H.D., Bremer, R.E.,"(1989) Improved initial water saturation distribution for a three dimensional model", 6th SPE Middle East Oil Show, SPE 17958.

Proehl, T., and Sabins, F. (2006). Drilling and Completion Gaps for HP/HT Wells in Deep Water. Society of Petroleum Engineers.

Protacio, J.A., Naina, Gupta, Raj, Yadav, Bastia, R., Sangvai, P. (2011). Interpretation of Gravity and Magnetic Data at the Krishna-Godavari basin, East Coast India. In: The 2nd South Asian Geoscience Conference and Exhibition, GEO India, 12-14th January, Greater Noida, New Delhi, India.

Purnachandra Rao, V. and Kessarkar, M. P. (2001). Geomorphology and Geology of the Bay of Bengal and the Andaman Sea. In: The Indian ocean – a perspective. R. S. Gupta and E. Desa (eds.), V. 2, pp. 815-868.

Quereshy, MN; Brahman, NK; Garde, SC and Mathur, BK (1968). Gravity anomalies and the Gondwana rift, India. Bull. Geol. Soc. Amer., 79, 1221-1230.

Quirein, J., Kimminau, S., Vigne, J. L., Singer, J. and Wendel, F. (1986,) "A coherent framework for developing and applying multiple formation evaluation models", Trans. SPWLA Twenty-seventh Annual logging Symposium.

Qureshi, M. N., Krishnabhramam, N., Aravamadhu, P. S. and Naqvi, S. M. (1968) Role of granitic intrusions in reducing the density of the crust, and other related problems as illustrated from a gravity study of the Cuddapah Basin, India. Proc. R. Soc. London, Ser. 304, 449–464.

Qureshi, M. N., Krishnabhramam, N., Garde, S. C. and Mathur, B. (1968). Gravity anomalies and the Godavari rift, India. Bull. Geol. Soc. Am, 79, 1221–1230.

Qureshi, M.N. (1964). A Gravity anomalies as related to the regional tectonics of Peninsular India. Report, International Geological Congress, 22nd Session, pp. 52-53.

Qureshi, M.N., Venkatachalam, S., and Subrahmanyam, C. (1974). Vertical tectonics in the middle Himalayas: An appraisal from recent gravity data, Geological Soc. America Bull., v.85, pp. 921-926.

R. Bhattacharyya and T. J. Majumdar (2013). Basin and sub-basin crustal structure of a part of the western offshore, India– J. Ind. Geophys. Union (April) Vol.17, No. 2, pp. 179-186.

Rabia, Hussain (2002). Well Engineering and Construction, Entrac Consulting, Modern well design/Bernt S. AadnØy. – 2nd edition.

Radhakrishn, B.P. (1987). Introduction on Purana basins of Peninsular India. GSI Mem-6, 1-15, (Ed.,B.P. Radhakrishna) GSI publication, Bangalore.

Radhakrishna B.P. (1984). Petroleum prospects in Precambrian. Current Sci., 53, 671-674.

Radhakrishna B.P. and Naqvi, S.M. (1986). Precambrian continental crust of India and its evolution, Jour. Geol., 94, 145-166.

Radhakrishna, B.P. (1989). Suspect tectono-stratigraphic terrane elements in the Indian subcontinent. Journal of the Geological Society of India 34 (1), 1-24.

Radhakrishna, I. (2000). Saline and Fresh water interface structure in Mahanadi delta region, Orissa, India, Environmental Geology, 40(3), 369–380.

Radhakrishna, M., Chand, S., Subrahmanyam, C. (2000). Gravity anomalies, sediment loading and lithospheric flexure associated with the Krishna- Godavari basin, eastern continental margin of India. Earth and Planetary Sciences Letters 175, 223-232.

Radhakrishna, M., Srinivasa Rao, G., Nayak, S., Bastia, R., Twinkle, D. (2012). Early Cretaceous fracture zones in the Bay of Bengal and their tectonic implications: constraints from multi-channel seismic reflection and potential field data. Tectonophysics 522, 187-197.

Raghunath H.K.and J. Abraham(2006). Formation Evaluation in Complex Lithology Using NMR data A Case Study of Western Onshore Basin, India.6th International Conference and Exposition on Petroleum Geophysics.

Raha, P. (1987). Stromatolites and correlation of the Purana (Middle to Late Proterozoic) basins of Peninsular India. GSI mem-6 (Ed., B.P. Radhakrishna) pp. 393-398, GSI publication, Bangalore.

Raha, P.K., and Sastry, M.V.A. (1973). Algal stromatoliies from the Jammu limestone and their geochronologic, Paleogeographic and environmental significance, Wadia. Inst. Him. Geol. (Third symposium) Abstracts.

Raina A, Pande HC, Saxena RK, Singh H, Singh RR (2010). Genetic characterization of shallow hydrocarbon gases by High Resolution Carbon Isotopic Studies: A case study of South Cambay Basin in India. Petrotech conference, New Delhi, India.

Raiveman, V., Rao, M.R. and Pal, D. (1985). Stratigraphy and structure of Pranhita-Godavari Graben. Pet. Asia Journal, v. 8(91), pp. 174-181.

Raiverman, V, M R Rao and D Pal (1983). Stratigraphy and Structure of the Pranhita Godavari Graben-Petroleum Asia Journal pp-174-189.

Raiverman, V. (1986). Sedimentation pattern and its controlling factors in the Cenozoic Basins of the Indian Subcontinent Bull. ONGC 23, No.1, 1-16.

Raiverman, V., Kunte, S.V., and Mukherjea, A. (1983). Basin geometry Cenozoic sedimentation and hydrocarbon prospects in northwestern Himalaya and Indo-gangetic plains, Pet. Asia J., v. 6, pp. 67-92.

Raiverman,V,S V Kunte and A. Mukherjea (1983). Basin Geometry Cenozoic Sedimentation and Hydrocarbon prospects in North-Western Himalaya and Indo-Gangetic Plains.- Petroleum Asia Journal SPEB-I Vol.I,pp 67-92.

Rajaram, Mita, Anand, S.P., Erram, V.C. (2000). Magnetic studies over Krishna-Godavari basin in eastern continental margin of India. Gondwana Research 3, 385-393.

Rajeev Kumar (2003). Under Balance Drilling- an Approach for Faster Drilling and higher roductivity– Proceedings of PETROTECH-2003, New Delhi.

Rajesh Kumar, P. K. Cherukupalli, B. L. Lohar, Dinesh Chandra (2002). "Saturation Modeling in a Multi layered Carbonate Reservoir Using Log derived Saturation- Height Function", SPE/DOE Thirteenth Symposium June 2002, Tulsa, Oklahoma, USA.

Rajesh Sarma (2002). Abnormal High pressure occurrence in the Himalayan Foreland Thurst fold belt in Northwest Himalaya of India- proc 1st conference of APG, India vol.1 pp 373-379.

Rajiv Nischal (2005). Factors effecting concept selection criteria for Deep-water Field Development, PetroTech.

Raju A.T.R et al. (1982). Depositional environment of oil bearing Lower and Middle Eocene sands of North Cambay Basin, India: Bull ONGC, V – 19. no: 2, pp: 55 - 72.

Raju A.T.R. et al. (1985). Middle Eocene Environments in Cambay Basin: Petroleum Asia Journal, V-VIII, no: II pp: 86-106.

Raju A.T.R., Hardas M.G., Srivastava D.C. and Maindarkar M.M. (1982). Depositional environment of oil bearing Lower and Middle Eocene sands of North Cambay Basin; ONGC Bull. Vol.10 No. 2.

Raju ATR, Srinivasan S (1993). Cambay basin–Petroleum habitat. In: Biswas SK, Dave A,Garg P, Pandey J, Maithani A, Thomas NJ (eds) Proceedings of the second seminar on petroliferous basins of India, vol 2. Indian Petroleum Publishers, pp 33–78.

Raju D S N et al. (2008). Stratigraphy of India- ONGC Bulletin V01.43 pp-10-39.

Raju, A.T.R. (1966). Geological evolution of Assam and Cambay Tertiary basins with particular reference to the associated lithological suites, Proceedings of the symposium on "Processes and products of sedimentation", Andhra University. Waltair (abstract).

Raju, A.T.R. (1968) Geological evolution of Assam and Cambay Tertiary basin of India. Amer. Assoc. Pet. Geol. Bull., v. 52, pp. 242-2437.

Raju, A.T.R. (1976). New global tectonics and petroleum exploration on continental margins-a review: July-Dec., Geonews, p. 24-40. 1979.

Raju, A.T.R. and Srinivasan, S. (1983) More hydrocarbon from well explored Cambay basin. Petroleum Asia Journal, Petroliferous Basins of India, v. 1, pp. 25-35.

Raju, D.S.N., Guha D.K., Bedi, T.S., Kumar, P., and Bhatt, D.K. (1970). Micrafauna, Biostratigraphy and Palaeoecology of the Middle Eocene to Oligocene sediments in Western India; Publication of the Centre of Advanced Study in Geology, Punjab University. Chandigarh.

Raju, D.S.N., Jaiprakash, B.C., Ravindran, C.N., Kalyanasunder, R., Ramesh, P. (2005). The magnitude of hiatus and sea level changes across K/T Boundary in Cauvery and Krishna-Godavari Basins. In: Indian Association of Petroleum Geologists,\Special Publication, vol. 1, 104-113 pp.

Raju, D.S.N., Ravindran, C.N., Kalyansundar, R. (1993). Cretaceous cycles of sea level changes in the Cauvery Basin, India – a first revision. Bulletin of the Oil and Natural Gas Commission 30, 101-113.

Raju, K.V.S., S.S. Bakshi and Dev Singh (1992). Study of Low Resistivity Pay Sands of Gandhar-An Approach.,ONGC Bull.,Vol. 29, No.2.

Raju, P. S. R. (1986). Geology and hydrocarbon prospects of Pranhita–Godavari Graben. J. Assoc. Explor. Geophys., VII, 131–146.

Raju,KVS., S.S. Bakshi and Dev Singh (1992). Study of Low Resistivity Pay Sands of Gandhar-An Approach Vol. 29, No.2.

Raju., A.T.R. (1974). Sedimentary Basins: Tectonic and Structural Framework -a Reviews-Bulletin of the Oil and Natural Gas Commission.,Vol. 11, No. 1.

Raju.,ATR and s. Srinivasan.(1983). More Hydrocarbons from well explored Cambay Basin- Petroliferous Basins of India.,Petroleum Asia Journal April 1984 SPEB-1 Publication,pp25-35.

Rakesh Dhir, K. Vijaya Kumar, P.K. Gupta, Piyush Panwar, S.K. Mattey, M.S. Gauma and K.A. Shah (2003). An Integrated Exploitation strategy of south Kadi field, Cambay Basin, India – a case study., Proceedings of PETROTECH.

Rakesh Dhir, K. Vijaya Kumar, P.K. Gupta, Piyush Panwar, S.K. Mattey, M.S. Gauma and K.A. Shah (2003). An Integrated exploitation strategy of South Kadi field, Cambay Basin,India – A case study– Proceedings of Petrotech-2003 January 9-12, New Delhi.

Rama Rao, L. (1956). Recent contribution to our knowledge of the Cretaceous rocks of south India. Proceedings of the Indian Academy of Science B.V.44, 185-245.

Rama Rao, L. (1968). The problems of Createaceous Tertiary boundary; In "Createaceous Tertiary Formations of South India"; Geol. Soc. Ind. Mem. No. 2.

Ramakrishna, MJ (1987). Applicability of equilibrium compaction depth in respect of geo-pressure generation, K.G. Basin. (ONGC Report-Unpublished).

Ramakrishna, T. S. and Chayanulu, A. Y. S. R. (1988). A geophysical appraisal of the Purana basins of India. J. Geol. Soc. India, 32,48–60. Research communications current science, vol. 88, NO. 7, 10 APRIL 2005 1175.

Ramakrishnan, M. (1987). Stratigraphy, sedimentary environment and evolution of the Late Proterozoic Indravati basin, central India. Purana Basins of Peninsular India, Geological Society of India Memoir 6, 139-160.

Ramam, P.K., Murthy, V.N. (1997). Geology of Andhra Pradesh. Geological Society of India, Bangalore, 98 p.

Raman, C. V. and Reddy, K. S. N. (2001). Sediment dispersal pattern off the Mahanadi-Nagavali continental shelf, Northwest Bay of Bengal, Jour. Geol. Soc. India, V. 58(2), pp. 123-133.

Raman, K.S., Kumar, Sukhdarshan, Neogi, B.B. (1986). Exploration in Bengal Basin India - An overview.,–Offshore South East Asia Show, 28-31 January 1986, Singapore.

Raman, K.S., Kumar, Sukhdarshan, Neogi, B.B. and A.K. Parakh (1986 a). New Approach towards the Exploration of Pre-Eocene Sediments in Bengal Basin. Bull. ONGC Vol.23.No. 1 p35-49.

Raman. K.S., B.B. Neogi, A.Saran and A.K. Parakh (1986). Carbonate reservoirs and their distribution in west Bengal –ONGC Bull vol.19 no.2 p.

Raman. K.S., B.B. Neogi,(1986 b). Exploration Leads from the study of Geochemical data., Bull. ONGC vol 23 no.1 p 163-174.

Ramana, M. V., V. Subrahmanyam, A. K. Chaubey, T. Ramprasad, and K.V. L. N. S. Sarma,(1997). Structure and origin of the 850 E RidgeJ. Geophys. Res., 102, 17995–18012.

Ramana, M.V., Krishna, K.S., Ramprasad, T., Desa, M., Subrahmanyam, V.Sarma, K.V.L.N.S. (2001). Structure and tectonic evolution of the north eastern Indian Ocean. In: Sen Gupta, R., Desa, E. (Eds.), 2001. The Indian Ocean e a Perspective, vol. 2. Oxford and IBH publishing Co., New Delhi, pp. 731-816.

Ramanathan, R.M. (1977). In: Venkatarengan, R. (Ed.), Depositional Systems and Thrust Area for Exploration in Cauvery Basin. ONGC Bulletin 24 (1), 53-69. 1989.

Ramanathan, S. (1964). Exploration for oil in the Cauvery basin, ONGC Reporter, 2, Nos. 2 and 3, pp. 5-8.

Ramanathan, S. (1968). Stratigraphy of Cauvery Basin with reference to its oil prospects. Geol Soc. Ind. Mem. No. 2. P. 153-167.

Ramanathan, S. and Rao, V. Raghavendra, (1965). Geology, Tectonics and Petroleum possibilities of Cauvery basin, south India, Third ECAFE symposium on the Development of Petroleum Resources of Asia and the Far East.

Ramanathan, S., et al. (1986). Exploration strategy of ONGC. Researches in sedimentary Basins - implications in the exploration of natural resources. Indian Petroleum Pub., pp1-16.

Ramani, K.K.V. (1985) Geology of Gondwana basins of peninsular India and their hydrocarbon potential-Status report. ONGC Report (Unpublished).

Ramasamy S.M., Kumanan C.J., Selvakumar R. and Saravanavel J (2011). Remote sensing revealed drainage anomalies and related tectonics of South India, Tectonophysics, Vol 501, pp 41-51.

Ramasamy, S., Banerji, R.K. (1991). Geology, petrography and systematic stratigraphy of the pre-Ariyalur sequence in Trichinopoly district, Tamil Nadu, India. Journal of the Geological Society of India 37, 577-594.

Ramaswamy, G., and K. L. N. Rao, (1978). Geology of continental shelf of the west coast of India in Facts and principles of world oil occurrence: Canadian Assoc. Petroleum Geologists, Calgary, Canada.

Rammohan Rao,T. (1987). Structure and tectonics of the East Coast of India in relating to delta setting: Field Seminar on Recent Deltas, Geology Deptt. Andhra University, p.1-6.

Ramsayer, G.R., 1979. Seismic strigraphy, a fundamental exploration tool. Proceed. Offshore Tech. Confr., pp. 1859-1866.

Ranga Rao A. (1983). Geology and Hydrocarbon potential of a part of Assam Arakan basin and it's adjoining region, Petrol, Asia, Jour. V.6(4), 127-158.

Ranga Rao A. and Samanta M.K. (1987). Structural style of the Naga overthrust belt and its implication on exploration, Bull ONGC v. 24(1) pp. 69-111.

Ranga Rao, A. (1972). New memmalian genera and species from the Kalakot Zone of Himalayan foot hills near Kalakot, Jammu and Kashmir State, India, Special paper No. 1, Directorate of Geology, ONGC, Dehradun.

Rangaraju M.K et al. (1991). Tectono-stratigraphy, structural styles Evolutionary model and hydrocarbon prospects of Cauvery and Palar basins Proc. SPEB-2

Rangaraju MK, Abinash Agrawal and K N Prabhakar (1993). Tectono Stratigraphy structural styles Evolutionary Model and Hydrocarbon Habitat Cauvery and Palar Basins - Proc Second Seminar Petroliferous Basins- Indian Petroleum Publications.,Vol.2 pp-371-388.

Rangaraju, M. K. (1987). Turbidites in petroleum exploration, ONGC Bulletin, Vol. 24, No. I, pp31-51.

Rangaraju, M.K., Agarwal, A., Prabhakar, K.N. (1993). Tectno-stratigraphy, Structural Style, Evolutionary Model and Hydrocarbon Prospects of the Cauvery Palar Basins of India, vol. 1. Indian Petroleum Publishers, Dehradun, 371-398 pp.

Ranjit K. Shaw, Mrinal K Sen, Ankur B. Chatterjee (2006). Estimation of Dip of Oblique Fractures Using AVOA Analysis 6th International Conference and Exposition on Petroleum Geophysics, Kolkata.

Rao D.R. and A.C. Seshadri (2003).,Identification of Gas Liquid contacts in old wells with the help of cased hole neutron logs-case study.,Proceedings of PETROTECH.

Rao G N. (1990). Subsurface stratigraphic nomenclature of Krishna-Godavari Basin: ONGC Report-Unpublished.

Rao G.N. (2001). Sedimentation, Stratigraphy and Petroleum Potential of Krishna Godavari Basin, Bull. AAPG, Vol., 85,pp1623-1643.

Rao K. L. N. (1969). Lithostratigraphy of paleogene succession of southern Cambay Basin Bulletin ONGC.,Vol. 6, No. 1.

Rao K.G.C.,D. Das and S.D. Deshpande (1968). Numerical Coefficients for the Reduction of Vertical Field to the Magnetic pole.,–ONGC Bull Vol. V, No. 2.

Rao N.D.J. (1967). An Interpretation of Gravity Data of Cambay Basin.,ONGC Bull Vol. 4, No. 1.

Rao V K. (1994). Paleogence Sedimentation Rates in Bombay – Ratnagiri Offshore of Western Indian Shelf and Their Implications in Drift History Events of Indian Plate Volume 31 Number 2 (December, Special Issue).

Rao V.K. (1991). Biofacies, Paleoecology and geological history of deep continental shelf area; Proceedings Conference IERAP, ONGC, Dehradun, p. 271-280.

Rao, C. G (1978a). Carbonate clay mineral relation and the dolomite problem in the carbonate rocks of Bombay High oil field: 10th Int. Cong. Sedimentology (Jerusalem), Proc.

Rao, C. G (1978b). Diagenesis of larger foraminifer (Lepidocyclina sp.) and its effect on porosity in the limestone reservoir rocks of Bombay Offshore area: Geol. Soc. India Jour., v. 13, no. 4, p. 165-168.

Rao, C. G and J. Peters, (1978). Application of Kotmogorov Semirnov 'D' statistic for comparison of two porosity samples distribution from Bombay offshore: Geol. Sec. India Jour.

Rao, C. G. (1977). Biogenic protodolomite from the limestone reservoir rocks of Bombay High oil field: Current Sci., v. 46, no. 3, p. 78-79.

Rao, G. N. (1991). Exploration results and leads in Krishna Godavari basin: Bulletin of Oil and Natural Gas Corporation, v. 28, no. 1, p. 69–87.

Rao, G. N. (1994). Sedimentation and strati-structural control for hydrocarbon accumulations in Krishna Godavari basin: Indian Journal of Earth Sciences, v. 21, no. 3, p. 153–163.

Rao, G. N. (2001). Sedimentation, stratigraphy, and petroleum potential of Krishna-Godavari basin. East Coast of India. Bull. Amer. Assoc. Petrol. Geol. V. 85(9), pp. 1625-1643.

Rao, G. N., and K. S. Mani, (1993). A study on generation of abnormal pressures in Krishna Godavari basin, India: Indian Journal of Petroleum Geology, v. 2, no. 1, p. 20–30.

Rao, G. N., K. Satyanarayana, and K. S. Mani, (1993). Seismo-stratigraphic analysis of Lower Gondwana (pre-Cretaceous) hydrocarbon producing horizons of Mandapeta area of Krishna Godavari basin: Journal of Association of Exploration Geophysicists, v. 14, no. 4, p. 165–172.

Rao, G. N., M. Man Mohan, and B. Prasad, (1996). A study of Paleocene depositional environments in Krishna Godavari basin: Indian Journal of Petroleum Geology, v. 5, no. 2, p. 1–20.

Rao, G.N (1984). Geology and Hydrocarbon Prospects of West Bengal Basin, Dissertation submitted to ISM for M.Tech degree.

Rao, G.N (1985). Geological History and Hydrocarbon Prospects of West Bengal Basin. Proceedings of Workshop on status and line of exploration in West Bengal by ONGC.

Rao, G.N (1999). Pre-Drilling Intelligentia to predict Abnormal Formation Pressures for safe Drilling in hydrocarbon Exploration., Proc.6th Oil and Gas Symposium.,p 86-103.

Rao, G.N. (1986). Hydrocarbon Generation potential of Tertiary, K.G. Basin, ONGC Report (Unpublished).

Rao, G.N. (1993a). Geology and hydrocarbon prospects of East coast sedimentary basins of India with special references to Krishna-Godavari Basin: Journal of the Geological Society of India, v.41,no.5,p. 444-454.

Rao, G.N. (1993b). Sedimentation and subsurface stratigraphy of Krishna- Godavari Basin, east coast of India: Unpublished Ph.D. thesis -Andhra University.

Rao, G.N. and D.S.A.N.Raju. (1992). Krishna Godavari Basin - Geological history, tectonics sedimentation and controls for hydrocarbon accumulation: ONGC Report - Unpublished.

Rao, G.N., Mallick, S., Mandal, N.C., Bandopadhay, S., Nayak, A.K., Dutta, S. (2002). Geology and hydrocarbon prospects of Mahanadi basin, proc. 1st conference and exhibition of APG, p 215-221.

Rao, K. M., Rajamanickam, G. V. and Rao, T. C. S. (1989). Holocene marine transgression as interpreted from bathymetry and sand grain size parameters off Gopalpur. Proc. Indian. Acad. Sci., V. 98, pp. 173-181.

Rao, K.L.N. (1966). Lithostratigraphy of Paleogene succession of southern Cambay basin; Bull. of Oil Natl.ral Gas Commission, Vol. 6, No 1.

Rao, N D J. (1968). An Interpretation of Gravity data of Cambay Basin Vol. IV, No. 2, January.

Rao, R. P. and D. C. Srivastava, (1979). Seismic fades of parts of Bombay offshore basin between Bombay and Ratnagiri; AEG Symposium held at Roorkee, India (in press).

Rao, R. P. and Talukdar, S.N. (1980). Petroleum Geology of Bombay High Field, India. In gaint Oil and Gas fields of the past decade. AAPG,. Mem. 30 pp.487-506.

Rao, R.P. and S. N. Talukdar (1980). Petroleum geology of Bombay High Field, India: AAPG Memoir 30, p,487-505.

Rao, R.P. and Talukdar S.N. (1980). Petroleum geology of Bombay high field, India. In. M.T. Halbouty (ed.) Giant Oil and Gas Fields of the Decade 1968-1978. Amar. ASSO Pet. Geol. Men. 30. pp-387-506.

Rao, T. Srinivasan (1987). The Pakhal Basin - A Perspective. In GSI Mem-6 (Ed., B.P. Radhakrishna) pp.161-187, GSI publication, Bangalore.

Rao, V P., and R sudhakar (1968). Appearance of oil and gas shows in shallow waters and in Ankleshvar oil field Bull ONGC Vol. V, No. 2, December.

Rao, V.K. and Gupta, A. (1985). Studies on fractures identification in Bassein wells Bull. ONGC, Vol.22(2).

Rao, V.K. and V.K. Mittal, (1990). Paleogene Neogene boundary in Bombay Offshore Basin: Geological Society of Indian Memoir Number 16, p. 91-96.

Rao, V.K. and V.S. Rao, (1987). Environmental analysis and facies studies on Tertiary clastic sediments of Bombay offshore area: Symposium Volume, IAS, Hyderabad, p. 220-232.

Rao, Y.S.N., D.P. Rao and A.A. Rahman. Re-appraisal of earlier reported wrench faulting in the Sub-Himalayan region south-east of Kotdwara, Uttar Pradesh., ONGC Bulletin.

Rao,G.N., P Mukhopadhyay and S.Bandopadhaya (2003). Subsurface Geology of Offshore Mahanadi Basin. Jou. Ind Assn. of Sedimentologists Vol.22 no. 1 and 2 p 57-66.

Rao., G. N. (1993). Electro-Facies Analysis of Reservoirs in Krishna-Godavari Basin. Jour Geological Society of India,vol.41, No.1 pp 74-82.

Rao., V K and K.N. Pandey (1989). Fracture evaluation and feasibility of oil exploitation in Bassein Field –Bulletin of Oil and Natural Gas Commission Vol. 26 No. 1.

Rao., V.K. and K.N. Pandey (1989). Fracture Evaluation and Ffeasibility of oil exploitation in Bassein field., ONGC Bulletin.,Vol. 26 No. 1.

Rao.,G. N. (1999). Pre-Drilling Intelligentia for safe drilling. Proceedings of 6th annual Oil and Gas conference conducted at Mumbai, India.

Rasheed, D.A., Ravindran, C.N. (1978). Proceedings of the VII Indian Colloquium Micropaleontology and Stratigraphy. In: Rasheed, D.A. (Ed.), Foraminiferal Biostratigraphic Studies of the Ariyalur Group of Trichirapalli Cretaceous Rocks of Tamil Nadu State, pp. 321-336.

Rathore SS, AR Vijan Krishna BN Parbhu and K N Misra (1999). Dating of glauconites fron Shirbhu Shales of Vindhyan Super Group- Proc PetroTech 99 pp-191-196.

Rathore, S.S., Venkatesan T.R., Srivastva R.K. (1996). Mundwara alkali igneous complex, Rajasthan: Chronology and Sr Isotope systematic. Journal Geological Society of India.v. 48(5) p.517-528.

Raval, U. (2001). Earthquakes over Kutch: A region of 'Trident' Space–Time Geodynamics. Curr. Sci., 81, 809–815.

Ravi Kumar M. and Mohan G. (2005). Mantle discontinuities beneath the Deccan volcanic province Earth and Planetary Science Letters, v. 237, p, 252– 263.

Ravi, K., Vargo, R., and amp; Lasley, B. (2008). Successful Cementing Case Study In Tuscaaloosa HPHT Well. Paper SPE 115643 Presented at the SPE Russian Oil and Gas Technical conference and Exhibition, Moscow, Russia, 28-30 October.

Ravindran, C.N., Kalyanasundar, R. (1995). Status of surface and subsurface stratigraphy of Cretaceous Sequence in the Trichinopoly Type Area, Cauvery Basin, India. Proceedings of Petrotech, New Delhi, pp. 53-64.

Rawat, R.S., S.N. Swamy, N.P. Juyal and P.N. Kapoor (1996). Palynostratigraphic and Paleoenvironmental Interpretation of Panna Formation in the East of Bombay High Bull ONGC Volume33 Number 2 December.

Ray D S., and R. S. Mishra. Hydrocarbon prospectivity of Mesozoics in western margin of Cambay Basin and adjoining areas, India – An exploration challenge.

Ray D S., and S. Mayor (2006). Syn-Rift Sequences Through Transfer Zones– Thrust Area for Future Exploration in Cambay Basin, Western India 6th International Conference and Exposition on Petroleum Geophysics "Kolkata 2006".

Ray, S., S.K. Deb, S.C. samanta, D.C. Gupta and S. Prasad (1982). Enhancement of Seismic data in the Mansa-Pethapur area of Cambay Basin.,ONGC Bull Vol. 19, No. 2.

Ray, S.B. and Mohanti, M. (1989). Sedimentary processes in the Mahanadi River estuary, East Coast of India. Workshop on Coastal processes and coastal Quaternaries of Eastern India. Geological Survey of India, Eastern Region, Calcutta, pp. 28-29.

Raymer, L.L. (1981). "Elevation and Hydrocarbon Density Correction for Log-Derived Permeability Relationship," The Log Amdyst.

Raymer, L.L., Hunt, E.R., and Gardner, (1980). "An Improved Sonic Transit Time-to-Porosity Transform," Trms., SPWLA Annual Logging Symposium.

Raynolds, R.C., and Johnson, G.D. (1985). Rates of Neogene depositional and deformational processes, northwest Himalayan foredeep margin, Pakistan, Geol. Soc. London, Mem.10., pp. 297-311.

Raza Khan, M S.,M.C. Srivastava and S.K. Sharma (2003). Genesis and Entrapment of Hydrocarbons in Mehsana block, north Gujarat, India –Proceedings of PETROTECH-2003 January 9-12, New Delhi.

Raza Khan, M S.,R. Kumar, A.T. Vushim, B.N. Prabhu and B.K. Sharma (2000). "Genesis and Entrapment of Hydrocarbons within Multi-layered Reservoirs of Geleki Field", AAPG Proceedings, 373-379.

Reading, H.R. (1982). Sedimentary basins Global tectonics – Proceedings of the Geological Association,– Vol. 93, Part 4., p. 321-347.

Reddy, A.N., Yadagiri, K. (2001). Benthic Foraminifera and clay mineralogy across the cretaceous tertiary boundary in Cauvery Basin, India. Journal of Geological Society of India 57, 339-351.

Reddy, B.S.R. and Prasad, K.V.S.R. (1982). The sand spit near Kakinada - Further studies. India J. Earth Sci., 2, No.2, pp.167-173.

Reddy, P.R., Venkateswarulu, N., Prasad, A.S.S.S.R.S., Koteswar Rao, P. (2002). Basement structure below the coastal belt of Krishna-Godavari basin: correlation between seismic structure and well information. Gondwana Research 5, 513-518.

Reid, P.I., et al., March, (1996), "Mechanism of Differential Sticking and a Simple Wellsite Test for Monitoring and Optimizing Drilling Mud Properties", SPE 35100.

Reineck, H.E. (1972). Tidal flats. (In) Recognition of Ancient sedimentary environments. Ed., by J.K. Rigby and W.K. Hamblin, pp.146-159. Spec. Publ. Soc. Econ. Paleont. Miner., 16, Tulsa.

Reynolds, E. B. (1970), Predicting over-pressured zones with seismic data. World Oil, Oct- 70, pp. 78–82.

Riazi, M. R. and Daubert, T. E.(1987), "Characterization Parameters for Petroleum Fractions," Ind. Eng. Chem. Res, Vol. 26, No. 24, pp. 755–759.

Richardson, J.G. and Blackwell, R.J. (1971). Use of Simple Mathematical Models for Predicting Reservoir Behaviour. J.Pet.Tech., September: 1145-1154. Trans. AIME.

Rima Chatterjee, K. Manoharan, and Manoj Mukhopadhyay (2013). Petrophysical and mechanical properties of Cretaceous sedimentary rocks of Cauvery basin, Eastern continental Margin of India-J. Ind. Geophys. Union-v.17,no.4, pp,349-359.

Robinson, E. A. (1983), Seismic Velocity Analysis and the Convolutional Model. IHRDC, Boston, USA.290p.

Robinson, E. S., and C. Coruh (1988). Basic Exploration Geophysics, John Wiley.

Rodolfo, K.S. (1969 a). Bathymetry and marine geology of Andaman Basin and tectonic implications for S.E. Asia. Bull. Geol. America, Vol. 80, pp 1203-1230.

Rodolfo, K.S. (1969b). Sediments of the Andaman basin N.E. Indian Ocean, Marine Goel. Vol. 7 pp 371-402.

Rodolfo, K.S. (1970). Geology of Andaman Sea and Burma. Further comments, Bull. Geol. Soc. America Vol. 81, pp 1851-1853.

Roland, N. Horne (1990). " Modern Well Test Analysis- A Computer Aided Approach", Stranford University, copy right (C), Petroway Inc.

Rommetveit R. (2003). HPHT Well Control; An Integrated Approach, Offshore Technology Conference, OTC 15322, Houston, Texas USA.

Ron Bland, Greg Mullen, Yohnny Gonzalez, Floyd Harvey, and Marvin Pless (2006). Baker Hughes Drilling Fluids, HPHT Drilling Fluid Challenges, IADC/SPE 103731.

Rouse, B., F. Hiebert, and L.W.Lake, (1992).Laboratory testing of a microbial enhanced oil recovery process under anaerobic conditions, SPE 24819, 547-555.

Roy A. B., and Alokesh Chatterjee (2015). Tectonic framework and evolutionary history of the Bengal Basin in the Indian subcontinent., current science, VOL. 109, NO. 2.

Roy Burman, A (1983). Geology and Hydrocarbon Prospects of West Bengal, Pet. Asia Jour. Nov.83, pp. 51-56.

Roy Burman, A (1992). Geological History and Hydrocarbon Exploration in Bengal Basin.,Ind Jour.Geol.V.64.No.3., P 235-258.

Roy Burman, A and K.V.S. Murty (1982). Exploration Targets in West Bengal Basin Bull.ONGC., Vol.19.no. 2 p119-126.

Roy Moulik, S.K.,H.J. Singh, R.K. Singh Rawat, Md. S.Akhtar, S. Mayor, and M. Asthana (2009). Sand Distribution Pattern and Depositional Model of Kopili Formation (Eocene) with Special Reference to Sequence Stratigraphic Framework from North Assam Shelf, Assam-Arakan Basin, India Posted July 28, AAPG Annual Convention, Denver, Colorado.

Roy, A.B. and Jakhar, S.R. (2002). Geology of Rajasthan (Northwest India) Precambrian to recent,. Scientific Publishers (India) Box 91, Jodhpur, p. 1-421.

Roy, A.K., D.P. Paine, S. Sett, and H.N. Bhattacharya (2012). Gravity-Magnetic and Magneto-Telluric surveys in Purnea Onland Basin,India - A case history. SPG-2012.

Roy, M.M and Vashist, N, (1994). " Environment of deposition and sand geometry of Daman Offshore." ONGC-Schlumberger Report-Unpublished, 1994.

Roy,TK (1983). Geology and Hydrocarbon Prospects of Andaman Nicobar Basin – Petroleum Asia Journal pp-37-50.

Roy., K.K.and A. Saha.(1976), SP. Log Interpretation should preferably be qualitative a debatable proposition., ONGC Bull., Vol.13 No. 1and 2.

Roy., T.K. (1990). Structural styles in southern cambay asin india and role of Narmada Geofracture in formation of giant Hydrocarbon Accumulations– Bull ONGC Vol. 27 No. 1.

Roychoudhury, S.C. Deshpande (1982). Regional distribution of carbonate facies, Bombay offshore region, India: AAPG Bulletin, v 66, no.10, p.1483-1696.

Roychowdhury, S.C and S.V. Deshpande, (1982). Regional distribution of carbonate facies, Bombay offsore region, India; AAPG Bulletin v.66, no.10, p.1483-1496.

Rybach, L. (1986). Amount and significance of radioactive heat sources in sediments. –In BURRUS, J. (ed.): Thermal modellingin sedimentarybasins. EditionsTechnique, Paris.

S chnorr DR (1996). "More Answers from Production Logging Than Just Flow Profiles," Transactions of the SPWLA 37th Annual Logging Symposium, New Orleans, Louisiana, USA, June 16-19, 1996, paper KK.

Sah S.L. and M.A. Ganapathy (1979). Three Dimensional Seismic Data acquisition and sign-bit recording., ONGC.Bull., 16 No.2.

Sah, S.L., J.K. Bhatnagar and Philip d'souza-(1985)- A Method of Measuring the Relative sensitivity of hydrophones of an active section of marine streamer., ONGC Bulletin Vol. 22 No.1.

Saha D. (2011). Integrated Analysis of Gravity and Magnetic Data in the Upper Assam Shelf and Adjoining Schupen Belt Area - A Critical Review- GEO-India, Greater Noida, New Delhi.

Saha, D., Ghosh, G. (1997). Tectonic setting of Proterozoic sediments around Somanpalli, Godavari Valley. Journal of Indian Association of Sedimentologists 7, 29-45.

Saha, D., Ghosh, G. (1998). Litho-stratigraphy of deformed Proterozoic rocks from around the confluence of Godavari and Indravati Rivers, South India. Indian Journal of Geology 70,217-230.

Sahay, B. (1978). Geology of Bombay High and adjoining structures (preprint). Offshore South-East Asia Conference, pp.1-11.

Sahay, Bhagwan and Fertl, W.H. (1988). Origin and evaluation of formation pressures. Allied Publishers Private ltd.pp-1-292.

Sahni, MR and Srivastava, RN (1954). New organic remains from the Vindhyan System of the probable systematic position of Fermoria Chapman. Curr. Sci., 23, 39.

Sahoo, M. and K.D, Gogoi, (2008). Paleocene to Holocene Structure, Stratigraphic Development and Hydrocarbon Accumulation in Upper Assam Foreland Basin, India: Proceedings of 2008 Geo India Convention and Exhibition, New Delhi, India.

Sahu. B.K., and Chandu Satyanarayana (1986). Multivariate factor model for well-log Analysis of Bombay High area, india., ONGC Bulletin.,Vol. 23 No. 1.

Saikia, M. M. and Dutta T.K. (1980). Depositional Environment of Source beds of High wax oils in Assam Basin, India. AAPG, Vol.64,No.3, pp.427-429.

Sain M. and Chopra A.K. (1991). Contribution of geochemical studies to exploration in Himalayan Foothills area and Ganga basin. In: proceedings of Conference on Integrated Exploration Research-Achievements and Perspectives, J. Pandey and V. Banerjie (eds.), KDMIPE, ONGC, Dehradun, pp. 461-469.

Sambasiva Rao, M., Nageswara Rao, K. and Vaidyanadhan, R. (1978). Morphology and evolution of Mahanadi and Brahmani Baitarani Deltas. Proceedings of Symposium on Morphology and Evolution of Landforms, Dept. Geology, Delhi University, India, pp. 241-248.

Samanta, U. and R.K. Shukla (1985). Trends of organic matter maturation in Krishna –Godavari Basin- Proc of Petroleum geochemistry and exploration in Afro-Asian Region, Dehradun., Ed-Ruby Kumar et. al., pp-297-300.

Sams, M.S., Atkins, D., Said, N., Parwito, E.,van Riel, P. (1999). Stochastic inversion for high resolution reservoir characterisation in the central Sumatra Basin, SPE Asia Pacific Improved Oil Recovery, Kuala Lumpur, SPE 57620.

Sander, N.J. and Humphrey, W.E. and Mason J.F. (1975). Tectonic framework of SE Asian and Australia-its significance in the occurrence of petroleum IX World Petroleum Congress, pp 83-106.

Sandwell, D.T., Smith, W.H.F. (1997). Marine gravity anomaly from Geosat and ERS I satellite altimetry. J. Geophys. Res. 102, 10039-10054.

Sanjeev Thakur, Anil Tyagi, Anish T.K and Kamlesh Saxena (2006). An Integrated Petrophysical Study for deep water turbidite reservoir to improve Hydrocarbon Saturation., 6th International Conference and Exposition on Petroleum Geophysics "Kolkata".

Sar, D.; M.K. Maheshwari, S. Rangarajan, C. S. Bahuguna (2009). "Eighty five degrees east ridge and its hydrocarbon potential". Geohorizons (Society of Petroleum Geophysicists (SPG), India) 15.

Sarin, L.N., I R. Haider and S. Sridhar (1986). Optimization of Drilling Parameters for Bombay offshore Basin., ONGC Bulletin., Vol. 23 No.1.

Sarin, L.N., I.R. Haider and S. Sridhar (1995). Optimization of Drilling parameters for the current wells of Krishna Godavari and Cauvery Basins., ONGC Bulletin Vol. 36 no.

Sarkar. A.K., Mohan.R. (2004). "Sub Basalt Imaging Using Low Frequency Processing and Angle stack In Saurashtra Region, Western India". 5th Conference and Exposition on Petroleum Geophysics, Hyderabad, India PP 90-94.

Sarma B. S. P and M. V. R. Krishna Rao. (2005). Basement structure of Godavari basin, India – Geophysical modelling current science, VOL. 88, NO. 7, 10.

Sarma, P. L. N., R. L. Basak, and A. K. Naithani, (2008). Exploration for Hydrocarbon Below Trap in Krishna-Godavari Basin- – A Seismic Imaging Issue. SPG proceedings.

Sarvesam, G, S. Choudhury, G.B.Ramamurthy, Ranbir Singh (2006). Advanced 3D Seismic Modeling Studies as a Tool for Effectve Seismic Campaign for Exploring Strati-structural Features in Developing Field., 6th International Conference and Exposition on Petroleum Geophysics "Kolkata".

Sastri V. V. and B. S. Venkatachala. Organic remains, age and environment of the pre-Siwalik sediments encountered in some deep wells drilled in southern part of Ganga Valley., ONGC Bulletin.

Sastri, V. V., and L. L. Bhandari (1973). Petroleum prospect of the Continental Shelves of India: Inst. Pet. Expin. ONGC, India (unpublished).

Sastri, V. V., Sinha, R. N., Singh, G., and Murti, K. V. S. (1973). Stratigraphy and tectonics of sedimentary basins on the east coast of peninsular India, AAPG Bulletin, v.57, pp: 655-678.

Sastri, V.V. and Bedi, T.S. (1962). On the occurrence of Niogypsina, Cycloclypena and orbulina in the Miocene of Andaman Islands, Current Science Vol. 31 pp. 20-21.

Sastri, V.V., Sinha, R.N., Gurucharan Singh and Murti,K.V.S. (1973). Stratigraphy and tectonics of sedimentary basins of east coast of peninsular India, Bull. Am.Asso.Pet.Geol.,57, 655-678.

Sastri, V.V., Venkatachala, B.S., Narayanan, V. (1981). The evolution of the east coast of India. Palaeogeography, Palaeoclimatology, Palaeoecology 36, 23-54.

Sastri,V V, and Raiverman. V. (1968). On the basin study programme of the cretaceous tertiary sediments of the Cauvery basin; in, Cretaceous-Tertiary formations of South India: Mem. Geol.Soc. India, Vol. 2.

Sastri., V. V., L. L. Bhandari, A. T. R. Raju, A. K. Datta (1971). Tectonic Framework and Subsurface Stratigraphy of the Ganga Basin., J GSI Volume 12, Issue 3.

Sastry, C.V.S., D.R. Ghosh, Shipra S. Bhagwani and S. Mahapatra (1984). Geometry of iii and iv sands of Kalol Formation in Kalol-Jhalora-Ahmedabad area, Cambay Basin–Bull ONGC VOL. 21 NO. 2.

Sastry, V.V., Madan Mohan and Guha, D.K. (1964). Tertiary Biofacies of Kutch, Cambay and Anklesvar areas, India; International Geological Congress, 22nd session, New Delhi.

Sattwati Dey, Chandramani Shrivastva, Sreemanti Gijare; Rajeev Kumar, Sanchita Ganguly (2010). Integrated Fracture Identification with Z-VSP and Borehole Images: A study from Cambay Basin., SPG.

Satyanarayana, K., G. N. Rao, P. R. Bhavana, H. R. Sinha, and A. N. Banerjee (1996). Miocene sedimentation in subsurface of Godavari delta: Bulletin of Oil and Natural Gas Corporation, v. 33, no. 1, p. 77–92.

Saxena K. and Klimentos T. (2004). Field study of integrated formation evaluation in thinly laminated reservoirs, SPWLA 45th Annual logging symposium.

Saxena, I.P. C.K. Singh and Sukhadarsan Kumar (1982). Neogene Geological model of a part of Bengal Basin. Bull ONGC. Vol.19.no.2 p101-110.

Saxena, I.P. C.K. Singh and Sukhadarsan Kumar (1984). Palaeo-Environmental Analysis of Post-Eocene Sequence Bengal Basin., Pet. Asia Jour.p 158-166.

Saxena, R K.,Shishir K. Saxena, Rashmi Anand, Rajeev Sharma and G.C. Datta (2003). Fatty Acid distribution in source rock of Cambay-Hazad Petroleum System of Cambay Basin, India –Proceedings of Petrotech-2003 January 9-12, 2003, New Delhi.

Saxena,R., M.M. Panigrahi, U. Goyal, V. Sharma, U.B. Singh, K.S. Prasad and S. Ramanan (2005).,Use of Logging while Drilling (lwd) – Geo-steering Technology in Drilling Horizontal well under hostile well bore environment -a case study of Mumbai High Field.,Petrotech.

Seeber, L., Armbruster, J.G., and Quitmeyer, R.C.(1981). Seismicity and continental subduction in the Himalayan arc, in Zagros, Hindu Kush, Himalaya, Geodynamic Evolution, Geodyn. Ser., vol. 3, pp. 215-242.

Seifert, W.K., J.M. Moldowan, and R.W. Jones (1980). Application of biological marker chemistry to petroleum exploration: Proceedings of the 10th World Petroleum Congress, Bucharest, p. 425–440.

Sekhar Srivastava,V J Pandey, D C Sarma and N.P Singh (1985). Paleo-environmental Reconstruction for sub surfsce Sedimentary Sequence around Shahgarh

Sen, G, V K Rao, P K Gupta, DVR Murty, S Dutta, AK Bansal, RP Singh and Rajiv Kumar (1993). Porasity distribution in Bassaim Limestone Sequence of Neelam Field West Coast offshore India, Proc Second Seminar Petroliferous Basins Indian Petroleum Publications Vol.2 pp-413-419.

Senapati,R B.,N.K. Singh, Ashish Kumar and C.L. Tikku (1993). Hydrocarbon Prospects of Patan Area, Patan-Sanchor Block, North Cambay Basin, India Vol. 30, No. 2.

Sengupta, S. N (1972). Status of oil exploration in the offshore areas adjoining India: 4th Symposium on Development of Asia and Far East, Proc., p. 112-114.

Sengupta, SN (1962). Basement configuration of the Indo-Gangetic plains shown by the aeromagnetic surveys. In Proceedings Seminar on Oil Prospecting in the Ganga Willey, Tech. Publ. No. 1, ONGC.

Serra,O. (1985). Sedimentary Environment from Wireline Logs: Schlumberger Document.

Shadravan, A. and amp; Amani, M. (2012). HPHT 101 - What Every Engineer or Geoscientist Should Know About High Pressure High Temperature Wells. Presented at the SPE Kuwait International Petroleum Conference and Exhibition, Kuwait.

Shah. J., Fainstein. R., Broetz. R.J. and Wygrala.B. (2009). "Hydrocarbons in Sub-basalt Sediment-Basin Modeling of Kerala-Konkan Basin, Offshore West India". 71st EAGE Conference and Exhibition-Amsterdam, the Netherlands, 8-11 June 2009.

Sham V. Rao. (2006). Discovering Medium-Giant Fields- Perspectives and Challenges., 6th International Conference and Exposition on Petroleum Geophysics "Kolkata".

Shanmugasamy,S., R. V. S. Murthy, R. L. Basak and A. K. Roy (2006). Removal of High Amplitude Noise and Preservation of Relative Amplitudes – A Case Study. 6th International Conference and Exposition on Petroleum Geophysics "Kolkata".

Sharma N.C (2002). History of Seismic Prospecting In ONGC - A Chronological Sketch of Events Oil and Natural Gas Corporation Limited. GEOHORIZONS.

Sharma, K. K. (2005). Malani Magmatism: An extensional lithospheric tectonic orign, in Plates, Plumes and Paradigms, edited by G.R. Foulger, J.H. Natland, D.C. Presnall and D.L. Anderson, Geological Society of America 388, p. 463-476.

Sharma, K.K. (2006). Intraplate Seismicity of the northwestern Indian shield: Implication for the reactivation of palaeo-tectonic elements. Geophysical Research Abstracts, V. 8, 03253, 2 European Geosciences Union 2006SRef-ID: 1607-7962/gra/EGU06-A-03253.

Sharma, O.P. Pardeep Kumar and R.P. Verma (2005). Identification of Natural Gas Hydrates with Nuclear Magnetic Resonance (NMR) log data., PetroTech Proceedings-article- 384.

Sharma, S.C., Beena Jhaldiyal, D.K. Ghosh and R.P. Verma (2005). Identification and Estimation of High Formation pressure by log data - a case study-Proceedings-article-296.

Sharma, S.K. U.V.S. Nohwar, J.Koithara.(1977),Analysis of Anomalous SP in Tipams of upper Assam valley., ONGC Bulletin Vol. 14 No. 1 and 2.

Sharma, V P. (1973). An account of the stratigraphy and structure of Doda-Bhadarwah-Basantgarh region, Jammu Himalaya, Seminar on Géo-dynamics of the Himalayan Region, Hyderabad, India.

Sharma,S. K., H.L. Kak, H.L. Meena and V. Pratap (2003). EOR process in Balol-Santhal Fields, India's honeymoon with in-situ combustion an overview–Proceedings of PETROTECH-2003 January 9-12, 2003, New Delhi.

Sharma., O.P. and O.P.Arya. (2006). Formation Strength Estimation From Well Log Data For Sand Cut Analysis in Tapti – Daman Area, Western Offshore Basin, India., 6th International Conference and Exposition on Petroleum Geophysics

Sharma., S.C. (2006). Evaluation of Shaly Sand Reservoirs with Anomalous Log Response In Kamlapuram Field -A Case Study/.6th International Conference and Exposition on Petroleum Geophysics.

Shashi Shankar and Kallole Ghosh (1985). Hydraulic Factors affecting the Rate of Penetration., ONGC Bullettin Vol 32 no.2.

Shaughnessy, J. and amp; Helweg, J. (2002). Optimizing HPHT Cementing Operations. Presented at the SPE Drilling Conference, Dallas, Texas.

Shaughnessy, J. M., RomoL. A. (2003). Problems of Ultra-Deep High Temperature High Pressure Drilling. SPE 84555, SPE ATCE, Denver Coloradp.

Shaw J.C. and Chee T. (1996). "Laboratory Evaluation Of Drilling Mud Systems For Formation Damage Prevention In Horizontal Wells" SPE-37121,.

Sheehy, A.J. (1990. Field studies of microbial EOR, SPE/DOE 20254, 785-790).

Shenai and M.S. Rao.(1984). Exploration offshore Godavari-Krishna Basin ONGC Bulletin Vol 20 no 2.

Shenai, K.R. and Rao, M. S. (1982). Offshore exploration of Godavari - Krishna Basin, Bull. ONGC., 19, no.2,pp-293 -300.

Sherbon Hills, E. (1963). Elements of Structural Geology- Asia Publishing House-pp 1-483.

Sheriff, R. E. (1975). Factors affecting amplitudes. Geophysical Prospecting,Vol. 23, P.125–138.

Shree Prakash Singh, Nikhil Agnihotri, Praveen Dhiman and Dr. G. K.Ghosh (2006). Crooked line seismic survey in thrust-belt and mountainous area of Mizoram, North East INDIA: A Case Study., 6th International Conference and Exposition on Petroleum Geophysics "Kolkata".

Shukla S.N., Katiyar G.C., Mudiar B., Murlikrishna A., Misra M.M. (1993). Lithostratigraphy of Indian Petroliferous Basins, Document-I, Ganga, Punjab and Purnea Basins. KDMIPE, ONGC, Dehradun, internal document.

Shukla, S.N. (1983). Petroleum Geology of India. In Schlumberger well Evaluation Conference, India, pp.23-25.

Shyam Chand, Subrahmanyam, C. (2001). Subsidence and isostacy along a Sheared margin-cauvery Basin, Eastern continental margin of India. Geophysical Research Letters 28 (No. 11), 2273-2276.

Shyam Mohan and Pranaya Sangai-(1995). ONGC Through Thar Desert-In Exploration View – Proc PetroTech 95 pp-319-326.

Shyam Mohan and S Mangal, (1997). Prospect Evaluation and Model Prediction fot Future oil Exploration towards western part of Bombay High Field- Proceedings of Second International National Petroleum conference - PetroTech -97 pp-445-352.

Singh NP- (1984). Additional to the Tertiary Bio Stratigraphy of Jaisalmer Basin –First Petrilferous Basins of India (SPEB) pp-106-128.

Singh, Dr N. P.,N. Ganesh, S. Prasad, Bachan Singh (2006). Offset VSP Surveys Going Beyond Traditional Boundaries: A Case Study.,6th International Conference and Exposition on Petroleum Geophysics.

Singh, I.B. (1973). Depositional environment of Vindhyan sediments in Son Valley area, Recent Res. Geol., 1, 146-152.

Singh, M.P., Singh, P.K. (1994). Indications of hydrocarbon generation in coal deposits of Rajmahal Basin, Bihar: Revelation of fluorescence microscopy, Jour. Of Geol. Soc.Ind. V. 43, pg. 647-658.

Singh, Paramjit (2000). Inversion of Gravity Data Using Effective Density Contrast in South Rewa Basin, 3rd Conference and Exposition on Petroleum Geophysics, New Delhi, pp405-408.

Singh, Paramjit, Kumar, J. and Ram, Jokhan. (2003). Concealed sub-surface structures below Punjab Alluvium: A geophysical evidence, Journal on Paleontologica Indica, India, Special issue on "Himalayan Orogen-Foreland Interaction".

Singh, T.P.,M. Pal, G. Mehra And S.K. Laskar (1991). Basement Mapping of A Part of Andaman Offshore Basin From Gravity-Magnetic Data - A Case History.,ONGC Bull Vol.28 JUNE, No.1.

Singh, V.K. Baid, S.P.S. Negi, D. Subrahmanyam and S. Biswal (2005). Seismic to Simulation of Porosity for Reservoir Characterisation- a case study.– PetroTech.

Singh,N K and N K Lal,(1993). Geology and Petroleum Prospects of Konkan Kerala Basin- Proc Second Seminar Petroliferous Basins Indian Petroleum Publications Vol.2 pp-462-469.

Singh,N.P., Boruah,R.M. and Dave, A. (1986). "Biostratigraphy of the Eocene sequence of Upper Assam Shelf", Bull. ONGC, vol.-23, no.2, pp 45-66.

Singh,N.P., Boruah,R.M. and Dave,A. (1986). "Biostratigraphy of the Eocene sequence of Upper Assam Shelf", Bull. ONGC, vol.-23, no.2, pp 45-66.2. Goel, S.M., Sharma, P.C.

Sinha, A.K. (1997). "The concept of terrane and its application in Himalayan and adjoining region", In A. K. Sinha, F. P. Sassi, and D. Papnikolaou (Eds.) Geodynamic domains in the Alpine- Himalayan Tethys (IGCP 276). A. A. Balkema, Rotterdam,1-43.

Sinha, B K. (2003). Successful utilization of Hydraulic Fracturing Technique to increase productivity and recoverable reserve of Kalol oil field in Gujarat – Proceedings of Petrotech-2003 January 9-12, 2003, New Delhi.

Sinha,A K., R K Yadav and S M Qureshi (1993). Status of Exploration in South Shahgarh Sub Basin of Jaiselmar Basin Rajasthan -Second Seminar on Peroliferous Basins of India –Vol-2 Indian Petroleum Publishers Dehradun.,pp-285-333.

Sisodia, M. S., Singh, U. K., Lashkari, G., Shukla, P.N., Shukla, A. D.,and N Bhandari (2005). Mineralogy and trace element chemistry of the Siliceous Earth of Barmer basin, Rajasthan: Evidence for a volcanic origin: Journal Earth System Science, v. 114 (2), p.111-124.

Sivaraman, T.V. and Raval, U. (1995). U-Pb isotope study of zircons from a few granitoids of Delhi-Aravalli belt: Journal Geological Society of India v.46 p. 461-475.

Skelt, Christoper, Pickford, Scott, and Harrison, Bob,(1995). "An integrated approach to saturation height analysis", SPWLA 36th annual logging symposium, Paris, France.

Sohoni, P. S.(2001). Structural studies on central Kachchh Mainland with special reference to Quaternary tectonism, Unpublished Ph D thesis, M. S. University of Baroda, 2001.

Soodan K.S. Mathur Y.K. Anjani Kumar and D.S.Mitra (1990). Integrated Geological Research in Bengal Basin. Conf. Integrted Exploration Research Proc. Editors Jagdish Pandey and V. Banerjee pp 329-342.

Soumen Paul, B S Bisht, Vinay C Raturi, Sapna Jaiswal (2013). Exploring frontiers through sequence stratigraphic approach: Vindhyan basin case study -10th Biennial International Conference and Exposition Kochin.

Soumen Paul, B S Bisht, Vinay C Raturi, Sapna Jaiswal. (2013). Exploring frontiers through sequence stratigraphic approach: Vindhyan basin case study.,SPG 10th Biennial International Conference and Exposition.

SreenivasaRao, T. (1985). A note on the stratigraphy of the upper Precambrian sediments around Ramagundam, Andhra Pradesh. Indian Minerals..

SreenivasaRao, T. (1987). The Pakhal Basin a perspective. Geological Society of India Memoir 6, 161-187.

Sridhar, S. and Nanda, J.S., 1981. Optimization of weight on bit and rotary speed, ONGC Bulletin, Vol.18 No. 1,June, pp 95-101.

Sridhar. S, (1983). Optimization of Drilling Parameters, ONGC Bullettin., Vol 20 No,2.

Srikantia S. V. and O. N. Bhargava (1998). "Geology of Himachal Pradesh", Geol. Soc.India, Banglore, 406.

Srikantia S. V. and R. P. Sharma, (1969). "The Geology of the Shali Belt and the adjoining area in the Simla Himalaya", Mem. Geol. Surv. India, 106, 31-161.

Srinivasan S and BM Khar (1995). Frontier Basin Exploration in India- Prospectives and Challenges- Proc PetroTech 95 pp-1-19.

Srinivasan, M.S. and Azmi,R.J. (1976). New Developments in the late cenozoic lithostratigraphy of Andaman-Nicobar Islands, Bay of Bengal. Proc. VI Colloguim of Indian Micropalentology and Stratigraphy, Varanasi.

Srinivasan, M.S., and Sharma, Vol. (1973). Stratigraphy and microfauna of Car Nicobar Island, Bay of Bengal. Jour. Geol. Soc. India. Vol. 1a, No. 1, pp. 1-11, pls. 1-3.

Srinivasan, S. and Khar, B.M., 1995. Status of oil exploration in west coast basins – a review. Proceedings on National Seminar on Recent Researches in Geology of Western India, Baroda, India, 29-31 October 1995, pp. 7-58.

SrinivasaRao, K., SreenivasaRao, T., RajagopalanNair, S. (1979). Stratigraphy of the upper Precambrian Albaka belt, east of the Godavari River in Andhra Pradesh and Madhya Pradesh. Journal of Geological Society of India 20, 205-213.

Srinivasarao, K., Subbaraju, M., Sreenivasarao, T., Imam Alikhan, Md. and Silekar,

Srivastava B N., M S Rana, Narendra K. Varma (1983): Geology and Hydrocarbon Prospects of Vindhyan Basin – Petroleum Asia Journal SPEB-I Vol 3 pp-179-189.

Srivastava D.C.,and P.N Mathur (1984). Seismic Facies of the Carbonate sequence of the Mahim area, Bombay offshore Basin., ONGC Bull VOL. 21 NO. 2.

Srivastava RP, Vedanti N, Akervoll I, Bergmo P, Yerramilli RC, Yerramilli SS, Dimri VP (2015). Study of CO2 EOR in a sector model from mature oil field, Cambay basin, India. In: Mukherjee S (ed) Petroleum geosciences: Indian contexts. Springer International Publishing, pp 87–98.

Srivastava, D K.,S.K. Suri, B.P. Dhulia, D.R. Saklani and S.N. Mohanty (2003). Accreting more reserves in a mature field - Acase study of South Kadi Field, north Cambay Basin, India– Proceedings of PETROTECH-2003 January 9-12, 2003, New Delhi.

Srivastava, D.K., Sharma, S., Goel, S.M., and Husain, A. (1999). Geological setting and exploitation strategy for heavy oil in north Cambay Basin, India, Proceedings PETROTECH-99, Third International Petroleum Conference, New Delhi, p. 123-130.

Srivastava, R.K. (1988). Magmatism in the Aravalli mountain range and its environs, Memoir.,Geological Society of India, v. 7, p.77- 94.

Srivastava, R.K.,S. Srivastava and S. Bhattacharya (2003). Integrated Reservoir Characterisation of Barail Main sand, Demulgaon Field by Geocellular Modelling – a new approach., Proceedings of PETROTECH.

Srivastava, R.P., Vedanti, N., and Dimri, V.P. (2007). Optimum Design of a Gravity Survey Network and its Application to Delineate the Jabera–Damoh Structure in the Vindhyan Basin, Central India, Pure Appl. Geophys, v.164, pp.1–14, doi: 10.1007/s00024-007-0252-1.

Srivastava, R.P.,Vedanti, N., Pandey, O.P., and Dimri,V.P. (2009). Detailed gravity studies over Jabera-Damoh region of the Vindhyan Basin, central India and crustal evolution, Journal of Geological Society of India, Vol. 73, 715-723.

Srivastava. M.C. and V. Pratap (2005). Occurrence of Gas Cap in under-saturated reservoir of North Kadi oil Field, Cambay Basin, India: a new insight., PetroTech.

Staggs, H.M. and Herbeck, E.F. (1971). Reservoir Simulation Models – An Engineering Overview. J.Pet.Tech., December: 1428-1435. Trans. AIME.

Stewart, G. and Wittman M.(1964). Control Through Dip Computations,"Bull., AAPG 48, No. 12.

Stone, H.L.: (1991). "Rigorous Black Oil Pseudo Functions", SPE 21207, SPE Symposium on Reservoir Simulation, Anaheim, California.

Subba Rao Dr.V.V., Dr.Y.S.K.Reddy, A.K.Agrawal (2006). Identification of Deep-water Depositional Features on Dipmeter and Borehole Image Logs- A Case Study from Cauvery Basin, India,6th International Conference and Exposition on Petroleum Geophysics.

Subba Rao, M. (1958). Studies on the Kakinada bay on the east coast of India. Qt. J. Geol. Min. and Metallurgical Soc. Ind., 39, pp.75-91.

Subir Kumar, Sunil Kumar and A.K.S. Rathore (2003). Role of saturates, aromatics, resins and asphaltenes in insitu combustion of Santhal Field– Proceedings of PETROTECH-2003 January 9-12, 2003, New Delhi.

Subrahmanyam, A. S., K. S. R. Murthy, T. C. S. Rao, M. Malleswara Rao, A. S. Lakshminarayan, and K. Venkateswarlu, (1994). A two-stage evolution of Visakhapatnam-Paradip shelf, East Coast of India, from magnetic study, Proceedings of Indian Academy of Science (Earth Planet. Sci.), 103, 17–26.

Subrahmanyam, A.S., Subrahmanyam, V., Sarma, K.V.L.N.S., Murty, G.P.S., Murthy, K.S.R. (2010). Landeocean tectonic signatures of the Krishna-Godavari basin, a qualitative evaluation. Current Science 98 (4), 485-487.

Subrahmanyam, C., Thakur, N.K., Gangadhara Rao, T., Ramana, M.V., Subrahmanyam, V. (1999). Tectonics of the Bay of Bengal, northeastern Indian Ocean: new insights from satellite-derived gravity and shipborne geophysical data. Earth and Planetary Science Letters 171, 237-251.

Subrahmanyam, V., Murthy, K.S.R., Subrahmanyam, A.S., Mohana Rao, K., Murty, G.P.S., Reddy, N.P.C., Sarma, K.V.L.N.S. (2006). Imprints of Chilika Lake in the offshore region – a geomorphologic evidence. Current Science,90, No. 9, 1180-1182.

Subramanyam V, et al. (1995). "Structure and Tectonics off the western continental margin of India",Tectonophysics, 249, 267-282.

Sucheta Dotiwala, Pangtey K K S and B K Dey (1995). Tectonics of South Rewa Gondwana Basin and Its Bearing on Petroleum Prospects –Based on Remote sensing Data- Proc PetroTech 95 pp-301-308.

Sudhakar R., Khan, S., Elias Ahmed, A. (1970). Deformation of Tertiary and Quaternary sediments of Cambay basin south of Mahisagar River; Bull. ONGC, Vol, 7, No. 1.

Sudhakar, R. and Basu D.N. (1973). A reappraisal of the Paleogene stratigraphy of southern Cambay Basin. Bull. ONGC, v. 10 (1-2), pp. 55-76.

Sudhir Sarma (1994). Analysis and Mapping of High Pressures in Cachar Area of Assam Arakan Basin. ONGC Bulletin Vol 31. No. 1 pp 99-114.

Sudipto Datta, P.Mukhopadhyay, D.Das, P.K.Das. (2012). Hydrocarbon Occurrences in Mahanadi Basin, ONGC., SPG Seminar, Hyderabad.

Summers, G.C. and Broding, R.A. (1952). Continuous Velocity Logging. Geophysics, Vol. 17.

Sundaram, R., Henderson, R.A., Ayyasami, K., Stilwell, J.D. (2001). A litho-stratigraphic revision and palaeo-environmental assessment of the Cretaceous System exposed in the onshore Cauvery Basin, southern India. Cretaceous Research 22, 743-762.

Sundaram, R., Rao, P.S. (1979). Litho-stratigraphic classification of Uttatur and Trichinopoly Group of rocks in Trichinopoly district, Tamilnadu. Geological Survey of India, Misc. Publication 45, 111-119.

Sunjay (2010). A New Wireline Tool: Temperature Logging by DTS and DTSS Techniques, 8th SPG.

Supriya Sengupta.(1960), Geological and Geophysical studies in western Part of Bengal Basin. Bull AAPG Vol 50.no.5. pp 1000-1017.

Surajit Gohain (2012). Mesozoic Prospectivity of Kerala Konkan Offshore Basin, Hindustan Petroleum Corporation Limited– SPG 2012

Surajit Gorain, (2012). Mesozoic Prospectivity of Kerala Konkan Offshore Basin -SPG (Hyderabad).

Suri. N.K. (2003). Exploitation strategy of oil Eeservoirs with large gas cap gas – deep water G-1 structure, Krishna Godavari Basin.,Proceedings of PETROTECH.

Sutko, A.A. and Myers, G. M.-(1971): "The Effect of Nozzle Size, Number and Extension on the Pressure Distribution Under a TriCone Bit". Journal of Petroleum Technology, November 1299.

Swamy.P, and K.David (2016). An Overview of Pranhita-Godavari sedimentary basin, Telangana, India International Research Journal of Natural and Applied Sciences Vol. 3, Issue 11, November 2016.

Swanson, B. (1981). A simple correlation between Permeabilities and Mercury Capillary Pressures: Journal of Petroleum Technology, v-33, p.2498-2504.

Sweeney, J. J., and A. K. Burnham (1990). Evaluation of a simple model of vitrinite reflectance base on chemical kinetics: AAPG Bulletin, v. 74, p. 1559–1570.

Takacs, G. (1976). Comparisons made for Computer Z–Factor Calculations. The Oil and Gas Journal, Dec. 20th: 64–66.

Talukdar, B. N., U. C. Das and N. Baruah (2005). Exploitation of Hydrocarbon by High Technology logging tools in Mature Eocene fields of upper Assam Basin in oil's areas of operations-Proceedings-article-235.

Talukdar, B. N., U. C. Das and N. Baruah (2006). Exploitation of Hydrocarbon by high technology logging tools in mature Eocene fields of Upper Assam basin in oil's areas of operations-6th International Conference and Exposition on Petroleum Geophysics "Kolkata".

Talukdar, R., J.S. Saluja, S.K. Shrivastava, B.K. Panda and S. Bhattacharya (2003). Development of Demulgaon field – a Historical perspective., PetroTech proceedings.

Talukdar, S.N.(1982) Geology And Hydrocarbon Prospects Of East Coast Basins Of India And Their Relationship To Evolution Of The Bay Of Bengal, Offshore South East Asia Show, Singapore

Talwani, P. and Gangopadhyay, A.(2001) Tectonic Framework of the Kachchh Earthquake of January 26,scsn.seis.sc.edu/Publications/sr101/SRLMSW.doc.

Tandon A.N., Srivastava A.K., and Verma N.K. (1998). Comprehensive study of geoscientific data to evaluate tectonic style and hydrocarbon prospects of Inner tectonic belt (Paror-Mandi-Sundernagar area, H.P.), NRBC, ONGC, Dehradun, internal report.

Taner, M. T., Koehler, F., and Sheriff, R. E. (1979). Complex seismic trace analysis. Geophysics, vol. 44,P1041–1063.

Taron, P.B. (1979). "Basic volcanic rocks of Uttar Pradesh and Himachal Pradesh, Lesser Himalaya – A review", GSI. Misc. Publ. 41(1), 315-336.

Tattancherry, BJ, D.Srinivas, V Vairavan and R. Rajababu (1999)-Kamalapuram Faormation,Model for subtle traps ecploration in Cauvery Basin- Proc. Workshop integrated Exploration fpr statigraphic and subtle traps-Bull. ONGC V 35 no.2, pp 63-67.

Tewari, H.C., Murthy, P.R.K., Dixit, M.M. (1996). Extension of the Pranhita–Godavari graben towards south-east under the coastal Godavari basin: an evidence from refraction seismics. Gondwana Nine 2, Geological Survey of India, Calcutta, pp. 953–962.

Tewari, S., G. Lahiri, P. K. Nandi, A.K. Sen, B. Bhaumik and A.D Singh (1985). Prediction of Abnormal sub-surface Formation Pressure and formation fracture pressure with special reference to Bodra well no. 2., ONGC Bull.,VOL. 22 NO.2.

Tewari, S, G. Lahiri, P K. Nandi,A K.Sen, B Bhaumik and A.D.Singh. Prediction of abnormal sub-surface formation pressure and formation fracture pressure with special reference to Bodra well no. 2 Bull. ONGC, vol.no.2.

Tewari R.R. and Dolly Pal Rana., (2003). Geochemical Prospecting in Offshorea Case Study-Proceedings of PETROTECH-2003.

Thakur and B S Rawat, (1992). Geologied Map of the Western Himalaya, Published under the authority of Surveyor General of India. p.10.

Thomas J. and Thomas N.J. (1988). Use of TTI to Evaluate Hydrocarbon Prospects in West Bengal Basin. Petro. Geo-chemistry and Expl. In Afro-Asian region.

Thomas N.J. and Sharma V.N. (1991). "Maturation modeling in Cauvery basin". Proc. of SPEB-2, Dec, ONGC Seminar.

Thomas, N. J. and Sharma, V. N. (1993). Thermal evolution of source rocks in Cauvery Basin. In: Proceedings of the 2nd Seminar on Petroliferous Basins of India (Eds S. K. Biswas et al.), Vol. 1, Indian Petroleum Publishers, Dehradun.

Thomeer, J. (1983). Air permeability as a function of three pore network parameters: Journal of Petroleum Technology, v.35,.p 809-814.

Thurber, C.H, and Latson, B.E. (1960). Acoustic Log cheeks casing cement jobs. The Petroleum Engineer reprint distributed by P.G.A.C.

Timur, A. (1968)"An investigation of permeability, porosity and residual water saturation relationship", The Log Analyst, Vol 9,no.,4,pp 8-17.

Tipper, G.H. (1911). The Geology of the Andaman Islands with special reference to the Nicobar. Mem. Geol. Survey India, Vol. 35, pt. 3, pp. 135-145.

Tissot, B. and Walte, D.H. (1978). Petroleum formation and occurrence, Springer-Verlag, Berlin, 538p.

Tiwari G., Samanta U., Kumar A., Upadhyay P., Gupta A., Arora H.K., Shah V.R., and Baruah A.K. (1991). Geochemical evaluation of source organics of sediments of Himalayan foothills. KDMIPE, ONGC, Dehradun, internal report.

Tiwari. S and S K. Banerjee (1985). Seismic Startigraphy of Onshore West Bengal basin Bull.ONGC. Vol.22.no. 2. P 97-122.

Tiwary, D.N., Birbal Singh, R.Rahaman, P.Saha and R.T.Arasu (2005). Delineation of Hydrocarbon pool in Jambusar area: An Innovative Approach.,PetroTech Proceedings.

Tixier, M P., Alger, R.P. and Doh. C.H. (1958). Sonic Logging. AMIE Meeting. Schlumberger Well Surveying Corporation; Houston Texas.

Todd, M.R., O'Dell, P.M. and Hirasaki, G.J. (1971). Methods for Increased Accuracy in Numerical Reservoir Simulators. Soc.Pet.Eng.J., December: 515- 530. Trans. AIME.

Torsvik, T. H., Carter, L.M., Ashwal, L.D., Bhushan. S.K., Pandit M.K. and Jamtvit B. (2001). Rodinia redefined or obscured: Palaeomagnetism of the Malani Igneous Suite (NW India). Precambrian Research, v. 108, p.319-333.

Torsvik, T. H., Pandit, M. K, Redfield, T. F., Ashwal, L. D. and Webb, S. J. (2005). Re-Magnetization of Mesozoic lime stones from the Jaisalmer basin, NW India: Geophys.J. Int. 161, p 57–64.

Trappe, H. and Hellmich, C. (1998)- A real Predication of Porosity thickness from 3D seismic Data by means of neural Networks, EAGE, 59th. Cong. And Exhibition Extended Abstracts.

Trilok Chand, A.K. Sethu Madhavan and B.S. Negi (2003). Geochemical Evaluation for Hydrocarbon Accumulation in Miocene sequences in Mehsana Block, Cambay Basin–Proceedings of PETROTECH-2003 January 9-12, 2003, New Delhi.

Udintsev, G.B. et al. (1975). Geological-Geophysical Atlas of the Indian Ocean, Academy of Sciences of USSR. 151 pp.

Uniyal AK., U Chandra, K Nain R Dhawan V Banerji and Kuldeep Chandra (1994). Environmental Studies in Ganga Basin Through Stable isotope ratios.- Proc second petroliferous Basins of India Indian petroleum Publishres, Dehradun Vol.3 pp173-184.

Uniyal, N.P.,A. Baruah, Sonal T. Shah, Y.N. Trivedi, and Jasbir Singh (2005). Implementation of MEOR Technique in reservoirs having BHT up to 750c.,PetroTech.

Upadhyay H.,Dhand A.K.,Choudhary P.K.,Dave A.,Thakur R.A. (2000). Lithofacies,Depositional Environment and reservoir characteristics of early Oligocene sands of Mahuva Formation, C-39 structure, Tapti-Daman Block.

Upadhyaya, R., Srivastava, R. K., and Agarwal, V., 1988, A statistical approach to the study of an igneous suite- a case history of Tavidar volcanics. N. Jb. Mier. Abh., 159, v.3, p.311-324.

V. S. (1977). Tectonic evolution of the Godavari graben. In Proc. IV Int. Gondwana Symp., Calcutta, vol. II, pp. 889–900.

Vail, P.R. and Mitchum, P.M. (1980). Global Cycles of sea level changes and their role in exploration. Proc. Tenth World Petroleum congress, 2, Expl. Supply and Demand, pp. 95-104.

Vail, P.R., et al. (1987). Seismic stratigraphy interpretation using sequence stratigraphy. In: Bally, A.W. (ed.) Atlas of Seismic Stratigraphy. Volume 1. Am. Ass. Petrol. Geol., Studies in Geology, 27: 1-10.

Vail, P.R., Mitchum, R.M. Jr., Thompson, S. III. (1977). Seismic stratigraphy and

Vairavan V., (1993). Tectonic Hisyory and Hydrocarbon prospects of Palar and Pennar Basins, India, Proceedings of second seminar on Petroliferous Basins of India, Vol-I, pp 308-396.

Vairavan, V, K.G. Vijayalakshmi, Y.E.R. Sastri, M.S.N. Sarma, P.N. Ananthanarayanan (1997). Cauvery Basin - Southern India: Petroleum Geology and Hydrocarbon Habitat AAPG International Conference and Exhibition, Vienna, Austria.

Vaithilingam, K. (2005). structural analysis of Adamtila Field using well logs with special emphasis to dip meter data- Proceedings-article- 261.

Valdia K.S. (1968). The stromatolites of the carbonate formations of the lesser formations and the Vindhyan. Jour. Geol, Soc. Of India, 10(1), pp. 1-25.

Valmiki., D and T.K.N.Gopala swamy (1988). Stratigraphic prospects in Bombay Offshore. JEGP; VOL.IX.No.4.

Van Barren, J.P. (1979). "Quick-look permeability estimates using sidewall samples and porosity logs", in 6 Th Annual European Logging Symposium Transactions: SPWLA, 19.

Van Daalen, F. and van Domselaar, H.R. (1972). Water Drive in Inhomogeneous Reservoirs - Permeability Variations Perpendicular to the Layer. Soc.Pet.Eng.J., June: 211 -219. Trans. AIME.

Venkanna, P., M. Purkait, K.L. Mehrotra and B. Guha-(2003). Source Rock Distribution and Hydrocarbon Potential Evaluation in the West Godavari Sub Basin, K.G. Basin, India--Proceedings of PETROTECH-2003.

Venkalachala, B.S. and Rawat, M.S. (1973). Organic remains from Kaladgi Basin. Geophytology, 3, 1, 26-35.

Venkata Raju D.Ch., R.S. Rajesh, D.C. Mishra (2002). Bouguer anomaly of the Godavari basin, India and magnetic characteristics of rocks along its coastal margin and continental shelf Journal of Asian Earth Sciences 21 (2002) 111–117.

Venkatachala, B.S. and J.Mukherjee. (1982). Source Rock Palynology of Bengal Basin., Bull ONGC. Vol 19.no.2 p-111-118.

Venkataragan, R., Rao, G. N., Prabhaikar, K. N.,Singh, D. N., Awasthi, A. K, Misra, P. K., Reddy, P. K., Roy, S.K. (1993). Lithostratigraphyof Krishna- Godavari Basin, ONGC internal Document.

Venkataraman.D.(1984). Petroleum prospects and Review and Reinterpretation of the Geology of a part of West Bengal Pet. Asia Jour. Vol. II pp 147-157.

Venkataramana, P. (2000). Late Cretaceous India-Madagascar fit and timing of break-up related magmatism: Terra Nova, v. 12, p. 220–224.10.

Venkataramnan, D, (1984). Petroleum Prospects and a review and reinterpretation of the Geology of part of West Bengal- Petroleum Asia Journal pp-147-157.

Venkatarangan, R., Prabhakar, K.N.Singh, D.N. Awasti, A.K.Reddy, P.K. Mishra, P.K.Roy, S.K. and Palakshi, K. (1993). Litho-stratigraphy of Indian Petroliferous basins, Document – VII, Cauvery Basin ONGC published report.

Venkateswarlu, P.D., Rao, A.M.V.R., Bose, P.K. (1992b). Marine magnetic anomalies and associated tectonics along the East Coast of India. Geological Survey of India Special Publication 29, 61-70.

Venkateswarlu, P.D., Sengupta, B.J., Rao, A.M.V.R., Bose, P.K. (1992a). Marine magnetic indication of a submerged volcano off Machilipatnam in Bay of Bengal. Journal of Geological Society of India 39, 197-203.

Verma, R. K. (1991). Geodynamics of the Indian peninsula and the Indian plate margin, Oxford and IBH Pub.Co. Pvt.Ltd.

Verma, R.K.,U.S. Singh, and Naresh Chand (1976). Bbble Point Pressure correlation for Indian oilfields., Bull.ONGC.Vol.13 No. 1and 2.

Vernik, L. (1994). "Predicting lithology and transport properties from acoustic velocities based on petrophysical classification of siliclastics", Geophysics, 59, 420 – 427.

Vernik, L. (2000)" Permeability prediction in poorly consolidated sliclastics based on porosity and clay volume logs", Petrophysics, 41, 2, 138 -147.

Vidrine, Dennis and Benitt, Emmett, (1967). Field verification of the effect of differential pressure on drilling rate, SPE1959.

Vij, S.K., Narasaiah, S.L., Walia, A. and Singh, G. (1998). "Multilaterals: An overview and issues involved in adopting this technology", SPE 39509.

Vijayakumar, K., et al., Reservoir delineation through multi-offset VSP, South Kadi Field, Cambay basin, India, paper published in 4th Conference and Exposition of Society of Petroleum Geophysics, India.

Vijayalakshmi, K.G. (1988). Miocene depositional systems in thrust areas in Krishna - Godavari Basin: ONGC Bulletin, v.25, no.2.p.1-10.

Vincent, Ph., Gardner, J-E., and Attali, G. (1977). "GEODIP -An Approach to Detailed Dip Determination Using Correlation by Pattern Recognition," paper SPE 6823 presented at the 1977 SPE Annual Technical Conference and Exhibition.

Virdi, N.S. (1995). "Proterozoic Sedimentalogical cycles and Geotectonic Evolution of the southern margin of the Proto-tethys: Evidence from the Western Himalaya", Ind.Jour. Pet. Geol., v.4(2), 45-73.

Viswanath., S.N (1997). A hundred Years of Oil, A narrative account of the search for oil in India, Vikas Publishing House PVT LTD., pp1-175.

Vogel, C.B. (1952). A seismic Velocity Logging Method. Geophysics, Vol. 17.

Wadhwan, Saurashtra, Gujarat State, Thingran com. Vol., Delhi University, Delhi, P. 229-239.

Wadia D.N. (1928). The Geology of Poonch State (Kashmir) and adjacent portions of the Punjab. Mem Geol Surv. India, 51 Pt. 2.

Wadia, D.N, (1938). Permo-Carboniferous limestone inliers in the Sub-Himalayan Tertiary Zone of Jammu-Kashmir Himalaya, Rec. Geol. Surv India, 72, Pt. 2.

Wadia, D.N. (1932). The Syntaxis of the north-west Himalaya. Its rocks, Tectonics and Orogeny. Rec. Geol. Surv. 65, Pt 2.

Wadia, D.N., and West, W,D, (1964). Structure of the Himalayas, Int. Geo1. Cong. 22nd Session, India.

Wadsle Y A. W (2011). Markov Chain "Methods for Reserves Estimation". International Petroleum Technology Conference. Doha, 2005.

Wandrey, C.J. (2004). Bombay geologic province Eocene to Miocene composite total petroleum system, India: U.S. Geological Survey Bulletin 2208-F, 26p.

Wandrey, C.J. (2004a). Sylhet–Kopili/Barail–Tipam Composite Total Petroleum System, Assam Geologic Province, India: U.S. Geological Survey Bulletin 2208-D, 19 p.

Wandrey, C.J. (2004b). Bombay geologic province Eocene to Miocene composite total petroleum system, India: U.S. Geological Survey Bulletin 2208-F, 20 p.

Wani, M R, R Hussain RP Singh and I Sarala (1995). Sequence Stratigraphy and Hydrocarbon Prospects of Lower Miocene Sediments in South western part of Cambay Basin India, - Proc Petro Tech, New Delhi, 1995 pp-175-187.

Waples, D.W. (1985). "Geochemistry in Petroleum Exploration", International Human Resources Development Corporation, Boston, 232.

Wardlaw. N.C., and R.P.Taylor (1976). Mercury capillary pressure curves and the interpretation of pore structure/capillary behaviour in reservoir rocks, Bulletin of Canadian petroleum Geology-V-24 No2-June.

Wardroup, W.R. and Cannon,G.E.(1956). "Some Factors Contributing to Increased Drilling Rate", Oil and Gas Journal.

Warren, T.M. and Winters,W.J. (1982). "The effect of Nozzle Diameter and Jet impact for a Tri Cone Bit", SPE Paper No. 11059, September 1982.

Warsi, W.E.K., and Molnar, P. (1977). Gravity anomalies and plate tectonics in the Himalaya, in Colloques Internationaux du CNRS, Himalaya, Science De La Terra, pp. 463-473.

Waxman, M. H., and L. J. M. Smits (1968). Electrical conductivities in oil-bearing shaly sands: Society of Petroleum Engineers Journal, v. 8, no. 2, p. 107-122, SPE-1863-A.

Weeks LG-(1952). "Factors of Sedimentary Basin Development That Control Oil Occurrence," Bulletin of the AAPG 36,no. 11,2071–2124.

Weeks, L.A., Harbison, R.N. and Peter, G. (1967). Island arc system in Andaman Sea. Bull. Amer. Assoc. Petrol Geol. Vol. 51 No. 9 pp. 1803-1851, Figure 1.

Welge, H.J. (1952). A Simplified Method for Computing Oil Recovery by Gas or Water Drive. Trans. AIME. 195: 91-98.

Welte DH and Yukler M A (1981). "Petroleum Origin and Accumulation in Basin Evolution–A Quantitative Model," AAPG Bulletin 65, no. 8: 1387–1396.

Welte DH: (1972)"Petroleum Exploration and Organic Geochemistry," Journal of Geochemical Exploration 1, no. 1: 117–136.

Welte, D.H. (1965). Relation between Petroleum and Source Rock; AAPG, Vol. 49, No. 12; pp. 2246-2268.

Wendt, W.A., Sukurai, S. and P.H. Nelson, (1986). Permeability prediction from well logs using Multiple Regression, in L.W. Lake and H.B. Carrol Jr. eds., Reservoir Characterization: Academic Press, Orlando, p, 181-212.

Wharton, R.P. and Delano, J.M. (1981). "An EPT Interpretation Procedure and Application in Fresh Water, Shaly, Oil Sands,"Trm., SPWLA Annual logging Symposium.

Wharton, R.P., Hazen, G.A., Rau, R.N., and Best, D.L. (1980). "Electromagnetic Propagation Logging Advances in Technique and Interpretation," paper SPE 9267 presented at the SPE Annual Technical Conference and Exhibition.

Whitmarh R.B., Weser, O.E. and D.A., Ross, (1974). Initial reports of DSDP Site, 219 v.23, p. 35-115.

Williams, i.e., "Development of 15,000 psig Well Testing Equipment," SPE paper presented at the Petroleum Engineering Conference, Aberdeen, 1986.

Winsaver, W.O., Sheain, KM., Jr., Masson, P.H., and Williams, M. (1952). "Resistivity of Brine-Saturated Sands in Relation to Pore Geometry," Bull,., AAPG 36, No. 2.

Wm. C. Gussow (1980). Differential Entrapment–Bulletin ONGC-Vol. 17, No. 2, December.

World Oil Outlook-WOO_2014.

World Oil Supply Scenarios- The Future is neither as bleak nor rosy as some assert.

Worthington, P.F. (1985). "The Evaluation of Shaly-Sand Concepts in Reservoir Evaluation," The Log Analyst.

Wuenschel, P. C. (1976). The vertical array in reflection seismology, Geophysics, vol. 41, p.219–232.

Wyllie, M. R. J., A. R. Gregory, and G. H. F. Gardner,(1958). An Experimental Investigation of Factors Affecting Elastic Wave Velocities in Porous Media, Geophysics, vol. 23, pp. 459–493.

Wyllie, M.R.J. (1963) 3rd edition. The Fundamentals of Well Log interpretation. Academic Press.

Wynne-Edwards, H.R. (1976). Proterozoic ensialic orogenesis: the millipede model of platetectonics. Am. J. Sci., 276, 927-953.

Xie, X. (1991). A formulation for the capillary pressure relationship and a statistical description of pore distribution", Academia, SPE 21890.

Yadav, L. and Sud, V. K. (2001). Proximate Analysis of Coals through Well Log Evaluation (PACTWEL): A Software Package, Proceedings of 4th International Petroleum Conference and exhibition, PETROTECH.

Yadav, L., Bhatt, U.C. and Prasad, K. K. (2002). Extrapolation of Matrix Parameters for Inversion of Open Hole Logs in Complex Lithological Reservoirs of Upper Assam,SPG-2002, Mumbai, India.

Yadav, L., Kumar, R., Singh, M and Sud, V. K. (2000). Pseudo mineral concept of secondary porosity and variable "m"technique for the evaluation of low resistivity carbonates: A Software Package, Proceedings of third Conference and exposition on petroleum Geophysics, SPG.

Yadav, Lalaji., Narender Pal and R.K. Bhardwaj (2003). Evolution of a Generalized Porosity –Permeability Transform for upper Assam oil fields., –Proceedings of PETROTECH.

Yeats, R.S., and Lawrence, R.D.(1984). Tectonics of the Himalayan thrust belt in northern Pakistan, in marine Geology and Oceanography of Arabian Sea and Coastal Pakistan, pp- 177-198, New york.

Yetunde, S., and amp; Ogbonna, J. (2011). Challenges and Remedy for Cementing of HPHT Wells In Nigerian Operation. Paper SPE 150751 Presented at the Nigerian Annual International Conference and Exhibition, Abuja, Nigeria.

Yi Xie, Kefeng Xin, James Sun, Carl Notfors, CGGVeritas (2010). 3D pre-stack depth migration with compensation for frequency dependent absorption and dispersion,., SPG 8th Hyderabad.

Yoshida, M., Funaki, M., Vitanage, P.W. (1992). Proterozoic to Mesozoic East Gondwana: The juxtaposition of India, Sri Lanka, and Antarctica. Tectonics, V.11, 381 – 391.

Yoshida, M., Rajesh, H.M., Santosh, M. (1999). Juxtaposition of India and Madagascar: a perspective. Gondwana Research 2, 449e462.

Yuan, Z., Schubert, J., and Teodoriu, C., et al. (2012). HPHT Gas Well Cementing Complications and its Effect on Casing Collapse Resistance. Paper SPE-153986-MS presented at SPE Oil and Gas India Conference and Exhibition.

Zaigham N. A. and K. Mallick, (2000). A Prospect of Hydrocarbon Associated with Fossil-Rift Structures of the Southern Indus Basin, Pakistan: AAPG Bulletin. v. 84; p. 11.

Zhaoguang, Y., Schubert, J. and Teodoriu, C. (2012). The Effect of Hole Angle and Cementing Complications On HPHT Well Integrity. Paper SPE 162839 Presented at the Canadian Unconventional Resources Conference. Calgary, Canada.

Ziegler, P. A. (1992). Geodynamics of rifting and implications in Hydrocarbon Habitat. Tectonophysics, 215, 221–253.

Zubov, I.P., Naugoloy, I.K., Zapivalov, N.P. and Chandra, P.K, (1966). Problem of correlation and distribution of hydrocarbon bearing horizons in the Eocene of Cambay basin; Bull. of Oil and Natural Gas Commission, Vol. 3, No. 2.

Zutshi P.L.,Jain M.M.,Mahajan P.K.,Mishra M.M.,Kumar Ram (1986). Oligo-Miocene formation of Surat Depression-A regional study ONGC Bull. Vol.23 No.1 pp 81-100.

Zutshi P.L.,Sood A.,Mohapatra P.,Ramani KKV, Dwivedi A.K. and Srivastava S.C. (1993). Lithostratigraphy of Indian Petroliferous Basins: Bombay Offshore Basin, Document-V, ONGC, Dehradun.

Zutshi., P. L. (1985). A Study of sands within the Barail Group in Rudrasagar field Assam.– Bulletin ONGC. Vol.22 no. 1.pp.

Index

R

S

T

U

V

W